New National Framework

MATHEMATICS 8+

M. J. Tipler K. M. Vickers

J. Douglas

Published in 2003 by:
Nelson Thornes Ltd
Delta Place
27 Bath Road
CHELTENHAM
GL53 7TH
United Kingdom

03 04 05 06 07 / 10 9 8 7 6 5 4 3 2 1

A catalogue record for this book is available from the British Library

ISBN 0 7487 6754 1

Illustrations by Angela Lumley, Oxford Designers and Illustrators, Harry Venning
Page make-up by Mathematical Composition Setters Ltd

Printed and bound in Spain by Graficas Estella

Acknowledgements
The authors and publishers would like to thank Jocelyn Douglas for her contribution to the development of this book.

The publishers thank the following for permission to reproduce copyright material.

Casio: 6, 36, 85; Corel 1 (NT): 108, 187; Corel 423 (NT): 364; Corel 433 (NT): 119, 177; Corel 624 (NT): 411; Corel 691 (NT): 389; Corel 772 (NT): 403; Corel 783 (NT): 242; Digital Stock 1 (NT): 398; Digital Stock 6 (NT): 315; Digital Stock 75 (NT): 90; Digital Vision 8 (NT); 16, 17, 32; Digital Vision 11 (NT): 137; Digital Vision SC (NT): 386; Photodisc 22 (NT): 244; Photodisc 32 (NT): 280; Photodisc 40 (NT): 376; Photodisc 44 (NT): 118; Photodisc 45 (NT): 153; Photodisc 61 (NT): 131; Stephen Frink/Digital Vision AF (NT): 9.

The publishers have made every effort to contact copyright holders but apologise if any have been overlooked.

Contents

Contents

Introduction

We hope that you enjoy using this book. There are some characters you will see in the chapters that are designed to help you work through the materials.

These are    

 This is used when you are working with information.

 This is used where there are hints and tips for particular exercises.

 This is used where there are cross references.

 This is used where it is useful for you to remember something.

 These are blue in the section on number.

 These are green in the section on algebra.

 These are red in the section on shape, space and measures.

 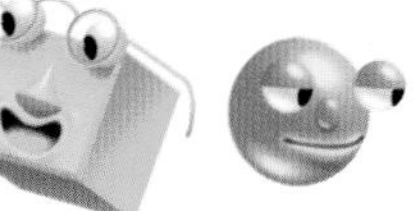 These are yellow in the section on handling data.

New National Framework Mathematics 8 and the Casio *fx-9750 GPLUS*

Throughout this book, you will find a range of activities that support the use of the Casio *fx-9750 GPLUS*. Each of these activities has been written so that they are fully linked to the Framework for teaching mathematics. The graphic calculator is often referred to in the Framework as an important resource for developing mathematical understanding using this ICT. This includes working with actual data and looking at relationships in numbers and number sequences for example. All the activities are fully integrated as part of this book and additional support for using graphic calculators can be found in the Teacher Support Files that accompany the Pupil Books.

Number Support

Place value, × and ÷ by multiples of 10, 100, 1000, ...

This chart shows **place value**.

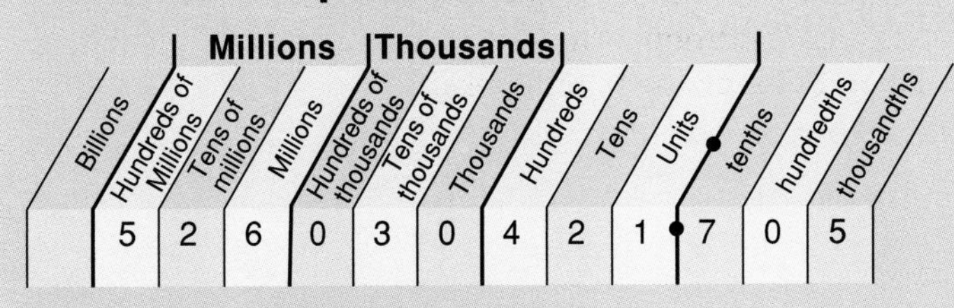

We use zeros rather than spaces as place holders.

A billion is a thousand million.
We read the number on the chart as 'five hundred and twenty-six million, thirty thousand, four hundred and twenty-one point seven zero five'.

We use place value to **add and subtract by multiples of 0·1, 0·01 and 0·001**.

Examples $8{\cdot}586 + 0{\cdot}01 = 8{\cdot}596$ — **Add 1 to the hundredths.**

$27{\cdot}723 - 0{\cdot}006 = 27{\cdot}717$ — **Subtract 6 from the thousandths.**

To **multiply by 10, 100, 1000,** ... move each digit one place to the left for each zero in 10, 100, 1000, ...

Examples $86 \times 10 = 860$ $\quad 0{\cdot}7 \times 100 = 70$ $\quad 0{\cdot}36 \times 1000 = 360$

To **divide by 10, 100, 1000,** ... move each digit one place to the right for each zero in 10, 100, 1000, ...

Examples $720 \div 10 = 72$ $\quad 7{\cdot}9 \div 100 = 0{\cdot}079$ $\quad 834 \div 1000 = 0{\cdot}834$

We can use multiplication and division by 10, 100, 1000, ... to **multiply and divide by multiples of 10, 100, 1000,** ...

Examples

$$\begin{aligned} 70 \times 500 &= 7 \times 10 \times 5 \times 100 \\ &= 7 \times 5 \times 10 \times 100 \\ &= 35\,000 \end{aligned}$$

$4{\cdot}8 \div 300 = 0{\cdot}016$
because $\quad 4{\cdot}8 \div 3 = 1{\cdot}6$
$\quad 1{\cdot}6 \div 100 = 0{\cdot}016$

Practice Questions 1, 2, 10, 14, 20, 30

Putting decimals in order. Rounding

To put **decimals in order**, compare digits with the same place value. Work from left to right to find the first digits that are different.

Example $7{\cdot}3682 > 7.3658$ Starting at the left, the digits are the same until we get to the digits 8 and 5.
8 is greater than 5 so $7{\cdot}3682 > 7{\cdot}3658$.

We can round to the nearest 10, 100, 1000, ... or nearest whole number.

3 653 872 to the nearest ten thousand is 3 650 000
to the nearest hundred thousand is 3 700 000
to the nearest million is 4 million.

When a number is halfway between two numbers we round up.

Example 450 000 to the nearest hundred thousand is 500 000.

To **round to the nearest whole number** we look at the tenths digit.
If the tenths digit is 5 or more, we round the whole number up. Otherwise, it stays the same.

Example 27·63
↑ The tenths digit is 5 or more so 27 becomes 28

27·63 to the nearest whole number is 28.

Practice Questions 8, 11, 24, 31, 48

Calculation with integers

The **integers** are ..., $^{-}4$, $^{-}3$, $^{-}2$, $^{-}1$, 0, 1, 2, 3, 4, ...
We can show these on a number line.

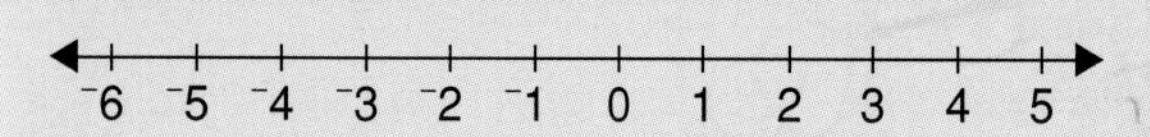

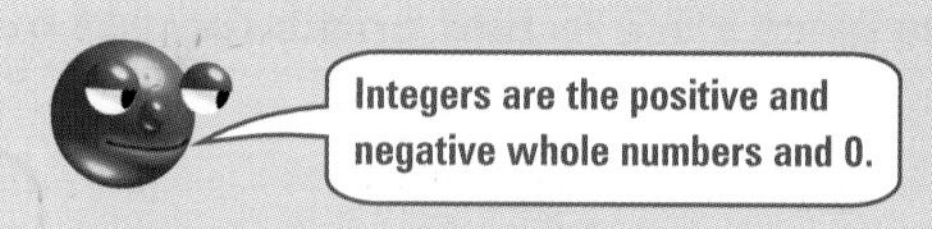

We can **add and subtract integers**. Sometimes we use a number line to help.

Examples $6 + {}^{-}2 = 4$ $8 + {}^{-}5 = 3$ ${}^{-}2 + {}^{-}7 = {}^{-}9$ ${}^{-}3 + 6 = 3$
$1 - {}^{-}4 = 5$ ${}^{-}6 - {}^{-}8 = 2$ ${}^{-}7 - {}^{-}8 = 1$

Multiplying or dividing two negative numbers gives a positive answer.
Multiplying or dividing one negative and one positive number gives a negative answer.

Examples ${}^{-}5 \times {}^{-}4 = 20$ ${}^{-}7 \times 6 = {}^{-}42$ $5 \times {}^{-}8 = {}^{-}40$ ${}^{-}30 \div {}^{-}6 = 5$

Practice Questions 3, 4, 6, 45, 54a,b,c, 57

Divisibility

A number is **divisible** by **2** if it is an even number
by **3** if the sum of its digits is divisible by 3
by **4** if the last two digits are divisible by 4
by **5** if the last digit is 0 or 5
by **6** if it is divisible by both 2 and 3
by **8** if half of it is divisible by 4
by **9** if the sum of its digits is divisible by 9
by **10** if the last digit is 0.

To check if a number is divisible by 24, check for divisibility by 3 and 8, not 4 and 6.
We check using numbers that multiply to 24 but have no common factors other than 1.

Practice Questions 25, 42

Prime numbers, factors

Prime numbers are only divisible by themselves and 1. 1 is not a prime number.
The prime numbers less than 30 are 2, 3, 5, 7, 11, 13, 17, 19, 23 and 29.
To test if a number is prime, try to divide it by each of the prime numbers in turn.

60 written as a **product of prime factors** is $2 \times 2 \times 3 \times 5$ or $2^2 \times 3 \times 5$.
We can use a table or a factor tree to write 90 as a product of primes.

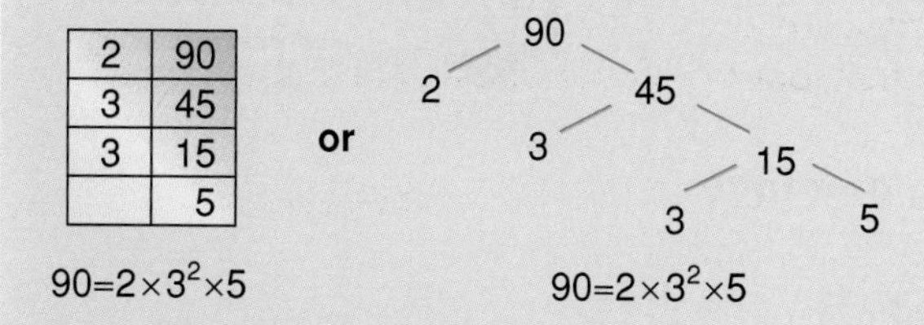

$90 = 2 \times 3^2 \times 5$ $90 = 2 \times 3^2 \times 5$

The **highest common factor (HCF)** of 36 and 54 is 18.
This is the largest number that is a factor of both 36 and 54.

The **lowest common multiple (LCM)** of 8 and 12 is 24.
This is the smallest number that is a multiple of both 8 and 12.
The lowest common multiple of 36 and 54 is 108.

$36 = 2 \times 2 \times 3 \times 3$
$54 = 2 \times 3 \times 3 \times 3$
$\text{HCF} = 2 \times 3 \times 3$
$= 18$
$\text{LCM} = 2 \times 2 \times 3 \times 3 \times 3$
$= 108$

We can find the HCF and LCM using a Venn diagram.

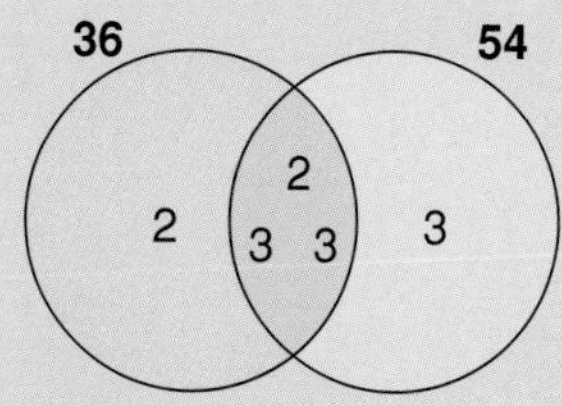

HCF is the product of numbers in the purple shaded area.
LCM is the product of all the numbers on the diagram.

Practice Questions 56, 60, 61, 63, 66, 69

Squares, square roots, cubes, cube roots, powers

6×6 can be written as 6^2 and is said as '**6 squared**'. $6^2 = 36$.
$5 \times 5 \times 5$ can be written as 5^3 and is said as '**5 cubed**'. $5^3 = 125$.

6 is the **square root** of 36 and is written as $\sqrt{36} = 6$.
4 is the **cube root** of 64 and is written as $\sqrt[3]{64} = 4$.

Squaring and finding the square root are **inverse operations.**
Cubing and finding the cube root are **inverse operations**.

To square a number using a calculator, use the $\boxed{x^2}$ key.
To find the square root of a number using a calculator, use the $\boxed{\sqrt{}}$ key.

Examples $8{\cdot}4^2$ is keyed as $\boxed{8{\cdot}4}$ $\boxed{x^2}$ $\boxed{=}$ to get 70·56.
$\sqrt{4{\cdot}9}$ is keyed as $\boxed{\sqrt{}}$ $\boxed{4{\cdot}9}$ $\boxed{=}$ to get 2·2 (1 d.p.).

2^5 is read as 'two to the power of 5'.
5 is called an **index**.

Practice Questions 32, 41, 44, 47, 54d, 67, 70

Mental calculation

Here are some ways of **adding and subtracting mentally**.

Counting up *Example* $6{\cdot}3 + 2{\cdot}9 = \mathbf{9{\cdot}2}$ →

+2 +0·9
6·3 8·3 9·2

Nearly numbers *Example* $7{\cdot}9 + 11{\cdot}2 = 8 - 0{\cdot}1 + 11 + 0{\cdot}2$
$= 19 - 0{\cdot}1 + 0{\cdot}2$
$= \mathbf{19{\cdot}1}$

Compensating *Example* $16{\cdot}4 - 5{\cdot}8 = 16{\cdot}4 - 6 + 0{\cdot}2$
$= 10{\cdot}4 + 0{\cdot}2$
$= \mathbf{10{\cdot}6}$

Always try to work out answers using a mental method.

Here are some ways of **multiplying and dividing mentally**.

Partitioning *Example* $5{\cdot}3 \times 15 = 5{\cdot}3 \times (10 + 5)$
$= 5{\cdot}3 \times 10 + 5{\cdot}3 \times 5$
$= 53 + 5 \times 5 + 0{\cdot}3 \times 5$
$= 53 + 25 + 1{\cdot}5$
$= \mathbf{79.5}$

Factors *Examples* $8{\cdot}2 \times 12 = 8{\cdot}2 \times 3 \times 4$
$= 24{\cdot}6 \times 4$
$= 24{\cdot}6 \times 2 \times 2$
$= 49{\cdot}2 \times 2$
$= \mathbf{98.4}$

$420 \div 15 = \mathbf{28}$ because $420 \div 5 = 84$
$84 \div 3 = 28$

Practice Questions 15, 16, 19, 23, 26

Order of operations

The **order in which we do operations** is from left to right doing
Brackets **then I**ndices **then D**ivision and **M**ultiplication **then A**ddition and **S**ubtraction (BIDMAS).

Examples $8 + 24 \div 6 = 8 + 4$
$= \mathbf{12}$

Do first

$5 \times 9 + 3 \times 7 = 45 + 21$
$= \mathbf{66}$

$\frac{72}{2 \times 3} = \frac{72}{6}$
$= \mathbf{12}$

The horizontal line acts as a bracket.

$\frac{5(8-4)}{12-2} = 5(8 - 4) \div (12 - 2) = 5(4) \div (10)$
$= 20 \div 10$
$= 2$

Do first

Practice Questions 7, 47

Estimating

Sometimes **an estimate** is a good enough answer to a question.

Example The number of people at a football match.

To **estimate the answer to a calculation** we round the numbers.
There is often more than one possible estimate. $\approx$ means 'is approximately equal to'.

Examples $393 \times 84 \approx 400 \times 80 = 32\,000$
or $393 \times 84 \approx 400 \times 100 = 40\,000$
$\frac{647}{78} \approx \frac{640}{80} = 8$

We try to round to 'nice numbers'. We round to $\frac{640}{80}$ rather than $\frac{650}{80}$ because 64 is a multiple of 8.

Practice Question 17

Written calculation

We **add and subtract** by lining up the units column or, with decimals, by lining up the decimal points.

Examples

$$\begin{array}{r} 58{\cdot}6 \\ +\ 4{\cdot}384 \\ \hline 62{\cdot}984 \\ {\scriptstyle 1} \end{array} \qquad \begin{array}{r} {\scriptstyle 4\ 9\,12\ 1} \\ 8503{\cdot}0 \\ -\ 27{\cdot}6 \\ \hline 8475{\cdot}4 \end{array}$$

Always try to use a mental method or written method before using a calculator.

Multiplying

Example $5{\cdot}87 \times 9$

$5{\cdot}87 \times 9 \approx 6 \times 9 = 54$

	5	0·8	0·07
9	45	7·2	0·63

Answer $45 + 7{\cdot}2 + 0{\cdot}63 = \mathbf{52{\cdot}83}$

or

$5{\cdot}87 \times 9$ is equivalent to

$5{\cdot}87 \times 100 \times 9 \div 100 = 587 \times 9 \div 100$

$$\begin{array}{r} 587 \\ \times 9 \\ \hline 5283 \\ {\scriptstyle 7\,6} \end{array} \qquad 5283 \div 100 = \mathbf{52{\cdot}83}$$

Example $27{\cdot}3 \times 5{\cdot}7$

$27{\cdot}3 \times 5{\cdot}7 \approx 30 \times 6 = 180$

	20	7	0·3
5	100	35	1·5
0·7	14	4·9	0·21

Answer $100 + 35 + 1{\cdot}5 + 14 + 4{\cdot}9 + 0{\cdot}21$
$= \mathbf{155{\cdot}61}$

or

$27{\cdot}3 \times 5{\cdot}7$ is equivalent to

$27{\cdot}3 \times 10 \times 5{\cdot}7 \times 10 \div 100 = 273 \times 57 \div 100$

$$\begin{array}{r} 273 \\ \times 57 \\ \hline 13650 \\ 1911 \\ \hline 15561 \\ {\scriptstyle 1} \end{array} \qquad 15\,561 \div 100 = \mathbf{155{\cdot}61}$$

Division

Examples $158{\cdot}7 \div 8 \approx 160 \div 8 = 20$

$$\begin{array}{rl} 8\ )\ 158{\cdot}64 & \\ -\ 80{\cdot}00 & 8 \times \mathbf{10} \\ \hline 78{\cdot}64 & \\ -\ 72{\cdot}00 & 8 \times \mathbf{9} \\ \hline 6{\cdot}64 & \\ -\ 6{\cdot}40 & 8 \times \mathbf{0{\cdot}8} \\ \hline 0{\cdot}24 & \\ 0{\cdot}24 & 8 \times \mathbf{0{\cdot}03} \\ \hline 0{\cdot}00 & \end{array}$$

Answer **19.83**

$89{\cdot}6 \div 14 \approx 90 \div 15 = 6$

$$\begin{array}{rl} 14\ )\ 89{\cdot}6 & \\ -\ 84{\cdot}0 & 14 \times \mathbf{6} \\ \hline 5{\cdot}6 & \\ -\ 5{\cdot}6 & 14 + \mathbf{0{\cdot}4} \\ \hline 0{\cdot}0 & \end{array}$$

Answer **6·4**

When we **divide by a decimal** we do an equivalent division. We make the divisor a whole number.

Example $72 \div 0{\cdot}9$ is equivalent to $720 \div 9$.
$50 \div 1{\cdot}3$ is equivalent to $500 \div 13$.

Always estimate first.
Check your answer against the estimate. You can also check your answer using **inverse operations** or by doing an **equivalent calculation**.

Example Check the answer to $196{\cdot}4 \times 5{\cdot}2 = 1021{\cdot}28$ by

- estimating $196{\cdot}4 \times 5{\cdot}2 \approx 200 \times 5 = 1000$
- inverse operations $1021{\cdot}28 \div 5{\cdot}2 = 196{\cdot}4$ or $1021{\cdot}28 \div 196{\cdot}4 = 5{\cdot}2$
- equivalent calculation $196{\cdot}4 \times 5{\cdot}2 = 196{\cdot}4 \times 5 + 196{\cdot}4 \times 0{\cdot}2$.

Practice Questions 5, 9, 18, 22, 43, 74, 75

Using a calculator

Sometimes we need to **interpret** the answer given by a calculator.

Examples [8.2] might be £8·20 or 8 hours 12 minutes or 8 metres 20 centimetres and so on.

To find the **remainder** when dividing using a calculator, subtract the whole number part of the answer and then multiply by the divisor.

Examples 872 ÷ 23

Key [872] [÷] [23] [=] to get 37·91304348

Key [−] [37] to get 0·91304348

Key [×] [23] to give the remainder 21.

Sometimes we use **brackets** on the calculator.

Example 8·2 − (7·1 − 2·5)

Key [8·3] [−] [(] [7·1] [−] [2·5] [)] [=] to get 3·7

In a fraction, the horizontal line acts as a bracket.

Example $\frac{5 \cdot 2 + 3 \cdot 6}{4 \cdot 7 - 2 \cdot 9}$

Key [(] [5·2] [+] [3·6] [)] [÷] [(] [4·7] [−] [2·9] [)] [=]

to get **4·9 (1 d.p.)**

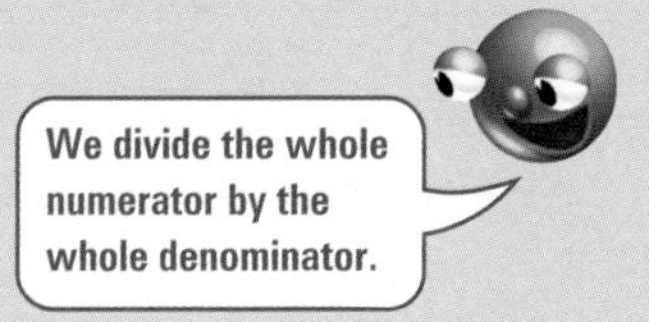

Practice Questions 71, 73, 77, 80, 82, 83

Fractions, decimals and percentages

Fractions

Unit fractions have a numerator of 1. $\frac{1}{8}$, $\frac{1}{3}$ and $\frac{1}{24}$ are all unit fractions.

When writing **one number as a fraction of another**, make sure the numerator and denominator have the same units.

Example 37 cm as a fraction of 1 m = $\frac{37}{100}$ ← Both in cm.

Mixed numbers to improper fractions

$1\frac{3}{4} = \frac{4}{4} + \frac{3}{4} = \frac{7}{4}$

$\frac{16}{5} = 3\frac{1}{5}$

Equivalent fractions

We make equivalent fractions by multiplying or dividing the numerator and denominator by the same number.

Examples

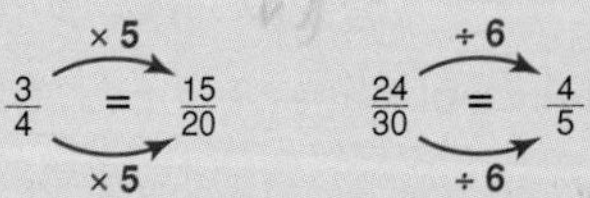
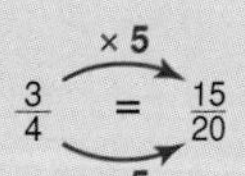

To cancel a fraction to its **lowest terms** we divide the numerator and denominator by the highest common factor. This is called **cancelling**.

Example 64 as a fraction of 124 = $\frac{64}{124} = \frac{16}{31}$

Numerator and denominator have been divided by 4.

If it is hard to find the HCF, you can cancel in steps.

Adding and subtracting fractions

It is easy to add fractions with the same denominator.

Example $\frac{3}{5} + \frac{4}{5} = \frac{3+4}{5}$ **Add the numerators.**

$= \frac{7}{5}$

$= 1\frac{2}{5}$

Fraction of

Example $\frac{3}{8}$ of 96

$\frac{1}{8}$ of $96 = 96 \div 8$

$= 12$

$\frac{3}{8}$ of $96 = 3 \times 12$

$= \mathbf{36}$

Fractions and decimals

To write a **decimal as a fraction**, write it with denominator 10, 100 or 1000.

Examples $0{\cdot}34 = \frac{34}{100} = \frac{17}{50}$ $\qquad 0{\cdot}375 = \frac{375}{1000} = \frac{3}{8}$

Always cancel the fraction to its simplest form.

To write a **fraction as a decimal** we can

1 use known facts. *Example* $\frac{5}{8} = 5 \times \frac{1}{8} = 5 \times 0{\cdot}125 = 0{\cdot}625$

2 make the denominator 10 or 100. *Example* $\frac{19}{25} = \frac{76}{100} = 0{\cdot}76$

3 divide the numerator by the denominator. *Example* $\frac{4}{9}$ $\qquad 9\overline{)4{\cdot}0^40^40}$ = $0{\cdot}4\,4\,4$

4 use a calculator

Example $\frac{41}{87}$

[41] [÷] [87] [=] to get 0·47 (2 d.p.)

A **terminating** decimal has a finite number of digits.

Example $\frac{3}{8} = 0{\cdot}375$

A **recurring decimal** is a decimal that has one or more repeated digits.

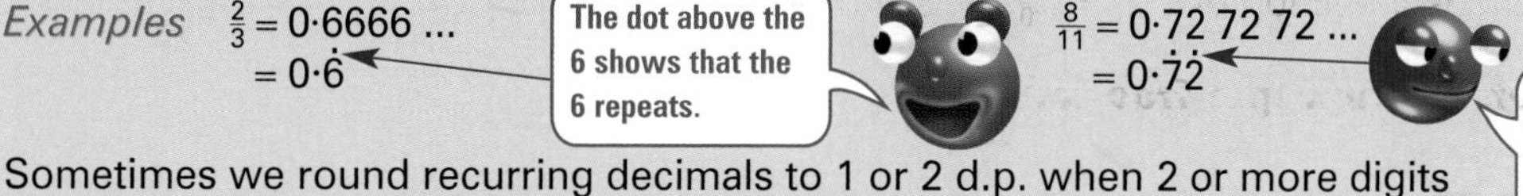

Examples $\frac{2}{3} = 0{\cdot}6666\ldots = 0{\cdot}\dot{6}$ $\qquad \frac{8}{11} = 0{\cdot}72\,72\,72\ldots = 0{\cdot}\dot{7}\dot{2}$

The dots above the 7 and 2 show that these repeat.

Sometimes we round recurring decimals to 1 or 2 d.p. when 2 or more digits repeat.

Percentages, fractions and decimals

To write a **percentage as a fraction or decimal**, write the percentage as the number of parts per hundred.

Examples $82\% = \frac{82}{100} = 0{\cdot}82$ $\qquad 65\% = \frac{65}{100} = \frac{13}{20}$

Cancel if you can.

$33\frac{1}{3}\% = \frac{1}{3} = 0{\cdot}\dot{3}$ $\qquad 66\frac{2}{3}\% = \frac{2}{3} = 0{\cdot}\dot{6}$ $\qquad 12\frac{1}{2}\% = \frac{1}{8} = 0{\cdot}125$

To write a **fraction or decimal as a percentage**, write it with a denominator of 100 or multiply by 100%.

Examples $0{\cdot}96 = \frac{96}{100} = 96\%$ $\qquad \frac{8}{15} = (8 \div 15) \times 100\%$

$= 53\%$ to the nearest whole number or $53{\cdot}\dot{3}\%$.

Percentage of

$10\% = \frac{1}{10}$ 5% is half of 10% $25\% = \frac{1}{4}$

We sometimes use these facts **to find the percentage of a quantity mentally**.

Example 15% of 120 km can be calculated by finding 10%, then 5% and adding these.
15% of 120 km is 18 km.

10% of 120 = 12
5% of 120 = $\frac{1}{2}$ of 12 = 6

We sometimes need to use a calculator to find the **percentage of a quantity**.
83% of £24 is keyed as [83] [÷] [100] [×] [24] [=] or [0·83] [×] [24] [=] or [83] [×] [0·24] [=]

using fractions — using decimals — finding 1% first

To find an **increase of** 12% we can find 12% and add it on or we can find 112%.
To find a **decrease of** 25% we can find 25% and subtract it or we can find 75%.
100% – 25% = 75%

Example Increase 60 by 35%.
35% of 60 = 21
Answer = 60 + 21
= 81

or 135% of 60 = 1·35 × 60
= 81

Practice Questions **12, 13, 21, 29, 33, 34, 35, 37, 38, 39, 40, 46, 52, 53, 55, 62, 65, 68, 72, 76, 78, 79, 81**

Ratio and proportion

8 : 11 and 7 : 2 : 3 are both **ratios**.

Pairs of numbers are in **direct proportion** if the ratio of the pairs is always the same.

Example

Amount of red paint (ℓ)	1	2	3	4	5
Amount of blue paint (ℓ)	2	4	6	8	10

The ratio of amount of red paint to amount of blue paint is always 1 : 2 or $\frac{1}{2}$.

Example If five packets of crisps cost £6·25.
We can find the cost of eight packets by finding the cost of one packet first.
Cost of one packet $= \frac{£6·25}{5} =$ £1·25.
Cost of eight $= 8 \times \frac{6·25}{5} =$ £10·00.

We can find an **equivalent ratio** by multiplying or dividing both parts of the ratio by the same number.

Examples 3 : 7 = 15 : 35 (× 5) 16 : 28 = 4 : 7 (÷ 4)

A ratio is in its **simplest form** when both numbers in the ratio have no common factors other than 1.

Example In its simplest form, the ratio 16 : 28 is 4 : 7.

The numbers in a ratio in its simplest form must not be fractions or decimals.

Examples $4\frac{1}{2} : 5 = 9 : 10$ Both parts have been multiplied by 2 to get whole numbers.
6·2 : 3·4 = 62 : 34 Both parts have been multiplied by 10 to get whole numbers.
= 31 : 17 Now both parts have been divided by 2.

Proportion compares part to whole.
Ratio compares part to part.

Example

The ratio of red to blue is 2 parts to 6 parts or 2 : 6 = 1 : 3.
The proportion that is red is 2 parts out of 8 parts or $\frac{2}{8} = \frac{1}{4}$ or 25% or 0·25.

We can **divide in a given ratio.**

Example Divide £8320 in the ratio 1 : 3 : 4.
There are 1 + 3 + 4 = 8 shares altogether.
One share is $\frac{£8320}{8}$ = £1040.
Three shares is 3 × £1040 = £3120.
Four shares is 4 × £1040 = £4160.
£8320 divided in the ratio 1 : 3 : 4 is £1040, £3120 and £4160.

Practice Questions 27, 28, 36, 49, 50, 51, 58, 59, 64

Practice Questions

Except for questions 68–83.

1 What is the value of the digit 6 in these?
a 8642·31 **b** 6·532 **c** 17·163 **d** 5004·604 **e** 198.346

2 Write these as decimal numbers.
a Three hundred and eighty-six thousand and five and seven tenths.
b Four hundred and sixty-two million, eight hundred and four point seven zero six.
c $18\frac{367}{1000}$.

3 A sea lion gives birth to her cub at ⁻60 m. She then swims up 15 m towards the surface. At what depth is she now?

4 Put these numbers in ascending order.
⁻5·2, 4·7, ⁻2·5, ⁻5·6, ⁻5, 3·4

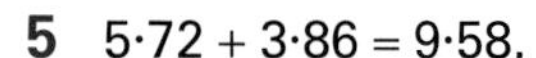

5 5·72 + 3·86 = 9·58.
What is the answer to 9·58 – 3·86?

6 The temperature was 3 °C. It dropped 7 °C then rose 4 °C.
What was the temperature then?

7 a Write the answers. (4 + 2) × 3 =
4 + (2 × 3) =
b Work out the answer to (2 + 4) × (6 + 3 + 1)
c Put brackets in the calculation to make the answer 50.
4 + 5 + 1 × 5
d Now put brackets in the calculation to make the answer 34.
4 + 5 + 1 × 5

8 This table gives the area of some of the world's largest islands.
a Round each area to the nearest 10 000.
b Put the islands in order from largest to smallest.

Island	Area in miles^2
Honshu	87 799
Great Britain	84 195
Sumatra	104 990
Greenland	839 852
Baffin Island	195 916

9 a A football club is planning a trip. **[SATs 2001 paper 1]**
The club hires 234 coaches. Each coach holds 52 passengers.
How many passengers is that altogether?

b The club wants to put one first aid kit into each of the 234 coaches.
The first aid kits are sold in **boxes of 18**.
How many boxes does the club need?

10

a	$11{\cdot}9 \times 100$	**b**	$36 \div 10$	**c**	$5{\cdot}8 \div 100$	**d**	$53{\cdot}6 \times 1000$	**e**	$472 \div 1000$
f	$0{\cdot}06 \times 100$	**g**	$11 \div 1000$	**h**	$0{\cdot}632 \times 10$	**i**	$49{\cdot}63 \times 1000$	**j**	200×60
k	$1600 \div 400$	**l**	$1{\cdot}4 \times 500$	**m**	$0{\cdot}54 \div 90$	**n**	$200 \div 4000$		

11 a A class was divided into seven groups to do an experiment. Each group weighed some powder at the beginning of the experiment. The masses were

1·38 kg, 1·34 kg, 1·287 kg, 1·491 kg, 1·83 kg, 1·8 kg, 1·3 kg

Put these masses in order from heaviest to lightest.

b Melissa's class found the masses of some pieces of rock they found on a geography field trip.

2·0431 kg, 2·08 kg, 2·1346 kg, 2·143 kg, 2·081 kg, 2·09 kg, 2·1 kg

Put these masses in order from heaviest to lightest.

c Find the mean of the masses given in **b**.
Round your answer to the nearest whole number.

12 a Look at these fractions. **[SATs 2001 paper 1]**

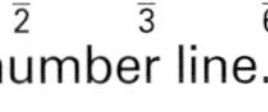

Mark each fraction on the number line.
The first one is done for you

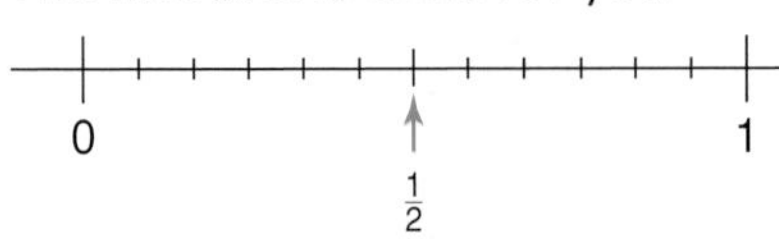

b *Copy these*. Fill in the missing numbers in the boxes.

$\frac{2}{12} = \frac{\square}{6}$ $\frac{1}{2} = \frac{12}{\square}$ $\frac{1}{\square} = \frac{6}{24}$

13 What fraction of
a 1 metre is 57 centimetres **b** 1 hour is 48 minutes?

14 a Adam filled ten test tubes with 16·5 mℓ of copper sulphate solution. How much copper sulphate solution did he need altogether?

b Praiwan measured the thickness of 100 sheets of art card as 568 mm.
How thick is each sheet?

15 Find the answers mentally.

a	$5 + 8 + 16 + 7$	**b**	$18 - 9 + 3 - 4$	**c**	$80 - 20 - 10$	**d**	$90 + 40 - 40$
e	$40 + 50 + 12$	**f**	$80 + 30 - 7$	**g**	$42 + 30$	**h**	$94 - 40$
i	$83 - 29$	**j**	$67 + 19$	**k**	$47 + 17$	**l**	$46 + 8 + 31 + 9$
m	$41 + 26 - 19$	**n**	$175 + 26$	**o**	$196 - 43$	**p**	$295 - 32$
q	$488 + 98$	**r**	$537 - 89$	**s**	$180 + 260$	**t**	$340 - 170$
u	$427 + 166$	**v**	$4073 + 1436$	**w**	$6013 - 2885$		

Mentally means in your head.

16 Rachel filled each of six test tubes with 1·3 mℓ of acid. How much acid did she need altogether?

17 Write down a way to find an estimate for these. Give an alternative way for **a–c**.

a $351 + 297$ **b** $4{\cdot}48 \times 5{\cdot}87$ **c** $1042 - 351$ **d** $362 \div 5{\cdot}2$

e $\dfrac{9{\cdot}3 \times 7{\cdot}2}{6{\cdot}8}$ ***f** $\dfrac{6{\cdot}4 + 3{\cdot}7}{5{\cdot}8 - 3{\cdot}6}$ ***g** $8{\cdot}2 \div (10{\cdot}3 - 6{\cdot}1)$

18 Do these using a written method. Estimate first.

a 58321 + 4386 **b** 8·93 + 4·7 **c** 13804 − 327 **d** 40·81 − 3·2
e 516·3 − 81·97 **f** 5·7 × 9 **g** 1·39 × 7 **h** 15·83 × 4
i 152·6 × 8 **j** 869 × 54 **k** 732 × 51 **l** 7·2 × 19
m 31·6 × 27 **n** 468 ÷ 18 **o** 852 ÷ 24 **p** 416·7 ÷ 9
q 69·4 ÷ 5 **r** 279 ÷ 0·3

19 Find the answers to these mentally.

a 0·8 + 0·6 **b** 0·9 + 0·7 **c** 0·5 + 0·9 **d** 1·2 − 0·4 **e** 1·6 − 0·8
f 1·8 − 0·5 **g** 4·3 + 0·8 **h** 7·8 − 1·2 **i** 14·6 + 7·9 **j** 17·3 − 5·7

20 What must be added or subtracted to or from 19·75 g of salt to get

a 19·95 g **b** 19·71 g **c** 19·68 g **d** 20·25 g **e** 18·95 g?

21 What goes in the gaps?

Fraction	Decimal	Percentage	Fraction	Decimal	Percentage
a	b	45%	$1\frac{7}{20}$	g	h
c	0·375	d	i	j	$12\frac{1}{2}$%
$\frac{43}{50}$	e	f	k	0·045	l

22 A school wants to divide 873 students into groups of 16.

a How many groups of 16 are there?
b How many students are in the last group?

23 Copy this magic square
Fill it in.

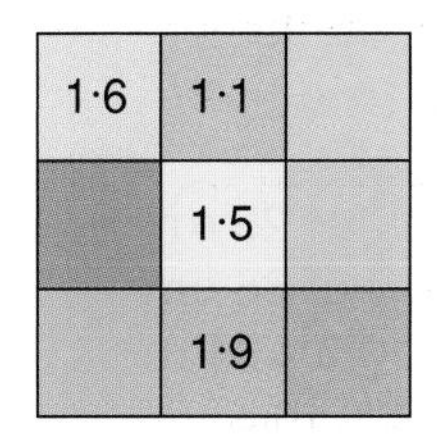

24 Which of <, > or = goes in the box?

a 550 mℓ □ 5 ℓ **b** 4500 m □ 4·5 km **c** 8420 g □ 8·4 kg **d** 7·23 cm □ 723 mm

25 a Which of these is divisible by 18? Explain how you can tell.

1872 5463 1478

b Which of these is divisible by 24?

2320 3024 2274

26 Find the answers to these mentally

a 390 ÷ 3 **b** 1950 ÷ 2 **c** 7·2 ÷ 9 **d** 5·6 ÷ 8 **e** 3·6 ÷ 6

27 Three oranges cost £1·10.
How much do nine oranges cost?

28 a What is the ratio of fertiliser to water in the mix?
b What proportion of the mix is fertiliser?
Give your answer as a percentage.
c What proportion of the mix is water?
Give your answer as a fraction.

Plant Food Mix
2 litres of fertiliser
8 litres of water

29 a Copy and complete this. $\frac{3}{4} = \frac{\square}{8} = \frac{9}{\square} = \frac{\square}{24} = \frac{\square}{60}$.

b Simplify these to their lowest terms by cancelling.

i $\frac{5}{10}$ **ii** $\frac{8}{12}$ **iii** $\frac{16}{20}$ **iv** $\frac{21}{28}$ **v** $\frac{27}{45}$ **vi** $\frac{32}{40}$

30 Cam was building a set for a play. He wrote down one of the measurements he needed as 4·93 m.
What should it have been if the measurement he wrote down is

a 0·1 m too long **b** 0·01 m too short
c 0·1 m too short **d** 0·01 m too long?

31 a Round 68 923 417 to the nearest **(i)** hundred thousand **(ii)** million.

b Round 796 429 to the nearest **(i)** hundred thousand **(ii)** million.

32 Find the answers to these mentally.

a $\sqrt{4}$ **b** $\sqrt{9}$ **c** $\sqrt{1}$ **d** $\sqrt{100}$ **e** $\sqrt{64}$ **f** $\sqrt{144}$

33 Write these as fractions in their lowest terms. **a** 0·655 **b** 0·045

34 Write these as decimals. Round **c** to 2 d.p.
a $\frac{17}{25}$ **b** $\frac{15}{30}$ **c** $\frac{2}{7}$ **d** $\frac{33}{6}$ **e** $4\frac{3}{8}$ **f** $\frac{1}{3}$

35 Max got 83%. Paulo got 19 out of 23. Who got the higher mark?

36 On a class trip there were 30 pupils, 6 adults and 4 pre-school children.
What is the ratio, in its simplest form, of pupils to adults to pre-school children?

37 Do these mentally.

a $\frac{1}{10}$ of £320 **b** $\frac{1}{5}$ of 25 m **c** $\frac{2}{7}$ of 140 cm **d** $\frac{3}{4}$ of 16 ℓ
e $\frac{2}{3}$ of £240 **f** $\frac{5}{6}$ of 36 cm **g** $\frac{7}{8}$ of 240 g **h** $\frac{1}{4}$ of £22
i $1\frac{1}{2}$ of 12 ℓ **j** 0·2 × 24 m **k** 0·7 × 50 kg

38 a What **fraction** of this shape is shaded?
Write your fraction as simply as possible. **[SATs 2001 paper 1]**

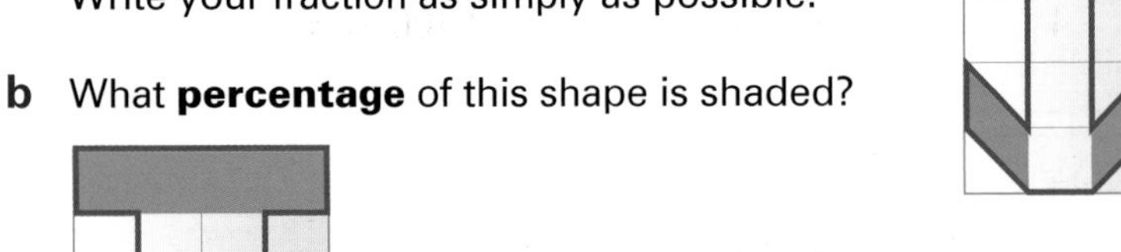

b What **percentage** of this shape is shaded?

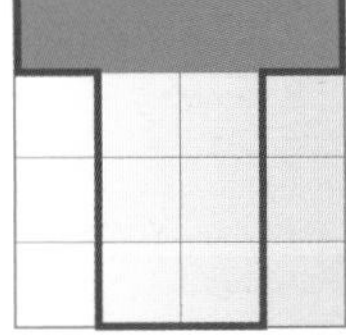

c Which shape has the **greater percentage** shaded?

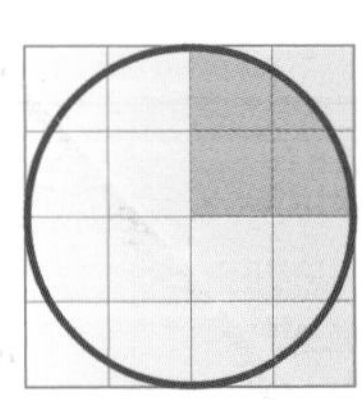

Shape A

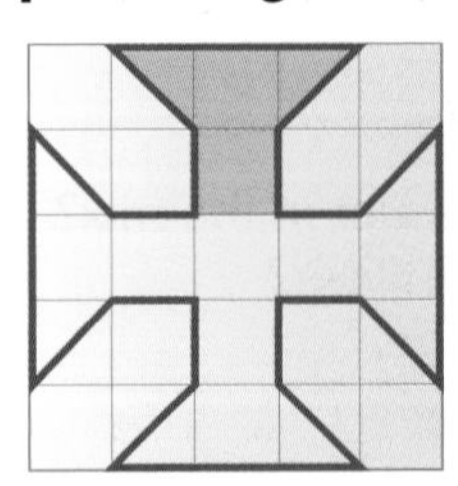

Shape B

Explain how you know.

39 This table shows some percentages of amounts of money. **[SATs 2000 paper 1]**
You can use the table to help you work out the missing numbers.

	£10	£30	£45
5%	50p	£1·50	£2·25
10%	£1	£3	£4·50

15% of £30 = £ ____

£6·75 = 15% of £ ____

£3·50 = ____ % of £10

25p = 5% of £____

40 Write the equivalent fraction with a denominator of 48.
a $\frac{1}{2}$ **b** $\frac{3}{4}$ **c** $\frac{2}{3}$ **d** $\frac{5}{6}$ **e** $\frac{7}{12}$

41 What is 30^2?
A 300 **B** 3000 **C** 3030 **D** 60 **E** 900

T

42 Use a copy of this.
Shade triangles with numbers that are
a divisible by 10 **b** divisible by 15
c divisible by 24 **d** divisible by 18.
What shape does the shading make?

2376 / 7970	6840 / 2470
7938 / 1275	3864 / 5830
7305 / 4872	8304 / 4824
1031 / 3587	5177 / 2869

43 **a** 240 people paid the entrance fee on Monday. **[SATs 2000 paper 1]**
How much money is this altogether?
Show your working.
b The museum took £600 in entrance fees on Friday.
How many people paid to visit the museum on Friday?
Show your working.

Museum
entrance fee
£1·20 per person

44 Calculate these mentally.
a $\sqrt{101+20}$ **b** $\sqrt{160-16}$ **c** $\sqrt{10^2-19}$ **d** $\sqrt{40-2^2}$
e $\sqrt{50-5^2}$ **f** $\sqrt{3^2+4^2}$ **g** $\sqrt[3]{20+7}$ **h** $\sqrt[3]{100-36}$

45 4 | $^-3$ | 0 | 9 | $^-6$ | 2 | $^-5$ | 10

a Choose a number card to give the answer 7. $^-5$ + 3 + ☐ = 7

b Choose two cards to give the lowest possible total. ☐ + ☐ = ____
What is this total?

c Choose two number cards to multiply to give 30. ☐ × ☐ = 30

d Choose two number cards to divide to give $^-2$. ☐ ÷ ☐ = $^-2$

46 Write these as percentages.
a $\frac{1}{4}$ **b** $\frac{7}{10}$ **c** $\frac{3}{5}$ **d** $\frac{17}{20}$ **e** 4 **f** $1\frac{3}{4}$ **g** 0·73
h 0·86 **i** 0·04 **j** 0·19 **k** 1·14

47 Calculate these mentally.
a $4 \times 7 - 3$ **b** $8 + 3 \times 4$ **c** $16 - 2 \times 7$ **d** $\frac{9+7}{8}$ **e** $\frac{36}{3 \times 2}$
f $4^2 - 1$ **g** $5(5^2 - 16)$ **h** 3×4^2 **i** $8^2 - 10$ **j** $\frac{4^2+9}{5}$
k $(2+9)^2$ **l** $8^2 - 3^2$ **m** $12^2 - 4^3$ **n** $\sqrt[3]{64} + 2$ **o** $3^3 - 3$

48 The population of Guildford to the nearest thousand is 127 000.

a What is the smallest number of people who might live there?

b What is the largest number of people?

49 Screenwash is used to clean car windows. **[SATs 2002 paper 2]**

To use Screenwash you mix it with water.

Winter mixture
Mix 1 part Screenwash with 4 parts water.

Summer mixture
Mix 1 part Screenwash with 9 parts water.

a In **winter**, how much water should I mix with **150 mℓ of Screenwash**?

b In **summer**, how much Screenwash should I mix with **450 mℓ of water**?

c Is this statement correct? Explain your answer.

25% of **winter** mixture is **Screenwash**

50 Matthew and Olivia share a 60 hour job in the ratio 7 : 3.
How many hours does Matthew work?

51 The ratio of sugar to flour in a recipe is 2 : 3.
How much flour is needed with this amount of sugar?

a 200 g **b** 100 g **c** 400 g **d** 500 g

52 **a** $\frac{1}{3} \times 18$ **b** $20 \times \frac{3}{5}$ **c** $\frac{3}{8} \times 32$ **d** $1\frac{1}{4} \times 16$ **e** $5 \times \frac{1}{3}$
f $\frac{2}{3} \times 8$ **g** $1\frac{1}{2} \times 5$ **h** $3{\cdot}1 \times 20$ **i** $0{\cdot}6 \times 60$

53 **a** $\frac{1}{2} + \frac{3}{4}$ **b** $\frac{3}{8} + \frac{3}{8}$ **c** $\frac{11}{12} - \frac{7}{12}$ **d** $\frac{3}{4} + \frac{1}{8}$ **e** $1 - \frac{2}{3}$
f $\frac{2}{3} + 1\frac{1}{3}$ **g** $\frac{3}{12} + \frac{8}{12} + \frac{5}{12}$ **h** $\frac{5}{6} - \frac{3}{6} + \frac{4}{6}$

i Olivia mixed $\frac{11}{12}$ ℓ of red paint with $\frac{5}{12}$ ℓ of blue paint.
How much paint did she have then?

54 **a** Two numbers **multiply** together to make $^-15$. **[SATs 2000 paper 1]**
They **add** together to make **2**.
What are the two numbers?

b Two numbers **multiply** together to make $^-15$,
but **add** together to make $^-$**2**.
What are the two numbers?

c Two numbers **multiply** together to make **8**,
but **add** together to make $^-$**6**.
What are the two numbers?

55 On a farm **80** sheep gave birth. **[SATs 2001 paper 1]**
30% of the sheep gave birth to two lambs.
The rest of the sheep gave birth to just one lamb.
In total, how many lambs were born?
Show your working.

56 What are the missing numbers in this factor tree?

136
2 A
B C
D E

57 Place the numbers from $^{-}3$ to 5 in the circles, so that each line has the total given.
a 3 **b** 6 **c** 0

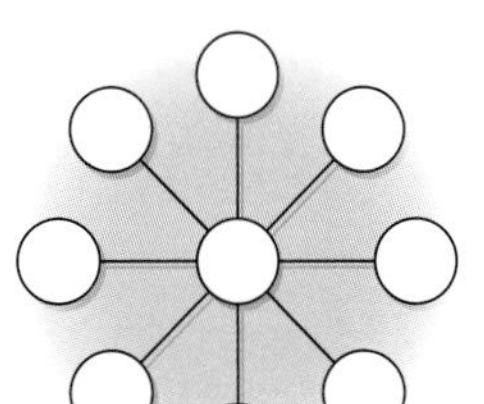
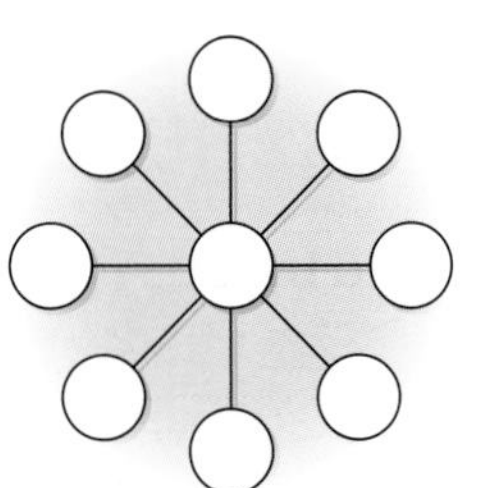
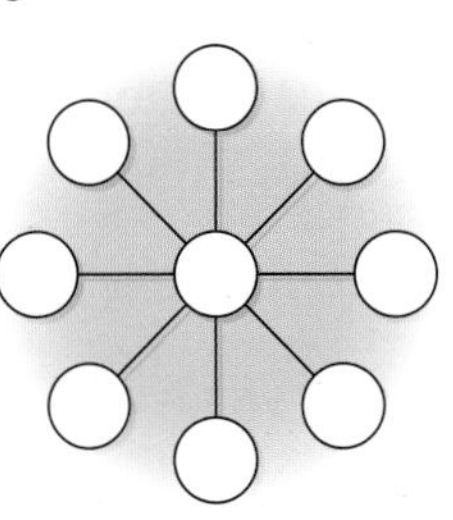

58 A recipe for mushroom pie uses 850 g of mushrooms for 4 people.
What mass of mushrooms are needed for 7 people?

59 Divide £750 in the ratio 2 : 2 : 1.

60 Use a table or factor tree to write these as products of prime factors in index notation.
a 40 **b** 225 **c** 1225

61 a What is the LCM (lowest common multiple) of 54 and 81?
b What is the HCF (highest common factor) of 54 and 81?

62 At both Direct Lighting and Lamplighter a lava lamp was originally £89.
a Which shop is cheaper now and by how much?
b Lamplighter reduces their prices by a **further** 15%.
How much is the lava lamp then?

63 Cancel these by finding the HCF.
a $\frac{45}{108}$ **b** $\frac{39}{65}$ **c** $\frac{28}{72}$ **d** $\frac{112}{180}$.

64 A salad is made from potatoes, pumpkin and carrots in the ratio 7 : 2 : 4.
Samir used 1000 g of carrots to make the salad.
What was the total mass of Samir's salad?

65 a How many eighths in 6?
b How many two-thirds in 6?
c $24 \times \frac{3}{4} = 18$ so $18 \div \frac{3}{4} = \square$
d Find the answer to $12 \div \frac{4}{5}$. Use a diagram to help.

66 Two prime numbers are subtracted.
The answer is 4.
What could the numbers be?
Find as many pairs as possible using prime numbers less than 50.

67 A four-digit square number has a units digit of 4.

 4

What must the square root of this number have as its units digit? Explain your choice.
A 4 **B** 2 **C** 2 or 8 **D** can't tell.

You may use a calculator for questions 68 to 83.

68 Write these as recurring decimals. **a** $\frac{2}{3}$ **b** $\frac{4}{9}$ **c** $\frac{5}{11}$.

69 Test whether these numbers are prime. Write Yes or No.
a 47 **b** 57 **c** 61 **d** 87 **e** 97 **f** 117

70 Use your calculator to find these.
a 13^2 **b** $\sqrt{196}$ **c** $1{\cdot}2^2$ **d** $\sqrt{1{\cdot}96}$

71 Find the remainder when 597 is divided by 37.

72 a 8% of £26.50 = £ [SATs 2000 paper 2]
b $12\frac{1}{2}$% of £98 = £

73 Calculate these. Give the answer to **d** to 1 d.p.
a $(42 + 63) \times 42$ **b** $5{\cdot}1 + (4{\cdot}7 - 0{\cdot}96)$ **c** $\frac{14{\cdot}7}{3{\cdot}7 + 1{\cdot}2}$ **d** $\frac{5{\cdot}6 + 7{\cdot}3}{7{\cdot}4 - 4{\cdot}8}$

74 Maria did these calculations. Check the answers using inverse operations. Show how you did this.
a $426 \times 72 = 30\,672$ **b** $8{\cdot}36 + 7{\cdot}94 = 16{\cdot}3$ **c** $\sqrt{7} = 2{\cdot}645751311$.

75 Check the answer to question **74a** using an equivalent calculation.
Show how you did this.

76 a 27% of £65 **b** 34% of 82 ℓ **c** 76% of £2573 **d** 17·5% of £27.

77 Convert
a 197 hours to days and hours **b** 572 seconds to minutes and seconds.

78 a The label on yoghurt A shows this information [SATs 2001 paper 2]
How many grams of **protein** does **100 g** of yoghurt A provide?
Show your working.

Yoghurt A **125g**	
Each 125 g provides	
Energy	430 kJ
Protein	4·5 g
Carbohydrate	11·1 g
Fat	4·5 g

b The label on yoghurt B shows different information.
A boy eats the same amount of yoghurt A and yoghurt B.
Which yoghurt provides him with more **carbohydrate**?
Show your working.

Yoghurt B **150g**	
Each 150 g provides	
Energy	339 kJ
Protein	6·6 g
Carbohydrate	13·1 g
Fat	0·2 g

79 Janna bought a stereo for £385. She sold it a year later at a loss of 19%. For what price did she sell it?

80 Hannah went on a cycling holiday. [SATs 2002 paper 2]
The table shows how far she cycled each day.

Monday	Tuesday	Wednesday	Thursday
32·3 km	38·7 km	43·5 km	45·1 km

Hannah says: 'On average I cycled **over 40 km** a day'.
Show that Hannah is wrong.

81 A garden centre sells plants for hedges.
The table shows what they sold in one week.

[SATs 2000 paper 2]

Plants	Number of plants sold	Takings
Beech	125	£212·50
Leylandii	650	£2437·50
Privet	35	£45·50
Hawthorn	18	£23·40
Laurel	5	£32·25
Total	**833**	**£2751·15**

a What percentage of the total number of plants sold was **Leylandii**?
Show your working.

b What percentage of the **total takings** was for Leylandii?
Show your working.

c Which is the **cheaper** plant, Beech or Privet?
Show working to explain how you know.

82 A drink from a machine costs 55p

[SATs 2001 paper 2]

The table shows the coins that were put into the machine one day.

Coins	Number of coins
50p	31
20p	22
10p	41
5p	59

How many cans of drink were sold that day?
Show your working.

83 A company sells and processes films of two different sizes.
The tables show how much the company charges.

[SATs 2002 paper 2]

Film size: **24** photos	
Cost to **buy** each film	£2·15
Postage	free
Cost to **print** each film	£0·99
Postage for each film	60p

Film size: **36** photos	
Cost to **buy** each film	£2·65
Postage	free
Cost to **print** each film	£2·89
Postage for each film	60p

I want to take **360** photos.
I need to buy the film, pay for the film to be printed, and pay for the postage.
Is it cheaper to use all films of size 24 photos, or all films of size 36 photos?
How much cheaper is it? Show your working.

1 Place Value, Ordering and Rounding

You need to know

✓ place value — page 1
✓ multiplying and dividing by 10, 100, 1000 ... — page 1
✓ putting decimals in order — page 1
✓ rounding to the nearest 10, 100 and 1000, and to the nearest whole number — page 2

Key vocabulary

billion, descending, exponent, index, power of ten, recurring decimal

In days gone by

- The Mayans lived in Central America. They were one of the first people to use place value. They used sticks and stones to show numbers.

0 1 2 3 4 5

6 was written as (five and one)

7 was written as (five and two)

We count in lots of 10. The Mayans counted in lots of 20. Place value was shown by putting one place above the other.

52 was written as 2 lots of 20 (40)

and

2 lots of 5 and 2 ones (12)

60 was written as

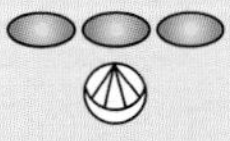

47 was written as

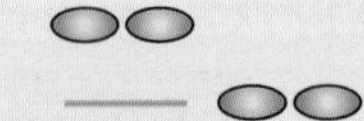

Write these numbers using Mayan symbols.

9 16 23 57 34 68 92

How might the Mayans add and subtract these?

$8 + 4$ $20 - 3$ $25 + 36$ $32 + 54$ $57 - 32$

Now try some with bigger numbers.

- Find out about another number system. You could choose

  Greek or *Chinese* or *Roman*.

Powers of ten

$10 = 10^1$
$100 = 10 \times 10 = 10^2$
$1000 = 10 \times 10 \times 10 = 10^3$
$10\,000 = 10 \times 10 \times 10 \times 10 = 10^4$

$10^1, 10^2, 10^3, 10^4, \ldots$ are called **powers of ten**.

The little $^1, ^2, ^3, ^4$ of $10^1, 10^2, 10^3, 10^4 \ldots$ are called **indices**. Indices is the plural of **index**.

Discussion

$10^2 = 10 \times 10$ so $\frac{1}{10^2} = \frac{1}{10 \times 10}$
$= \frac{1}{100}$
$= 0{\cdot}01$

Use this keying sequence to find 10^{-2}.

Compare the answers for $\frac{1}{10^2}$ and 10^{-2}.

What if you compare $\frac{1}{10^3}$ with 10^{-3}?
What if you compare $\frac{1}{10^4}$ with 10^{-4}?
What if ...

Make a statement about $\frac{1}{10^n}$ and 10^{-n}. **Discuss.**

This **place value chart** shows the *base 10* number system.
Each place to the left in a number is one power of ten bigger.

	Millions			Thousands			Hundreds					
(Thousands of millions) Billions	Hundreds of millions	Tens of millions	Millions	Hundreds of thousands	Tens of thousands	Thousands	Hundreds	Tens	Units	• tenths	hundredths	thousandths
10^9	10^8	10^7	10^6	10^5	10^4	10^3	10^2	10^1	10^0	• 10^{-1}	10^{-2}	10^{-3}

Note: $10^1 = 10$ $\quad 10^0 = 1$ $\quad 10^{-1} = \frac{1}{10^1} = \frac{1}{10}$ $\quad 10^{-2} = \frac{1}{10^2} = \frac{1}{100}$ $\quad 10^{-3} = \frac{1}{10^3} = \frac{1}{1000} \ldots$

The Earth is about one hundred and forty-nine million, six hundred thousand kilometres from the Sun.
We write this distance as 149 600 000 km or as $1{\cdot}496 \times 10^8$ km.

We say $1{\cdot}496 \times 10^8$ as 'one point four nine six times ten to the power of eight'.

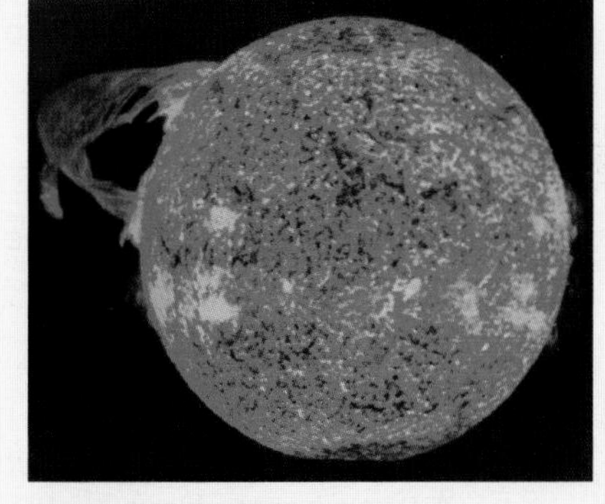

The thickness of a piece of thick paper is 0·0089 mm. We can write this as $8{\cdot}9 \times 10^{-3}$ mm.
We say $8{\cdot}9 \times 10^{-3}$ as 'eight point nine times ten to the power of negative three'.

A **kilo**metre is 1000 metres. **Kilo** is 10^3.
A **centi**litre is $\frac{1}{100}$ of a litre. **Centi** is 10^{-2}.

These **prefixes** are associated with the given powers of 10.

power	prefix		power	prefix
10^9	giga		10^{-2}	centi
10^6	mega		10^{-3}	milli
10^3	kilo		10^{-6}	micro
			10^{-9}	nano
			10^{-12}	pico

The prefixes are usually used with the following units.

metre (m) gram (g) litre (ℓ) second (s)

Practical

A computer had 20 gigabytes of memory.

Find some other examples of measurements which use one of the prefixes given in the table above.

Exercise 1

1 Write these as a whole number or a decimal.

a 10^2 **b** 10^4 **c** 10^1 **d** 10^3 **e** 10^8 **f** 10^9 **g** 10^{11}
h 10^0 **i** 10^{-2} **j** 10^{-1} **k** 10^{-4} **l** 10^{-6} **m** 10^{-9}

2 Write these as a power of ten.

a a thousand **b** a million **c** a hundred **d** a hundred thousand
e ten million **f** a billion **g** ten **h** ten billion
i one tenth **j** one **k** one thousandth *l one hundred-thousandth
*m one millionth

3 Copy and complete these. The first one has been done for you.

a $10^{-1} = \frac{1}{10^1} = \frac{1}{10} = 0{\cdot}1$ **b** $10^{-2} = \frac{1}{\_\_} = \frac{1}{100} = \_\_\_$ **c** $\_\_ = \frac{1}{10^4} = \_\_\_ = \_\_\_$
d $\_\_\_ = \_\_\_ = \frac{1}{1000} = \_\_\_$

4 Write the numbers in red in words.

a A heart beats about **37 000 000** times each year.
b The average person will blink about **5 625 000** times each year.
c One megabyte is **1 048 576** bytes.
d The African continent has an area of **30 271 000** km^2.
e The Galaxy NGC2207 is **114 100 000** light years from Earth.

5 Write the numbers in red in words.

- **a** The world's highest observatory is $\mathbf{4{\cdot}3 \times 10^{3}}$ m above sea level.
- **b** Saturn's rings are $\mathbf{2{\cdot}7 \times 10^{5}}$ km in diameter.
- **c** Jupiter's largest moon is $\mathbf{5{\cdot}262 \times 10^{3}}$ km in diameter.
- **d** Jupiter's largest moon is $\mathbf{1{\cdot}07 \times 10^{6}}$ km from Jupiter.
- **e** Saturn is $\mathbf{1{\cdot}427 \times 10^{9}}$ km from the Sun.
- **f** The thickness of a piece of paper is about $\mathbf{6 \times 10^{-2}}$ cm.
- **g** The thickness of a hair is about $\mathbf{1{\cdot}5 \times 10^{-5}}$ m.
- **h** The Sun is about $\mathbf{1{\cdot}6 \times 10^{-6}}$ light years from Earth.
- **i** The diameter of an atom is about $\mathbf{1{\cdot}1 \times 10^{-10}}$ mm.

This is linked to science.

T

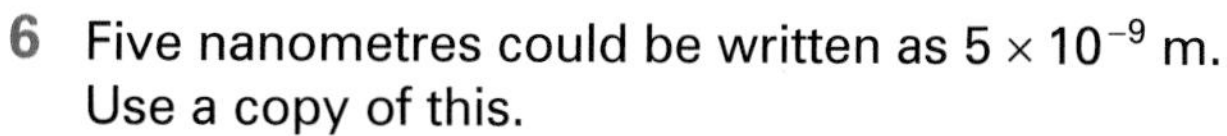

6 Five nanometres could be written as 5×10^{-9} m.
Use a copy of this.
Find an equivalent in the square for each of these measurements. Shade the box it is in.
The shading makes a letter. Which letter?

- **a** 2 gigabytes
- **b** 2 megabytes
- **c** 2 kilobytes
- **d** 8 centimetres
- **e** 8 millimetres
- **f** 8 micrometres
- **g** 8 picometres
- **h** 8 nanometres
- ***i** 20 gigabytes
- ***j** 200 kilobytes
- ***k** 800 nanometres
- ***l** 80 micrometres
- ***m** 2 bytes

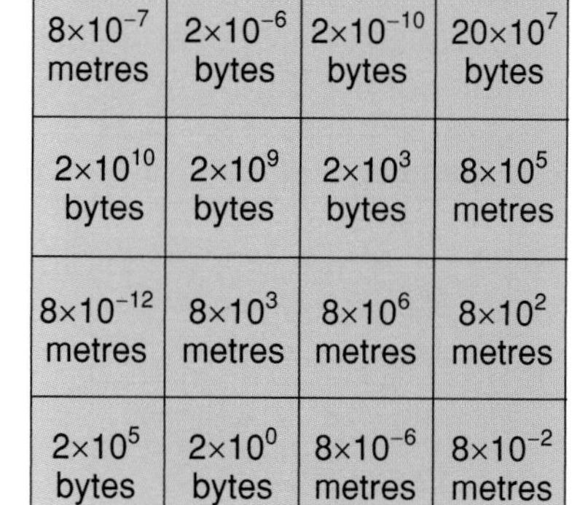

8×10^{-3} metres	8×10^{-5} metres	2×10^{6} bytes	8×10^{-9} metres
8×10^{-7} metres	2×10^{-6} bytes	2×10^{-10} bytes	20×10^{7} bytes
2×10^{10} bytes	2×10^{9} bytes	2×10^{3} bytes	8×10^{5} metres
8×10^{-12} metres	8×10^{3} metres	8×10^{6} metres	8×10^{2} metres
2×10^{5} bytes	2×10^{0} bytes	8×10^{-6} metres	8×10^{-2} metres

Review 1 Write these as powers of ten.

- **a** ten thousand
- **b** a hundred million
- **c** a thousand million
- **d** one ten-thousandth
- ***e** one thousand thousandth

Review 2 Write the numbers in red in words.

- **a** Neptune takes a bit more than $\mathbf{1{\cdot}6 \times 10^{2}}$ Earth years to orbit the Sun.
- **b** Neptune's largest moon is $\mathbf{2{\cdot}71 \times 10^{3}}$ km in diameter.
- **c** Neptune is $\mathbf{4{\cdot}497 \times 10^{9}}$ km from the Sun.
- **d** The wavelength of light we can see is $\mathbf{5{\cdot}0 \times 10^{-5}}$ cm.
- **e** The mass of a magnesium atom is $\mathbf{4 \times 10^{-23}}$ g.

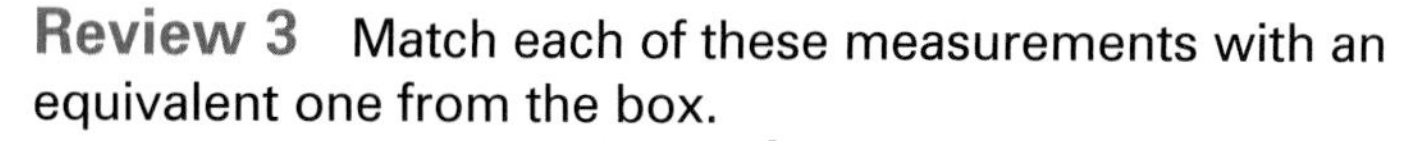

Review 3 Match each of these measurements with an equivalent one from the box.

- **a** 9 nanoseconds
- **b** 9 milliseconds
- **c** 9 microseconds
- **d** 9 picoseconds
- **e** 900 nanoseconds

A 9×10^{-3} seconds
B 9×10^{-12} seconds
C 9×10^{-7} seconds
D 9×10^{-9} seconds
E 9×10^{-6} seconds

Puzzle

I am a 6-digit number between 400 and 500.
All of my digits are different.
The difference between my units digit and my thousandths digit is the same as the difference between my tens digit and my hundredths digit.
My tenths digit is three times my hundredths digit.
My units digit is twice my tenths digit.
The sum of my digits is 21.
What number am I?

Multiplying and dividing by powers of ten

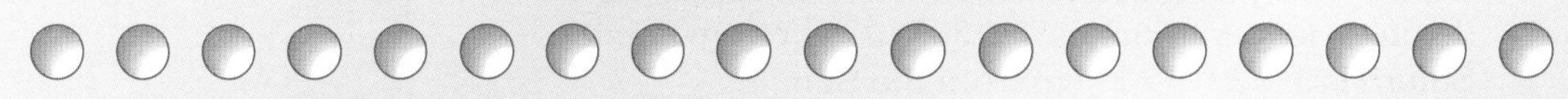

Investigation

× and ÷ by 0·1 and 0·01

You will need a calculator or spreadsheet package.

T

Use a copy of the table.
Fill it in.
Use a calculator or spreadsheet to help. If you use a spreadsheet ask your teacher for a copy of the Bigger or Smaller ICT Worksheet.

Number	× 10	÷ 0·1	× 100	÷ 0·01	÷ 10	× 0·1	÷ 100	× 0·01
4·7	47	47	470	470				
0·8	8	8						
0·92	9.2	9.2						
16	160							
12·3	123							
572	5720							
46·23	462.3							
5271	52710							

If you use a spreadsheet, you can change the numbers in this column.

1 What do you notice about multiplying by 10 and dividing by 0·1?
What do you notice about multiplying by 100 and dividing by 0·01?
What about dividing by 10 and multiplying by 0·1?
What about dividing by 100 and multiplying by 0·01?

2 From what you have found out in **part 1** above, what can you deduce about
multiplying by 0·001 **and** dividing by 0·001
multiplying by 0·0001 **and** dividing by 0·0001?

3 Notice that 4·7 × 10 = 47 and 4·7 × 0·1 = 0·47.

Multiplying by 10 gives a larger answer than 4·7.
Multiplying by 0·1 gives a smaller answer than 4·7.

Is it true that all positive numbers,
when multiplied by 10 get larger
when multiplied by 0·1 get smaller?

What if we multiply by 0·01?
What if we divide by 0·1?
What if we divide by 0·01?

Which of **smaller** or **bigger** goes in the gap?

Multiplying by 0·1 or 0·01 makes a positive number _______ .
Dividing by 0·1 or 0·01 makes a positive number _______ .

Discussion

- 

Robert Nishi

Nishi wrote

$5{\cdot}6 \times 0{\cdot}1 = 5{\cdot}6 \times \frac{1}{10}$
$= 5{\cdot}6 \div 10$
so $5{\cdot}6 \times 0{\cdot}1$ is the same as $5{\cdot}6 \div 10$.

Use Nishi's way to explain why multiplying by 0·01 is the same as dividing by 100.

- 

Robert

Nishi began like this.

$18 \div 0{\cdot}1$
$0{\cdot}1$ is the same as $\frac{1}{10}$. So $18 \div 0{\cdot}1 = 18 \div \frac{1}{10}$.
This is read as 'how many 0·1 s (or tenths) in 18?'
There are 10 tenths in 1, so in 18 there are 18×10 tenths.
So $18 \div 0{\cdot}1 = 18 \times 10$.

Use Nishi's way to explain why dividing by 0·01 is the same as multiplying by 100.

- Nishi wanted to know the answer to $5 \times 0{\cdot}6$.
Robert wrote

$5 \times 0{\cdot}6 = 5 \times 6 \times 0{\cdot}1$
$= 5 \times 6 \div 10$
$= 30 \div 10$
$= 3$

Is Robert right?
Use Robert's way to find the answer to **0·6** × **0·7**.

* ● Nishi wanted to know the answer to $0{\cdot}4 \times 0{\cdot}3$.
She used this diagram.
There are 100 squares and each small square has side 0·1.
She wrote

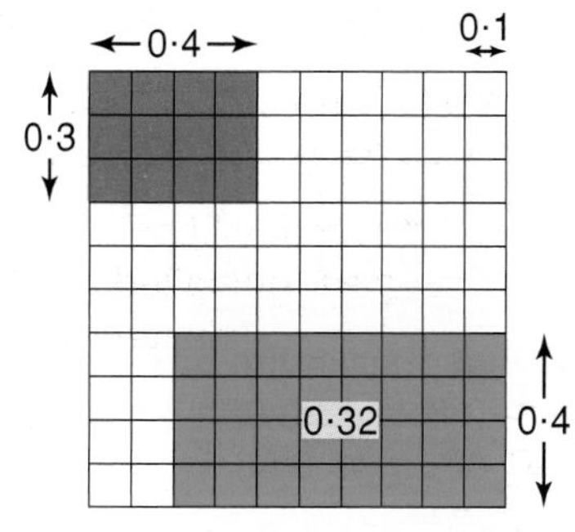

The red rectangle is $0{\cdot}4 \times 0{\cdot}3$.
It has 12 squares.
12 out of 100 is $\frac{12}{100}$ or 0·12.
$0{\cdot}4 \times 0{\cdot}3 = 0{\cdot}12$.

To find the answer to $0{\cdot}32 \div 0{\cdot}4$, she wrote

The blue rectangle is $0{\cdot}4 \times ?$
It has 32 squares (out of 100). This is 0·32.
The missing length is 8 squares long.
Each square has side 0·1, so 8 squares is 0·8.
$0{\cdot}32 \div 0{\cdot}4 = 0{\cdot}8$

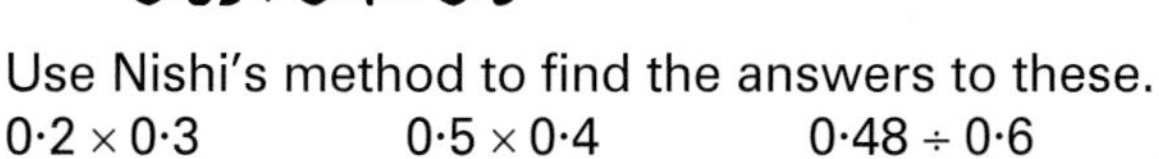

Use Nishi's method to find the answers to these.
$0{\cdot}2 \times 0{\cdot}3$ $0{\cdot}5 \times 0{\cdot}4$ $0{\cdot}48 \div 0{\cdot}6$ $0{\cdot}28 \div 0{\cdot}7$

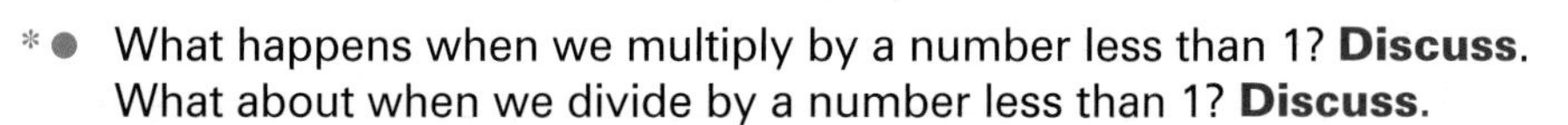

* ● What happens when we multiply by a number less than 1? **Discuss**.
What about when we divide by a number less than 1? **Discuss**.

Multiplying by 0·1 is the same as dividing by 10.
Multiplying by 0·01 is the same as dividing by 100.
Dividing by 0·1 is the same as multiplying by 10.
Dividing by 0·01 is the same as multiplying by 100.

To remind yourself about multiplying by 10, 100, 1000, ... see page 1.

Worked Example

a $34{\cdot}2 \div 0{\cdot}01$ b $0{\cdot}86 \times 0{\cdot}1$ c $^{-}7 \times 0{\cdot}4$ d $0{\cdot}3 \times 0{\cdot}1$

Answer

a $34{\cdot}2 \div 0{\cdot}01 = 34{\cdot}2 \times 100$
$= \mathbf{3420}$

b $0{\cdot}86 \times 0{\cdot}1 = 0{\cdot}86 \div 10$
$= \mathbf{0{\cdot}086}$

c $^{-}7 \times 0{\cdot}4 = {}^{-}7 \times 4 \times 0{\cdot}1$
$= {}^{-}28 \div 10$
$= \mathbf{{}^{-}2{\cdot}8}$

d $0{\cdot}3 \times 0{\cdot}1 = 0{\cdot}3 \div 10$
$= \mathbf{0{\cdot}03}$

or $0{\cdot}3 \times 0{\cdot}1 = \frac{3}{10} \times \frac{1}{10}$
$= \frac{3}{100}$
$= \mathbf{0{\cdot}03}$

Worked Example

Convert 5 mm^2 to cm^2.

See the area conversions on page 336.
1 mm^2 = 0·01 cm^2.

Answer

5 mm^2 = $5 \times 0{\cdot}01$ cm^2
= **0·05 cm^2**

because 1 cm^2 = 10 mm × 10 mm
= 100 mm^2
So 1 mm^2 = $\frac{1}{100}$ cm^2
= 0·01 cm^2

Exercise 2

1 a $6{\cdot}42 \times 100$ b $86{\cdot}2 \div 100$ c $56 \times 10\,000$ d $5{\cdot}9 \div 1000$
e $9062 \times 10\,000$ f $5872 \div 10\,000$ g $0{\cdot}68 \div 1000$ h $5{\cdot}4 \times 1\,000\,000$
i $0{\cdot}068 \div 1000$ j $5{\cdot}68327 \times 1\,000\,000$ k $5{\cdot}6 \times 400$ l $8{\cdot}6 \div 2000$
m $0{\cdot}48 \times 700$ n $560 \div 80\,000$ o $49{\cdot}7 \div 70\,000$

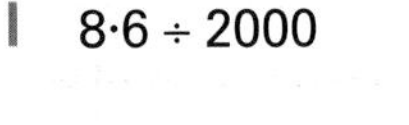

Let's start with some practice at × and ÷ by 10, 100, ...

2 a Which of these is equivalent to $4{\cdot}8 \div 0{\cdot}01$?
A $4{\cdot}8 \times 10$ B $4{\cdot}8 \div 10$ C $4{\cdot}8 \div 100$ D $4{\cdot}8 \times 100$
b Which of these is equivalent to $864 \times 0{\cdot}01$?
A 864×100 B $864 \div 100$ C 864×10 D $864 \div 0{\cdot}01$
c Which of these is equivalent to $0{\cdot}024 \times 0{\cdot}1$?
A $0{\cdot}024 \times 10$ B $0{\cdot}024 \div 100$ C $0{\cdot}024 \div 10$ D $0{\cdot}024 \div 0{\cdot}1$

3 a $8 \times 0{\cdot}1$ b $32 \times 0{\cdot}1$ c $5 \div 0{\cdot}1$ d $17 \times 0{\cdot}1$ e $832 \times 0{\cdot}01$ f $0{\cdot}86 \div 0{\cdot}1$
g $4{\cdot}8 \times 0{\cdot}01$ h $7{\cdot}653 \div 0{\cdot}1$ i $82 \div 0{\cdot}01$ j $5 \times 0{\cdot}4$ k $6 \times 0{\cdot}2$ l $4 \times 0{\cdot}7$
m $7 \times 0{\cdot}5$ n $3 \times 0{\cdot}9$ o $5 \times 0{\cdot}6$ p $^{-}8 \times 0{\cdot}4$ q $^{-}6 \times 0{\cdot}8$ r $^{-}7 \times 0{\cdot}9$

4 What goes in the box?
a $0{\cdot}8 \times \square = 0{\cdot}08$ b $5{\cdot}6 \times \square = 56\,000$ c $7{\cdot}2 \times \square = 0{\cdot}072$ d $0{\cdot}861 \div \square = 86{\cdot}1$
e $56 \times \square = 0{\cdot}56$ *f $0{\cdot}7 \times \square = 0{\cdot}14$ *g $42 \times \square = 8{\cdot}4$ *h $1{\cdot}6 \times \square = 0{\cdot}064$

5 In a science experiment Lyle filled each of six beakers with 0·03 ℓ of a salty solution. He then added 0·006 ℓ of acid solution to each beaker. How much solution did he use in total?

6 Write down the next 3 lines of these patterns.

a $0{\cdot}5 \times 7 = 3{\cdot}5$
$0{\cdot}5 \times 0{\cdot}7 = 0{\cdot}35$
$0{\cdot}5 \times 0{\cdot}07 = 0{\cdot}035$

b $0{\cdot}4 \times 9 = 3{\cdot}6$
$0{\cdot}4 \times 0{\cdot}9 = 0{\cdot}36$
$0{\cdot}4 \times 0{\cdot}09 = 0{\cdot}036$

c $0{\cdot}8 \times 0{\cdot}7 = 0{\cdot}56$
$0{\cdot}08 \times 0{\cdot}7 = 0{\cdot}056$
$0{\cdot}008 \times 0{\cdot}7 = 0{\cdot}0056$

d $6 \times 1{\cdot}2 = 7{\cdot}2$
$0{\cdot}6 \times 1{\cdot}2 = 0{\cdot}72$
$0{\cdot}06 \times 1{\cdot}2 = 0{\cdot}072$

e $8 \div 2 = 4$
$8 \div 0{\cdot}2 = 40$
$8 \div 0{\cdot}02 = 400$

f $1{\cdot}6 \div 0{\cdot}4 = 4$
$0{\cdot}16 \div 0{\cdot}4 = 0{\cdot}4$
$0{\cdot}016 \div 0{\cdot}4 = 0{\cdot}04$

g $0{\cdot}48 \div 6 = 0{\cdot}08$
$0{\cdot}48 \div 0{\cdot}6 = 0{\cdot}8$
$0{\cdot}48 \div 0{\cdot}06 = 8$

7 Write down a pattern like the ones in question **6** to work these out.

a $4 \times 0{\cdot}009$ **b** $0{\cdot}5 \times 0{\cdot}08$ **c** $0{\cdot}6 \times 0{\cdot}007$ **d** $0{\cdot}004 \times 0{\cdot}3$ **e** $0{\cdot}04 \times 1{\cdot}5$
f $0{\cdot}12 \times 0{\cdot}11$ **g** $0{\cdot}64 \div 0{\cdot}008$ **h** $8{\cdot}1 \div 0{\cdot}000\,09$ **i** $0{\cdot}48 \div 0{\cdot}016$

8 Melissa measured a distance on a map as 0·6 cm. To find the actual distance she had to multiply this by 20 000.
What was the actual distance in metres?

9 Explain how this diagram can be used to show that these are true.

a $0{\cdot}6 \times 0{\cdot}4 = 0{\cdot}24$ **b** $0{\cdot}5 \times 0{\cdot}3 = 0{\cdot}15$ **c** $0{\cdot}18 \div 0{\cdot}2 = 0{\cdot}9$

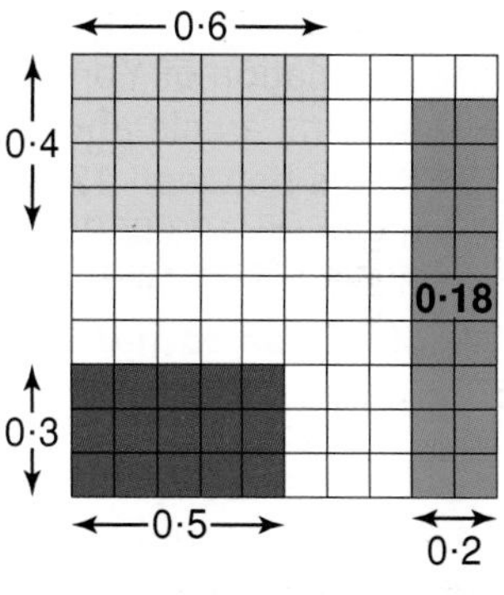

***10** Use the facts in the box to help do these conversions.

a 7 cm^2 to mm^2 **b** 9·4 m^2 to cm^2 **c** 4 cm^3 to mm^3
d 0·46 m^3 to cm^3 **e** 90 000 mm^2 to cm^2 **f** 84 000 mm^3 to cm^3
g 8·9 mm^2 to cm^2 **h** 72 cm^2 to m^2 **i** 764 mm^3 to cm^3
j 9256 cm^3 to m^3

1 cm^2 = 100 mm^2
1 m^2 = 10 000 cm^2
1 cm^3 = 1000 mm^3
1 m^3 = 1 000 000 cm^3

See page 336 for more about these measures.

***11** A rectangular stamp has an area of 600 mm^2.

a What is its area in cm^2?
b One side is 20 mm. What is the length of the other side?

T

Review 1

56·3 0·42 0·00087   56·3 0·56 4·8 0·00087 0·563   0·56 13 700

420 0·56 0·0563 0·00021 0·56 0·0563   4·8 13·7 870

0·56 0·0563 0·137 0·00087   U: 0·08 870 0·00087 0·00021   13·7 870

13·7   5·6 0·56 0·56

Use a copy of this box. Put the letter beside each question above its answer in the box.

U $0{\cdot}8 \times 0{\cdot}1 = 0{\cdot}08$ **C** $13{\cdot}7 \div 100$ **E** $8{\cdot}7 \div 10\,000$ **F** $137 \div 0{\cdot}01$ **H** $42 \times 0{\cdot}01$
A $0{\cdot}137 \div 0{\cdot}01$ **S** $8{\cdot}7 \div 0{\cdot}01$ **D** $0{\cdot}42 \div 2000$ **N** $5{\cdot}63 \times 0{\cdot}01$ **L** $42 \div 0{\cdot}1$
R $5{\cdot}63 \times 0{\cdot}1$ **T** $5{\cdot}63 \div 0{\cdot}1$ **W** $8 \times 0{\cdot}6$ **O** $0{\cdot}8 \times 0{\cdot}7$ **Z** $8 \times 0{\cdot}7$

Review 2 Sebastian bought 8 packets of special sweets for his birthday party. Each packet had a mass of 0·03 kg. One quarter of these were left at the end of the party. What mass of sweets was left?

Review 3 Write down a pattern like the ones in question **6** to work out the answers to these.

a $5 \times 0{\cdot}03$ **b** $0{\cdot}4 \times 0{\cdot}07$ **c** $0{\cdot}08 \times 0{\cdot}9$ **d** $0{\cdot}06 \times 1{\cdot}1$

e $0{\cdot}021 \div 0{\cdot}007$ **f** $0{\cdot}024 \div 0{\cdot}08$ ***g** $0{\cdot}0088 \div 0{\cdot}011$

Review 4 Use the facts in the box in question **10** on the previous page to do these conversions.

a 9 cm^2 to mm^2 **b** 8·6 m^2 to cm^2 **c** 5·3 cm^3 to mm^3

d 500 mm^3 to cm^3 **e** 96 820 cm^3 to m^3.

Rounding to powers of ten

The population of York is 124 609.
A magazine article about York rounded this to the nearest 10 000.
124 609 is between 120 000 and 130 000.
It is nearer to 120 000 than to 130 000.
124 609 to the nearest 10 000 is 120 000.
We often **use rounding when estimating**.

120 000 130 000
124 609
York

Worked Example

The population of England is about 60 million.
The populations, to the nearest hundred thousand, of the five largest cities are:

London	6·4 million
Birmingham	1·0 million
Liverpool	0·5 million
Sheffield	0·4 million
Leeds	0·4 million

Remember: when a number is exactly half way between we round up.

a What is the greatest population London could have?

***b** The tenth largest city is Coventry with a population of 0·3 million.
Estimate the percentage of the population of England who live in the ten largest cities.

Answer

a 6·4 million is 6 400 000 and has been rounded to the nearest hundred thousand.
The greatest number that would be rounded to 640 000 is **6 449 999**. If it was one more, **6 450 000**, it would be rounded to 6·5 million.

***b** The sixth to the ninth largest cities must have populations between 0·3 million and 0·4 million. We estimate that the mean population of these cities is 0·35 million.
Total millions in ten largest cities $= 6{\cdot}4 + 1 + 0{\cdot}5 + 0{\cdot}4 + 4 \times 0{\cdot}35 + 0{\cdot}3 = 10{\cdot}4$ million
Percentage in ten largest cities $= \frac{10{\cdot}4}{60} \times 100\%$
$=$ **17%** to the nearest percent.

Exercise 3

1 This chart gives the estimated number of people in the world who follow five minor religions. Write, in figures, the estimated number who follow each religion.

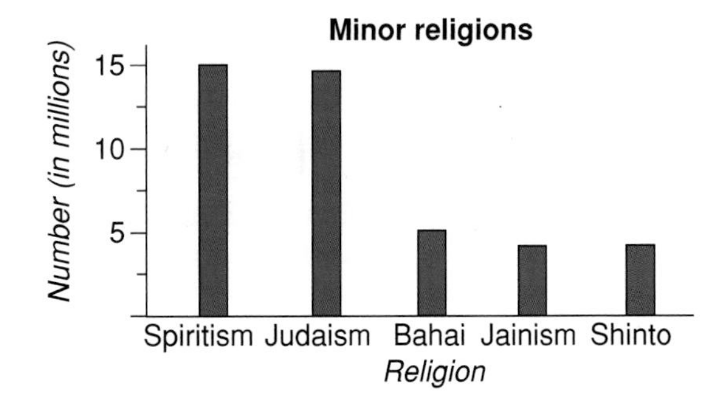

2 A newspaper reported that in one area of Britain 180 000 sheep had been destroyed because of foot and mouth disease. The figure was to the nearest ten thousand.
   a What is the smallest number of sheep that could have been destroyed?
   b What is the largest number?

3 One year, 352 760 motoring offences were picked up by automatic cameras.
   a A report said 'Nearly 400 000 offences were picked up by automatic cameras'. Was this given to the nearest 10 000, nearest 100 000 or nearest million?
   b Another year, the report said 'About 350 000 offences were picked up by automatic cameras'. The number had been rounded to the nearest 50 000.
   What is the greatest and smallest number of offences that could have been picked up?

4 James read this headline and said 'We have just over nine million school pupils in Britain'.
Hunter read the same headline and said 'We have nearly nine and a half million school pupils in Britain'.
Is James or Hunter more accurate? Explain your answer.

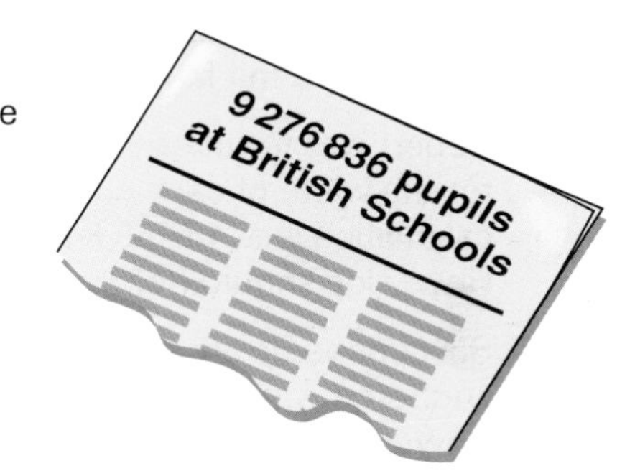

5 Val and Maya both have £2500 to the nearest £100.
Val has £3000 to the nearest £1000.
Maya has £2000 to the nearest £1000.
They both have a whole number of pounds.
   a Write down three amounts that Val could have.
   b Write down three amounts that Maya could have.

6 This table gives the population of the five largest countries in Europe.

Population	
Russian Federation	105 984 000
Germany	81 912 000
United Kingdom	58 144 000
France	58 375 000
Italy	57 193 000

   a Round each population to the nearest million then add these rounded figures.
   b Add the five populations, then round the answer to the nearest million.
   c Explain why the answers to **a** and **b** are different.
   d The estimated population of Europe is 688 million.
   Estimate the percentage of people in Europe who do **not** live in the five largest countries.
   *e The population of the tenth largest country, the Netherlands, is 15 517 000.
   Estimate the percentage of people in Europe who live in the ten largest countries.

Review About 74·5 million barrels of oil was produced one year.
a This figure has been rounded to the nearest hundred thousand.
What is the greatest and smallest number of barrels that could have been produced?
b One report said 'Nearly seventy-five million barrels were produced'.
Another said 'Just over seventy million barrels were produced'.
Which report is more accurate? Explain.
*c This table gives the amount of oil produced by the five top oil-producing countries.
The eighth top producer is Venezuela and it produced 3·2 million barrels.
Estimate the percentage of the world's oil produced by the top eight oil-producing countries.

Million barrels	
Saudi Arabia	9·1
USA	7·7
Russian Fed	6·5
Iran	3·8
Mexico	3·5

Rounding to decimal places

Discussion

- The 12 office staff at The Tannery won a prize of £34 859 on the pools. This prize was to be divided equally between the 12 people.
  How should they share this prize? **Discuss**.
- Three families picked a total mass of 12·4 kg of raspberries. These raspberries were to be shared equally between the three families. When they calculated 124 ÷ 3, their calculator gave an answer of 4·1333333.
  How should the families share the raspberries? **Discuss**.
- Robert was to rule an A4 piece of paper into 13 equally spaced columns. His calculation for the width of each column was 210 mm ÷ 13. The calculator display for this calculation was 16·153 846.
  How wide should Robert make each column? **Discuss**.
- Six cupboards are to be built along a 4 metre wall in a kitchen. The width of each cupboard was worked out on the calculator. The calculator display was 0·6666666.
  How wide should each cupboard be made? **Discuss**.

To give a sensible answer to a calculation we often need to **round the answer**.
Rounding gives an approximate answer.

Example Seven friends share a 250 g bag of popcorn.
To find how many grams each gets we key 250 ÷ 7 = to get 35.71428571
We could give the answer as 35·7 g to the nearest tenth of a gram.
We have rounded to 1 decimal place. There is 1 digit after the decimal point.

Follow these steps to **round to a given number of decimal places**, for example, to round 8·74862579 to 2 decimal places.

1 Keep the number of digits you want.
2 Before 'throwing the rest away', look at the next decimal place.
3 If this digit is 5 or greater, increase the last digit you are keeping by 1. Otherwise leave it as it is.
Throw the unwanted digits away.

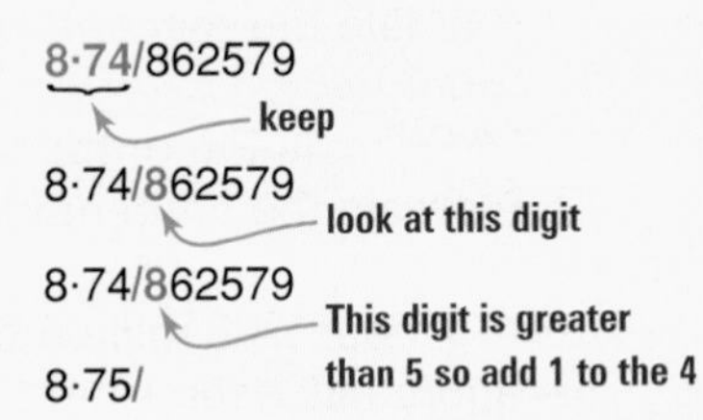

Worked Example
Round 3·647 253 to 2 d.p.

Answer
a We want 2 digits after the decimal point.

3·647253 = **3·65** (2 d.p.)

Discussion

Never round until the final answer is found.

Suppose $A = \frac{0{\cdot}34}{26{\cdot}1}$ and $B = 5{\cdot}9A$. You are to find the answer to B to 2 decimal places.

What answer do you get for B if you round the answer for A?
What answer do you get for B if you do not round the answer for A? **Discuss.**

Rounding to **the nearest whole number** is the same as rounding to **0 d.p.**

Example 3·6472 rounded to the nearest whole number is 4.

Sometimes the answer to a calculation has one or more repeating digits.

Example 5489 ÷ 3 = 1829·666666 ...

Instead of rounding, we often write these as **recurring decimals**.
1829·666666 ... is written as $1829{\cdot}\dot{6}$.
The dot above the 6 shows the digit repeats.

Example 489 ÷ 11 = 44·45454545 ...
$= 44{\cdot}\dot{4}\dot{5}$

The dots above 4 and 5 show that these digits repeat.

There is more about recurring decimals on page 94.

Exercise 4

1 Approximate these to **i** the nearest whole number **ii** 2 d.p. **iii** the nearest tenth

a 2·8237 **b** 6·827 **c** 0·7348 **d** 1·292 **e** 1·835
f 2·896 **g** 12·995 **h** 4·0038 **i** 1·0037 **j** 22·881
k 0·859 **l** 0·895 **m** 0·985 **n** 325·092

2 Round these to the given number of decimal places.

a 3·4215 (2 d.p.) **b** 24·01 (1 d.p.) **c** 0·004 (2 d.p.) **d** 13·995 (2 d.p.)
e 64·58 (0 d.p.) **f** 2·0997 (1 d.p.) **g** 125·704 (2 d.p.) **h** 4·98 (0 d.p.)
i 14·003 (1 d.p.) **j** 1·638 (2 d.p.) **k** 13·6492 (2 d.p.) **l** 0·0507 (1 d.p.)
m 6·98 (1 d.p.) **n** 7·997 (2 d.p.) **o** 0·096 (2 d.p.)

3 Trudy bought the following for her design and technology class.
Give each amount to the nearest tenth of a kilogram.

a Meat

2840 g

b Potatoes

1356 g

c Tomatoes

1720 g

4 Do these calculations on your calculator. Give the answers as recurring decimals or round them to 2 d.p.

a 8 ÷ 3 **b** 43 ÷ 9 **c** 7·593 ÷ 6 **d** 82·1 ÷ 4 **e** 0·57 ÷ 12 **f** 2·68 ÷ 11
g 22·9 ÷ 3 **h** $\frac{86{\cdot}473}{5}$ ***i** $\frac{72{\cdot}791}{11}$ ***j** $\frac{568{\cdot}4}{1{\cdot}68 \times 2{\cdot}37}$ ***k** $\frac{0{\cdot}0683}{6} + 8{\cdot}63$

5 Round $\frac{1}{680}$ to one decimal place.

6 In art, half of a roll of 6·8 m of paper was shared equally by 3 students. To the nearest tenth of a metre, how much did each get?

7 a The area of a cupboard door is $2{\cdot}11\ m^2$.
To the nearest centimetre, how high is the door?
Give the answer in metres.
b How many posters 26 cm high can fit up the height of the door?

0·9 m
$2{\cdot}11\ m^2$?

Link to area.

You will need to use the unrounded answer to **a**.

8 Round the answers to these sensibly.
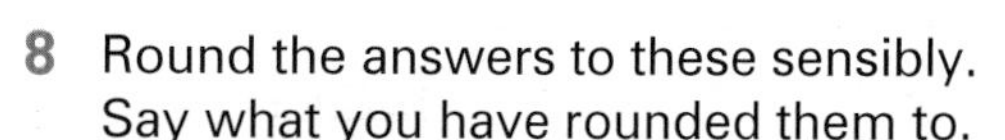
Say what you have rounded them to.
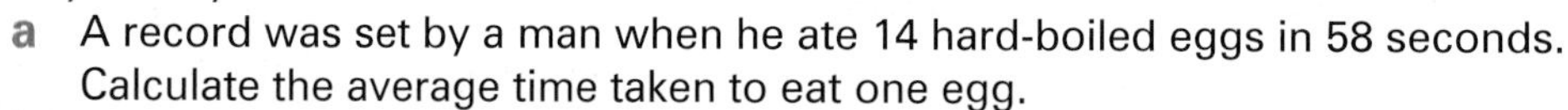
a A record was set by a man when he ate 14 hard-boiled eggs in 58 seconds.
Calculate the average time taken to eat one egg.
b In a science experiment, Joshua had to find the mean height that a pendulum swung to.
These are the heights it swung to.

8·3 cm, 6·4 cm, 7·2 cm, 8·4 cm, 7·8 cm, 7·7 cm

Find the mean height.
c A tap that drips once each second wastes 30 ℓ of water each day.
How much water is wasted in each drip?
d A crate is 1·68 m wide by 1·73 m long.
What is the area of the floor of the crate?
The volume is found by multiplying the floor area by the height.
What is the volume of the crate if it is 72 cm high?
*e The formula for finding the area of a circle is $A = \pi r^2$.
The area of a circular pond is $8{\cdot}04\ m^2$.
What is the radius of the pond?
Use $\pi = 3{\cdot}14$.

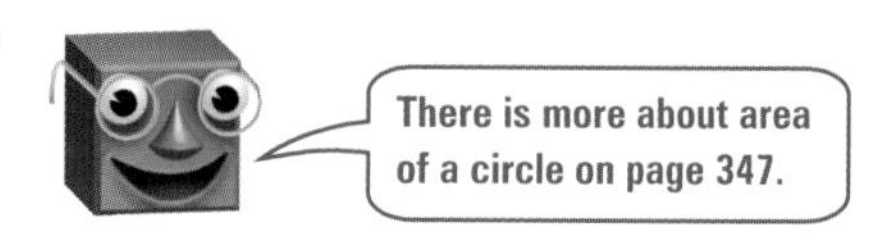

Review 1
a Round 3562 m to the nearest tenth of a kilometre.
b Round 7314 g to the nearest tenth of a kilogram.
c Round 852 cm to the nearest tenth of a metre.

Review 2 Use your calculator to find the answers to these.
Give the answers as recurring decimals or round them to 2 d.p.
a $25 \div 6$ b $31 \div 3$ c $1{\cdot}24 \div 9$ d $\frac{25}{15}$ e $\frac{7}{3}$

Review 3 Three friends spent £159·65 on food for a party. They shared the cost.
How much did each pay? Round your answer sensibly.

Review 4 Round the answers to these sensibly.
Say what you have rounded them to.
In one of the heaviest rainfalls ever recorded, 1880 mm fell in 24 hours.
a What was the average rainfall each hour?
b What was the average rainfall each minute?

Review 5 Ten customers chosen at random spent these amounts at a hardware store.

£23·65	£142·71	£5·36	£18·24	£29·63
£87·64	£97·60	£7·82	£35·64	£64·36

a Find the average amount spent by the ten customers.
b On average, the store has 751 customers each week.
About how much is spent by customers each week?

Practical

A **You will need** a spreadsheet package.
Ask your teacher for the Rounding Numbers ICT worksheet.

B **You will need** a graphical calculator.
Use the table facility on your graphical calculator.
Set your graphical calculator to round to 2 decimal places.

Enter Ran# as the function in the table and then as each number appears, round it to two decimal places.
Check your rounding on the table.

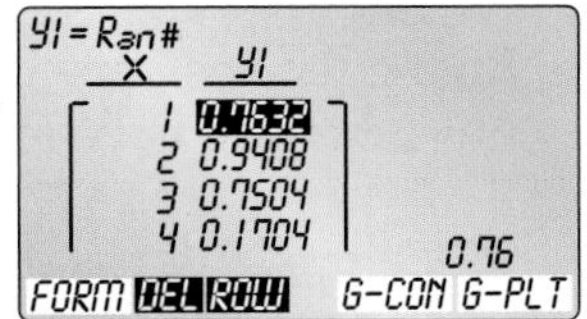

C Repeat B but set your graphical calculator to round to 3 d.p.

Summary of key points

This chart shows the **powers of ten**.

	Millions			Thousands			Hundreds					
(Thousands of millions) Billions	Hundreds of millions	Tens of millions	Millions	Hundreds of thousands	Tens of thousands	Thousands	Hundreds	Tens	Units •	tenths	hundredths	thousandths
10^9	10^8	10^7	10^6	10^5	10^4	10^3	10^2	10^1	10^0 •	10^{-1}	10^{-2}	10^{-3}

$6{\cdot}3 \times 10^4$ is read as 'six point three times ten to the power of four'.
$8{\cdot}7 \times 10^{-3}$ is read as 'eight point seven times ten to the power of negative three'.

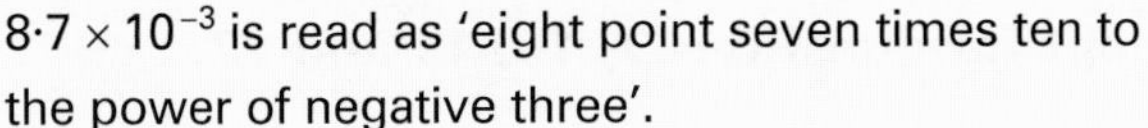

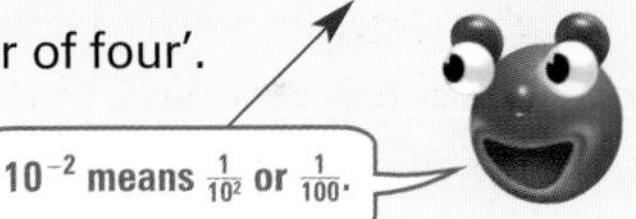

Multiplying and dividing by powers of 10.

Multiplying by 0·1 is the same as dividing by 10.
Multiplying by 0·01 is the same as dividing by 100.
Dividing by 0·1 is the same as multiplying by 10.
Dividing by 0·01 is the same as multiplying by 100.

Examples $3{\cdot}6 \times 0{\cdot}1 = 3{\cdot}6 \div 10 = 0{\cdot}36$ $\quad 0{\cdot}52 \div 0{\cdot}01 = 0{\cdot}52 \times 100 = 52$

Rounding to powers of 10.

We often use rounding to make estimates.

Example The population of the world is about 5300 million.
The populations of the four largest cities are:
Mexico City 21·5 million, Sao Paulo 19·9 million, Tokyo 19·5 million, New York 15·7 million.
The tenth largest city is Rio de Janeiro with a population of 11·9 million.
We can estimate the populations of the 5th, 6th, 7th, 8th and 9th largest cities as having an average population of $\frac{15{\cdot}7 + 11{\cdot}9}{2} = 13{\cdot}8$ million.
The total population of the ten largest cities is then
$21{\cdot}5 + 19{\cdot}9 + 19{\cdot}5 + 15{\cdot}7 + 5 \times 13{\cdot}8 + 11{\cdot}9 = 157{\cdot}5$ million.
The approximate percentage of people in the world who live in the ten largest cities is $\frac{157{\cdot}5}{5300} \times 100\% = 3{\cdot}0\%$ (1 d.p.)

To round to a given number of **decimal places**:

1 keep the number of digits asked for after the decimal point.

2 delete any following digits. If the first digit to be deleted is 5 or more, increase the last digit kept by 1.

Examples 8·6832 = 8·7 (1 d.p.)
53·0535 = 53·05 (2 d.p.)

Rounding to the nearest whole number is the same as rounding to 0 d.p.

Example 81·235 to the nearest whole number is 81.

Sometimes the answer to a calculation is written as a **recurring decimal** and not rounded.

Example $35 \div 6 = 5{\cdot}833333\ldots$
$= 5{\cdot}8\dot{3}$

Note: Never round until the final answer is found.

Example If we are finding the answer to $\frac{8{\cdot}64}{3{\cdot}1} \times 5{\cdot}8$, we don't round the answer to $\frac{8{\cdot}64}{3{\cdot}1}$ before multiplying by 5·8.

Test yourself

Except for questions 7, 10, 11 and 12.

1 Write these as a power of ten.
a a hundred thousand **b** ten thousand **c** a million **d** a billion

2 **a** The Sun is about $1{\cdot}49 \times 10^8$ km from Earth.
Write $1{\cdot}49 \times 10^8$ in words.
b The mass of an atom is about $2{\cdot}0 \times 10^{-23}$ g.
Write $2{\cdot}0 \times 10^{-23}$ in words.

3 What power of ten goes in these gaps?
a Mount Everest is 8·848 kilometres or 8·848 × ___ metres high.
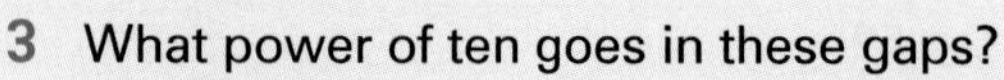
b Michael's computer has 9 gigabytes or 9·0 × ___ bytes of memory.
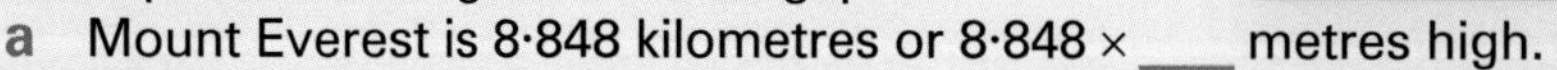
c Two hundred nanoseconds or 200 × ___ seconds is the time taken to send a signal via GPS.

4 Which of these is the same as $1{\cdot}08 \div 0{\cdot}01$?
A $1{\cdot}08 \div 10$ **B** $1{\cdot}08 \times 10$ **C** $1{\cdot}08 \times 100$ **D** $1{\cdot}08 \div 100$

5 Find the answers to these
a $5{\cdot}7 \times 0{\cdot}1$ **b** $4{\cdot}73 \div 0{\cdot}1$ **c** $83{\cdot}5 \times 0{\cdot}01$ **d** $0{\cdot}68 \div 0{\cdot}01$

6 Paulo paid 0·7 times as much as his brother to hire a suit for five days.
If his brother paid £9·36 per day for the suit, how much did Paulo pay?

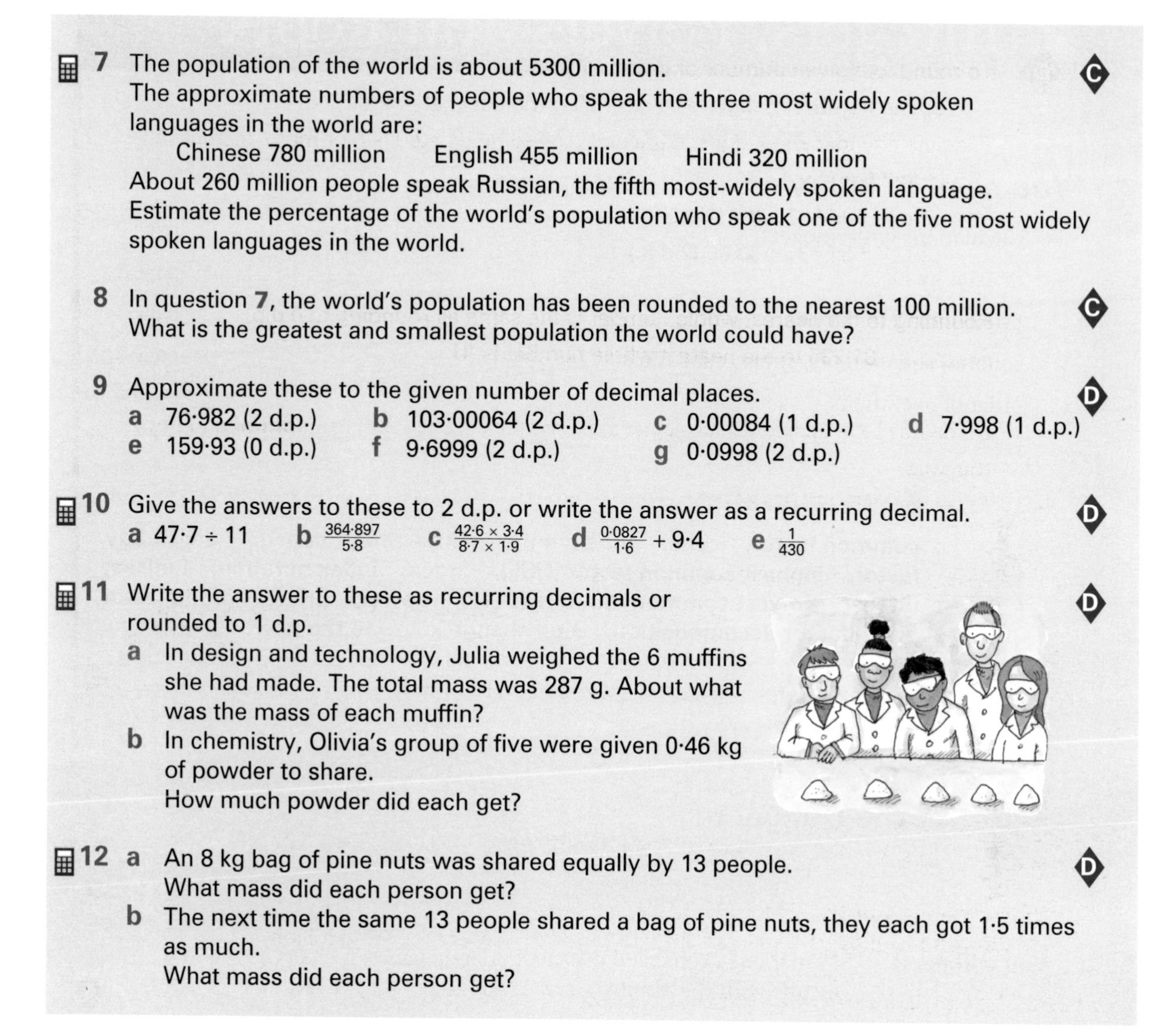

7 The population of the world is about 5300 million.
The approximate numbers of people who speak the three most widely spoken languages in the world are:

Chinese 780 million English 455 million Hindi 320 million

About 260 million people speak Russian, the fifth most-widely spoken language.
Estimate the percentage of the world's population who speak one of the five most widely spoken languages in the world.

8 In question **7**, the world's population has been rounded to the nearest 100 million.
What is the greatest and smallest population the world could have?

9 Approximate these to the given number of decimal places.

a 76·982 (2 d.p.) **b** 103·00064 (2 d.p.) **c** 0·00084 (1 d.p.) **d** 7·998 (1 d.p.)
e 159·93 (0 d.p.) **f** 9·6999 (2 d.p.) **g** 0·0998 (2 d.p.)

10 Give the answers to these to 2 d.p. or write the answer as a recurring decimal.

a $47{\cdot}7 \div 11$ **b** $\frac{364{\cdot}897}{5{\cdot}8}$ **c** $\frac{42{\cdot}6 \times 3{\cdot}4}{8{\cdot}7 \times 1{\cdot}9}$ **d** $\frac{0{\cdot}0827}{1{\cdot}6} + 9{\cdot}4$ **e** $\frac{1}{430}$

11 Write the answer to these as recurring decimals or rounded to 1 d.p.

a In design and technology, Julia weighed the 6 muffins she had made. The total mass was 287 g. About what was the mass of each muffin?

b In chemistry, Olivia's group of five were given 0·46 kg of powder to share.
How much powder did each get?

12 a An 8 kg bag of pine nuts was shared equally by 13 people.
What mass did each person get?

b The next time the same 13 people shared a bag of pine nuts, they each got 1·5 times as much.
What mass did each person get?

2 Integers, Powers and Roots

You need to know

Key vocabulary

common factor, cube, cube number, cube root, cubed, divisibility, factor, highest common factor (HCF), index, index notation, indices, integer, lowest common multiple (LCM), power, prime, prime factor, prime factor decomposition, sign change key, to the power of

To the Dungeon!

Game for 2 players

You will need 13 markers (You could use chalk marks, paper or chairs.)
A cube with the numbers 1, 2, 3, $^{-}1$, $^{-}2$, $^{-}3$ on the faces.

To Play: In an old fairy story, a rich King played a game with his people.
Each person who played had a chance to win gold.
The King had a staircase with 13 steps. He put a person on the middle step, then tossed a die with 1, 2, 3 and $^{-}1$, $^{-}2$, $^{-}3$ on it.
Each time the die was tossed the person would move up or down the staircase this number of steps.
If the person reached the bottom the game was over and he or she was thrown in the dungeon.
If the person reached the top, he or she was given gold.
Make your own staircase with markers or chairs.
Play this game.

Adding, subtracting, multiplying and dividing integers

Remember

We can **add and subtract integers** using a number line.

Adding

Examples $^{-}2 + 3 = 1$

$1 + {}^{-}3 = {}^{-}2$

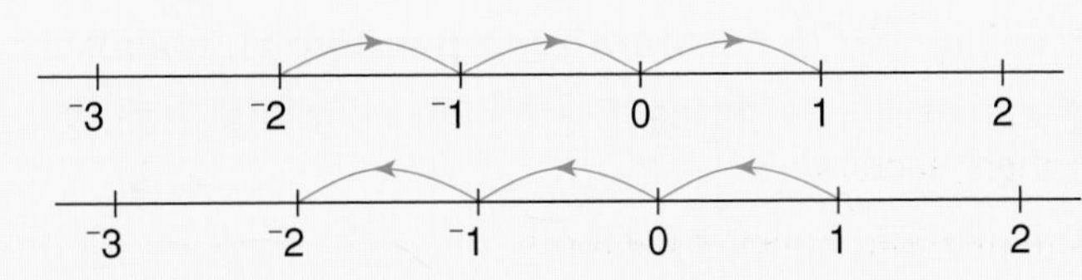

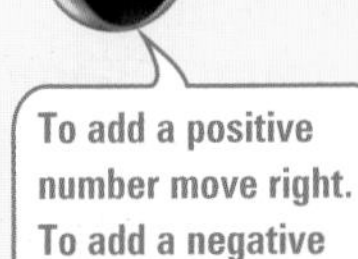

Subtracting

Adding and subtracting are inverse operations so we move in the opposite direction when we subtract.

Examples $^{-}1 - 2 = {}^{-}3$

$^{-}2 - {}^{-}5 = 3$

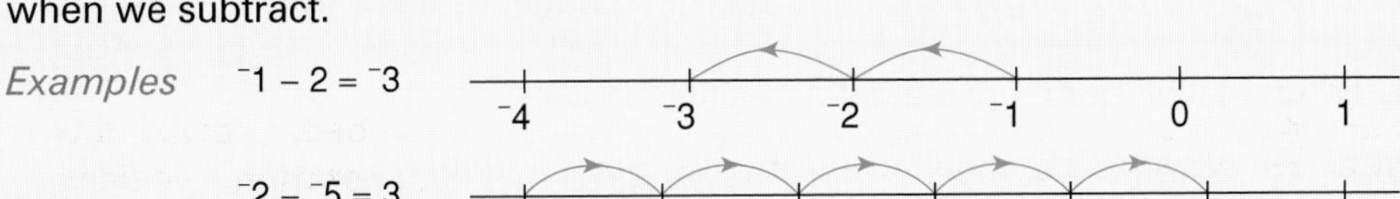

To subtract a positive number, move left.
To subtract a negative number, move right.

Discussion

- We can add and subtract integers without using a number line.
  We know that $1 + {}^{-}1 = 0$, $2 + {}^{-}2 = 0$, $3 + {}^{-}3 = 0$, $4 + {}^{-}4 = 0, \ldots$
  How could you use this to work these out? **Discuss**.
  $^{-}27 + 27$ $562 + {}^{-}562$ $1 + 1 + {}^{-}1 + 1 + {}^{-}1$

- Daniel worked out some answers like this.

  $56 + {}^{-}47 = 56 + {}^{-}56 + 9$
  $= 0 + 9$
  $= 9$

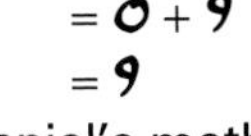

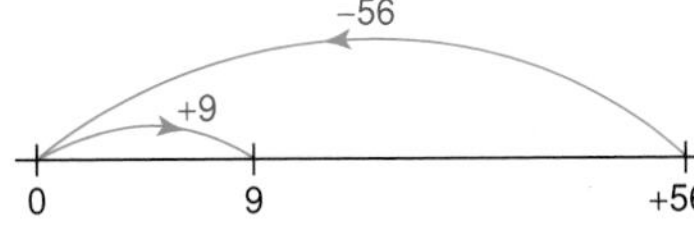

  $^{-}53 + 60 = {}^{-}53 + 53 + 7$
  $= 0 + 7$
  $= 7$

  Discuss Daniel's method.
  Use his method to work these out.
  $68 + {}^{-}57$ $^{-}72 + 80$ $50 + {}^{-}20 + 10 + {}^{-}30$

- Simon wanted to know what $0 - {}^{-}1$ was equal to.
  He wrote

  I know that $0 = 1 + {}^{-}1$.
  So $0 - {}^{-}1 = (1 + {}^{-}1) - {}^{-}1$ substituting $1 + {}^{-}1$ for 0
  $= 1 + {}^{-}1 - {}^{-}1$
  $= 1 + 0$ adding $^{-}1$ then taking $^{-}1$ away again leaves 0
  $= 1$

  How could you show that these are true? **Discuss**.
  $0 - {}^{-}2 = 2$, $0 - {}^{-}3 = 3$, $0 - {}^{-}4 = 4, \ldots$

- * Simon used the facts $0 - {}^{-}1 = 1$, $0 - 1 = {}^{-}1, \ldots$ to work this out.

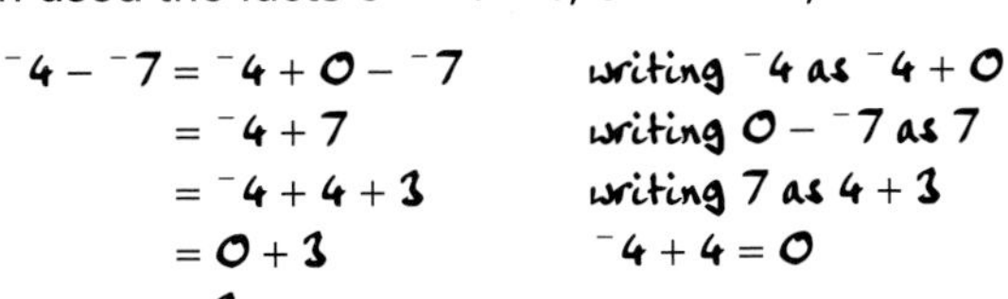

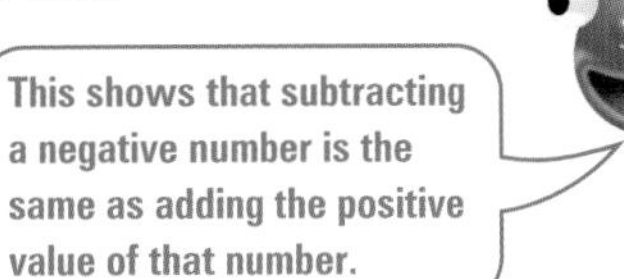

  Discuss Simon's method.
  How might Simon find the answer to these?
  $^{-}6 - 7 = \square$ $8 - {}^{-}5 = \square$ $^{-}3 - {}^{-}8 = \square$ $^{-}7 - \square = 3$

Remember
Adding a positive number or **subtracting a negative** number is the same as **adding**.

Examples $^{-}8 + {}^{+}2 = {}^{-}8 + 2$, $= {}^{-}6$ $\quad$ $5 - {}^{-}2 = 5 + 2$, $= 7$ $\quad$ $^{-}4 - {}^{-}3 = {}^{-}4 + 3$ $= {}^{-}1$

Remember
Adding a negative number or **subtracting a positive** number is the same as **subtracting**.

Examples $4 + {}^{-}2 = 4 - 2$, $= 2$ $\quad$ $^{-}3 + {}^{-}2 = {}^{-}3 - 2$ $= {}^{-}5$ $\quad$ $^{-}5 - {}^{+}3 = {}^{-}5 - 3$ $= {}^{-}8$

Remember
When we multiply a positive and a negative number we get a negative answer. $\ominus \times \oplus = \ominus$ or $\oplus \times \ominus = \ominus$

When we multiply two negative numbers we get a positive answer. $\ominus \times \ominus = \oplus$

When we divide with a positive and a negative number we get a negative answer. $\oplus \div \ominus = \ominus$
When we divide with two negative numbers we get a positive answer. $\ominus \div \ominus = \oplus$

Worked Examples
a $^{-}7 \times 3$ $\quad$ b $^{-}5 \times {}^{-}7$ $\quad$ c $^{-}12 \div 3$ $\quad$ d $\frac{^{-}30}{^{-}5}$

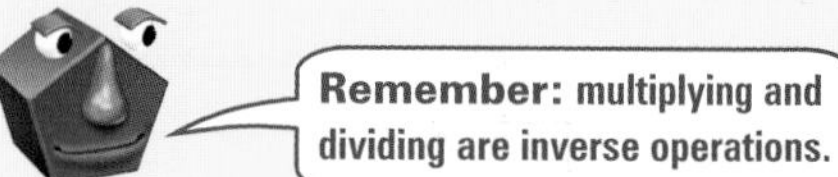

Answers
a $^{-}7 \times 3 = {}^{-}21$ $\quad$ $7 \times 3 = 21$ and $\ominus \times \oplus = \ominus$
b $^{-}5 \times {}^{-}7 = 35$ $\quad$ $5 \times 7 = 35$ and $\ominus \times \ominus = \oplus$
c $^{-}12 \div 3 = {}^{-}4$ $\quad$ $12 \div 3 = 4$ and $\ominus \div \oplus = \ominus$
d $\frac{^{-}30}{^{-}5} = 6$ $\quad$ $30 \div 5 = 6$ and $\ominus \div \ominus = \oplus$

The **sign change key** on the calculator is (−).

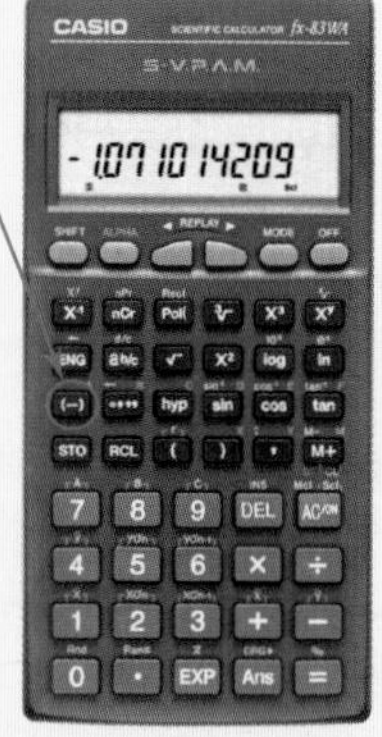

Example $^{-}5$ is keyed as (−) 5 .

Worked Examples
Use your calculator to find
a $485 - {}^{-}292$ $\quad$ b $37 \times {}^{-}4{\cdot}8$

Answers
a key 485 − (−) 292 = to get 777.
b key 37 × (−) 4·8 = to get $^{-}177{\cdot}6$.

Always follow the rules for order of operations when calculating mentally.

Brackets, **I**ndices, **D**ivision and **M**ultiplication, **A**ddition and **S**ubtraction.

Worked Examples
a $5 - {}^{-}2 \times 4$ $\quad$ b $^{-}3 \times 5 - {}^{-}4 \times {}^{-}2$ $\quad$ c $^{-}2(4 + {}^{-}7)$ $\quad$ d $\frac{^{-}4 + {}^{-}8}{^{-}2}$

Answers
a $5 - {}^{-}2 \times 4 = 5 - {}^{-}8$
$= \mathbf{13}$

b $^{-}3 \times 5 - {}^{-}4 \times {}^{-}2 = {}^{-}15 - 8$
$= \mathbf{{}^{-}23}$

Multiply first
$^{-}3 \times 5 = {}^{-}15$
$^{-}4 \times {}^{-}2 = 8$

c $^{-}2(4 + {}^{-}7) = {}^{-}2 \times {}^{-}3$
$= \mathbf{6}$
or $^{-}2(4 + {}^{-}7) = {}^{-}2 \times 4 + {}^{-}2 \times {}^{-}7$
$= {}^{-}8 + 14$
$= \mathbf{6}$

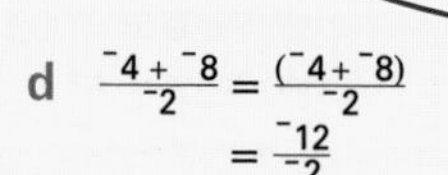

d $\frac{^{-}4 + {}^{-}8}{^{-}2} = \frac{(^{-}4 + {}^{-}8)}{^{-}2}$
$= \frac{^{-}12}{^{-}2}$
$= \mathbf{6}$

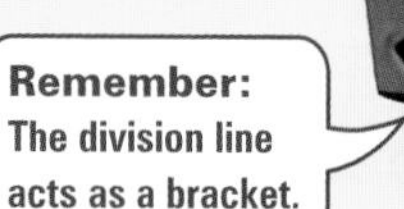

Exercise 1

Except questions 4, 9 and Review 3.

T

1 Use a copy of this. Shade the answers.
What does the shading make?

$^-8$	$^-10$	0	10	9	3	25	$^-85$
4	48	11	103	$^-4$	$^-6$	51	28
4	211	$^-120$	42	42	0	$^-4$	640
1	$^-28$	0	19	$^-4$	4	12	$^-15$
0	641	$^-12$	20	$^-25$	$^-26$	126	89
652	100	36	81	13	47	$^-3$	101

a $5 + {}^-5$ **b** ${}^-24 + 24$ **c** $326 + {}^-326$
d ${}^-879 + 879$ **e** $42 + {}^-39$ **f** $52 + {}^-48$
g ${}^-76 + 80$ **h** ${}^-52 + 46$ **i** ${}^-86 + 90$
j $28 + {}^-36$ **k** $7 - {}^-4$ **l** ${}^-11 - {}^-12$
m ${}^-20 - {}^-8$ **n** ${}^-55 - {}^-30$ **o** $42 - {}^-84$
p ${}^-47 - {}^-56$ **q** ${}^-99 - {}^-73$ **r** $426 + {}^-215$
s ${}^-380 - {}^-260$ **t** ${}^-810 - {}^-800$

2 What goes in the box?
a ${}^-3 + \square = 5$ **b** $2 + \square = {}^-1$ **c** $4 - \square = {}^-1$ **d** ${}^-2 + \square = {}^-3$ **e** ${}^-4 - \square = {}^-2$
f ${}^-7 - \square = {}^-9$ **g** ${}^-10 - \square = {}^-6$ **h** $8 - \square = 12$ **i** ${}^-7 + \square = {}^-16$ **j** $3 + \square = {}^-13$
k ${}^-15 - \square = {}^-21$ **l** ${}^-7 - \square = 4$ **m** $23 + \square = {}^-11$ **n** ${}^-15 - \square = 21$
o $50 + {}^-20 + {}^-10 + 40 = \square$ **p** $80 + {}^-20 + 10 + {}^-30 = \square$

T

3 Use a copy of this.

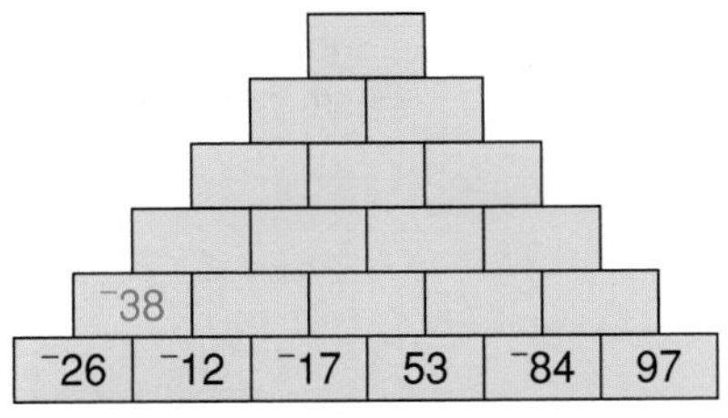

On this green pyramid each pair of numbers is added to get the number above.

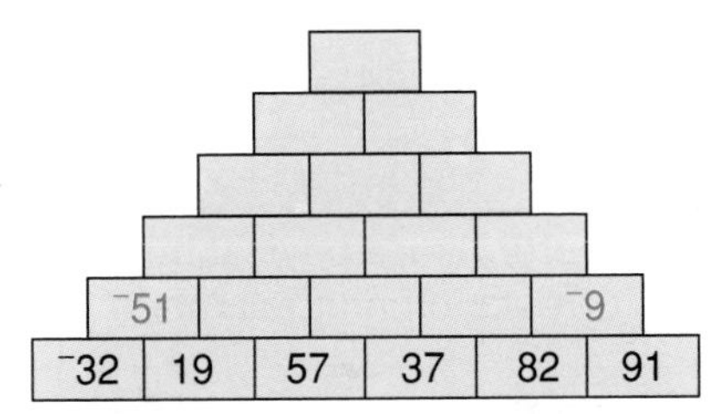

On this blue pyramid the number on the right is subtracted from the number on the left to get the number above.

a What number is at the top of the green pyramid?
b What number is at the top of the blue pyramid?

T

4 Use a copy of this
Find a route from the blue to the pink box by shading the answers to these.

	282	$^-41·6$	19·89	$^-19·89$	
$^-282$	$^-104$	84·2	$^-4·04$	249·6	4·04
41·6	594	32·9	13·16	$^-23·8$	$^-249·6$
$^-594$	$^-32·9$	$^-87·54$	–3008	42·33	104
$^-5·7$	87·54	100·7	$^-13·16$	8·6	$^-5·7$

a ${}^-84 + {}^-198$ **b** $51·4 - 9·8$ **c** $253 - {}^-341$
d ${}^-41·6 - {}^-8·7$ **e** ${}^-87·2 - 0·34$ **f** $47 \times {}^-64$
g ${}^-2·8 \times {}^-4·7$ **h** ${}^-14 \times 1·7$ **i** ${}^-8·3 \times {}^-5·1$
j ${}^-61·92 \div {}^-7·2$ **k** $11·97 \div {}^-2·1$ **l** $\frac{{}^-52 \times 14}{{}^-7}$
m $\frac{{}^-6·4 \times {}^-7·8}{{}^-0·2}$ **n** $\frac{{}^-8·5 + {}^-11·7}{{}^-5}$

5 Write down the next four lines of these.

a
$2 \times 6 = 12$
$1 \times 6 = 6$
$0 \times 6 = 0$
${}^-1 \times 6 = {}^-6$

b
$1 \times {}^-3 = {}^-3$
$0 \times {}^-3 = 0$
${}^-1 \times {}^-3 = 3$
${}^-2 \times {}^-3 = 6$

c
${}^-6 \times 7 = {}^-42$
${}^-7 \times 7 = {}^-49$
${}^-8 \times 7 = {}^-56$
${}^-9 \times 7 = {}^-63$

d
${}^-12 \times 4 = {}^-48$
${}^-13 \times 4 = {}^-52$
${}^-14 \times 4 = {}^-56$
${}^-15 \times 4 = {}^-60$

6 **a** How many negative threes make negative twelve?
b How many negative eights make sixty-four?
c How many fours make negative sixty-four?

7
a $3 \times {}^-7$ b ${}^-2 \times 5$ c ${}^-8 \times 2$ d ${}^-4 \times {}^-5$ e $3 \times {}^-5$
f $8 \div {}^-2$ g ${}^-24 \div {}^-4$ h ${}^-12 \div 3$ i ${}^-18 \div {}^-6$ j $20 \div {}^-5$
k $\frac{{}^-16}{4}$ l $\frac{{}^-30}{{}^-5}$ m ${}^-2 \times {}^-3 \times {}^-5$ n $2 \times {}^-3 \times {}^-1$ o $\frac{{}^-6 \times {}^-3}{2}$
p $\frac{8 \times {}^-5}{4}$ q $\frac{{}^-2 \times {}^-6 \times 3}{{}^-3 \times {}^-3}$ r $\frac{{}^-3 \times 8}{{}^-2 \times {}^-6}$ s $2 \times {}^-1{\cdot}5$ t ${}^-4 \times {}^-2{\cdot}2$
u $2{\cdot}5 \times {}^-3$ v $4{\cdot}8 \div {}^-3$ w $\frac{{}^-8{\cdot}2}{{}^-2}$ x $\frac{{}^-16{\cdot}8}{8}$

T **8** Use two copies of this grid.
Find two different ways to fill it in.

×		5	⁻7	
		⁻10	14	
⁻4		⁻20		
	45			⁻15
				⁻24

9 Use a calculator.
Round the answers to 1 d.p.
a $15{\cdot}7 \div {}^-8{\cdot}4$ b ${}^-23{\cdot}6 \div 4{\cdot}7$ c ${}^-568 \div {}^-53$
d $4{\cdot}73 \div {}^-0{\cdot}93$ e ${}^-5{\cdot}86 \div {}^-3{\cdot}4$

10 Find the answers to these mentally. Use jottings if you need to.
a ${}^-2 + 3 \times {}^-4$ b $3 + {}^-4 \times {}^-2$ c $4 - 3 \times {}^-2$ d $4 - {}^-3 \times {}^-2$
e ${}^-1 - 2 \times 3$ f ${}^-3 - {}^-1 \times {}^-4$ g ${}^-2 \times {}^-1 + {}^-3 \times 2$ h ${}^-6 \times 2 + 3 \times {}^-4$
i ${}^-5 \times {}^-3 - 2 \times {}^-3$ j $3(4 + {}^-5)$ k $5(3 + {}^-4)$ l $4({}^-2 + {}^-3)$
m ${}^-3(2 + {}^-6)$ n ${}^-1(8 + {}^-3) - 4$ o ${}^-8 + 3({}^-2 + {}^-1)$ p $\frac{6 + {}^-4}{{}^-2}$
q $\frac{12 + {}^-3}{3}$ r $\frac{8 - {}^-4}{{}^-3}$ s $\frac{10 - {}^-10}{5}$ *t $\frac{{}^-3 - {}^-4 + 9}{{}^-5}$
*u $\frac{2({}^-4 - 16)}{8}$ *v $\frac{{}^-8 + 2(3 + {}^-4)}{{}^-5}$ *w $\frac{{}^-5({}^-4 + 5) - 10}{{}^-2}$ *x $\frac{8(3 + {}^-7) + 8}{4 + {}^-5}$

11 Barry has an overdraft. He pays some off each month.
He uses this formula to find the amount still owing, N.

$$N = D + mb$$

D is the initial overdraft, m is the number of months and b is the amount he pays each month.
Find N if
a $D = {}^-£160$, $m = 2$, $b = £30$
b $D = {}^-£500$, $m = 6$, $b = £40$
c $D = {}^-£1080$, $m = 11$, $b = £24$

12 a Two integers are added.
The answer is ${}^-124$.
What could the integers be?
b $\square - \square = {}^-321$
What integers might go in the boxes?

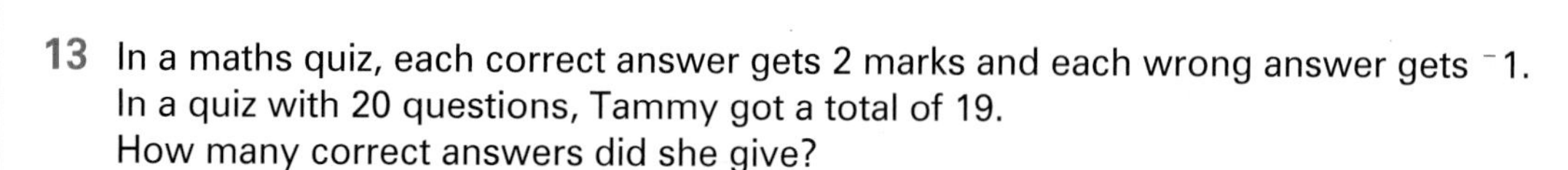

13 In a maths quiz, each correct answer gets 2 marks and each wrong answer gets ${}^-1$.
In a quiz with 20 questions, Tammy got a total of 19.
How many correct answers did she give?

***14** **⁻5 ⁻3 ⁻2 ⁻2 ⁻7**
With these five numbers we can make ${}^-5 \times {}^-3 - {}^-2 \times {}^-7 + {}^-2 = {}^-1$.
What is the largest answer you can make?

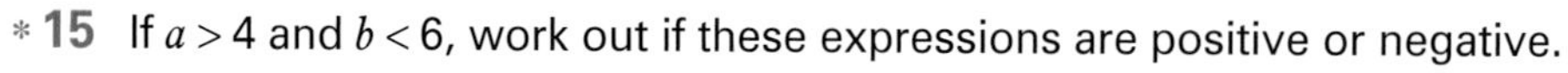

***15** If $a > 4$ and $b < 6$, work out if these expressions are positive or negative.
a $4 - a$ b $\frac{b - 6}{4 - a}$

Review 1 Find the answers to these mentally.

a $^-16 + ^-12$ **b** $^-22 + ^-36$ **c** $17 + ^-29$ **d** $^-13 - ^-24$
e $^-29 - ^-14$ **f** $^-80 + 50 + 30 - 10$ **g** $^-70 - 30 + 40 + 30$ **h** $^-36 - 17 + 21 + 15$

Review 2 $\square - \square = ^-352$.
What integers could go in the boxes?

T **Review 3**

_	_	_		_	_	_	_	_		_	F	
$^-140$	$^-102$	475		24·9	$^-140$	8·25	25·62	$^-65$		$^-26$	$^-326$	

24·9	$^-140$	8·25	1620	$^-1827$	25·62	$^-1827$	$^-140$	$^-65$		$^-1827$	24·9

3844	$^-24·9$	338·4	338·4	475	25·62

$^-8·25$	$^-26$	182·7	$^-26$	338·4	$^-26$	$^-2·6$	$^-65$

Use a copy of this box. Write the letter beside each question above its answer in the box.

F $^-84 + ^-242 = ^-326$ **H** $247 + ^-349$ **E** $348 - ^-127$ **I** $^-29 \times 63$
C $^-31 \times ^-124$ **L** $^-94 \times ^-3·6$ **O** $1456 \div ^-56$ **M** $^-792 \div 96$
N $^-29 \times ^-6·3$ **G** $145·6 \div ^-56$ **U** $^-792 \div ^-96$ **Y** $^-416 \div 6·4$
T $^-84 + ^-98 - ^-42$ **P** $144 \times ^-25 \times ^-0·45$ **S** $^-8·3 \times 3·9 \div ^-1·3$ **A** $^-8·3 \div 1·3 \times 3·9$
D $^-4·2 \times ^-6·1$

T **Review 4**

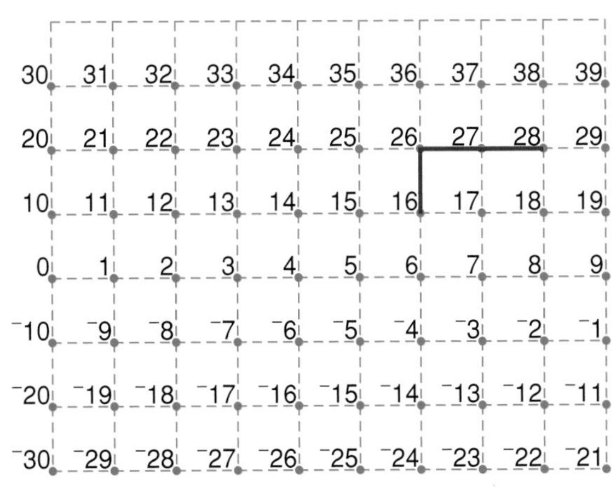

Use a copy of this.
Do the calculations in order from **a** to **q**. Join the answers on the grid.
What picture do you get?
a, **b** and **c** are done for you.

a $^-6 \times ^-5 + ^-2$ **b** $^-2(^-14 + 1)$ **c** $4 + ^-3 \times ^-4$ **d** $9 + ^-1 \times 2$
e $4 \times ^-5 + 7$ **f** $^-2(4 - ^-8)$ **g** $^-4 \times 8 + 5$ **h** $3(^-5 + 2) + 1$
i $2 \times ^-1 - 3 \times ^-5$ **j** $^-1 \times 3 - 4 \times ^-2$ **k** $^-6 \times ^-2 + 4 \times ^-4$ **l** $\frac{3(6 - ^-2)}{4}$
m $\frac{6 \times ^-5}{^-2}$ **n** $^-8 \times ^-2 - ^-9$ **o** $40 - ^-1 \times ^-4$ **p** $^-4 \times ^-9 - ^-1$
q $^-7 \times ^-5 + ^-4 \times 2$

Puzzle

1 $9 = \frac{^{-}4}{^{-}4} - [^{-}4 + (^{-}4)]$

Write each of the other 1-digit numbers (0, 1, 2, 3, 4, 5, 6, 7, 8) using $^{-}4$ *exactly* four times.

2

$^{-}4$	$^{-}5$	1	$^{-}4$	$^{-}3$	2	$^{-}2$	$^{-}1$	2	5	$^{-}5$	0	$^{-}4$	$^{-}4$	2	$^{-}2$
1	6	–2	$^{-}4$	$^{-}5$	$^{-}6$	1	3	$^{-}2$	$^{-}5$	$^{-}6$	1	0	1	4	1
$^{-}4$	1	$^{-}1$	1	4	$^{-}1$	0	2	$^{-}4$	$^{-}1$	2	$^{-}1$	5	$^{-}5$	$^{-}2$	$^{-}1$
$^{-}1$	$^{-}5$	4	3	$^{-}1$	4	$^{-}2$	$^{-}5$	1	$^{-}4$	$^{-}5$	1	$^{-}3$	0	3	$^{-}6$
$^{-}2$	$^{-}3$	$^{-}1$	$^{-}6$	0	3	$^{-}2$	$^{-}1$	4	$^{-}3$	$^{-}2$	$^{-}2$	3	$^{-}2$	$^{-}3$	5
2	1	6	2	$^{-}4$	$^{-}1$	4	5	$^{-}2$	$^{-}6$	1	1	1	$^{-}4$	$^{-}1$	$^{-}4$
0	$^{-}4$	$^{-}2$	$^{-}1$	$^{-}2$	1	$^{-}2$	4	$^{-}6$	$^{-}5$	0	$^{-}1$	4	$^{-}2$	$^{-}2$	1
3	1	4	1	$^{-}6$	2	$^{-}8$	$^{-}3$	2	5	$^{-}1$	$^{-}4$	$^{-}7$	$^{-}2$	$^{-}5$	0
3	$^{-}8$	2	0	$^{-}3$	$^{-}2$	3	$^{-}3$	$^{-}6$	2	$^{-}1$	$^{-}2$	7	5	2	$^{-}7$
$^{-}4$	0	$^{-}3$	$^{-}5$	2	2	0	$^{-}1$	$^{-}3$	3	$^{-}9$	2	$^{-}1$	0	$^{-}3$	0

Score:		
	100 or more	Excellent
	60–99	Very good
	20–59	Good
	fewer than 20	Keep trying

Find as many 'triple statements' as you can. A 'triple statement' is a true statement formed from 3 adjacent integers, either horizontally or vertically or diagonally.

For instance, the ringed numbers are a triple statement since $^{-}1 - (^{-}2) = 1$. You may use the operations +, −, ×, ÷.

T

Practical

You will need a graphical calculator.

Use the list facility to practise adding and subtracting integers.

Use the random number generator to fill the first two lists.

The command is Int(21 Ran # – 10).

	List1	List2	List3	List4
1	2	-4	-2	
2	-7	3	-4	
3	0	6		
4	-4	-1		
5	0	-9		

SRT-A SRT-D DEL DEL-A INS

*

Divisibility

Remember

The **rules for divisibility** are on page 2.

Worked Example

Prove that any two-digit number in which the tens digit is twice the units digit is always divisible by 7.

Prove means it must be true for every case. We usually use algebra when we prove something.

Answer

If the number of units is t, then the number of tens must be $10 \times 2t = 20t$.

The number has the form $20t + t = 21t$.

$21t$ is always divisible by 7 because 21 is divisible by 7.

Exercise 2

1 Adele started with 59.
She reversed the digits to get 95.
She found the difference, $95 - 59 = 36$.
She did this with some other two-digit numbers.
Adele thinks the difference is always divisible by 9.

a **Show** using three examples that Adele could be correct.

b Adele started to **prove** that the difference is always divisible by 9. Copy and finish Adele's proof.

Call the number you chose $10t + u$
When you reverse the digits, the number is $10u + t$.
The difference between these two expressions is ...

*c Prove that this is also true for a 4-digit number.
Hint: Call the number you choose $1000s + 100h + 10t + u$.

2 583, 264, 781 and 374 are all divisible by 11.
Boris noticed that in all of these numbers, the sum of the hundreds digit and the units digit equals the tens digit.
He started to prove that this was true for all 3-digit numbers.

Let the number be $100h + 10(h + u) + u$.
$100h + 10(h + u) + u = \ldots$

***3** Prove that any three-digit number, when repeated to give a six-digit number, is divisible by 7, 11 and 13.

Example If we start with 864 and repeat the digits to give 864 864, then 864 864 is divisible by 7, 11 and 13.

***4** Prove this statement.

Any three-digit number is divisible by 9 if the sum of the digits is divisible by 9.

Review Carlota noticed that all of the two-digit numbers where the tens digit is equal to the units digit are divisible by 11.
11, 22, 33, 44, 55, 66, 77, 88 and 99 are all divisible by 11.
She wanted to prove this, using algebra.
Show how she could do this.

Using prime factor decomposition

Remember
We can use a diagram to find the **highest common factor (HCF)** and the **lowest common multiple (LCM)**

Example 72 and 60

We find the prime factors of 72 and 60 using a table or tree diagram.

2	72
2	36
2	18
3	9
	3

$72 = 2^3 \times 3^2$

60
2 30
2 15
3 5

$60 = 2^2 \times 3 \times 5$

Find the HCF and LCM using a Venn diagram.
HCF $= 2 \times 2 \times 3 = 12$
LCM $= 2 \times 3 \times 2 \times 2 \times 3 \times 5 = 2^3 \times 3^2 \times 5 = 360$

72 60
2 3 | 2 2 3 | 5
HCF (purple section)
LCM (dotted ring)

When cancelling fractions, the HCF of the denominator and numerator is the largest number that will divide into both.

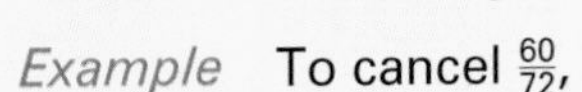

Example To cancel $\frac{60}{72}$, **divide by the HCF, 12.**

$\frac{^5\not{60}}{_6\not{72}} = \frac{5}{6}$ **dividing both numerator and denominator by 12**

When we add and subtract fractions, we find the LCM of the denominators. This LCM is the **lowest common denominator**.

Example $\frac{3}{72} + \frac{9}{60} = \frac{15+54}{360}$ $\leftarrow \frac{3}{72} = \frac{15}{360}$ **and** $\frac{9}{60} = \frac{54}{360}$

$= \frac{69}{360}$

$= \frac{23}{120}$

Exercise 3

1 Cancel these fractions by finding the HCF of the numerator and denominator.

a $\frac{36}{80}$ **b** $\frac{45}{125}$ **c** $\frac{72}{120}$ **d** $\frac{144}{180}$ **e** $\frac{90}{144}$ **f** $\frac{105}{240}$
g $\frac{120}{288}$ **h** $\frac{164}{280}$ ***i** $\frac{4nm}{2n}$ ***j** $\frac{9a^2b}{3ab}$ ***k** $\frac{3x^2yz}{6wx}$

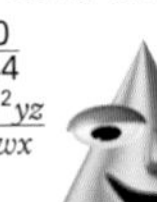

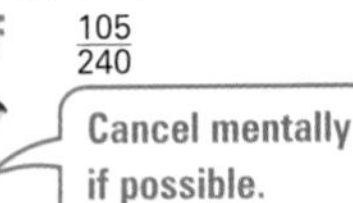

2 Add these fractions by finding the LCM of the denominators.

a $\frac{1}{24} + \frac{7}{36}$ **b** $\frac{7}{32} + \frac{11}{48}$ **c** $\frac{23}{56} + \frac{39}{70}$ **d** $\frac{19}{28} - \frac{7}{38}$ **e** $\frac{48}{72} - \frac{3}{20}$
f $\frac{134}{150} - \frac{79}{120}$ **g** $\frac{57}{132} + \frac{9}{48}$ **h** $\frac{5}{6} - \frac{4}{9} + \frac{7}{15}$ ***i** $\frac{5}{12} + \frac{7}{18} - \frac{9}{24}$

Find the LCM mentally if possible.

3 Tristram found this method of finding the HCF and LCM in his sister's maths book.

To find the LCM and HCF of 30 and 48:

Divide both by the same smallest prime possible	2	30 48	$30 \div 2 = 15$, $48 \div 2 = 24$
Divide both again by the smallest prime possible	3	15 24	$15 \div 3 = 5$, $24 \div 3 = 8$
		5 8	

It is not possible to divide both 5 and 8 by the same prime number so stop.
HCF $= 2 \times 3 = 6$
LCM $= 2 \times 3 \times 5 \times 8 = 240$ LCM $= 2 \times 3 \times 5 \times 8 = 240$

Use this method to find the HCF and LCM of these.

a 12 and 30 **b** 24 and 30 **c** 16 and 20 **d** 20 and 30
e 20 and 48 **f** 48 and 100 **g** 96 and 144

4 The HCF of 20 and another number is 5.
What might the other number be? Write down more than one answer.

***5** You are given the HCF and LCM of two numbers.
What might the two numbers be?

a HCF = 5, LCM = 60 **b** HCF = 12, LCM = 84
c HCF = 24, LCM = 720 **d** HCF = $4ab$, LCM = $16a^2b^2$

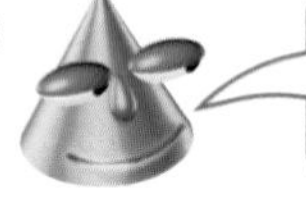

There is more than one answer for some of these. Try to find them all.

***6** When is the LCM of two numbers
a one of the numbers **b** the product of the two numbers?

Review 1 Cancel these fractions by finding the HCF of the numerator and denominator.

a $\frac{36}{90}$ **b** $\frac{52}{84}$ **c** $\frac{28}{120}$ **d** $\frac{138}{195}$ ***e** $\frac{12x^2y}{4xy}$

Review 2 Add these fractions by finding the LCM of the denominators.

a $\frac{3}{14}+\frac{7}{21}$ **b** $\frac{5}{36}+\frac{9}{40}$ **c** $\frac{4}{27}+\frac{9}{45}$ **d** $\frac{21}{28}-\frac{14}{38}$ **e** $\frac{83}{96}-\frac{17}{28}$

***Review 3** When is the HCF of two numbers one of the numbers?

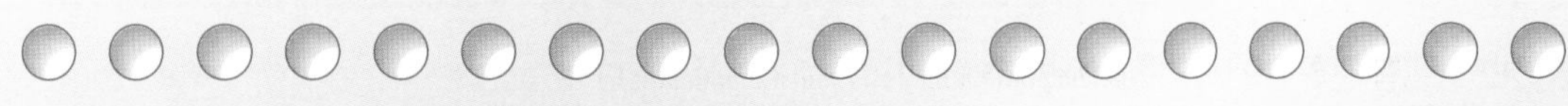

Investigation

Number of factors

1 $36 = 2^2 \times 3^2$ has $3 \times 3 = 9$ factors (1, 2, 3, 4, 6, 9, 12, 18, 36)
$45 = 3^2 \times 5^1$ has $3 \times 2 = 6$ factors (1, 3, 5, 9, 15, 45)
$200 = 2^3 \times 5^2$ has $4 \times 3 = 12$ factors

How has the number of factors been worked out from the indices? **Investigate.**

Write the numbers below as a product of prime numbers in index notation.
Use this to write down the number of factors each has.
60 125 750 3200

2 Find the smallest number that has 18 factors.

3 Find the smallest number greater than 50 that has the same number of factors as 50.
Justify your answer.

***Puzzle**

I am a multiple of 21 and 35.
I have four digits.
What is the smallest number I could be?

Powers and roots

Indices

Remember
3^4 is read as 'three to the power of 4'.
The 4 in 3^4 is called an **index**. The plural of index is **indices**.
On a calculator, squares are keyed using x^2 .
cubes are keyed using x^3 .

Example $(^-8{\cdot}1)^3$ is keyed as
((–) 8·1) x^3 = to get 531.441 .

We need the brackets around $^-8{\cdot}1$ so that we find $^-8{\cdot}1 \times {}^-8{\cdot}1 \times {}^-8{\cdot}1$ and not $^-(8{\cdot}1 \times 8{\cdot}1 \times 8{\cdot}1)$.

On a calculator, indices are keyed using x^y .

Example 21^5 is keyed as 21 x^y 5 = to get 4 084 101.

Discussion

Find the value of these.

10^0 18^0 24^0 116^0 256^0 $(^-3)^0$ $5{\cdot}7^0$

Make a statement about the value of x^0 for any value of x. **Discuss.**

Any number to the power of zero equals 1.

$x^0 = 1$

Exercise 4

1 Find the value of these. Give the answers to **f**, **g**, **h** and **k** to 1 d.p.

a 14^3 **b** 17^4 **c** 6^6 **d** 7^8 **e** 23^4 **f** $5{\cdot}6^4$
g $8{\cdot}72^5$ **h** $19{\cdot}63^3$ **i** $(^-5)^4$ **j** $(^-7)^3$ **k** $(^-2{\cdot}3)^5$

2 The number of bacteria cells doubled each hour, as shown in this table.
How many cells will there be after
a 6 hours **b** 12 hours **c** a day?

Hour	Number of cells
0	1 = 2^0
1	2 = 2^1
2	4 = 2^2
3	8 = 2^3
.	. .
.	. .
.	. .

There is more about powers of ten and indices on page 19.

3 0·1 written using indices is 10^{-1}.
Write these using indices.
a 10 **b** 1 **c** 1000 **d** 0·01 **e** $\frac{1}{100}$ **f** $\frac{1}{1000}$

4 The difference of the squares of two consecutive even numbers is 20.
What are these even numbers?

5

2	9	16	28	35	54	65	72	91

$16 = 8 + 8$
$= 2^3 + 2^3$

$91 = 64 + 27$
$= 4^3 + 3^3$

All of the numbers in the box can be written as the sum of two cubes.
Show how.

6 **a** Write 60 as a product of prime factors.
b Use your answer to **a** to find the smallest number that 60 must be multiplied by to get a square number.
c What is this square number?

7 8^3 is 512. Without using a calculator work out the units digit of 8^{12}.

8 I am thinking of a six-digit square number with a units digit of 4. □□□□□4.
Could its square root be a prime number?
Explain your answer.

***9** Find the smallest number that can be written as the sum of two cubes in two different ways.

Review 1 Find the value of these. Give the answers to **c** and **e** to 1 d.p.
a 8^4 **b** 7^6 **c** $8{\cdot}1^5$ **d** $(^-3)^7$ **e** $(^-2{\cdot}4)^4$

Review 2
a If $9 \times x$ is a square number, what is the smallest value of x?
*__b__ If 3PQ9 is a square number, what digits do the letters P and Q stand for? Is there more than one answer?

Puzzle

1 The sum of the squares of six consecutive whole numbers is 1111.
What are the six numbers?

*2 Find the missing digits. ∗ can stand for any digit.
a $(∗8)^2 = ∗8∗$ b $(∗∗)^2 = 5∗∗5$ c $(∗∗)^2 = ∗∗25$
d $(∗∗∗)^2 = ∗88∗44$ e $(6 \times ∗∗)^2 = 54∗56$
Is there more than one answer for some?

*3 Find the missing digits. ∗ can stand for any digit.
a $(∗∗)^3 = ∗∗∗7$
Explain why there is just one answer.
b $(∗∗)^3 = ∗∗∗2$
Explain why there is just one answer.
c $(∗∗)^3 = ∗∗∗5$
Explain why there is just one answer.
d $(∗∗)^3 = ∗∗∗∗9$
Find the two possible answers.
Explain why there are no more than two answers.

Hint: What single digit cubed ends in 7?

*4 Great Great Grandmother wouldn't tell when she was born.
She did say that she was A years old in the year A^2.
What year was she born? (**Hint:** A is between 40 and 50.)

Investigations

1 Sums and Differences

Do this investigation mentally.

Numbers	1 and 2	2 and 3	3 and 4	4 and 5	5 and 6	6 and 7
Squares	1 and 4	4 and 9	9 and 16	16 and 25	25 and 36	36 and 49
Difference of squares	3	5	7	9	11	13

This table gives the differences of the squares of pairs of numbers which differ by 1.
The differences of the squares give a sequence of odd numbers.
What other pattern do you notice? **Hint:** Add each pair of numbers.
Use this pattern to predict the difference of the squares of
a 9 and 10 **b** 25 and 26 **c** 82 and 83 **d** 107 and 108.

Numbers	1 and 3	2 and 4	3 and 5	4 and 6	5 and 7	6 and 8
Squares	1 and 9	4 and 16	9 and 25	16 and 36	25 and 49	36 and 64
Difference of squares	8	12	16	20	24	28

This table gives the differences of the squares of pairs of numbers which differ by 2.
The differences of the squares give the sequence 8, 12, 16, 20, ...
What else do you notice? **Hint:** Begin by adding each pair of numbers.

Predict the difference of the squares of
a 16 and 18 **b** 20 and 22 **c** 78 and 80 **d** 135 and 137.

Repeat for pairs of numbers that differ by 3.
Predict the difference of the squares of **a** 27 and 30 **b** 92 and 95.

What patterns do you think you would get in the differences of the squares of pairs of numbers which differ by 4? by 5? by 6? ...
Use your results to write down the answers for
a $18^2 - 12^2$ **b** $36^2 - 24^2$ **c** $174^2 - 154^2$.

2 True or False

Is it true that the sum of n consecutive odd numbers, starting at 1, is n^2?

3 End Digits

a **You could use a graphical calculator**.

$9^1 = \mathbf{9}$
$9^2 = 9 \times 9 = 8\mathbf{1}$
$9^3 = 9 \times 9 \times 9 = 81 \times 9 = 72\mathbf{9}$
$9^4 = 9 \times 9 \times 9 \times 9 = 729 \times 9 = 656\mathbf{1}$
$9^5 = 9 \times 9 \times 9 \times 9 \times 9 = 6561 \times 9 = 59\ 04\mathbf{9}$
$9^6 = 9 \times 9 \times 9 \times 9 \times 9 \times 9 = 59\ 049 \times 9 = 531\ 44\mathbf{1}$

The last digits in $9^1, 9^2, 9^3, 9^4, 9^5, 9^6$ are 9, 1, 9, 1, 9, 1.
Without doing any more calculations, predict the last digits of 9^{12} and 9^{15}.

We can use a graphical calculator to repeatedly multiply by the same number.
Key 9 EXE Ans × 9 EXE, then continue to press EXE to get the display shown.

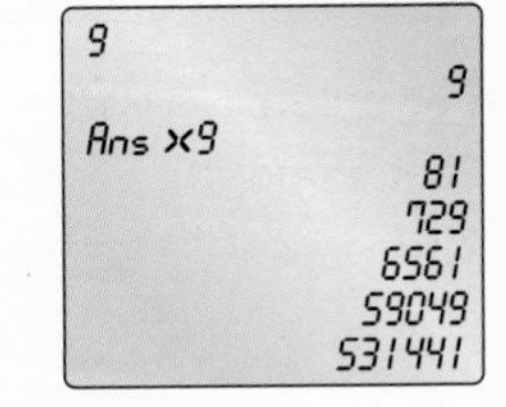

b Work out the answers to these mentally first. Check using a calculator.

Investigate to find the pattern in the last digits of $4^1, 4^2, 4^3, 4^4, 4^5, 4^6, \ldots$
Use this pattern to write down the last digits of 4^{21} and 4^{30}.

* **Investigate** to find the pattern in the last digits of $2^1, 2^2, 2^3, 2^4, 2^5, 2^6, \ldots$ and $3^1, 3^2, 3^3, 3^4, 3^5, 3^6, \ldots$

Predict the last digits of $2^{11}, 2^{12}, 2^{13}$ and 2^{14}.
Predict the last digits of $3^9, 3^{12}, 3^{15}$ and 3^{22}.

* Predict the last digits of 7^{41} and 8^{41} by **investigating** the patterns in $7^1, 7^2, 7^3, 7^4, \ldots$ and $8^1, 8^2, 8^3, \ldots$

Square roots and cube roots

Remember
$\sqrt{81}$ is read as 'the square root of 81'.
We must find what number squared equals 81.

Squaring and finding the square root are inverse operations. So are cubing and finding the cube root.

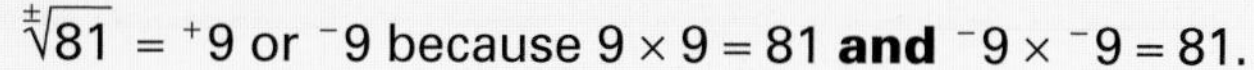

$\sqrt[\pm]{81} = {}^{+}9$ or ${}^{-}9$ because $9 \times 9 = 81$ **and** ${}^{-}9 \times {}^{-}9 = 81$.
9 is the positive square root and ${}^{-}9$ is the negative square root.

$\sqrt{81}$ means the positive square root, 9. $\sqrt[\pm]{81}$ means the **positive** and **negative** square roots, 9 and ${}^{-}9$.

$\sqrt[3]{27}$ is read as 'the cube root of 27'.
We must find what number cubed equals 27.
$\sqrt[3]{27} = 3$ because $3 \times 3 \times 3 = 27$

On a calculator we use ∛ to find cube roots.

Example $\sqrt[3]{91}$ is keyed as ∛ 91 = to get 4·5 (1 d.p.).

Discussion

- 

  Discuss Jack's and Chelsea's statements. Are they correct?

  Find these. ${}^{-}3 \times {}^{-}3 \times {}^{-}3 = ({}^{-}3)^3 = \square$
  ${}^{-}4 \times {}^{-}4 \times {}^{-}4 = ({}^{-}4)^3 = \square$
  ${}^{-}1{\cdot}2 \times {}^{-}1{\cdot}2 \times {}^{-}1{\cdot}2 = ({}^{-}1{\cdot}2)^3 = \square$

  What can you say about the cube root of a negative number? **Discuss**.

- Calculate $\sqrt{4} + \sqrt{1}$.
  Calculate $\sqrt{5}$.
  Does $\sqrt{4} + \sqrt{1} = \sqrt{5}$?

  Does $\sqrt{9} + \sqrt{16} = \sqrt{25}$?
  Does $\sqrt{25} + \sqrt{49} = \sqrt{74}$?

  Make a statement about whether $\sqrt{a} + \sqrt{b}$ is equal to $\sqrt{a+b}$. **Discuss**.

We can find the **square root** of a number by **factorising**.

Examples

$\sqrt{36} = \sqrt{9 \times 4}$
$= \sqrt{9} \times \sqrt{4}$
$= 3 \times 2$
$= \mathbf{6}$

$\sqrt{225} = \sqrt{9 \times 25}$
$= \sqrt{9} \times \sqrt{25}$
$= 3 \times 5$
$= \mathbf{15}$

Factorising means writing as a product of factors.
$225 = 9 \times 25$
9 and 25 are factors of 225.

When finding square roots, estimate the answer by finding an integer upper and lower bound for the answer. Use the square numbers you know.

Example $\sqrt{4} < \sqrt{8} < \sqrt{9}$ so $2 < \sqrt{8} < 3$

known square numbers — lower bound — upper bound

Exercise 5

Except for question 8 and Review 5.

1 Find the answers to these mentally.
a $\sqrt{24-8}$ **b** $\sqrt{23+13}$ **c** $\sqrt{89-25}$ **d** $\sqrt{36+16-3}$ **e** $\sqrt{101-36-1}$

2 Give the positive *and* negative answer to these.
a $\pm\sqrt{9}$ **b** $\pm\sqrt{25}$ **c** $\pm\sqrt{49}$ **d** $\pm\sqrt{144}$ **e** $\pm\sqrt{121}$

3 Find these by factorising.
a $\sqrt{256}$ **b** $\sqrt{400}$ **c** $\sqrt{324}$ **d** $\sqrt{441}$ ***e** $\sqrt{484}$

4 Find an upper and lower bound for these using the square numbers you know.
a $\sqrt{3}$ **b** $\sqrt{7}$ **c** $\sqrt{10}$ **d** $\sqrt{11}$ **e** $\sqrt{5}$

5 Round the answer to these to 1 d.p.
a $\sqrt[3]{7}$ **b** $\sqrt[3]{11}$ **c** $\sqrt[3]{100}$ **d** $\sqrt[3]{6\cdot4}$ **e** $\sqrt[3]{8\cdot95}$ **f** $\sqrt[3]{0\cdot8}$
g $\sqrt[3]{1\cdot34}$ **h** $\sqrt[3]{25\cdot6}$ **i** $\sqrt[3]{21\cdot43}$

6 Write true or false for these.
a $\sqrt{12}+\sqrt{4}=\sqrt{16}$ **b** $\sqrt{29}-\sqrt{4}=\sqrt{25}$ **c** $\sqrt{4}\times\sqrt{4}=\sqrt{16}$ **d** $\sqrt{5}\times\sqrt{5}=\sqrt{25}$

7 What goes in the gap, **positive** or **negative**?
a There are two square roots of a _______ number. One is _______ and the other _______.
b The cube root of a positive number is _______.
c The cube root of a negative number is _______.

8 Use your calculator to find these. Give the answers to 2 d.p.
a $\sqrt{187-46}$ **b** $\sqrt[3]{364-197}$ **c** $\sqrt[3]{1064-21^2}$ **d** $\sqrt{4^2+7^2}$ **e** $\sqrt{26^2-8^2}$
f $\sqrt[3]{32^2-17^2}$ **g** $\sqrt{18\cdot6+6\cdot88}$ **h** $\sqrt[3]{5\times17^2}$ **i** $\sqrt{16\cdot94-4\cdot1^2}$ **j** $\sqrt{9\cdot8^2-2\times0\cdot68^2}$
k $\sqrt[3]{186-235}$ **l** $\sqrt{14^2-5^3}$ **m** $\sqrt{12}+\sqrt{13}$ **n** $\sqrt{27}+\sqrt{39}$ ***o** $\sqrt[3]{3}+\sqrt[3]{5}$
***p** $\sqrt[3]{8\cdot6^3-5\cdot2^2}$

9 Paul estimated $\sqrt{11}$ like this.

$\sqrt{9} < \sqrt{11} < \sqrt{16}$
$3 < \sqrt{11} < 4$
11 is closer to 9 than to 16 so $\sqrt{11}$ will be closer to 3 than to 4.
$\sqrt{11} \approx 3\cdot3$

9 and 16 are the closest square numbers.

Find a rough estimate for these using Paul's method.
a $\sqrt{19}$ **b** $\sqrt{39}$ **c** $\sqrt{72}$ **d** $\sqrt{94}$ **e** $\sqrt{130}$

Review 1 Find the answers to these mentally.
a $\sqrt{36-11}$ **b** $\sqrt{99+22}$ **c** $\sqrt{84+21-5}$ **d** $\sqrt{84-35}$

Review 2 Give the positive and negative square roots of **a** 81 **b** 100.

Review 3 Find these by factorising. **a** $\sqrt{196}$ **b** $\sqrt{729}$

Review 4 $\sqrt{5}$ lies between 2 and 3. Explain how you know this without doing the calculation.

Review 5 Use your calculator to find these. Give the answers to 1 d.p.
a $\sqrt[3]{86}$ **b** $\sqrt{16-3}$ **c** $\sqrt[3]{152-61}$ **d** $\sqrt{8^2+5^2}$ *__e__ $\sqrt[3]{5^3+3^3}$

Discussion

Sharyn was using a calculator which didn't have a $\sqrt{\ }$ or $\sqrt[3]{\ }$ key.

She decided to use trial and improvement to find $\sqrt{19}$ to 2 decimal places. She began as follows.

19 lies between the two square numbers 16 and 25.
$\sqrt{19}$ lies between $\sqrt{16}$ and $\sqrt{25}$.
$\sqrt{19}$ lies between 4 and 5.

Try 4·5 $4{\cdot}5 \times 4{\cdot}5 = 20{\cdot}25$ too big
Try 4·3 $4{\cdot}3 \times 4{\cdot}3 = 18{\cdot}49$ too small
Try 4·4 $4{\cdot}4 \times 4{\cdot}4 = 19{\cdot}36$ too big

I now know that $\sqrt{19}$ lies between 4·3 and 4·4

Try 4·32 $4{\cdot}32 \times 4{\cdot}32 = 18{\cdot}6624$ too small

Discuss Sharyn's method. How could Sharyn continue?
Find $\sqrt{19}$ to 2 decimal places. **Discuss**.

Exercise 6

1 Use trial and improvement to find these, to 2 d.p. Do not use the $\sqrt{\ }$ or $\sqrt[3]{\ }$ key.
a $\sqrt{8}$ **b** $\sqrt{24}$ **c** $\sqrt[3]{50}$ **d** $\sqrt[3]{81}$

Review Use trial and improvement to find these, to 2 d.p. Do not use the $\sqrt[3]{\ }$ or $\sqrt{\ }$ key.
a $\sqrt{35}$ **b** $\sqrt[3]{16}$

Investigation

T

Painted Cubes

The outside of a cube made from 27 smaller cubes is painted pink.

Use a copy of this table. Fill it in to show the number of small cubes with 0, 1, 2 or 3 faces painted pink.

	Faces painted pink			
Total no. of small cubes	No. with 0 faces	No. with 1 face	No. with 2 faces	No. with 3 faces
27				
64				
125				
216				

What if the cube was made from 64 small cubes? What about 125, 216, ... ? **Investigate**.

Try to find a pattern or rule.

Multiplying and dividing numbers with indices

Discussion

- $2^5 = 2 \times 2 \times 2 \times 2 \times 2 = 32$ $\quad 2^3 = 2 \times 2 \times 2 = 8$ $\quad 2^8 = 2 \times 2 \times 2 \times 2 \times 2 \times 2 \times 2 \times 2 = 256$ $\quad 2^5 \times 2^3 = 32 \times 8 = 256$

  It seems that $2^5 \times 2^3 = 2^{(5+3)} = 2^8$.

  Is it true that $2^4 \times 2^3 = 2^7$?

  Try multiplying other powers of 2. **Discuss** the results.

  What if the 2 was replaced by 3 or some other number?
  What if we began with two different 'base' numbers?

  Examples $\quad 2^3 \times 5^3 \quad 3^4 \times 2^2$

- $2^6 = 2 \times 2 \times 2 \times 2 \times 2 \times 2 = 64$ $\quad 2^2 = 2 \times 2 = 4$ $\quad 2^4 = 2 \times 2 \times 2 \times 2 = 16$ $\quad 2^6 \div 2^2 = 64 \div 4 = 16$

  It seems that $2^6 \div 2^2 = 2^{6-2} = 2^4$.

  Is it true that $2^8 \div 2^5 = 2^3$?
  Investigate for dividing other powers of 2.

  What if the 2 was replaced by 3 or some other number? **Discuss.**

The **index laws** are:
Indices are **added** when **multiplying**.
Indices are **subtracted** when **dividing**.

Worked Example
Write each of these as a number with a single index.
a $7^4 \times 7^9$ **b** $7^8 \div 7^2$

Answer
a $7^4 \times 7^9 = 7^{(4+9)} = 7^{13}$ **b** $7^8 \div 7^2 = 7^{(8-2)} = 7^6$

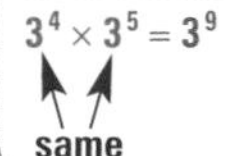

Exercise 7

Except for question 3 and Review 1.

1 Write these as single powers of 5.
a $5^3 \times 5^6$ **b** $5^4 \times 5^3$ **c** $5^4 \times 5^4$ **d** $5^7 \times 5^{10}$ **e** $5^{16} \div 5^4$ **f** $\frac{5^8}{5^4}$

2 Simplify.
a $3^4 \times 3^6$ **b** $6^3 \times 6^2$ **c** $4^7 \div 4^3$ **d** $\frac{10^{11}}{10^9}$ **e** $\frac{8^6}{8^5}$ **f** $7^5 \times 7^5$
g $27^3 \times 27^4$ **h** $\frac{16^9}{16^7}$ **i** $11^8 \times 11^4$ **j** $24^{24} \div 24^{16}$

3 Use the index laws to simplify these. Then find the answers using a calculator.
a $2^4 \times 2^3$ **b** $3^2 \times 3^3$ **c** $\frac{5^7}{5^5}$ **d** $\frac{3^9}{3^9}$ **e** $4^2 \times 4^3 \times 4^0$ **f** $2^2 \times 2^3 \times 2^2$
g $\frac{4^3 \times 4^6}{4^7}$ ***h** $\frac{2^{13}}{2^4 \times 2^5}$ ***i** $\frac{7^{11} \times 7^4}{7^9 \times 7^6}$ ***j** $\frac{5^7 \times 5^3}{5^2 \times 5^5}$

4 Which of the following statements are correct?

a $4^3 \times 4^2 = 4^6$ **b** $4^3 + 4^2 = 4^5$ **c** $4^3 \times 2^2 = 8^5$ **d** $4^8 \div 2^3 = 2^5$ **e** $4^5 - 4^3 = 4^2$

5 Sarah thinks that the index laws are true for algebra as well as arithmetic.

She wrote $a^4 \times a^2 = (a \times a \times a \times a) \times (a \times a)$

$= a^6$

So $a^4 \times a^2 = a^{(4+2)} = a^6$

a How could she show that the index laws were true for $m^3 \div m^2$?

b Show that the index laws are true for these.

i $n^4 \times n^3 = n^7$ **ii** $p^8 \div p^3 = p^5$

6 Use the index laws to simplify these.

a $a^4 \times a^5$ **b** $n^7 \times n^5$ **c** $p^8 \div p^2$ **d** $m^6 \times m^4$ **e** $b^9 \div b^7$

f $r^6 \times r^{11}$ **g** $d^{12} \div d^4$ **h** $x^7 \times x^{12}$ **i** $\frac{n^{10}}{n^4}$ **j** $a^0 \times a^4$

k $\frac{y^6}{y^2}$ **l** $\frac{b^{20}}{b^5}$ **m** $\frac{a^{10}}{a^{10}}$ **n** $b^2 \times b^4 \times b^5$

Review 1 Simplify these, then find the answer using a calculator.

a $4^3 \times 4^1$ **b** $3^2 \times 3^2$ **c** $\frac{4^6}{4^4}$

d $3^2 \times 3^1 \times 3^2$ **e** $\frac{5^2 \times 5^3}{5^5}$ **f** $\frac{2^{11}}{2^2 \times 2^4}$ **g** $\frac{8^5 \times 8^6}{8^3 \times 8^7}$

Review 2 Use the index laws to simplify these.

a $y^5 \times y^8$ **b** $m^{20} \div m^4$ **c** $a^3 \times a^0$ **d** $\frac{p^9}{p^3}$ **e** $b^3 \times b^4 \times b^5$

Summary of key points

We **add and subtract integers** using a number line, patterns or using facts we already know such as $1 + {}^{-}1 = 0$, $0 - 1 = {}^{-}1$ and $0 - {}^{-}1 = 1$.

Example $45 + {}^{-}36 = 9 + 36 + {}^{-}36 = 9 + 0 = 9$

Multiplying or dividing two negative numbers gives a positive number.
Multiplying or dividing one negative and one positive number gives a negative number.

Examples ${}^{-}3 \times {}^{-}4 = 12$ ${}^{-}5 \times 7 = {}^{-}35$ ${}^{-}32 \div 8 = {}^{-}4$ ${}^{-}80 \div {}^{-}10 = 8$

B We can use the **sign change key**, (–), to add and subtract integers.

Always follow the rules for **order of operations**.

Brackets, **I**ndices, **D**ivision and **M**ultiplication, **A**ddition and **S**ubtraction (BIDMAS)

Examples ${}^{-}7 - 2 \times {}^{-}3 = {}^{-}7 - {}^{-}6$ ${}^{-}4({}^{-}6 + 1) = {}^{-}4 \times {}^{-}5$

$= {}^{-}1$ $= 20$

D We can use algebra to **prove** some divisibility rules.

Example Any two-digit number in which the tens and units digits are the same is divisible by 11.

If the number of units is b, then the number of tens is $10b$.

The number has the form $10b + b = 11b$ and $11b$ is always divisible by 11.

We use **prime factor decomposition** to find the **highest common factor, HCF**, and the **lowest common multiple, LCM**.

The LCM can be used to find a common denominator when adding and subtracting fractions.

Example $\frac{7}{60} + \frac{9}{140} = \frac{49+27}{420}$

$= \frac{76}{420}$

$= \frac{19}{105}$

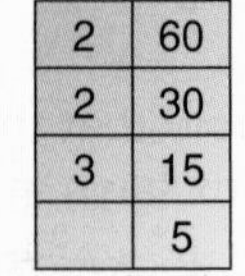

2	60
2	30
3	15
	5

$60 = 2^2 \times 3 \times 5$

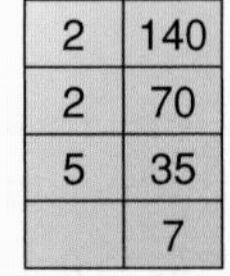

2	140
2	70
5	35
	7

$140 = 2^2 \times 5 \times 7$

HCF = 20
LCM = 420

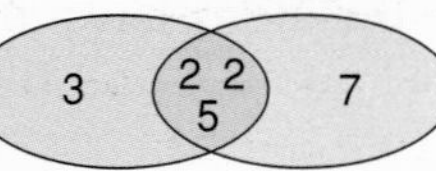

The HCF can be used to cancel fractions.

Example $\frac{60}{140} = \frac{3}{7}$ **dividing numerator and denominator by the HCF of 20.**

On a calculator, **cubes** are keyed using the [x^3] key, and **indices** are keyed using the [x^y] key.

Example 5^4 is keyed as [5] [x^y] [4] [=] to get 625.

Note: $a^0 = 1$ Any number to the power of 0 is equal to 1.

On a calculator we use [$\sqrt[3]{\ }$] to find **cube roots**.

Example $\sqrt[3]{97}$ is keyed as [$\sqrt[3]{\ }$] [97] [=] to get $32.\dot{3}$.

Note: All positive numbers have two square roots, a positive and a negative one.

Example $\sqrt[\pm]{25} = {}^{+}5$ or ${}^{-}5$.

The cube root of a positive number is positive.

The cube root of a negative number is negative.

We can **estimate square roots and cube roots**.

Example a rough estimate for $\sqrt{30}$ is $5 < \sqrt{30} < 6$

because $\sqrt{25} < \sqrt{30} < \sqrt{36}$

known square numbers.

Example We can estimate the value of $\sqrt[3]{16}$ using **trial and improvement**.

See page 49.

The index laws

When multiplying, add the indices. *Example* $8^6 \times 8^7 = 8^{(6+7)} = 8^{13}$

When dividing, subtract the indices. *Example* $\frac{12^{11}}{12^5} = 12^{(11-5)} = 12^6$

The index laws also apply to algebra.

Examples $m^5 \times m^7 = m^{(5+7)}$
$= m^{12}$

$p^9 \div p^4 = p^{(9-4)}$
$= p^5$

The 'base' number must be the same.

Test yourself

Except for questions 2, 7, 9, 11 and 12.

T

1 Use a copy of these.

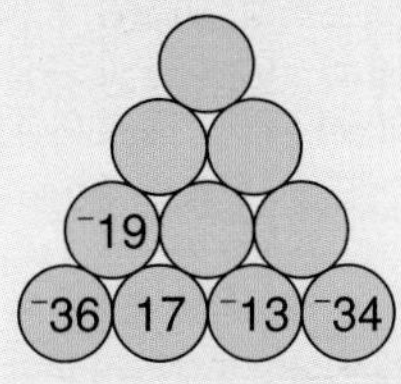

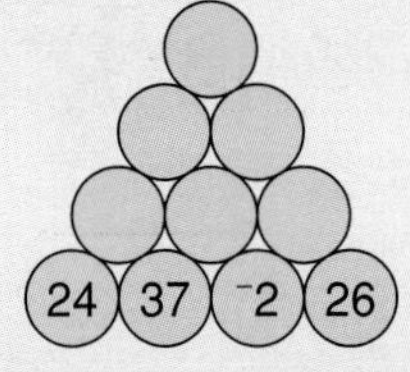

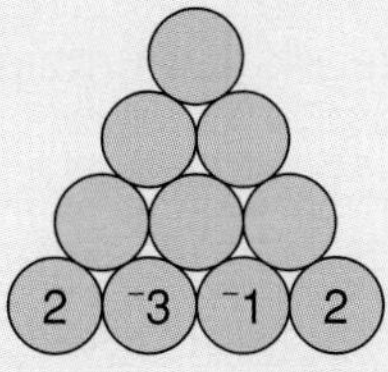

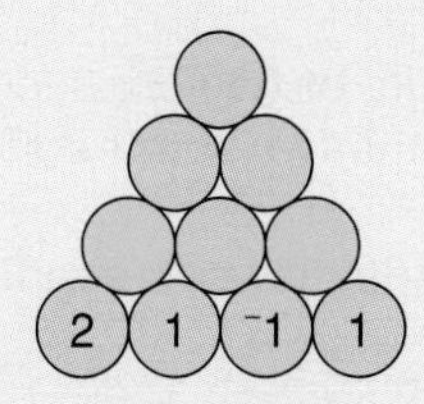

What number goes in the top circles?

a In the green circles, two numbers are added to get the number above.

b In the blue circles, the number on the right is subtracted from the number on the left to get the number above.

c In the purple circles, two numbers are multiplied to get the number above.

d In the pink circles, the number on the left is divided by the number on the right to get the number above.

T

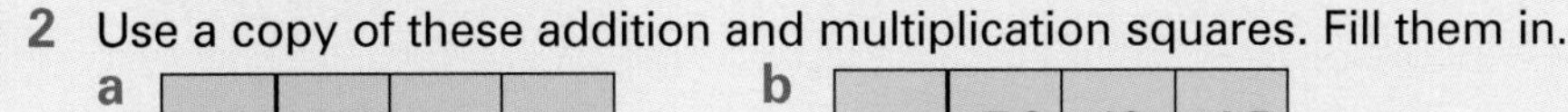

2 Use a copy of these addition and multiplication squares. Fill them in.

a

+	$^-109$	79	$^-212$
346			
$^-127$			
$^-449$			

b

×	$^-7{\cdot}8$	42	$^-4{\cdot}5$
$^-34$			
8·6			
$^-3{\cdot}8$			

3 **a** $2(^-4 + ^-7)$ **b** $^-3 + ^-1(5 - ^-4)$ **c** $2 \times ^-3 + ^-4 \times 7$ **d** $\frac{^-2(8 - ^-7)}{^-6}$

4 What number goes in the box?

a $9 + ^-9 = \square$ **b** $64 + ^-59 = \square$ **c** $18 - ^-3 = \square$ **d** $^-14 - ^-8 = \square$

e $^-105 + 131 = \square$ **f** $^-15 + \square = 18$ **g** $^-9 - \square = 12$

5 Cancel these fractions by finding the HCF of the numerator and denominator.

a $\frac{72}{168}$ **b** $\frac{88}{198}$ **c** $\frac{56}{98}$ **d** $\frac{80}{236}$ **e** $\frac{96}{184}$ **f** $\frac{92}{161}$

6 Use the LCM to add and subtract these.

a $\frac{1}{18} + \frac{3}{20}$ **b** $\frac{7}{24} + \frac{9}{30}$ **c** $\frac{8}{45} + \frac{7}{60}$ **d** $\frac{7}{120} + \frac{129}{200}$

7 Find the value of these. Give the answers to **c**, **d** and **e** to 2 d.p.

a 5^7 **b** 6^5 **c** $1{\cdot}4^4$ **d** $0{\cdot}42^5$ **e** $(^-8{\cdot}4)^3$ **f** 114^0

8 Find these answers mentally.

a $(6 + ^-2)^2$ **b** $(9 - 13)^2$ **c** $\frac{(5+1)^2}{(7-5)^2}$ **d** $\sqrt{4^2 + 20}$ **e** $\pm\sqrt{196}$

f 4^3 **g** $\sqrt[3]{27}$

9 Give the answers to 2 d.p. if you need to round.

a $7{\cdot}4^2 + 6$ **b** $2{\cdot}8^2 + 3 \times 4{\cdot}6^2$ **c** $\pm\sqrt{9{\cdot}5}$ **d** $\sqrt{7 \times 18^2}$ **e** 13^3

f $(^-8)^3$ **g** $\sqrt[3]{2197}$ **h** $17^3 - 12^2$

10 Is $\sqrt[3]{^{-}84}$ positive or negative? 

11 Use trial and improvement to find these to 2 d.p. Do not use the $\sqrt{\ }$ or $\sqrt[3]{\ }$ keys. 

a $\sqrt{45}$ **b** $\sqrt[3]{35}$

12 The product of two consecutive whole numbers is 2862.
What are the two numbers? 

Hint: Use $\sqrt{\ }$ key to get an estimate.

13 Use the index laws to simplify these to a single power. 

a $8^3 \times 8^4$ **b** $\frac{9^{12}}{9^7}$ **c** $\frac{5^2 \times 5^4}{5^3}$ **d** $a^5 \times a^3$ **e** $\frac{m^8}{m^3}$

*****14** Prove that all 4-digit numbers where the thousands and units digits are the same *and* the hundreds and the tens digits are the same are divisible by 11. 

3 Mental Calculation

You need to know

Key vocabulary

best estimate, complements, degree of accuracy, difference, estimate, operation, order of operations, product, quotient

A Star is Born

Using these two charts you can work out the day of the week someone was born.

Month	Jan	Jan (leap yr)	Feb	Feb (leap yr)	Mar	Apr	May	Jun	Jul	Aug	Sep	Oct	Nov	Dec
Month number	8	7	11	10	11	7	9	5	7	10	6	8	11	6

Remainder	0	1	2	3	4	5	6
Day	Saturday	Sunday	Monday	Tuesday	Wednesday	Thursday	Friday

Follow these steps: *Example* **Robbie Williams 13 February 1974**

1 Write down the last two digits of the year the person was born. **74**
2 Divide this number by 4. **18**
Ignore the remainder.
3 Write down the date of the day the person was born. **13**
4 Write down the number for the month the person was born. Use the chart above. **11**
5 Add up the numbers you wrote for 1–4. **74 + 18 + 13 + 11 = 116**
6 Divide the answer to 5 by 7. **116 ÷ 7 = 70 ÷ 7 + 46 ÷ 7 = 10 + 6 R 4**
= 16 R 4
7 Write down the remainder. **4**
8 Find the day using the chart above. **Wednesday**

Robbie Williams was born on a Wednesday.

1800s – subtract 5 from the total in **step 5**.
1700s – subtract 3 from the total in **step 5**.
2000s – subtract 1 from the total in **step 5**.

Find what day some other people were born. Do the calculations mentally using jottings if you need to.

Here are the birthdays of some famous people:

Michael Jackson 29 August 1958
Steven Spielberg 18 December 1947

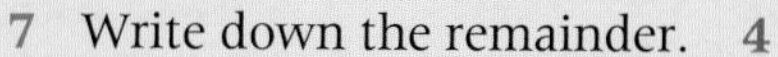

Adding, subtracting, multiplying and dividing

Adding and subtracting

These strategies can be used to **add and subtract mentally**.

1 Complements in 1, 10, 50, 100 and 1000

Example $8{\cdot}63 + 5{\cdot}37 = 8 + 0{\cdot}63 + 5 + 0{\cdot}37$
$= 8 + 5 + 1$
$= \mathbf{14}$

0·63 and 0·37 are **complements** in 1 and we can add in any order.

2 Partitioning

Examples $6{\cdot}83 - 0{\cdot}68 = 6{\cdot}83 - 0{\cdot}6 - 0{\cdot}08$
$= 6{\cdot}23 - 0{\cdot}08$
$= \mathbf{6{\cdot}15}$

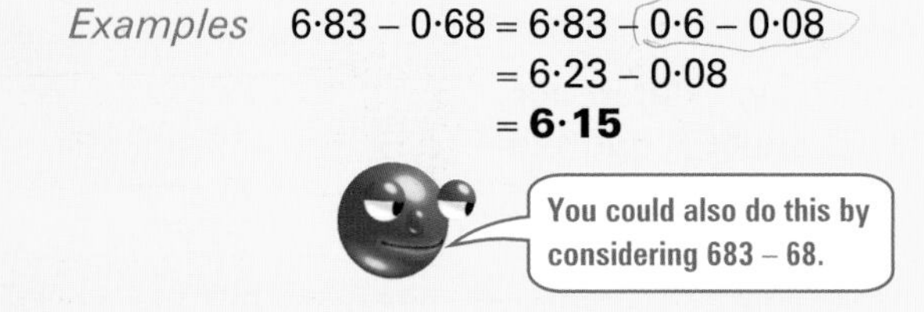

To find the length of wood to make this box kite frame,
$4(12{\cdot}2) + 4(18{\cdot}4) + 4(13{\cdot}4)$
$= 4(12{\cdot}2 + 18{\cdot}4 + 13{\cdot}4)$
$= 4(44)$
$= 4 \times 40 + 4 \times 4$
$= \mathbf{176\ cm}$

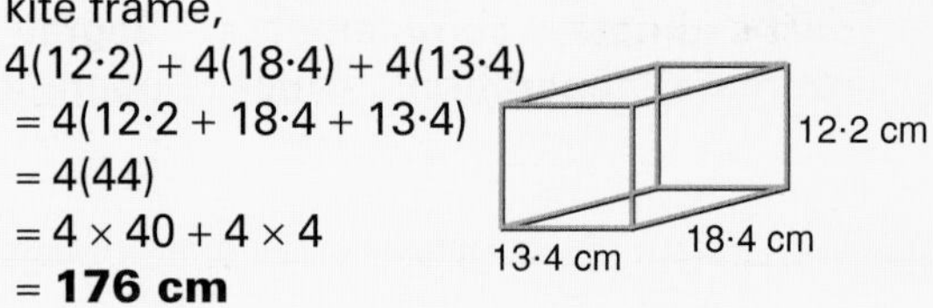

3 Using facts you already know

Example $0{\cdot}28 + 0{\cdot}43$
We know that $28 + 43 = 71$.
so $0{\cdot}28 + 0{\cdot}43 = \mathbf{0{\cdot}71}$.

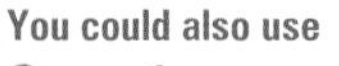
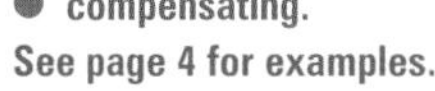

Multiplying and dividing

These strategies can be used to **multiply and divide mentally**.

1 Multiplying and dividing by multiples of 10, 100 and 1000

Examples $500 \times 9000 = 5 \times 100 \times 9 \times 1000$
$= 45 \times 100\,000$
$= \mathbf{4\,500\,000}$

$200 \div 4000 = 200 \div 4 \div 1000$
$= 50 \div 1000$
$= \mathbf{0{\cdot}05}$

2 Multiplying by near 10s, 100s

Example $56 \times 29 = 56 \times 30 - 56 \times 1$
$= \mathbf{1624}$

3 Using knowledge of place value

Examples $8{\cdot}7 \times 0{\cdot}01 = 8{\cdot}7 \times \frac{1}{100}$
$= 8{\cdot}7 \div 100$
$= \mathbf{0{\cdot}087}$

$900 \times 0{\cdot}6 = 90 \times 6$
$= \mathbf{540}$

$920 \div 0{\cdot}04 = 92\,000 \div 4$
$= \mathbf{23\,000}$

$0{\cdot}16 \times 0{\cdot}4 = 16 \div 100 \times 4 \div 10$
$= 16 \times 4 \div 1000$
$= 64 \div 1000$
$= \mathbf{0{\cdot}064}$

$48 \div 0{\cdot}8 = 48 \div (8 \times 0{\cdot}1)$
$= 48 \div 8 \div 0{\cdot}1$
$= 48 \div 8 \times 10$
$= 6 \times 10$
$= \mathbf{60}$

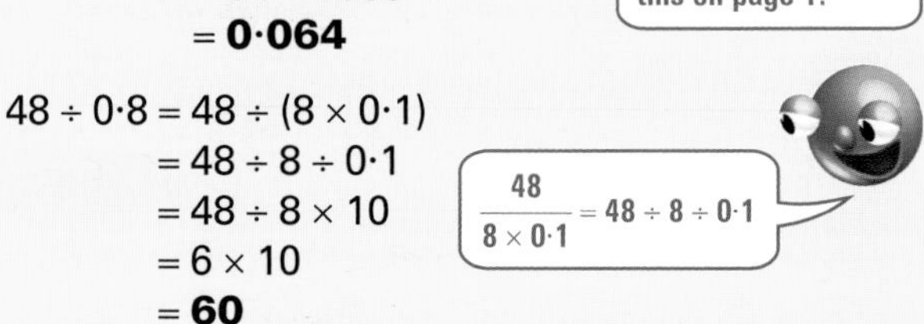

4 Using facts you already know

Example If we know that $32 \times 25 = 800$ we can work out 32×24 from $800 - 32 = \mathbf{768}$.

We know that $5 = 10 \div 2$, $25 = 100 \div 4$, $50 = 100 \div 2$ and $25 = 5 \times 5$.

Examples

$468 \times 50 = 468 \times 100 \div 2$
$= 46\,800 \div 2$
$= \mathbf{23\,400}$

$3{\cdot}24 \times 25 = 3{\cdot}24 \times 100 \div 4$
$= 324 \div 4$
$= \mathbf{81}$

$475 \div 25 = 475 \div 5 \div 5$
$= 95 \div 5$
$= \mathbf{19}$

5 Doubling and halving

We can double one number and halve the other.

Examples

Half of 16 is 8 and double 7·5 is 15.

$16 \times 7{\cdot}5 = 8 \times 15$
$= 4 \times 30$
$= \mathbf{120}$

$2{\cdot}4 \times 4{\cdot}5 = 1{\cdot}2 \times 9$
$= 1 \times 9 + 0{\cdot}2 \times 9$
$= 9 + 1{\cdot}8$
$= \mathbf{10{\cdot}8}$

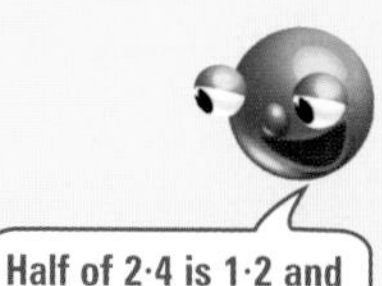

Half of 2·4 is 1·2 and double 4·5 is 9.

$^{-}11 \times 1{\cdot}5 = {}^{-}5{\cdot}5 \times 3$
$= \mathbf{^{-}16{\cdot}5}$

$^{-}12{\cdot}8 \times {}^{-}1{\cdot}25 = {}^{-}6{\cdot}4 \times {}^{-}2{\cdot}5$
$= {}^{-}3{\cdot}2 \times {}^{-}5$
$= \mathbf{16}$

You could also use partitioning or factors – see page 4.

Exercise 1

Find the answers mentally. You may use jottings.

1 Is it possible to get these totals by throwing four darts at this dart board? If so, how? Assume no darts miss the board.

a 110 **b** 135 **c** 130 **d** 145 **e** 140

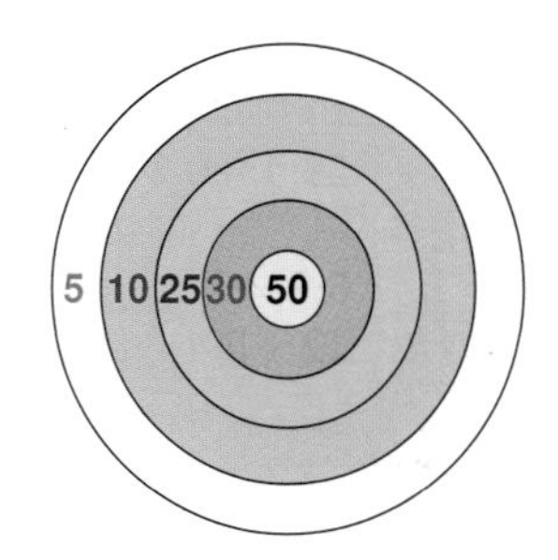

T **2** Use a copy of this.

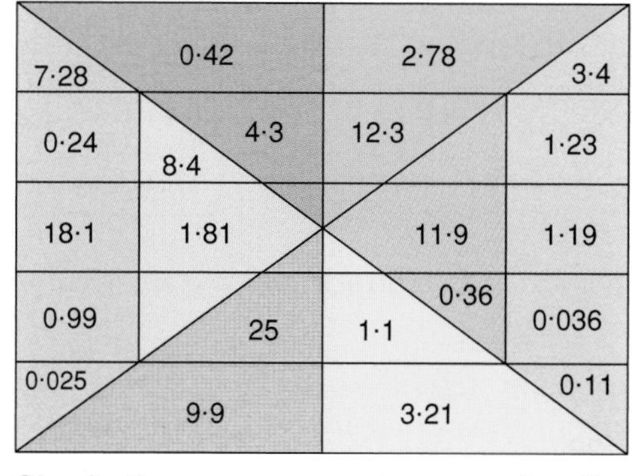

Shade the answers to these on the diagram.

a $9{\cdot}1 - 5{\cdot}7$ **b** $8{\cdot}3 + 9{\cdot}8$ **c** $0{\cdot}62 + 0{\cdot}57$ **d** $0{\cdot}34 + 0{\cdot}65$
e $0{\cdot}72 - 0{\cdot}48$ **f** $0{\cdot}047 + 0{\cdot}063$ **g** $0{\cdot}091 - 0{\cdot}066$ **h** $0{\cdot}083 - 0{\cdot}047$
***i** $8{\cdot}13 - 6{\cdot}9$ ***j** $8{\cdot}17 - 0{\cdot}89$

Which two sections of the diagram are shaded?

3 **a** $0{\cdot}3 + 0{\cdot}4 - 0{\cdot}1 + {}^{-}0{\cdot}6$ **b** $^{-}1{\cdot}2 - 0{\cdot}6 + 0{\cdot}4 - {}^{-}0{\cdot}7 + {}^{-}0{\cdot}3$
c $^{-}2{\cdot}1 + 3{\cdot}6 + {}^{-}3{\cdot}7$ **d** $4{\cdot}26 - 4{\cdot}1 + {}^{-}3{\cdot}2$

4 **a** $420 + 360$ **b** $320 - 260$ **c** $5800 + 6300$ **d** $3200 - 1800$ **e** $5200 - 3500$
f $384 + 119$ **g** $492 - 149$ **h** $237 + 197$ **i** $393 - 157$ **j** $3597 - 307$
k $4423 + 274$ **l** $3285 - 1106$ ***m** $11\,043 - 687$ ***n** $10\,630 - 479$

5 **a** 50×60 **b** 200×40 **c** $900 \div 30$ **d** $80 \div 400$ **e** 500×6000
f $400 \div 8000$ **g** $600 \times 80\,000$ **h** $48\,000 \div 800$ *__i__ $60 \div 90\,000$

T

6 Use a copy of these number chains.
Fill in the missing numbers.

a 1·4 → ×6 → ◯ → ÷4 → 2·1 → ×9 → ◯ → ÷0·3 → ◯ → ×0·5 → ◯

b 1·2 → ×3 → ◯ → ÷0·9 → ◯ → ×0·6 → ◯ → ÷8 → ◯ → ×1·2 → ◯

T

7 Use a copy of this.
Find the answers on the dot grid.
Join the dots in order. The first two are done.

a $4·5 \times 0·1 =$ **0·45** **b** $0·68 \times 0·1 =$ **0·068** **c** $5·7 \times 0·01$
d $0·5 \div 0·1$ **e** $0·052 \div 0·1$ **f** $0·3 \times 0·02$
g $5·7 \div 0·01$ **h** $5 \times 0·6$ **i** $5 \times 0·4$
j $0·5 \times 0·9$ **k** $0·04 \times 0·6$ **l** $^{-}0·07 \times 4$
m $^{-}0·09 \times {}^{-}8$ **n** $0·2 \div 4$ **o** $0·04 \div {}^{-}5$
p $0·6 \div 0·4$ **q** $0·07 \div 2$ **r** $^{-}0·09 \div {}^{-}2$
s $0·036 \div 6$ **t** $0·26 \times 0·4$ **u** $500 \times 0·8$
v $81 \div 0·9$ **w** $0·052 \div 0·1$

What shape have you made?

38	5·7	0·57	0·057	5	90	400
104	40	0·068	50	9	0·52	0·104
6	0·45	3·5	2	3	570	0·006
24	0·024	57	35	20	15	0·045
2·8	$^{-}$0·28	0·72	0·05	$^{-}$0·008	1·5	0·035

8 Given that $48·4 \times 1·7 = 82·28$, find the answer to these.
a $48·4 \times 0·17$ **b** $82·28 \div 17$ **c** $82·28 \div 0·17$ **d** $4·84 \times 0·17$ **e** $8·228 \div 17$

9 **a** $6·2 \times 2$ **b** $2·4 \times 3$ **c** $2·6 \times 9$ **d** $3·7 \times {}^{-}5$ **e** $2·9 \times 11$
f $12 \times 3·4$ **g** $4·5 \times {}^{-}19$ **h** $8·6 \times 21$ **i** $^{-}9·3 \times 11$ **j** $12 \times 1·01$
k $16 \times 2·02$ **l** $24 \times 2·01$

10 **a** $2·4 \times 4·5$ **b** $3·4 \times 2·5$ **c** $0·24 \times 2·5$ **d** $3·4 \times 1·2$
e $6·12 \times 2·5$ **f** $8·4 \times 1·5$ **g** $^{-}1·7 \times 4·5$ **h** $3·4 \times {}^{-}3·5$
i $^{-}3·6 \times 1·25$ **j** $^{-}8·8 \times {}^{-}1·25$

Try doubling and halving.

11 **a** $270 \div 6$ **b** $846 \div 4$ **c** $2·68 \div 4$ **d** $186 \div 12$ **e** $13·8 \div 6$
f $112 \div 16$ **g** $918 \div 18$

12 Given that $40 \times 25 = 1000$, find these.
a 40×24 **b** 39×25 **c** 41×25 **d** 40×26

13 Given that $36 \times 35 = 1260$, find these.
a 36×34 **b** 35^2 **c** 72×35

14 **a** $15 \times 49 = 735$. What is $15 \times 4·9$?
b Decrease 19·54 by 12·61.
c $8·634 - 7·98 = 0·654$. What is $8·634 - 0·654$?
d How much must I add to the sum of 3·82 and 2·79 to get 8·94?
e Find the **quotient** when 12·8 is divided by 8.
f What number divided by 8 gives the answer 0·12?
g Find the sum of 18·6 and 3·4 and then add this to the product of 24 and 1·6.

15 Find four consecutive odd numbers with a total of 80.

16 Find ways of filling in the box, the circle and the triangle.

a □ × ○ = 0·06 **b** □ ÷ ○ = 0·4 **c** □ × ○ = 0·32

d □ × ○ = 1·44 ***e** □ × ○ × △ = 0·08 ***f** □ + △ × ○ = 0·66

***17** □ × □·□ = □□·□

Put one of the digits 2, 3, 4, 5, 6 and 8 in each box to make a true statement.

***18** Choose numbers from the box to make these true.
Use each number only once in each question.

1·6	2·5	8
3·4	4	3

a □ × □ + □ = 7 **b** (□ + □) × □ = 20

c (□ + □) ÷ (□ − □) = 1 **d** □ × □ − □ × □ = ⁻9·6

T

Review 1

	H						
2418	1154	394		4199	7664	11·92	2418

0·027 7664 4199 4199 7664 5·92   7·51 7664 11·3 4199

7664 7·51   0·027 0·38 5·92 0·027 394 11·3   906 11·92

11·92 0·89 906 5·92   0·027 0·38 5·92 0·027 394 11·3

Use a copy of this box. Write the letter beside each calculation above its answer in the box.

H 779 + 375 = **1154**
I 526 + 380
E 690 − 239 − 57
R 8·6 + 3·2 − 4·1 − ⁻3·6
T 7324 − 4906
F 8·37 − 0·86
N 8·42 − 2·5
S 8·25 + 3·67
O 4681 + 2983
K 4·61 − 3·72
A 0·97 − 0·59
M 5206 − 1007
C 0·084 − 0·057

T

Review 2

18 000 ⁻1   0·02 60 ⁻1

9·3 ⁻1 84 18 000 12·5 64·8 16·5 20·3 0·078 0·13

64·8 ⁻1 ⁻1 20·3   0·02 **1 600 000** (H) 0·078   0·13 16·5 1 200 000 0·078

Use a copy of this box. Write the letter beside each calculation above its answer in the box.

H 800 × 2000 = **1 600 000**
W 48 000 ÷ 800
N 540 ÷ 0·03
F 225 ÷ 18
M 300 × 4000
E 7·8 × 0·01
L 12 × 5·4
K 8·12 × 2·5
T 60 ÷ 3000
R 0·84 ÷ 0·01
C 6·2 × 1·5
S 1·3 × 0·1
O ⁻2·5 × 0·4
A 198 ÷ 12

Review 3 Given that 35 × 43 = 1505, find these.

a 34 × 43 **b** 36 × 43 **c** 35 × 44 **d** 35 × 0·43 **e** 150·5 ÷ 4·3

Review 4

a Find the product of 320 and 0·06.
b Find the product of 3·4 and 8. Add 9·6.
c Find the quotient when 372 is divided by 1·2.
d What is the answer when the difference between 960 and 390 is added to the quotient of 420 and 15?

* **Review 5** The product of two numbers is 675.
One number is three times the other.
What are the two numbers?

Puzzle

$$\begin{array}{r} A\cdot AB \\ \times\ A\cdot\ \ A \\ \hline A\cdot BCB \end{array}$$

Replace A, B and C with digits to make this correct.

Order of operations mentally

Worked Example
Calculate these mentally.

a $4(7-2)-3\times 8$ **b** $(29-16+5-14)^3$ **c** $\sqrt[3]{25+39}$ **d** $\frac{(7\times 5)^2}{15-8}$

e $48\div(7+5)-6+4\times(12\div 4)^3$ **f** $^-8^2+4$ **g** $(\frac{5}{2})^2$

Answer

a $4(7-2)-3\times 8 = 4\times 5-3\times 8 = 20-24 = \mathbf{^-4}$

b $(29-16+5-14)^3 = 4^3 = \mathbf{64}$

c $\sqrt[3]{25+39} = \sqrt[3]{64} = \mathbf{4}$

d $\frac{(7\times 5)^2}{15-8} = \frac{{}^{1}\cancel{7}\times 5\times 7\times 5}{\cancel{7}^{1}} = 5\times 7\times 5 = \mathbf{175}$

e $48\div(7+5)-6+4\times(12\div 4)^3 = 48\div 12-6+4\times 3^3$
$= 48\div 12-6+4\times 27$
$= 4-6+4\times 27$
$= 4-6+108$
$= \mathbf{106}$

Remember BIDMAS:
Brackets
Indices (powers and roots)
Division and **M**ultiplication
Addition and **S**ubtraction.

f $^-8^2+4 = {^-64}+4 = \mathbf{^-60}$

g $(\frac{5}{2})^2 = 5^2\div 2^2 = \frac{25}{4} = \mathbf{6\frac{1}{4}}$

Exercise 2

Find the answers mentally.
You may need to use jottings for some.

1 **a** $(2+1)^3$ **b** $(5+4)^2$ **c** $(14-9)^3$ **d** $(7-3)^3$
e $(^-4)^2+3$ **f** $^-4^2+3$ **g** $(^-8)^2-8$ **h** $(^-11)^2-3(5+2)$
i $(25-17+2-5)^3$ **j** $(18+9-13+6-10)^3$

2 **a** $3(7-2)-5\times 3$ **b** $4(5-3)^3-2$ **c** $18-3\times 2^2$ **d** $20+4\times 5^2$
e $2\times 4^3+4$ **f** $(8-3\times 2)^3$ **g** $(1+2\times 5)\div 11$ **h** $(16-3)\times(8+2)^3$
i $(5\times 3-10)^3$ **j** $7(7-3\times 2)^3$ **k** $36\div(8+4)-11+4\times(16\div 8)^3$
l $(\frac{3}{4})^2$ **m** $(\frac{4}{3})^2$

3 **a** $\frac{(3+2)^2}{8+4}$ **b** $\frac{8\times 2^2}{5^2-1}$ **c** $\frac{(4\times 2-3)^3}{5}$ **d** $\frac{(3+1)^2}{(3-1)^2}$
e $\frac{(1+2\times 4)^2}{(2+1)^2}$ **f** $\frac{5\times 6^2}{5\times 2}$ **g** $\frac{(5\times 6)^2}{5\times 2}$ **h** $\frac{{}^-8^2+4}{7^2+1}$
i $\frac{({}^-8)^2+4}{7^2+1}$ **j** $\frac{{}^-5^2+({}^-5)^2}{10^2}$

Give the answer as a fraction or whole number.

4 **a** $\sqrt{37-12}$ **b** $\sqrt[3]{136-11}$ **c** $\sqrt{4+5\times 9}$ **d** $\sqrt{5^2+24}$ **e** $\sqrt[3]{2^2-3}$
f $\sqrt[3]{7^2+15}$ **g** $\sqrt[3]{5^2+10^2}$

***5** **a** $\sqrt{900}$ **b** $\sqrt{324}$ **c** $\sqrt{576}$ **d** $\sqrt{225}$ **e** $\sqrt{441}$

Write each number in question 5 as the product of two square numbers first.

T

Review

$\frac{1}{9}$	16	112	$\frac{1}{9}$	10	5	6	6	9	$\frac{1}{9}$	N 1000

49	88	$2\frac{7}{9}$	N 1000	300	16	$\frac{1}{9}$	6	$^-74$	125	$\frac{1}{9}$	$^-74$	112	88

Use a copy of this box. Write the letter beside each calculation above its answer in the box.

N $(7+3)^3 = 1000$ **D** $(15-4\times 2)^2$ **C** $(11-14)^2$ **T** ${}^-9^2+7$ **R** $({}^-9)^2+7$
W $(36-21-7+3-6)^3$ **K** $3(22-6\times 2)^2$ **U** $\frac{(3+2)^2}{3\times 2-1}$ **A** $\frac{5-2^2}{(2+1)^2}$ **I** $(\frac{5}{3})^2$
E $\frac{(7\times 8)^2}{7\times 4}$ **S** $\frac{7\times 8^2}{7\times 4}$ **G** $\sqrt[3]{635+365}$ **L** $\sqrt{10^2-8^2}$

Fractions, decimals, percentages mentally

Discussion

- 50% = 0·5
  5% = 0·05
  How could you write 0·005 as a percentage? **Discuss**.
  How could you write $67\frac{1}{2}$% as a decimal? **Discuss**.
  How could you write 10·5 as a percentage? **Discuss**.
- Ruby found $\frac{3}{5}$ of 16·5 like this.
  $\frac{1}{5}$ of 16·5 = 16·5 ÷ 5
  = 15 ÷ 5 + 1·5 ÷ 5
  = 3 + 0·3
  = 3·3
  $\frac{3}{5}$ of 16·5 = 3 × 3·3
  = 9·9
  Discuss Ruby's method. How else could she have found $\frac{3}{5}$ of 16·5?

  How could you find these? **Discuss**.
  $\frac{2}{5}$ of £6.50 $\frac{7}{20}$ of £15 $\frac{3}{8}$ of 24·8 ℓ $17\frac{1}{2}$% of 60 m

Patty bought a watch in this sale.
Its original price was £58·60
The new price can be worked out mentally in two ways.

Either 10% of £58·60 = £5·86
20% of £58·60 = 2 × £5·86
= £11·72
£58·60 − £11·72 = £58·60 − £11 − £0·72
= £47·60 − £0·72
= **£46·88**

or A 20% reduction in price is the same as 80% of the original price.
So we need to find 80% of £58·60.
10% of £58·60 = £5·86
80% of £58·60 = 8 × £5·86
= 2 × 2 × 2 × £5·86
= 2 × 2 × £11·72
= 2 × £23·44
= **£46·88**.

Worked Example Find $12\frac{1}{2}$% of 80.

Answer 10% of 80 = 8
1% of 80 = 0·8
2% of 80 = 1·6
$\frac{1}{2}$% of 80 = 0·4
so $12\frac{1}{2}$% of 80 = 8 + 1·6 + 0·4
= **10**

or $12\frac{1}{2}\% = \frac{1}{8}$
$\frac{1}{8}$ of 80 = **10**

Exercise 3

This exercise is to be done mentally.

1 What goes in the gaps?

Fraction	Percentage	Decimal
$\frac{1}{8}$		
	$0·\dot{6}$	
		175%
		$33\frac{1}{3}$%
$\frac{3}{15}$		
	0·03	
		37·5%
	0·625	
$12\frac{1}{2}$		

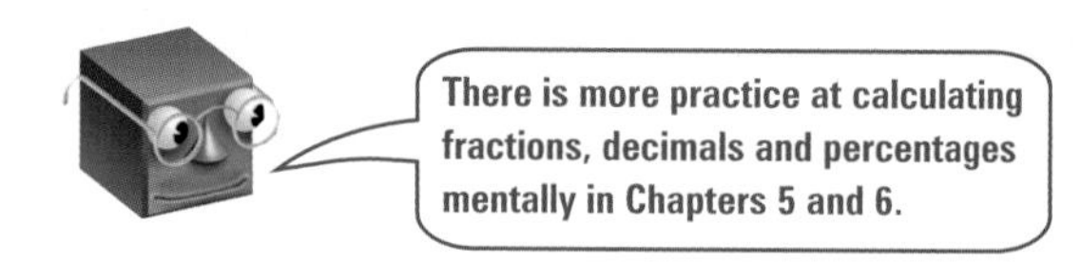

2 Find the answers mentally.

a $\frac{2}{5}$ of 40 **b** $\frac{7}{12}$ of 84 **c** $\frac{3}{8}$ of 120 **d** $\frac{2}{3}$ of 1·2 **e** $\frac{3}{5}$ of 12·5
f $1\frac{1}{2}$ of 18 **g** $1\frac{1}{4}$ of 12 **h** $2\frac{3}{4}$ of 16 **i** 35% of 20 **j** 65% of 40
k 125% of 360 **l** 175% of 240 **m** $\frac{3}{5}$ of 10·5 **n** $\frac{3}{8}$ of 5600 **o** 20% of £2·50
p 35% of £5 **q** 85% of 15 mm **r** $1\frac{3}{5}$ of £20·50 **s** $2\frac{5}{12}$ of 3600 kg

3 There is a discount of 15% on a £65 fishing rod in a sale.
By how much is the rod's price reduced?

4 Robert bought a shirt, a pair of shoes and a coat.
How much did he save on each?

a **b** **c**

5 **a** Increase 360 by 30%. **b** Decrease 420 by 65%.
c Decrease 800 by 12%. **d** Increase 7 by 150%.
e Find the sale price of a hat that was £45 and is reduced by 30% in the sale.
f Find 25% of 10% of 800 mℓ.

6 **a** 20% of a number is 15. What is the number?
*__b__ 0·9 is 15% of what number?

***7** A tank, when $\frac{3}{4}$ full, holds 84 litres. How much does it hold when full?

***8** When half-full, a container holds 7·2 litres. How many more litres are needed to make it $\frac{2}{3}$ full?

Review

a Write $\frac{2}{16}$ as a decimal and as a percentage.
b Write $17\frac{1}{2}$% as a decimal.
c Write 18·5 as a percentage.
d Find $\frac{3}{5}$ of 15·5.
e Find 30% of £10·50.
f Find 15% of 5 m.
g Decrease 600 ℓ by 22%.

Solving problems mentally

Worked Example
Twelve square tiles of side 5·5 cm were used to tile around a basin. What is the total area of the tiles?

Answer
Area of 1 tile = 5·5 × 5·5
Area of 12 tiles = 5·5 × 5·5 × 12
$= \frac{11}{{}_1 2} \times \frac{11}{{}_1 2} \times 12^3$
= 11 × 11 × 3
= **363 cm²**

or

5·5 × 5·5 × 12
= 5·5 × 12 × 5·5
= 66 × 5·5
= 66 × 5 + 66 × 0·5
= 60 × 5 + 6 × 5 + 66 × 0·5
= 300 + 30 + 66 × 0·5
= 330 + 33
= **363 cm²**

5·5 × 12 = 5 × 12 + 0·5 × 12
= 60 + 6
= 66

Exercise 4

This exercise is to be done mentally.
You may use jottings.

There are lots of links to other areas of maths in this exercise. Try to find them.

1 Write the answers to these as quickly as possible.

- **a** What is one less than six and a half million?
- **b** Last year Tim grew from 1·56 m to 1·8 m. How much did he grow in centimetres?
- **c** How many minutes are there in 2·25 hours?
- **d** How many days are there in 60 hours?
- **e** How many hours and minutes are there in 155 minutes?
- **f** Two angles of a triangle are 37° and 95°. What is the other angle?
- **g** The area of a square is 25 m^2. What is the length of one side?
- **h** You get A$100 for £80. How much do you get for £200?
- **i** What is the perimeter of a rectangle with sides 5 cm and 8 cm?
- **j** What is the mean of 9, 11 and 16?
- **k** How many centimetres are there in 1·864 m?
- **l** Five pears cost 65p. What do 8 pears cost?
- **m** Bathroom tiles are 20 cm by 20 cm. How many are needed to cover one square metre?
- **n** Three angles make a full circle. If two of the angles are 136° each, what is the size of the third angle?
- **o** The probability of Lucy winning a tennis match against Natasha is 0·95. What is the probability of her not winning?
- **p** Use the assumed mean method to find the mean of 2·1, 2·3, 2·8, 2·9 and 2·4.
- **q** Find the LCM of 45 and 54.
- **r** Find the HCF of 48 and 60.
- **s** Find the area of a triangle with base 4·5 cm and height 12 cm.
- **t** 75 miles per hour is about 33 metres per second. About how many metres per second is 50 miles per hour?
- **u** Find the volume of a cuboid with dimensions 8 cm by 3·5 cm by 6 cm.
- **v** Find the area of a parallelogram of base 16 mm and height 2·1 cm.
- **w** Half the tickets for the school play were bought by parents, one-third by siblings and the rest by relatives. What fraction were bought by relatives?
- ***x** The ratio of sugar to flour in a recipe is 5 : 8. There are 280 g of sugar. How many grams of flour are there?
- ***y** $a = 2$ and $b = 3$. What is the value of a to the power of b plus b to the power of a?
- ***z** Solve **i** $137 + x = 189$ **ii** $(4 + x)^2 = 36$ **iii** $(20 - x)^2 = 64$.

2

? ?
? ?

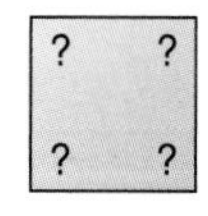

Arrange the numbers 1 to 12 in the corners of these squares. Use the following rules:

1. There must be just one number in each corner.
2. The four numbers in each square must add to 26.

Is there more than one answer?

3 Sarah measured the distance between two towns on a map as 3·8 cm. The scale on the map is 1 : 150 000. What is the actual distance between the two towns?

4 This table gives the 'human years' for animals. For example a 1-year-old dog is 5·7 'human years' old and a 3-year-old dog is 3 × 5·7 = 17·1 'human years' old.

Animal	Human Years
Dog	5·7
Cat	6·3
Rabbit	11·7
Horse	3·8

Animal	Human Years
Guinea pig	16·9
Mouse	25
Elephant	1·1
Giant Tortoise	0·6

a How old are these in 'human years'? Round your answer to the nearest year.

i	a 2-year-old dog	**ii**	a 7-year-old cat	**iii**	a 4-year-old rabbit
iv	a 2-year-old guinea pig	**v**	a 49-year-old elephant	**vi**	a $2\frac{1}{2}$-year-old mouse
vii	a 29-year-old giant tortoise	**viii**	a 19-year-old horse		

b Find the age in 'human years' of any pets you have.

*__c__ If a human is expected to live to 75, what is the life expectancy of a giant tortoise?

5 A clock chimes once at 1 o'clock, twice at 2 o'clock, three times at 3 o'clock and so on. Every half hour it chimes once. How many chimes does it make between 6·45 a.m. and 8·15 p.m.?

6 In a science experiment, Clare measured the heights of four plants. They were 122 cm, 116 cm, 109 cm and 137 cm tall. What is their mean height?

7 Find a and b.

a The sum of a and b is 24.
The product of a and b is 108.

b The sum of a and b is 1·2.
The product of a and b is 0·32.

c The sum of a and b is 0·7.
The product of a and b is 0·12.

d The sum of a and b is $^{-}10$.
The product of a and b is 24.

e The difference between a and b is 3.
The quotient of a and b is 2.

f The difference between a and b is 4.
The quotient of a and b is $^{-}1$.

8 These two cards have got numbers on the back as well.
When I put the two cards on the table, the sum of the numbers showing is either 14, 15, 16 or 17.
What two numbers are written on the back?
Is there more than one possible answer?

9 In science, Rebecca got five different amounts of powder left when she repeated an experiment five times.
She got 11·7 g, 11·4 g, 12·1 g, 11·3 g and 12·5 g.
Using the assumed mean method, find the mean of the five amounts.

There is more about the mean on page 386.

10 It costs 40p to send a letter to someone in England and 60p to send a letter abroad.
John sent 8 letters and the stamps cost him £3·80.
How many letters did he send abroad?

11 The probability that Jodie will beat her brother at tennis is 0·06. Out of 40 matches, how many would you expect her to win?

12 Find 2% of 5% of 1000 mm.

13 **a** Gabriel bought 37 balls altogether.
How many packs of each did she buy? Is there more than one answer?

b Winston spent £20.80 altogether on tennis and golf balls.
How many packs of each did he buy?

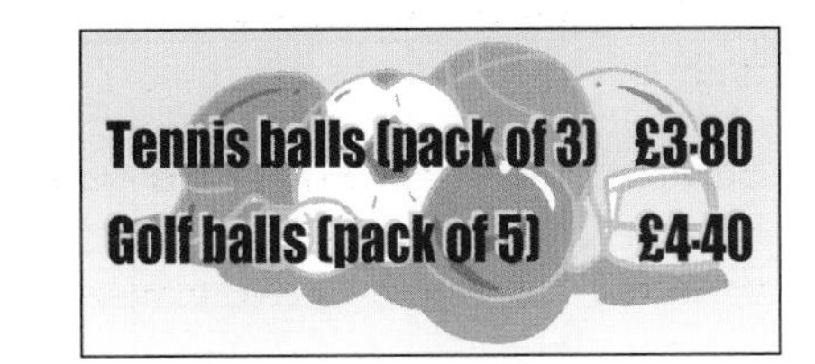

14 **a** Use 3, 4, 5 and 6 once each with some of ×, ÷, +, − and brackets to make 30.

b $4 \times 3 + 6 - 5 \times 3 + 1$
Put brackets into this to make as many different answers as you can.

*__c__ Use any of +, −, ×, ÷, √ and brackets, and four 4s to make
i 17 **ii** 9 **iii** 20

Remember BIDMAS page 4.

*15 Three bananas and two oranges cost £3·99.
Two bananas and three oranges cost £3·81.
How much does a banana cost?

Review 1

a Eight people bought tickets for an animal park. Each paid £5·60.
What was the total cost?

b A group paid £61·60 for their tickets to the animal park.
How many were in the group?

Review 2 Six consecutive numbers are put on the six faces of a cube.
Numbers on opposite faces have the same sum.
What number is opposite **a** 15, **b** 16 **c** 12?

Review 3 Five friends each put in some money for lunch.
The amounts put in are £6·90, £7·20, £7·50, £6·80 and £7·10.
What is the mean amount put in?

Review 4 A group of friends bought 9 drinks.
The total cost was £10.
What did they buy?

DRINKS	
Milkshake	£1·20
Thickshake	£1·80
Juice	55p

Review 5 Thirty years ago, Britain used pounds, shillings and pence.
There were 20 shillings in £1.
A gallon of petrol cost 7 shillings thirty years ago.
Today it costs about £3·90.
By how much has the cost of petrol risen in the last thirty years?

T

Puzzle

Use a copy of the table.
Rewrite each number given in each row within the table using the numbers at the left of that row.
You may use +, −, ×, ÷ and brackets.
As an example, the first row has been completed.

	Use 2 of the numbers	Use 3 of the numbers	Use 4 of the numbers	Use all of the numbers
2 3 16 8 5	**4** $= \frac{8}{2}$	**9** $= 5^2 - 16$	**1** $= \frac{16}{8} + 2 - 3$	**18** $= 2(16 - 8) - 3 + 5$
1 2 7 3 4	10 =	8 =	42 =	17 =
4 5 12 1 9	7 =	8 =	9 =	10 =
6 5 13 7 4	28 =	27 =	22 =	24 =
3 4 15 8 2	7 =	0 =	7 =	12 =
3 9 18 1 4	36 =	13 =	1 =	4 =

Estimating

Discussion

- Robbie is painting his bedroom.
  He needs to calculate the area of the walls to work out how much paint to buy.
  He measures the lengths and heights to the nearest 10 cm, then calculates the area.
  Will his estimate for the area be accurate enough? **Discuss**.
- **8348** We could give a number of approximations to this.
  We could give 8000 or 8300 or 8350 or 8500 or 10 000.
  Think of situations where each of these approximations for 8348 might be used.
  Discuss.
- What does '**degree of accuracy**' mean? **Discuss**.

Exercise 5

1 Decide the degree of accuracy needed for these.

  a A vet uses the mass of a cat to calculate how much medicine to inject.
  b A builder calculates the amount of wood he needs to buy to build a shed.
  c A chef calculates how much flour and sugar are needed for a cake.
  d A reporter calculates the number of people who have attended an art exhibition in the last week.

A. As accurate as possible.
B. A rough estimate will do.
C. An estimate is fine but it must be reasonably accurate.

2 **a** Maria rounded the length and width of her bedroom to the nearest metre. She then calculated the area so that she could work out the cost of recarpeting. Would her estimate for the area be accurate enough? Explain.

  b Thomas bought the same book for all of his 10 cousins for Christmas. He worked out the length of paper, to the nearest cm, needed to wrap one book. He used this to calculate the total length of paper he would need. Will his estimate for the length for one book be accurate enough?

Review 1 Write down two calculations where the answer must be as accurate as possible and two where a rough estimate is good enough.

Review 2 Rose rounded the length of one curtain to the nearest 10 cm and then worked out how much material she would need for 24 curtains of this length.
Will her estimate be accurate enough?

Discussion

- Is $8 \div 2$ or $9 \div 2$ a better approximation for $8{\cdot}59 \div 2{\cdot}37$? **Discuss**.
- Is 5×4 or 5×5 a better approximation for $4{\cdot}5 \times 4{\cdot}5$? **Discuss**.

Remember
$\approx$ means '**is approximately equal to**'.

Guidelines for estimating

- Look for '**nice' numbers** that enable you to do the calculation mentally.

Example $200 \div 5{\cdot}7 \approx 200 \div 5$ rather than $200 \div 6$.

Example $\frac{72{\cdot}6 \times 347{\cdot}05}{0{\cdot}89} \approx \frac{100 \times 350}{1}$

- Look for **numbers that will cancel.**

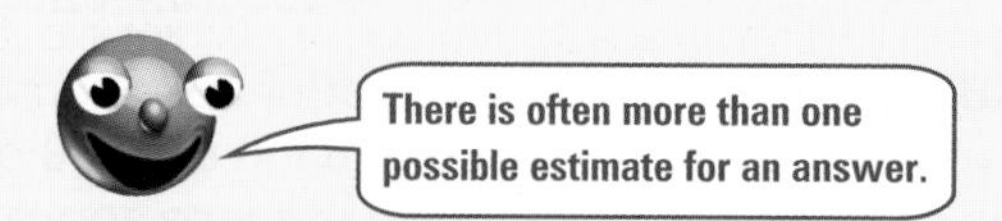

Example $\frac{12{\cdot}48 \times 487{\cdot}31}{3{\cdot}69} \approx \frac{\cancel{12}^3 \times 500}{\cancel{4}^1} = 1500$

- When multiplying or dividing **never approximate a number to 0**.
Use 0·1, 0·01 or 0·001 etc.

Example $205{\cdot}7 \times 0{\cdot}012$ should not be approximated as 200×0. It is better to use $200 \times 0{\cdot}01$ or $200 \times \frac{1}{100}$, which gives an estimate of 2.

- When **multiplying** two numbers, try to **round one up and one down.**
When **dividing** two numbers, try to **round both numbers up** or **both numbers down.**

Example It is better to estimate $2{\cdot}5 \times 3{\cdot}5$ as 2×4 or 3×3 rather than 3×4.

$2{\cdot}5 \times 3{\cdot}5 = 8{\cdot}75$ so $2 \times 4 = 8$ or $3 \times 3 = 9$ both give a closer estimate than $3 \times 4 = 12$.

Example It is better to estimate $\frac{83{\cdot}2}{8{\cdot}5}$ as $\frac{80}{8}$ rather than $\frac{81}{9}$.

$\frac{83{\cdot}2}{8{\cdot}5} = 9{\cdot}79$(2d.p.) so $\frac{80}{8} = 10$ gives a closer estimate than $\frac{81}{9} = 9$.

Worked Example
Estimate the area of this kitchen unit.

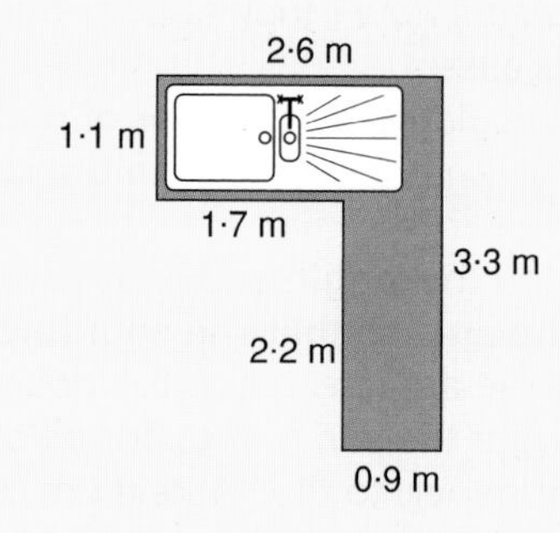

Answer
Area = area A + area B
$= 1{\cdot}1 \times 2{\cdot}6 + 2{\cdot}2 \times 0{\cdot}9$
$\approx 1 \times 3 + 2 \times 1$
$= 3 + 2$
$= 5\ m^2$

Worked Example
Estimate the answers to these.

a $8{\cdot}98 \times 24{\cdot}6$ **b** $(6{\cdot}35)^2$ **c** $\frac{198 \times 71{\cdot}6}{11{\cdot}3 \times 0{\cdot}83}$ **d** $0{\cdot}09 \times 59{\cdot}6$.

Answer

a $8{\cdot}98 \approx 10$, $24{\cdot}6 \approx 25$
An estimate is $10 \times 25 = 250$.

b $(6{\cdot}35)^2$ is more than 6^2 but less than 7^2.
An estimate for $(6{\cdot}35)^2$ is: between 36 and 49.

c $198 \approx 200$, $71{\cdot}6 \approx 70$, $11{\cdot}3 \approx 10$, $0{\cdot}83 \approx 1$

An estimate is $\frac{200 \times \cancel{70}^7}{{}_1\cancel{10} \times 1} = 1400$.

d $0{\cdot}09 \approx 0{\cdot}1 = \frac{1}{10}$, $59{\cdot}6 \approx 60$
An estimate is $\frac{1}{10} \times 60 = 6$.

Exercise 6

Except for questions 3, 6 and Review 2.

1 Choose the best estimate for these.

a	$4{\cdot}5 \times 5{\cdot}5$	A 5×6	B 5×5	C 4×6
b	$41{\cdot}3 \div 6{\cdot}6$	A $42 \div 7$	B $42 \div 6$	C $48 \div 6$
c	$8{\cdot}4 \times 7{\cdot}8$	A 7×7	B 8×8	C 8×7
d	$96{\cdot}4 \div 12$	A $96 \div 15$	B $90 \div 10$	C $100 \div 10$

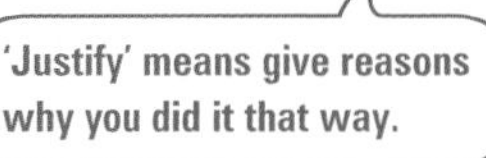

2 Estimate the answers to these. Justify your estimates.

a $7{\cdot}6 \times 4{\cdot}123$ b $67{\cdot}34 \div 9{\cdot}3$ c $7{\cdot}24 \times 18{\cdot}07$ d $(10{\cdot}14)^2$ e $\frac{19{\cdot}6 \times 34{\cdot}7}{4{\cdot}35}$

f $\frac{7{\cdot}62 + 2{\cdot}21}{5{\cdot}23}$ g $81{\cdot}2 \times 0{\cdot}27$ h $\frac{27{\cdot}8 \times 3{\cdot}67}{7{\cdot}64}$ i $\frac{28{\cdot}6 \times 24{\cdot}4}{5{\cdot}67 \times 4{\cdot}02}$ j $\frac{18{\cdot}3 + 11{\cdot}1}{57{\cdot}03}$

3 Use the calculator to find the answers to these. Round the answers sensibly. Check that the answer is reasonable by making an estimate.

a $37{\cdot}64 \times 23{\cdot}1$ b $44{\cdot}9 \div 8{\cdot}76$ c $0{\cdot}47 \times 19{\cdot}1$ d $\frac{38{\cdot}4 + 22{\cdot}5}{18{\cdot}4}$

e $\frac{274 \times 31{\cdot}4}{49{\cdot}3}$ f $\frac{31{\cdot}2}{0{\cdot}24}$ g $(3{\cdot}24)^2$ h $\frac{(7{\cdot}05)^2}{4{\cdot}68}$

i $\frac{87{\cdot}9}{1{\cdot}3 + 5{\cdot}01}$ j $\frac{4{\cdot}7 \times 49{\cdot}2}{0{\cdot}18}$ k $\frac{51{\cdot}6 \times 0{\cdot}12}{9{\cdot}8}$ l $\frac{24{\cdot}4 \times 8{\cdot}2}{\sqrt{65}}$

4 A nautical mile is approximately 1·853 km.
Estimate how many km are in 214 nautical miles.

5 An ounce is about 28·35 grams.
Estimate the number of ounces in 600 grams.

6 Estimate, then use your calculator, to find the answers to the following. If rounding is required, round your answers sensibly.

a Find the cost of 9·7 m of material at £8·19 per metre.

b Find the perimeter of this triangle.

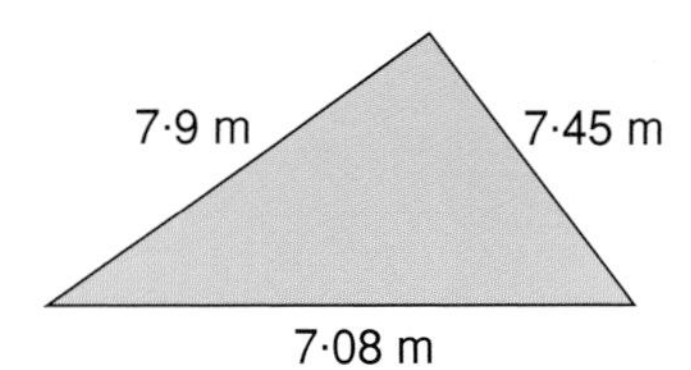

Remember, when calculating the accurate answer, don't round until the final answer is to be found.

c Shirts in a sale were priced at £8·85, £7·95, £11·15 and £5·45.
During the first day of the sale, the following numbers were sold.
27 at £8·85 15 at £7·95 24 at £11·15 47 at £5·45
How much were all these shirts sold for?

d A formula for finding the area of a trapezium is
$A = \frac{1}{2}(a + b) \times h$.
a and b are the lengths of the parallel sides and h is the distance between these sides.
Find the area of this trapezium.

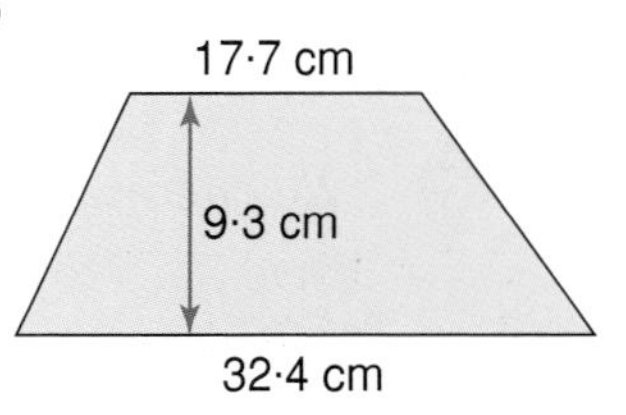

e Kareema is reading this book.
She takes an average of 2 minutes 5 seconds to read a page.
How long will it take her to read the book?

*f Jim bought 213 feet of decking timber.
This was in planks, each 5′11″ long.
How many planks did Jim buy?

*g A formula for the volume of a pyramid is $V = \frac{1}{3}$(base area) $\times$ height.
Find the volume of a pyramid which has a rectangular base, measuring 81·2 mm by 58·6 mm, and height 19·7 mm.

*h Amanda has been visiting relatives in Australia.
If the exchange rate is A$1 for 37·02p, how much British money would Amanda get for the A$48·65 she brought back to England with her?

*i A formula for finding the area of a circle is $A = \pi r^2$

i Using $\pi = 3{\cdot}14$, find the area of a circle with radius 26·7 cm.

ii Using $\pi = \frac{22}{7}$, find an estimate for the area of a circle with radius 26·7 cm.

iii Did you get the same estimate for **i** and **ii**?
Did you get the same answer, to the nearest cm^2, for the calculation?

There is more about areas of circles on page 351.

*7 Write down 10 different calculations that could have an estimated answer of 15.
Use at least two of the operations +, −, ×, ÷ and squaring in each calculation.
Use decimals in all of the calculations.

Review 1 Estimate the answers for these.

a $38{\cdot}2 \times 4{\cdot}67$ **b** $\frac{28{\cdot}7}{0{\cdot}44}$ **c** $\frac{21{\cdot}4 \times 38{\cdot}7}{3{\cdot}68 \times 4{\cdot}71}$ **d** $\frac{7{\cdot}204 + 2{\cdot}63}{1{\cdot}934}$ **e** $4{\cdot}9 \times (3{\cdot}14)^2$

Review 2 Estimate, then use your calculator to find the answers to the following. Round your answers sensibly.

a $\frac{36{\cdot}7 \times 72{\cdot}6}{6{\cdot}94}$ **b** $48{\cdot}6 \times 0{\cdot}098$ **c** $\frac{7{\cdot}64 + 14{\cdot}1}{3{\cdot}84}$ **d** $\frac{(9{\cdot}63)^2}{4{\cdot}78}$

Summary of key points

There are many strategies you can use to **calculate mentally**. These include

adding and subtracting
- complements in 1, 10, 50, 100 and 1000
- partitioning
- using facts you already know
- counting up
- nearly numbers
- compensating

multiplying and dividing
- multiplying and dividing by multiples of 10, 100 and 1000
- multiplying by nearly 10s, 100s, ...
- using knowledge of place value
- using facts you already know
- doubling and halving
- partitioning
- factors

We carry out operations in this order:

Brackets
Indices
Division and **M**ultiplication
Addition and **S**ubtraction

We often **calculate mentally with fractions, decimals and percentages.**

Example $17\frac{1}{2}$% of 50 can be found by calculating

10% of 50 = 5
1% of 50 = 0·5
7% of 50 = 3·5
$\frac{1}{2}$% of 50 = 0·25
so $17\frac{1}{2}$% of 50 = 5 + 3·5 + 0·25
= 8·75

We can use mental strategies to **solve problems.**

Sometimes an **exact answer** is not always needed and an **estimate** is sufficient. Other times an exact answer is needed and an estimate is not sufficient.

Example An estimate is good enough when calculating the amount of cola needed for a party.
An exact answer is needed when calculating the area of a room for carpet laying.

When **estimating**:

- Look for numbers so you can do the calculation mentally.
- Look for numbers that will cancel.
- Never approximate a number to 0. Use 0·1, 0·01, ...
- When multiplying two numbers, try to round one up and one down.
- When dividing two numbers, try to round both up or both down.

Examples $5{\cdot}36 \div 0{\cdot}8 \approx 5{\cdot}4 \div 0{\cdot}9$ $\quad$ $850 \div 2{\cdot}5 \approx 900 \div 3$

$= 6$ $\quad$ $= 30$

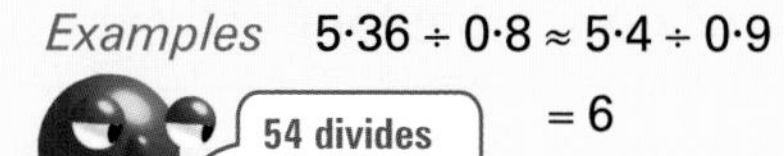

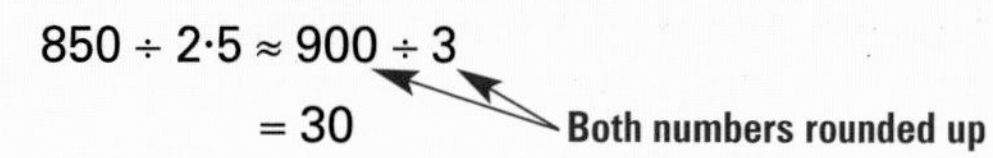

Test yourself

Except for questions 18 and 19b.

1 a 0·86 + 0·37 b 0·92 − 0·49 c 0·091 − 0·077 d 5·21 − 0·93 A

2 a 14 − ⁻3 − 8 + ⁻4 + 2 b 0·6 + 0·8 − ⁻0·2 − 1·3 − 0·7 A

3 In this pyramid, each number is the sum of the two numbers below it.
What number goes in the blue square? A

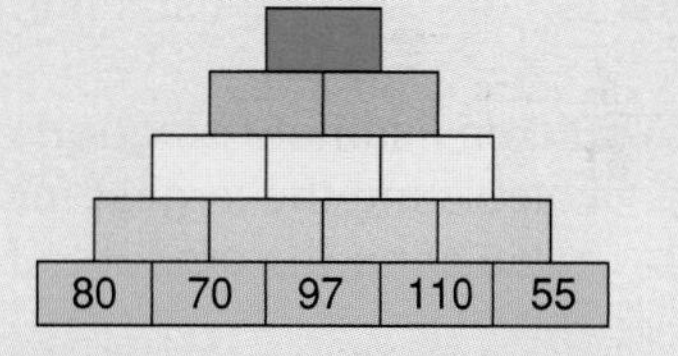

4 a 700 × 6000 b 72 000 ÷ 900 c 50 ÷ 80 000 d 8·6 × 0·01 e 0·062 ÷ 0·01 A
f 8·6 × 0·7 g ⁻0·09 × 8 h ⁻0·4 ÷ 8 i 0·28 × 6 j 700 × 0·9
k 8·6 × 4 l 5·7 × ⁻4 m 12 × 3·2 n 25 × 2·02

5 Given that 46 × 25 = 1150, find a 46 × 24 b 45 × 25. A

6 Find ways to fill in the boxes. □ × □ × □ = 0·06 A

T 7 Use a copy of this number chain.
Fill in the missing numbers. A

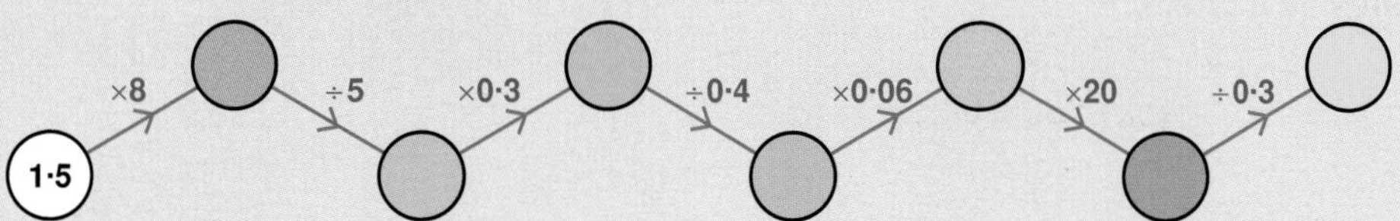

8 Find the answers to these. B

a $(^-3)^2 - 4$ b $4(8-5) - 2 \times 3$ c $2(12-6)^2$ d $\frac{(8-4)^2}{2^2}$ e $\sqrt[3]{10^2 + 5^2}$

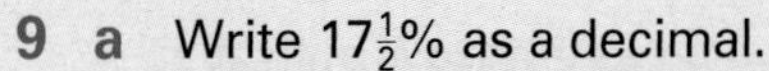
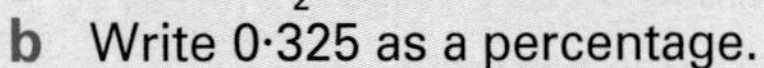

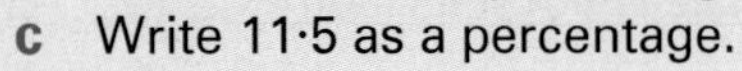

9 a Write $17\frac{1}{2}\%$ as a decimal.
b Write 0·325 as a percentage.
c Write 11·5 as a percentage.

10 Find the answers mentally.
a $\frac{2}{5}$ of 55·5 b $\frac{7}{8}$ of 7200 c 30% of £8·50 d 45% of £5

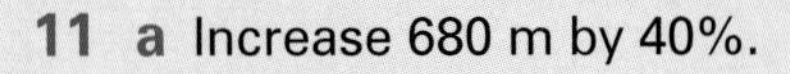

11 a Increase 680 m by 40%. b Decrease 500 kg by 12%.

12 10% of a number is 15. What is the number?

13 Find the answers to these as quickly as possible.
a The area of a square is 121 cm^2. What is the length of one side?
b Four notepads cost £2·60. How much do twelve cost?
c Two angles join to make a straight line. One angle is 96°. What is the other angle?
d What is the mean of 17, 23 and 41?

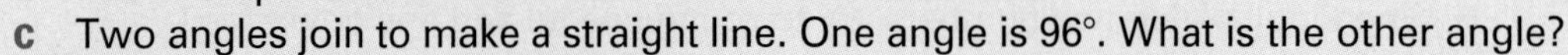

e You get about €1·6 for £1. About how many euros do you get for £12?
f 80 km is about the same as 50 miles. About how many miles is 60 km?
g In a contest, a quarter of the entries are from England, a third from Scotland and the rest from Wales. What fraction of the entries are from Wales?
h The ratio of correct to incorrect answers Joanna got in a test was 3 to 5. She got 45 correct answers. How many incorrect answers did she get?

14 In science, Becky had 50 mℓ of solution to put into 6 test tubes.
She put 7·5 mℓ in the first test tube then shared the rest equally between the other test tubes. How much solution did she put in each of the five test tubes?

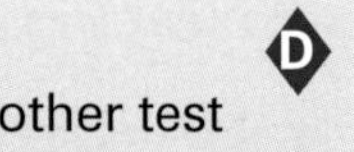

15 Write **one** of the prime numbers 5, 7, 11, 13, 17, 19 and 23 in each circle so that the three primes in each line add to the same number. Use each number only once.

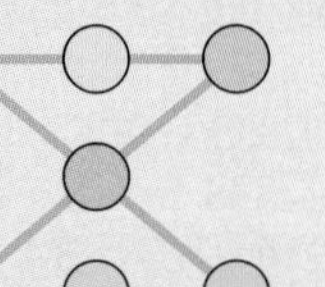

16 Would an estimate be good enough when
a calculating the largest mass a bridge will hold
b calculating the amount of flour needed to mix with 25·5 g of butter to make scones?

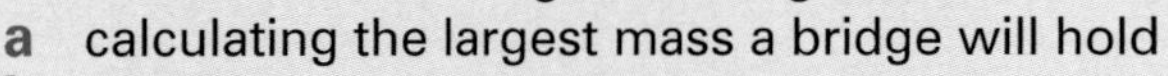

17 Estimate the answers. a $0{\cdot}25 \times 83{\cdot}4$ b $\frac{7{\cdot}8 \times 21{\cdot}4}{0{\cdot}23}$

18 Estimate the answer to these and then use your calculator to find the accurate answer.
a $44{\cdot}7\,(5{\cdot}82 + 2{\cdot}15)$ b $\frac{0{\cdot}59 \times 135{\cdot}09}{6{\cdot}8 \times 1{\cdot}04}$

19 a Estimate the area and perimeter of this triangle.
b Calculate the area and perimeter, using your estimates as a check.

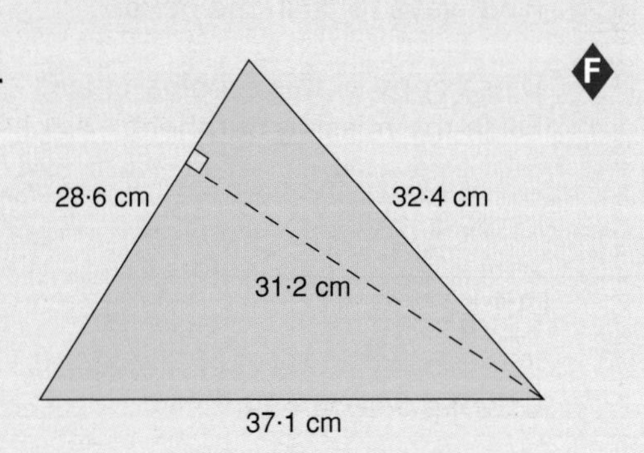

4 Written and Calculator Calculation

You need to know

Key vocabulary

brackets, divisor

Back to front

A number such as 373, which reads the same from left to right as it does from right to left, is a **palindromic number**.
Other examples are 88, 1441 and 36 763.

Many palindromic numbers can be made as follows.

Begin with any number, say	168
Reverse the digits	+861
Add	1029
Reverse the digits	+9201
Add	10230
Reverse the digits	+03201
Add	13431

PALINDROMES
13431
2286822
5943495
721127

It took 3 reversals of the digits to make the palindromic number 13431.

Find out the number of reversals needed to make palindromic numbers from other starting numbers.

Adding and subtracting

Remember
We **add and subtract decimals** by lining up the decimal points.
Always estimate the answer first.

Example These are the areas of three small islands.

1583 km^2 890·006 km^2 8·0035 km^2

The total area of all three is 2481·009 km^2.

```
  1583
   890·006
+    8·0035
  2481·0095
   111
```

Worked Example
Calculate 8614·16 − 52·9 − 0·7

Answer

```
 8614·16             8561·26          52·9           8614·16
−  52·9     and   −     0·7    or   + 0·7    and   −   53·6
 8561·26             8560·56          53·6           8560·56
```

The answer is **8560·56**.

Exercise 1

1 **a** 19 − 4·97 − 4·6 **b** 21 − 5·72 − 6·8 **c** 4936·27 − 83·9 − 0·06
d 96 + 4·38 − 24·7 + 3·72 **e** 9683 + 294·308 + 0·0067 **f** 79 632·4 + 83·096 + 5·007
g 8326 − 4·094 − 38·62

2 Evelyn bought a stereo for £1086, a CD rack for £58·70 and a CD cleaning tissue for 85p. How much did this cost altogether?

3 The largest mammal ever measured was a blue whale, 31·996 m long.
The smallest mammal ever measured was a pygmy white-toothed shrew, 5·98 cm long.
What is the difference in length between these two mammals?

***4** Put the digits 1 to 9 in the boxes to make each true.
a □□□□ + 0·000□□ − □·0□□ = 8722·94439
b □0□□·□□ − □□·□ − 0·□ = 972·22

Review 1
a 83·72 − 68·4 − 0·89 − 5·72 **b** 8672 + 36·4 + 0·0073 − 5·802

Review 2 Marie weighs a container of igneous rock. It has a mass of 0·839 kg.
The container has a mass of 62 g.
What is the mass of the rock in grams?

Multiplying and dividing by numbers between 0 and 1

Discussion

$40 \times 5 = 200$
$40 \times 50 = 2000$
$40 \times 500 = 20\,000$
$40 \times 5000 = 200\,000$
$40 \times 50000 = 2\,000\,000$

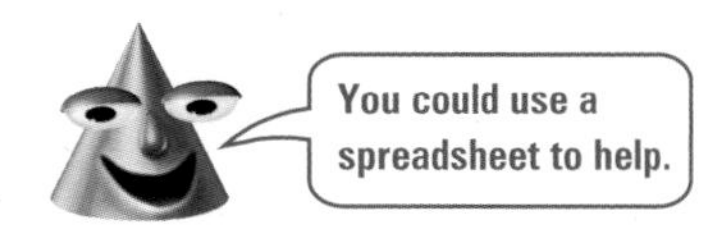

- **Discuss** how to use the number pattern in the box to find the answer to the following multiplications.

  $40 \times 0{\cdot}5$ $\quad 40 \times 0{\cdot}05$ $\quad 40 \times 0{\cdot}005$

  'Multiplying a positive number, n, by a number that is greater than 1, gives an answer greater than n'. Make a similar statement about multiplying a positive number, n, by a number that is between 0 and 1. **Discuss**.

- **Discuss** how to use the number pattern in the box to find the answers to these divisions.

  $\frac{8000}{4000} = 2$
  $\frac{8000}{400} = 20$
  $\frac{8000}{40} = 200$
  $\frac{8000}{4} = 2000$

  $\frac{8000}{0{\cdot}4}$ $\quad \frac{8000}{0{\cdot}04}$ $\quad \frac{8000}{0{\cdot}004}$

  'Dividing a positive number, m, by a number that is greater than 1, gives an answer smaller than m.
  Make a similar statement about dividing a positive number, m, by a number that is between 0 and 1. **Discuss.**

- * Multiplying and dividing are inverse operations. **Discuss** how this fact could be used to justify the statements you made earlier in this discussion.

Worked Example

Which of these will have an answer greater than 0·72? Which will have an answer less than 0·72? Explain.

a $0{\cdot}72 \times 1{\cdot}8$ **b** $0{\cdot}72 \div 1{\cdot}8$ **c** $0{\cdot}72 \times 0{\cdot}27$ **d** $0{\cdot}72 \div 0{\cdot}27$

Answer

a The answer will be greater than 0·72 because we are multiplying by a number greater than 1.
b The answer will be less than 0·72 because we are dividing by a number greater than 1.
c The answer will be less than 0·72 because we are multiplying by a number between 0 and 1.
d The answer will be greater than 0·72 because we are dividing by a number between 0 and 1.

Exercise 2

1 Which of the following calculations will have an answer less than 200?

a $200 \times 0{\cdot}4$ **b** $200 \div 0{\cdot}2$ **c** $200 \times 0{\cdot}5$ **d** $200 \div 0{\cdot}004$ **e** $200 \div 0{\cdot}05$
f $200 \times 0{\cdot}004$ **g** $200 \times 4{\cdot}5$ **h** $200 \div 4{\cdot}5$ **i** $200 \div 54$ **j** 200×54

2 Which of the following calculations will have an answer greater than 0·4?

a $0{\cdot}4 \div 0{\cdot}1$ **b** $0{\cdot}4 \times 0{\cdot}1$ **c** $0{\cdot}4 \div 0{\cdot}02$ **d** $0{\cdot}4 \div 0{\cdot}004$ **e** $0{\cdot}4 \times 0{\cdot}002$
f $0{\cdot}4 \times 0{\cdot}04$ **g** $0{\cdot}4 \times 2{\cdot}4$ **h** $0{\cdot}4 \div 24$ **i** $0{\cdot}4 \div 2{\cdot}4$ **j** $0{\cdot}4 \times 24$

3 Choose the calculation which has an answer less than 1·04.

	A	B	C
a	$1{\cdot}04 \times 2{\cdot}6$	$1{\cdot}04 \div 2{\cdot}6$	$1{\cdot}04 \div 0{\cdot}26$
b	$1{\cdot}04 \div 0{\cdot}58$	$1{\cdot}04 \times 5{\cdot}8$	$1{\cdot}04 \times 0{\cdot}58$
c	$1{\cdot}04 \div 29$	$1{\cdot}04 \div 0{\cdot}29$	$1{\cdot}04 \times 29$
d	$1{\cdot}04 \times 1{\cdot}04$	$1{\cdot}04 \div 0{\cdot}104$	$1{\cdot}04 \div 1{\cdot}04$
e	$1{\cdot}04 \div 0{\cdot}9$	$1{\cdot}04 \times 0{\cdot}99$	$1{\cdot}04 \times 9{\cdot}9$

4 What goes in the gap, **greater** or **less**?

a When we multiply a positive number, n, by a number between 0 and 1, the answer will be ______ than n.

b When we divide a positive number m, by a number between 0 and 1, the answer will be ______ than m.

Review Which of these will have an answer greater than 1·83? Explain.

a $1{\cdot}83 \div 2{\cdot}7$ **b** $1{\cdot}83 \times 0{\cdot}27$ **c** $1{\cdot}83 \times 2{\cdot}7$

d $1{\cdot}83 \div 0{\cdot}27$ **e** $1{\cdot}83 \div 0{\cdot}027$ **f** $1{\cdot}83 \times 0{\cdot}027$

Multiplying

A medicine is to be injected at the rate of 0·45 mℓ per kilogram of body weight.
Emmalene is 64·2 kg.
The nurse needs to inject $64{\cdot}2 \times 0{\cdot}45$ mℓ.

$64{\cdot}2 \times 0{\cdot}45 \approx 60 \times 0{\cdot}5 = 6 \times 5 = 30,$

and is equivalent to $642 \times 45 \div 1000$.

≈ means is approximately equal to.

$64{\cdot}2 \times 0{\cdot}45 = 642 \div 10 \times 45 \div 100$

$$\begin{array}{r} 642 \\ \times \quad 45 \\ \hline 25680 \\ 3210 \\ \hline 28890 \\ \hline \end{array}$$

and $28\,890 \div 1000 = 28{\cdot}890$.

Find an equivalent calculation without decimals.

Use your estimate to check the decimal point is in the right place.

Check this against the estimate of 30. It is the right **order of magnitude**.

The answer is 29 mℓ to the nearest millilitre.

Always round your answer sensibly.

Exercise 3

1 **a** $68{\cdot}7 \times 2{\cdot}3$ **b** $5{\cdot}9 \times 6{\cdot}42$ **c** $5{\cdot}7 \times 8{\cdot}33$ **d** $72{\cdot}6 \times 0{\cdot}39$ **e** $0{\cdot}72 \times 8{\cdot}64$

2 Calculate how much a nurse must inject over 24 hours for these body masses. Give the answer to the nearest mℓ.

a 85 kg **b** 77 kg **c** 53·2 kg **d** 68·7 kg

Dose over 24 hours

0·65 mℓ per kilogram of body mass

3 Oil costs £1·82 per litre. What is the cost of 2·7 litres?

4

Stir-fry vegetables for one	
0·35 kg	carrots
0·25 kg	broccoli
0·55 kg	cauliflower

Cost of vegetables	
carrots	£0·89 kg
broccoli	£1·06 kg
cauliflower	£1·65 kg

How much do the vegetables for 'Stir-Fry Vegetables for one' cost in total?

T

Review 1

						I	
0·6256	0·9672	20·13	7·6941	4·7652	0·8736	133·92	14·872

0·8736	5·172	20·13	3·318	5·208	0·2912	33·934

				I		
20·13	5·208	2·2568	0·2912	133·92	14·872	5·172

10·325	3·318	0·9672	0·6256	0·8736	5·172	7·6941	0·8736

				I				
20·13	5·208	0·6256	14·872	133·92	5·208	0·8736	5·172	20·13

0·2912	20·13	0·8736	0·8736	20·13	0·9672	14·872	4·7652	0·8736

Use a copy of this box. Write the letter beside each calculation above its answer in the box.

I $18·6 \times 7·2 =$ **133·92**	**Y** $3·61 \times 9·4$	**N** $1·4 \times 3·72$	**H** $4·31 \times 1·2$	**S** $2·6 \times 5·72$
E $18·3 \times 1·1$	**O** $1·58 \times 2·1$	**D** $2·72 \times 0·23$	**W** $41·3 \times 0·25$	**L** $0·16 \times 1·82$
G $4·03 \times 0·56$	**A** $0·927 \times 8·3$	**R** $0·186 \times 5·2$	**T** $0·364 \times 2·4$	**M** $0·836 \times 5·7$

Review 2 Find the area of this rectangular park.
Give your answer to 2 d.p.

1·02 km

0·097 km

Investigation

Theories

Investigate to see if these are always true or sometimes true.

1 When you multiply an even number by 0·6 the last digit of the even number is the same as the last digit in the answer.

Example $0·6 \times$ **28** $=$ 16·**8**

2 To multiply a 2-digit number by 1·1 follow these steps.

Step 1 Write the 2-digit number with the tens digit in the hundreds place. Leave a space for the tens digit.

Example $63 \times 1·1$ 6__3

Step 2 Add the digits of the 2-digit number

Example $6 + 3 =$ **9**

Step 3 Put the answer to Step 2 in the tens digit space and insert the decimal point.

Example $63 \times 1·1 = 6$**9**$·3$

Puzzle

Find the value of each letter.

a A·B × B·A = B·CB **b** L·M × L·L = ML·LM

Dividing

To find the current in an electrical circuit, Dylan had to divide the voltage of 86·7 volts by 16.

He needed to know the answer to 1 decimal place.

86·7 ÷ 16 is approximately 100 ÷ 20 = 5.

```
16 ) 86·7
    −80·0     16 × 5
      6·7
     −6·4     16 × 0·4
      0·30
     −0·16    16 × 0·01
      0·14
```

Dylan wants the answer to 1 d.p., so he needs to work out the answer to 2 d.p. then round to 1 d.p.

Answer 5·41 R 0·14
5·4 to 1 d.p.

To **divide by a decimal** we do an **equivalent calculation** that has a whole number as the **divisor**.

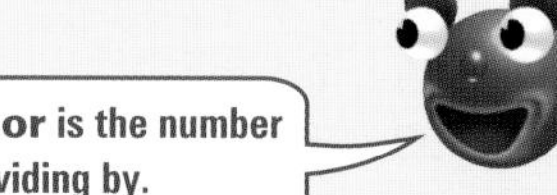

Worked Examples

Calculate these. **a** 352 ÷ 2·6 **b** 5·24 ÷ 0·045

Answers

a 352 ÷ 2·6 is approximately 360 ÷ 3 = 120.
352 ÷ 2·6 is equivalent to 3520 ÷ 26.

```
26 ) 3520
    −2600     26 × 100
      920
     −780     26 × 30
      140
     −130     26 × 5
      10·0
      −7·8    26 × 0·3
       2·20
      −2·08   26 × 0·08
       0·12
```

Answer 135·38 R 0·12
135·4 (1 d.p.)

b 5·24 ÷ 0·045 is approximately
$5 \div 0{\cdot}05 = \frac{5}{5 \times 0{\cdot}01}$
$= 5 \div 5 \div 0{\cdot}01$
$= 5 \div 5 \times 100$
$= 1 \times 100$
$= 100.$

5·24 ÷ 0·045 is equivalent to 5240 ÷ 45.

```
45 ) 5240
    −4500     45 × 100
      740
     −450     45 × 10
      290
     −270     45 × 6
       20
      −18     45 × 0·4
       2·0
      −1·8    45 × 0·04
       0·20
```

Answer 116·44 R 0·2
116·4 (1 d.p.)

Exercise 4

1 Give the answers to these to 1 d.p.

a 487 ÷ 14 **b** 321 ÷ 17 **c** 843 ÷ 23 **d** 560 ÷ 19 **e** 57·1 ÷ 16
f 64·8 ÷ 17 **g** 93·4 ÷ 14 **h** 96·9 ÷ 21 **i** 87·9 ÷ 36 **j** 65·2 ÷ 39
k 94·1 ÷ 43 **l** 87·4 ÷ 28 **m** 53·4 ÷ 24

2 A supermarket sells small soaps in packets.

15 for 56p
24 for 86p
36 for £1·32

Which packet is best value for money?
Why do you think the supermarket sells one packet more cheaply, per soap, than others?

3 Which is equivalent to the calculation given?

a	94 ÷ 3·4	**A**	94 ÷ 34	**B**	940 ÷ 3·4	**C**	940 ÷ 34	**D**	9400 ÷ 34
b	386 ÷ 7·9	**A**	386 ÷ 79	**B**	3860 ÷ 7·9	**C**	3860 ÷ 79	**D**	38 600 ÷ 79
c	472 ÷ 4·6	**A**	4720 ÷ 46	**B**	472 ÷ 46	**C**	47 200 ÷ 46	**D**	4720 ÷ 4·6
d	842·6 ÷ 0·9	**A**	84·26 ÷ 9	**B**	8426 ÷ 9	**C**	8426 ÷ 90	**D**	84 260 ÷ 9
e	364·2 ÷ 0·06	**A**	3642 ÷ 6	**B**	3642 ÷ 60	**C**	36 420 ÷ 6	**D**	364 200 ÷ 6
f	0·058 ÷ 0·0045	**A**	5·8 ÷ 45	**B**	5800 ÷ 45	**C**	58 ÷ 45	**D**	580 ÷ 45
g	0·872 ÷ 7·3	**A**	8·72 ÷ 73	**B**	872 ÷ 73	**C**	8720 ÷ 73	**D**	8720 ÷ 73

4 **a** 378 ÷ 4·2 **b** 481 ÷ 7·4 **c** 234 ÷ 2·6 **d** 195 ÷ 3·9 **e** 425 ÷ 1·7
f 238 ÷ 2·8 **g** 164 ÷ 1·6 **h** 198 ÷ 4·8 **i** 512 ÷ 2·5 **j** 456 ÷ 3·2
k 396 ÷ 4·8 **l** 286 ÷ 1·6 **m** 684 ÷ 2·5 ***n** 386 ÷ 6·4

Think about whether the answer should be bigger or smaller than the dividend.

5 Give the answers to these to 1 d.p.

a 524 ÷ 0·6 **b** 388 ÷ 0·7 **c** 298·3 ÷ 0·3 **d** 426·3 ÷ 0·09 **e** 342 ÷ 2·6
f 47·9 ÷ 3·1 **g** 46·3 ÷ 2·7 **h** 683 ÷ 3·5 **i** 0·87 ÷ 0·69 **j** 0·874 ÷ 0·0045
k 0·583 ÷ 0·0096 **l** 0·832 ÷ 5·7 ***m** 86·3 ÷ 2·7 ***n** 0·142 ÷ 0·048

***6** The currency exchange rate between British pounds and euro is €1 = £0·63.
How many euro would you get for

a £55 **b** £196·52 **c** £0·65 **d** £37·42?

***7** Find the missing measurements. Give them to the nearest cm.

a area = 0·012 m² ; ? ; 0·08 m

b 0·85 m ; ? ; area of triangle = 0·64 m²

c ? ; 0·44 m ; regular hexagon area = 0·84 m²

Review 1 Give the answers to 1 d.p.

a 582 ÷ 17 **b** 924 ÷ 26 **c** 87·6 ÷ 14 **d** 59·3 ÷ 18

Review 2 A gym is selling package deals.

12 visits for £41
25 visits for £79
48 visits for £140

Which package is the best value for money?
Why do you think the gym sells package deals at different prices per visit?

Review 3 Give the answers to these to 1 d.p.

a $389 \div 0{\cdot}9$ **b** $423 \div 0{\cdot}07$ **c** $521 \div 3{\cdot}6$ **d** $0{\cdot}479 \div 6{\cdot}4$ ***e** $0{\cdot}96 \div 0{\cdot}27$

***Review 4** This plan shows some areas in Andrea's bedroom.
What are the missing measurements?
Give them to the nearest cm.

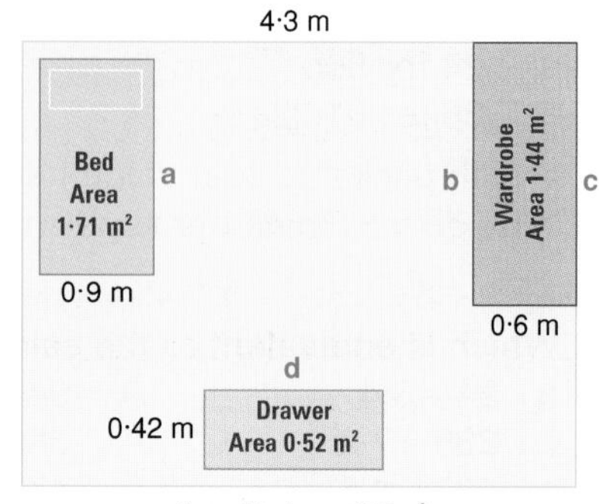

Area of bedroom 15·2 m²

Checking answers

We can **check an answer** using one of these.

1 Check that the answer is sensible.

Example 142 168 151 133 147

Jan found the mean of this data and got the answer 18·6.
This is not a sensible answer because the data values are all much higher than 18·6.

Example Bill calculated that the floor area of his garden shed was 186 m^2.
This is not sensible as it is about the area of a house.

2 Check the answer is the right order of magnitude.

Example The answer to $72 \times 1{\cdot}8$ must be bigger than 72 because we are multiplying by a number bigger than 1.
The answer to $72 \div 1{\cdot}8$ must be smaller than 72 because we are dividing by a number bigger than 1.
The answer to $72 \times 0{\cdot}8$ must be smaller than 72 because we are multiplying by a number smaller than 1.
The answer to $72 \div 0{\cdot}8$ must be bigger than 72 because we are dividing by a number smaller than 1.

3 Estimate first then check if the answer is the right order of magnitude.

Examples $\sqrt{11}$ lies between $\sqrt{9}$ and $\sqrt{16}$ or between 3 and 4.

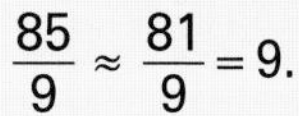
$\frac{85}{9} \approx \frac{81}{9} = 9.$

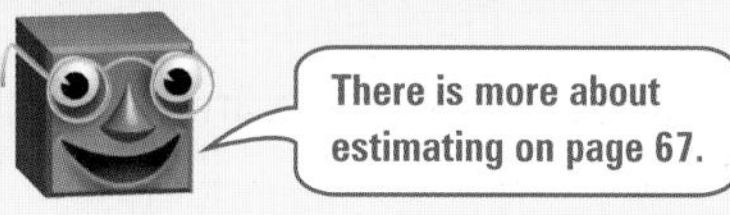

4 Check using inverse operations.

Examples Mark calculated 105·8 ÷ 46 as 2·3.
He could check this answer by calculating 2·3 × 46.
The answer should be 105·8.

The area of a square tile is 127 cm^2.
Alisia calculated $\sqrt{127}$ as 11·26942767. This is the length of each side. She could check this by finding (11·26942767)^2. She should get 127.

Some calculators show the answer as 126·999 999 99.

5 Check using an equivalent calculation.

Example 48 × 12 = 576 can be checked by calculating 96 × 6 or 48 × 4 × 3 or 12 × 4 × 12.

6 Check the last digits.

Example 78 × 3·**2** = 257·**4** is wrong because 8 × 2 = 16.
The last digit should be 6.

Exercise 5

Except for questions 3, 6, 8 and Review 2.

1 Veronica worked out the answers to these problems.
Her answers are given in the blue boxes.
Are her answers sensible? If not, explain why not.

a Mary took 125 g of vitamin C powder each day.
How many grams of vitamin C does she take each week? **17·9 g**

b Samuel saved £4·80 each week.
By the end of June he had saved £57·60.
How many weeks did he take to save this? **12**

c A packet of 4 small cakes cost £3·85.
How much does each cake cost? **£15·40**

d These are the heights of Veronica's four brothers.
Calculate the mean of these heights. **159·75 m**
1·46 m   1·63 m   1·58 m   1·72 m

2 Will the answers to these be bigger or smaller than 52?
a 52 × 0·6   **b** 52 × 1·6   **c** 52 ÷ 5   **d** 52 ÷ 0·5   **e** 52 × 2·7
f 52 ÷ 2·7   **g** 52 × 14·6   **h** $\frac{52}{0\cdot4}$ **i** $\sqrt{52}$

3 Check the answers to these using inverse operations.
Which ones are wrong?
a 73 × 1·2 = 87·6   **b** 578 ÷ 32 = 19·2   **c** 8·64 + 23·7 = 32·34   **d** 5·7 × 9·6 = 547·2
e $\frac{2}{3}$ of 86·4 = 57·6   **f** 4·7^2 = 29·02   **g** $\sqrt{8}$ = 2·8284271

4 Janet entered the number 8653 on her calculator. She then added 89 and then multiplied by 7. What should Janet then key to get 8653 displayed again?

5 Which of these is not equivalent to 24 × 36?
A 12 × 2 × 12 × 3   **B** 48 × 18   **C** 48 × 12 × 2   **D** 24 × 6 × 6

6 Check these by doing an equivalent calculation.
Write down the calculations you did.
a 512 ÷ 16 = 32   **b** 8·3 × 1·1 = 9·13   **c** 586 × 99 = 58 014

7 Check these calculations by looking at the last digits.
Which ones are wrong?
a $81 \times 64 = 5148$ **b** $7{\cdot}2 \times 3{\cdot}8 = 27{\cdot}36$ **c** $4{\cdot}5 \times 37 = 165{\cdot}6$

8 Check the answers to these using a method of your choice.
Show how you did it.
a $8{\cdot}7 \times 3{\cdot}9 = 33{\cdot}93$ **b** $19{\cdot}63 + 52{\cdot}71 = 72{\cdot}34$ **c** $867 \div 17 = 51$ **d** $96 \times 28 = 2688$
e $\sqrt{17} = 4{\cdot}1231056$

***9** Manisha bought a car in this sale.
Two months after she bought the car, its value was 10% more than she paid for it.
The value of the car then was not the same as it was the day before the sale. Is this possible? Explain.

***10** Without using a calculator, pick a possible answer to the calculation.
Explain your choice.

		A	B	C
a	48×89	4862	4272	4383
b	53×32	1696	1486	1782
c	$79 \times 1{\cdot}5$	118·5	142·5	120
d	$539 \times 0{\cdot}72$	321·6	632	388·08
e	$488 \times 1{\cdot}47$	432·86	768·7	717·36
f	526^2	26 384	276 676	10 527
g	$426 \div 0{\cdot}52$	256	81·92	819
h	$58 \div 0{\cdot}072$	805·6	80·6	8·056

Review 1 Paula did a science experiment. She worked out the total mass of four pieces of sodium chloride, each with a mass of 21·7 g. She got an answer of 18·8 g.
Is Paula's answer sensible? Explain.

Review 2 Check the answers to these using a method of your choice.
Show how you did it.
a $4{\cdot}6 \times 7{\cdot}3 = 33{\cdot}58$ **b** $960 \div 16 = 60$ **c** $87 \times 39 = 3393$ **d** $57 \times 1{\cdot}86 = 106{\cdot}02$

* **Review 3** Without using a calculator, choose a possible answer for the calculation.
Explain your choice.

		A	B	C
a	27×48	1686	1296	1312
b	$5{\cdot}2 \times 3{\cdot}1$	18·13	14·32	16·12
c	$\sqrt{864}$	20·7	400	29·4
d	$834 \div 0{\cdot}79$	1055·7	106·7	10·6

Discussion

Paul keyed a number into his calculator and cubed it.
What must he key to get back to his starting number?

Will his exact starting number always be shown on the screen by doing this? **Discuss.**
If not, explain why not.

Brackets on a calculator

Scientific calculators carry out operations according to the rules for order of operations, as long as you only key equals at the end.
When **brackets** are part of the calculation, key them as you come to them.

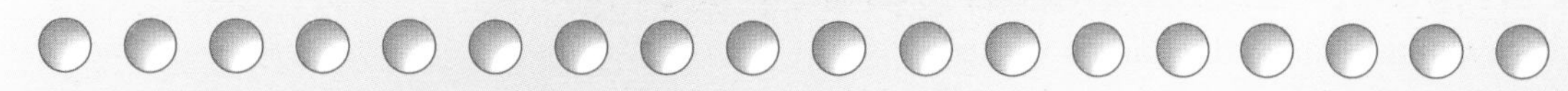

Investigation

Brackets on a calculator

Investigate keying sequences for these.

$\frac{8+12}{2}$ $\frac{144}{2\times 3}$ $\frac{8+12}{2+3}$ $\frac{8+12(11-7)}{4\times 7}$ $\frac{8+12(11-7)}{16-2(3+4)}$

Worked Example

a $\sqrt{(11^2-5^2)}$ **b** $\frac{^{-}7{\cdot}3-2\times 4{\cdot}1}{2\times 3}$ **c** $\frac{(12-7)^2(19-8)^2}{5(15-12)^4}$

Answer

a $\sqrt{(11^2-5^2)} \approx \sqrt{10^2-5^2} = \sqrt{100-25} = \sqrt{75} \approx \sqrt{81} = 9$

Key √ (11 x^2 − 5 x^2) = to **get 9·8 (1d.p.)**

b $\frac{^{-}7{\cdot}3-2\times 4{\cdot}1}{2\times 3} \approx \frac{^{-}7-8}{6} \approx \frac{^{-}15}{5} = {^{-}3}$

Key ((−) 7·3 − 2 × 4·1) ÷ (2 × 3) = to get $^{-}2{\cdot}6$ (1d.p.)

Note: We need to add brackets to the calculation so that the whole numerator is divided by the whole denominator.
An alternative keying sequence is

Key (−) 7·3 − 2 × 4·1 = ÷ 2 ÷ 3 =

$\frac{200}{4\times 5} = 200 \div 4 \div 5$

c $\frac{(12-7)^2(19-8)^2}{5(15-12)^4} = \frac{5^2\times 11^2}{5\times 3^4} = \frac{25\times 121}{5\times 81} \approx \frac{25\times 100}{400} \approx \frac{2400}{400} = 6$

Key (12 − 7) x^2 (19 − 8) x^2 ÷ (5 (15 − 12) x^y 4) =
to get 7·5 (1 d.p.)

Note: There are two sets of brackets keyed for the denominator. One set is around the whole denominator.

Sometimes there are two sets of brackets given in a calculation.

Example 140 ÷ {20 − (13 − 8)}
We key this as

140 ÷ (20 − (13 − 8)) = to get 9·3 (1d.p.)

Exercise 6

1 a $3{\cdot}6 - (4{\cdot}9 + 2{\cdot}3) + (4{\cdot}6 + 7)$ b $4{\cdot}7 + (8{\cdot}6 - 3{\cdot}2) - (9{\cdot}7 - 7{\cdot}63)$
c $3{\cdot}2 \times (9{\cdot}7 - 2{\cdot}3) - (4{\cdot}6 - 3{\cdot}8)$ d $19{\cdot}3 - 5{\cdot}7(6{\cdot}8 - 2{\cdot}3)$
e $(16.4 - 3{\cdot}2) \times (6{\cdot}8 - 5{\cdot}92)$ f $^{-}(362 \times 4 + 582) + 4 \times 361 - (1 - 382)$
g $18{\cdot}7 - {}^{-}(8{\cdot}36 \times 4{\cdot}2 + 3{\cdot}6) + 5 \times 8{\cdot}9 - (3 - 4{\cdot}6)$

2 a James keyed $\frac{13 + 5}{10 - 1}$ as

(13 + 5 = ÷ (10 − 1) =

What mistake did James make?

b Janus keyed $\frac{3 + 5 \times 9}{5 - 2}$ as

(3 + 5 × 9) ÷ (5 − 2 =

What mistake did Janus make?

c Olivia keyed $\frac{10 + 12}{4 \times 2}$ as

(10 + 12) ÷ 4 × 2 =

What mistake did Olivia make?

d Kiran keyed $\frac{15 \times 10 + 12}{4 \times 2}$ as

(15 × 10 + 12) ÷ 4 × 2 =

What is wrong with his keying sequence?

e Michaela keyed $\frac{15 + 64 - 2 \times 4}{3(2 + 5) - 1}$ as

15 + 64 − 2 × 4 = ÷ (3 × (2 + 5) − 1 =

What is wrong with her keying sequence?

3 Give the answers to 2 d.p. if you need to round.

a $\frac{91 - 17}{9 - 6}$ b $\frac{69}{5 \times 4}$ c $\frac{4{\cdot}6 + 2{\cdot}3}{5{\cdot}8}$ d $\frac{8 + 6}{2 \times 3}$

e $\frac{3 \times 7{\cdot}2}{4 + 3}$ f $\frac{1{\cdot}4 + 3 \times 8}{6}$ g $\sqrt{16^2 - 5^2}$ h $\sqrt{25^2 - 24^2}$

i $\sqrt{8^2 + 15^2}$ j $5 \times (4{\cdot}68)^2$ k $\frac{(4 + 6)^2}{(16 - 5)^2}$ l $\frac{(15 - 9)^2(8 - 4)^2}{(9 - 4)^2}$

m $\frac{(11 - 4)^2(6 - 2)^2}{(7 - 4)^3}$ n $\frac{(17 - 10)^2(18 - 16)^2}{3(17 - 12)^3}$ *o $\frac{(5{\cdot}2 - 3{\cdot}6)^2(4{\cdot}1 - 2{\cdot}3)^3}{18 - (3{\cdot}2 - 1{\cdot}4)^2}$

4 Give the answers to 1 d.p. if you need to round.
a $140 \div \{20 - (13 - 6)\}$ b $16 \times \{24 - (8 - 3)\}$
c $5 \times \{29 - (20 - 8)\}$ d $\{41 + (13 - 7)\} \div 6$
e $56 \div \{(13 + 3) \times 5\}$

***5** $4{\cdot}2 \times 6 + 3 - 3 \times 4 + 1{\cdot}8$
Put one set of brackets in this calculation to make
a the largest possible answer b the smallest possible answer.

***6** Put decimal points, +, −, ×, ÷ and brackets in these to make them true.
a 19 36 2 = 2·62 b 128 69 3 = 10·5
c 124 4 18 = 1·3 d 84 4 16 2 = 36·8
e 4 116 5 22 = 14·4 f 24 4 68 27 = 18·8
g 69 24 3 18 = 9·5 h 34 2 66 3 = 4·6
i 27 11 5 7 = 0·7

It may take a long time to do these.

T

Review 1

					A										
37·68	7	7	16	16·74	1·5	15	7	484·6	3·5	15	7	7	1	7	16

Use a copy of this box. Put the letter that is beside each calculation above its answer in the box.

A $4 \cdot 7 - (8 \cdot 3 + 3 \cdot 7) + (5 \cdot 8 + 3) = 1 \cdot 5$ H $(18 \cdot 7 - 9 \cdot 4) \times (8 \cdot 7 - 6 \cdot 9)$ I $\frac{15 + 24}{17 - 6}$ to 1 d.p.

V $\sqrt{(17^2 - 8^2)}$ F $7 \times 8 \cdot 32^2$ to 1 d.p. Y $\frac{(5 + 7)^2}{(21 - 9)^2}$

S $160 \div \{18 - (12 - 4)\}$ E $140 \div \{4 \times (3 + 2)\}$ B $(14 \cdot 6 - 8 \cdot 32) \times \{18 \div (9 - 6)\}$

Review 2 Put one set of brackets in $180 \div 6 \cdot 39 + 4 \cdot 3 \times 2 \cdot 7 - 6 \cdot 8$ to make the smallest possible answer.

Using the calculator memory

We can use the **memory** on the calculator to help us with calculations.

0 STO M+ clears the memory.
Always clear the memory before beginning a calculation.

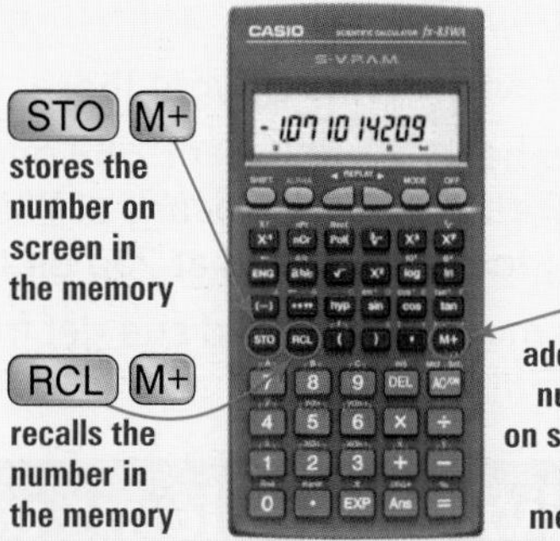

Example Jane bought 23 bars of chocolate at 42p each and 29 cans of drink at 35p each for her class party.
How much change did she get from £20?

We could key

0 STO M+ 23 × ·42 + 29 × ·35 = STO M+ 20 − RCL M+ = to get 0·19.

This puts the cost of the bars and cans into memory. This recalls the cost of the bars and cans from memory.

We could have used the M+ key by keying

0 STO M+ 23 × ·42 STO M+ 29 × ·35 M+ 20 − RCL M+ = to get 0·19.

Cost of bars put into memory. Cost of cans added to memory.

The keying might be different on other calculators.

Jane got **£0·19** or **19p** change.

Exercise 7  **Use the memory key on the calculator to do these.**

1 For a design and technology project Marcel bought 8 kg of mixed vegetables at £1·89 per kg, 1·4 kg of mixed nuts at £8·70 per kg and 0·6 ℓ of sauce at £7·83 per ℓ.
How much change did he get from £50?

2 Annie set herself a goal of running 50 km each week. She ran 3·8 km on day 1, twice as far on day 2 and on each of days 3 to 7, she ran three-quarters of the distance she ran on day 2.
How many kilometres short of her goal was she?

3 Sam bought 3 pairs of track pants at £21·86 each, 4 pairs of socks at £3·54 each and a coat at 10% off its normal price of £136·50.
How much change did he get from £500?

4 £1 = 175 yen.
How many yen would Rosalie get for
a £50 **b** £75.80 **c** £120·64 **d** £0·96?

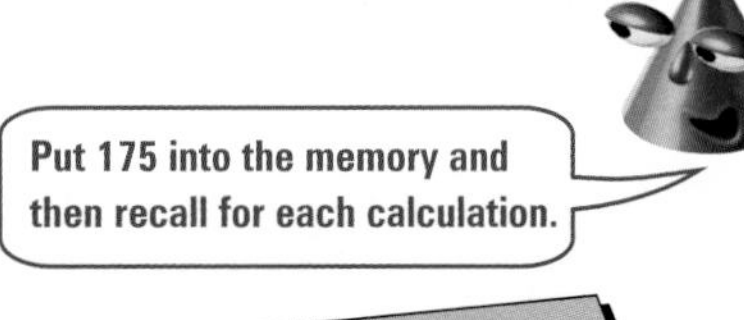

5 What is the sale price of a shirt that was
a £76·75 **b** £93·40 **c** £47·95 **d** £50·20?

Review Penny bought these.
5 bags of apples at £2·89 per bag
3 cans of baked beans at 78p each
4 loaves of bread at 20p off their normal price of £1·59.
How much change did she get from £25?

Solving problems using a calculator

Exercise 8 

1 Seadown Youth Club hired a bus for a day outing. The total cost of this hire was £124·20, which consisted of a fixed charge of £45 and 45p per kilometre travelled. How many kilometres did the Seadown Youth Club travel on this outing?

2 Amanda put 56·5 ℓ of petrol in her car. It cost 68p per litre.
How much did this petrol cost altogether?

3 The petrol consumption rate for Amanda's car is 13·4 km per litre.
- **a** Amanda travelled from Aberdeen to Birmingham, a distance of 711 kilometres. How many litres of petrol did her car use on this journey?
- **b** Amanda used the conversion rate '1 mile is about 1·609 km' to work out the distance she had travelled in miles. To the nearest mile, she calculated this distance to be 442 miles. Is she correct?
- **c** What distance could Amanda travel on 28 litres of petrol?
- **d** On a 113 km section of the journey, Amanda travelled at a constant speed. This section of the journey took Amanda 1 hour 21 minutes. About how many kilometres did Amanda travel each 15 minutes?

4 One day, a factory made 73 skirts and 22 jackets. The sizes made and the material used for each size are given in the table.

	Size	Material (metres)	Number
Skirt	10	1·2	18
Skirt	12	1·3	32
Skirt	14	1·5	23
Jacket	10–12	2·5	10
Jacket	12–14	2·9	12

a How much material was used on this day?

b The next day, the same amount of material was used to make a quantity of size 14 skirts. How many skirts were made?

c On the day that 73 skirts were made, one person did all the finishing work on them. This person was paid a daily wage of £30 with an extra 75p for every skirt in excess of 50 that were finished. How much did this person earn on this day?

*__d__ One of the tailors was paid an hourly rate of £6·20 for the first $7\frac{1}{2}$ hours worked, then £9·30 per hour after that. How many hours did this tailor work on a day in which he earned £60·45?

Review Keung measured the length of his desk as 60 cm. How many desks of this length could be placed side by side along a classroom wall that is 26 feet long? (Use 1 foot = 0·3048 metres.)

Puzzle

1 Two families went to a wildlife park.
Rani's family bought tickets for one adult and four children and paid £25.
Becky's family bought tickets for two adults and two children and paid £23.
What was the cost of a child's ticket?

2 At a Bring and Buy sale table, slices of apple pie were priced at 40p per slice. When none had been sold at this price, a decision was made to reduce them to less than half price. Once this had been done, they all sold quickly for a total of £3·91.
By how much was the price of each slice reduced?

Summary of key points

We **add and subtract decimals** by lining up the decimal points.
Always estimate first.

When we multiply **a positive number, a, by a number between 0 and 1**, the answer is **smaller** than a.

When we **divide a positive number, b, by a number between 0 and 1**, the answer is **larger** than b.

Example 58·6 × 0·72 = 42·192 which is smaller than 58·6
187·2 ÷ 0·31 = 603·9 (1 d.p.) which is larger than 187·2.

Multiplying

When **multiplying with decimals**, we can do an equivalent calculation without decimals.

Example 59·7 × 0·68

59·7 × 0·68 is approximately equal to 60 × 0·7 = 42

and is equivalent to 597 × 68 ÷ 1000.

597 × 68 ÷ 1000 = 40 596 ÷ 1000

= 40·596

```
   597
 ×  68
 35820
  4776
 40596
 11
```

Dividing

When the **divisor is a decimal** we do an equivalent calculation where the divisor is not a decimal.

Example 8·83 ÷ 0·32

8·83 ÷ 0·32 is approximately equal to 9 ÷ 0·3 = 30

and is equivalent to 883 ÷ 32.

```
32)883
  -640     32 × 20
   243
  -224     32 × 7
    19
   -16     32 × 0·5
    3·0
  -2·88    32 × 0·09
   0·12
```

If we want the final answer to 1 d.p. we must find the answer to 2 d.p. and then round to 1 d.p.

Answer 27·59 R 0·12

27·6 to 1 d.p.

We can **check the answer to a calculation** in one of these ways.

- Check the answer is sensible.
- Check the answer is about the right order of magnitude.
- Estimate first then check the order of magnitude of the answer is the same as the estimate.
- Check using inverse operations.
- Check using an equivalent calculation.
- Check using last digits.

When **brackets** are part of a calculation, we key them as we come to them.

Example (5·6 + 2·3) × (8·4 − 6·9) is keyed as

(5·6 + 2·3) × (8·4 − 6·9) = to get 11·85.

Sometimes we need to **add brackets to a calculation.**

The whole numerator must be divided by the whole denominator.

Example $\frac{5+11}{18-7}$ is keyed as

to get 1·45. (2 d.p.)

We use the **calculator memory** to do a calculation.

STO M+	stores the number on the screen in memory.
M+	adds the number on the screen to the number already in memory.
RCL M+	recalls the number that is in the memory back to the screen.

Test yourself

Except for questions 12, 13 and 14.

1 Todd needed to add these masses in a science experiment.
What total should he get?
8·2 kg 523 g 0·0067 kg

2 Without doing any calculation, decide which of these has an answer greater than 0·26. Explain your answer.
a $0{\cdot}26 \div 1{\cdot}4$ **b** $0{\cdot}26 \times 0{\cdot}05$ **c** $0{\cdot}26 \times 11{\cdot}81$ **d** $\dfrac{0{\cdot}26}{0{\cdot}89}$

3 **a** $52{\cdot}7 \times 0{\cdot}8$ **b** $38 \times 4{\cdot}63$ **c** $8{\cdot}74 \times 0{\cdot}54$ **d** $6{\cdot}7 \times 0{\cdot}114$ **e** $0{\cdot}897 \times 0{\cdot}047$

4 A rectangular plate on a space shuttle was 1·76 m by 0·085 m.
What is the area of the plate?

5 Give the answers to these to 1 d.p.
a $587 \div 16$ **b** $429 \div 28$

6 Renee times a pendulum with a stopwatch.
Seventeen swings of the pendulum take 31·62 seconds.
What is the mean time taken for one swing?

7 Is it cheaper per slice to buy 36 pizza slices or 24 pizza slices?

8 Give the answers to 1 d.p.
a $126 \div 3{\cdot}3$ **b** $83 \div 0{\cdot}7$ **c** $526{\cdot}7 \div 0{\cdot}06$ **d** $58{\cdot}6 \div 5{\cdot}3$ **e** $8{\cdot}62 \div 0{\cdot}025$ **f** $0{\cdot}896 \div 0{\cdot}037$

9 Marilyn ran round a 3·6 kilometre track 18 times.
She worked out she had run 648 km.
Is this answer sensible?

10 Check each calculation by looking at the size of the answer. If the answer is wrong, explain how you can tell.
a $5{\cdot}2 \times 6{\cdot}3 = 2{\cdot}9$ **b** $8{\cdot}3 \div 0{\cdot}4 = 2{\cdot}075$ **c** $89 \div 6{\cdot}8 = 197$

11 Write down an equivalent calculation you could do to check the answer to $4{\cdot}6 \times 2{\cdot}1 = 9{\cdot}66$. Is the answer correct?

12 Check the answers to these using inverse operations.
Which ones are wrong?
a $4{\cdot}6 \times 3{\cdot}9 = 19{\cdot}74$ **b** $582 \div 6 = 97$ **c** $\sqrt{7} = 2{\cdot}6457513$ **d** $68 \div 0{\cdot}0035 = 194$

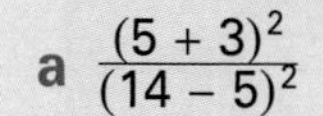

13 Give the answers to 2 d.p. if you need to round.
a $\dfrac{(5+3)^2}{(14-5)^2}$ **b** $\dfrac{(18-11)^2(9-4)^2}{(7-3)^2}$ **c** $\dfrac{4 \times 3{\cdot}6 - {}^-2{\cdot}7}{3(8{\cdot}9 - 6{\cdot}7)}$ **d** $\dfrac{1{\cdot}2 - 4{\cdot}7 - {}^-1}{1{\cdot}4 + 2 \times 0{\cdot}8}$
e ${}^-(386 \times 4 + 583) + 4 \times 261 - (2 - 387)$

14 Use the calculator memory to find the answers.
At a book and magazine sale, Dianne buys
6 paperbacks at £0·55 each
4 books at £1·19 each
8 books at £2·35 each
and 2 magazines at £0·80 each.
How much change does Dianne get from £30?

5 Fractions

You need to know

✓ fractions, decimals and percentages page 6

— one number as a fraction of another

— mixed numbers, improper fractions

— equivalent fractions

— cancelling a fraction to its lowest terms

— adding and subtracting fractions with the same denominator

— finding a fraction of

Key vocabulary

cancel, convert, lowest terms, mixed number, recurring decimal, simplest form, terminating decimal, unit fraction

The Sound of Music

Music is written in 'bars'.
In a bar the note values often add up to 1 whole semibreve (called 1 whole note in this exercise).
The possible notes are

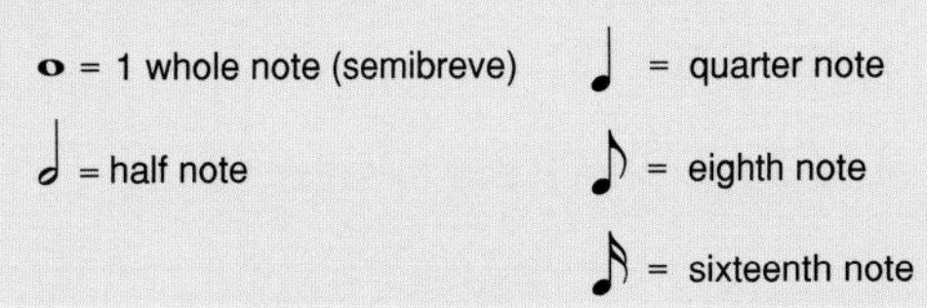

1 Which of these bars do *not* have notes which add to one whole note?

2 Write the bars for these notes.

a $\frac{1}{2}, \frac{1}{4}, \frac{1}{4}$ **b** $\frac{1}{4}, \frac{1}{8}, \frac{1}{8}, \frac{1}{4}, \frac{1}{4}$ **c** $\frac{1}{16}, \frac{1}{16}, \frac{1}{16}, \frac{1}{16}, \frac{1}{8}, \frac{1}{8}, \frac{1}{4}, \frac{1}{8}, \frac{1}{8}$

3 What goes in the box, > or < or = ?

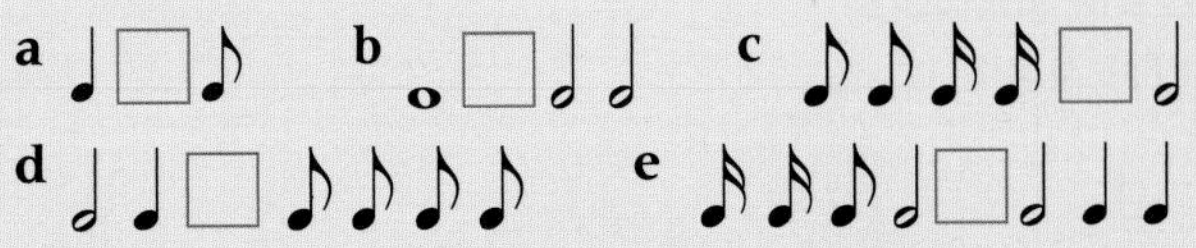

4 Write four bars of your own music. Make each bar add to one whole note.

5 Investigate more about music and fractions.

Fractions of shapes

Worked Example

Find the fraction of this square that is shaded.

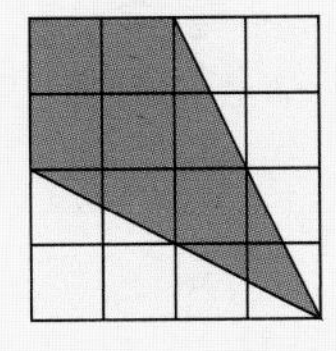

Answer

Count the whole squares shaded, 5.

Fit shaded pieces together to make whole squares.

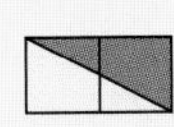 = 1 shaded square, 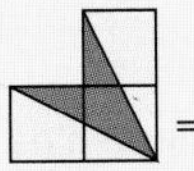 = 1 shaded square, 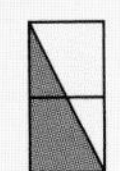 = 1 shaded square,

Altogether 8 whole squares out of 16 are shaded.

So $\frac{1}{2}$ of the square is shaded.

Exercise 1

1 Find the fraction of each shape that is red.

a

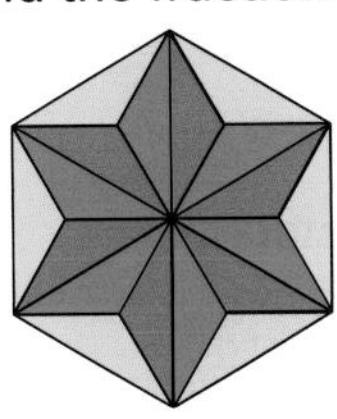

b

c

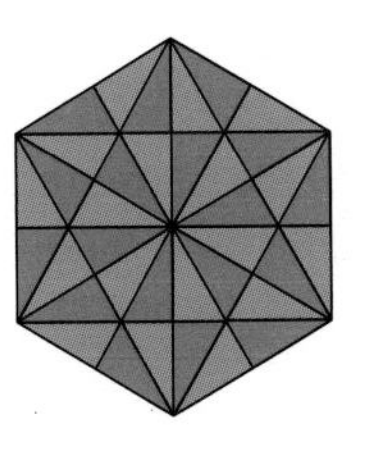

Remember to give the answer in its simplest form.

2 Find the fraction of each square that is coloured.

a

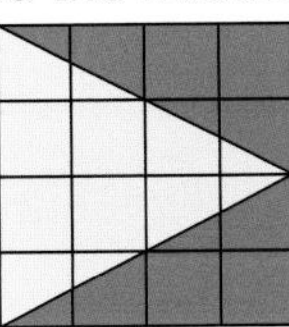

b

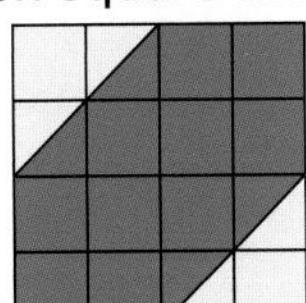

c

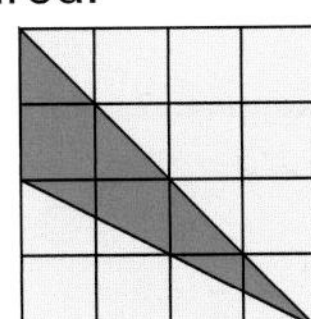

***3** Express the shaded shape as a fraction of the large rectangle.
Note: The curves are all semi-circles or quarter-circles.

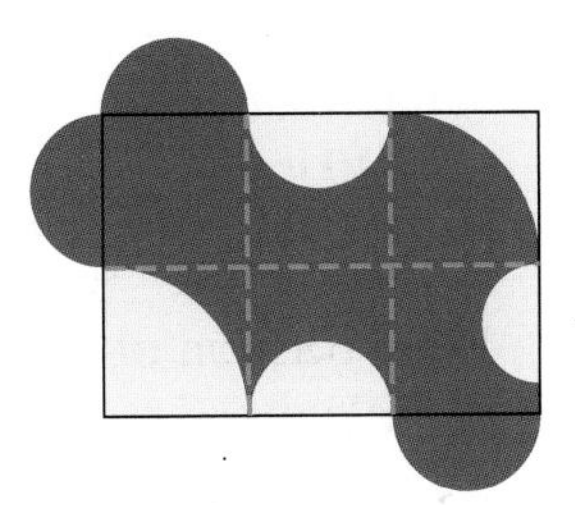

Review Find the fraction of each shape that is shaded purple.

a

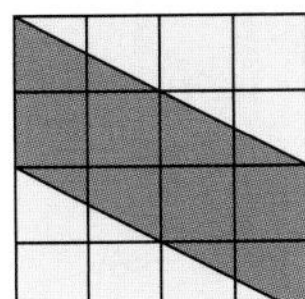

b

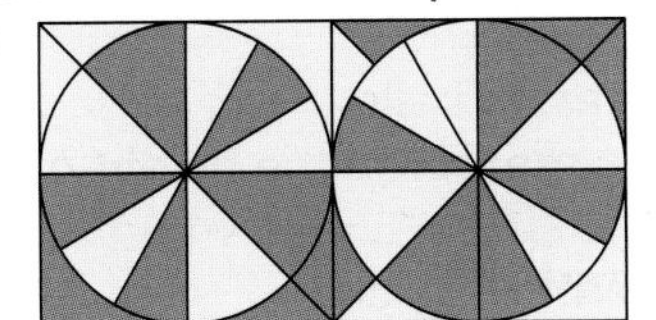

One number as a fraction of another

Worked Example

What fraction of **a** 240 is 160 **b** 60 is 80?

Answer

a We write this as $\frac{160}{240} = \frac{2}{3}$.
160 is $\frac{2}{3}$ of 240.

b We write this as $\frac{80}{60} = \frac{4}{3}$ or $1\frac{1}{3}$
80 is $\mathbf{1\frac{1}{3}}$ of 60.

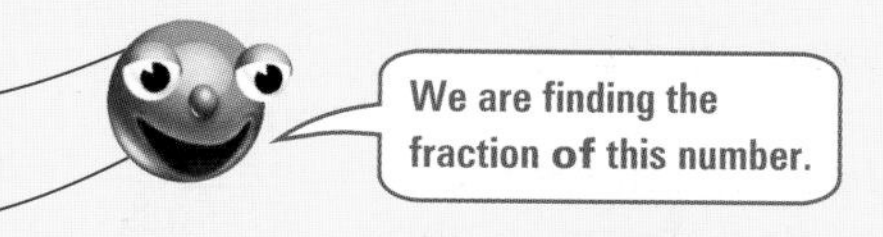

Worked Example

This frequency diagram shows the sunshine hours for UK weather stations one day in May.
The sunshine hours were put into class intervals
$0 \leqslant s < 1$, $1 \leqslant s < 2$, etc.

What fraction of stations had

a between 2 and 4 hours of sunshine recorded

b 4 or more hours sunshine?

Sunshine hours at weather stations

Number of stations

Number of hours

Answer

a We must add the number of stations that had between 2 and 3 and 3 and 4 hours of sunshine.

$7 + 13 = 20$

20 stations had between 2 and 4 hours of sunshine.
We must also add up the heights of the bars to find the total number of stations.

$2 + 2 + 7 + 13 + 9 + 8 + 15 + 9 = 65$

Fraction of stations that had between 2 and 4 hours of sunshine $= \frac{20}{65} = \frac{4}{13}$.

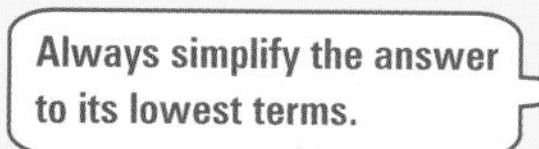

b We must add the number of stations that had between 4 and 5, 5 and 6, 6 and 7 and 7 and 8 hours of sunshine.

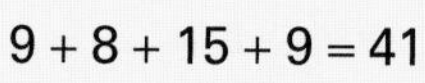

$9 + 8 + 15 + 9 = 41$

Fraction of stations that had 4 or more hours $= \mathbf{\frac{41}{65}}$

Exercise 2

1 What fraction of

a 240 is 200 **b** 300 is 280 **c** 10 is 15 **d** 6 is 10
e 150 is 200 **f** 100 is 180 **g** 180 is 240 **h** 240 is 360?

2 What fraction of 10 m is

a 100 cm **b** 150 cm **c** 1250 cm **d** 2500 cm **e** 2250 cm?

3 What fraction of an hour is

a 100 minutes **b** 160 minutes **c** 320 minutes?

4 Bernadette did an experiment with three different balls.
She dropped each from a height of 2 m. She measured the height of the first bounce. These are her results

Ball 1 180 cm **Ball 2** 155 cm **Ball 3** 85 cm

What fraction of the height from which they were dropped did each bounce?

5 The red shape has been enlarged to get the blue shape.
On the red shape, AB is 16 mm.
On the blue shape A′B′ is 40 mm.

a What fraction of AB is A′B′?

*__b__ What is the scale factor for the enlargement?

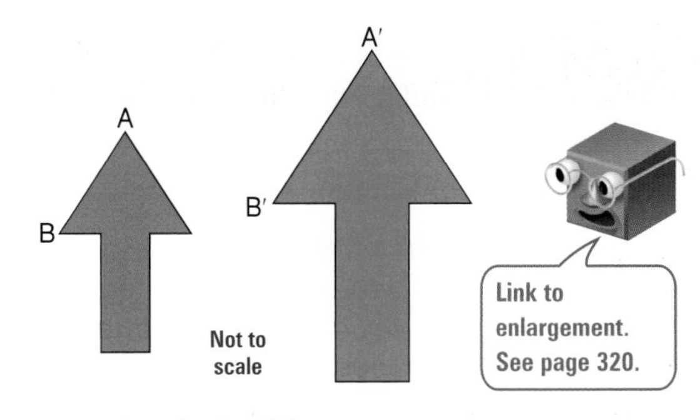

6 This frequency diagram shows the lengths of the vehicles parked in one section of a supermarket car park.
What fraction of vehicles were

a 3·5 m or longer but less than 4·5 m

b 5 m or longer

c shorter than 4 m?

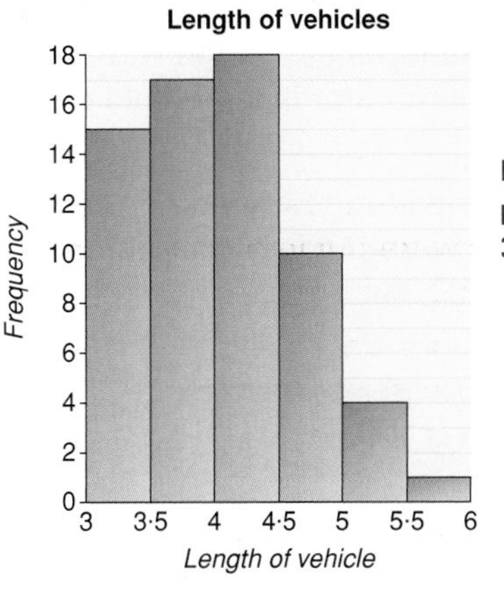

Note: The lengths were put into class intervals $3 \leqslant l < 3{\cdot}5$, etc.

7 What fraction of the small shape is the large one?

a 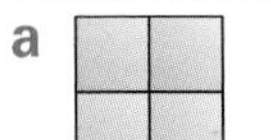 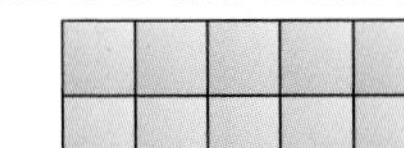

b 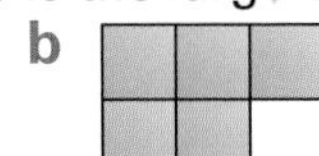 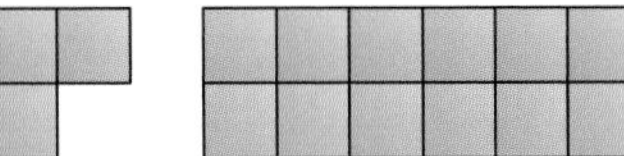

*__8__ Jenni 1·50 m Lucy 1·45 m Jesse 1·65 m
Write

a Jenni's height as a fraction of Jesse's

b Jenni's height as a fraction of Lucy's

c Jesse's height as a fraction of Jenni's.

Review 1 Mary had three hours for an exam. She began at 9:15 am and finished at 11:45 a.m. What fraction of the time allowed did she not use?

Review 2 This frequency diagram shows the distance of the longest tee-shot taken by students in a golf tournament.
What fraction of students had a tee-shot

a 140 m or longer but less than 170 m

b 160 m or more

c less than 150 m?

Note: The distances were put into class intervals of $120 \leqslant d < 130$, etc.

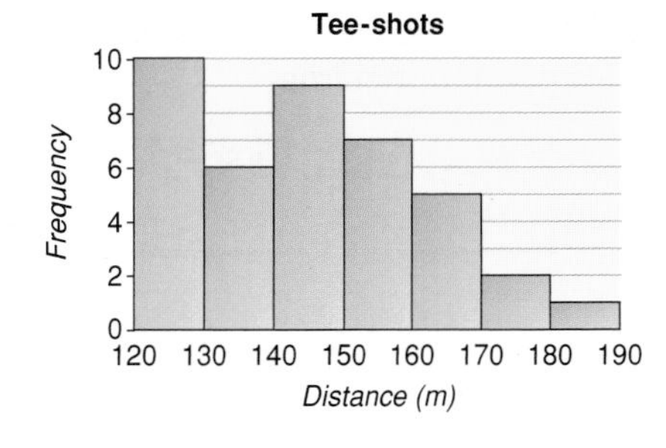

Review 3 What fraction of the small angle is the large one?

a

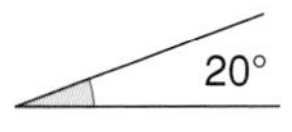

90°

b

100°

270°

Puzzle

What fraction am I?

a I am a unit fraction.
My denominator is a multiple of 3.
My denominator is double the number of factors it has.
I am between $\frac{1}{8}$ and $\frac{1}{20}$.

b I am an improper fraction.
My denominator is the HCF of 6 and 18.
My numerator is a multiple of 5.
I am between 1 and 2.

***c** If 1 is added to my numerator I am equivalent to $\frac{1}{3}$.
If 1 is added to my denominator I am equivalent to $\frac{1}{4}$.

Fractions and decimals

Remember

To **convert a decimal to a fraction** write with denominator 10, 100 or 1000 and then cancel to its simplest form.

Example $0{\cdot}682 = \frac{682}{1000}$
$= \frac{341}{500}$

To **convert a fraction to a decimal** we can
1 use known facts **or**
2 make the denominator 10, 100 or 1000 **or**
3 divide the numerator by the denominator.

We can also use a calculator.

Example $\frac{27}{39}$ key 27 ÷ 39 = to get 0·69 (2 d.p.)
or key 27 a b/c 39 = a b/c to get 0·69 (2 d.p.)

Keying this again gives the answer as a decimal.

Sometimes we get a **recurring** (repeating) decimal.

Example $\frac{2}{11} = 0{\cdot}181\ 818\ 18\ ...$
$= 0{\cdot}\dot{1}\dot{8}$

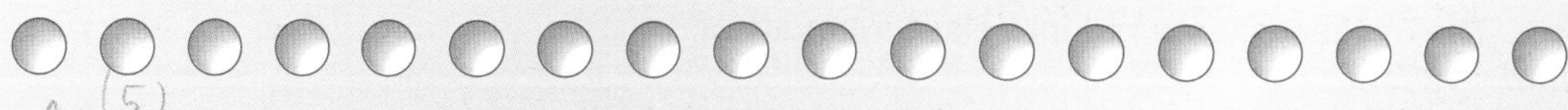

Investigation

Prime Factors and Recurring Decimals

$\frac{1}{2} = 0{\cdot}5$ $\quad \frac{1}{8} = 0{\cdot}125$ $\quad \frac{1}{10} = 0{\cdot}1$ $\quad \frac{1}{16} = 0{\cdot}0625$ $\quad \frac{1}{20} = 0{\cdot}05$

These fractions give terminating decimals.
What are the prime factors of each of the denominators?

Example The prime factors of 20 are 2 and 5.

$\frac{1}{3} = 0{\cdot}\dot{3}$ $\frac{3}{11} = 0{\cdot}\dot{2}\dot{7}$ $\frac{7}{15} = 0{\cdot}4\dot{6}$ $\frac{2}{7} = 0{\cdot}\dot{2}8571\dot{4}$

These fractions give recurring decimals.
What are the prime factors of each of the denominators?

The string of digits 285714 repeats so we put a dot above the first and last digit of the string.

Investigate to find which denominators give terminating decimals and which give recurring decimals.

You could use the $a^{b/c}$ key on your calculator.

Can all fractions be written as either terminating or recurring decimals? **Investigate**.

Investigate what goes in this gap.

Fractions with denominators that have prime factors other than ________ will recur if written in decimal form.

You should learn these **recurring decimal conversions**.

$0{\cdot}333333 \cdots = \frac{1}{3}\ (= \frac{3}{9})$ $0{\cdot}666666 \cdots = \frac{2}{3}\ (= \frac{6}{9})$

$0{\cdot}111111 \cdots = \frac{1}{9}$ $0{\cdot}999999 \cdots = \frac{9}{9} = 1$

$\frac{1}{7} = 0{\cdot}142\,857\,142\,857\,142\,857$. We write this as $0{\cdot}\dot{1}4285\dot{7}$. The digits 142857 repeat so we put a dot above the first and last digit of the string.

Exercise 3

Except for questions 3b, 4d, 5, 7 and Review 1.

1 Write these as fractions in their lowest terms.
a 0·96 **b** 0·06 **c** 0·875 **d** 0·005 **e** 1·68 **f** 3·725

2 Write these as decimal fractions.
a $\frac{3}{5}$ **b** $\frac{1}{20}$ **c** $\frac{15}{40}$ **d** $\frac{3}{20}$ **e** $\frac{4}{25}$ **f** $\frac{19}{20}$
g $\frac{17}{25}$ **h** $\frac{36}{40}$ **i** $\frac{18}{60}$ **j** $\frac{27}{50}$ **k** $\frac{23}{25}$ **l** $1\frac{3}{5}$
m $2\frac{3}{4}$ **n** $4\frac{5}{8}$ **o** $\frac{33}{30}$ **p** $\frac{104}{50}$ **q** $\frac{48}{40}$ **r** $\frac{175}{50}$

3 Which of these, when written as a decimal, gives a terminating decimal?
a **A** $\frac{3}{7}$ **B** $\frac{5}{9}$ **C** $1\frac{1}{3}$ **D** $\frac{7}{8}$
b **A** $\frac{9}{16}$ **B** $\frac{5}{14}$ **C** $1\frac{3}{11}$ **D** $\frac{4}{13}$

4 Convert the answers to these to a decimal.
a A chemical reaction takes 20 seconds.
After 18 seconds, 95% of the reaction is complete.
What fraction of the total time is this?
b Molly ran 24 km and walked 40 km each week.
What fraction of the total distance did she run each week?
c The brass used to make a 500 g statue is made of 325 g of copper and the rest is zinc.
What fraction of the brass is copper?
d On its first swing a pendulum reaches a height of 45 mm above its resting position. On the second swing it reaches 41 mm.
What fraction of the height of the first swing did the second swing reach?

5 Use your calculator to convert these to decimals. Round to 2 d.p. if you need to round.

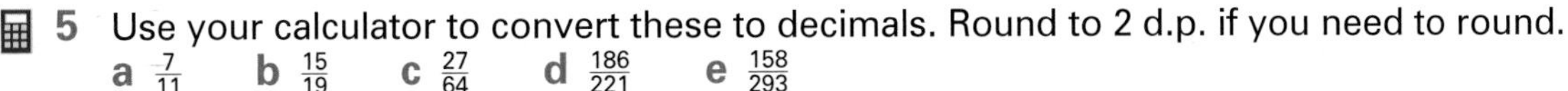

a $\frac{7}{11}$ **b** $\frac{15}{19}$ **c** $\frac{27}{64}$ **d** $\frac{186}{221}$ **e** $\frac{158}{293}$

6 Write these as fractions.

a 0·44444 ... **b** 0·77777 ... **c** 0·33333 ... **d** 0·66666 ...

e 0·999999 ... ***f** 0·18181818 ...

***7** Write $\frac{1}{7}$, $\frac{2}{7}$, $\frac{3}{7}$, $\frac{4}{7}$ and $\frac{5}{7}$ as recurring decimals.
Is it possible to predict what $\frac{6}{7}$ will be?

T

Review 1

1·26	0·75	0·35	0·988	$1\cdot\dot{4}$	0·8	0·96	0·95		0·45	0·8	$1\cdot\dot{2}\dot{7}$	0·84

$1\cdot\dot{2}$	$1\cdot\dot{6}$	S 0·625	1·41	0·96	0·8	0·35	0·95	1·41	0·75	0·96

S 0·625	1·2	0·8	0·35	0·95	0·45

Use a copy of this box.
Convert these to decimals. Use a calculator for the last row of letters and round the answers to 2 d.p.

S $\frac{5}{8} = 0\cdot625$	I $\frac{7}{20}$	W $\frac{21}{25}$	R $\frac{32}{40}$	G $\frac{27}{60}$	A $1\frac{2}{3}$
F $1\frac{2}{9}$	D $1\frac{4}{9}$	L $\frac{247}{250}$	H $\frac{150}{200}$	P $\frac{36}{30}$	
E $\frac{26}{27}$	N $\frac{20}{21}$	T $\frac{38}{27}$	C $\frac{107}{85}$	O $1\frac{3}{11}$	

Review 2 Which of these are recurring decimals?

a $\frac{5}{6}$ **b** $\frac{19}{20}$ **c** $\frac{24}{27}$ **d** $\frac{8}{122}$

If the denominator has prime factors other than 2 or 5 they give recurring decimals.

Ordering fractions

We can **compare fractions** by

1 writing them with a common denominator **or**

2 writing them as decimals.

Worked Example

Which is larger $\frac{3}{5}$ or $\frac{5}{8}$?

Answer

The two ways are shown.

Remember: LCM is lowest common multiple.

The denominators are 8 and 5. The multiples of 8 are 8, 16, 24, 32, 40, 48, ... The multiples of 5 are 5, 10, 15, 20, 25, 30, 35, 40, ... The LCM is 40. Write both fractions with a denominator of 40. 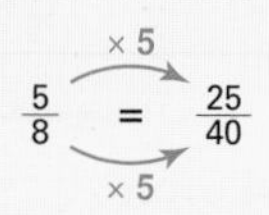 $\frac{5}{8} = \frac{25}{40}$ (× 5) $\frac{3}{5} = \frac{24}{40}$ (× 8) $\frac{25}{40} > \frac{24}{40}$ so $\frac{\mathbf{5}}{\mathbf{8}} > \frac{\mathbf{3}}{\mathbf{5}}$	$\frac{5}{8} = 5 \times 0\cdot125$ $= 0\cdot625$ $\frac{3}{5} = 3 \times 0\cdot2$ $= 0\cdot6$ $0\cdot625 > 0\cdot6$ so $\frac{\mathbf{5}}{\mathbf{8}} > \frac{\mathbf{3}}{\mathbf{5}}$ 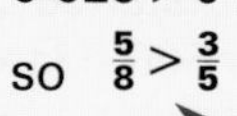

Use this method when it is hard to find a common denominator.

Worked Example

Which is larger $3\frac{5}{8}$ or $26 \div 7$?

Answer

$3\frac{5}{8} = \frac{29}{8} = 29 \div 8 = 3{\cdot}625$

$26 \div 7 = 3{\cdot}714$

26 ÷ 7 is larger.

Exercise 4

Except for questions 2, 5, 7 and Review 1.

1 Compare these fractions by writing them with a common denominator.
Which of < or > goes in the box?

a $\frac{1}{4} \square \frac{1}{3}$ **b** $\frac{3}{4} \square \frac{2}{3}$ **c** $\frac{1}{4} \square \frac{4}{12}$ **d** $\frac{3}{4} \square \frac{17}{20}$ **e** $\frac{2}{3} \square \frac{7}{9}$

f $\frac{2}{5} \square \frac{3}{10}$ **g** $\frac{7}{8} \square \frac{4}{5}$ **h** $\frac{1}{3} \square \frac{2}{5}$ **i** $\frac{7}{8} \square \frac{13}{16}$ **j** $\frac{7}{9} \square \frac{5}{8}$

k $\frac{5}{6} \square \frac{4}{5}$ **l** $\frac{5}{8} \square \frac{2}{3}$ **m** $\frac{2}{3} \square \frac{4}{5}$ **n** $\frac{1}{3} \square \frac{3}{8}$

2 In Hanover Street 16 out of 45 houses use electric heating.
In Merlin Street 6 out of 15 houses use electric heating.
Which street has the greater fraction of houses using electric heating?

3 Paula and Rani each bought a quiche.
Paula ate $\frac{4}{5}$ of hers. Rani ate $\frac{7}{8}$ of hers.
Who had more left?

4 Peter, Brendon and Kieran are in different classes for science.
In the last test Peter got 16 out of 20, Brendon got 7 out of 10 and Kieran got 33 out of 40.
Who got the best mark?

5 Which is larger?

a $1\frac{4}{5}$ or $11 \div 6$ **b** $3\frac{5}{9}$ or $29 \div 8$ **c** $2\frac{3}{7}$ or $\frac{22}{9}$ **d** $56 \div 9$ or $6\frac{3}{16}$

e 7 out of 11 or 9 out of 13 **f** 6 out of 17 or 9 out of 21

6 Put these in order from smallest to largest.

a $\frac{1}{3}, \frac{1}{4}, \frac{5}{6}, \frac{5}{12}$ **b** $\frac{7}{16}, \frac{3}{8}, \frac{1}{4}, \frac{15}{32}$ **c** $\frac{1}{2}, \frac{7}{10}, \frac{3}{5}, \frac{9}{10}, \frac{4}{5}$

d $\frac{1}{6}, \frac{1}{3}, \frac{1}{4}, \frac{5}{12}, \frac{5}{24}, \frac{1}{8}$ **e** $\frac{5}{6}, \frac{4}{5}, \frac{27}{30}, \frac{17}{30}, \frac{11}{15}$ **f** $\frac{5}{6}, \frac{2}{3}, \frac{3}{4}, \frac{7}{8}, \frac{5}{8}, \frac{19}{24}, \frac{11}{12}$

7 Use your calculator to put these in order from largest to smallest.

a $\frac{3}{7}, \frac{5}{8}, \frac{4}{9}, \frac{8}{17}$ **b** $\frac{8}{11}, \frac{2}{3}, \frac{5}{8}, \frac{7}{9}$ **c** $\frac{15}{19}, \frac{4}{5}, \frac{7}{8}, \frac{22}{27}$

8 Maddison used a graphing method to compare fractions.
She plotted the denominator on one axis and the numerator on the other.
This example shows $\frac{3}{4}$ and $\frac{4}{5}$.

a How could Maddison use the graph to tell if $\frac{3}{4}$ or $\frac{4}{5}$ is larger?

b Draw a set of axes with denominator and numerator from 0 to 20.
Plot these fractions on your graph. Use your graph to put them in order from largest to smallest.

$\frac{7}{8}$ $\frac{3}{4}$ $\frac{13}{20}$ $\frac{9}{13}$ $\frac{7}{12}$

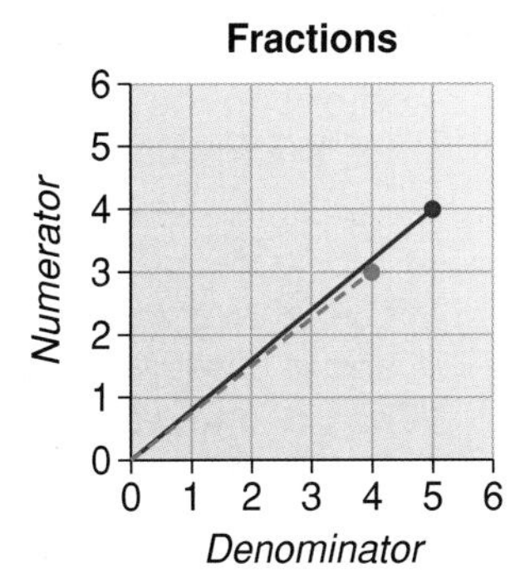

T

9 Use a copy of these number lines.

a $\frac{2}{5}$ is shown with an arrow.
Show these fractions with an arrow. $\frac{16}{20}, \frac{9}{15}, \frac{9}{6}, \frac{5}{25}, \frac{30}{25}$

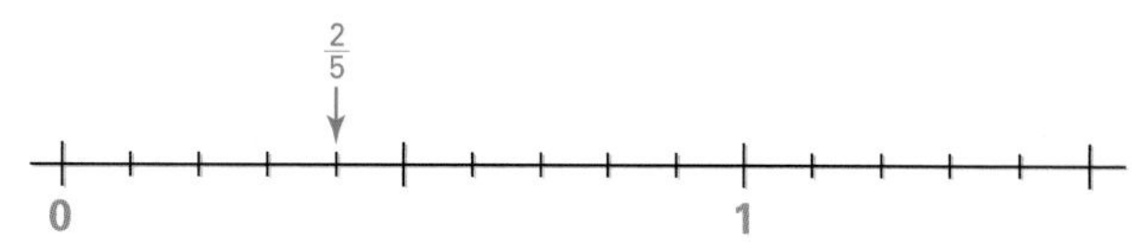

b Show these fractions with an arrow. $\frac{17}{25}, \frac{17}{20}, \frac{12}{15}, \frac{11}{10}$

*** 10** Find the fraction exactly halfway between these.

a $\frac{3}{4}$ and $\frac{11}{12}$ **b** $\frac{1}{3}$ and $\frac{5}{9}$ **c** $\frac{3}{5}$ and $\frac{7}{15}$

d $\frac{2}{3}$ and $\frac{4}{5}$ **e** $\frac{2}{5}$ and $\frac{4}{7}$.

*** 11** x is a decimal with one decimal place. Write down some values x might be if both of these are true.

$\frac{1}{3} < x < \frac{3}{4}$ **and** $\frac{1}{4} < x < \frac{5}{8}$

Review 1 Which of < or > goes in the box?

a $\frac{3}{5} \square \frac{11}{15}$ **b** $\frac{3}{8} \square \frac{11}{24}$ **c** $\frac{3}{4} \square \frac{5}{8}$ **d** $\frac{7}{11} \square 0{\cdot}7$

e $\frac{5}{9} \square 0{\cdot}56$ **f** $\frac{4}{7} \square \frac{8}{15}$ **g** $\frac{9}{31} \square \frac{7}{27}$

Review 2 Paula had three dogs, Runner, Rabbit and Ratty.
Runner ate $\frac{3}{4}$ of a can of food each day. Rabbit ate $\frac{5}{7}$ of a can. Ratty ate $\frac{11}{15}$ of a can.
Who ate the most?

Review 3 Put the fractions in order from largest to smallest.

a $\frac{1}{2}, \frac{13}{16}, \frac{7}{8}, \frac{3}{4}, \frac{7}{16}$ **b** $\frac{4}{5}, \frac{3}{4}, \frac{1}{2}, \frac{7}{10}, \frac{13}{20}, \frac{17}{20}$

You could use Maddison's graphing method in question 8.

T

Review 4 Use a copy of this number line.
Show these fractions with an arrow. $\frac{13}{25}, \frac{19}{20}, \frac{24}{30}, \frac{6}{5}$

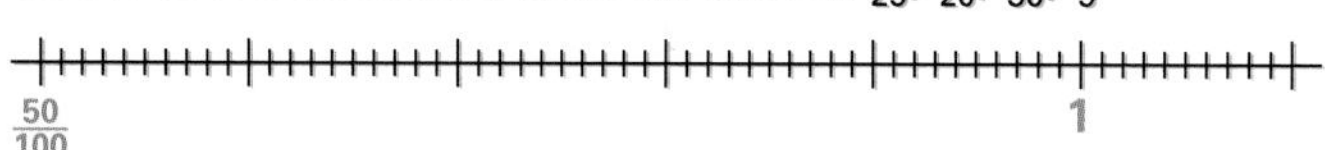

Investigation

Adding One

Add one to the numerator and one to the denominator of each of these fractions.
Does the fraction get bigger or smaller? **Investigate**.

$\frac{3}{4}$ $\frac{2}{3}$ $\frac{1}{8}$ $\frac{9}{16}$ $\frac{25}{27}$ $\frac{182}{187}$ $\frac{36}{23}$ $\frac{104}{89}$

What if two was added to both the numerator and denominator?
What if three was added to both the numerator and denominator?
What if one was subtracted from both the numerator and denominator?
What if ...

Adding and subtracting fractions

Remember

Fractions can be **added and subtracted** easily if they have the same denominators.

Example $\frac{7}{24} + \frac{23}{24} = \frac{7+23}{24}$

$= \frac{30}{24}$

$= 1\frac{6}{24}$

$= \mathbf{1\frac{1}{4}}$

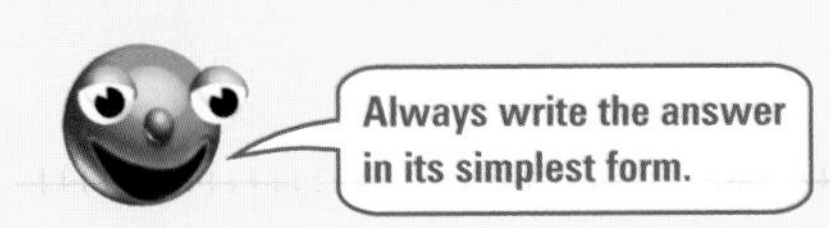

You could use these three steps to **add or subtract fractions with different denominators**.

1. Find the LCM (lowest common multiple) of the denominators.
2. Write equivalent fractions with this LCM as the common denominator.
3. Add or subtract using the equivalent fractions. Give the answer in its simplest form.

Example $\frac{4}{5}$ and $\frac{3}{4}$ have different denominators.

1. Find the lowest common multiple of 4 and 5.
   LCM of 4 and 5 is 20.
2. Write both fractions with a denominator of 20.

$5 = 5 \quad 4 = 2^2$

LCM $= 5 \times 2^2 = 20$

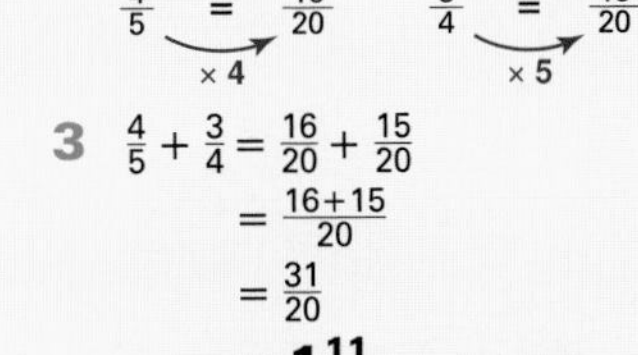

3. $\frac{4}{5} + \frac{3}{4} = \frac{16}{20} + \frac{15}{20}$

$= \frac{16+15}{20}$

$= \frac{31}{20}$

$= \mathbf{1\frac{11}{20}}$

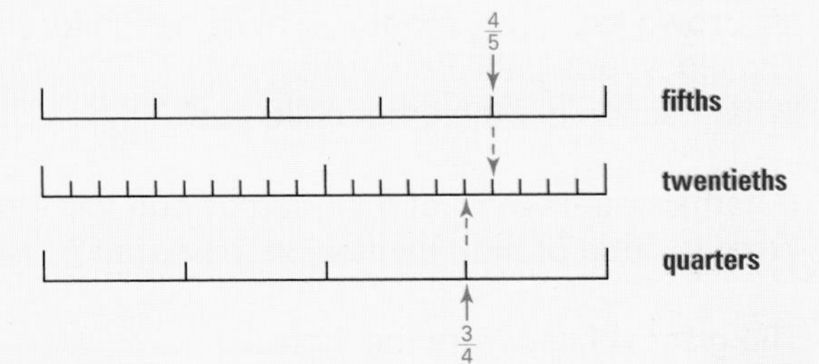

We can show this using **a diagram**.

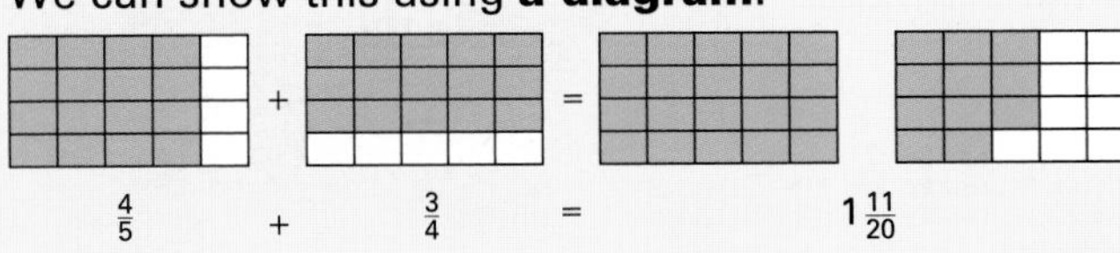

Find the LCM to find the number of parts to have in each diagram.

We can **add and subtract fractions** using the $a^{b/c}$ button on a calculator.

$\frac{4}{5}$ is keyed as [4] [$a^{b/c}$] [5] to get 4⌟5.

Example $\frac{4}{5} + \frac{3}{4}$ is found by keying

[4] [$a^{b/c}$] [5] [+] [3] [$a^{b/c}$] [4] [=] to get 1⌟11⌟20.

We read this as $1\frac{11}{20}$.

Worked Example

In Baysdown, $\frac{1}{8}$ of the population is under 5 and $\frac{1}{3}$ is aged from 5 to 16. What fraction of the population is over 16?

Answer

Fraction aged 16 or under $= \frac{3}{8} + \frac{1}{3}$ LCM of 8 and 3 is 24.

$= \frac{9}{24} + \frac{8}{24}$

$= \frac{9+8}{24}$

$= \frac{17}{24}$

Fraction aged 16 or over $= 1 - \frac{17}{24}$

$= \frac{24}{24} - \frac{17}{24}$

$= \frac{24-17}{24}$

$= \mathbf{\frac{7}{24}}$

Exercise 5

Except for question 3 and Review 3.

1 a $\frac{3}{8}+\frac{1}{4}$ b $\frac{3}{7}+\frac{5}{14}$ c $\frac{7}{8}-\frac{1}{4}$ d $\frac{11}{12}-\frac{3}{4}$ e $\frac{1}{12}+\frac{2}{3}$
f $\frac{5}{6}-\frac{1}{3}$ g $\frac{7}{10}-\frac{3}{5}$ h $\frac{1}{4}+\frac{1}{3}$ i $\frac{5}{8}-\frac{1}{3}$ j $\frac{4}{5}-\frac{3}{4}$
k $\frac{1}{2}+\frac{2}{5}$ l $\frac{7}{8}-\frac{5}{6}$ m $\frac{2}{5}+\frac{2}{3}$ n $\frac{3}{4}+\frac{3}{5}$ o $\frac{5}{6}-\frac{1}{8}$

2 a $\frac{1}{3}+\frac{1}{4}+\frac{5}{12}$ b $\frac{9}{10}-\frac{2}{5}-\frac{1}{2}$ c $\frac{3}{5}+\frac{7}{10}-\frac{3}{20}$ d $\frac{5}{8}+\frac{3}{4}-\frac{13}{24}$
e $\frac{2}{3}+\frac{1}{8}+\frac{3}{4}$ f $\frac{3}{4}+\frac{5}{6}-\frac{1}{8}$ g $\frac{2}{3}-\frac{1}{4}+\frac{1}{2}$ h $\frac{3}{4}+\frac{5}{6}-\frac{1}{3}$
i $\frac{7}{10}-\frac{1}{5}+\frac{2}{3}$ j $\frac{5}{8}+\frac{3}{4}-\frac{13}{20}$

3 Use your calculator to find these.
a $\frac{3}{8}+\frac{4}{7}$ b $\frac{5}{7}-\frac{7}{12}$ c $\frac{8}{9}+\frac{4}{5}+\frac{3}{4}$ d $2-\frac{2}{3}-\frac{3}{4}$
e $\frac{7}{8}+\frac{4}{7}+\frac{3}{4}$ f $\frac{7}{8}-\frac{1}{3}-\frac{4}{9}$ g $\frac{14}{15}+\frac{1}{3}-\frac{2}{9}$ h $3-\frac{2}{7}-\frac{11}{15}$

4 In a horse race, there was $\frac{1}{3}$ of a length between the first and second horses and $\frac{1}{2}$ of a length between the first and third horses.
What fraction of a length was there between the second and third horses?

5 Mr Brown left all his money to his three children.
The eldest got $\frac{1}{2}$, the youngest got $\frac{1}{5}$.
What fraction did the other child get?

6 A netball team won $\frac{2}{5}$ of their games and lost $\frac{1}{8}$ of them.
What fraction of their games did they draw?

7 The glass is filled from the thermos.
Which now contains more, the glass or the thermos?

8 Find the next term in these sequences.
a $\frac{1}{2}, \frac{2}{3}, \frac{5}{6}, 1, \ldots$ b $\frac{1}{4}, \frac{3}{8}, \frac{1}{2}, \frac{5}{8}, \ldots$ c $1, \frac{13}{16}, \frac{5}{8}, \frac{7}{16}, \ldots$

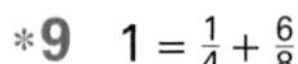

***9** $1=\frac{1}{4}+\frac{6}{8}$
1 is written as the sum of two fractions. Each of the four digits in the fractions is different.
a Find some other ways of writing 1 using two fractions and four different digits.
b Find some ways of writing 1 using two fractions and five different digits.

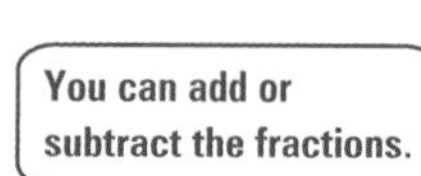

***10** The answer is $\frac{5}{8}$. What might the question be if I am
a adding two fractions b subtracting two fractions?

***11** These sequences were made by adding or subtracting a constant number each time.
What is the value of n?
a $\frac{1}{4}, m, n, \frac{5}{8}, \ldots$ b $1\frac{3}{4}, m, n, 1\frac{1}{2}, \ldots$

Review 1

a $\frac{3}{4}-\frac{5}{8}$ b $\frac{7}{8}+\frac{3}{4}$ c $\frac{5}{9}+\frac{2}{3}$ d $\frac{5}{6}-\frac{2}{3}$
e $\frac{2}{3}+\frac{3}{4}$ f $\frac{3}{5}-\frac{1}{3}$ g $\frac{7}{8}+\frac{3}{5}-\frac{1}{4}$ h $\frac{3}{8}+\frac{5}{6}+\frac{3}{12}$

Review 2 The Earth has three main layers: crust, mantle and core. The mantle is about $\frac{4}{5}$ of the Earth and the core is about $\frac{1}{6}$.
What fraction of the Earth is crust?

Review 3 Use your calculator to find $\frac{7}{9} + \frac{5}{8} - \frac{3}{4}$.

* **Review 4** $2 = \frac{5}{4} + \frac{6}{8}$
Find some other ways of writing 2 using two fractions and four different digits.

* **Review 5** This sequence was made by subtracting a constant number each time.
What is the value of b? $\frac{2}{3}, a, b, \frac{1}{3}$

Puzzle

What digits could ▲ be to make these true?

a $\frac{▲}{4} + \frac{▲}{▲} = \frac{3}{▲}$ b $\frac{▲}{5} + \frac{▲}{▲▲} = \frac{7}{▲▲}$

c $\frac{▲}{5} + \frac{▲}{▲▲} = \frac{3}{▲▲}$ d $\frac{▲}{▲▲} + \frac{▲}{8} = \frac{11}{▲▲}$

Find as many ways as you can for each.

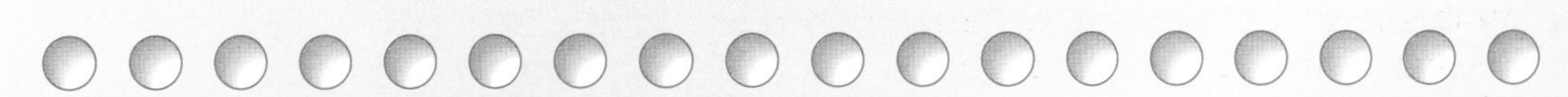

Investigation

Egyptian Fractions

The ancient Egyptian number system used these symbols.

one stick	two sticks								heel bone	coiled rope
1	2	3	4	5	6	7	8	9	10	100

They had a symbol, ⬭, called 'ro' which was used to show fractions.

$= \frac{1}{2}$ $= \frac{1}{5}$ $= \frac{1}{10}$ $= \frac{1}{103}$

All Egyptian fractions are unit fractions because they all have a numerator of 1.
Other fractions such as $\frac{3}{4}$ were written as the sum of two different unit fractions.

$\frac{3}{4} = \frac{1}{2} + \frac{1}{4}$. This was written as

$\frac{2}{3} = \frac{1}{2} + \frac{1}{6}$. This was written as

$\frac{2}{5}$ could not be written as because the fractions had to be different.

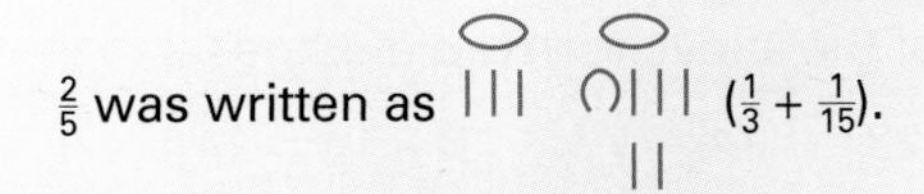

$\frac{2}{5}$ was written as $(\frac{1}{3} + \frac{1}{15})$.

1 What fraction does each of these show?

a  **b** **c** **d**

2 How would the Egyptians have written these?

a $\frac{5}{6}$ **b** $\frac{3}{10}$ **c** $\frac{5}{8}$ **d** $\frac{2}{9}$ **e** $\frac{7}{12}$ **f** $\frac{8}{15}$

3 Some fractions had to be written as the sum of more than two unit fractions.

Example $\frac{7}{8} = \frac{1}{2} + \frac{1}{4} + \frac{1}{8}$ This was written as  

Write these as the sum of three unit fractions.

a $\frac{11}{12}$ **b** $\frac{9}{20}$ **c** $\frac{3}{5}$

4 Find some more fractions that can be written as the sum of unit fractions. **Investigate**.

Puzzle

1 A tub was $\frac{1}{4}$ full of water. Another 5 ℓ was added. Then the tub was $\frac{1}{3}$ full. How much does the tub hold altogether?

* 2 At Pene's party, $\frac{1}{8}$ of the guests arrived at 7 p.m. 38 arrived after 7 p.m. but before 8.30 p.m. and $\frac{2}{5}$ were an hour late.
How many people were at the party?

Invitation to a Party for Pene
at ________
on ________
Time 7.30 p.m.

To **add or subtract mixed numbers**, write each as an improper fraction **or** add the whole numbers first.

Examples

$$3\frac{2}{3} + 1\frac{3}{4} = \frac{11}{3} + \frac{7}{4} = \frac{44}{12} + \frac{21}{12} = \frac{65}{12} = 5\frac{5}{12}$$

or

$$3\frac{2}{3} + 1\frac{3}{4} = 3\frac{8}{12} + 1\frac{9}{12} = 4\frac{17}{12} = 5\frac{5}{12}$$

$$3\frac{1}{6} - 1\frac{1}{2} = \frac{19}{6} - \frac{3}{2} = \frac{19}{6} - \frac{9}{6} = \frac{10}{6} = \frac{5}{3} = 1\frac{2}{3}$$

On a calculator we would key $3\frac{2}{3} + 1\frac{3}{4}$ as

3 a b/c 2 a b/c 3 + 1 a b/c 3 a b/c 4 = to get 5⌟5⌟12

Discussion

Discuss the advantages and disadvantages of both of the above **written** methods.

Discuss how you could find these. $3\frac{2}{3} + 1\frac{3}{4}$ $15\frac{1}{2} + 23\frac{3}{8}$ $14\frac{3}{5} - 7\frac{3}{10}$ $3\frac{1}{6} - 1\frac{1}{2}$ $14\frac{3}{10} - 7\frac{3}{5}$

Exercise 6

Except question 2 and Review 2.

1 a $1\frac{1}{2} + 2\frac{3}{4}$ b $3\frac{2}{3} + 1\frac{1}{4}$ c $2\frac{4}{5} - 1\frac{1}{2}$ d $5\frac{1}{2} + \frac{7}{10}$
e $4\frac{5}{8} - 2\frac{3}{4}$ f $2\frac{1}{3} - \frac{4}{5}$ g $2\frac{1}{3} - 1\frac{1}{2}$

2 Use your calculator to find these.
a $3\frac{5}{8} + 2\frac{4}{9}$ b $5\frac{1}{4} - 2\frac{7}{12}$ c $8\frac{3}{4} + 3\frac{7}{15} - 2\frac{1}{8}$ d $5\frac{1}{6} + 7\frac{4}{7} - 3\frac{5}{8}$

3 Sam works in a restaurant for $3\frac{1}{2}$ hours each Saturday and $2\frac{3}{4}$ hours each Sunday.
How long does Sam work each weekend?

4 The last 8 pages of a magazine are for advertisements.
Two days before publication, $6\frac{7}{8}$ pages of advertising had been sold. Then $2\frac{1}{4}$ pages were cancelled. What fraction of pages still has to be sold?

5 Find the next term in the sequence $\frac{2}{3}, 2\frac{1}{2}, 4\frac{1}{3}, 6\frac{1}{6}, \ldots$

6 What might ■ and ● be in each of the following?
a ■ + ● = $3\frac{5}{6}$ b ■ − $2\frac{1}{2}$ = ●

T

***7** Use a copy of these magic squares.
Remember: The numbers in each row, column and diagonal must add to the same total.

$\frac{1}{2}$		
$1\frac{2}{3}$	1	
$\frac{5}{6}$		

a

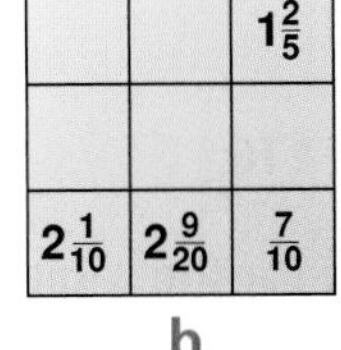

		$1\frac{2}{5}$
$2\frac{1}{10}$	$2\frac{9}{20}$	$\frac{7}{10}$

b

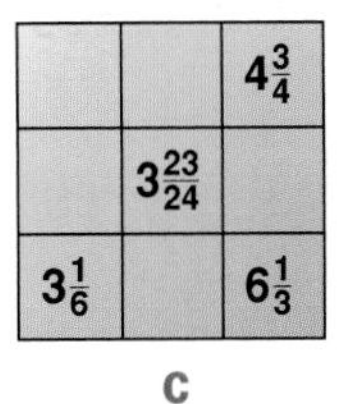

		$4\frac{3}{4}$
	$3\frac{23}{24}$	
$3\frac{1}{6}$		$6\frac{1}{3}$

c

Review 1 Find the answers to these.
a $3\frac{1}{4} + 1\frac{7}{10}$ b $2\frac{2}{3} - 1\frac{1}{4}$ c $2\frac{3}{8} - 1\frac{2}{3}$

Review 2 Use your calculator to find these.
a $3\frac{7}{8} + 8\frac{5}{12}$ b $5\frac{5}{6} - 2\frac{3}{4} + 4\frac{3}{7}$

Review 3 Liz made plum sauce. She filled these two bottles to the top.
a How much plum sauce did Liz make altogether?
b How much more did Liz pour into the large bottle than into the small bottle?

Fraction of

Remember
In mathematics 'of' means multiply.
We can often find a **fraction of a quantity** mentally.

Example To find $\frac{2}{3}$ of 27, find $\frac{1}{3}$ of 27 first.
$\frac{1}{3}$ of 27 = 9
$\frac{2}{3}$ of 27 = 2 × 9
= 18

Worked Example

About $\frac{3}{8}$ of Mr Health's 300 g patty is protein.
About $\frac{5}{12}$ of Home Burgers 276 g patty is protein.
Which patty has more protein?

Answer

$\frac{3}{8}$ of $300 = \frac{3}{8} \times 300$
$= 3 \times \frac{1}{8} \times 300$
$= \frac{900}{8}$
$= 112\frac{1}{2}$ g

$8 \overline{)9^1 0^2 0}$ = $112\frac{4}{8}$

$\frac{5}{12}$ of $276 = \frac{5}{12} \times 276$
$= 5 \times \frac{1}{12} \times 276$
$= \frac{1380}{12}$
$= 115$ g

$12 \overline{)13^1 8^6 0}$ = 115

There is more protein in a Home Burger patty.

Exercise 7

1 Find the answers to these mentally.
a $\frac{2}{5}$ of 40 **b** $\frac{3}{8}$ of 64 **c** $\frac{7}{25}$ of 125 **d** $\frac{5}{9}$ of 81 **e** $1\frac{1}{2}$ of 18
f $1\frac{3}{4}$ of 60 **g** $2\frac{1}{5}$ of 45 **h** $1\frac{5}{8}$ of 72 **i** $3\frac{1}{3}$ of 48

2 Find the answers to these mentally.
Which is greater?
a $\frac{3}{4}$ of 16 or $\frac{1}{3}$ of 33 **b** $\frac{5}{8}$ of 56 or $\frac{2}{3}$ of 54 **c** $\frac{4}{7}$ of 42 or $\frac{3}{8}$ of 40 **d** $\frac{5}{16}$ of 64 or $\frac{7}{8}$ of 24

3 Which is greater?
a 24% of 65 or $\frac{4}{7}$ of 45 **b** 68% of 120 or $\frac{2}{3}$ of 116.

4 A test took 120 minutes. Josie took $\frac{1}{3}$ of the time for the multi-choice section, $\frac{1}{4}$ of the time for an essay and $\frac{3}{8}$ for a problem solving section. She spent the rest of the time on the short answers. How much time did she spend on the short answers?

5 In a science laboratory, Arshad's group used $\frac{1}{5}$ of a 750 mℓ bottle of copper sulphate solution. Bea's group used $\frac{2}{3}$ of the solution that was left. How many millilitres were left in the bottle?

6 In a bag of 24 sweets, $\frac{1}{8}$ are red, $\frac{1}{4}$ are green, $\frac{1}{3}$ are yellow, 4 are blue and the rest are orange.
What fraction are orange?

This is linked to mutually exclusive events – see page 418.

7 In a class of 36 pupils, $\frac{1}{3}$ were from England, $\frac{2}{9}$ were from India, $\frac{1}{12}$ were from Asia, 6 were from Europe and the rest were from other places.
What fraction were from other places?

***8** If $\frac{2}{5}$ of what I spent is £8, how much did I spend?

Review 1 Find the answers to these mentally.
a $\frac{7}{8}$ of 56 **b** $\frac{9}{25}$ of 250 **c** $1\frac{1}{2}$ of 24 **d** $2\frac{3}{5}$ of 85 **e** $3\frac{1}{5}$ of 45

Review 2 Which is greater?
a $\frac{3}{4}$ of 80 or $\frac{3}{8}$ of 168 **b** 16% of 60 or $\frac{1}{12}$ of 80 **c** 80% of 112 or $\frac{7}{8}$ of 96

Review 3 A country exported £48 million worth of produce. Of this, $\frac{1}{3}$ was meat exports, $\frac{1}{6}$ was dairy produce, $\frac{1}{8}$ was timber, £5 million worth was wool and the rest was other products. What fraction was other products?

Puzzle

1 Ben spent $\frac{3}{4}$ of his money.
Half of what he had left was 40p.
How much did he have to start with?

2 The head of a fish is 4 cm long.
The body is as long as the head and tail together.
The tail is as long as the head and half the body.
How long is the fish?

3 Half a number and twice half the number is two more than twice half the number.
What is the number?

4 Jamilah spent one third of her savings on her plane fare and one quarter of the remainder on her accommodation. She then had £120 left for spending money.
What was Jamilah's plane fare?

*5 To escape, a spy has to cross three borders.
The spy agrees to pay one half of her money to someone at each border in order to be escorted to the next border.
The spy needs to have at least £200 left after crossing the last border.
What is the least amount of money this spy needs in order to escape?

Multiplying fractions

Discussion

- This diagram can be used to find the answer to $\frac{1}{4} \times \frac{1}{3}$.
Discuss.

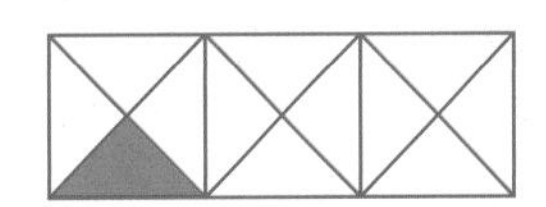

- Draw diagrams to find the answers to $\frac{1}{2} \times \frac{1}{5}$, $\frac{3}{4} \times \frac{1}{3}$, $\frac{1}{4} \times \frac{2}{3}$, $\frac{5}{8} \times \frac{3}{4}$, $\frac{3}{8} \times \frac{4}{5}$ and $\frac{3}{4} \times \frac{2}{3}$.
Discuss how the answers could be found without using diagrams.

- 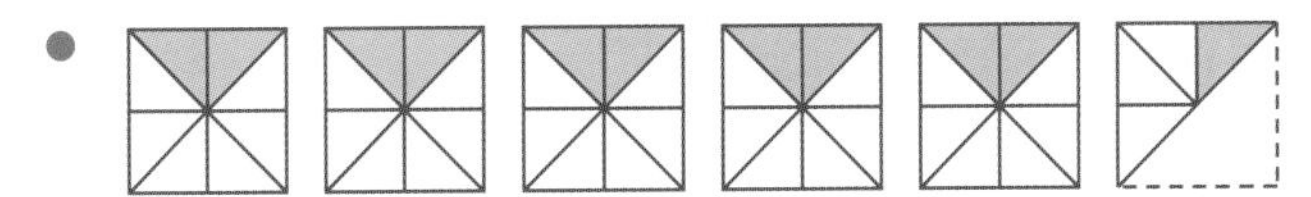

This diagram shows that the answer to $\frac{1}{4} \times 5\frac{1}{2}$ is $\frac{11}{8}$ or $1\frac{3}{8}$.

Discuss how to draw diagrams to find the answers to $\frac{1}{2} \times 3\frac{1}{4}$, $\frac{3}{4} \times 2\frac{1}{2}$, $\frac{2}{3} \times 3\frac{3}{4}$ and $\frac{3}{5} \times 4\frac{1}{3}$.

Discuss how to find the answers without drawing diagrams.

- **Discuss** how the following diagrams could be used to show that the answer to $2\frac{1}{2} \times 3\frac{2}{5}$ is $8\frac{1}{2}$.

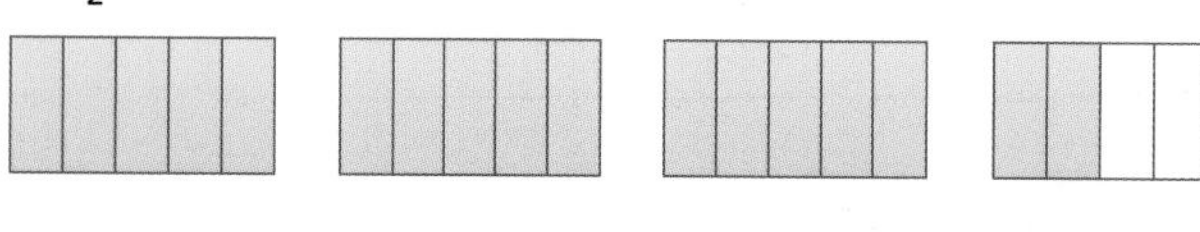

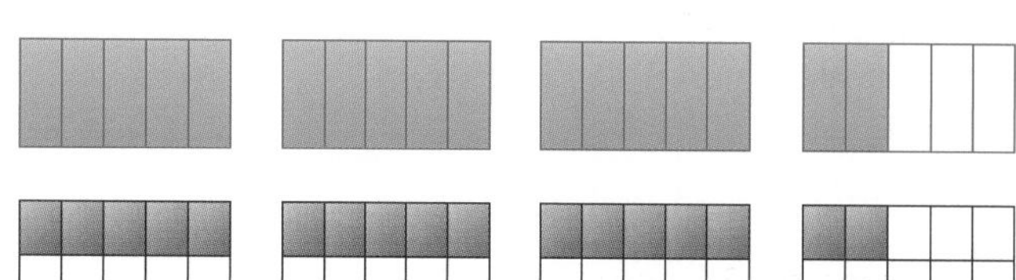

Discuss how the answer to $2\frac{1}{2} \times 3\frac{2}{5}$ could be found without using diagrams. As part of your discussion, you might like to find the answer to other multiplications involving mixed numbers.

- Make and test statements about how to multiply fractions. **Discuss.**

Fraction multiplications can be made easier by **cancelling** first.

Examples $\frac{3}{4} \times \frac{6}{5} = \frac{3}{{}_{2}\cancel{4}} \times \frac{\cancel{6}^{3}}{5}$ $= \frac{9}{10}$

$\frac{5}{8} \times \frac{2}{15} = \frac{{}^{1}\cancel{5}}{{}_{4}\cancel{8}} \times \frac{\cancel{2}^{1}}{\cancel{15}_{3}}$ $= \frac{1}{12}$

To multiply fractions:

Step 1 Write whole numbers or mixed numbers as improper fractions.
Step 2 Cancel if possible.
Step 3 Multiply the numerators; multiply the denominators.

Examples $\frac{5}{8} \times \frac{16}{25} \times \frac{15}{8} = \frac{{}^{1}\cancel{5}}{{}_{1}\cancel{8}} \times \frac{{}^{2}\cancel{16}}{{}_{5}\cancel{25}} \times \frac{15}{8}$ $= \frac{1}{1} \times \frac{{}^{1}\cancel{2}}{{}_{1}\cancel{5}} \times \frac{{}^{3}\cancel{15}}{{}_{4}\cancel{8}}$ $= \mathbf{\frac{3}{4}}$

$\frac{1}{4}(3 - \frac{1}{3}) = \frac{1}{4} \times 2\frac{2}{3}$ $= \frac{1}{{}_{1}\cancel{4}} \times \frac{{}^{2}\cancel{8}}{3}$ $= \mathbf{\frac{2}{3}}$

$3\frac{3}{4} \times 1\frac{1}{5} = \frac{{}^{3}\cancel{15}}{{}_{2}\cancel{4}} \times \frac{{}^{3}\cancel{6}}{{}_{1}\cancel{5}}$ $= \frac{9}{2}$ $= \mathbf{4\frac{1}{2}}$

$(2\frac{3}{4})^2 = (\frac{11}{4})^2$ $= \frac{121}{16}$ $= \mathbf{7\frac{9}{16}}$ $\quad (\frac{11}{4})^2 = \frac{11^2}{4^2}$

Exercise 8

1 Find the answers to these.

a $\frac{2}{5} \times \frac{2}{7}$ **b** $\frac{2}{3} \times \frac{3}{5}$ **c** $\frac{5}{12} \times \frac{8}{9}$ **d** $\frac{2}{3} \times \frac{4}{5}$ **e** $\frac{5}{6} \times \frac{9}{11}$
f $\frac{3}{4} \times \frac{10}{9} \times \frac{2}{5}$ **g** $\frac{3}{5} \times \frac{20}{33} \times \frac{22}{14}$ **h** $\frac{7}{9} \times \frac{27}{14} \times \frac{18}{5}$ **i** $\frac{5}{9} \times \frac{3}{10} \times \frac{45}{12}$ **j** $6 \times 6 \times \frac{2}{5}$
k $3 \times 3 \times \frac{4}{7}$ **l** $\frac{5}{6} \times 8 \times 4$ **m** $(\frac{2}{3})^2$ **n** $(\frac{4}{3})^2$

2 Calculate.

a $2\frac{1}{4} \times 3\frac{1}{3}$ **b** $2\frac{3}{5} \times 1\frac{2}{3}$ **c** $3\frac{3}{4} \times 1\frac{3}{5}$ **d** $3\frac{1}{3} \times 1\frac{2}{5}$ **e** $4\frac{1}{6} \times 2\frac{4}{5}$
f $4\frac{2}{7} \times 2\frac{1}{3}$ **g** $2 \times 1\frac{5}{8}$ **h** $4 \times 3\frac{1}{3}$ **i** $\frac{1}{4}(3 - \frac{1}{2})$ **j** $\frac{2}{3}(2 - \frac{1}{4})$
k $\frac{3}{4}(\frac{3}{8} + 1\frac{1}{2})$ **l** $(2\frac{1}{4})^2$ **m** $(1\frac{3}{8})^2$ **n** $(4\frac{2}{5})^2$

3 Susan claims that three-fifths of one-quarter is the same as three-quarters of one-fifth. Is she correct?

4 Two-fifths of a garden is used to grow vegetables. One-quarter of this is planted in potatoes.
What fraction of the garden is planted in potatoes?

5 A train travels at an average speed of 96 mph.
How far does it travel in $2\frac{1}{4}$ hours?

6 David's bedroom floor measures $3\frac{1}{4}$ metres by $3\frac{1}{2}$ metres.
What is the area of this floor?

7 A frame for a picture is $5\frac{1}{4}$ inches by $9\frac{5}{8}$ inches. Calculate its area.

8 Find the answers to these.
a $(1\frac{1}{2})^3$ **b** $\frac{2}{3} \times 1\frac{2}{5} \times 2\frac{1}{7}$ **c** $2\frac{1}{2} \times 3\frac{3}{10} \times 1\frac{2}{3}$ *__d__ $(2\frac{2}{3})^3$

***9** **a** Find three different pairs of fractions which multiply to $\frac{25}{32}$.
b $\frac{a}{3} \times \frac{2}{b} = \frac{8}{15}$ What values could a and b have?
Is there more than one answer?

***10** What number is one-third of the way from $\frac{1}{3}$ up to $\frac{5}{12}$?

***11** Use three of these four fractions to make each statement true. $\frac{1}{4}, \frac{1}{3}, \frac{1}{5}, \frac{1}{8}$
a $\square + \square \times \square = \frac{7}{24}$ **b** $\square \times \square + \square = \frac{23}{60}$ **c** $\square - \square \times \square = \frac{17}{60}$

Review 1 Find the answer to these.
a $\frac{2}{3} \times \frac{9}{16}$ **b** $\frac{5}{8} \times \frac{4}{15} \times \frac{2}{8}$ **c** $6\frac{1}{2} \times 2\frac{2}{3}$ **d** $3\frac{1}{4} \times 1\frac{3}{5}$ **e** $(\frac{4}{5})^3$
f $8 \times 8 \times \frac{3}{16}$ **g** $(4\frac{2}{3})^2$ **h** $(2\frac{1}{3})^2$

Review 2 Find the area of the top of this square table.

Dividing fractions

Discussion

- Sirah drew this number line.

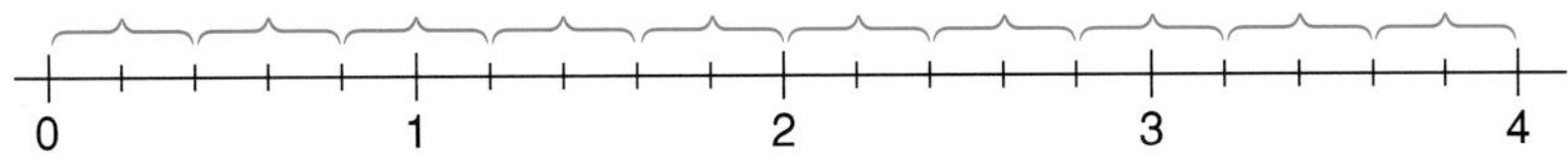

How could she use this to find the answer to $4 \div \frac{2}{5}$? **Discuss.**

Tim knew that $10 \times \frac{2}{5} = 4$.
How could he use this to find the answer to $4 \div \frac{2}{5}$? **Discuss.**

Is there any other way you could find the answer to $4 \div \frac{2}{5}$? **Discuss.**

How could you find the answers to these? **Discuss.**
$3 \div \frac{1}{7}$ $4 \div \frac{1}{5}$ $5 \div \frac{1}{6}$ $6 \div \frac{3}{5}$ $9 \div \frac{3}{4}$ $8 \div \frac{2}{3}$

- How many eighths are there in the whole diagram?
  How many eighths are there in one half of the diagram?
  What does $\frac{1}{2} \div \frac{1}{8}$ equal? **Discuss.**
  What does $\frac{1}{2} \times \frac{8}{1}$ equal? **Discuss.**

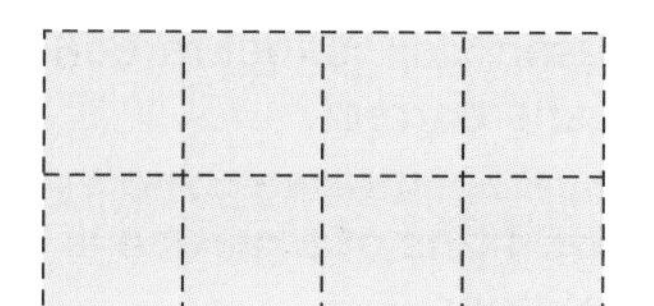

  Use diagrams to find these.
  $\frac{1}{2} \div \frac{1}{3}$ $\quad \frac{1}{4} \div \frac{1}{5}$

  Does $\frac{1}{2} \div \frac{1}{3} = \frac{1}{2} \times \frac{3}{1}$?
  Does $\frac{1}{4} \div \frac{1}{5} = \frac{1}{4} \times \frac{5}{1}$? **Discuss.**

We use the **inverse rule** to **divide by fractions**.
To divide by a fraction, multiply by the inverse.

Turn the fraction you are dividing by upside down and multiply.

Example $\frac{3}{4} \div 6 = \frac{3}{4} \div \frac{6}{1}$
$= \frac{{}^{1}3}{4} \times \frac{1}{6_{2}}$
$= \frac{1}{8}$

Cancel once you've got the multiplication written down.

Example $6 \div \frac{3}{4} = \frac{6}{1} \div \frac{3}{4}$
$= \frac{{}^{2}6}{1} \times \frac{4}{3_{1}}$
$= 8$

Example $\frac{3}{4} \div \frac{5}{12} = \frac{3}{{}_{1}4} \times \frac{12^{3}}{5}$
$= \frac{9}{5}$
$= 1\frac{4}{5}$

Example $1\frac{2}{3} \div 4\frac{1}{6} = \frac{5}{3} \div \frac{25}{6}$
$= \frac{{}^{1}5}{3_{1}} \times \frac{{}^{2}6}{25_{5}}$
$= \frac{2}{5}$

Exercise 9

1 **a** $\frac{3}{4} \div 3$ **b** $6 \div \frac{2}{5}$ **c** $\frac{9}{10} \div 5$ **d** $12 \div \frac{2}{3}$ **e** $\frac{7}{10} \div 3$
f $8 \div \frac{3}{4}$ **g** $2 \div \frac{4}{5}$ **h** $9 \div \frac{3}{10}$ **i** $7 \div \frac{2}{3}$ **j** $3 \div \frac{2}{5}$
k $\frac{2}{3} \div \frac{1}{2}$ **l** $\frac{7}{10} \div \frac{4}{5}$ **m** $\frac{9}{10} \div \frac{3}{4}$ **n** $\frac{5}{6} \div \frac{2}{3}$

2 **a** $3\frac{1}{2} \div \frac{3}{4}$ **b** $2\frac{3}{4} \div \frac{5}{8}$ **c** $1\frac{2}{3} \div \frac{1}{3}$ **d** $1\frac{1}{3} \div \frac{5}{9}$ **e** $2\frac{3}{8} \div \frac{3}{4}$
f $10 \div 1\frac{2}{3}$ **g** $8 \div 2\frac{2}{3}$ **h** $4\frac{1}{2} \div 3$ **i** $2\frac{1}{4} \div \frac{5}{8}$ **j** $3\frac{2}{5} \div \frac{3}{10}$

3 **a** $(1 - \frac{1}{4})(1 - \frac{3}{8})$ **b** $(2 - 1\frac{1}{3})(2 - 1\frac{5}{8})$ **c** $(3 - 1\frac{1}{2})(4 - 2\frac{1}{4})$

4 A recipe for a Christmas cake needs $\frac{1}{4}$ kg of flour.
How many of these cakes can be made from a $2\frac{1}{2}$ kg bag of flour?

5 1 gallon (8 pints) is about $4\frac{1}{2}$ litres.
About how many pints is 1 litre?

6 A yacht is sailing at an average speed of $7\frac{1}{2}$ km/h.
How long will it take to travel 60 km?

7 If $2\frac{1}{2}$ kg of fruit costs, £3, what is the cost per kg?

8 A hovercraft completes one crossing in three-quarters of an hour.
What is the greatest number of crossings that it could make in 18 hours?

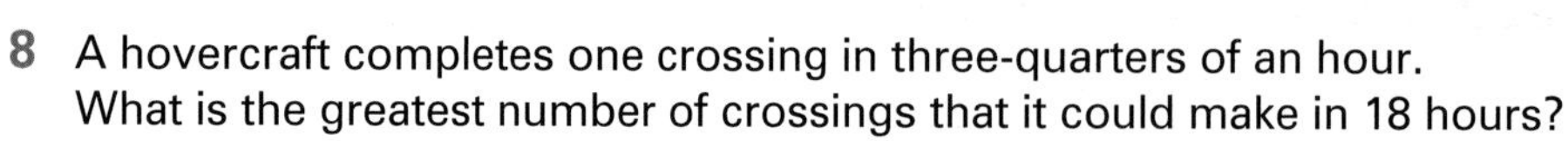

9 **a** Find three different pairs of fractions which give the answer $\frac{4}{9}$ when they are divided.
***b** $\frac{4}{a} \div \frac{b}{3} = \frac{6}{7}$
What values could a and b have? Is there more than one answer?

Review 1 **a** $\frac{3}{5} \div \frac{9}{10}$ **b** $6\frac{1}{2} \div \frac{2}{3}$ **c** $3\frac{1}{8} \div 1\frac{2}{3}$ **d** $(1 - \frac{3}{4})(1 - \frac{1}{3})$

Review 2 It takes Saad $1\frac{1}{2}$ minutes to make a milkshake.
How many could he make in 12 minutes?

Review 3 A short skirt can be made from $\frac{3}{4}$ m of fabric.
How many of these skirts could be made from $10\frac{1}{2}$ m of this fabric?

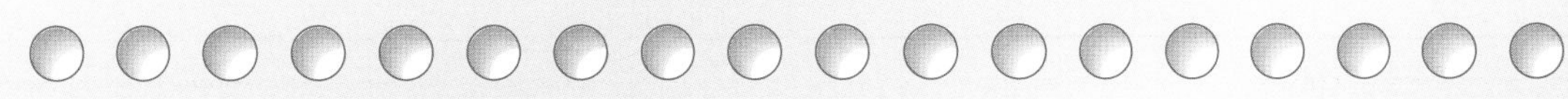

Investigation

One-ninth

$\frac{1}{9} = \frac{10\,638}{95\,742}$ Each of the ten digits 0 to 9 has been used once to form the fraction $\frac{1}{9}$.

Can these ten digits be rearranged in other ways to form the fraction $\frac{1}{9}$? **Investigate**.

Summary of key points

A We can **find the fraction** of a shape that is shaded.

Example If we assume this shape is made from identical triangles, then $\frac{21}{36}$ triangles are shaded.

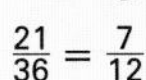

$\frac{21}{36} = \frac{7}{12}$

$\frac{7}{12}$ of this shape is shaded.

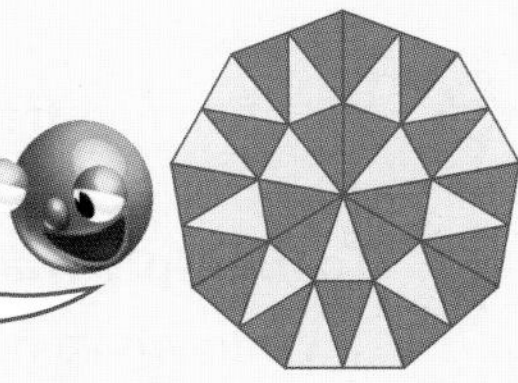

B We can write **one number as a fraction of another**.

Example 150 as a fraction of 120 = $\frac{150}{120}$

$= \mathbf{1\frac{1}{4}}$

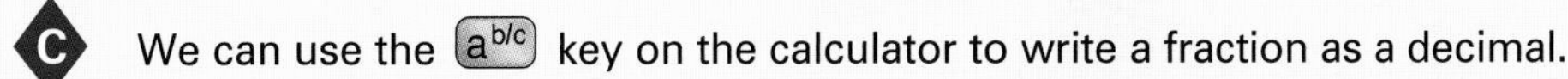

C We can use the [$a^{b/c}$] key on the calculator to write a fraction as a decimal.

Example $\frac{37}{63}$

Key [37] [$a^{b/c}$] [63] [=] [$a^{b/c}$] to get 0·59 (2 d.p.)

Keying this again gives the answer as a decimal.

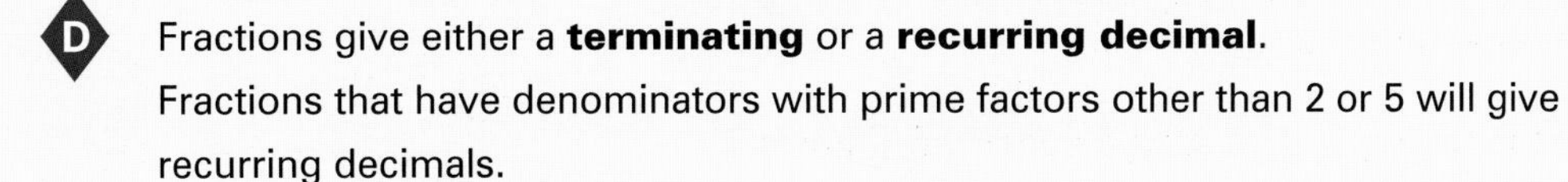

D Fractions give either a **terminating** or a **recurring decimal**.

Fractions that have denominators with prime factors other than 2 or 5 will give recurring decimals.

Example $\frac{2}{5} = 0{\cdot}4$ which is a terminating decimal

$\frac{3}{11} = 0{\cdot}\dot{2}\dot{7}$ which is a recurring decimal

You should know these fraction/decimal conversions.

0·333333 ... = $\frac{1}{3}$ ($\frac{3}{9}$) 0·666666 ... = $\frac{2}{3}$

0·111111 ... = $\frac{1}{9}$ 0·999999 ... = 1 ($\frac{9}{9}$)

We can **compare fractions** by

1 writing them with a common denominator **or**

2 writing them as decimals.

See page 96 for an example.

Example Compare $\frac{3}{4}$ and $\frac{5}{7}$.

1 The LCM of 4 and 7 is 28.

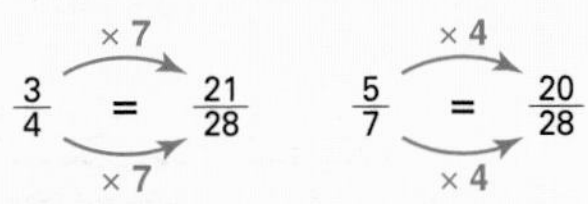

$\frac{3}{4} = \frac{21}{28}$ (×7) $\quad \frac{5}{7} = \frac{20}{28}$ (×4)

$\frac{21}{28} > \frac{20}{28}$

$\frac{3}{4} > \frac{5}{7}$

2 $\frac{3}{4} = 0{\cdot}75$

$\frac{5}{7} = 0{\cdot}71$ (2 d.p.)

$0{\cdot}75 > 0{\cdot}71$

$\frac{3}{4} > \frac{5}{7}$

To **add and subtract fractions** that do not have the same denominator we find the LCM and write both fractions with this as the denominator.

Examples $\frac{5}{8} + \frac{2}{3} = \frac{15}{24} + \frac{16}{24} = \frac{15 + 16}{24} = \frac{31}{24} = 1\frac{7}{24}$

$2\frac{5}{8} - \frac{3}{5} = \frac{21}{8} - \frac{3}{5} = \frac{105 - 24}{40} = \frac{81}{40} = 2\frac{1}{40}$

We can use the a b/c key on a calculator.

Examples $3\frac{5}{8} + 2\frac{2}{3}$ **key** 3 a b/c 5 a b/c 8 + 2 a b/c 2 a b/c 3 =

to get 6⌟7⌟24

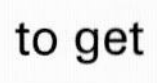

We often find **fractions of quantities**.

To **multiply fractions**

1 Write whole numbers or mixed numbers as improper fractions.

2 Cancel if possible.

3 Multiply numerators; multiply denominators.

Examples $\frac{^1\cancel{7}}{_2\cancel{8}} \times \frac{\cancel{4}^1}{\cancel{21}_3} = \frac{1}{6}$

$1\frac{1}{2} \times 2\frac{2}{3} = \frac{^1\cancel{3}}{_1\cancel{2}} \times \frac{\cancel{8}^4}{\cancel{3}_1} = \frac{4}{1} = 4$

To **divide by a fraction**, multiply by the inverse.

Examples $\frac{2}{3} \div \frac{4}{9} = \frac{^1\cancel{2}}{_1\cancel{3}} \times \frac{\cancel{9}^3}{\cancel{4}_2} = \frac{3}{2} = 1\frac{1}{2}$

$1\frac{3}{4} \div \frac{7}{8} = \frac{7}{4} \div \frac{7}{8} = \frac{^1\cancel{7}}{_1\cancel{4}} \times \frac{\cancel{8}^2}{\cancel{7}_1} = 2$

Test yourself

Except for questions 5, 7 and 13.

1 Express the shaded part as a fraction of the green rectangle.

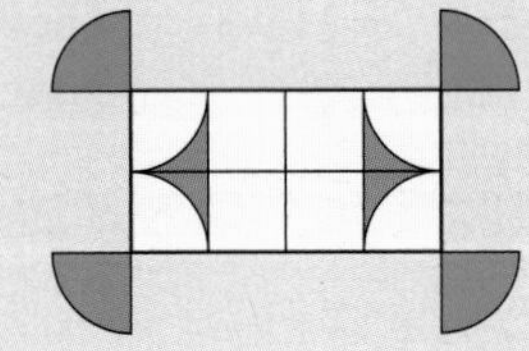

2 What fraction of 240 is 360?

3 This frequency diagram shows the heights of Mr Yin's piano pupils.
He put the heights into class intervals of $150 \leqslant h < 155$, ...
What fraction of Mr Yin's pupils were
a 160 cm or taller but less than 175 cm
b shorter than 160 cm?

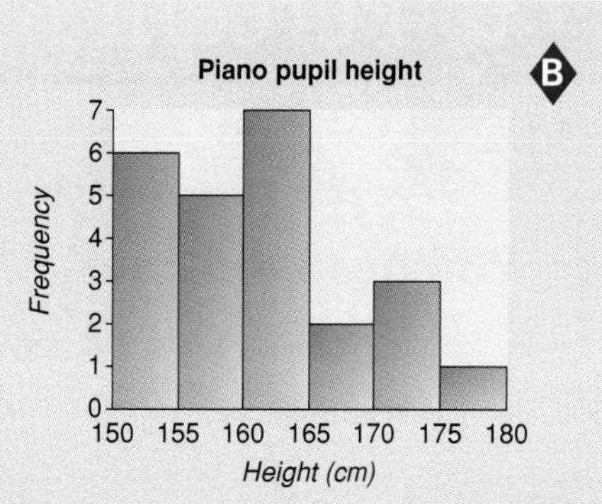

4 Convert these mentally to terminating or recurring decimals.

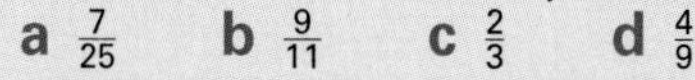
a $\frac{7}{25}$ **b** $\frac{9}{11}$ **c** $\frac{2}{3}$ **d** $\frac{4}{9}$

5 Use your calculator to write these as decimals.

a $\frac{17}{31}$ **b** $\frac{28}{37}$ **c** $\frac{118}{236}$ **d** $\frac{5}{365}$ **e** $\frac{17}{840}$

6 Write these as fractions.

a 0·5555 ... **b** $0{\cdot}\dot{7}$ **c** $0{\cdot}\dot{6}$

7 7 out of 16 pupils from the lunchtime computer club went on a camp. 19 out of 48 from the after school computer club also went. Which computer club had the greater fraction of pupils that went?

8 Put these in order from smallest to largest.

$\frac{1}{2}, \frac{2}{5}, \frac{7}{25}, \frac{17}{50}, \frac{3}{10}$

9 Copy this number line.

Show these fractions with an arrow. $\frac{12}{20}, \frac{13}{10}, \frac{35}{25}, \frac{36}{40}$

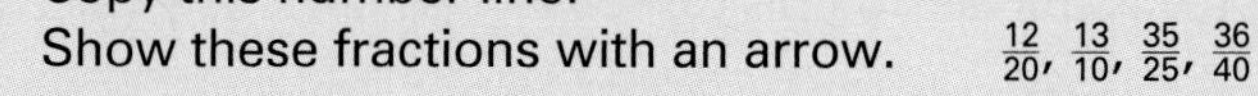

10 Find the fraction halfway between $\frac{3}{8}$ and $\frac{11}{24}$.

11 **a** $\frac{2}{5} + \frac{4}{5}$ **b** $\frac{7}{12} + \frac{2}{3}$ **c** $\frac{5}{8} + \frac{2}{5} + \frac{7}{10}$ **d** $\frac{7}{12} - \frac{3}{8}$
e $\frac{14}{15} - \frac{7}{10}$ **f** $\frac{11}{12} + \frac{5}{6} - \frac{3}{8}$ **g** $\frac{1}{4} + \frac{3}{5} - \frac{3}{10}$

12 **a** $\frac{3}{4} \times \frac{2}{15}$ **b** $8 \times 3\frac{3}{4}$ **c** $2\frac{2}{5} \times 3\frac{1}{8}$ **d** $\frac{5}{9} \div \frac{2}{3}$
e $3\frac{3}{5} \div 2\frac{1}{10}$ **f** $\frac{5}{6} \times \frac{1}{4} \times \frac{12}{25}$ **g** $\frac{2}{5} \times \frac{3}{4} \times \frac{1}{2}$ **h** $2\frac{2}{3} \times 5\frac{3}{5}$
i $3\frac{1}{4} \div 1\frac{5}{6}$ **j** $(1\frac{1}{2})^3$ **k** $\frac{3}{5}(\frac{2}{3} + 2\frac{1}{4})$

13 Find these using your calculator.

a $\frac{3}{7} + \frac{4}{9} - \frac{2}{3}$ **b** $\frac{5}{8} - \frac{3}{20} + \frac{4}{5}$ **c** $1\frac{3}{4} \times 3\frac{5}{12}$ **d** $3\frac{1}{3} \div 2\frac{7}{11}$

14 Which is greater?

a $\frac{3}{8}$ of 64 or $\frac{7}{9}$ of 45 **b** 72% of 36 or $\frac{3}{5}$ of 41

15 In a 240 g bag of mixed nuts, $\frac{1}{3}$ are brazil nuts, $\frac{1}{4}$ are almonds, $\frac{1}{8}$ are cashews, 30 g are peanuts and the rest are other sorts of nuts. What fraction are other sorts of nuts?

*16 A shelf of books is half full. Four more books are put on the shelf and then it is two-thirds full. How many books will fit on the shelf?

6 Percentages, Fractions, Decimals

You need to know

Key vocabulary

interest, loss, profit, service charge, tax, unitary method, value added tax (VAT)

How Much?!

- We often estimate percentages.

  About how full, as a percentage, is the petrol tank of this car?

  About what percentage of this file has downloaded?

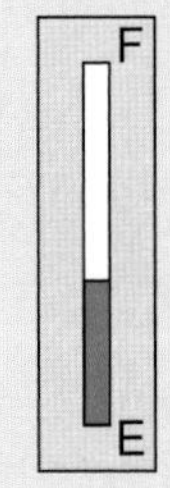

  Think about other times we estimate percentages.

- We often give proportions of a whole as percentages.

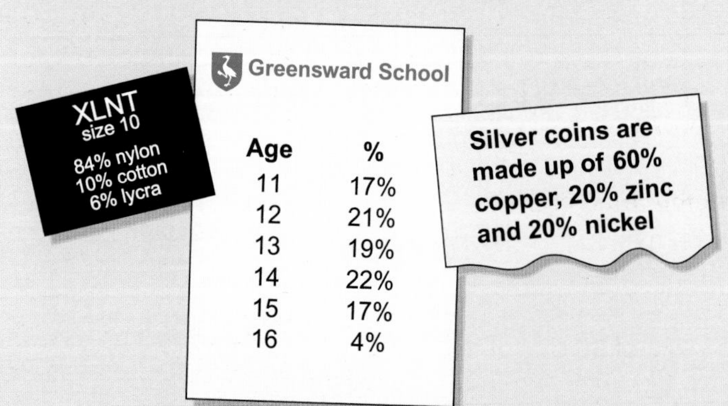

  What should the total add to in each case?

  Think about other times we divide a whole into percentages.

- How are percentages used to help sell goods?

Percentage of

We can find the **percentage of a quantity** using a written method or using a calculator.

To find 17% of 64·2 we can use one of these **written methods**.

Using fractions	Using decimals	Finding 1% first
17% of 64·2 = $\frac{17}{100} \times 64{\cdot}2$ $= \frac{17 \times 64{\cdot}2}{100}$ $= \frac{1091{\cdot}4}{100}$ = **10·914** 642 × 17: 6420 + 4494 = 10914 64·2 × 17 = 1091·4	17% of 64·2 = 0·17 × 64·2 = 17 × 642 ÷ 1000 = 10914 ÷ 1000 = **10·914** 642 × 17: 6420 + 4494 = 10914	1% of 64·2 = 0·642 17% of 64·2 = 17 × 0·642 = 17 × 642 ÷ 1000 = 10914 ÷ 1000 = **10·914** 642 × 17: 6420 + 4494 = 10914

Using a calculator instead we would key 17 ÷ 100 × 64·2 = **or** 0·17 × 64·2 = **or** 17 × 0·642 = to get **10·914**.

Note: $33\frac{1}{3}$% can be keyed as 33 a b/c 1 a b/c 3 ÷ 100 or as 0.33333333333

Some calculators give the answer to a calculation like 0.33333333333 × 18 as 5·999999999. We interpret this answer as 6.

Always try to use a mental method or written method first.

Exercise 1

Except for question 10 and Review 2.

Do some using fractions, some using decimals and some by finding 1% first.

1 Find the answers to these mentally.

a 85% of 60 mℓ b 35% of 140 ℓ c 45% of 160 km d 125% of £60
e 160% of 50 m f 21% of 300 cm g 51% of 80 g h 39% of £1500

2 Find these.

a 16% of 20 m b 26% of 80 ℓ c 37% of 200 cm d 13% of £68
e 24% of 84 km f 19% of 120 g g $33\frac{1}{3}$% of 195 mm h $66\frac{2}{3}$% of 123 m
i 181% of £52 j 177% of 110 km k 117% of 840 cm l $17\frac{1}{2}$% of £68
m $12\frac{1}{2}$% of £83·50 n $5\frac{1}{2}$% of £164 o $33\frac{1}{3}$% of £54

3 a How much was taken off the price of the TV in the sale?
b William bought a stereo in the sale. How much did he save?

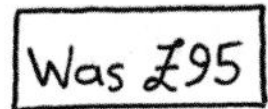

4 Last year Bill earned £26 524. He was taxed at 24% on this. How much did Bill pay in tax?

5 Samantha borrowed £8500 from her sister. Her sister charged her 4·5% interest each year. How much interest did she owe after 3 years?

6 

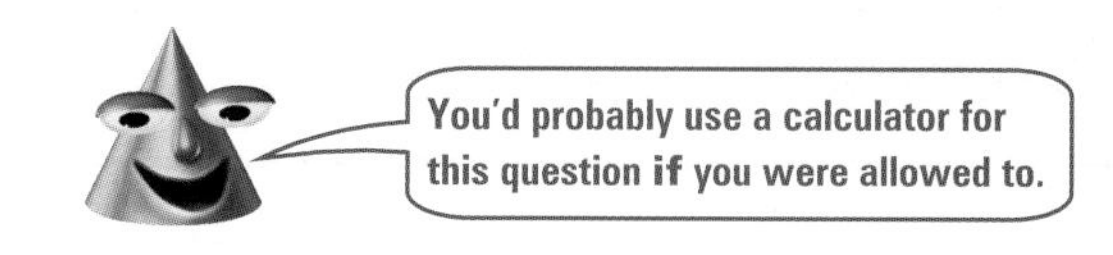

At these shops the full price of a bag was £10·20.
Which shop is now cheaper and by how much?

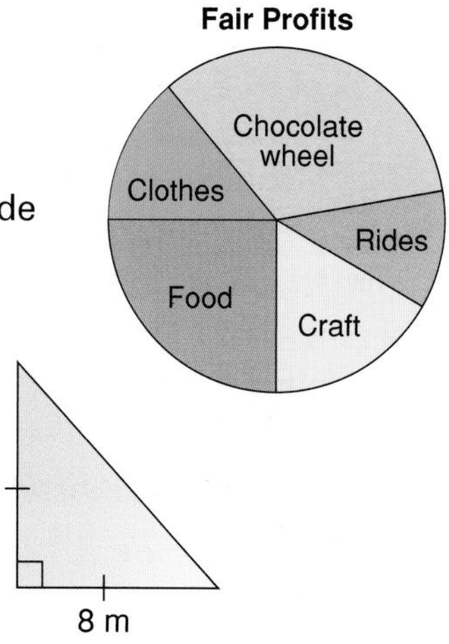

7 This pie chart shows the profit made by the stalls at Ravensdale School fair. The total profit was £3600. The chocolate wheel made about $33\frac{1}{3}\%$ of this profit.
Estimate the profit in £ made from these stalls.
a rides **b** craft **c** food.

8 Marita has a triangular garden. 17% of it is planted in carrots.
a What area of the garden is planted in carrots?
b 35% of the area planted in carrots needs weeding. What area needs weeding? Round your answer sensibly.

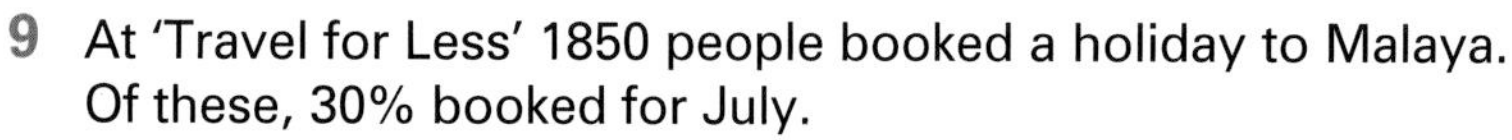

9 At 'Travel for Less' 1850 people booked a holiday to Malaya. Of these, 30% booked for July.
a How many was this?
b 20% fewer people booked a holiday to Malaya for August than for July. How many booked for August?
*__c__ 30% fewer people booked a holiday to Malaya for September than for August. How many booked for September?

10 27% of the export earnings of a country come from farming.
a If the total export earnings are £296 843 500, how much is earned from farming?
*__b__ 36% of the earnings from farming come from cheese. How much comes from cheese?

Review 1 Find these.

a 15% of 30 m **b** 47% of 400 m **c** 53% of £38 **d** 34% of 68 km
e 119% of 210 cm **f** $33\frac{1}{3}\%$ of £870 **g** $3\frac{1}{2}\%$ of 55 km.

Review 2 Only use a calculator if you need to.

19% of the population in a town is aged under 20.
How many of the 23 500 people in the town are under 20?

Review 3 A stereo is on sale in two shops. Echo music have 22% off and Music Maker have £30 off. Which shop is cheaper?

Review 4 'Contacts 4U' did a survey of 1240 people.
15% of those surveyed wore glasses.
63% of those who wear glasses said they would like to try contact lenses.
a How many wear glasses?
*__b__ How many would like to try contact lenses? Round your answer sensibly.

Percentage increase and decrease

Discussion

- Isia bought £1000 worth of shares.
  They increased in value by 100%.
  Which of these is true about Isia's shares? **Discuss**.
  They have doubled in value.
  They have trebled in value.
  They are worth $1\frac{1}{2}$ times as much as she paid.
- If Isia's shares had increased in value by 500% instead, which of these would be true? **Discuss**.
  They would be worth 5 times the original amount.
  They would be worth 6 times the original amount.
- If Isia's shares had decreased in value by 100% instead, how much would they be worth? **Discuss**.
- If Isia's shares had increased in value by 50% instead, which of these would be true? **Discuss**.
  An increase of 50% will give 150% of the original value.
  An increase of 50% will give 200% of the original value.
- On 31 March Isia's shares were worth £1200. They increased in value by 25% in the next 3 months. To find the new value of Isia's shares, her brother found 25% of £1200 and added it to £1200.
  Isia got the same answer by multiplying 1200 by a single number.
  What was the number? **Discuss**.

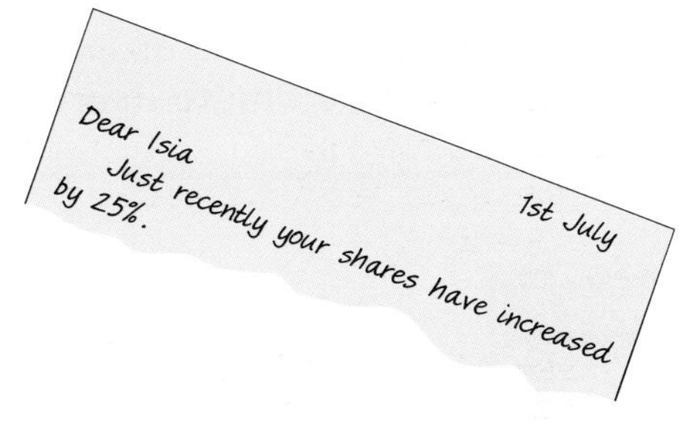

- What would Isia's number be if the shares had
  increased by 35%   increased by 85%   decreased by 5%?

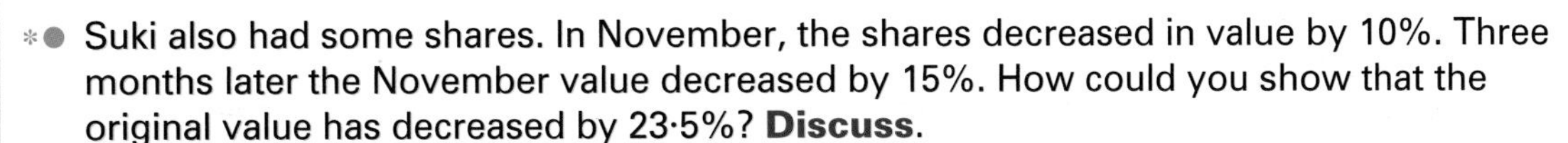

- * Suki also had some shares. In November, the shares decreased in value by 10%. Three months later the November value decreased by 15%. How could you show that the original value has decreased by 23·5%? **Discuss**.

Practical

This label says you get 40% more fruit.

Does this give a false impression?

Collect lots of similar labels.

Decide which ones give the customer a false impression.

Worked Example
Donna is a buyer for 'Fashion Warehouse'. One line of dresses she buys cost £12·50. A 30% mark up is put on these.

a What price does the Fashion Warehouse sell these dresses for?
b At the end of the summer these dresses are reduced by 25%. What profit or loss does the Fashion Warehouse make on them?

Answers
a They sell for 130% of £12·50 = 1·3 × 12·50
= £16·25

b Reduced price = 75% of £16·25
= 0·75 × 16·25
= £12·19 (to the nearest penny)

This reduced price is less than the cost. Loss = £12·50 − £12·19
= **31p** on each dress

Worked Example
Mark is enlarging a 10 cm drawing. He first enlarges it by setting the photocopier to 144%. He puts this enlarged drawing on the photocopier and sets it to 144%. How long is the final copy of the drawing?

Answer
1st enlargement = 144% of 10 cm
2nd enlargement = 144% of 1st enlargement
= 144% of (144% of 10 cm)
= 1·44 × 1·44 × 10 cm
= **20·7 cm (to nearest mm)**

Exercise 2

Only use a calculator if you need to.

1 Samuel bought a model aeroplane for £15.
a He painted and repaired it and sold it for a profit of 38%.
How much did he sell it for?
b The next owner crashed the plane and then sold it at a 55% loss.
How much did it sell for?

2 Aisling bought an Ace Computer Special.
a She paid an extra 15% to have some added parts installed.
How much did she pay altogether?
b She sold the computer two years later and made a 35% loss on the total price paid.
What price did she sell it for?

3 Which is the better buy?
a A 600 g block of cheese for £3·40 or a 500 g block plus 20% extra free for £3·45
b

4 Janita wanted to know what total percentage saving she would make on red dot items.

She worked out that on red dot items, the original price has been reduced by 32%. She is correct.
Show how she might have done this.

5 The salesperson multiplied all the original prices by 0·85 to find the sale price.

She then multiplied the sale price by 0·85 to find the last day sale price.

a What single number could she multiply the original prices by to get the last day sale price?

b By what percentage have the original prices been reduced on the last day? Explain your answer.

6 Bridget owns a boutique. She prices all the clothes so she makes a 40% profit. When her friend Anna bought a dress from her, she gave Anna a 40% discount. Did Bridget make any profit on this dress? Explain.

7 Neil claims than an increase of 10% followed by a decrease of 10% is less than an increase of 20% followed by a decrease of 20%. Is Neil correct?

8 Emalia sold her bike to Anna and made 25% profit. Anna sold the bike to Tara for a further 5% profit. Anna told Emalia that the bike had been sold to Tara for 30% more than Emalia had paid for it. Is Anna correct? Explain.

9 A manufacturer adds 60% to the cost of materials to cover labour. She then adds a further 25% to get the selling price. At what price would she sell an article made from materials which cost 75p?

10 Kimberley's had a sale on all jumpers. The marked price of all jumpers was reduced by 25% and a cash customer was given a further 5% off. What price would a cash customer pay for a jumper originally marked at £26?

11 Didier enlarged a drawing in a book by setting the photocopier to 130%. He decided this was too large. He put the enlarged drawing on the photocopier and set it to 90%. If the original drawing was 14 cm by 18 cm, what are the dimensions of the final copy?

12 Ashley embroidered tablecloths. She added 50% to the cost of the tablecloths and sold them to Linen For Everyone. They sold them for 25% profit. If the tablecloths cost Ashley £15, how much did Linen For Everyone sell them for? Give your answer to the nearest pound.

***13** Fitness House wanted to increase its membership by 15% each year for 5 years. In the first year membership increased by only 10% to 847.

a If the 15% target is reached for the following 4 years, how many members will Fitness House have? (Round to the nearest person after each year's increase.)

b If the 15% target had been reached for all 5 years, how many members would they have had?

***14** A woman pays 5% of her salary on life insurance and saves 9% of the remainder. What percentage of her full salary is she saving?

Review 1 The population in a town increased 20% in the last 5 years. It is predicted to increase a further 25% in the next 5 years. If the population five years ago was 18 420, what is it predicted to be in 5 years time?

Review 2 A car worth £15 000 is insured for 20% more than its value. The car is stolen and the insurance company pays out the sum insured less 15%. What did the insurance company pay out?

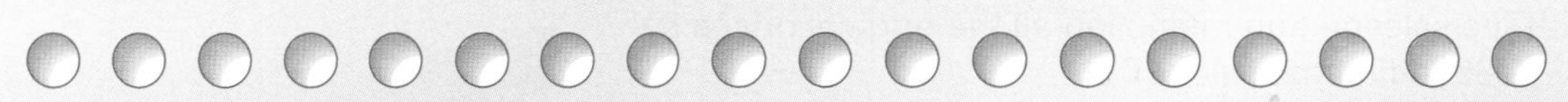

Investigation

Population Increase

You will need a spreadsheet or graphical calculator.

The rabbit population on an island was increasing at the rate of 4% per year.
At the start of year 1 the rabbit population is 1000.
How long will it take to double?

Ask your teacher for a copy of the Population Increase ICT Worksheet or use a graphical calculator.

What if the rabbit population was 2500 at the start?
What if it was 5000 or 7500 or ...?
What if the rate at which it increased was 6% or 8% or 10% or ...?

Increase or decrease as a percentage

Discussion

- James scored 2 goals in last week's match.
  In this week's match he scored 4 goals.

  Did James get 50% more goals this week than last week?
  Did James get 100% more goals this week than last week? **Discuss**.

  The school roll increased from 250 in 2000 to 750 in 2004.
  Is this an increase of 200% or 300%? **Discuss**.

- Think of other situations where increases in numbers might be given as a % increase. **Discuss**.

  Think of situations where decreases in numbers might be described as % decreases. **Discuss**.

Noah's trainer told him to increase his exercise time by 20%.
Noah increased his daily exercise time from 40 to 48 minutes.
He checked to see what percentage increase this is.

Fractional increase $= \frac{8}{40}$ ← **actual increase** / ← **original amount**

Percentage increase $= \frac{8}{40} \times 100\%$

$= 20\%$

% increase $= \frac{\textbf{actual increase}}{\textbf{original amount}} \times$ **100%**

% decrease $= \frac{\textbf{actual decrease}}{\textbf{original amount}} \times$ **100%**

Worked Example

A jersey which was priced at £39 is reduced to £28 in a sale.
What percentage reduction is this?

Answer

Actual decrease in price = £39 − £28
= £11

% decrease $= \frac{\text{actual decrease}}{\text{original amount}} \times 100\%$
$= \frac{11}{39} \times 100\%$
= 28% (to the nearest percentage).

Exercise 3

Round your answers sensibly.

1 Over a two year period, a house increases in value from £145 000 to £205 000.
  a What actual increase in value is this?
  b What percentage increase is this?

2 Rebecca increased her scoring rate from 20 runs per match last season to 25 runs per match this season.
What was the percentage increase in Rebecca's scoring rate?

3 A motorbike depreciated in value from £4500 to £3200.
What percentage depreciation is this?

4 A car appreciated in value from £17 500 to £25 800.
What percentage appreciation is this?

5 Dale bought some shares for £50 and later sold them for £177.
What percentage profit did Dale make on these shares?

6 The prices of five items P, Q, R, S and T five years ago and this year are shown on these scales.

Price (£) five years ago

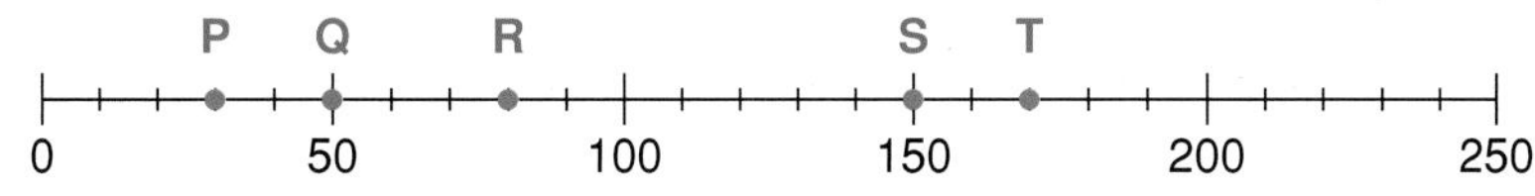

Price (£) this year

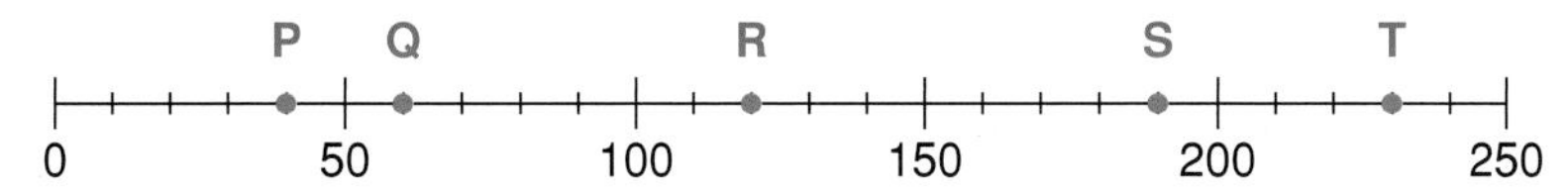

  a What is the percentage increase of each item?
  b Which item showed the greatest percentage increase over the five years?

7 Thomas was studying expansion in science. He heated a metal bar and noticed it increased in length from 1·16 m to 1·173 m.
Calculate the percentage increase to one decimal place.

*8 The Atkinsons thought their living room was too small. The dashed lines show how they extended it. Calculate the percentage increase in the area of the living room.

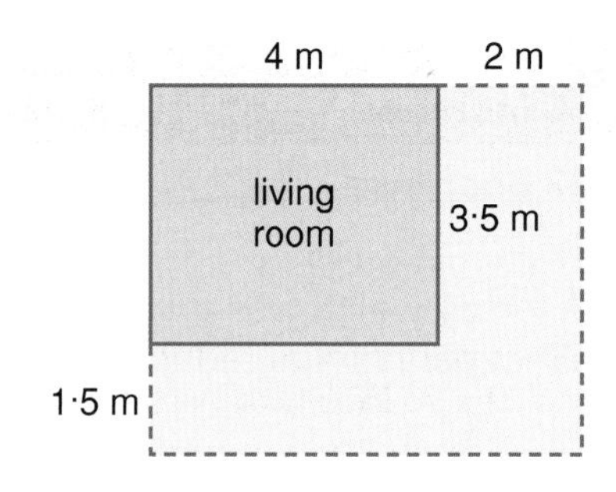

*9

Energy: oil production			
		barrels produced daily	
rank	country	1990	2000
1.	Saudi Arabia	7 105 000	9 145 000
2.	USA	8 915 000	7 745 000
3.	Russian Federation	10 405 000	6 535 000
4.	Iran	3 255 000	3 770 000
5.	Mexico	2 975 000	3 450 000
6.	Norway	1 740 000	3 365 000
7.	China	2 775 000	3 245 000
8.	Venezuela	2 245 000	3 235 000
9.	Canada	1 965 000	2 710 000
10.	United Kingdom	1 915 000	2 660 000

a Which countries given in the table had an increase in the number of barrels produced daily from 1990 to 2000?
b Which country had the greatest increase in barrels produced daily?
c Which country had the greatest percentage increase in barrels produced daily? What was the percentage increase?
d Which country had the greatest percentage decrease in barrels produced daily? What was the percentage decrease?

Review 1 **Marriages in United Kingdom**

Year	1979	1989	1999
Number	369 000	345 000	264 000

This table gives the number of marriages, to the nearest thousand. Find the percentage decrease
a between 1979 and 1989 **b** between 1989 and 1999 **c** between 1979 and 1999.

Review 2 Calculate the percentage increase in these prices.
a £1·65 to £1·85 **b** £1560 to £3425.

Practical

You will need a spreadsheet package.

Compare the percentage savings on ten or more items that are on special offer. Use a spreadsheet to help. Ask your teacher for the Savings ICT sheet.

Calculating the original amount

Discussion

- Blake read that the town where he lived had increased in population by 10% this year to 8635. Blake wanted to know what the population was last year. He decided that to get the new population, the old population must have been *multiplied* by 110% or 1·1. To find the old population he *divided* the new population by 110% or 1·1. Was Blake correct? **Discuss.**
- Blake read that the population of the next town had also increased by 10% to 8950. He worked out that the population the previous year was $8136{\cdot}\dot{3}\dot{6}$. In fact the population was 8136. Why are these figures different? **Discuss.**
- Jasmine wanted to know the original price of the shirt.
  She wrote

  £65 represents 85% (100% − 15%)
  £65 ÷ 85 represents 1%
  £65 ÷ 85 × 100 represents 100%

  Is Jasmine correct? **Discuss.**
  What is the original price of a shirt?

Worked Example
This year Wolver Farm has 345 sheep. This is 15% more than last year. How many did they have last year?

Answer
Three different ways to find the answer are shown.

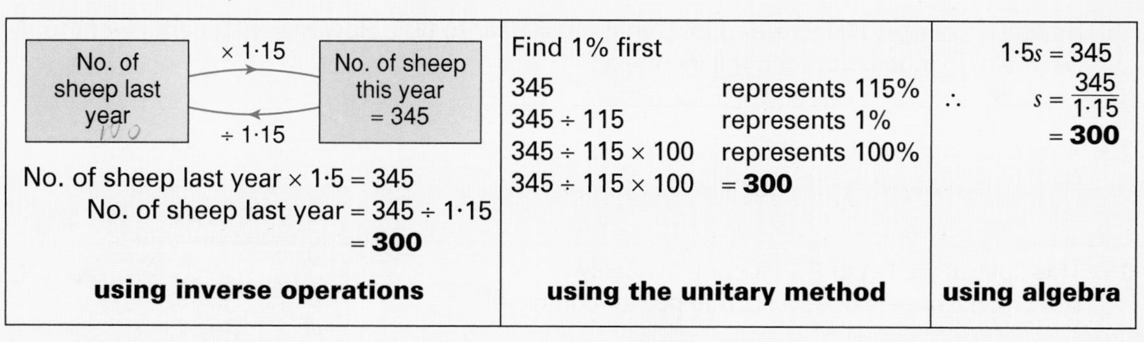

using inverse operations	using the unitary method	using algebra
No. of sheep last year —× 1·15→ No. of sheep this year = 345 (← ÷ 1·15) No. of sheep last year × 1·5 = 345 No. of sheep last year = 345 ÷ 1·15 = **300**	Find 1% first 345 represents 115% 345 ÷ 115 represents 1% 345 ÷ 115 × 100 represents 100% 345 ÷ 115 × 100 = **300**	$1{\cdot}5s = 345$ $\therefore\ s = \frac{345}{1{\cdot}15}$ = **300**

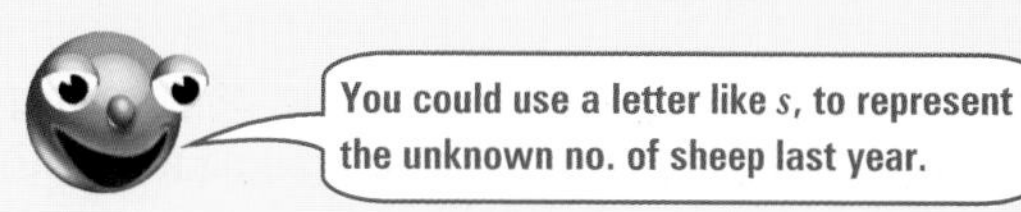

Exercise 4

Only use a calculator if you need to.

1 Find the answers to these mentally.
- **a** 25% of a number is 8. What is the number?
- **b** 30% of a number is 3. What is the number?
- **c** 75% of a number is 15. What is the number?

There are more mental percentage calculations on page 61.

2 A coat is on sale for £100, which is 80% of its original price. What was its original price?

3 This year 16 500 people visited a memorial site. This was an increase of 20% from the previous year. How many visitors were there the previous year?

4 An antique was sold for £1400. This was a profit of 40%. What price was paid for the antique originally?

5 Mrs Schmidt's salary was increased by 6% to £47 700. What was her salary before the increase?

6 Find the original price of a painting sold for £1729 at a 5% loss.

7 When 9% interest was added to Julia's savings she had £163·50. How much did Julia have originally?

8 Denise surveyed the clubs at her school. She worked out the percentage increase or decrease in membership in the last year, giving the percentages to the nearest per cent. What was the membership of each club last year?

Club	Current membership	Increase or decrease
Drama	62	Increase 17%
Chess	18	Decrease 14%
Painting	67	Decrease 3%
Cricket	70	Increase 23%
Craft	101	Increase 1%
Athletics	45	Decrease 8%

9 Newfield College roll increased by $\frac{1}{10}$ in the last year to 605. How many students went to Newfield College before the roll increased?

*10 A plant increased its height by $\frac{2}{3}$ to 28 cm. How high was the plant originally?

*11 This special packet of P + Ps contains 80 sweets. How many sweets does a normal packet contain?

Review 1 A model was made of an antique chest so that a replica could be produced. The model was increased 75% in size to make the replica. If the height of the replica was 87·5 cm, how high was the model?

Review 2 'Accessories Unlimited' had 25% off all belts and handbags. The new prices were rounded to the nearest penny. In the sale Julia bought a belt for £7·46 and a handbag for £26·25. What was the original price of these items?

Review 3 The number of telephone calls Tait Electronics made this month increased on last month by $12\frac{1}{2}$%. This month Tait Electronics made 765 calls. How many calls did the company make last month?

Mixed percentage calculations

Exercise 5

Only use a calculator if you need to.

1 Sophia bought this ring.
  a How much did it cost?
  b Her friend bought the same ring at another shop and paid 15% less. How much more did Sophia pay? Give your answer to the nearest penny.

2 Rachel put £7583 in the bank. She earned interest of $8\frac{1}{2}$% per year. How much did she have in total after one year? Give your answer to the nearest penny.

3 Last Thursday 530 people came to a parents' evening. This was $\frac{1}{3}$ of all parents. At the evening 480 people bought a school magazine. What percentage of all parents is this? Give your answer to 1 d.p.

4 Which of these fruit bars has the highest proportion of fruit? What percentage of fruit does it have?

Fruity Tuti
Total mass 50 g
Fruit mass 33 g

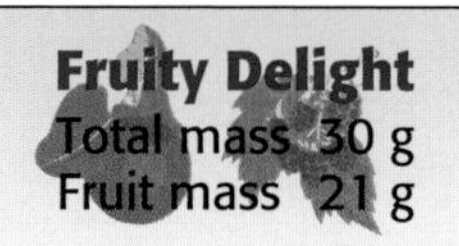

Fruito
Total mass 20 g
Fruit mass 15 g

5 Brandy bought a mobile phone. He paid a deposit and monthly payments for six months.
How much *more* than the cash price did he pay?

6 **a** 20% of this shape is to be red. How many *more* squares must be shaded red?
  b $87\frac{1}{2}$% of the remaining squares are to be black. What percentage of the whole shape is to be black?

7 This pie chart shows how Jessica got her Vitamin A requirements last week. She worked out she got 8 g of Vitamin A from meat.
Estimate how much she got from vegetables.

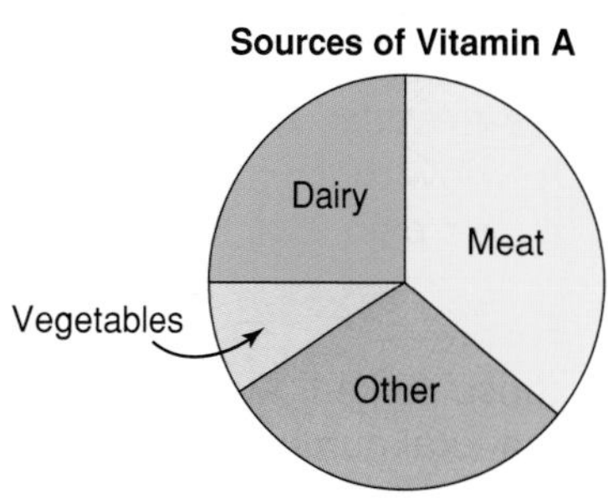

8 A marketing agency asked 2400 people which sport out of netball, hockey and horse riding they would like to see more of on TV. 1000 said netball, 900 said hockey and 500 said horse riding.
What percentage said
  a netball
  b hockey
  c horse riding?
Round your answers to the nearest per cent.
Add up the percentages you got for **a**, **b** and **c**.
Explain your answer.

9 Alan and Paul both bought identical sets of headphones from this shop. Alan bought his set on Monday. Paul bought his on Tuesday. Alan paid 40p more than Paul.
What was the original price of the headphones?
(**Hint:** Alan paid 5% more than Paul, so 5% of the cost = 40p.)

Monday 15% off everything	Tuesday 20% off everything
CLOSING DOWN SALE	

10 Mel sold her yacht for £1500 at a 20% profit.
What did she pay for the yacht?

11 Paul and Lucy went out to dinner. When they went to pay, the waiter added VAT of $17\frac{1}{2}$% and then a $12\frac{1}{2}$% service charge. Paul argued that the service charge should have been added first.
Who was correct? Give a mathematical reason for your answer.

12 Craig Cycles ran a sales promotion. At 9 a.m. on 13 July the price of cycles dropped by 5%. At the beginning of each hour after this, until 5 p.m., the price dropped a further 5%. The price was rounded to the nearest pound each time it dropped. How much would a cycle priced at £299 at 8:30 a.m. cost at 4:30 p.m.?

***13** This diagram shows two overlapping identical rectangles.
20% of each rectangle is shaded.
What fraction of the whole diagram is shaded?

***14** A photocopier enlarges an image. The ratio of lengths of the enlarged image to lengths on the original image is 5 : 2.
What percentage increase in length is this?

***15** If Oren sold his stereo for £179·55 he would make 5% profit. How much does Oren need to sell it for to make 10% profit? Give your answer to the nearest pound.

***16** Kayla made 10 ℓ of orange drink. It was 50% water. How much water must she add to it to make it 75% water?

***17** **a** Both sides of this square are increased by 20%.
What is the percentage increase in the area of the square?
b The area of the square shown increases by 21%. The new shape is still a square.
What is the length of each side?

5 m

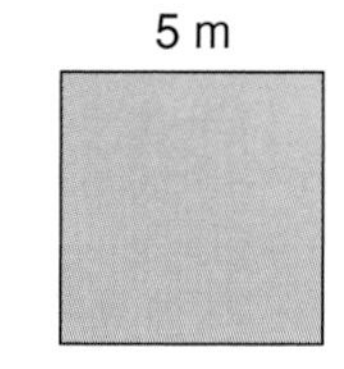

Review 1 Robert earned £8520. He got a bonus of 5% at the end of the year.
a How much did he earn in total?
b He paid 23% tax on his total earnings. How much did he have after tax?

Review 2 160 wildlife supporters went to a fund-raising dinner. This was one-quarter of those invited. At the dinner, 90 people bought a 'Save the Tiger' car sticker. What percentage of the total number invited, bought a car sticker? Give your answer to 1 d.p.

Review 3 This pie chart shows the export earnings of a country.

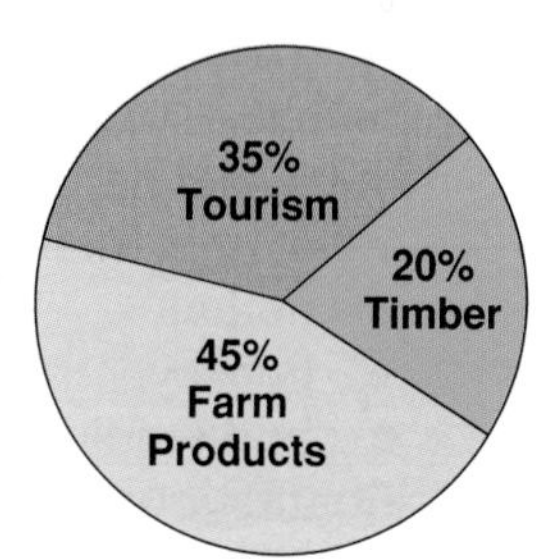

If last year it earned £7 million from tourism, how much did it earn from
a timber **b** farm products?

Review 4 When 15% discount was given, the price of a shirt was £13·60. What was the original price?

Review 5 A company's profit in 2003 was 12% higher than in 2002. In 2004 the profit decreased by 25%. If the profit in 2002 was £1 836 420, what was it in 2004? Give your answer to the nearest pound.

Practical

Percentages are all around us. We find them on food labels, clothing labels, in government statistics, in finance, in geographical data, ...

Either

- Make a poster displaying as many examples of where percentages are used as possible.

or

- Choose one area where percentages are used and make a poster about this.

Examples

Matthew chose food.
On his poster he displayed ten biscuit labels.
He compared the percentages of fat and sugar in each.

Melissa chose the weather.
On her poster she displayed 12 pie charts, one for each month of last year.
Each pie chart showed the proportion of rainy, cloudy and sunny days.
She got the information from the Internet.

Javed chose Africa.
On his poster he displayed percentages for crops, land use, minerals, ethnic groups, ...

Summary of key points

We can find **'percentage of' using a written method.**

Example 16% of 78

Using fractions

$16\% \text{ of } 78 = \frac{16}{100} \times 78$

$= \frac{16 \times 78}{100}$

$= \frac{1248}{100}$

$= \mathbf{12{\cdot}48}$

$$\begin{array}{r} 78 \\ \times\ \underline{16} \\ 780 \\ \underline{468} \\ \underline{1248} \end{array}$$

Using decimals

$16\% \text{ of } 78 = 0{\cdot}16 \times 78$

$= 16 \times 78 \div 100$

$= 1248 \div 100$

$= \mathbf{12{\cdot}48}$

Finding 1% first

$1\% \text{ of } 78 = 0{\cdot}78$

$16\% \text{ of } 78 = 16 \times 0{\cdot}78$

$= 16 \times 78 \div 100$

$= 1248 \div 100$

$= \mathbf{12{\cdot}48}$

To **increase an amount** by 20% we multiply by 120%.

To **decrease an amount** by 15% we multiply by 85% (100% − 15%).

To increase by 30% and then 15%, multiply by $1{\cdot}3 \times 1{\cdot}15$.

To increase by 40% and then decrease by 10% multiply by $1{\cdot}4 \times 0{\cdot}9$.

% increase = $\frac{\textbf{actual increase}}{\textbf{original amount}}$ × **100%**

% decrease = $\frac{\textbf{actual decrease}}{\textbf{original amount}}$ × **100%**

Example The price of petrol increased from 69p to 98p per litre.

% increase = $\frac{29}{69}$ × 100% **98 − 69 = 29p**

= 42% to nearest whole number.

We can use the **unitary method** or **inverse operations** to **find the original amount** if we know the amount **after** the **percentage change**.

Example After a 25% price increase, the packet of coffee was £3·40.

We can find the original price in one of these ways.

Using inverse operations

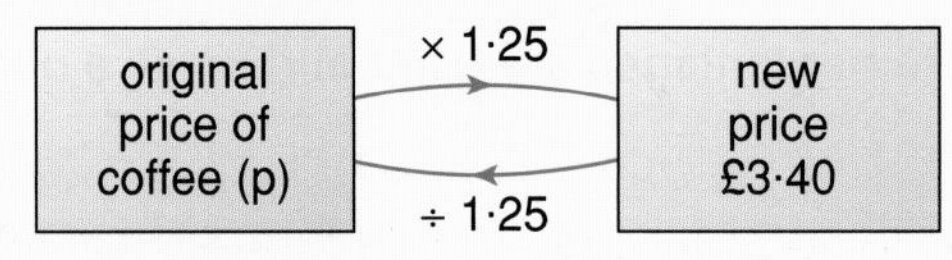

or Using algebra

$p \times 1{\cdot}25 = 3{\cdot}40$

$p = \frac{3{\cdot}40}{1{\cdot}25}$

= **£2·72**

or Using the unitary method

£3·40	represents 125%
£3·40 ÷ 125	represents 1%
£3·40 ÷ 125 × 100	represents 100%

£3·40 ÷ 125 × 100 = **£2·72**

Test yourself **Only use a calculator if you need to.**

1 Use a written method to find these. Check using your calculator. A

a 17% of 40 m **b** 36% of 65 ℓ **c** 87% of 150 km **d** 182% of £84

2 This pie chart shows the colour choices made by 8NT for the new school tie. Ten pupils in 8NT wanted green ties. Estimate how many wanted A

a navy blue ties **b** red ties.

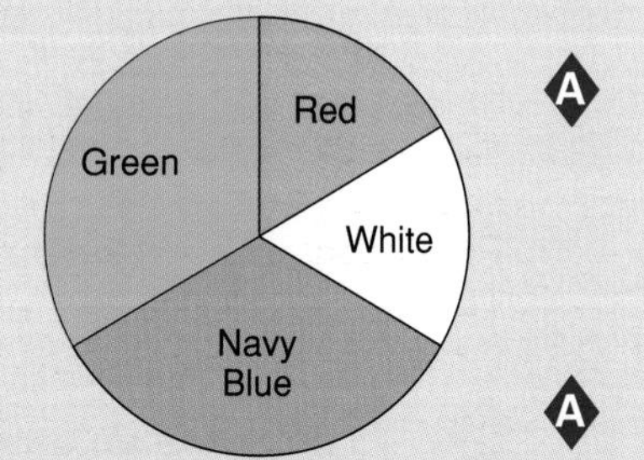

3 In an experiment a ball was dropped from a height of 1·5 m. In its first bounce it rose 85% of the height from which it was dropped. A

a To what height did it rise on its first bounce?

b On its second bounce it rose to 75% of the height it did on its first bounce. To what height did it rise on its second bounce? Round the answer sensibly.

4 A publisher increased the price of an atlas by 10% in each of two consecutive years.
The total increase in price was **A** 20% **B** less than 20% **C** more than 20%. 

5 Which is a better buy?  
a 200 g pack of crackers at £1·32 or
a pack of crackers with 200 g plus 25% extra free at £1·64

6 What is the percentage reduction on this camera?

7 In the last five years the amount spent on space transport has risen from 320 million dollars to 1550 million dollars.
What percentage increase is this?

8 This scale shows the distances five girls jumped on their first and second long jumps. 

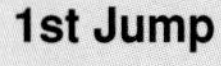
1st Jump

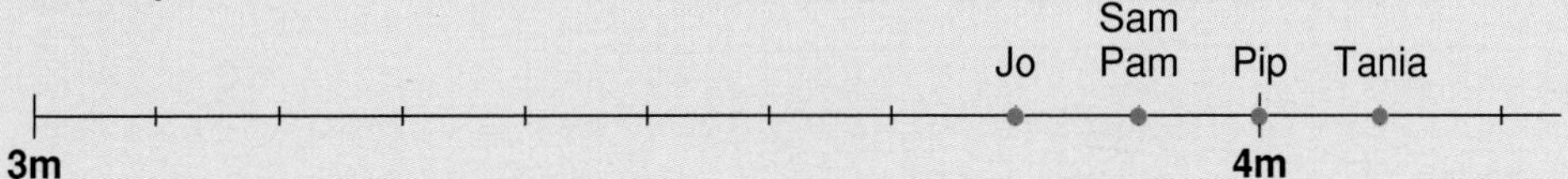

2nd Jump

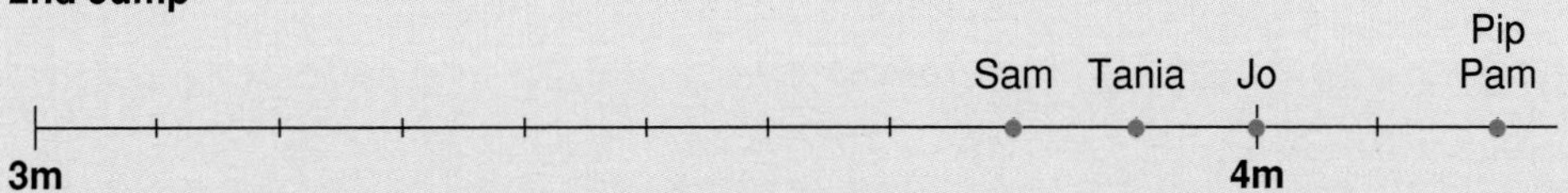

a Who had the greatest percentage increase from her first to her second jump?
b Who had the greatest percentage decrease from her first to her second jump?

9 This special pack contains 18 chocolate mints.
How many chocolate mints are in an ordinary pack?

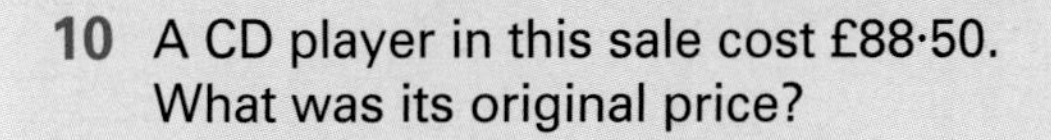
10 A CD player in this sale cost £88·50.
What was its original price?

11 At the end of the twentieth century about 50% of the world's tropical rain forest had been destroyed. 
At present, each year about 195 000 square kilometres are destroyed.
This is about 1·3% of the remainder.
Estimate the original area of the tropical rain forests.

7 Ratio and Proportion

You need to know

✓ ratio and proportion page 8

Key vocabulary

direct proportion, proportional to (∝)

Camping Out

The teacher–pupil ratio at Pam's school is 1 : 20.

What is the teacher–pupil ratio at your school?

The ratio of adults to pupils at Pam's school camp was 1 : 5.

What was the adult to pupil ratio at your last school camp or class outing?

Does your school have any policies about this?

Direct proportion

Sunil mixed 5 ℓ of purple paint with 2 ℓ of white paint.
This table shows the amounts of each colour she should mix together when she makes different amounts of paint.

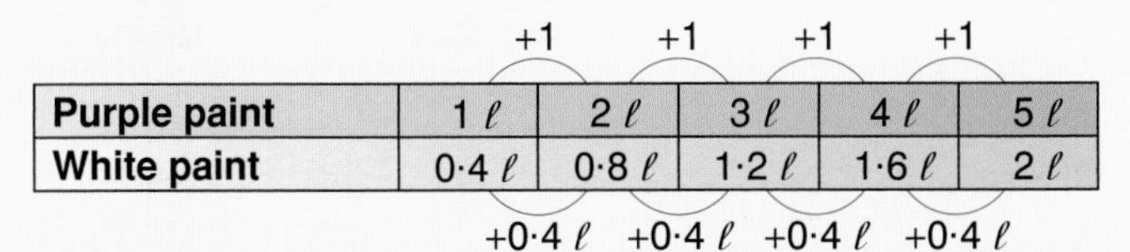

Purple paint	1 ℓ	2 ℓ	3 ℓ	4 ℓ	5 ℓ
White paint	0·4 ℓ	0·8 ℓ	1·2 ℓ	1·6 ℓ	2 ℓ

The amounts of purple and white paint are in **direct proportion**.
We say the amount of purple paint is **proportional to** the amount of white paint.

We can write this as
amount of purple paint ∝ amount of white paint.

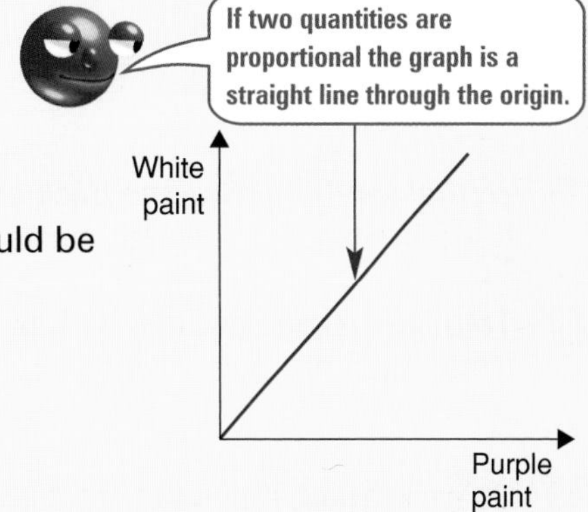

A graph of the amounts of white paint and purple paint would be a straight line through the origin.

Dana wants to make 8 gingerbread people.
She could use the **unitary method** and find the amounts for **one** gingerbread person by dividing by 6 first and then multiplying each amount by 8,

or

she could multiply each amount by a number, **the constant multiplier**, to get the amounts for 8.

Note: The constant multiplier method is the same as the unitary method done in one step.

6 Gingerbread people

100 g butter
100 g brown sugar
300 g flour
100 g syrup
2 tsp ginger
1 tsp bicarbonate of soda
currants to decorate

To find the **constant multiplier**

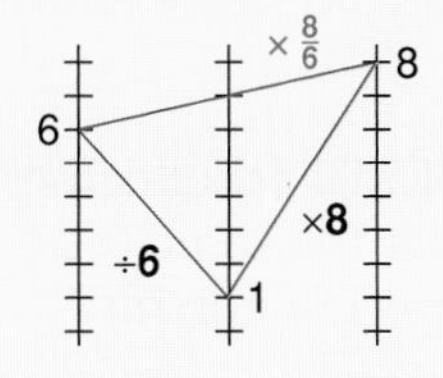

The constant multiplier to find the amounts for 8 gingerbread people is $\frac{8}{6}$.

The constant multiplier can also be found from the ratio of amount required : amount given $= 8 : 6 = \frac{8}{6}$.

Worked Example

Two horses are training on two different tracks.
For every 2 times Racer goes round track A, Mighty goes round track B 7 times.
If Racer goes round track A 120 times, how many times does Mighty go round track B?

Answer

1 round for Racer $= \frac{7}{2}$ rounds for Mighty
120 rounds for Racer $= \frac{7}{2} \times 120$ for Mighty
$=$ **420** rounds for Mighty.

or

ratio of rounds for Mighty to rounds for Racer is 7 : 2.
rounds for Mighty $= 120 \times \frac{7}{2}$
$=$ **420 rounds**

$\frac{7}{2}$ is the constant multiplier.

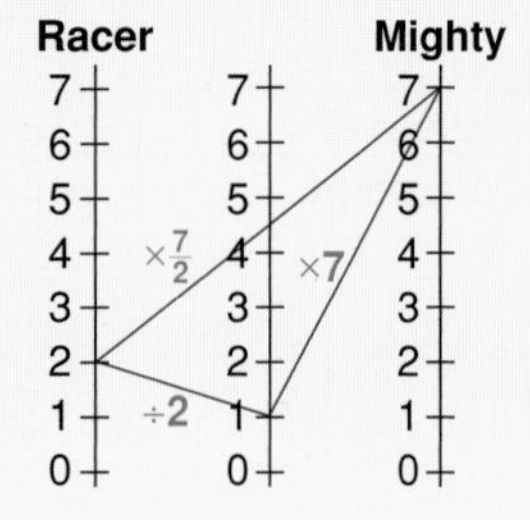

Exercise 1

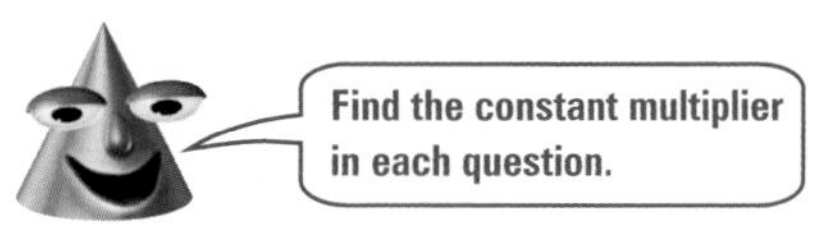

1 On Cassandra's bike, the back cog makes 10 turns for every 3 turns the front cog makes. If the front cog makes 180 turns, how many does the back cog make?

2 To make orange paint, mix 5 ℓ of yellow with 3 ℓ of red paint.
How much red paint would you mix with these amounts of yellow paint?
a 10 ℓ **b** 8 ℓ **c** 12 ℓ **d** 1·5 ℓ **e** 8·75 ℓ

3 To make glue, mix 6 mℓ of tube A with 14 mℓ of tube B.
How much of tube A would you need to mix with these amounts of tube B?
a 8 mℓ **b** 24 mℓ **c** 35 mℓ **d** 3 mℓ **e** 8·5 mℓ

4 £1 is worth NZ $3·20.
a Austin changes £1000 into NZ $.
How many NZ $ does he get?
b He spends NZ $2800 on his holiday.
He changes what he has left back to £.
About how much does Austin get?

5 5 miles is about 8 km.
a Manzoor cycled 17 miles to his friend's house.
How many kilometres is this?
b On the way home he got a puncture after $14\frac{1}{2}$ km.
How many miles from home was he?

6 **a** Adapt this recipe to feed 10 people.

Apple Crumble (for 4 people)

4 large cooking apples
50 g margarine (or butter)
100 g sugar
200 g flour

b Caleb was the camp cook at a scout camp.
He used 275 g of butter for the same recipe.
How many people was this enough for?

7 Sandy wanted to make this recipe for 10 people.

a Write the recipe Sandy should use.

b Gracie used 75 mℓ of soya sauce for the same recipe.
How many people was this enough for?

Egg Foo Yung for 6 people

8 eggs
200 g broccoli
100 g cauliflower
150 g peas
50 ml soya sauce

*__8__ Marcia wants to make a scale drawing of a shape.
The ratio to be used is 2 : 25.
One edge of the shape is measured as 125 mm.
How long should this edge be on the scale drawing?

*__9__ A photocopier is set to reduce in the ratio 2 : 5.

a What is the length of a reduced diagram if the length of the original is 4 cm?

b What length on an original will be reduced to 30 mm?

*__10__ The photo on the right to be enlarged in the ratio 5 : 2.
What are the dimensions of the enlargement?

Review 1 Two toy trains are going round their tracks.

For every 2 times train A goes round, train B goes round 5 times.
If train B goes round 155 times, how many times does train A go round?

Review 2

a Adapt this recipe to feed 16 people.

b Tina used 1·25 kg of baked beans for the same recipe.
How many people was this enough for?

Chilli Con Carne (for 12 people)

3 kg mince
6 small onions
600 g tomatoes
750 g baked beans
chilli powder to taste
3 bay leaves

Review 3 The scale of a map is 1 : 500 000.

a What length on the map is a distance of 10 km in real life.

b The length of a road on the map is measured as 52 mm.
Which of these is the actual length of the road?

A 2·6 km **B** 26 km **C** 260 m **D** 26 m

T

Practical

You will need a spreadsheet package

1 Find out the cost of some of your favourite foods.
Use a spreadsheet to find the cost of 1, 2, 3, 4, 5 ... of each.

2 Ask your teacher for the Constant Multiplier ICT sheet.

Simplifying ratios

Remember
We can simplify a ratio to its **simplest form** by **cancelling**.

Example

8 : 12 : 16 (each part $\div 4$)
= 2 : 3 : 4

All parts of the ratio have been divided by 4.

All parts of a ratio must have the same units.
Once they are the same and the ratio is simplified, we do not include the units.

Worked Example
Write the ratio *adult cost to child cost* in its simplest form.

Answer
The ratio is £1·60 : 70p.
We must write both parts either in pounds or pence.

Using pence, the ratio is

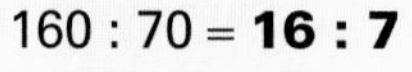
160 : 70 = **16 : 7**

Using pounds the ratio is

1·60 : 0·7 ($\times 10$, $\times 10$)

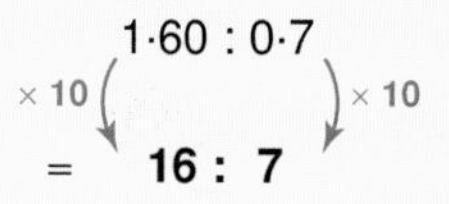
= **16 : 7**

Remember: A ratio is written using whole numbers rather than fractions or decimals.

Exercise 2

1 Write each ratio in its simplest form.

a 6 : 4 : 10	b 8 : 16 : 4	c 12 : 9 : 15	d 24 : 8 : 32
e 7 : 21 : 28	f 20 : 35 : 60	g 160 : 40 : 200	h 64 : 160 : 24
i 450 : 1000 : 800	j 1·2 : 3·6 : 5·4	*k $\frac{3}{4} : \frac{1}{2} : \frac{5}{8}$	*l 2·7 : $\frac{1}{3}$

2 There are 27 pupils, 6 teachers and 3 parents on a school camp.
Write the ratio of
a pupils to teachers to parents
b pupils to adults.

	No. of grams
Carbohydrate	24
Fat	8
Protein	16
Sugar	12

3 Write each of these as a ratio in its simplest form.
a carbohydrate to fat to sugar
b protein to carbohydrate to sugar.

4 Write these ratios in their simplest form.

a 2 mm : 5 mm	b 3 mm : 1 cm	c 53 cm : 2 m	d 13 min : 1 hour
e 60c : \$3	f \$4·50 : \$3	g 450 g : 2 kg	h 1 litre : 340 mℓ
i 25 min : 1 hour	j 3 m : 35 cm	*k 250 mℓ : 2 ℓ : 1250 mℓ	*l 3 cm : 90 mm : 0·6 m

Link question **4** to metric conversions page 253.

5 Sarah and Kate bought a raffle ticket which cost £2.
Sarah paid 89p and Kate paid the rest.
What was the ratio of the amount Kate paid to the amount Sarah paid?

6 Sasha braided her hair with 1·8 m of red ribbon and 95 cm of green.
What was the ratio of red to green ribbon?

7 A recipe uses 1·2 kg of flour and 400 g of sugar.
Write in its simplest form, the ratio of flour to sugar.

8 A farmer mixed 2.4 ℓ of weedkiller with 14·4 ℓ of water and 600 mℓ of oil.
Write, in its simplest form, the ratio of weedkiller to water to oil.

***9** Find the ratio of $x : y$ if
a $x = 12z$ and $y = 20z^2$ **b** $x = 16wz^2$ and $y = 10w^2z$ **c** $x = 3(z + 4)$ and $y = 6z + 24$

Review 1 Write these ratios in their simplest form.
a 8 : 48 : 12 **b** 136 : 120 : 64 **c** £3 : 40p **d** 35 min : 2 hours
e 8·2 mm : 4 cm ***f** 5·6 : 9·2 : 0·4 ***g** $\frac{5}{8} : \frac{5}{6} : \frac{1}{4}$

Review 2 What is the ratio of the price of a small pack of crisps to a family pack to a giant pack?

CRISPS	
Small	*35p*
Family	*£1.05*
Giant	*£2.65*

***Review 3** Find the ratio of p to q if $p = 5r^2$ and $q = 10r$.

Comparing ratios

The ratio of games won to games lost for Julian's team is 17 : 12 and for Tim's team is 23 : 15.
It is hard to tell from these ratios which team has the higher ratio of wins to losses.
To **compare ratios** we write them both in the form 1 : m or m : 1.

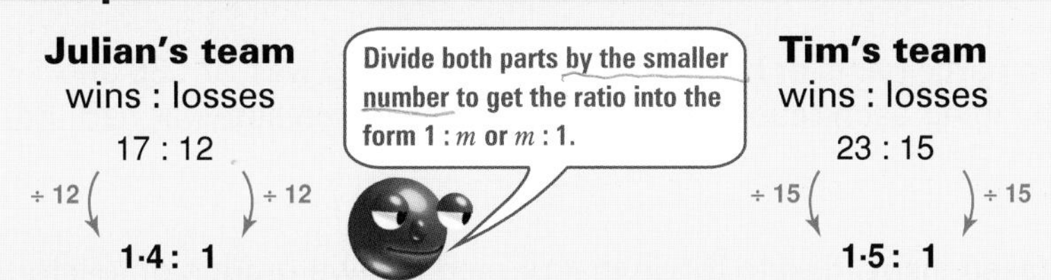

Tim's team has a higher ratio of wins to losses because $\frac{1·5}{1}$ (or 1·5 : 1) is greater than $\frac{1·4}{1}$ (or 1·4 : 1).

$\frac{1·5}{1} > \frac{1·4}{1}$

Worked Example
Which of these crackers has the smaller proportion of fat?

Answer

Crackerjacks
fat : other
3 : 50
÷ 3 ÷ 3
1 : 16·$\dot{6}$

Crispy
fat : other
4 : 62
÷ 4 ÷ 4
1 : 15·5

Ratio of fat to other ingredients given on packet

Crackerjacks has a smaller ratio of fat to other ingredients because
$\frac{1}{16·\dot{6}}$ ($\approx \frac{1}{17}$)
is less than $\frac{1}{15·5}$ ($\approx \frac{1}{16}$).

$\frac{1}{16·\dot{6}} < \frac{1}{15·5}$

We could say **Crackerjacks has a smaller proportion of fat.**

Exercise 3

1 Write these ratios in the form m : 1. Round to 1d.p. if you need to round.
a 8 : 3 **b** 17 : 3 **c** 40 : 19 **d** 83 : 25 **e** 124 : 87

2 Write these ratios in the form 1 : m. Round to 1d.p. if you need to round.
a 5 : 12 **b** 7 : 41 **c** 8 : 23 **d** 18 : 53 **e** 23 : 85

3 By changing each ratio to the form m : 1, say which pair of trousers has the greater proportion of wool. Explain.

4 The ratio of bread to meat in two brands of sausages is given.

By changing each ratio to the form 1 : m, decide which brand has the higher proportion of meat. Explain.

5 The ratio of trys converted to trys scored by two rugby teams, A and B, are respectively 18 : 5 and 15 : 4. By changing each ratio to the form m : 1, say which team has the better goal kicker. Explain.

6 The ratio of male to female participants at two seminars on parenting was 27 : 35 and 31 : 41 respectively. By changing each ratio to the form 1 : m say which seminar had the greater proportion of males. Explain.

***7** The ratio of toasters that failed inspection to those that passed for two brands of toaster, Popper and Browner, are 18 : 187 and 23 : 219 respectively.
Which brand had the higher proportion of toasters that failed inspection? Explain.

***8** Brad and Marlene were practising catching cricket balls. The ratio of balls batted to balls caught for Brad and Marlene was 17 : 4 and 13 : 3 respectively. Who is the better catcher?

Review 1 The ratio of elastine to other materials in two fabrics, A and B, are 3 : 26 and 4 : 39. By changing each ratio to the form 1 : m, say which fabric has the greater proportion of elastine. Explain.

***Review 2** The ratio of votes cast to votes received for two candidates is given.
Who got the greater proportion of the whole vote?

	Votes cast : votes received
B.Thomas	236 : 183
M.Grace	143 : 119

Solving ratio and proportion problems

Worked Example

500 people were at a musical.
This pie chart shows the number who bought the CD of the musical.
How many did not buy the CD?

Bought CD
Didn't buy CD
288°

Answer

$\frac{288}{360}$ bought the CD.

$\frac{288}{360} \times 500 = 400$

400 bought the CD.

$500 - 400 = 100$

100 people did not buy the CD.

Remember: There are 360° in a circle.

Worked Example

In a photo a 1·8 m tall man is 3 cm tall.
In the same photo, a child is 2·1 cm tall.
How tall is the child in real life?

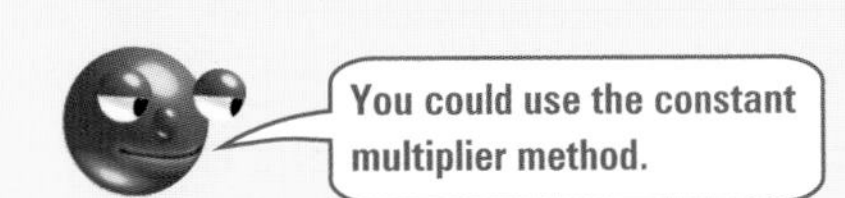

Answer

In real life the man is 1·8 m or 180 cm tall and in the photo he is 3 cm.

height in photo $\longrightarrow$ $\boxed{\times \frac{180}{3}}$ $\longrightarrow$ height in real life

2·1 cm $\longrightarrow$ $\boxed{\times \frac{180}{3}}$ $\longrightarrow$ 126 cm

The child is **1·26** m tall in real life.

Exercise 4

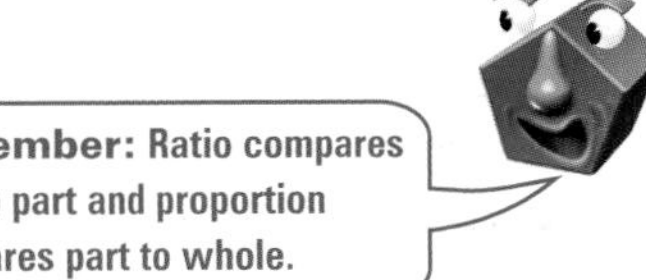
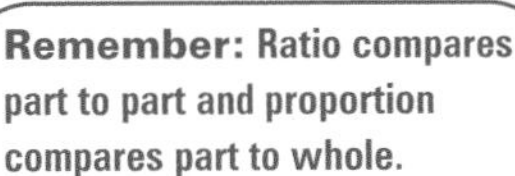

1 Adults normally have 32 teeth.

8 incisors 4 canines
8 bicuspids 12 molars

a Write these as ratios in their simplest form.

i incisors to molars **ii** bicuspids to canines  **iii** molars to all teeth

b What proportion of adult teeth are

i incisors **ii** canines **iii** molars?

Give each answer as a fraction, decimal and percentage.

2 When Mr Smythe died he left his farm to his two sons Jim and Steve. The farm was divided in the ratio 4 : 3. Steve got the smaller share, which was 420 hectares.
How big was Mr Smythe's farm?

3 Scott uses 12 onions to make $1\frac{1}{2}$ ℓ of chutney.

a How much chutney can be made with 48 onions?

b How many onions are needed to make $4\frac{1}{2}$ ℓ of chutney?

4 This is a recipe for 6 people.
How much of each ingredient would you need for 9 people?

Pancakes
3 cups flour
2 tsp baking powder
1 cup milk

5 Eight bars of chocolate cost £5·60.
Mick has £16. How many can he buy for this?

6 The ratio of the dimensions of the picture shown are 7 : 3.
Find the length, ℓ, of the picture.

35 cm

ℓ

7 Paula asked two groups of 100 people if they thought the age at which you can get a driver's licence should be lowered.

a How many of group A said No?

b How many of group B said No?

c One group were all aged under 18. Which group do you think this was?

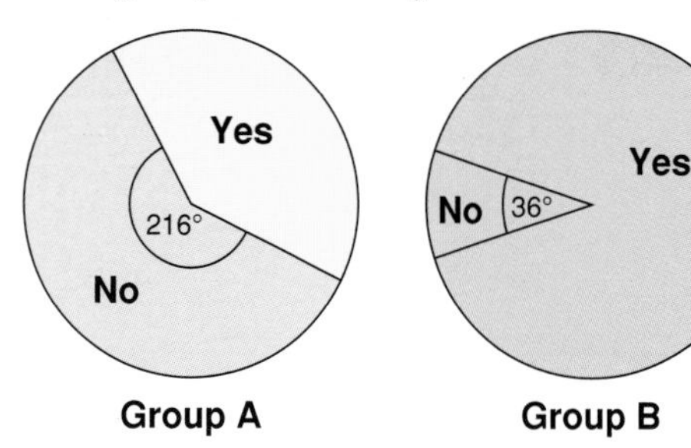

8 A 20 m high tree casts a 15 m shadow. At the same time a flagpole casts a 12 m shadow. How high is the flagpole?

9 Pie filling is made from meat, gravy, peas and carrots in the ratio 6 : 5 : 3 : 2. How much meat is needed to mix with 95 g of gravy?

10 Four people, Angela, Paula, Bryan and Tess, buy a national lottery ticket each week. They contribute money in the ratio 1 : 4 : 7 : 8. If they win they share the money in the same ratio as their contributions.

a One week they win £162 427. How much should each get?

b The next year they win another prize. Bryan gets £23 499·98.
How much do the others get?

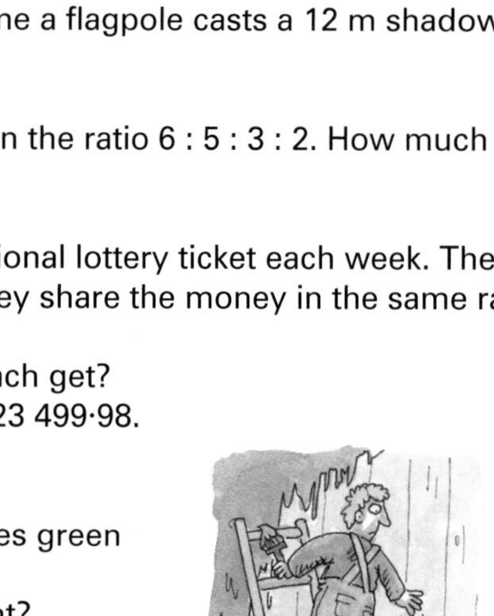

11 2 parts yellow paint mixed with 3 parts blue paint makes green paint.

a How much of each is needed to make a litre of paint?

b Henry has 100 mℓ of yellow paint and 200 mℓ of blue paint.
What is the maximum amount of green paint he can make?

c Lindy has $1\frac{1}{2}$ ℓ of yellow paint and 1125 mℓ of blue paint.
What is the maximum amount of green paint she can make?

12 Last week Emily earned £100.
This week Emily earned 25% more than she did last week.
What is the ratio of what she earned this week to what she earned last week?

13 A regular packet of mints weighs 750 g. A special pack weighs 30% more.
What is the ratio of the masses of a special pack to a regular pack?

14 Kate bought two cakes for a total of £8·75.
One cake cost two and a half times as much as the other.
How much did each cake cost?

15 A square and a rectangle have the same area.
The sides of the rectangle are in the ratio 4 : 1. Its perimeter is 200 cm.
What is the length of the side of the square?

*16 The ratio of mares to geldings in a cross country race is 4 to 3.
If two of the mares are withdrawn and replaced with two geldings there are the same number of each.
How many horses are in the cross country race?

*17 A and B are in the ratio 3 : 4. If A is reduced by 2 the ratio becomes 2 : 3.
Find A and B.

Review 1

a What is the ratio of cotton to polyester?
b What is the ratio of lycra to polyester to cotton?
c What proportion of the trousers are lycra?
d Give the proportion of cotton in the trousers as a percentage.

Trousers	
Material	**grams**
Cotton	20
Polyester	45
Lycra	15

Review 2 Garden fertiliser is made from manure, blood and bone, and loam in the ratio 5 : 3 : 2. How much of each is in a 15 kg bag?

Review 3 A recipe for pickle uses 85 g of cucumber for every 150 g jar of pickle. Jessica wants to make twenty 450 g jars of pickle for the school fair. How many kilograms of cucumber does she need?

Puzzle

If the front chain wheel (the wheel the chain goes around) on a bike has 40 teeth and the back cog has 10 teeth, the cog will go around four times faster than the chain wheel. The gear ratio is 4 : 1.

Front chain wheels		Rear cogs	
Big	48 teeth	1	28 teeth
Medium	38 teeth	2	24 teeth
Small	28 teeth	3	21 teeth
		4	18 teeth
		5	16 teeth
		6	14 teeth

This chart gives the number of teeth on the three front chain wheels and six rear cogs of an 18-speed bike.

1 Write down the 18 gear ratios possible.

2 One of the gear ratios is 28 : 21 or 4 : 3. This means for every 3 times the front chain wheel rotates, the back cog rotates 4 times.
Notice that the ratio of the number of turns is the opposite way round to the ratio of the number of teeth.
Work these out.
i How many turns does the big front chain wheel make when it is attached to cog 5 and cog 5 makes 288 turns?
ii How many turns does the rear cog 2 make if the small front chain wheel is attached to it and turns 1440 times?

*3 Each time the rear cog turns, the back wheel of the bike travels 210 cm. How many turns does the medium front chain wheel make if it is attached to cog 4 and the bike travels 4 km?

Summary of key points

+1 (1 tie costs £7·25) +7·25
+1 (2 ties costs £14·50) +7·25
3 ties costs £21·75

The number of ties and the cost of the ties are in **direct proportion**.

If 8 pizza slices cost £6·75 we can find the cost of 14 pizza slices using the **unitary method** or by finding the **constant multiplier**.

Unitary method

Cost of 1 slice of pizza $= \frac{£6·75}{8}$

Cost of 14 slices of pizza $= 14 \times \frac{£6·75}{8}$

= **£11·81 (nearest penny)**

Constant multiplier method

Ratio of no. of slices required to no. of slices given

= 14 : 8

So the constant multiplier is $\frac{14}{8}$.

Cost of 14 slices $= \frac{14}{8} \times £6·75$

= **£11·81 (nearest penny)**

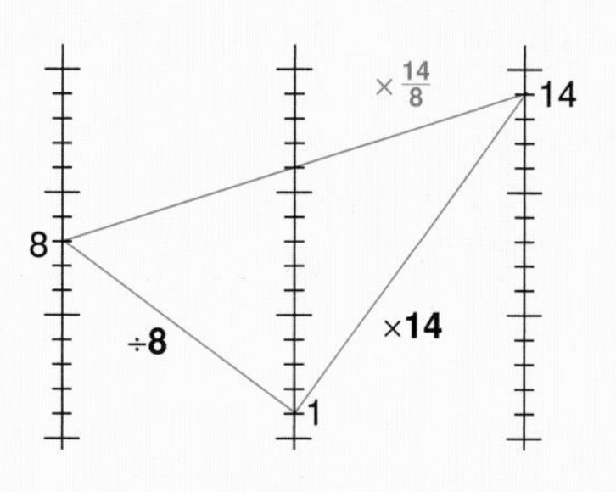

We **simplify a ratio** by cancelling.

All parts of a ratio must have the same units.

We usually multiply the parts of a ratio so they are all whole numbers.

Examples

8·5 : 3 (×2) → 17 : 6

1 mm : 5 m = 1 mm : 5000 mm
= 1 : 5000

To **compare ratios** we write each in the form 1 : m or m : 1

Example The ratios of butter to sugar in recipe A and recipe B are 2 : 3 and 7 : 11.
To compare these we write both in the form 1 : m.

Recipe A 2 : 3 (÷2) → 1 : 1·5

Recipe B 7 : 11 (÷7) → 1 : 1·6 (1d.p.)

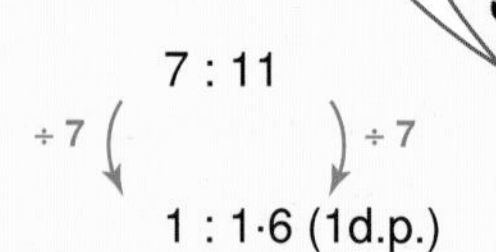

There is a greater proportion of butter in Recipe A because $\frac{1}{1·5} > \frac{1}{1·6}$.

We can **solve ratio and proportion problems**.

Test yourself **Only use a calculator if you need to.**

1 At a factory, a machine has two wheels, A and B.
Wheel A makes 5 turns for every 3 turns wheel B makes.
If wheel B makes 195 turns, how many turns does wheel A make? 

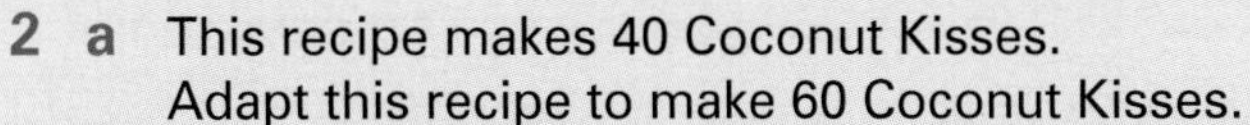
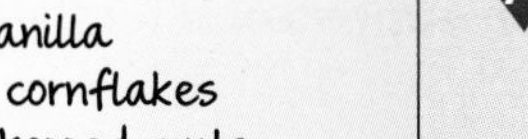

Coconut Kisses

4 egg whites
$\frac{1}{2}$ tsp vanilla
2 cups cornflakes
$\frac{1}{2}$ cup chopped nuts
1 cup sugar
$1\frac{1}{2}$ cups coconut

2 **a** This recipe makes 40 Coconut Kisses.
Adapt this recipe to make 60 Coconut Kisses. (A)
b Jeremy uses 5 cups of cornflakes for this recipe.
How many Coconut Kisses will this make?

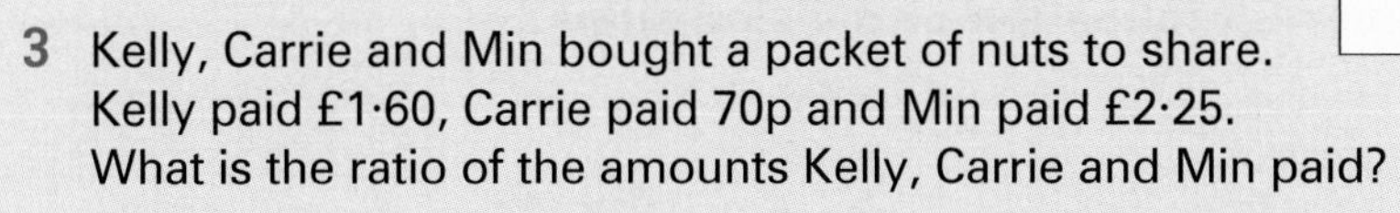

3 Kelly, Carrie and Min bought a packet of nuts to share.
Kelly paid £1·60, Carrie paid 70p and Min paid £2·25.
What is the ratio of the amounts Kelly, Carrie and Min paid? (B)

4 Write 55 cm : 1·4 m as a ratio in its simplest form. 

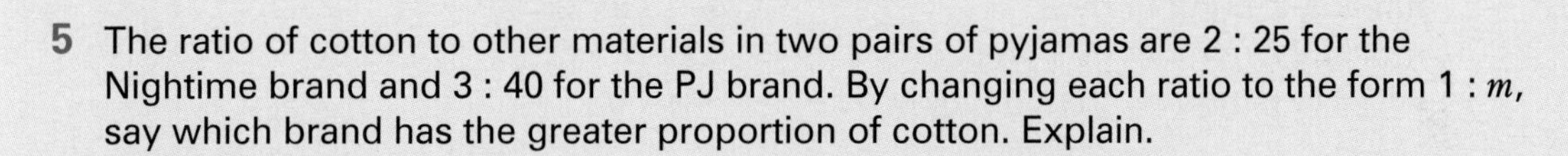

5 The ratio of cotton to other materials in two pairs of pyjamas are 2 : 25 for the Nightime brand and 3 : 40 for the PJ brand. By changing each ratio to the form 1 : m, say which brand has the greater proportion of cotton. Explain. 

6 Ginger ale is made from ginger beer and lemonade in the ratio 2 : 1.
How much ginger beer is needed to make 450 mℓ of ginger ale? 

7 This pie chart shows the proportion of 480 pupils at a school who are left-handed.
How many are left-handed? 

left-handed
126°
right-handed

8 A fabric is made from cotton, polyester, nylon and lycra in the ratio 9 : 4 : 2 : 1.
How much cotton is needed to blend with 64 g of polyester? (D)

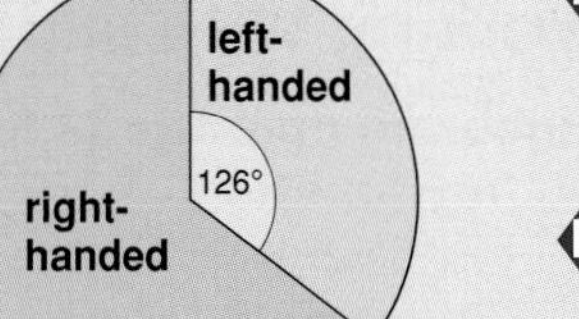

9 Fuel for a motor mower is mixed in the ratio shown on the label.
Hal has 1 litre of petrol and 30 mℓ of oil.
What is the maximum amount of motor mower fuel he can make? (D)

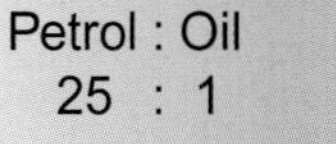

10 When she was eight, Ming was 1 m tall. By the time she was twelve, her height had increased 45%.
What was the ratio of her height at twelve to her height at eight? (D)

***11** Find the ratio of a to b if $a = 3x^2$ and $b = 6x$. 

***12** The ratio of attempts at goal to goals scored by the goal shooters in St John's and St Christopher's netball teams are 19 : 5 and 23 : 7 respectively. By changing each ratio to the form m : 1, decide which team has the better goal shooter. Explain. (C)

Algebra Support

Writing expressions

Theresa bought n pies for £2 each and a cake for £3.

An **expression** for the cost is is $\mathbf{2n + 3}$.

cost of n pies — cost of cake

We **write expressions without multiplication or division signs**.

Examples

add 6 to a number	$n + 6$
subtract 4 from a number	$n - 4$
subtract a number from 8	$8 - n$
divide a number by 5	$\frac{n}{5}$
multiply a number by 3	$n \times 3$ or $3n$
multiply a number by 2 then add 7	$2n + 7$
add 6 to a number then multiply by 4	$4(n + 6)$
$(a + b) \div c$	$\frac{a+b}{c}$
multiply n by itself	n^2

We write $n \times 3$ without a multiplication sign as $3n$.

Practice Questions 4, 6, 7

Working with expressions

Algebraic operations follow the same rules as arithmetic operations.

arithmetic	**algebra**
$4 + 5 = 5 + 4$	$a + b = b + a$
$3 + 6 + 9 = 6 + 9 + 3$	$p + q + r = q + r + p$
$5 \times 9 = 9 \times 5$	$xy = yx$
$5 \times 8 \times 3 = 8 \times 3 \times 5$	$cde = dec$

In arithmetic we can add in any order and multiply in any order. We can in algebra too.

Simplifying expressions

We can **simplify expressions by collecting like terms**.

Examples $a + a = \mathbf{2a}$ $3b + 2b = \mathbf{5b}$ $8y - 5y = \mathbf{3y}$

$4x + 3 - 2x - 4 = 4x - 2x + 3 - 4$ ← Write like terms next to each other.
$= \mathbf{2x - 1}$

We can simplify expressions by **cancelling**.

Examples $\frac{{}^1\cancel{x}}{{}^1\cancel{x}} = 1$ $\frac{{}^1\cancel{3}a}{{}^1\cancel{3}} = a$ $\frac{5\cancel{y}^1}{\cancel{y}^1} = 5$ $\frac{m^2}{m} = \frac{m \times {}^1\cancel{m}}{\cancel{m}^1}$
$= m$

Brackets

We use the rules of arithmetic to write an expression **without brackets**.

$8(10 + 6) = 8 \times 10 + 8 \times 6$

$7(x + 3) = 7 \times x + 7 \times 3$
$= \mathbf{7x + 21}$

	x	+3
7	$7x$	+21

$4(a - 2) = 4 \times a + 4 \times {}^-2$
$= \mathbf{4a - 8}$

	a	⁻2
4	$4a$	⁻8

Substituting into expressions

We can **evaluate an expression** by **substituting** values for the unknown. We follow the **order of operations** rules.

Example If $n = \mathbf{4}$ then

$3n = 3 \times \mathbf{4}$
$= 12$

$3n + 6 = 3 \times \mathbf{4} + 6$
$= 12 + 6$
$= 18$

$\frac{12}{n} = \frac{12}{\mathbf{4}}$
$= 3$

Practice Questions 3, 9, 12, 13, 19, 20, 26, 28, 31

Formulae

A **formula** is a rule for working something out.

Example Mr Gardener works out his charges using the formula

charge = call out fee + £5 × number of hours worked

If the call out fee is £35 and he works for 8 hours
then charge = £35 + £5 × 8
= £75

Example $\boldsymbol{P = 8T}$ is a formula for the pressure, P, in a container at temperature T.
If $T = \mathbf{15}$ then $P = 8 \times \mathbf{15}$
$= 120$.

Practice Questions 2, 33

Equations

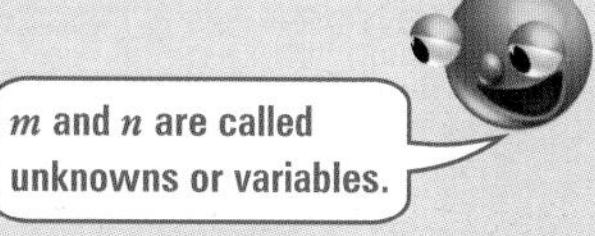

$n + 7 = 12$ is an **equation**. n has a particular value.
$m + n = 15$ is an equation. m and n can take any values as long as they add to 15.

We can **solve equations** with just one unknown **using inverse operations** or by **transforming both sides in the same way**.

Example **Using inverse operations**

$2y + 7 = 13$
$2y = 13 - 7$ The inverse of adding 7 is subtracting 7.
$2y = 6$
$y = \frac{6}{2}$ The inverse of multiplying by 2 is dividing by 2.
$\boldsymbol{y = 3}$

or

y → [multiply by 2] → $2y$ → [add 7] → $2y + 7$

3 ← [divide by 2] ← 6 ← [subtract 7] ← 13

$\boldsymbol{y = 3}$

Example **Transforming both sides**

$\frac{x}{5} + 4 = 7$
$\frac{x}{5} + 4 - \mathbf{4} = 7 - \mathbf{4}$ subtract 4 from both sides
$\frac{x}{5} = 3$
$\frac{x}{5} \times \mathbf{5} = 3 \times \mathbf{5}$ multiply both sides by 5
$\boldsymbol{x = 15}$

Practice Questions 14, 15, 17, 27, 32, 35

Sequences

A **sequence** is a set of numbers in a given order.
Each number is called a **term**.

Examples 3, 8, 13, 18, 23, ... ← **These dots mean the sequence continues forever. It is infinite.**

1st term (3) **5th term** (23)

8, 16, 24, 32, 40, ... 80 ← **This sequence is finite. It starts at 8 and ends at 80.**

We can **write a sequence** by **counting on** or **counting back**.

Examples Starting at 5 and counting on in steps of 4 we get 5, 9, 13, 17, 21, ...

Starting at 6 and counting back in steps of 3 we get 6, 3, 0, $^{-}3$, $^{-}6$, ...

Starting at 4 and counting on in steps of 1, 2, 3, 4, ... we get 4, 5, 7, 10, 14, ...

We can **write a sequence** if we know the first term and the **rule for finding the next term**.

Example **1st term** 3 **rule** add 10 gives 3, 13, 23, 33, 43, ...

We can write a sequence if we know the rule for the ***n*th term**.

Example If the rule for the nth term is $2n + 1$ the sequence is 3, 5, 7, 9, 11, ...

We find the terms of the sequence by substituting $n = 1$, $n = 2$, $n = 3$ and so on into $2n + 1$.

Sequences in practical situations

We can find a rule for the nth shape in a sequence of diagrams.

Example Fancy Fences makes these fences.

1 panel
4 pieces

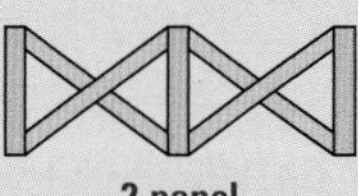

2 panel
7 pieces

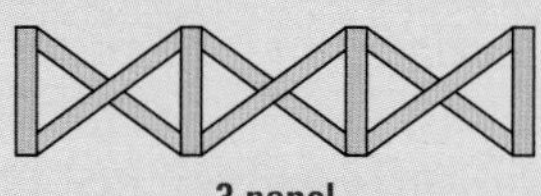

3 panel
10 pieces

The sequence for the number of pieces of wood is 4, 7, 10, ...
Each time a new panel is added, 3 more pieces of wood are needed. The first panel needs one more piece of wood than the added panels.
In an n panel fence there will be $3 \times n$ pieces of wood plus one extra for the first panel.
The rule for the nth panel is $\mathbf{3n + 1}$.

Finding the rule for the *n*th term using a difference table.

Example For the sequence 2, 5, 8, 11, 14, ...

Term number	1	2	3	4	5
Term	2	5	8	11	14
Difference	3	3	3	3	

The rule will be of the form $T(n) = 3n + ?$

We can see from the terms that $T(n) = 3n - 1$ because

$T(1) = 3 \times 1 - 1 = 2$ $T(2) = 3 \times 2 - 1 = 5$

Practice Questions 1, 5, 10, 18, 21, 24, 25, 30

Functions

$x \rightarrow$ [multiply by 2] $\rightarrow$ [add 4] $\rightarrow y$ This is a **function machine**.

The rule for the function machine is written as $y = 2x + 4$ or $x \rightarrow 2x + 4$.

If we are given the input we can find the output.

Example 2, 7, 3, 4 $\rightarrow$ [add 2] $\rightarrow$ [multiply by 3] $\rightarrow$ 12, 27, 15, 18

$(2 + 2) \times 3 = 12$
$(7 + 2) \times 3 = 27$
$(3 + 2) \times 3 = 15$
$(4 + 2) \times 3 = 18$

We can show the input and output of a function on a **mapping diagram**.

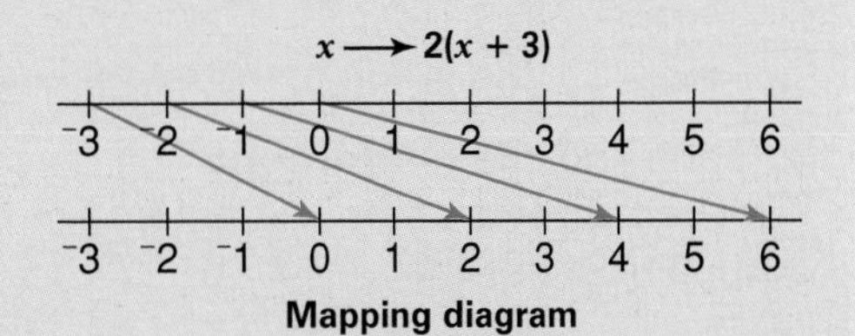

Mapping diagram

If we are given the input and output we can **find the rule**.

Example 5, 3, 0, 2 $\rightarrow$ [?] $\rightarrow$ 20, 12, 0, 8

The rule for this function machine is 'multiply by 4'.

Practice Questions 11, 29, 33, 35

Graphs

(5, 6) is a **coordinate pair**. (x-coordinate, y-coordinate)

Each of the **coordinate pairs** $(^{-}1, ^{-}2)$, $(0, ^{-}1)$, $(1, 0)$, $(2, 1)$, $(3, 2)$ satisfies the rule $y = x - 1$.

Each y-coordinate is found by subtracting 1 from the x-coordinate.

To draw a **straight-line graph** of $y = x + 2$:

1. Find 3 or more coordinate pairs that satisfy the equation $y = x + 2$.

Substitute some x-values into $y = x + 2$.

x	$^{-}4$	$^{-}2$	0	2
y				

Examples When $x = {^{-}4}, y = {^{-}4} + 2 = {^{-}2}$; $x = {^{-}2}, y = {^{-}2} + 2 = 0$; $x = 0, y = 0 + 2 = 2$; $x = 2, y = 2 + 2 = 4$

Now the table can be filled in.

x	$^{-}4$	$^{-}2$	0	2
y	$^{-}2$	0	2	4

2. Plot the coordinate pairs on a grid.
Draw a straight line through the points.
Label your line.

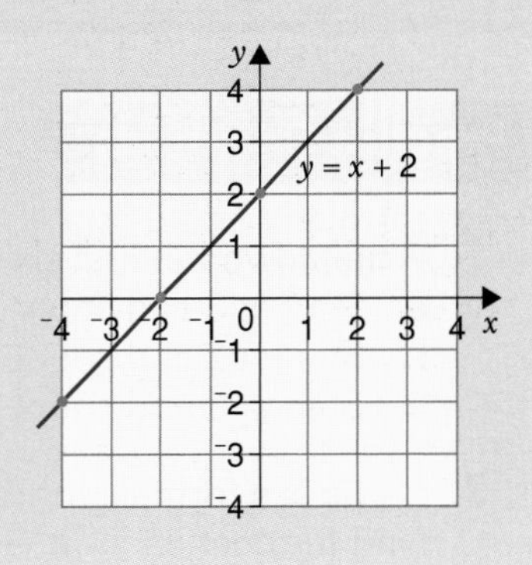

The **equation of the line** shown is $y = x + 2$.
The coordinate pairs of **all** points on the line satisfy the equation of the line.

$y = mx$ is the equation of a straight line through the origin.
m represents the **gradient** or slope and can have any value.
The greater the value of m, the steeper the slope.

If m is positive, the gradient is positive.
If m is negative, the gradient is negative.

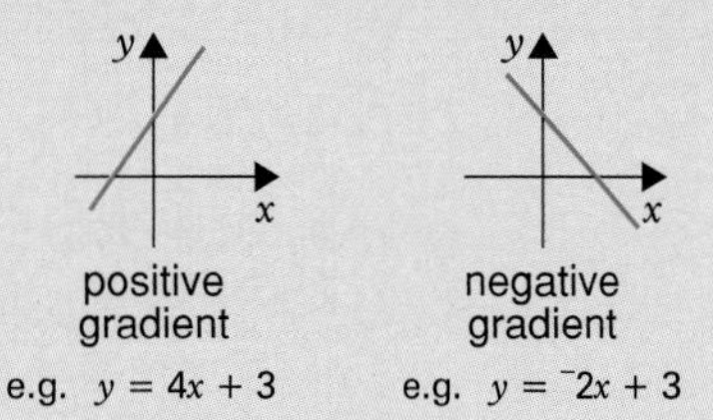

e.g. $y = 4x + 3$ e.g. $y = {}^{-}2x + 3$

The y-intercept is where the graph crosses the y-axis.

$y = mx + c$ is the graph of a straight line.

c represents the **y-intercept**.
The coordinates of the y-intercept are $(0, c)$ where c is given by the equation of the line.

Example $y = \mathbf{3}x - \mathbf{2}$ has a gradient of **3** and crosses the y-axis at $(0, {}^{-}\mathbf{2})$

Lines **parallel to the x-axis** all have equation $y = a$.
a is any number.

Examples $y = {}^{-}3$ $y = 2\frac{1}{2}$

Lines **parallel to the y-axis** all have equation $x = b$.
b is any number.

Examples $x = 5$ $x = {}^{-}2$.

Practice Questions 8, 16, 22

Graphs of real-life situations

We often draw straight-line graphs to represent **real-life situations**. We can often estimate other values from these graphs.

Example This graph shows the charges for Piper's Plumbing.

The points (0, £40), (10, £70) and (20, £100) have been plotted and joined with a straight line.

We can estimate other values from the graph.

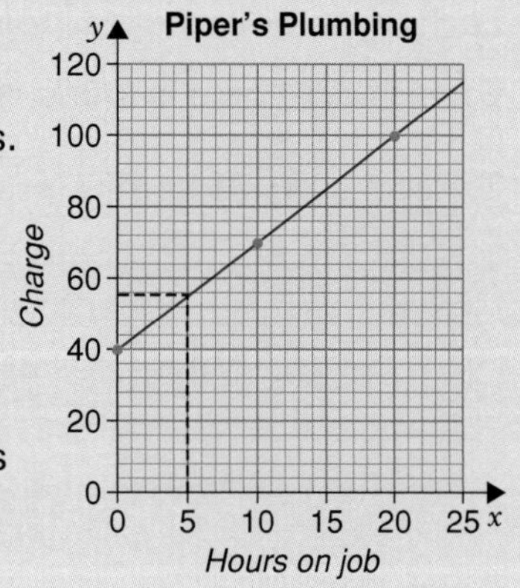

Example If Piper's Plumbing spend 5 hours on a job the charge is about £55.

Practice Questions 23, 36

Practice Questions

Except for questions 30, 32, 34.

1 Write down the first six terms of these sequences.
  a Start at 3 and count on in steps of 1·5.
  b Start at 2 and count back in steps of 3.
  c Start at 3 and count on in steps of 2, 4, 6, 8, ...

2 $C = 4n$ is a formula for the cost, in pence, of n jelly jubes.
Find the cost of **a** 5 jelly jubes **b** 20 jelly jubes **c** 35 jelly jubes.

3 For each of these, copy and finish the equivalent statement for algebra.

	Arithmetic	Algebra
a	$3 + 5 = 5 + 3$	$a + b =$
b	$7 \times 6 = 6 \times 7$	$ab =$
c	$7 \times 9 \times 2 = 9 \times 2 \times 7$	$pqr =$
d	$3 + (6 + 4) = (3 + 6) + 4$	$x + (y + z) =$

Write your expressions without a multiplication sign.

4 Write an expression for these. Let the unknown number be n.

a subtract 5 from a number
b add 6 to a number
c multiply a number by 7
d divide a number by 4
e multiply a number by 4 then add 3
f multiply a number by 5 then subtract 2
g add 4 to a number then multiply by 6
h subtract 3 from a number then multiply by 2
i multiply a number by itself
j divide a number by 4 and then add 3
*__k__ multiply a number by 3 and multiply this expression by itself
*__l__ add 5 to a number then divide the result by 3

5 Write down the first five terms of these. The first term and the rule for finding the next term is given.

a **1st term** 2, **rule** add 0·5
b **1st term** $^-1$, **rule** subtract 2
c **1st term** 2000, **rule** divide by 10
d **1st term** 1·5, **rule** multiply by 2

6 Write these without a multiplication or division sign.

a $3 \times n$ **b** $y \times 4$ **c** $4 \times (x + 2)$ **d** $(t - 4) \times 3$ **e** $a \times a$
f $(a + x) \div b$ **g** $x \times x \times 3$ **h** $5 \times a \times a$ **i** $4m \times 4m$ **j** $3 \times 6n$

7 A teacher has **5 full packets** of mints and **6 single** mints. **[SATs 2001 paper 2]**
The number of mints inside each packet is the same.

The teacher tells the class:

'**Write an expression** to show **how many mints** there are **altogether**.
Call the number of mints inside each packet y'

Here are some of the expressions that the pupils write:

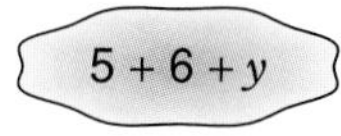

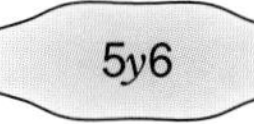

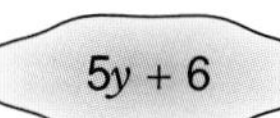

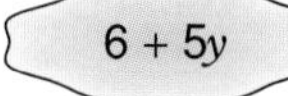

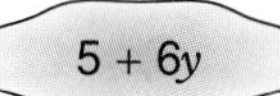

$(5 + 6) \times y$

a Write down **two** expressions that are correct.
b A pupil says: 'I think the teacher has a total of **56 mints**'.
Could the pupil be correct? Explain how you know.

T

8 Use a copy of this.

x	$^-3$	0	1
y			

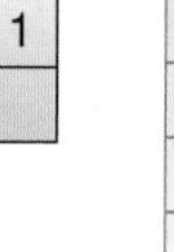

a Complete the table for $y = x + 2$.
b Write down the coordinate pairs for the points.
c On the grid, plot these three points.
Draw and label the line with equation $y = x + 2$.
d Does the point ($^-4$, $^-2$) lie on the line?
e Write down the coordinates of two other points that lie on the line $y = x + 2$.

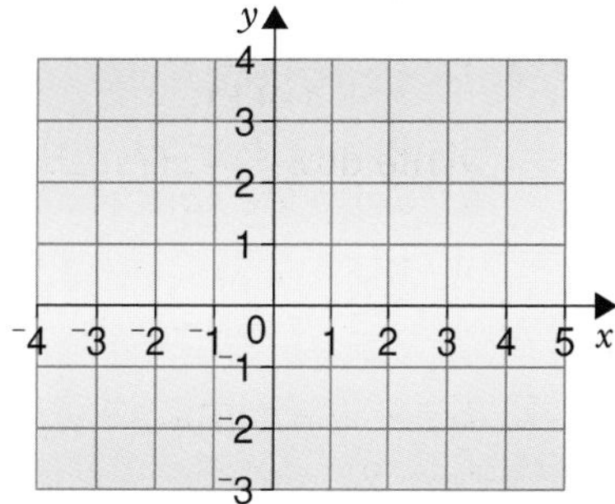

9 Simplify these.

a $3a + 4a$ **b** $7x + 3x$ **c** $8y - y$ **d** $3b + 2b + 4b$

e $5n + 3n - 2n$ **f** $4x + 3 + 7x - 2$ **g** $4a + 3b - 2a - 2b$ **h** $8y + 7 - 3y - 4$

10 Match the descriptions with the sequences given in the box.

a the even numbers from 4 to 14

b the sequence begins at 3 and increases in steps of 4

c each term is one more than a multiple of 5

d the sequence begins at 7 and decreases in steps of 1, 2, 3, 4, ...

A 4, 6, 8, 10, 12, 14
B 6, 11, 16, 21, ...
C 7, 6, 4, 1, $^{-}3$, ...
D 7, 10, 13, 16, ...
E 3, 7, 11, 15, 19, ...
F 3, 12, 48, 192, ...

11 Find the output for these.

a 3, 6, 2, 4 → [add 7] → __, __, __, __

b 7, 0, $\frac{1}{2}$, 2·5 → [multiply by 2] → [add 1] → __, __, __, __

12 Simplify these expressions by cancelling.

a $\frac{m}{m}$ **b** $\frac{y}{y}$ **c** $\frac{2n}{2}$ **d** $\frac{5x}{5}$ **e** $\frac{3a}{a}$ **f** $\frac{6p}{p}$

g $\frac{8m}{2}$ **h** $\frac{12x}{6}$ **i** $\frac{16a}{4}$ **j** $\frac{28a}{21}$ **k** $\frac{3b^2}{b}$

T

13 a Use a copy of this. [SATs 2001 paper 2]

Join pairs of algebraic expressions that have the **same value** when a = **3**, b = **2** and c = **6**.

One pair is joined for you.

***b** Repeat **a** when $a = b = c$

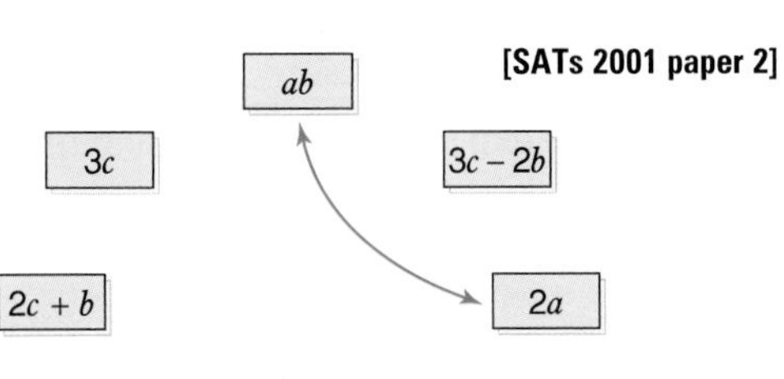

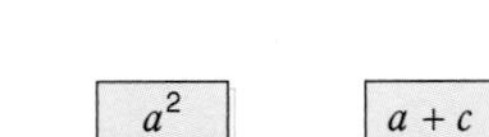

14 If $y = x + 6$ write down three other facts that must be true.

15 Look at this table. [SATs 2002 paper 1]

Write in words the meaning of each equation below.

The first one is done for you.

	Age (in years)
Ann	a
Ben	b
Cindy	c

a $b = 30$ **Ben is 30 years old**

b $a + b = 69$

c $b = 2c$

d $\frac{a+b+c}{3} = 28$

16 a $y =$ __x. What number could go in the gap to give the graph of a line which is steeper than $y = x$?

b Which of these graphs will have the steeper slope? $y = 5x + 4$, $y = 2x + 7$

c Which of these graphs will intersect the y-axis at (0, $^{-}2$)? $y = {}^{-}2x + 2$, $y = 2x - 2$

17 a A yard is 15 m long. A fence 50 m long is needed to go right around the perimeter of the yard. If w is the width of the yard, which of these is correct?

A $30 + 2w = 50$ **B** $15 + w = 50$ **C** $50 + 2w = 30$ **D** $w = 50 - 30$

b Solve the correct equation to find the width of the yard.

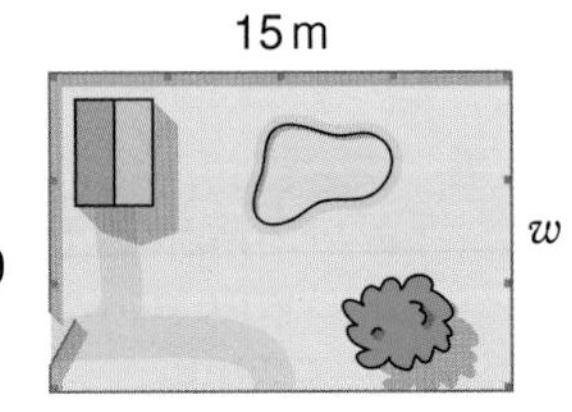

18 You can make 'huts' with matches. **[SATS 2000 paper 2]**

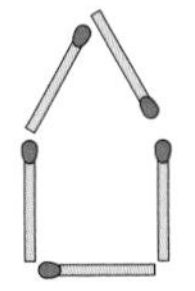
1 hut needs
5 matches

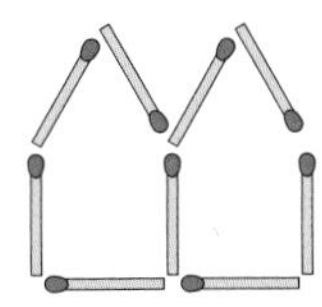
2 huts need
9 matches

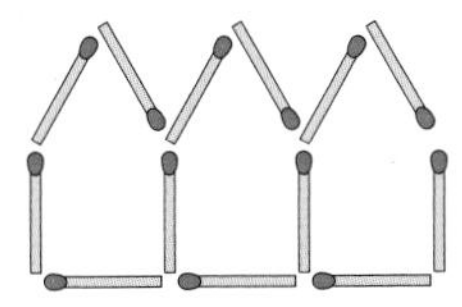
3 huts need
13 matches

A rule to find how many matches you need is

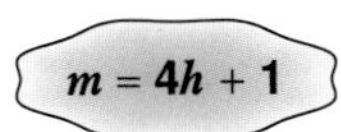

m stands for the number of matches.
h stands for the number of huts.

a **Use the rule** to find how many matches you need to make **8** huts.
Show your working.

b I use **81 matches** to make some huts.
How many huts do I make?
Show your working.

c Andy makes different 'huts' with matches.

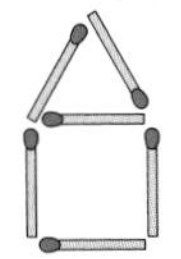
1 hut needs
6 matches

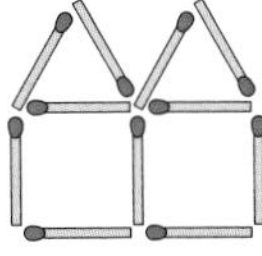
2 huts need
11 matches

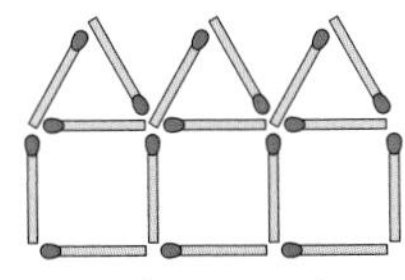
3 huts need
16 matches

Which rule below shows how many matches he needs?
Remember: m stands for the number of matches.
h stands for the number of huts.

$m = h + 5$ $m = 4h + 2$ $m = 4h + 3$

$m = 5h + 1$ $m = 5h + 2$ $m = h + 13$

19 A teacher has a large pile of cards. **[SATs 2002 paper 2]**
An expression for the **total** number of cards is $\mathbf{6n + 8}$

a The teacher puts the cards in two piles.
The number of cards in the first pile is $\mathbf{2n + 3}$.

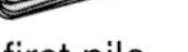
first pile

second pile

Write an expression to show the number of cards in the second pile.

b The teacher puts all the cards together.
Then he uses them to make **two equal piles**.

Write an expression to show the number of cards in one of the piles.

c The teacher puts all the cards together again, then he uses them to make two piles. There are **23** cards in the first pile.

How many cards are in the second pile? Show your working.

[SATs 2000 paper 1]

20 Write each expression in its simplest form.
a $7 + 2t + 3t$ **b** $b + 7 + 2b + 10$

21 Jackie made these patterns with strips of felt.
a Copy and fill in this table.
b Describe the sequence and how it continues.
c Explain how you could find the number of strips needed for the nth tree.

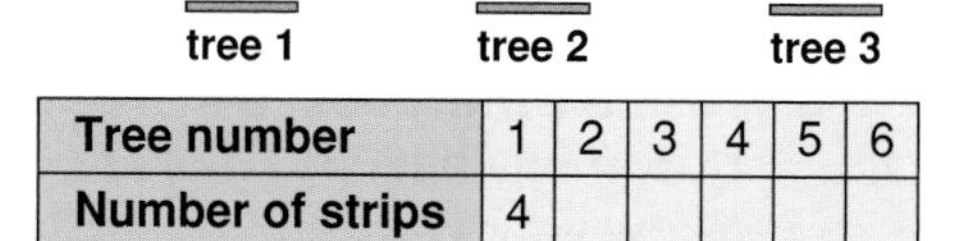

Tree number	1	2	3	4	5	6
Number of strips	4					

22 Which of these are the equations for a line parallel to the x-axis?
a $y = 3$ **b** $x = {}^{-}6$ **c** $y = {}^{-}2$ **d** $y = x$

T

23 The cost to hire a hall with supper provided is £150 + £20 per hour.
a Copy and complete this table.

Hours	1	4	8	10
Cost (£)				

b Use a copy of this grid.
Plot the points from the table on the grid.
Draw a straight line through the points.
c Explain what the symbol ⚡ on the vertical axis means.

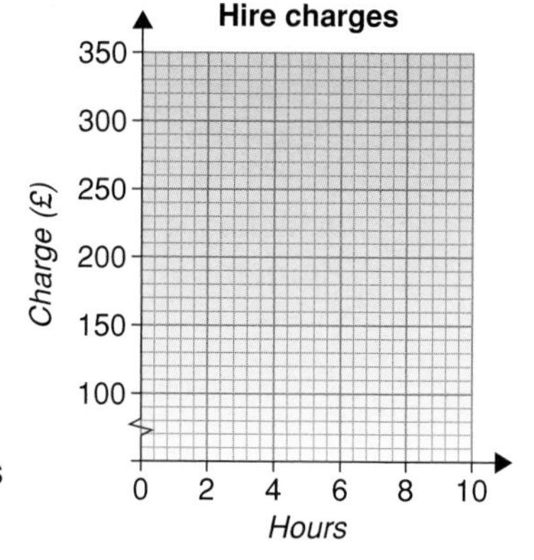

Show how to use your graph to answer these.
d How much does it cost to hire the hall for 9 hours?
e Sue hired the hall and it cost her £270. For how many hours did she hire it?

24 The nth term of a sequence is given. Find the first five terms.
a $n + 3$ **b** $n - 1$ **c** $2n + 1$ **d** $10 - n$ **e** $20 - 3n$ **f** $0{\cdot}6n$

25 Write down the 20th term of each of the sequences in question **24**.

26 If $x = 3$ and $y = 5$ find the value of these expressions.
a xy **b** $2x - 3$ **c** $4(x + y)$ **d** $\frac{x+2}{y}$ **e** $\frac{3y-3}{2x}$ **f** $3x^2 + 2y$

27 Write an equation for each of these. Use n for the unknown.
a I think of a number. I add 4. The answer is 10.
b I think of a number. I multiply by 3. The answer is 7.
c Three times the mass of Bessy, the cat, is 15 kg.
d When the length of an animal pen was doubled and then 3 metres added, the final length was 26 m.

28 Multiply out these brackets.
a $5(x + 4)$ **b** $7(n - 4)$ **c** $7(2a + 3)$ **d** $8(2y - 4)$ **e** $5(3y - 2x)$

29 **a** Copy and fill in this table for the function machine.

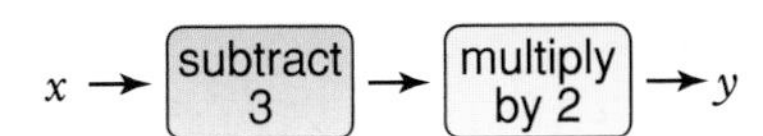

x	3	5	10	12
y				

b Write the rule for the function machine as $y =$ _____ and as $x \longrightarrow$ _____.

c Copy and fill in this mapping diagram for the function, for x-values from $^{-}1$ to 6.

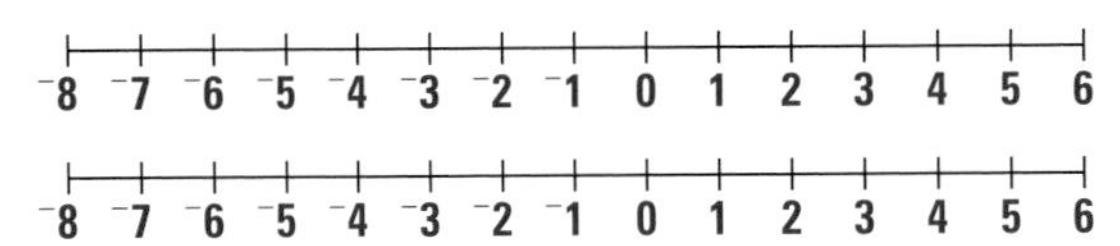

You could use a graphical calculator.

30 **a** The rule for finding the next term of a sequence is 'add 7'.
Write down the first five terms of two sequences that have this rule.

b Is it possible to find a sequence with this rule for which all terms are even?

31 **a** When $x = 5$, work out the values of the expressions below. **[SATs 2002 paper 1]**

$2x + 13$ $\quad$ $5x - 5$ $\quad$ $3 + 6x$

b When $2y + 11 = 17$ work out the value of y. Show your working.

32 Solve these equations.

a $a + 4 = 11$ **b** $y - 4 = 17$ **c** $5p = 45$ **d** $3n = 81$ **e** $\frac{m}{8} = 5$

f $\frac{c}{5} = 12$ **g** $2x + 3 = 5$ **h** $4m - 2 = 6$ **i** $5a - 3 = 47$

33 You can work out the cost of an advert in a newspaper by using this formula: **[SATs 2001 paper 2]**

C = 15*n* + 75	**C** is the cost in pounds ***n*** is the number of words in the advert

a An advert has **18 words**.
Work out the cost of the advert.
Show your working.

b The cost of an advert is **£615**.
How many words are in the advert?
Show your working.

34 Write down the function for each of these.

a 5, 3, 7, 2 → [?] → 12, 10, 14, 9

b 3, 8, $^{-}2$, 4 → [?] → [add 3] → 9, 19, $^{-}1$, 11

***c** 2, 1, 3, 4, 5 → [?] → [?] → 2, $^{-}1$, 5, 8, 11

35 **a** Solve this equation.
$7 + 5k = 8k + 1$

b Solve these equations. Show your working. **[SATs 2001 paper 1]**

i $10y + 23 = 4y + 26$

ii $\frac{3(2y + 4)}{14} = 1$

T

36 Use a copy of this. Helen took her toddler to the shops to buy an ice cream. They walked 800 m to the shop in 10 minutes, then stopped and talked to the shopkeeper for 5 minutes. They walked 400 m towards home in 5 minutes, then stopped at the park for 10 minutes, then walked home in 5 minutes. Finish drawing the graph.

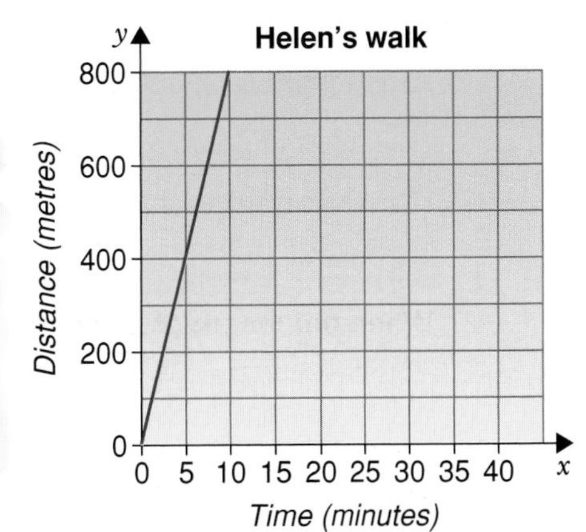

8 Expressions, Formulae and Equations

You need to know

Key vocabulary

algebraic expression, brackets, collect like terms, evaluate, expression, formula, formulae, function, inequality, linear equation, linear expression, multiply out, prove, solution, substitute, symbol, therefore (∴), transform, variable, verify

Stand ins

1 This square is on the door of a vault.

The letters A, B, C, D and E stand for whole numbers.

The red numbers show the sum of each row and column.

$E = 3$

To unlock the vault you must know what numbers the other letters stand for. What do they stand for?

2 This square is on another vault door.

The letters P, Q, R, S and T stand for whole numbers.

The red numbers show the sum of each row and column.

$T = 4$

What number do the other letters stand for?

Understanding algebra

Remember

$x + 8$, $2y - 7$, $\frac{3b + 2}{5}$, $9(p - q)$ and $5a + 7b$ are **linear expressions**.
An expression is made by combining letters and numbers with operations, +, −, ×, ÷, ( ).

$x + 8 = 7$, $2y - 7 = 3$, $9(p - q) = 3p$ and $5a + 7b = 4$ are all **linear equations**.

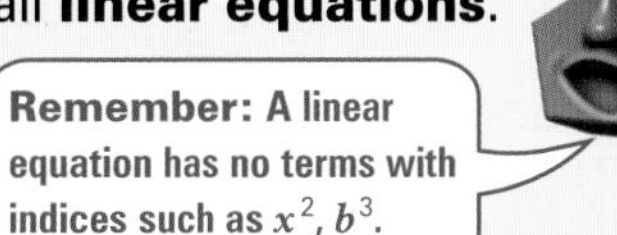

$t = 20k + 40$ is a **formula**. t is the time in minutes it takes to cook a chicken of mass k kilograms.

A formula is a rule which relates two or more variables.
The variables in a formula always stand for something specific.

Discussion

- What are the differences and similarities between expressions, equations and formulae? **Discuss**.
- In this expression $3x + 4$, x can have any value.
  Can x have any value in the equation $3x + 4 = 7$? **Discuss**.

Functions

Remember

$x \rightarrow$ [multiply by 2] $\rightarrow$ [subtract 5] $\rightarrow y$

A function is a special sort of equation.

We can write the rule for this function machine as $y = 2x - 5$.

$y = 2x - 5$ is called a **function**.

For **any** value we choose for x, we can calculate the value of y.

A function gives the relationship between the variables on the x- and y-axes of a graph.

Exercise 1

1 In each of the following, there is *one* equation, *one* expression, *one* formula and *one* function.
Which is which?

a $y = 2x - 4$, $a + b = 7$, $3x + 7$, $C = 4b - T$, where C is the cost, b is the number of bikes and T is the discount.

b $F = ma$, where F is the force, m the mass and a the acceleration,
$5x + 3 = 18$, $\frac{7n}{p}$, $y = \frac{1}{2}x - 3$

c $\frac{9y - 2}{x}$, $v = u^2 + 2as$ where v is the final velocity, u is the initial velocity, a is the acceleration and s is the distance, $y = \frac{x}{7} - 3$, $3a + 14 = a - 7$

d $\frac{5a + 21}{9} = 17$, $P = VI$ where P is the power, V is the voltage and I is the current,
$\frac{7n - 8}{2m}$, $y = \frac{3x - 2}{4}$

2 $y = 2x - 7$ is the equation of a straight line.
Is $y = 2x - 7$ a function?

***3** Write your own example of
a an expression **b** an equation **c** a formula **d** a function.

Review In each of the following, there is *one* equation, *one* expression, *one* formula and *one* function.
Which is which?

a $y = 3x + 7$, $2x + 4 = 7$, $5n - 3$, $v = u + at$ where v is the final velocity, u is the initial velocity, a is the acceleration and t is the time

b $P = nRT$ where P is the pressure, n and R are constant values and T is the temperature, $2p + n$, $2p + 7 = {}^-3$, $y = \frac{x+3}{7}$

More understanding algebra

An **'equals' sign** means 'is equal to'.
The left-hand side of the equals has exactly the same numerical value as the right-hand side.

Discussion

- Sanjay wrote this.

$$28 + 46 = 28 + 40$$
$$= 68 + 6$$
$$= 74$$

What is wrong with what Sanjay wrote? **Discuss.**

- 

Does 'equals' mean the same as 'makes'? **Discuss.**

- If $x + y = w + z$, what does the equals mean?
Would $w + z = x + y$? **Discuss.**

- Liam wrote these in his algebra test. What is wrong with each? **Discuss.**

a $a - b = 83 - 27$
$= 83 - 30$
$= 53 + 3$
$= 56$

b If $y = 4$, $6y - 3 = 6 \times 4 - 3$
$= 6 \times 1$
$= 6$

c $c + d + e = dec$

Inequalities

$n > {}^{-}5$ means 'n is greater than ${}^{-}5$'
$n \geqslant {}^{-}5$ means 'n is greater than or equal to ${}^{-}5$'
$n < 3$ means 'n is less than 3'
$n \leqslant 3$ means 'n is less than or equal to 3'

These are called 'inequalities'.

${}^{-}4 < n < 7$ means 'n is between ${}^{-}4$ and 7'
or 'n is greater than ${}^{-}4$ but less than 7'
${}^{-}4 \leqslant n \leqslant 7$ means 'n is greater than or equal to ${}^{-}4$ but less than or equal to 7'
${}^{-}4 \leqslant n < 7$ means 'n is greater than or equal to ${}^{-}4$ but less than 7'
${}^{-}4 < n \leqslant 7$ means 'n is greater than ${}^{-}4$ but less than or equal to 7'

Exercise 2

1 Decide which inequality best describes the statement.

a More than 250 mm of rain fell yesterday.
A $r < 250$ **B** $r \leqslant 250$ **C** $r > 250$ **D** $r \geqslant 250$

b The speed limit through a village is 60 km/h.
A $s < 60$ **B** $s \leqslant 60$ **C** $s > 60$ **D** $s \geqslant 60$

c T-bone steaks weigh between 100 g and 200 g.
A $100 < w < 200$ **B** $100 \leqslant w < 200$ **C** $100 < w \leqslant 200$ **D** $100 \leqslant w \leqslant 200$

d Hans never arrives at school earlier than 8:30 a.m.
A $t \leqslant 8{:}30$ **B** $t < 8{:}30$ **C** $t \geqslant 8{:}30$ **D** $t > 8{:}30$

e The typing speeds of the students in a class were all greater than 35 words per minute.
A $t < 35$ **B** $t \leqslant 35$ **C** $t > 35$ **D** $t \geqslant 35$

f In a test, every student gained at least 70%.
A $m < 70\%$ **B** $m \leqslant 70\%$ **C** $m > 70\%$ **D** $m \geqslant 70\%$

g Shane takes from 3 to 4 minutes to iron a shirt.
A $3 < t < 4$ **B** $3 \leqslant t < 4$ **C** $3 < t \leqslant 4$ **D** $3 \leqslant t \leqslant 4$

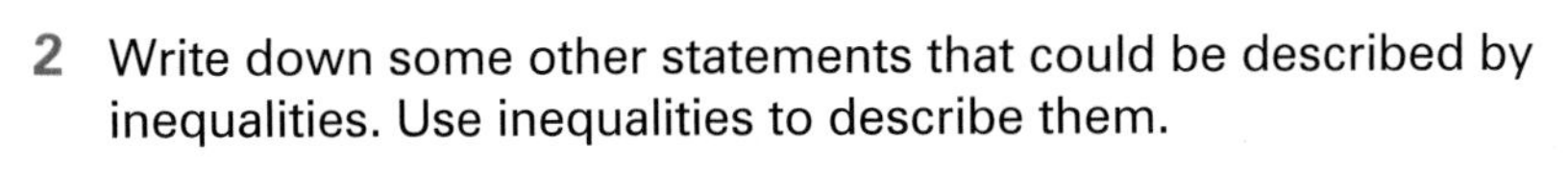

2 Write down some other statements that could be described by inequalities. Use inequalities to describe them.

Review Decide which inequality best describes the statement.

a 18 is the smallest number, and 27 is the greatest number of students in the classes in a school.
A $18 < s < 27$ **B** $18 \leqslant s < 27$ **C** $18 < s \leqslant 27$ **D** $18 \leqslant s \leqslant 27$

b Joanne never arrives at school later than 8:30 a.m.
A $t \leqslant 8{:}30$ **B** $t < 8{:}30$ **C** $t \geqslant 8{:}30$ **D** $t > 8{:}30$

Inverse operations

In arithmetic, using the **commutative rule** and **inverse operations**, we can make a 'family of facts'. We can do the same in algebra.

If	$6 + 8 = 14$	If	$a + b = c$	If	$3 \times 4 = 12$	If	$xy = z$
then	$8 + 6 = 14$	then	$b + a = c$	then	$4 \times 3 = 12$	then	$yx = z$
	$14 - 8 = 6$		$c - a = b$		$12 \div 4 = 3$		$x = \frac{z}{y}$
	$14 - 6 = 8$		$c - b = a$		$12 \div 3 = 4$		$y = \frac{z}{x}$

Remember
Algebraic operations follow the same rules as arithmetic operations.

In arithmetic if $22 = 3 \times \mathbf{6} + 4$ $\quad$ **6** → multiply by 3 →(18)→ add 4 → 22

then $\quad \mathbf{6} = \frac{22-4}{3}$ $\quad$ $\frac{22-4}{3}$ ← divide by 3 ←(18)← subtract 4 ← 22

The order of operations is **BIDMAS**:
Brackets
Indices
Division and **M**ultiplication
Addition and **S**ubtraction.

In algebra if $m = 3n + 4$ $\quad$ n → multiply by 3 →($3n$)→ add 4 → $3n + 4$

then $\quad n = \frac{m-4}{3}$ $\quad$ $\frac{m-4}{3}$ ← divide by 3 ←($m - 4$)← subtract 4 ← m

To find the inverse, work backwards doing inverse operations.

Examples $\quad$ If $\frac{p}{3} - 2 = q$ $\quad$ p → divide by 3 →($\frac{p}{3}$)→ subtract 2 → $\frac{p}{3} - 2$

then $p = 3(q + 2)$ $\quad$ $3(q+2)$ ← multiply by 3 ←($q+2$)← add 2 ← q

If $5(a - 3) = b$ $\quad$ a → subtract 3 →($a - 3$)→ multiply by 5 → $5(a-3)$

then $a = \frac{b}{5} + 3$ $\quad$ $\frac{b}{5} + 3$ ← add 3 ←($\frac{b}{5}$)← divide by 5 ← b

Exercise 3

1 Find two matching equations from the box for each of these.
You could check by choosing numbers to substitute for the unknowns.

a $l + m = 7$
b $3a = b$
c $\frac{x}{3} = y$
d $5a = b$
e $pq = 6$

$7 + m = l$	$lm = 7$	$m + l = 7$	$7 + l = m$	$7 - m = l$
$ab = 3$	$3b = a$	$\frac{b}{3} = a$	$\frac{b}{a} = 3$	$ab = \frac{1}{3}$
$3y = x$	$xy = 3$	$x + y = 3$	$y - 3 = x$	$3 = \frac{x}{y}$
$5ab = 1$	$\frac{b}{a} = 5$	$b = 5 \times a$	$5 + a = b$	$ab = 5$
$p + q = 6$	$\frac{6}{p} = q$	$6 - q = p$	$6p = q$	$6 = qp$

There is more about substituting on page 166.

2

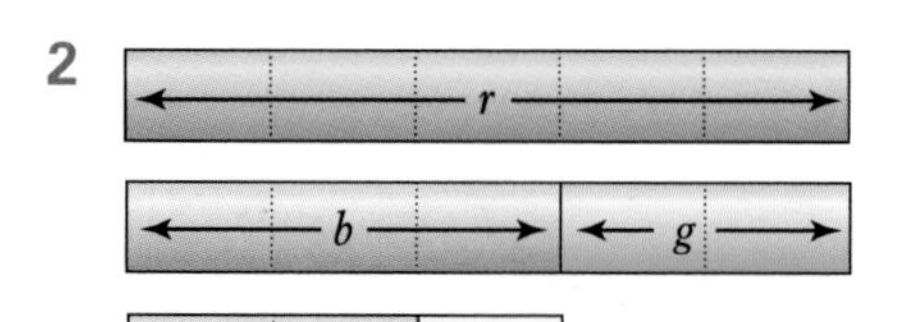

The length of the red rod is r.

The length of the blue rod is b and the green rod is g.

The length of the purple rod is p and the yellow rod is y.

For these coloured rods we could write the equation, $r = b + g$.

a Which of these are true?
$b = p + y \qquad r - g = b \qquad 2g = r \qquad 2p + y = r \qquad r - y = 2p$

b Write eight more equations that are true for these rods.
Use inverse operations to help.

3 What is the inverse of these?

a adding 8
b multiplying by 9
c multiplying by 6 then adding 2
d dividing by 3 then subtracting 4
e adding 2 then multiplying by 3
f subtracting 4 then dividing by 2.

You could draw function and inverse function machines to help.

4 Choose the correct answer. You could use substitution to help.

a If $3x + 2 = y$ then **A** $\frac{x}{3} - 2 = y$ **B** $\frac{y}{3} - 2 = x$ **C** $\frac{y-2}{3} = x$

b If $4(a + 2) = b$ then **A** $a = 4b + 2$ **B** $a = \frac{b}{4} - 2$ **C** $4(b + 2) = a$

c If $\frac{m}{3} - 2 = n$ then **A** $m = 2n + 3$ **B** $m = 3n + 2$ **C** $m = 3(n + 2)$

d If $\frac{x-3}{4} = p$ then **A** $x = \frac{p+3}{4}$ **B** $x = 4p + 3$ **C** $x = 4(p - 3)$

e If $5(n - 3) = m$ then **A** $n = \frac{m-3}{5}$ **B** $n = 5m + 3$ **C** $n = \frac{m}{5} + 3$

Remember: The operations must be undone in the opposite order.

Review 1

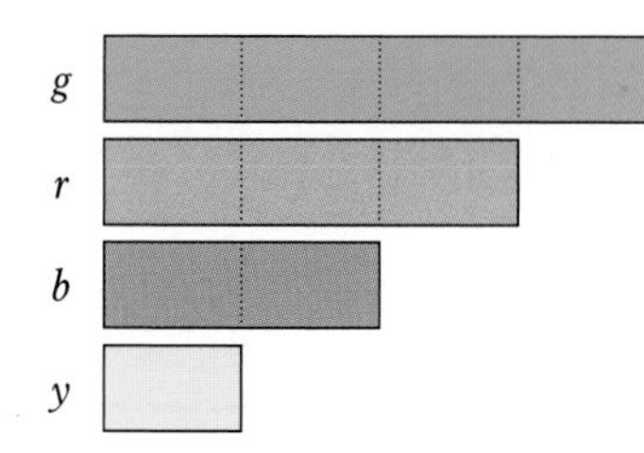

The total length of each rod is given by the letter beside it.

a Which of these are true?

$r + y = g$ $3y = g$ $y = \frac{r}{3}$ $g = 2b$

b Write two equations beginning $g =$ ____ .

c Repeat **b** for $r =$ ____ , $b =$ ____ and $y =$ ____ .

Review 2 Choose the correct answer.

a If $5y + 7 = x$ then **A** $y = \frac{x}{5} - 7$ **B** $y = \frac{x-7}{5}$ **C** $y = 5(x + 7)$

b If $3(x - 2) = y$ then **A** $x = \frac{y}{3} + 2$ **B** $x = \frac{y}{3} - 2$ **C** $x = \frac{y+2}{3}$

Practical

You will need a spreadsheet package.

Ask your teacher for the Equations and their Inverses ICT worksheet.

Simplifying expressions – multiplying and dividing

Worked Example

Simplify these. **a** $7 \times 6n$ **b** $5 \times {}^-2n$ **c** ${}^-3 \times {}^-4b$

Answer

a $7 \times 6n = 7 \times 6 \times n$
$= 42 \times n$
$= 42n$

7 lots of $6n = 42n$.

b $5 \times {}^-2n = 5 \times {}^-2 \times n$
$= {}^-10 \times n$
$= {}^-10n$

c ${}^-3 \times {}^-4b = {}^-3 \times {}^-4 \times b$
$= 12 \times b$
$= 12b$

We often use **cancelling** when dividing.

Examples $\frac{5p^2}{5} = \frac{{}^1 5p^2}{{}^1 5} = p^2$ $\frac{24x^2}{30x} = \frac{{}^4 24x \times {}^1 x}{{}^5 30 {}^1 x} = \frac{4x}{5}$

Discussion

- How could you show that $a^3 \times a^4 = a^{3+4} = a^7$? **Discuss.**
- How could you show that $n^6 \div n^4 = n^{6-4} = n^2$? **Discuss.**

Sometimes we use the **index laws** to help **simplify an expression**.

Worked Example

a $3m^3 \times 5m^2$ *b $\frac{9x^5}{12x^2}$

There is more about the index laws on page 50.

Answers

a $3m^3 \times 5m^2 = 3 \times 5 \times m^3 \times m^2$

$= 15 \times m^{3+2}$

$= \mathbf{15m^5}$

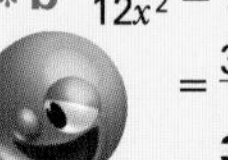

*b $\frac{9x^5}{12x^2} = \frac{{}^{3}9x^5}{{}^{4}12x^2}$

$= \frac{3x^{5-2}}{4}$

$= \mathbf{\frac{3x^3}{4}}$

Exercise 4

1 Simplify these.

a $2 \times 3a$ b $4 \times 2b$ c $3 \times 5n$ d $6 \times 2c$ e $7 \times 3m$
f $3 \times {}^{-}2x$ g $5 \times {}^{-}3p$ h $4 \times {}^{-}6m$ i ${}^{-}3 \times 5b$ j ${}^{-}4 \times 7h$
k ${}^{-}5 \times 3b$ l ${}^{-}2 \times {}^{-}2a$ m ${}^{-}3 \times {}^{-}4q$ n ${}^{-}6 \times {}^{-}3y$ o ${}^{-}3 \times 5e$
p $7 \times {}^{-}6m$ q $4x \times x$ r $5b \times b$ s $8p \times {}^{-}p$ t ${}^{-}4a \times a$
u $3b \times 5b$ v $6m \times 2m$ w $8b \times 7b$ *x $7a \times 6b$ *y $5n \times 3m$

2 Simplify these by cancelling.

a $\frac{3b}{3}$ b $\frac{8n}{4}$ c $\frac{15x}{3}$ d $\frac{3a}{a}$ e $\frac{7n}{n}$ f $\frac{12x}{8}$ g $\frac{25n}{15}$
h $\frac{56p}{32}$ i $\frac{49y}{21}$ j $\frac{81a}{45}$ k $\frac{125b}{75}$ l $\frac{112y}{80}$ m $\frac{117p}{65}$

3 Simplify these using the index laws.

a $a \times a^2$ b $p^2 \times p$ c $d^2 \times d^2$ d $n \times n^3$ e $m^2 \times m$
f $x^3 \times x^3$ g $p^2 \times 2p$ h $y^3 \times 4y$ *i $2y \times 3y^2$ *j $3x \times 2x^2$
*k $4a^2 \times a^3$ *l $3b^4 \times 5b^5$ *m $6m^2 \times 2m^3$ *n $8b^2 \times 2b^7$

4 Simplify these using the index laws.

a $\frac{a^2}{a}$ b $\frac{m^3}{m^2}$ c $\frac{b^3}{b}$ d $\frac{a^4}{a^2}$ e $\frac{r^4}{r^3}$ f $\frac{p^4}{p}$
g $\frac{x^5}{x}$ h $\frac{n^5}{n^3}$ i $\frac{q^5}{q^4}$ j $p^3 \div p^2$ k $x^4 \div x^2$ l $a^3 \div a^3$
m $x^4 \div x^4$ n $a^5 \div a^2$ o $y^2 \div y$ *p $\frac{a^2}{a^3}$ *q $n^4 \div n^5$

***5** Simplify these by cancelling and using the index laws.

a $\frac{4n^2}{4}$ b $\frac{3n^2}{3}$ c $\frac{12m^2}{4m}$ d $\frac{15b^2}{3b}$ e $\frac{4a^2}{2a}$ f $\frac{8x^3}{12x}$
g $\frac{3x^3}{9x^2}$ h $\frac{16y^5}{4y^2}$ i $\frac{20b^7}{24b^3}$ j $\frac{30g^5}{45g^3}$ *k $\frac{20a^4}{36a^7}$ *l $\frac{42x^3}{72x^8}$

*6 Write an expression for the length of these rectangles.

a 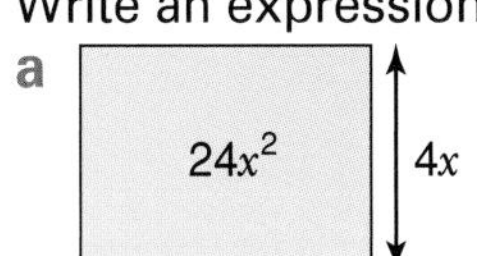

b 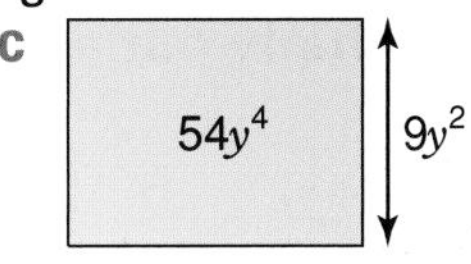

c $54y^4$ $9y^2$

*7 The area of a rectangle is $64x^3$. What might the lengths of the sides be?

Review 1 Simplify these.

a $4 \times 5b$ b $7 \times 3x$ c $^-2a \times 5$ d $^-6n \times {}^-4$ e $8 \times {}^-7p$
f $5p \times p$ g $5x \times x$ *h $3a \times 5a$ *i $7n \times 6n$

Review 2

$\overset{\text{I}}{\overline{a^3}}$ $\overline{a^4}$ $\overset{\text{I}}{\overline{a^3}}$ $\overline{\frac{5a}{3}}$ $\overline{\frac{4b}{3}}$ $\overline{2a}$ $\overset{\text{I}}{\overline{a^3}}$ $\overline{4a}$ $\overline{\frac{4b}{3}}$ $\overline{\frac{5a}{3}}$ $\overline{30}$ $\overline{5b^2}$ $\overline{12n^3}$ $\overline{4a}$

$\overline{1}$ $\overline{a^4}$ $\overline{b^2}$ $\overline{30}$ $\overline{\frac{5a}{3}}$ $\overline{\frac{3b^2}{4}}$ $\overline{5b^2}$ $\overline{a^4}$ $\overline{\frac{5a}{3}}$ $\overline{30}$ - $\overline{2a}$ $\overline{a^3}$ $\overline{a}$ $\overline{5b^2}$

$\overline{a^2}$ $\overline{1}$ $\overline{\frac{5a}{3}}$ $\overline{\frac{5a}{3}}$ $\overline{b^2}$ $\overline{5b^2}$ $\overline{\frac{4b}{3}}$ $\overline{1}$ $\overline{2a}$ $\overline{\frac{3b^2}{2}}$ $\overline{1}$ $\overline{\frac{3b^2}{2}}$ $\overline{12n^3}$ - $\overline{\frac{3b^2}{2}}$ $\overline{1}$ $\overline{b^2}$ $\overline{12n^3}$

$\overline{\frac{4b}{3}}$ $\overline{1}$ $\overline{b^2}$ $\overline{\frac{8a}{7}}$

Use a copy of this box. Simplify these expressions.
Write the letter beside each expression above its answer in the box.

I $a \times a^2 = a^3$	N $a^2 \times a^2$	F $\frac{8a}{4}$	R $\frac{20a}{5}$	Y $\frac{30b}{b}$	D $\frac{64a}{56}$
B $\frac{a^5}{a^3}$	V $a^6 \div a^5$	E $\frac{10b^2}{2}$	W $\frac{9b^2}{12}$	T $\frac{45a^2}{27a}$	
S $\frac{72b}{54}$	L $b^4 \div b^2$	O $b^3 \div b^3$	C $\frac{21b^3}{14b}$	A $3n \times 4n^2$	

***Review 3** Simplify these.

a $8n^3 \times 4n^2$ b $\frac{45a^6}{36a^4}$ c $\frac{72y^7}{48y^4}$ *d $\frac{36m^5}{56m^8}$

Collecting like terms

$3a$ and $4a$ are **like terms**. $3a$ and $4b$ are not.

$3a + 4m + 2a + 5m = 3a + 2a + 4m + 5m$
$= \mathbf{5a + 9m}$

$5a + 3m - 6a + 2m = 5a - 6a + 3m + 2m$
$= \mathbf{{}^-a + 5m}$

$5a - 4m - a + 6 - 6m = 5a - a - 4m - 6m + 6$
$= \mathbf{4a - 10m + 6}$

It is easier if you first move **like terms** next to one another. Always move the sign in front of the term, with the term.

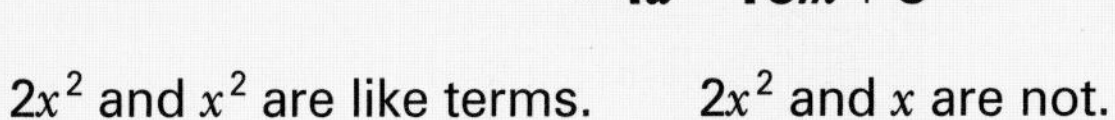

$2x^2$ and x^2 are like terms. $2x^2$ and x are not.

Examples $3x^2 + 2x^2 = 5x^2$
$4x^2 + 3x^2 + x = 7x^2 + x$

You cannot add $7x^2 + x$ to get $8x^2$ or $7x^3$. We can check this by substituting a number for x.
If $x = 2$ $7 \times x \times x + x = 7 \times 2 \times 2 + 2 = 30$ but $8x^2 = 8 \times 2 \times 2 = 32$ and $7x^3 = 7 \times 2 \times 2 \times 2 = 56$.

Sometimes we have to **multiply out a bracket** before we **simplify**.

Worked Example

Simplify **a** $8(x + 2y) - 4(2x + y)$ **b** $12 - (p - 4)$

Answers

a $8(x + 2y) - 4(2x + y) = 8x + 16y - 8x - 4y$ multiply out the brackets.
$= 8x - 8x + 16y - 4y$ write like terms together.
$= 0 + 12y$ add or subtract like terms.
$= \mathbf{12y}$

b $12 - (p - 4) = 12 - \mathbf{1}(p - 4)$
$= 12 - p + 4$ because $^-1 \times p = {}^-p$ and $^-1 \times {}^-4 = 4$
$= 16 - p$

Worked Example

Write an expression for the missing lengths, AB and BC.

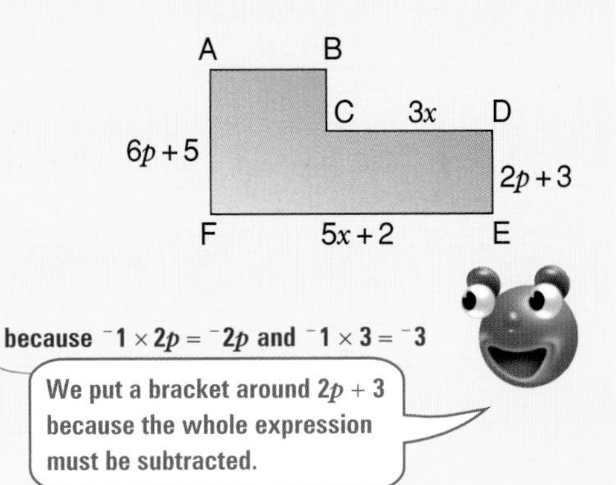

Answer

$AB + 3x = 5x + 2$
$AB = 5x + 2 - 3x$
$= \mathbf{2x + 2}$

$BC + (2p + 3) = 6p + 5$
$BC = 6p + 5 - (2p + 3)$
$= 6p + 5 - \mathbf{1}(2p + 3)$
$= 6p + 5 - 2p - 3$ because $^-1 \times 2p = {}^-2p$ and $^-1 \times 3 = {}^-3$
$= 6p - 2p + 5 - 3$
$= \mathbf{4p + 2}$

Exercise 5

1 **a** $5a + 6b + b - 2a$ **b** $6a + 2b - 3a - b$ **c** $4n + 4a - 2a - n$
d $5a + 6b - 3a + b$ **e** $10x - 3y - 9x + 5y$ **f** $14p - 5q - 7p - 3q$
g $12a - 7b - 6a - 9b$ **h** $15x - 3y - 7x - 4y$ **i** $5n + 3m - 6n - 2m$
j $7x + 3y - 8x + 4y$ **k** $3p - 5q - 5p + 3q$ **l** $15m - 17n - 23m + 25n$

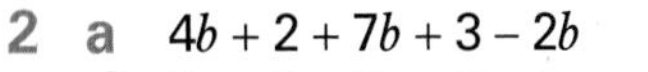

2 **a** $4b + 2 + 7b + 3 - 2b$ **b** $7a + 6 - 3a + 9$ **c** $5b + 2b - 3 - 2b + 7$
d $9y - 5 + 5y + 8 - 3y$ **e** $8y - 4 - 7y + 8$ **f** $9a - 8 - 8a - 4 - a$
g $14y - 10 - 5y - 3$ **h** $5a - 7 - 4a - 3$ **i** $3x - 5 - 5x + 6$
j $7b + 8 - 9b - 6$ **k** $5 + 2y - 7 - 5y$ **l** $64p - 37q - 13 - 82p + 13q + 8$
m $29y - 17x - 31 - 29x + 16y + 12$ **n** $16 - 17a + 3b - 11a - 24b - 35$

This is linked to adding and subtracting integers – see page 35.

3 **a** $10x^2 - 4x^2$ **b** $8x^2 + x^2$ **c** $5x^2 + 2x^2 + 3x^2$ **d** $4x^2 + 2x^2 + x^2$
e $3x^2 + 4x^2 - 2x^2$ **f** $7x^2 - 3x^2 - 4x^2$ **g** $3x^2 - x + 2x^2$ **h** $7x^2 - 3x + 5x$
i $6x^2 + 3x + 2x^2 - 4x$ **j** $5x^2 + 4x - 7x^2 - 8x$ **k** $6x^2 - 3x - 12x^2 + 4x$
l $9x^2 - 3x - 5x^2 - 2x$ ***m** $7x - 4x^2 - 3x - 5x^2$

4 Write an expression for the perimeter of these rectangles. Simplify your expression.

a

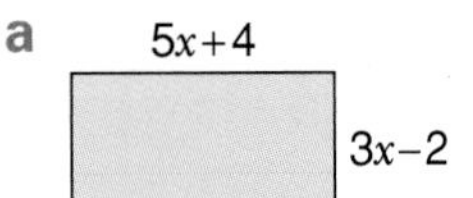

b

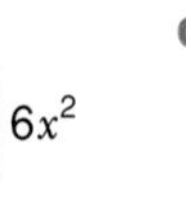

c

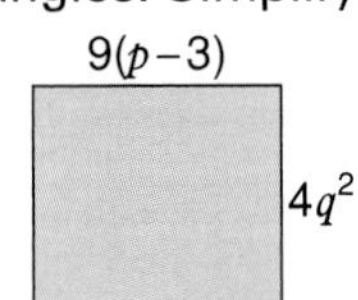

5 Jake made this rectangular garden.
Write an expression for the perimeter of the garden.
Simplify your expression.

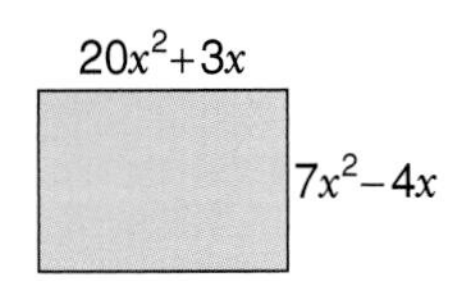

6 The expression in each box is found by adding the expressions in the two boxes beneath it.

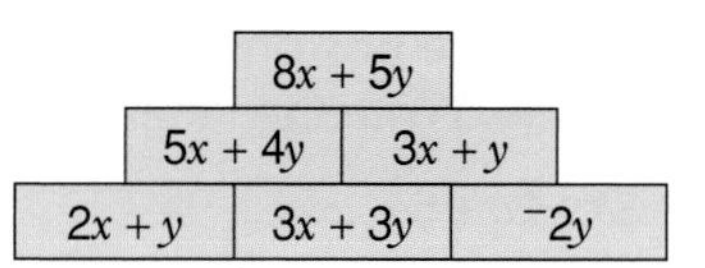

Find the missing expressions in these. Simplify them.

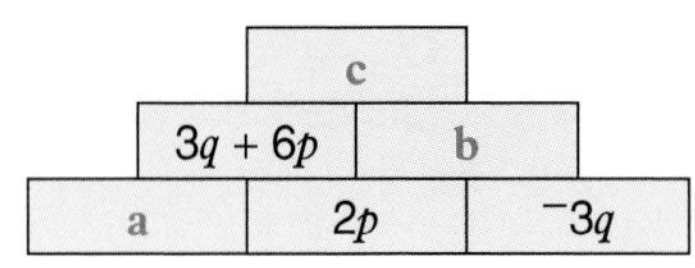

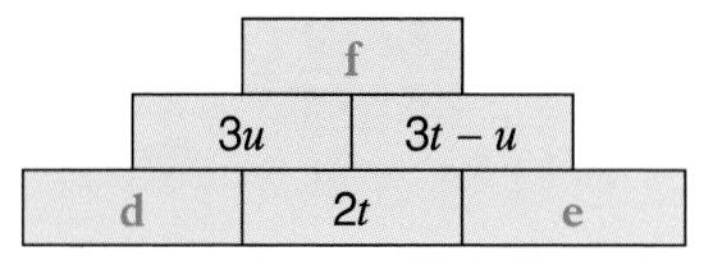

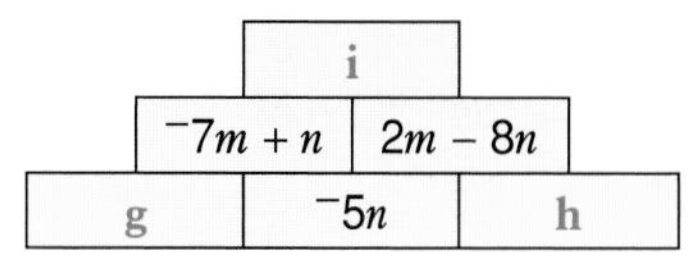

7 Multiply out the brackets and then simplify.

a $3(x + 4) + 2(x + 1)$ **b** $2(x + 3) + 3(x + 4)$ **c** $3(a - 4) + 2(a - 1)$
d $2(3x - 1) + 3(x + 4)$ **e** $5(3 - 2x) + 4(5x + 1)$ **f** $3(5y + 2) - 5$
g $4(1 - y) + 2y$ **h** $5(6 - 2a) + 3(7 - 4a)$ **i** $7(3 - 4m) + 3(5 - 7m)$

8 Write these without brackets and then simplify.

a $8 - 2(a + 1)$ **b** $10 - 3(b + 2)$ **c** $16 - 4(2x + 2)$
d $20 - 2(m + 1)$ **e** $15 - 3(b - 2)$ **f** $16 - 3(3x - 2)$
g $8 - (b - 1)$ **h** $10 - (2n - 2)$ **i** $8(a + 3) - 2(4 - 3a)$
j $7(n + 2) - 3(2n + 1)$ **k** $4(x + 2) - 3(2 - x)$ **l** $3(u - 3) - 3(4 - u)$
*__m__ $5(2m + n) - 3(m + n)$ *__n__ $4(2a + b) - 8(2b + a)$ *__o__ $6(2t + u) - (5t - 3u)$
*__p__ $7(4x + 3y) - (2x - y)$ *__q__ $5(4p - q) - (3p - 2q)$

***9** Write an expression for the missing lengths.

a

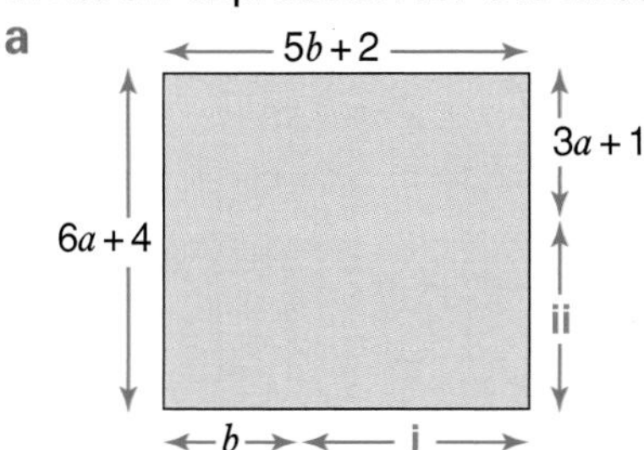

b

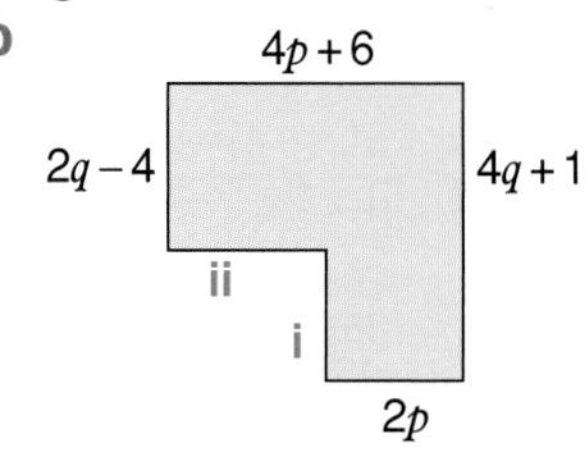

c

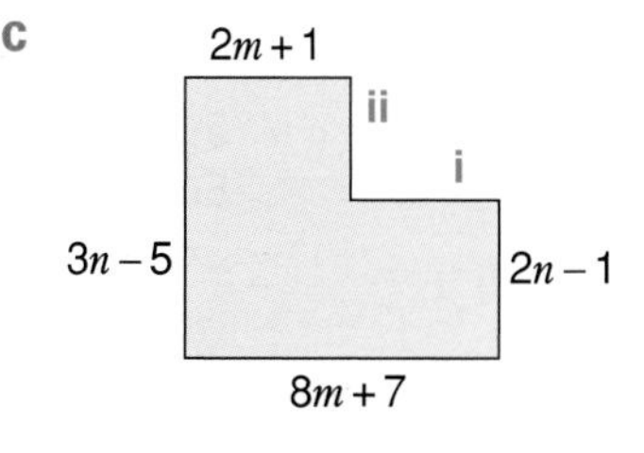

10 What might the missing lengths be?

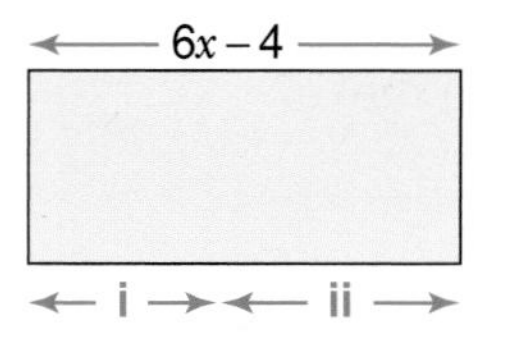

Review 1

$\overline{11a - 6}$ $\overline{3a + b}$ $\overline{^-a + b}$ $\overline{4a^2}$ $\overline{9b + 8}$ $\overline{^-a + b}$ $\overline{7a + 1}$ $\overline{4a^2}$ $\overline{^-a + 2}$ $\overline{^-a + b}$

$\overline{3a + b}$ $\overline{^-a + b}$ $\overline{8a^2 - 3a}$ $\overline{^-a + 2}$ $\overline{3a + b}$ $\overline{11a - 6}$ $\overline{6a + 4b}$ $\overline{10a + b}$ $\overline{4a^2}$

$\overline{3a^2 - 7a}$ $\overline{4a^2}$ N $\overline{6a + 9b}$ $\overline{8a^2 - 3a}$ N $\overline{6a + 9b}$ $\overline{11a - 6}$ $\overline{3a + b}$ $\overline{^-a + b}$

$\overline{3a^2 - 7a}$ $\overline{8a^2 - 3a}$ $\overline{^-4a - 2}$ $\overline{^-4a - 2}$ $\overline{15b + 1}$ $\overline{^-a + b}$ $\overline{4a^2}$ $\overline{^-a + 2}$ $\overline{^-a + b}$ $\overline{5a + 5b}$

$\overline{5a + 3b}$ $\overline{4a^2}$ $\overline{5a + 5b}$ $\overline{10a + b}$ $\overline{8a^2 - 3a}$ $\overline{9b + 8}$ $\overline{^-a + b}$ $\overline{10a + b}$ $\overline{^-a + b}$ $\overline{^-a + b}$ $\overline{11a - 6}$

$\overline{5a + 5b}$ $\overline{8a^2 - 3a}$ $\overline{^-2b - 1}$ $\overline{8a^2 - 3a}$ N $\overline{6a + 9b}$ $\overline{a + 11}$ $\overline{3a + b}$ $\overline{^-a + b}$ $\overline{5a + 5b}$

Use a copy of the box on the previous page.
Simplify these. Write the letter beside each, above its answer in the box.

N $4a + 3b + 2a + 6b = 6a + 9b$ **S** $9a + 3b - 4a + 2b$ **W** $3a + 4b + 2a - b$
F $8a + 4b + 2a - 3b$ **O** $5a + 6b + a - 2b$ **H** $7a + 2b - 4a - b$ **R** $5a + 3 + 2a - 2$
V $3b + 11 + 5b - 3 + b$ **C** $9a + 7 - 8a + 4$ **L** $12b - 7 + 3b + 8$ **G** $4a + 6 + 3a - 4 - 8a$
E $5a + 3b - 6a - 2b$ **T** $6a - 2 + 5a - 4$ **X** $4b - 10 - 6b + 9$ **D** $4a + 5 - 8a - 7$
A $9a^2 - 5a^2$ **I** $6a^2 + 2a^2 - 3a$ **M** $5a^2 - 3a - 2a^2 - 4a$

Review 2 Multiply out the brackets and then simplify.

a $4(x + 6) + 3(x + 2)$ **b** $4(a + 3) + 2(3a + 1)$ **c** $2(x - 3) + 4(2x + 7)$ **d** $4(3y - 2) - 7$
e $7(1 - n) + 12n$ **f** $10 - 3(1 + a)$ **g** $5 - (2 - b)$ **h** $3(2a + 3) - 3(1 - 2a)$

* **Review 3** The number in each box is found by adding the numbers in the two boxes below it.
Write possible expressions for the empty boxes.

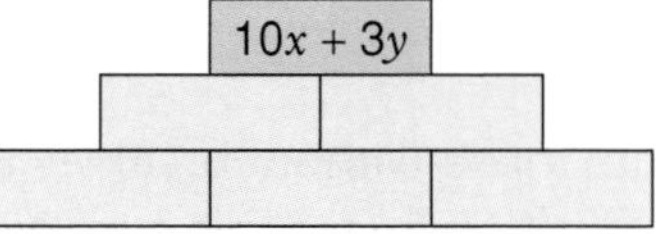

Puzzle

Use a copy of this.
Simplify these. Shade the answer on your diagram.

a $5(2x - 7)$ **b** $3(x - 4) - (x + 2)$
c $5(x - 4) - (x - 2)$ **d** $4 \times 2x$
e $3x \times 2x$ **f** $3x \times 5x^2$
g $x(x + 3)$ **h** $\frac{36xy}{18x}$
i $\frac{5x^3}{15x}$ **j** $\frac{12x^4}{10x^2}$

What does the shading make?

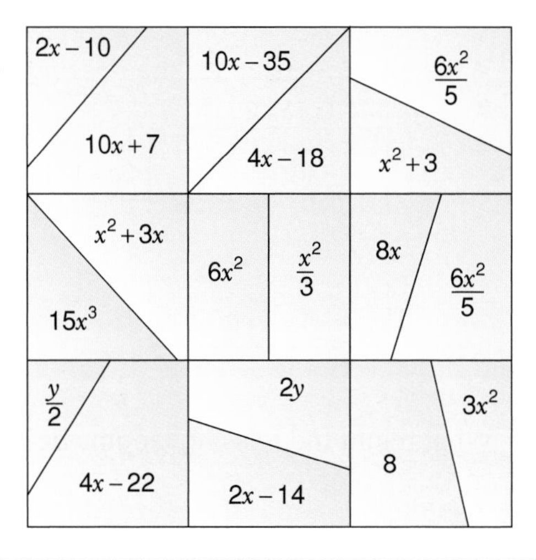

Practical

You will need a graphical calculator.

Use a graphical calculator to show that
$y = (x + 6) - (x - 2) = 8$ for any value of x.

You could do this using the table function.

x	y1	y2	y1–y2
1	7	-1	8
2	8	0	8
3	9	1	8
4	10	2	8
5	11	3	8

or

You could draw the graphs of the two straight lines

$y = x + 6$ and $y = x - 2$.

How could you use the vertical distance between the graphs to show that
$y = (x + 6) - (x - 2) = 8$?

Practical

You will need a spreadsheet package.

Ask your teacher for the Brackets ICT Worksheet.

T

Puzzle

Use a copy of this.

1 Fill in the expressions in the squares to make this a magic square.

$a+b$		
$a-b+c$		
$a-c$	$a+b+c$	

*2 Write an expression for each square to make this a magic square with a magic sum of $3x-3$.

		$x+2$
$x+1$		
		$x-4$

More writing and simplifying algebraic expressions

Discussion

A courtyard was built around a tree.
Jesamine wrote this expression for the area of the courtyard.

$$12 \times 8 - 3 \times x = 96 - 3x$$

Is her expression correct? **Discuss**.

Ronan wrote this expression for the area of the courtyard.

$$(12 - x) \times 3 + 5 \times 12 = 3(12 - x) + 60$$

Is Ronan's expression correct? **Discuss**.

Are Jesamine's and Ronan's expressions equivalent? **Discuss**.

Equivalent means they are the same for all values of x.

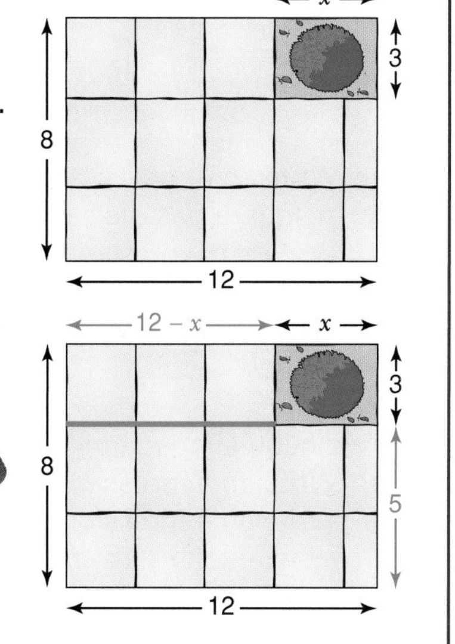

Algebra

Exercise 6

1 Rosie wrote down the perimeter of her bedroom as $a + a + b + b$.

a Write down a different expression for the perimeter of her room.

b Show that the expressions are equivalent.

2 Benjamin made this gate with lengths of steel.

a Write an expression for the total length of steel used.

b Write a different expression for the total length of steel used.

c Show that the expressions you wrote in **a** and **b** are equivalent.

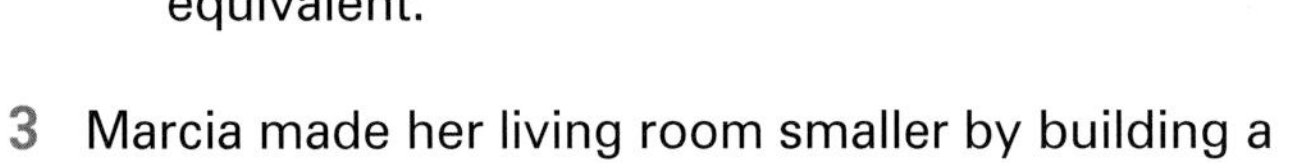

3 Marcia made her living room smaller by building a cupboard in the corner.
She said 'the area of the new living room is $48 - 2n$ metres^2'.

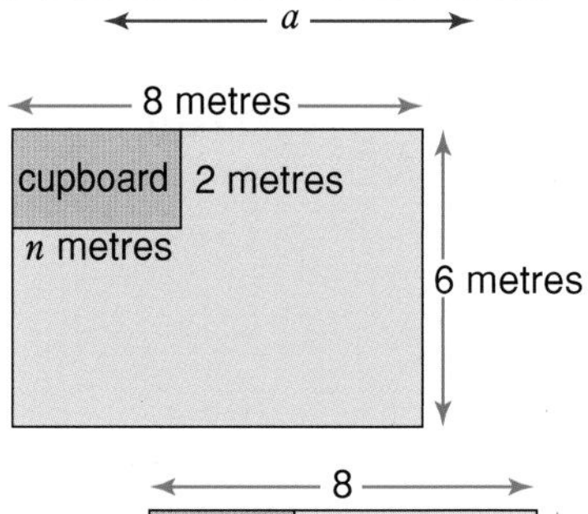

a Is Marcia's expression correct? Explain your answer.

b Marcia's brother found the area of the new living room by dividing it into two rectangles.

area of **A** $= 2(8 - n)$
area of **B** $= 4 \times 8$
total area $= 2(8 - n) + 32$

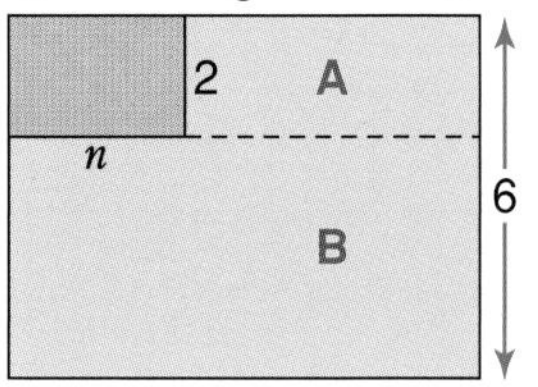

Show that $2(8 - n) + 32$ is equivalent to $48 - 2n$.

c Divide the new living room into two *different* rectangles.
Use this to write a different expression for the red shaded area.

d Show that the expression you wrote in **c** is equivalent to the other two.

4 Write three different expressions for the blue-shaded areas of each of these.
Show, in each case, that your three expressions are equivalent.

a

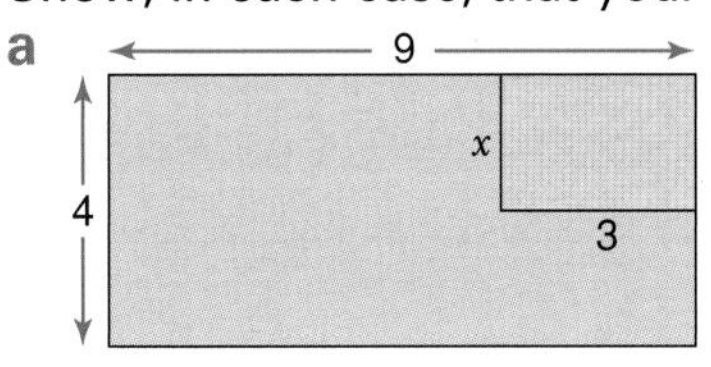

b

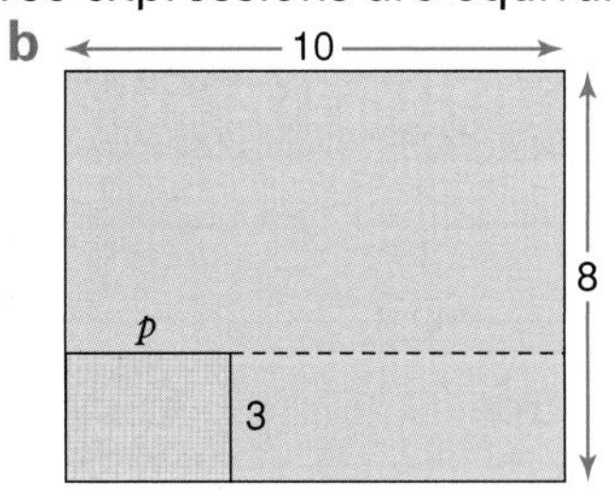

c

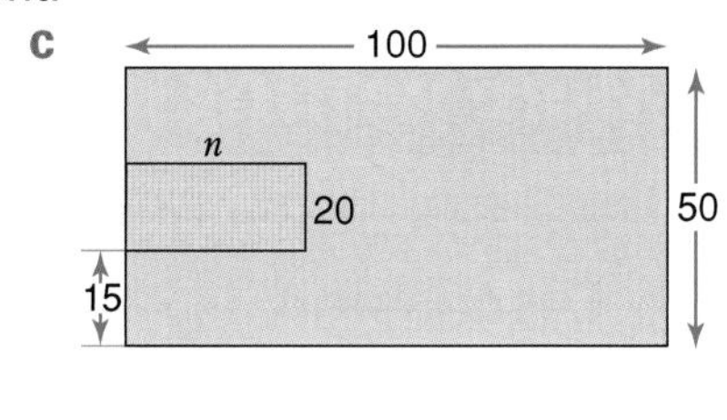

5 a Write an expression for the perimeter of this rectangle.

b A square has the same perimeter as this rectangle.
Write an expression for the length of the side of the square.

c Write an expression for the area of the square.

d Which has the larger area, the rectangle or the square?
By how much?

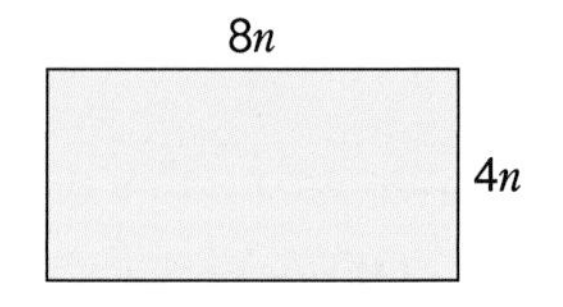

6 Two cube-shaped tanks are filled with water.

a Write an expression for the volume of the larger tank.

b Write an expression for the volume of the smaller tank.

c How much more water will the larger tank hold than the smaller one?

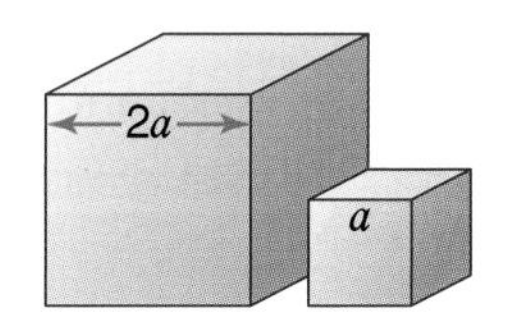

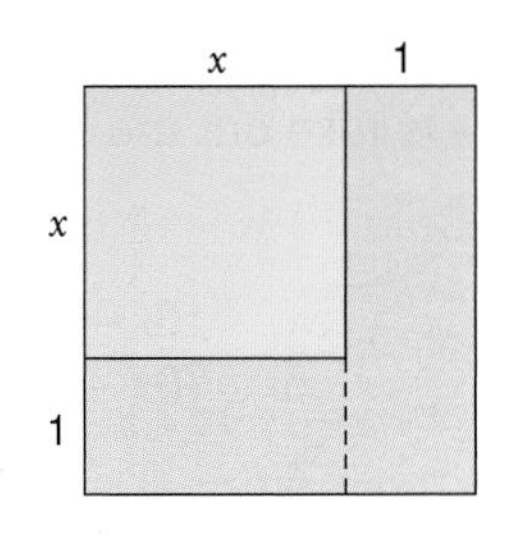

*7 **a** Write an expression for the area of the smaller square.
b Write an expression for the area of the larger square by dividing it into rectangles as shown.
c Write an expression for the difference between the larger square and the smaller square.
d If the difference in the areas is 37, what is the value of x?

Review Write three different expressions for the purple-shaded area.
Simplify your expressions to show they are equivalent.

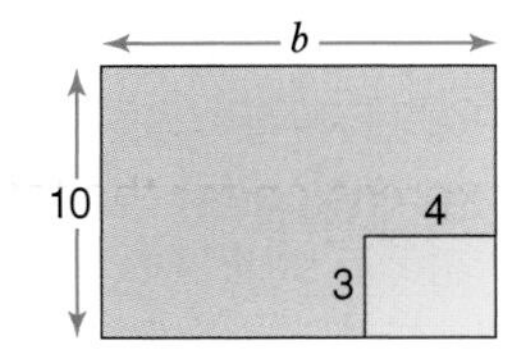

Factorising

Factorising an expression is the inverse of multiplying out a bracket.

Example $2(x + 3) = 2x + 6$
If we are asked to factorise $2x + 6$, we get $2(x + 3)$.

We factorise $2x + 6$ by finding the common factor of $2x$ and 6, which is 2. This is then put outside the bracket.

$2x + 6 = 2(x + 3)$

Discussion

- Karen factorised $12x - 8$ like this.

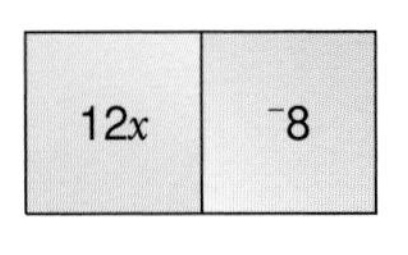

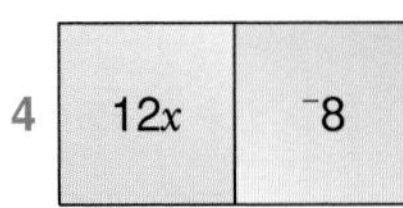

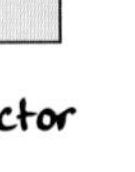

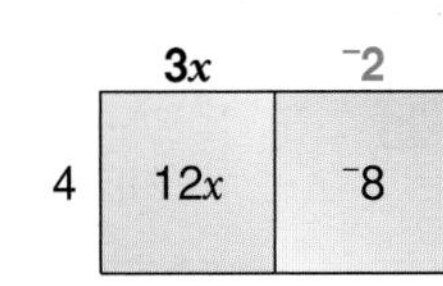

4 is a common factor of $12x$ and $^-8$.

$4 \times 3x = 12x$ or $\frac{12x}{4} = 3x$ or $4\overline{)12x}$ gives $3x$

$4 \times {}^-2 = {}^-8$ or $\frac{^-8}{4} = {}^-2$ or $4\overline{)^-8}$ gives $^-2$

So $12x - 8 = 4(3x - 2)$

Discuss Karen's method.

- Saad factorised $12x - 8$ like this.

$12x - 8 = 4(\quad\quad)$
$= 4(3x \quad)$ because $4 \times 3x + 12$
$= 4(3x - 2)$ because $4x - 2 = {}^-8$

Discuss Saad's method.

Always take out the **highest common factor** possible.

Example $16x - 12$ could be factorised as

$16x - 12 = 2(8x - 6)$

or $16x - 12 = 4(4x - 3)$

$4(4x - 3)$ is factorised completely because 4 is the HCF of $16x$ and $^{-}12$.

Always check your answer by multiplying out the brackets.

$4(4x - 3) = 4 \times 4x + 4 \times {}^{-}3 = 16x - 12$

Worked Example

Factorise these. **a** $12n + 6m$ **b** $y^3 + y^2 + 5y$

Answers

a $12n + 6m = \mathbf{6(2n + m)}$

check $6(2n + m) = 6 \times 2n + 6 \times m$
$= 12n + 6m$ ✓

	$2n$	m
6	$12n$	$6m$

b $y^3 + y^2 + 5y = \mathbf{y(y^2 + y + 5)}$

check $y(y^2 + y + 5) = y \times y^2 + y \times y + y \times 5$
$= y^3 + y^2 + 5y$ ✓

	y^2	y	5
y	y^3	y^2	$5y$

Exercise 7

T

1 Factorise these. Use a copy of the diagrams to help.

Remember to check your answers.

a $2x + 4$

2	$2x$	4

b $3y + 9$

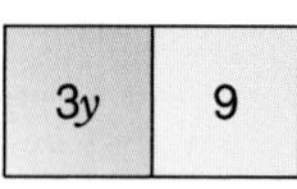

c $16p + 24$

	$16p$	24

d $8a - 16$

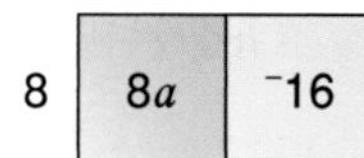

e $14n - 28$

	$16n$	$^{-}28$

f $5a + 5b$

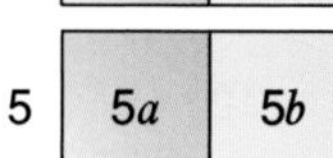

g $2n + 4m$

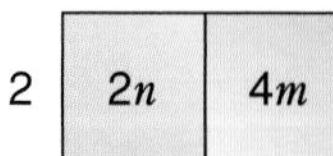

h $12x + 8y$

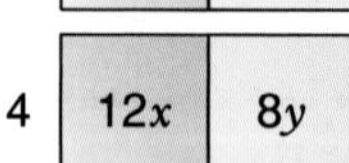

i $x^2 + x$

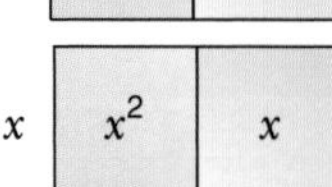

j $p^3 + p$

p	p^3	p

2 Copy and complete.

a $3x + 6 = 3\,(\_\_ + \_\_)$ **b** $5a - 10 = 5\,(\_\_ - \_\_)$ **c** $14x + 4 = 2\,(\_\_ + \_\_)$
d $16n - 12 = 4\,(\_\_ - \_\_)$ **e** $4x + 4 = 4\,(\_\_ + \_\_)$ **f** $12n - 4 = \_\_(3n - 1)$
g $15d - 25 = \_\_(3d - 5)$ **h** $18 + 3n = \_\_(6 + n)$ **i** $6 - 3a = \_\_(\_\_ - a)$
j $6 + 9x = \_\_(2 + \_\_)$ **k** $15x - 10 = \_\_(\_\_ - 2)$.

3 Factorise these.

a $2n + 2$ **b** $3 - 3a$ **c** $4x + 12$ **d** $6 + 12a$ **e** $14y - 7$
f $9x + 3$ **g** $8 - 12y$ **h** $10x + 25$ **i** $8n + 4$ **j** $11 - 22n$
k $10 + 15n$ **l** $9 - 21x$ **m** $12n + 8$ **n** $40 - 15n$ **o** $2x - 20$
p $20n + 16$ **q** $18 - 6a$ **r** $12 + 16n$ **s** $6x - 20$ **t** $21 - 6n$
u $32x - 24$ **v** $18n + 24$ **w** $16y - 24$ **x** $24 - 36n$ **y** $40 + 24a$
z $18a - 45$

4 Copy and complete.

a $2a + 2b = 2(\_\_\_ + \_\_\_)$ b $5x + 10y = 5(\_\_\_ + \_\_\_)$ c $4a + 8b = \_\_\_(a + \_\_\_)$
d $6n - 3m = 3(\_\_\_ - \_\_\_)$ e $7p - 14q = \_\_\_(\_\_\_ - 2q)$ f $12w - 8x = \_\_\_(\_\_\_ - \_\_\_)$

5 a $2n^2 + n = n(\_\_\_ + \_\_\_)$ b $ax - a = a(\_\_\_ - \_\_\_)$ c $4x + 3x^2 = x(\_\_\_ + \_\_\_)$
d $6x - x^2 = \_\_\_(6 - \_\_\_)$ e $10n^2 + 4 = 2(\_\_\_ + \_\_\_)$ f $n^2 + 3n = n(\_\_\_ + \_\_\_)$
g $p^3 + p^2 + 4p = p(\_\_\_ + \_\_\_ + \_\_\_)$ h $m^3 + m^2 + 5m = \_\_\_(\_\_\_ + \_\_\_ + 5)$

6 Factorise these.

a $x^2 + 5x$ b $a^2 + 9a$ c $p^2 - 3p$ d $5y - y^2$ e $x + x^2$
f $2y^2 - 5y$ g $a + 2a^2$ h $4n^2 - n$ i $2p - 5p^2$ j $5a + 6a^2$
k $2a + a^2$ l $5a - a^2$ m $5x^2 + 2x$ n $9n^2 + 4n$ o $2a^2 + 2$
p $5 + 5n^2$ q $8x^2 + 4$ r $12 - 3y^2$ s $a^3 + a^2 + 3a$ t $y^3 + y^2 + 7y$
*u $3y^2 + 3y + 3$ *v $8y^2 + 4y + 4$

***7** a The perimeter of a rectangle is $8x^2 + 16x$.
What might the lengths of its sides be?
b The area of a rectangle is $8x^2 + 16x$.
What might the lengths of its sides be?

***8** a Prove that the sum of three consecutive integers is always a multiple of 3.
b Prove that the sum of five consecutive integers is always a multiple of 5.

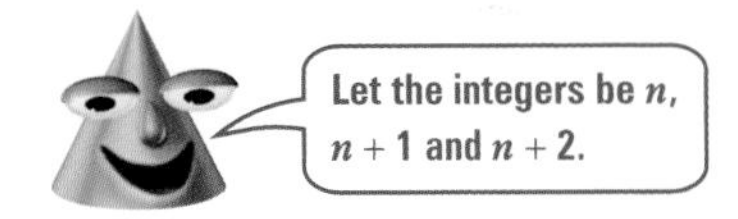

***9** I think of a number.
I multiply it by 4, add 20, divide by 4, subtract 5.
Try this for several starting numbers. What do you notice?
Use algebra to prove the result.

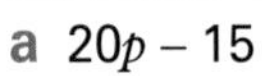

T **Review 1** Factorise these. Use a copy of the diagrams to help.

a $20p - 15$

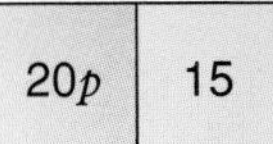

b $5x^2 + x$

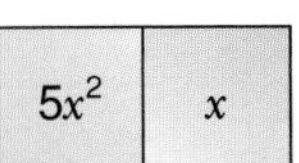

Review 2 Factorise these.

a $5y + 10$ b $20p - 10$ c $3n + 18$ d $20 - 5m$ e $4n + 12m$
f $8q - 16p$ g $3n^2 + n$ h $10a^2 + 5$ i $x^3 + x^2 + 4x$

* **Review 3** Suggest some possible lengths for the sides of this rectangle.

area = $4x^2 + 8x$

* **Review 4** Use algebra to show that the sum of two odd numbers is always an even number.
Hint: $2n$ is an even number.

Substituting

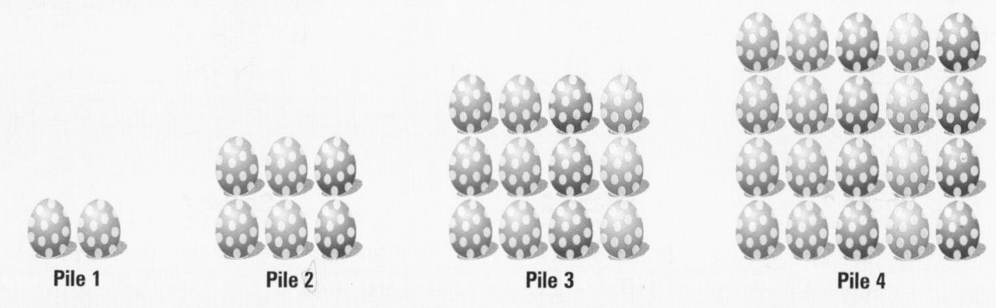

An expression for the number of Easter eggs in pile n is $n^2 + n$. To find the number of Easter eggs in pile **8** we **substitute 8** for n.

$$\begin{aligned} n^2 + n &= \mathbf{8}^2 + \mathbf{8} \\ &= 64 + 8 \\ &= 72 \end{aligned}$$

There are 72 eggs in Pile 8.

We **evaluate** an expression by **substituting** values for the unknown into the expression.

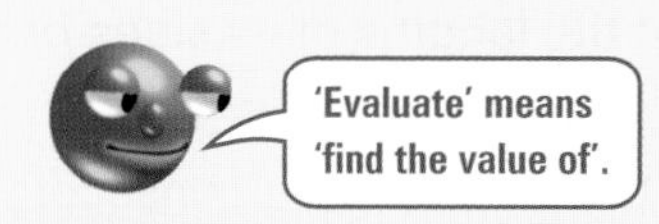

When we evaluate the expression the **order of operations** is the same as for arithmetic.

Brackets
Indices
Division and **M**ultiplication
Addition and **S**ubtraction

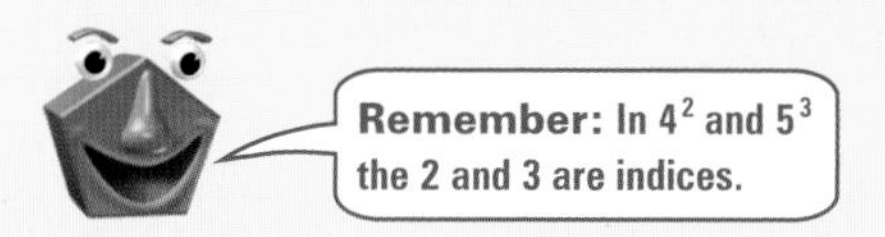

Use **BIDMAS** to help you remember this.

Worked Example

Find the value of

a $2a^2 + 6$ when $a = {}^-\mathbf{3}$ b $10 - 4b^2$ when $b = \mathbf{1{\cdot}5}$ c $\frac{3a^2(a-5)}{5a}$ when $a = \mathbf{8}$

Answer

a $$\begin{aligned} 2a^2 + 6 &= 2 \times a^2 + 6 \\ &= 2 \times ({}^-\mathbf{3})^2 + 6 \\ &= 2 \times 9 + 6 \\ &= 18 + 6 \\ &= \mathbf{24} \end{aligned}$$

b $$\begin{aligned} 10 - 4b^2 &= 10 - 4 \times b^2 \\ &= 10 - 4 \times \mathbf{1{\cdot}5}^2 \\ &= 10 - 4 \times 2{\cdot}25 \\ &= 10 - 9 \\ &= \mathbf{1} \end{aligned}$$

c $$\begin{aligned} \frac{3a^2(a-5)}{5a} &= \frac{3 \times \mathbf{8}^2 \times (\mathbf{8} - 5)}{5 \times \mathbf{8}} \\ &= \frac{3 \times 64 \times 3}{40} \\ &= \frac{576}{40} \\ &= \mathbf{14{\cdot}4} \end{aligned}$$

Exercise 8

Only use a calculator if you need to.

1 If $x = 10$ find the value of these expressions.

a $3x^2$ b $\frac{3x^2}{2}$ c $\sqrt{4x^2}$ d $\sqrt{325 - x^2}$ e $3(x^2 - 9)$

f $3x^2(x - 5)$ *g $\frac{2x^2(x-3)}{7x}$ *h $\frac{3x^2(2x-5)}{5x}$

2 Find the value of these when $x = {}^-4$.

a $2x + 12$ b $3x + 2$ c $20 - x$ d $3(x - 2)$ e $\frac{2x+4}{x}$ f $\frac{x-2}{x+2}$ g $3x^2 + 4$ *h $4x^3 - 3x$

3 Repeat question **2** for $x = 0{\cdot}1$.

4 Find the value of $y = \frac{4x + 2}{2}$ when

a $x = 3$ **b** $x = 0{\cdot}4$ **c** $x = {}^-2$ **d** $x = {}^-4$.

5

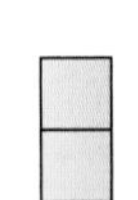
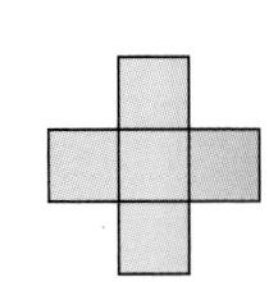
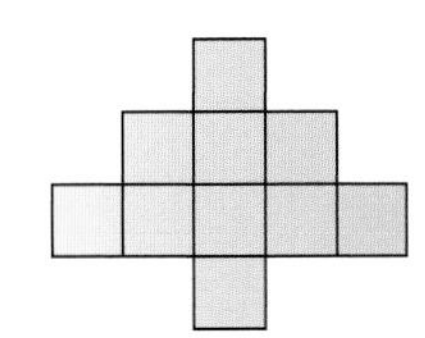
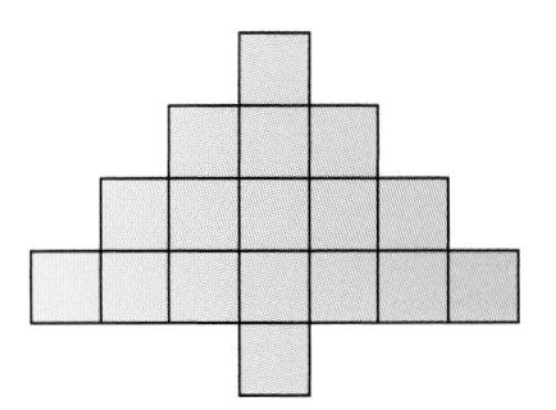

Shape 1 **Shape 2** **Shape 3** **Shape 4**

The number of squares in Shape n is given by $n^2 + 1$.
Find the number of squares in

a Shape 10 **b** Shape 9 **c** Shape 13 **d** Shape 24.

6 If $a = 3$, $b = {}^-2$, $c = 8$ and $d = {}^-4$, evaluate these.

a $a^2 + 2$ **b** b^3 **c** $3a(b - 2)$ **d** $3b^2 + b^3$ **e** $2a^2 - 2b^3$

f $3a^3 + 2$ **g** $\frac{b-2}{2-b}$ **h** $2c^3 + b^2$ ***i** $\sqrt{d^2 - b^2 - 3}$ ***j** $\sqrt{4d^2}$

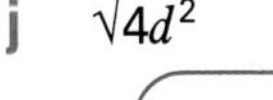

***k** $\frac{d^2 + 4}{b - 2}$ ***l** $\frac{d^2}{b^2}$ ***m** $\sqrt{4b^2}$

Remember: If you are squaring or cubing a negative number on the calculator key (− ...) x^2.

7 Cans of spaghetti are stacked in piles.

If the pile has p rows, the number of cans stacked is

$\frac{1}{2}(p^2 + p)$

How many cans are stacked in a pile with

a 12 rows **b** 16 rows **c** 24 rows?

Review 1

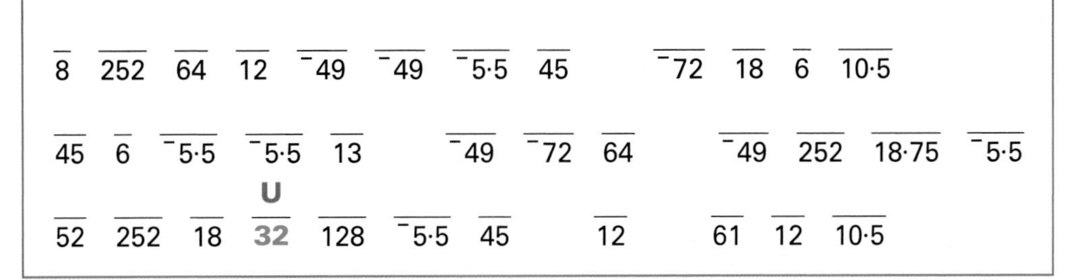

8 252 64 12 ⁻49 ⁻49 ⁻5·5 45 ⁻72 18 6 10·5

45 6 ⁻5·5 ⁻5·5 13 ⁻49 ⁻72 64 ⁻49 252 18·75 ⁻5·5

52 252 18 U 32 128 ⁻5·5 45 12 61 12 10·5

Use a copy of this box.

If $n = 4$, $m = 2{\cdot}5$ and $p = {}^-3$, evaluate these. Write the letter that is beside each above its answer in the box.

U $2n^2 = 32$ **M** $3n^2 + 4$ **T** $2n^3$ **D** $4n^2 - 3$ **L** $\frac{n^3 - 4}{10}$

V $3m^2$ **P** $4m + 3$ **E** $2 - 3m$ **Y** $7(m - 1)$ **N** $4(m + 2)$

O $3p^3 + 9$ **S** $3n^2 + p$ **F** $2m + 2p^3$ **A** $\frac{3n^2}{4}$ **R** $2n^2(n - 2)$

I $3n^2(m^2 - 1)$ **G** $3(n - 2) - 4(p + m)$

Review 2 If $x = 0{\cdot}1$, $y = 4$ and $z = {}^-2$, find the value of these.

a $\frac{24}{x}$ **b** $\frac{y}{8} + z$ **c** $2x^2$ **d** $y^2 - x^2$ **e** $\frac{4x}{y}$ **f** $\frac{y^2 - 2z}{x}$

g $20 - (y^2 - z^2)$ **h** $\sqrt{x^2 + z^2 + 0{\cdot}83}$ **i** $\frac{3z^2(y + 5)}{2y}$

Review 3

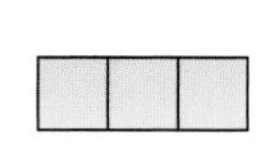
Pattern 1

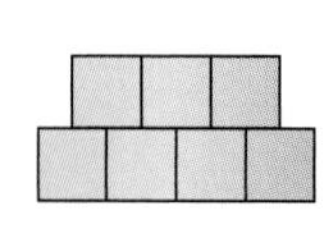
Pattern 2

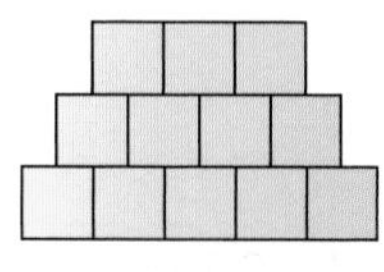
Pattern 3

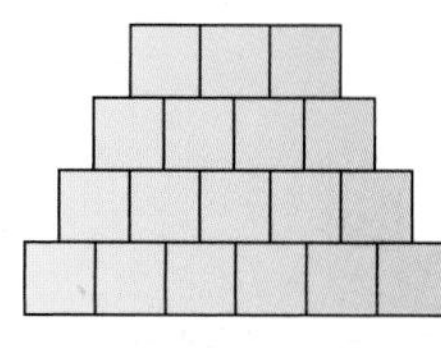
Pattern 4

Amri made these wall patterns.

If the pattern number is n, the number of bricks is $\frac{1}{2}(n^2 + 5n)$.

How many bricks are in pattern **a** 12 **b** 20 **c** 28?

T

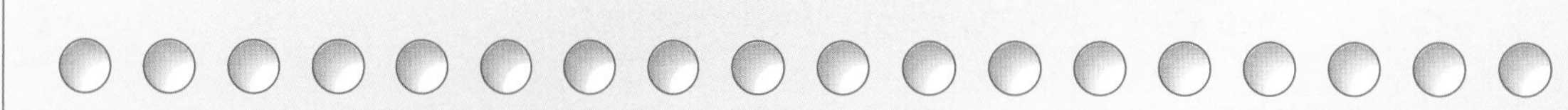

Investigation

Substituting

1 **You will need** a spreadsheet package.
Ask your teacher for the Formula Game ICT worksheet.

Substituting into formulae

The area of this sail is

$$A = \frac{b \times h}{2}$$

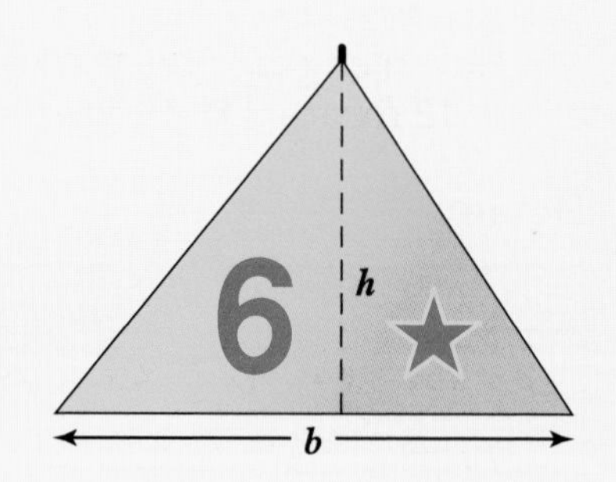

To find the area we need to know the value of b and h.
If $b = 1{\cdot}64$ m and $h = 2{\cdot}13$ m, we **substitute** these into the formula to find A.

$A = \frac{b \times h}{2}$ **Write the formula down first**

$= \frac{1{\cdot}64 \times 2{\cdot}13}{2}$ **Substitute the values**

$= \mathbf{1{\cdot}7466\ m^2}$ **Remember the units.**

Exercise 9

1 The perimeter of this trampoline is given by

$$P = 2l + 2w$$

Find the perimeter if

a $l = 3$ m, $w = 2$ m **b** $l = 3{\cdot}4$ m, $w = 3$ m **c** $l = 3{\cdot}5$ m, $w = 2{\cdot}5$ m.

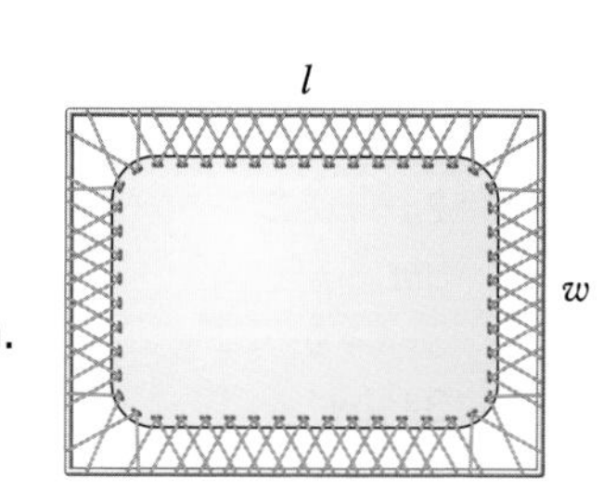

2 The formula for converting temperature in °C to K (Kelvin) is $K = °C + 273$.
Use the formula to convert these to Kelvins.
a 4 °C **b** 29 °C **c** $^-1$ °C **d** $^-9{\cdot}5$ °C **e** $^-19{\cdot}6$ °C

3 The surface area of this parcel is given by

$$S = 2lw + 2lh + 2wh$$

Find the surface area if
a $w = 60$ mm, $l = 88$ mm, $h = 19$ mm
b $w = 8{\cdot}6$ cm, $l = 10{\cdot}7$ cm, $h = 4$ cm
c $w = 0{\cdot}34$ m, $l = 0{\cdot}573$ m, $h = 0{\cdot}262$ m.

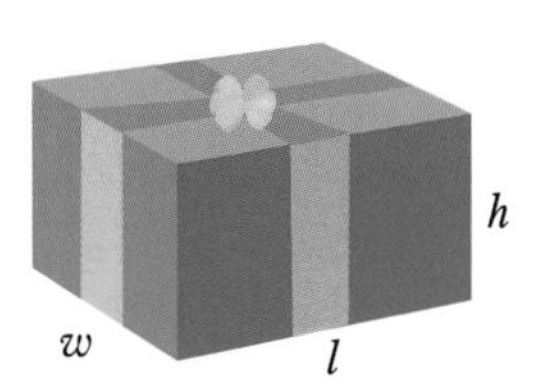

4 The formula for changing degrees Celsius to degrees Fahrenheit is

$$F = \tfrac{9}{5}C + 32 \quad \text{or} \quad F = \tfrac{9}{5}(C + 40) - 40.$$

Check that both of these formulae give the same Fahrenheit temperature if the Celsius temperature is
a 100 °C **b** 20 °C **c** 5·7 °C **d** $^-8{\cdot}4$ °C **e** $^-32$ °C

5 The mean of four numbers, a, b, c, d is given by $m = \frac{1}{4}(a + b + c + d)$
Find the mean if
a $a = 2, b = 6, c = 12, d = 1$
b $a = 6{\cdot}4, b = 3{\cdot}2, c = 1{\cdot}8, d = 9{\cdot}4$
c $a = \frac{1}{2}, b = 1\frac{3}{4}, c = 2\frac{1}{2}, d = 1\frac{1}{8}$.

6 The formula for the area of a trapezium is $A = \frac{h}{2}(a + b)$.

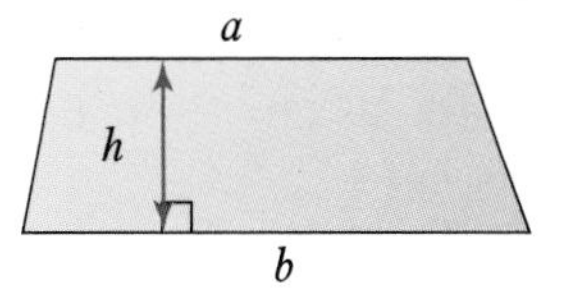

Find the areas of these trapeziums.

a
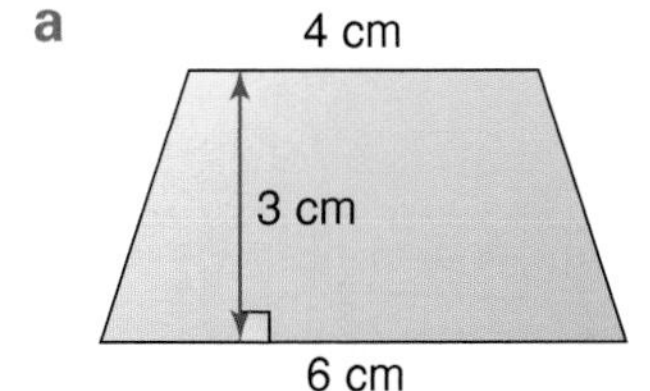

b
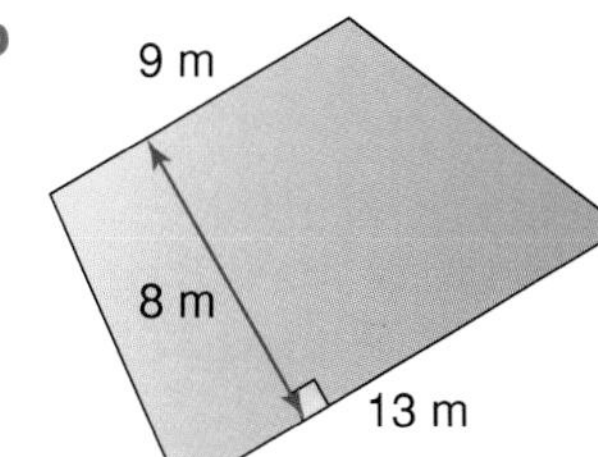

c
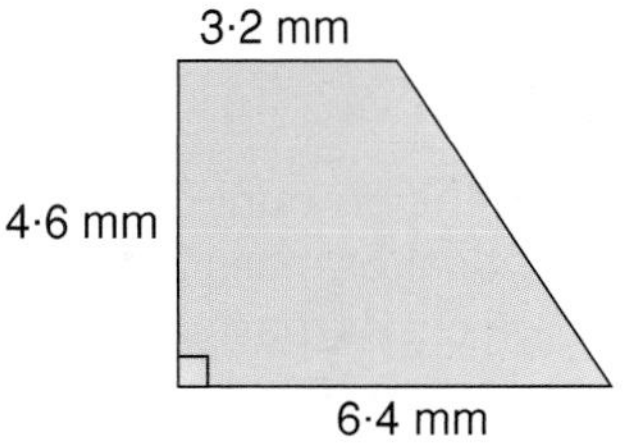

7 The final mark, F, given to a student is worked out using the formula

$$F = \tfrac{2}{5}E + \tfrac{3}{5}P$$

E is the mark the student got in the exam and P is the mark given for practical work during the year.
What is the final mark for these students?

a	**Allie**	Exam 40	Practical 50
b	**Brendon**	Exam 60	Practical 55
c	**Travis**	Exam 42	Practical 64

8 The speed of a van, v, in m/s is given by the formula $v^2 = 20 + 2as$. a is the acceleration in m/s^2 and s is the distance, in metres.
What is the speed if
a $a = 2$ m/s^2, $s = 11$ m **b** $a = 4$ m/s^2, $s = 20$ m
c $a = {}^-\frac{1}{2}$ m/s^2, $s = 17\frac{3}{4}$ m?

*9 Hero's formula for working out the area of a triangle is

$$A = \sqrt{s(s-a)(s-b)(s-c)}$$

where $s = \frac{1}{2}(a + b + c)$.

Use Hero's formula to find the areas of these triangles.
Round your answers sensibly.

a 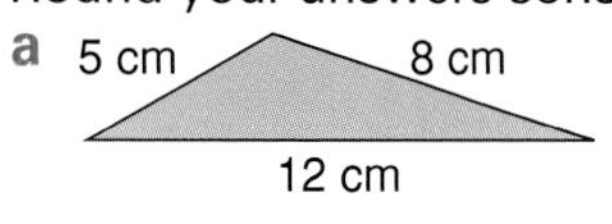

b 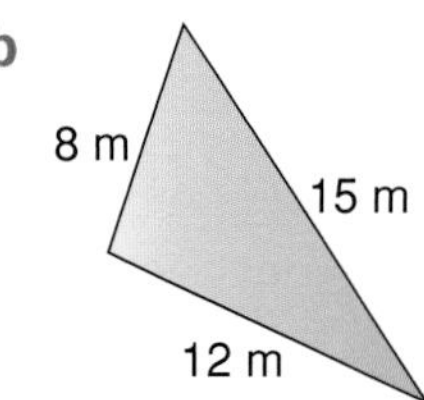

c 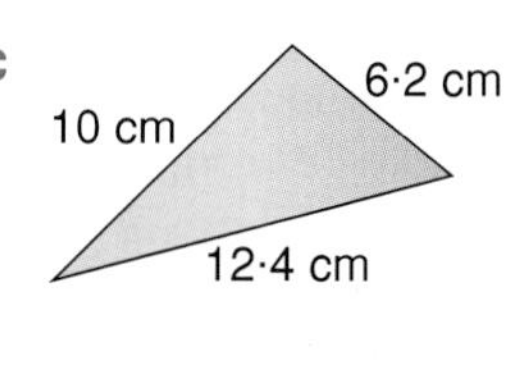

*10 Toby used this formula to work out the resistance, R ohms, in a circuit he set up in science.

$$R = \frac{R_1 R_2}{R_1 + R_2}$$

R_1 and R_2 are the resistances in two components of the circuit.
Find R. Round your answers sensibly.

a $R_1 = 5$ ohms, $R_2 = 12$ ohms **b** $R_1 = 3\frac{1}{2}$ ohms, $R_2 = 7\frac{1}{2}$ ohms

Review 1 The area of this poster is given by

$A = 25 + 3h$.

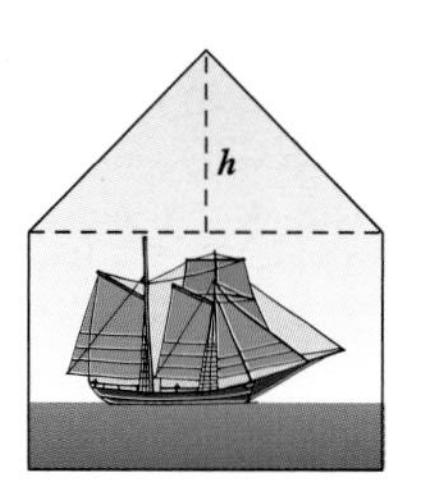

Calculate the area when h is

a 36 mm **b** 17 cm **c** 6·3 cm **d** 8·25 cm.

Review 2 The formulae give the dose, D in mℓ, of worming medication needed for cats and dogs of mass m kg are

cat $D = (10 \times m) - m$ **dog** $D = \frac{(9 \times m) + 10}{2}$

Find the amount of worming medication needed for

a a cat of mass 3 kg **b** a dog of mass 8 kg
c a cat of mass 4·5 kg **d** a dog of mass 27·6 kg
e a dog of mass 30·5 kg

*__Review 3__ $s = 4t + \frac{1}{2}at^2$ is a formula for finding the distance, s metres, that an object has moved after time, t seconds, and acceleration, a m/s^2. Find s if

a $t = 4$ seconds, $a = 3$ m/s^2 **b** $t = 1·5$ seconds, $a = {}^-3$ m/s^2 **c** $t = 1\frac{1}{2}$ seconds, $a = {}^-4\frac{1}{2}$ m/s^2.

In $F = ma$, F is called the **subject** of the formula.
Sometimes, once we substitute given values into a formula, the unknown is not the subject.
We must solve an equation to find the value of the unknown.

Worked Example

The voltage, V in volts, in a circuit is given by $V = IR$ where I is the current, in amps, and R is the resistance, in ohms.
Find I if $V = 20$ volts and $R = 4$ ohms.

Answer

$V = IR$
$V = I \times R$
$20 = I \times 4$ **substitute the known values**
$\frac{20}{4} = I$ **solve an equation to find I**
$\mathbf{I = 5}$

The current is **5 amps**.

Exercise 10 **Only use a calculator if you need to.**

1 The formula for finding force, F, in Newtons (N) from mass, m, in kilograms (kg) and acceleration, a, in metres per second² (m/s²) is $F = ma$.
Find
a a if $F = 60$ N and $m = 20$ kg
b m if $F = 2{\cdot}5$ N and $a = 5$ m/s²
c a if $F = 60{\cdot}4$ N and $m = 100$ kg.

2 The formula for finding speed, S, in metres per second (m/s) from distance, D, in metres and time, T, in seconds is $S = \frac{D}{T}$.
Find a D if $S = 30$ m and $T = 2{\cdot}5$ s b D if $S = 6{\cdot}4$ m and $T = 7$ s.

3 The formula for the change, £C, from £40 for n pieces of pie is $C = 40 - 5n$.
Find n if C is a £10 b £15 c £0 d £5.

4 $P = 2(l + w)$ gives the formula for the perimeter of this rectangle.
Find
a w if $l = 2$ m and $P = 6$ m
b l if $w = 5$ cm and $P = 24$ cm
c w if $l = 7{\cdot}5$ mm and $P = 50{\cdot}5$ mm.

5 The formula used to calculate mobile phone charges each month, £C, is $C = F + 0{\cdot}5n$ where F is the fixed monthly charge and n is the number of minutes used.
Find
a F if $C = £30$ and $n = 30$
b n if $C = £50$ and $F = £25$.
c F if $C = £56{\cdot}50$ and $n = 50$
d n if $C = £63{\cdot}80$ and $F = £24{\cdot}80$.

6 The cost, £C, to hire a vintage car is given by the formula $C = 39 + 15h$ where h is the number of hours hired.
a Roseanne hired a vintage car for her wedding. The cost was £136.50. For how many hours did she hire it?
b Pierre hired two cars for his birthday outing. It cost £183. For how many hours did he hire the cars?

7 The area, A cm², enclosed by an ellipse is given by
$$A = \pi ab.$$
Find the length of a if $b = 4{\cdot}6$ mm and $A = 82{\cdot}4$ mm.
Give your answer to 1 d.p.

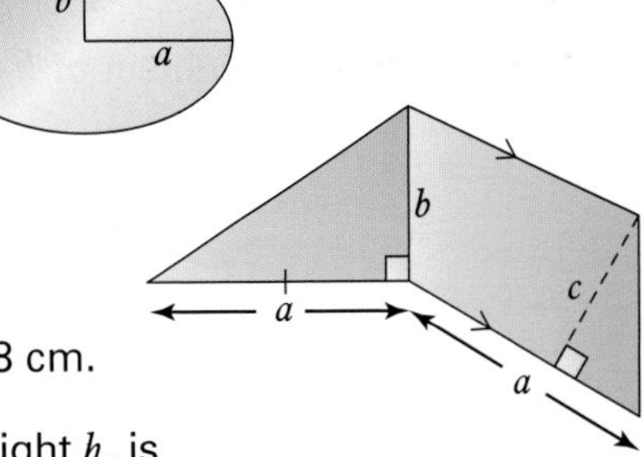

8 The area A cm² of a shape is given by
$$A = a(\tfrac{b}{2} + c).$$
Find the length of b if $A = 409{\cdot}5$ cm, $a = 19{\cdot}5$ cm, and $c = 12{\cdot}8$ cm.

*9 The surface area, S, of a cuboid of length l, width w and height h, is
$$S = 2lw + 2lh + 2hw$$
a Find h if $S = 528$ mm, $l = 18$ mm and $w = 10$ mm.
b Find w if $S = 258$ m, $l = 9$ m and $h = 6$ m.

Review 1 The voltage, V volts, in an electrical circuit, with current I amps and resistance, R ohms, is given by the formula $V = IR$.
Find
a R when $V = 6$ volts and $I = 2$ amps
b I when $V = 3{\cdot}6$ volts and $R = 4$ ohms
c R when $V = 13{\cdot}8$ volts and $I = 4{\cdot}6$ amps
d I when $V = 8{\cdot}1$ volts and $R = 3{\cdot}6$ ohms.

Review 2 The circumference of a circle of radius r is given by the formula

$C = 2\pi r$.

Find r to 2 d.p. if $C = 186{\cdot}4$ cm.

Changing the subject of a formula

Monique was making kites out of two triangular pieces of nylon. She knew the area and the length of the base of each piece. She wanted to know the height.

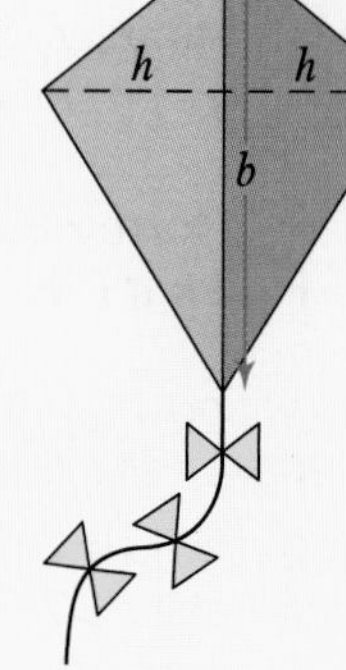

The formula for the area of a triangle is $A = \frac{b \times h}{2}$.
A is the **subject of the formula**.
Monique wants to change the formula so that h is the subject.
She uses inverse operations.

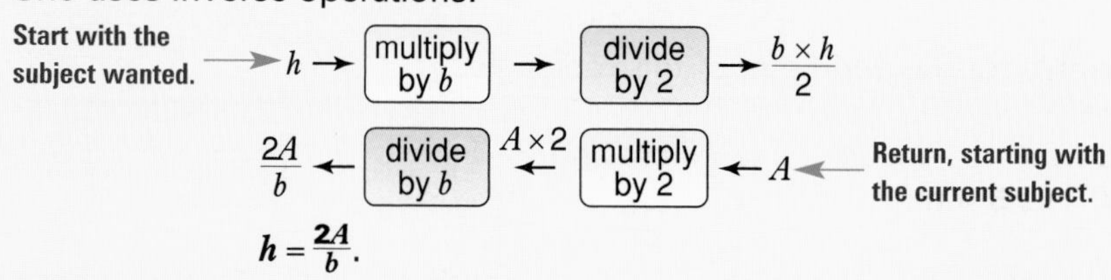

$h = \frac{2A}{b}$.

Worked Example

Make t the subject of $v = u + at$.

Answer

There is more about inverse functions on page 216.

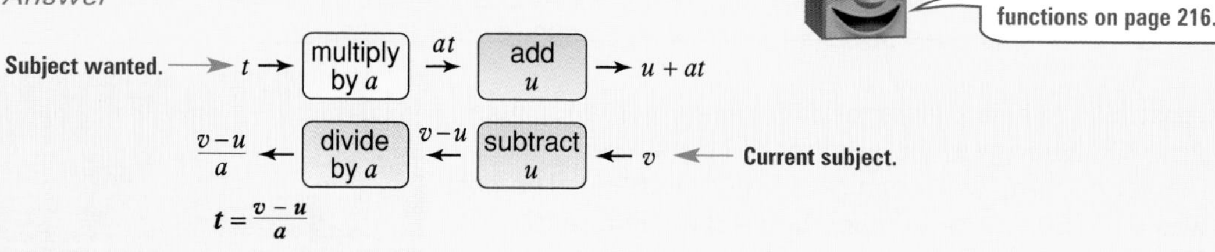

$t = \frac{v - u}{a}$

Exercise 11

1 **a** Make m the subject of $F = ma$. Use the function machines to help.

$m \rightarrow$ ☐ $\rightarrow$

$\leftarrow$ ☐ $\leftarrow F$

b Use your new formula to find m if $F = 20$ and $m = 2{\cdot}5$.

2 **a** Make a the subject of $F = ma$.

$a \rightarrow$ ☐ $\rightarrow$

$\leftarrow$ ☐ $\leftarrow F$

b Find a if $F = 16$ and $m = 3{\cdot}2$.

3 **a** Make l the subject of $A = lb$.
b Find l if $A = 168$ and $b = 32$.

4 **a** Make w the subject of $P = 2(l + w)$

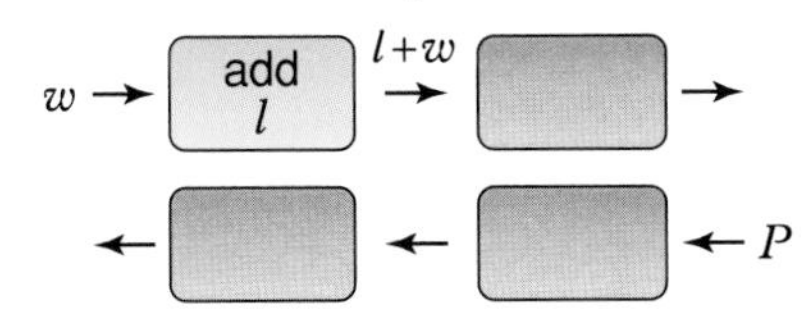

b Find w if $P = 68$ and $l = 7{\cdot}4$.

5 **a** Make l the subject of $P = 2(l + w)$.
b Find l if $P = 112$ and $w = 16{\cdot}75$.

6 **a** Make r the subject of $C = 2\pi r$.
b Find r to 1 d.p. if $C = 32{\cdot}4$ cm.

*__7__ **a** Make r the subject of $V = \pi r^2 h$.
b Find r if $V = 116$ and $h = 4{\cdot}2$.

T

Review 1 Use a copy of these diagrams to help.
a Make l the subject of $V = lbh$.

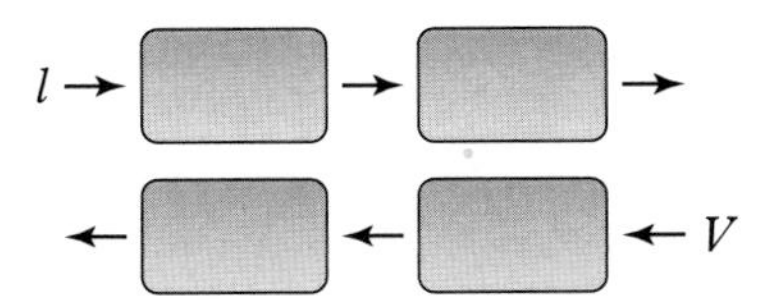

b Find l if $V = 60$, $b = 2{\cdot}5$, $h = 3$.

Review 2
a Make h the subject of $A = \frac{3h}{2}$.
b Find h if $A = 32\text{ cm}^2$.

* **Review 3**
a Make r the subject of $A = \pi r^2$.
b Find r if $A = 36{\cdot}4\text{ m}^2$.

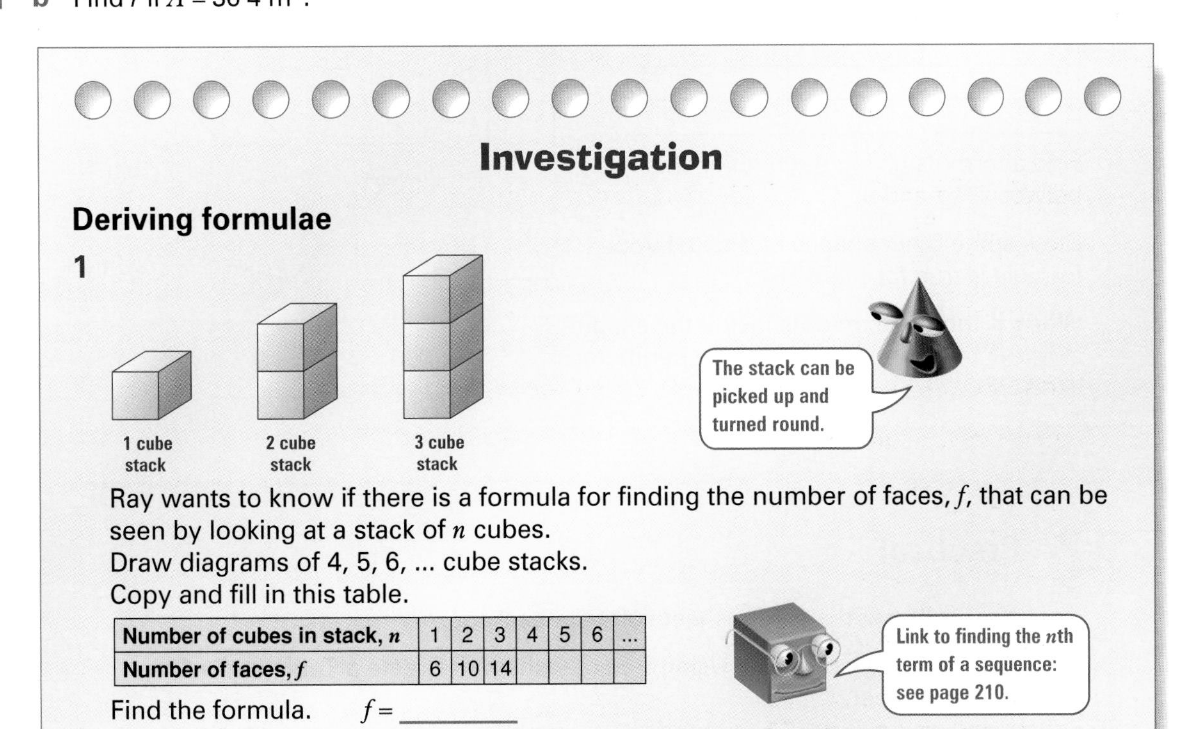

Investigation

Deriving formulae

1

Ray wants to know if there is a formula for finding the number of faces, f, that can be seen by looking at a stack of n cubes.
Draw diagrams of 4, 5, 6, ... cube stacks.
Copy and fill in this table.

Number of cubes in stack, n	1	2	3	4	5	6	...
Number of faces, f	6	10	14				

Find the formula. $f =$ ________

2

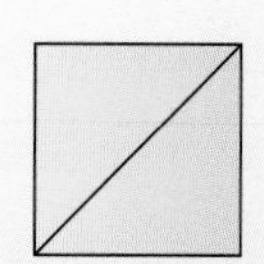

4 sides
sum of interior angles = 360°

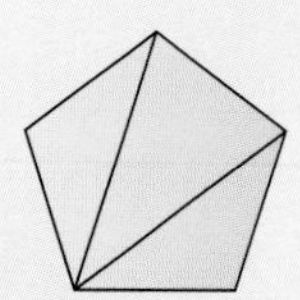

5 sides
sum of interior angles = 540°

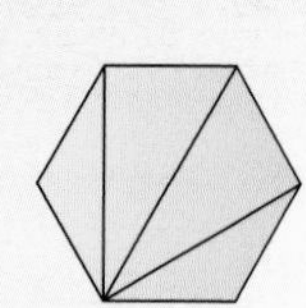

6 sides
sum of interior angles = 720°

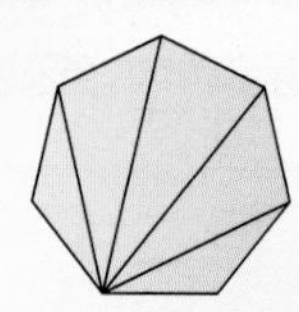

7 sides
sum of interior angles = 900°

Derive a formula for the sum of the interior angles, S, of an n-sided polygon.

*3 **a** Show, using algebra, that the perimeter P, of a semicircle of radius r is

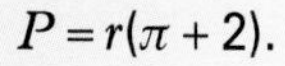

$$P = r(\pi + 2).$$

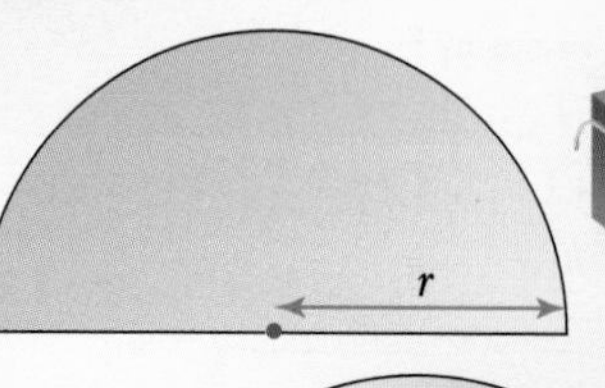

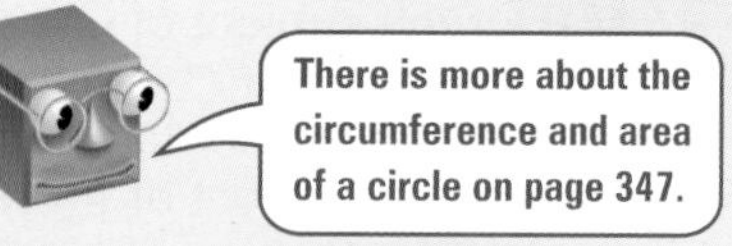

b Find a formula for the area of an annulus (ring) with outer radius r_1, and inner radius r_2.

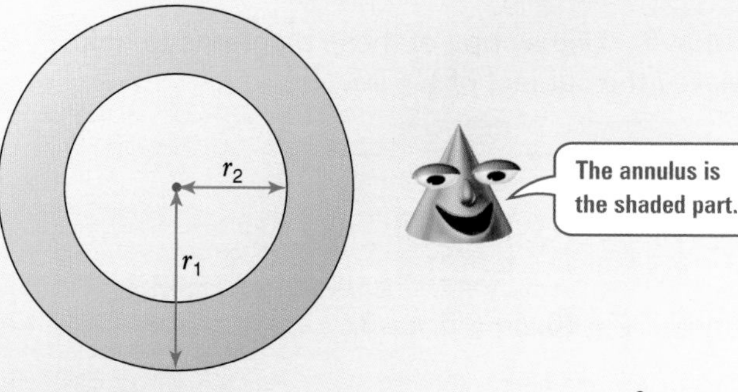

*4 The length of one side of a cube is a cm. The surface area of the cube is 5 cm^2. Prove, using algebra, that $s^3 = 216V^2$ where V is the volume in cm^3.

*5 Fill in a table like this for the shapes shown.

Number of dots on perimeter (P)	Number of dots inside shape (I)	Area of shape (units^2) (A)

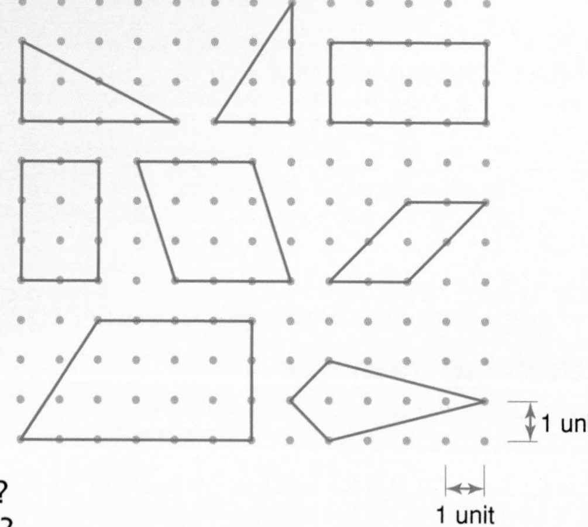

Find a formula giving the relationship between P, I and A.

Draw some other shapes and test if your formula is true for them also.

What if there are no dots inside the shape?
What if there are no dots on the perimeter?
What if ...

T

Practical

You will need a spreadsheet software package.

Ask your teacher for the Using a Spreadsheet to Create a Table and a Graph ICT worksheet.

Writing and solving equations

Discussion

- Kayla solved the equation $5d - 8 = {}^{-}22$ by transforming both sides.

$5d - 8 = {}^{-}22$

$5d - 8 + \mathbf{8} = {}^{-}22 + \mathbf{8}$ add 8 to both sides

$\therefore\ 5d = {}^{-}14$

$\therefore\ \frac{5d}{\mathbf{5}} = \frac{{}^{-}14}{\mathbf{5}}$ divide both sides by 5

$\therefore\ \mathbf{d = {}^{-}2{\cdot}8}$

Remember: ∴ means therefore.

How else could Kayla have solved this equation? **Discuss.**

- Jakob solved the equation $37x^2 = 29{\cdot}97$ using inverse operations.

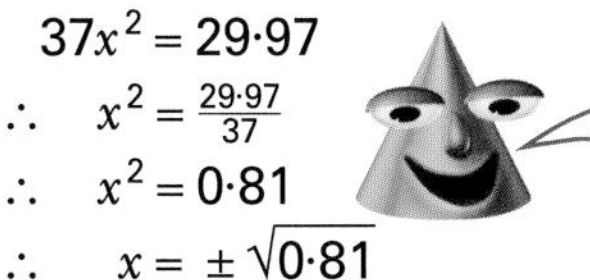

$37x^2 = 29{\cdot}97$

$\therefore\ x^2 = \frac{29{\cdot}97}{37}$

$\therefore\ x^2 = 0{\cdot}81$

$\therefore\ x = \pm\sqrt{0{\cdot}81}$

$= {}^{+}\mathbf{0{\cdot}9}$ **or** ${}^{-}\mathbf{0{\cdot}9}$

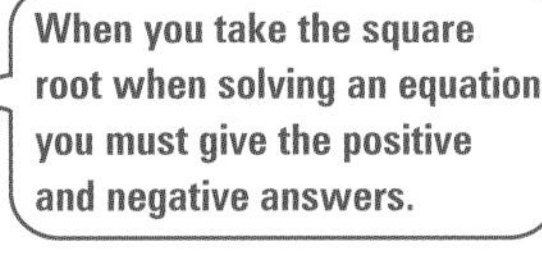

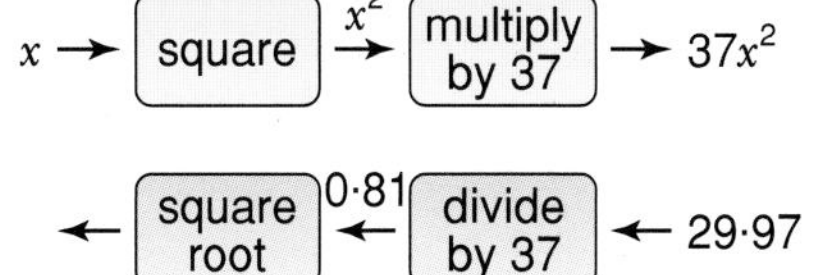

Discuss Jakob's method. How else could he have done it?

- Both Kayla and Jakob verified their answers.
  How might they have done this? **Discuss.**

- **Discuss** the most efficient way to solve this equation.
  Does the bracket need to be multiplied out first?

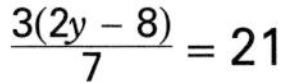

$\frac{3(2y - 8)}{7} = 21$

- **Discuss** how to write a single equation, with brackets, for this.
  How would you solve it?

  Supreme pizza cost £3 more than regular pizza.
  Regular pizza cost £x.
  Peter bought 5 Supreme pizzas. They cost £40 altogether.
  How much does a regular pizza cost?

Worked Example

Paula has £n. Rosa has £16 more than Paula. Ralph has twice as much as Rosa.
Altogether they have £144.
How much does each have?

Answer

Paula	**Rosa**	**Ralph**	**Total**
n	$n + 16$	$2(n + 16)$	144

$n + (n + 16) + 2(n + 16) = 144$

$n + n + 16 + 2n + 32 = 144$ multiply out brackets

$\therefore\ 4n + 48 = 144$ collect like terms

$\therefore\ 4n + 48 - \mathbf{48} = 144 - \mathbf{48}$ subtract 48 from both sides

$\therefore\ 4n = 96$

$\therefore\ \frac{4n}{\mathbf{4}} = \frac{96}{\mathbf{4}}$ divide both sides by 4

$n = 24$

Paula has £24, Rosa has £40 and Ralph has £80.

Worked Example
The length of a rectangle is four times its width. The perimeter is 66 cm.
Find its area.

Answer
Call the width x. Then the length is $4x$.

Perimeter $= 2(l + w)$.
$= 2(4x + x)$
$= 2 \times 5x$
$= 10x$
$10x = 66$ cm
$\therefore \quad x = \frac{66}{10}$ cm
$=$ **6·6 cm**

Area $= x \times 4x$
$= 6{\cdot}6 \times 4 \times 6{\cdot}6$
$=$ **174·24 cm²**

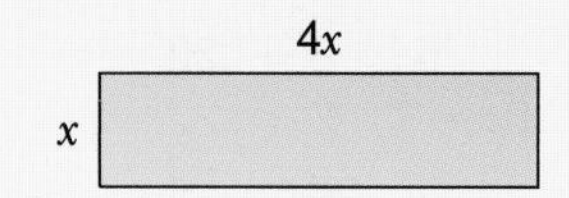

Exercise 12

Only use a calculator if you need to.

1 Solve these equations.

a $4d - 8 = {}^-12$ **b** $5m + 6 = {}^-9$ **c** $4x + 4 = {}^-13$ **d** ${}^-17 = 8a + 5$
e $\frac{2x + 5}{3} = 5$ **f** $\frac{4x + 2}{2} - 3 = {}^-10$ **g** $\frac{x - 2}{2} - 7 = {}^-11$ **h** $8{\cdot}4p + 2{\cdot}7 = 44{\cdot}7$
i $5{\cdot}1n - 3{\cdot}2 = {}^-34{\cdot}82$ **j** $34{\cdot}7 = 6{\cdot}4y - 3{\cdot}7$ **k** ${}^-3{\cdot}14 = 1{\cdot}4x + 8{\cdot}9$
l ${}^-21{\cdot}775 = 5{\cdot}3m - 7{\cdot}2$ **m** $12{\cdot}6 = {}^-1{\cdot}78y + 3{\cdot}7$ **n** $5(x - 3) = {}^-3$
o $4(5 - b) = 8$ ***p** $9(n - 2) + 4 = {}^-12$

2 Find the solution to these equations.

a $3x + 7 + 2x - 1 = 26$ **b** $4n - 3 + 5n = 15$ **c** $6p + 8 - 4p + 3p = 28$
d $7t + 3 + 6t = 68$ **e** $3(m - 2) + 2m = 24$ **f** $4(n + 6) - 3n = 25$
g $8(x - 3) + 2(x + 2) = 30$ **h** $4(n + 3) + 4(n - 1) = 20$ **i** $4(n - 7) - 6n = 0$
j $4(a - 1) - 5(a + 1) = 100$ **k** $3(x + 2) - 5(x - 1) = 7$ **l** $5(p + 3) - 7(p - 4) = 53$

3 **a** Write an expression for the perimeter of this glasshouse.
b The perimeter is 56 m.
Write an equation for this.
c Solve the equation to find the width of the glasshouse.

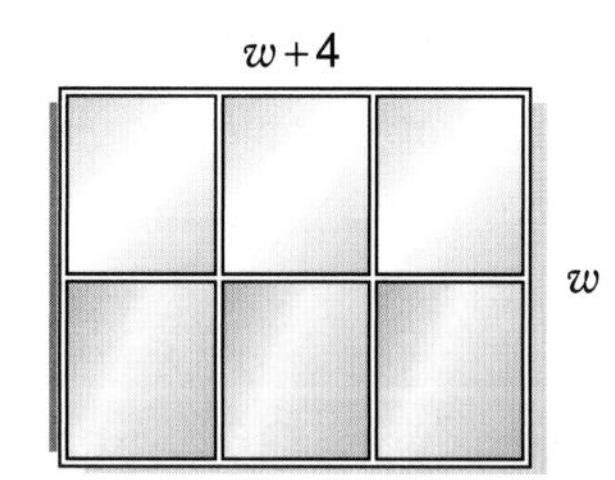

4 Marilyn made 80 pieces of fudge.
She put n pieces in one box. In the second box she put 12 more pieces than in the first box. In the third box she put 4 fewer pieces than in the second box.
Write and solve an equation to find the number of pieces of fudge in each box.

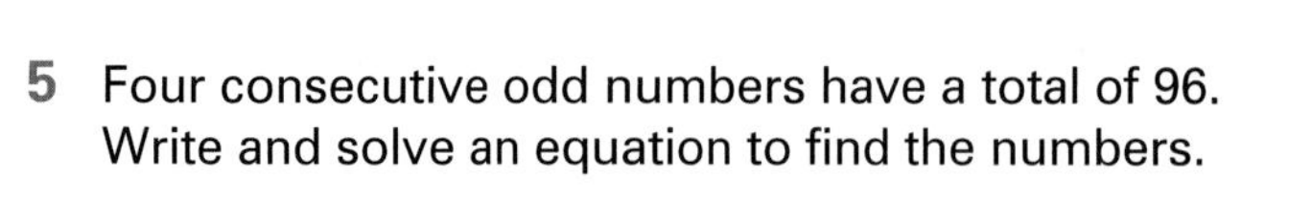

5 Four consecutive odd numbers have a total of 96.
Write and solve an equation to find the numbers.

6 I think of a number, add 7, multiply by 3, subtract 3, divide by 6, then multiply by 12.
The answer is 72.
What was the number I thought of?

7 **a** In this number pyramid, each number is the sum of the two numbers immediately above it.
Find the missing number, n, which makes the bottom number correct.

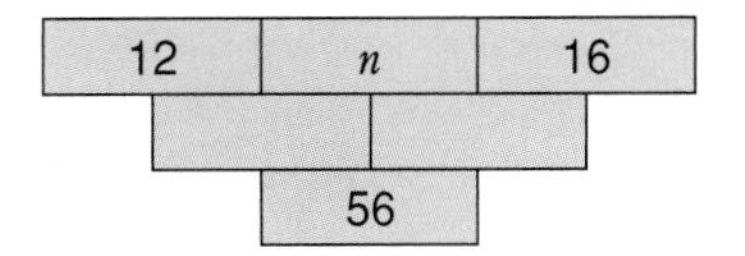

b Find n for each of these.

i

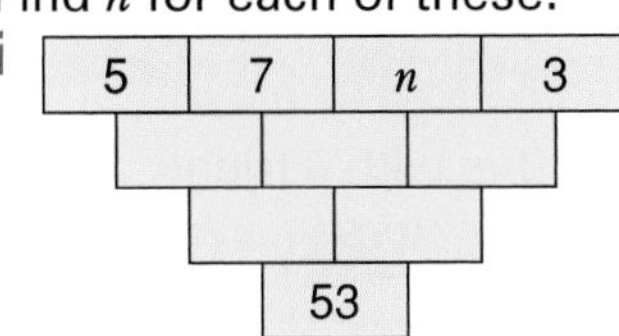

ii

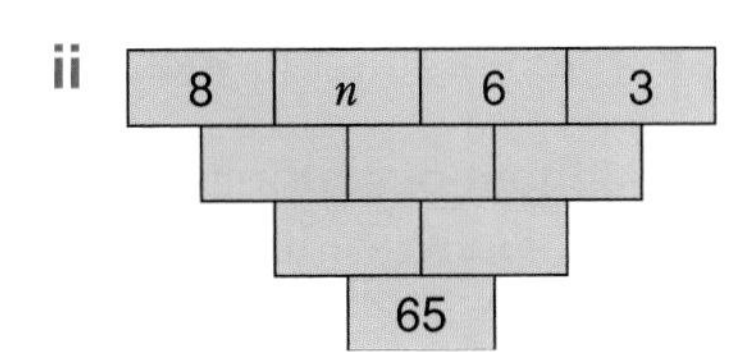

8 Write and solve an equation to find the answers to these.

a In a yacht race, the second leg is 6 km shorter than twice the first leg. The total distance for both legs is 204 km.
How long is the first leg?

b In Kent's class there are three times as many pupils who are right-handed as left-handed. There are 28 pupils altogether.
How many are left-handed?

c Two years ago, Adam was five times as old as his daughter. The sum of their ages now is 52.
How old is the daughter?

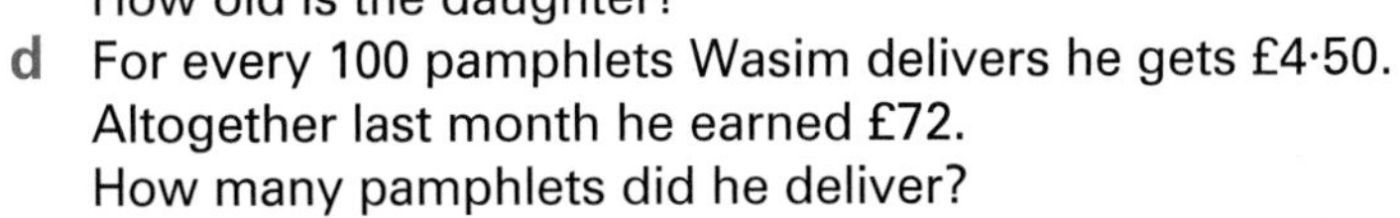

d For every 100 pamphlets Wasim delivers he gets £4·50.
Altogether last month he earned £72.
How many pamphlets did he deliver?

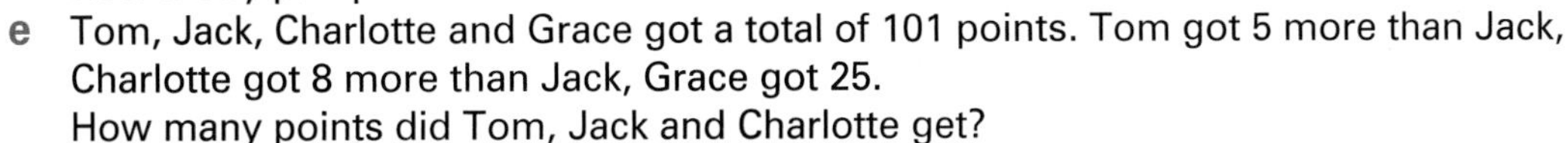

e Tom, Jack, Charlotte and Grace got a total of 101 points. Tom got 5 more than Jack, Charlotte got 8 more than Jack, Grace got 25.
How many points did Tom, Jack and Charlotte get?

f The length of a rectangle is three times its width. Its perimeter is 32 m.
Find its area.

9 The number in each square is the sum of the numbers in the two circles either side.
Find the missing numbers.
Write and solve an equation to help.

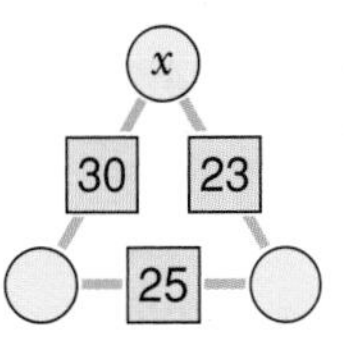

10 Solve these equations.

a $6{\cdot}3a^2 = 14{\cdot}175$ **b** $3{\cdot}2m^2 = 20$ **c** $1{\cdot}4b^2 = 24{\cdot}696$

***11** The solutions to $3x + 5 = {}^-4$ and $5x + 20 = 5$ are both $^-3$.
Write some more equations that have a solution of $^-3$.

***12** Find a pair of numbers that satisfy $7x - 2y = 38$ if one of the numbers is three times the other.
Is there more than one answer?

***13** Find two consecutive even numbers so that the difference between their squares is 100. You could use the diagram to help.

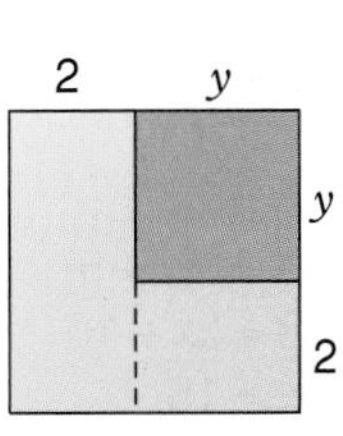

Review 1 Solve these equations.

a $5d - 6 = {}^-7$ **b** $\frac{2x + 4}{10} - 3 = {}^-7$ **c** $4{\cdot}9y - 2{\cdot}6 = {}^-8{\cdot}97$ **d** $5a - 7 = {}^-3$

e $\frac{8x - 3}{2} = {}^-27$ **f** $4(p + 4) - 5 = {}^-17$ **g** $5(q - 3) + 5 = {}^-25$ **h** $3{\cdot}2b^2 = 96{\cdot}8$

Review 2 Solve these equations.

a $2x + 3 + 4x - 7 = 20$ b $5n - 2 - 2n = 19$ c $7p - 3 - 4p + 6 = 27$

d $3(m + 1) + 2(m - 3) = 36$ e $2(y - 3) - 4(y - 1) = {}^{-}6$ *f $2(x + 3) - 5(x - 1) = 23$

Review 3 Subtracting 6 from 5 times a number then dividing by 2 and then adding 4 gives $^{-}6{\cdot}5$. Write and solve an equation to find the number.

Review 4 Jayne had a 16 cm length of liquorice. She shared this with a friend. The piece she gave her friend was 2 cm less than twice the length she kept for herself. Write and solve an equation to find the length of the piece Jayne kept for herself.

* **Review 5** Write and solve an equation to find the missing numbers.

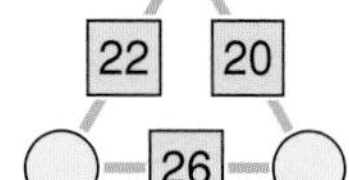

Equations with unknowns on both sides

Discussion

Discuss how to write and solve an equation for this.
Is it possible to solve the equation using inverse operations?

Multiplying a number by 3 and then adding 2 gives the same answer as subtracting the number from 30 and adding 4.

To solve an equation with **unknowns on both sides**, we must get the unknowns onto one side.

Worked Example

Solve these.

a $8(y + 4) = 100 - 4(20 - y)$

*b $\frac{8}{(x+5)} = \frac{3}{(x+1)}$

Answer

a Using transforming both sides:

$$8(y + 4) = 100 - 4(20 - y)$$

$\therefore\ 8y + 32 = 100 - 80 + 4y$ multiply out the brackets

$\therefore\ 8y - \mathbf{4y} + 32 = 100 - 80 + 4y - \mathbf{4y}$ subtract $4y$ from both sides

$\therefore\ 4y + 32 = 20$

$\therefore\ 4y + 32 - \mathbf{32} = 20 - \mathbf{32}$ subtract 32 from both sides

$\therefore\ 4y = {}^{-}12$

$\therefore\ \frac{4y}{\mathbf{4}} = \frac{{}^{-}12}{\mathbf{4}}$ divide both sides by 4

$\therefore\ \mathbf{y = {}^{-}3}$

*b $\frac{8}{(x+5)} = \frac{3}{(x+1)}$

$8 = \frac{3 \times (x+5)}{x+1}$ The inverse of dividing by $(x + 5)$ is multiplying by $(x + 5)$.

$8 \times (x + 1) = 3(x + 5)$ The inverse of dividing by $(x + 1)$ is multiplying by $(x + 1)$.

$8x + 8 = 3x + 15$ Multiplying out the brackets.

$\therefore\ 8x + 8 - 3x = 15$ The inverse of adding $3x$ is subtracting $3x$.

$\therefore\ 5x + 8 = 15$

$\therefore\ 5x = 15 - 8$ The inverse of adding 8 is subtracting 8.

$\therefore\ 5x = 7$

$\therefore\ x = \frac{7}{5}$ The inverse of multiplying by 5 is dividing by 5.

$= \mathbf{1{\cdot}4}$

Exercise 13

Only use a calculator if you need to.

1 Solve these equations.

a $a + 3(a + 1) = 2a$ **b** $2(3x - 1) = 4(x - 1)$ **c** $2p - 2(p - 1) = 5p$

d $4(y - 1) - 2 = 3y$ **e** $3m + 7 + 3(m - 1) = 2(2m + 6)$ **f** $4(b - 1) + 3(b + 2) = 5(b - 4)$

***g** $\frac{12}{(n+1)} = \frac{21}{(n+4)}$ ***h** $\frac{5}{(n+3)} = \frac{4}{(n+5)}$ ***i** $\frac{15}{(x+4)} = \frac{19}{(x+3)}$

2 Three times a number plus 6 is equal to twice the number plus 15. What is the number?

3 Multiplying a number by 3 and then adding 4 gives the same answer as subtracting the number from 20. Find the number.

4 To go from the start to the finish, begin with a number, p and follow one of the two paths to end with the number q.
Find p, so that whichever path you follow you end up with the *same* value of q.

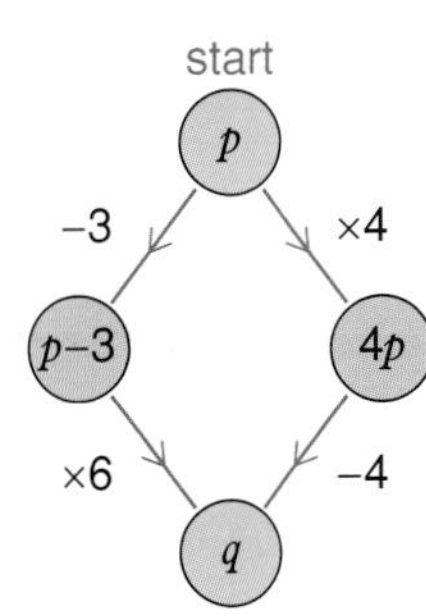

5 Repeat question **2** for these diagrams.

a

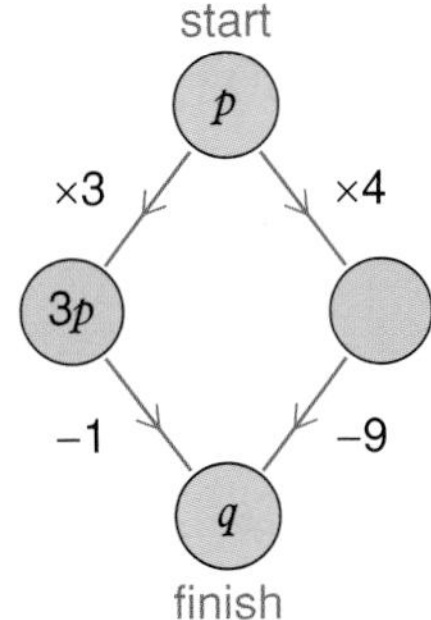

b

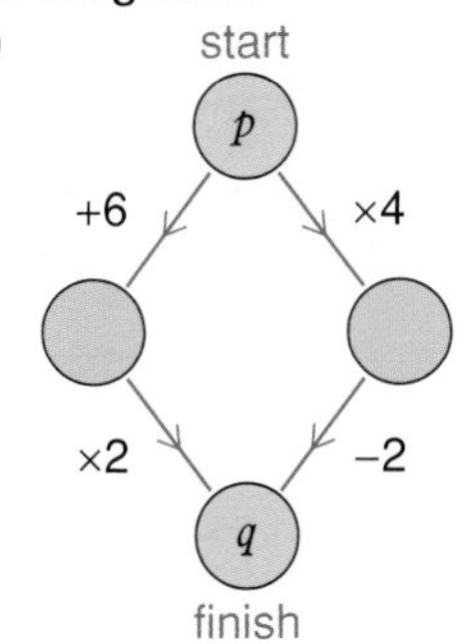

c

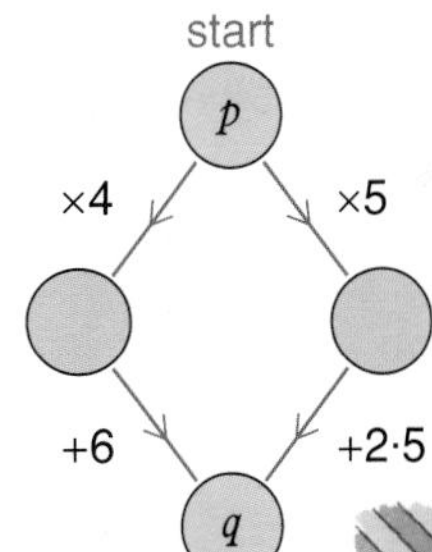

6 Meggie buys 4 ice creams and a drink and this costs the same as 2 ice creams and 4 drinks. If a drink costs 80p, how much does an ice cream cost?

7 For each of these, calculate the value of a and then use it to find the perimeter of the rectangle.

a **area = 115 mm²**

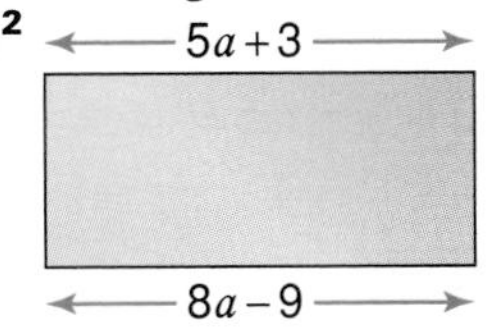

b **area = 7100 cm²**

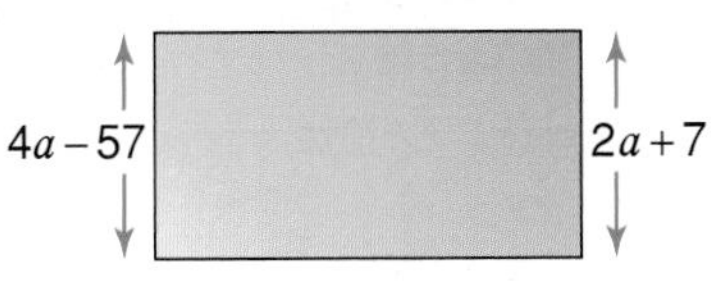

***8** Write and solve an equation to find the answers to these.

a Ben and Maddy are given the same amount of money for Christmas. Before Christmas, Ben had £205 and Maddy had £25. Now Ben has four times as much as Maddy.
How much were they each given for Christmas?

b Marshall and Gretchen have 200 tropical fish between them. Marshall gave Gretchen 10 of his fish. Now Gretchen has three times as many as Marshall.
How many fish does Marshall have now?

c Ray has twice as many euro as Julie. If Ray gave Julie 5 euro he would still have 10 more euro than Julie.
How many euro does Ray have?

Review 1 Solve these equations.

a $3 - 2(2x + 1) = x + 17$ **b** $x + (x + 2) = 792 - (x + 4)$ ***c** $\frac{5}{(x + 2)} = \frac{3}{(x + 5)}$

Review 2 Write and solve an equation for these.

a Five times a number plus 6 is the same as the number subtracted from 30.
What is the number?

***b** Adrienne has three times as many pounds as Karim. Adrienne gives Karim £20. She then has £5 more than Karim.
How much does Adrienne have now?

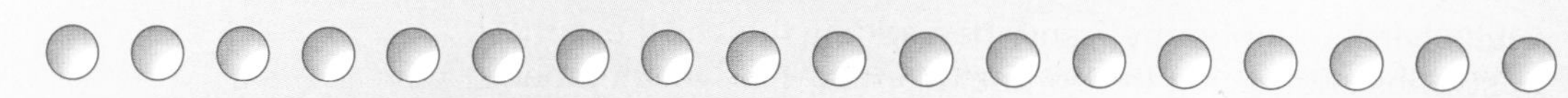

Investigation

Solving equations

Consider the equation $2x = 8$.

This equation may be solved as follows:

Step 1 Let x be equal to any number, say $x = 5$.

Step 2 Multiply the value you gave to x by 2, the number on the left-hand side of the equation. $5 \times 2 = 10$

Step 3 Divide the number on the right-hand side of the equation by the result of **step 2**.
$8 \div 10 = 0{\cdot}8$

Step 4 Multiply the result of **step 3** by the original guess in **step 1**. $0{\cdot}8 \times 5 = 4$

The result of all of the above steps is 4. $x = 4$ is the solution to the equation $2x = 8$!
It seems to be a way to solve $2x = 8$. Was putting $x = 5$ in **step 1** just lucky?

What if $x = 10$ in **step 1**?
What if $x = 2$ in **step 1**?
What if ...

Can you use a method similar to that shown above to solve $3x = 12$?
Investigate.

What if we began with the equation $\frac{x}{4} = 5$?
What if we began with the equation $2x + 3 = 8$?
What if ...

Solving equations using a graph

Discussion

- One bag of peanuts costs £3.
  How much do 2, 3, 4, 5, ... bags cost? **Discuss.**

  We could write a table for this.
  What is the ratio *cost : number of bags* for each number of bags?
  What do you notice? **Discuss.**

Number of bags	1	2	3	4	5	...
Cost (£)	£3					...

  Is the cost directly proportional to the number of packets bought? **Discuss.**

If variables are in direct proportion, corresponding values are always in the same ratio.

We could draw a graph of cost versus number of bags.

Will the graph be a straight line? **Discuss.**

Use a copy of this grid and plot the points to check.

What equation could you write for the relationship between number of bags (x) and cost (y)? **Discuss**

$y =$ ______

How could you use the graph to find the cost of 8 bags? **Discuss.**

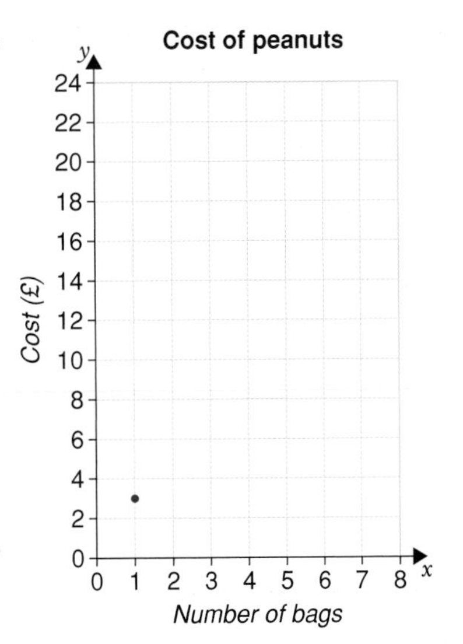

- Think of some other examples where one variable is directly proportional to another? **Discuss.**

  Is the graph of the two variables always a straight line? **Discuss.**

If two quantities x and y are directly proportional, the ratios of corresponding values of y and x are equal.

x	1	2	3	4	5	...
y	2·4	4·8	7·2	9·6	12	...

The ratios of corresponding values of y and x are

$\frac{y}{x} = \frac{2\cdot4}{1} = \frac{4\cdot8}{2} = \frac{7\cdot2}{3} = \frac{9\cdot6}{4} = \frac{12}{5} = \ldots = \mathbf{2\cdot4}$

∝ is the symbol for 'is directly proportional to'.

$\therefore$ y is **directly proportional to** x. $\quad y \propto x$.

The relationship between y and x is given by

$y = 2\cdot4x$ This is the ratio of $y : x$.

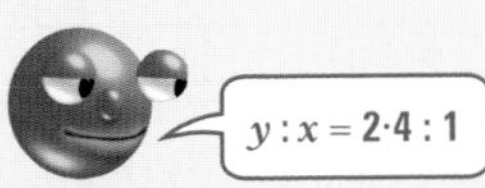

The **graph of y against x** will be a **straight line**.

Exercise 14

1 a One pizza costs £4.
Copy and fill in this table.

Number of pizzas	1	2	3	4
Cost (£)				

b Work out the ratio $\frac{\text{cost}}{\text{number bought}}$ for each pair of values on the table.
What do you notice?

c Is the cost directly proportional to the number of pizzas bought? Explain.

d Use a copy of this grid.
Plot the graph of cost versus number of pizzas bought.
Do the points lie in a straight line?

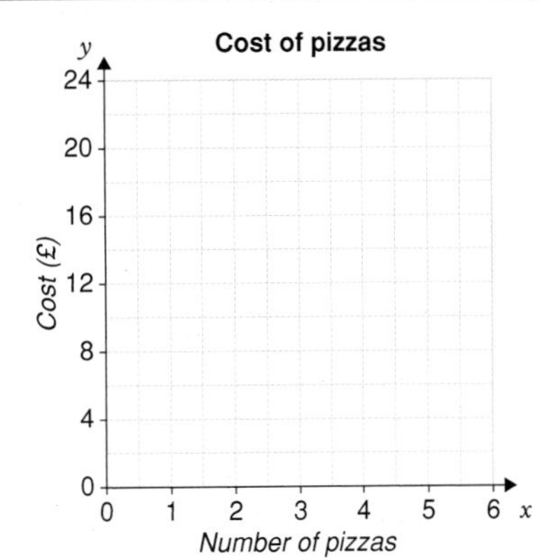

e Write a formula for the relationship between cost (y) and number of pizzas bought (x).
$y =$ ________.

f Find the cost of 8 pizzas.
Explain how you found this.

2 Mobile phone covers were reduced from £4 each to £3 each in a sale.

a Copy and fill in this table.

Number	1	2	3	4	5	...
Original price	£4	£8				...
Sale price	£3					...

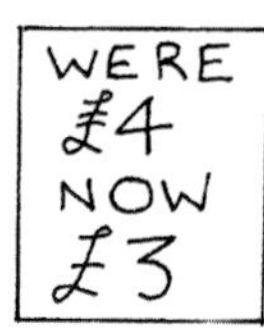

b Work out the ratio *sale price : original price* for each pair of values on the table.
What do you notice?

c Is the sale price directly proportional to the original price?

d Draw a graph of sale price versus original price.
Have original price on the horizontal axis.
Do the points lie in a straight line? Is this what you would expect? Explain.

e Write a formula for the relationship between the sale price (y) and the original cost (x).

f Find the sale price of phone covers that originally cost £48.

3 Rachel ran 250 metres every 2 minutes.
She ran at a constant speed.

a Write down five pairs for distance and time.
For example (2,250).

b Is the ratio *distance travelled : time taken* constant?
What is this ratio?

c Draw a distance versus time graph.

d Write a formula for the relationship between distance (y) and time (x).

$y =$ ________.

e Use this relationship to find the distance she would run in
i 18 minutes **ii** 9 minutes.

Assume she runs at the same speed.

4

x	1	2	3	4	5	6	7	...
y	4·5	9	13·5	18	22·5	27	31·5	...

a Are x and y directly proportional? Explain.

b Write an equation $y =$ ______ to show the relationship between x and y.

c Will the graph of y against x be a straight line?

T

Review

a There are 5 miles to every 8 kilometres.
Copy and fill in this table.

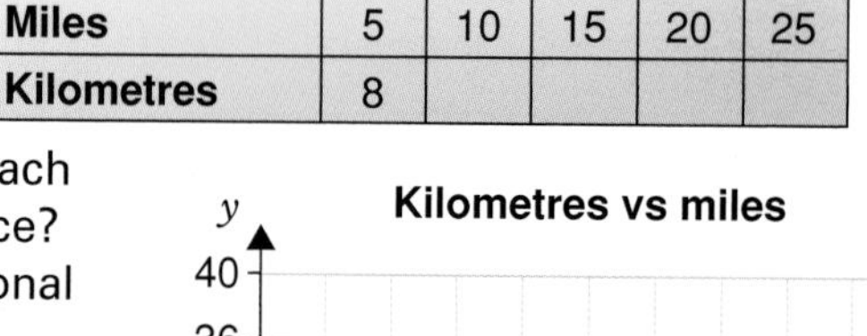

Miles	5	10	15	20	25
Kilometres	8				

b Work out the ratio *kilometres : miles* for each pair of values on the table. What do you notice?

c Is the number of kilometres directly proportional to the number of miles? Explain.

d Use a copy of this grid.
Plot the graph of kilometres versus miles.
Do the points lie in a straight line?

e Write a formula for the relationship between kilometres (y) and miles (x).

$y =$ ________.

f Find the number of kilometres in these numbers of miles.
i 45 miles **ii** 105 miles

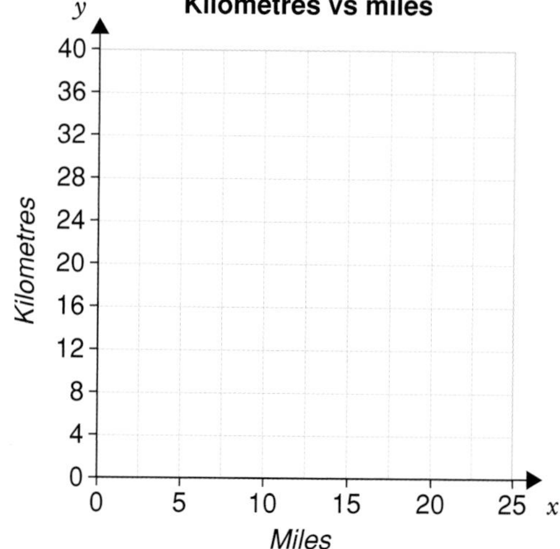

Solving non-linear equations

A non-linear equation has terms with indices greater than 1, such as x^2, x^3, … .
We can solve some **non-linear equations** to get an exact answer.

Worked Example
Solve these. **a** $x^2 - 10 = 246$ **b** $5 = \frac{125}{x^2}$

Answer

a
$x^2 - 10 = 246$
$x^2 = 246 + 10$
$x^2 = 256$
$\sqrt{x^2} = \pm\sqrt{256}$
$x =$ **16 or $^{-}$16**

We can only solve to get an **exact** answer if the number under the square root is a square number.

Whenever you take the square root when solving an equation, you must give the positive and negative solutions.

b
$5 = \frac{125}{x^2}$
$5x^2 = 125$
$x^2 = \frac{125}{5}$
$x^2 = 25$
$\sqrt{x^2} = \pm\sqrt{25}$
$x =$ **5 or $^{-}$5**

Exercise 15

1 Solve these to find the exact solutions.
a $m^2 + 20 = 84$ **b** $y^2 - 24 = 97$ **c** $p^2 - 124 = 200$ **d** $4 = \frac{100}{x^2}$ **e** $7 = \frac{112}{n^2}$

Review Solve these.
a $a^2 + 16 = 52$ **b** $n^2 - 35 = 365$ **c** $8 = \frac{288}{b^2}$

We can solve harder non-linear equations using **trial and improvement**, or using **a calculator, spreadsheet** or **graph plotting software**.
The answer we get is usually not exact.

Using a calculator

Example $m^3 + 3 = 53$
$m^3 = 50$ subtracting 3 from both sides
$\sqrt[3]{m} = \sqrt[3]{50}$
key [∛] [50] [=] to get 3·68 (2 d.p.)

50 is not a cube number so the answer will not be exact.

Using trial and improvement and a calculator

When using trial and improvement it is important to be systematic.

Example Solve $x^3 + x = 40$ giving the answer to 2 d.p.

It is no good trying 4 because $4^3 = 64$ so $4^3 + 4$ is going to be much too big.

Guess what x might be close to.
Try $x = 3$ If $x = 3$, $x^3 + x = 30$ too small
Try $x = 3·5$ If $x = 3·5$, $x^3 + x = 46·375$ too big
Try $x = 3·3$ If $x = 3·3$, $x^3 + x = 39·237$ too small but close
Try $x = 3·4$ If $x = 3·4$, $x^3 + x = 42·704$ too big

We can see that the answer is between 3·3 and 3·4. It is closer to 3·3 since 39·237 is closer to 40 than is 42·704.

Try $x = 3·33$ If $x = 3·33$, $x^3 + x = 40·256037$ too big
Try $x = 3·32$ If $x = 3·32$, $x^3 + x = 39·914368$ too small
The answer is between 3·33 and 3·32.
Because 39·914037 is closer to 40 than 40·256368, the answer is closer to 3·32.
The answer to 2 decimal places is **3·32**.

Using a spreadsheet

Example Solve $x^3 + x = 40$

Increase the number by 0·1 each time

	A	B
1	x	x³+x
2	3	=A2*A2*A2+A2
3	=A2+0.1	=A3*A3*A3+A3
4	=A3+0.1	
5	=A4+0.1	

	A	B
1	x	x³+x
2	3	30
3	3.1	32.891
4	3.2	35.968
5	3.3	39.237
6	3.4	42.704

Using this we can see that the solution is between 3·3 and 3·4.
Then we can change column A so that it starts at 3·3 and increases by 0·01 each time.

Exercise 16

1 Use a calculator to find the solutions to these.
Give the answers to 2 d.p.
a $a^3 = 50$ **b** $b^3 = 80$ **c** $x^3 = 110$ **d** $3a^3 = 50$ **e** $1{\cdot}2x^3 = 8{\cdot}4$

2 Use a calculator and trial and improvement to find an approximate solution to each of these. Give your answers to 2 d.p.
a $m^3 - m = 80$ **b** $x^3 + x = 97$ **c** $2y^3 + y = 6$

3 The area of this rectangle is 84·32 cm².

$a + 3$

a

a Write an equation for the area of the rectangle.
b Find the value of a to 2 d.p. Show your working.
You could use a table like this one.

a	$a + 3$	$a(a + 3)$	too big or too small

T

*4 Use a copy of the tables.
Fill in the table for the equation and values of x given.
Then continue and find the value of x, to 1 d.p., that gives the value of y closest to 0.
Use trial and improvement.

a $y = x^3 - x - 4$

x	0	1	2	3	4
y					

b $y = 2x^3 + 5x - 10$

x	0	1	2	3	4
y					

Review 1 Use a calculator to find the solutions to these to 1 d.p.
a $b^3 = 75$ **b** $4y^3 = 24$ **c** $7{\cdot}2a^3 = 16{\cdot}72$

Review 2 Use trial and improvement and a calculator to find a solution to 2 d.p.
a $p^3 - p = 10$ **b** $x^3 + x = 47$

Review 3 The area of this rectangle is 59 m².
a Write an equation for the area of the rectangle.
b Find the value of x to 2 d.p. Show your working.

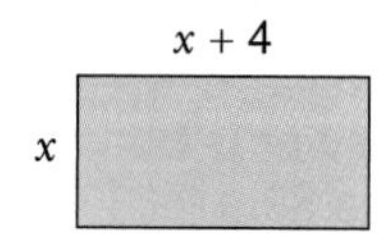

Summary of key points

$3x - 4$, $5p - 2$, $\frac{8y - 3}{2}$ and $4x + 3y$ are all **linear expressions**.

Expressions are made by using letter symbols (unknowns) and numbers together with operations such as +, −, ×, ÷, ( ).

$3x - 4 = 7$, $5p - 2 = {}^{-}18$, $\frac{8y - 3}{2} = 6$ and $4x + 3y = {}^{-}2$ are all **linear equations**.

A **formula** gives the relationship between unknowns that stand for something specific.

Example $V = IR$ is the voltage in an electrical circuit with current I and resistance R.

A **function** is a special sort of equation. It gives the relationship between the input x, and the output y, of a function machine.

This relationship is also the relationship between the variables on the x- and y-axes of a graph.

An **inequality** has <, >, ⩽ or ⩾.

Examples $x < 4$ $\quad 7 < y \leqslant 3$.

Algebraic operations follow the same rules as **arithmetic operations**.

Inverse operations

Examples If $f + g = h$ then $h - g = f$ and $h - f = g$.

If $pq = r$ then $\frac{r}{p} = q$ and $\frac{r}{q} = p$.

If $5x - 6 = y$ then $x = \frac{y + 6}{5}$.

We can **simplify expressions**.

Examples

$$\frac{18p^2}{12p} = \frac{{}^{3}\cancel{18} \times p \times \cancel{p}^{1}}{{}^{2}\cancel{12} \times \cancel{p}^{1}} = \frac{3p}{2}$$

We cancel by dividing both numerator and denominator by the same number.

$$\frac{y^4}{y^2} = y^{4-2} = y^2$$

$$3a \times 2a = 3 \times 2 \times a \times a = 6a^2$$

$$n^2 \times n^4 = n^{2+4} = n^6$$

We can simplify expressions by **collecting like terms**.

Examples

$$8p + 4q - 10p + 7q = 8p - 10p + 4q + 7q = {}^{-}2p + 11q$$

$$15 - 2(x - 3y) = 15 - 2x + 6y$$

$$8x^2 - 3x + 5x^2 + 2x = 8x^2 + 5x^2 - 3x + 2x = 13x^2 - x$$

We cannot subtract x from $13x^2$ because they are not like terms.

When we **factorise** an expression we put the highest common factor of the terms outside the bracket.

Examples

$12x - 16$ is factorised as $4(3x - 4)$ — 4 is the HCF of $12x$ and 16

$8y + 24w$ is factorised as $8(y + 3w)$ — 8 is the HCF of $8y$ and $24w$

$m^3 + m^2 + 3m$ is factorised as $m(m^2 + m + 3)$ — m is the HCF of m^3, m^2 and $3m$

Always check your factorising by multiplying out the brackets.

We **evaluate** an expression by **substituting** values for the unknown. We follow the rules for **order of operations**.

Example If $x = {}^-4$ and $y = 0{\cdot}1$, then

$$\begin{aligned} 2x^2(x+6) &= 2 \times ({}^-4)^2 \times ({}^-4 + 6) \\ &= 2 \times 16 \times (2) \\ &= \mathbf{64} \end{aligned}$$

$$\begin{aligned} 3y^3 - 2y &= 3 \times 0{\cdot}1^3 - 2 \times 0{\cdot}1 \\ &= 3 \times 0{\cdot}001 - 0{\cdot}2 \\ &= 0{\cdot}003 - 0{\cdot}2 \\ &= \mathbf{{}^-0{\cdot}197} \end{aligned}$$

When we substitute values for unknowns into a **formula** we sometimes need to solve an equation.

Example The formula for finding the speed, S, in km/h, is $S = \frac{D}{T}$ where D is the distance, in km, and T is the time taken in hours.

If $S = 80$ km/h and $T = 2$ hours, then

$80 = \frac{D}{2}$

$80 \times 2 = D$ The inverse of dividing by 2 is multiplying by 2.

$D = 160$ km

Sometimes we want to **change the subject** of a formula.

Example C is the subject of the formula $C = 2\pi r$ where C is the circumference and r is the radius of a circle.

To make r the subject, use inverse operations.

Start with the subject required → r → multiply by π → multiply by 2 → $2\pi r$

$\frac{C}{2\pi}$ ← divide by π ← $\frac{C}{2}$ ← divide by 2 ← C ← return with the current subject.

$\mathbf{r = \frac{C}{2\pi}}$

We can **solve equations** using **transforming both sides of the equation** or by **inverse operations**.

Examples

$4p - 9 = {}^-27$

$4p - 9 + 9 = {}^-27 + 9$ adding 9 to both sides

$4p = {}^-18$

$\frac{4p}{4} = \frac{{}^-18}{4}$ dividing both sides by 4

$p = \mathbf{{}^-4{\cdot}5}$

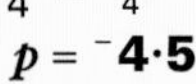

$5x^2 = 27{\cdot}63$

$x^2 = \frac{27{\cdot}63}{5}$ The inverse of multiplying by 5 is dividing by 5.

$x^2 = 5{\cdot}526$

$\sqrt{x^2} = \pm\sqrt{5{\cdot}526}$ The inverse of squaring is taking the square root.

$x =$ **2·35 or ⁻2·35 (2 d.p.)**

Use transforming both sides if there are unknowns on both sides.

Example

$4(a + 3) = 24 - 2(9 - a)$

$4a + 12 = 24 - 18 + 2a$ multiplying out the brackets

$4a + 12 - 2a = 6$ subtracting $2a$ from both sides

$2a = 6 - 12$ subtracting 12 from both sides

$2a = {}^-6$

$a = \frac{{}^-6}{2}$ dividing both sides by 2

$a = \mathbf{{}^-3}$

We can **solve some equations using a graph**. If the two variables are directly proportional, the ratios of corresponding values are always the same.

Example

x	1	2	3	4	5
y	1·5	3	4·5	6	7·5

$\frac{y}{x} = \frac{1·5}{1} = \frac{3}{2} = \frac{4·5}{3} = \frac{6}{4} = \frac{7·5}{5}$

$y = 1·5x$

Non-linear equations can sometimes be solved to get an exact answer.

Example $2x^2 - 12 = 60$

$2x^2 = 60 + 12$ adding 12 to both sides

$2x^2 = 72$

$x^2 = \frac{72}{2}$ dividing both sides by 2

$x^2 = 36$

$x = \sqrt[\pm]{36}$ taking the square root of both sides

$x = +$**6 or** $^-$**6**

Sometimes we can use **trial and improvement** and **ICT** tools such as a calculator, spreadsheet or graph plotting software. The answer is not usually exact.

Example $m^2 + 5 = 27$

$m^2 = 27 - 5$

$m^2 = 22$

$m = \sqrt[\pm]{22}$

$m = +$**4·69 or** $^-$**4·69**

See pages 183 and 184 for trial and improvement examples.

Test yourself

Only use a calculator if you need to.

1 Write true or false for each of these. (A)

- **a** $5x + 7 = 23$ is an equation. x has a particular value.
- **b** An expression always has an equals sign.
- **c** $F = \frac{9C}{5} + 32$ is a formula where F is temperature in degrees Fahrenheit and C is temperature in degrees Celsius.
- **d** A formula always has more than one variable.
- **e** A function gives the relationship between the input and output of a function machine.
- **f** $y \leqslant {}^-4$ is an inequality and means y is less than 4.
- **g** $1 < m \leqslant 3$ is an inequality and means m is greater than 1 but less than or equal to 3.

2 The *total* length of each rod is given by the letter written inside it.

- **a** Which of these are true?

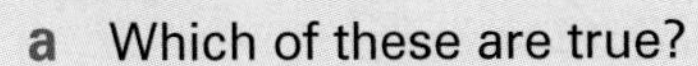

$4r = p$ $\quad g = \frac{r}{2}$ $\quad 2g + r = p$ $\quad p - 2g = r$ $\quad g = \frac{p - r}{2}$

- **b** Write two more equations that are true for these rods.

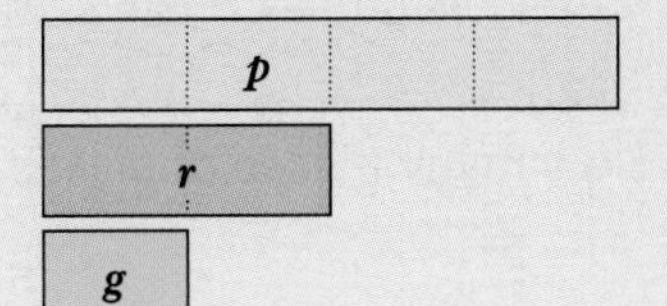

3 Choose the correct answer.

a If $5x - 7 = y$ then **A** $x = \frac{y}{5} + 7$ **B** $x = \frac{y+7}{5}$ **C** $x = \frac{y-7}{5}$

b If $4(a - 3) = y$ then **A** $a = 4y + 3$ **B** $a = \frac{y+3}{4}$ **C** $a = \frac{y}{4} + 3$

4 Simplify these expressions.

a $6 \times 7y$ **b** $8 \times 4x$ **c** $^{-}3b \times 4$ **d** $^{-}8m \times {}^{-}3m$ **e** $a^2 \times a^4$

f $\frac{9a^2}{3a}$ **g** $\frac{40q^4}{24q}$ **h** $\frac{5x^4}{25x^2}$ **i** $\frac{28n^3}{16n^2}$ **j** $p^6 \div p^2$

5 This is a magic square.
Copy it and fill in the missing expressions.

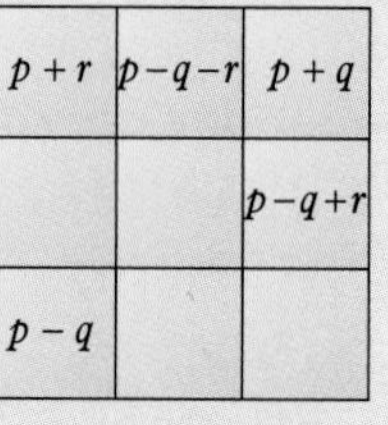

6 Simplify these.

a $5a + 4b - 3a + 7b$ **b** $3x + 12 + 4x - 9 - x$ **c** $7m - 8 - 4m - 4$ **d** $5q - 3 - 8q + 6$

e $3n + 4 - 2n - 5$ **f** $5a^2 + 3a^2$ **g** $3p^2 - 2p^2 + 6p$ **h** $9y^2 - 2y + 3y^2$

i $5(x + 2) + 2(x + 7)$ **j** $4(1 - p) + 8p$ **k** $20 - 4(n + 2)$ **l** $4 - (y - 1)$

m $5(3p + q) - (5p - 4q)$ **n** $6(3x - y) - (4x - 5y)$

7 Write expressions for the missing lengths, **a** and **b**.

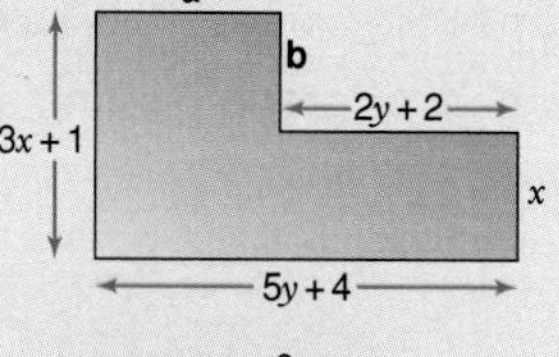

8 Write three different expressions for the area of the purple section.
Show that all three are equivalent.

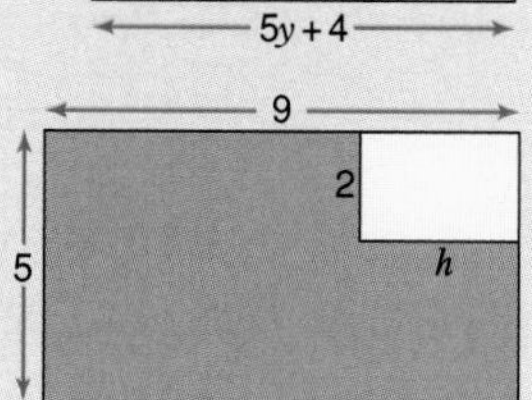

9 Factorise

a $9y + 18$ **b** $6p - 9$ **c** $6 - 8n$ **d** $5p + 15q$

e $5x^2 + x$ **f** $6n + n^2$ **g** $3a^2 + 6a + 6$ **h** $x^3 + x^2 + 2x$

10 Allanah started this proof to show that all numbers of the form ABBA are divisible by 11.

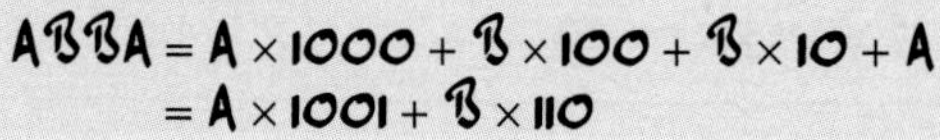

$ABBA = A \times 1000 + B \times 100 + B \times 10 + A$

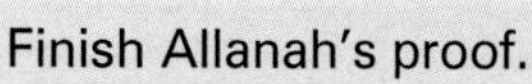

$= A \times 1001 + B \times 110$

Finish Allanah's proof.

11 Evaluate these expressions if **a** $n = 0{\cdot}2$ **b** $n = {}^{-}2$

i $3n^2 + 2$ **ii** $4n^3 - 3n$ **iii** $\frac{5n+2}{2}$ **iv** $\frac{4n^2+1}{5}$

12 Find the values of a and b when $x = 10$.

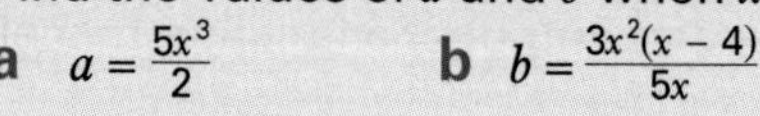

a $a = \frac{5x^3}{2}$ **b** $b = \frac{3x^2(x-4)}{5x}$

13 Ted made photo collages by joining strips of coloured plastic rod.
If the collage has P rows, the number of plastic rods needed is

$\frac{1}{2}(3P^2 + 3P)$.

How many pieces of plastic rod are needed for a collage with 15 rows?

14 Dietitians use this formula. 

BEE (for males) = 66 + 13·7m + 1·7h − 4·7a

where BEE is the Basal Energy Expenditure (energy required for basic life such as breathing, etc.), in kilojoules, m is the mass in kg, h is the height in cm, and a is the age in years.
What is the BEE for these males?

a **Rod** 75 kg, 182 cm, 37 years **b** **Brandon** 54 kg, 154 cm, 13 years

15 The formula for the pressure, P, inside a container of volume V and temperature T is given by 

$$P = 3VT$$

a Find V if $P = 24$ and $T = 2·5$.
b Find T if $P = 116$ and $V = 30·4$.
c Make T the subject of $P = 3VT$.

***16** **a** Make C the subject of $F = \frac{9C}{5} + 32$, where F is temperature in degrees Fahrenheit and C is temperature in degrees Celsius. 
b Find C if $F = 98$ °F.

17 This shape has 6 faces, 8 edges and 12 vertices.
There is a famous formula called *Euler's Rule* that gives the relationship between the number of faces (F), edges (E) and vertices (V) in any solid shape. Draw lots of solid shapes and complete this formula. $F + V =$ _____

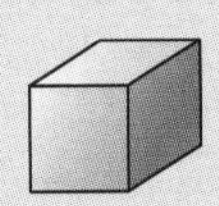

18 Write and solve an equation for these. 
a The length of a rectangular pool is five times its width. Its perimeter is 84 m. Find its area.
b In a triangle PQR, $\angle Q$ is $\frac{3}{4}$ of $\angle P$ and $\angle R$ is $\frac{1}{2}$ of $\angle P$.
Find all three angles of the triangle.

19 Solve these equations. 

a $5p = 7$ **b** $3(p + 2) = 21$ **c** $4 = \frac{20}{n}$
d $4·6a + 2·3 = 18·4$ **e** $5(m - 1) + 4(m + 2) = 75$ **f** $5a + 2 = 4a + 4$
g $5m - 7 = 13 - 3m$ **h** $4(q + 3) = 32 - 3(q + 2)$ **i** $5p - 8 = -13$
j $8·7x^2 = 235·248$ **k** $5(z - 6) - 8(z + 4) = -74$ ***l** $\frac{13}{(y + 3)} = \frac{9}{(y + 2)}$
***m** $3(2m - 1) = 5(4m - 1) - 4(3m - 2)$

20 Starting at B and ending at F gives the same answer no matter which route you follow.
Call the starting number n.
Write and solve an equation to find n.

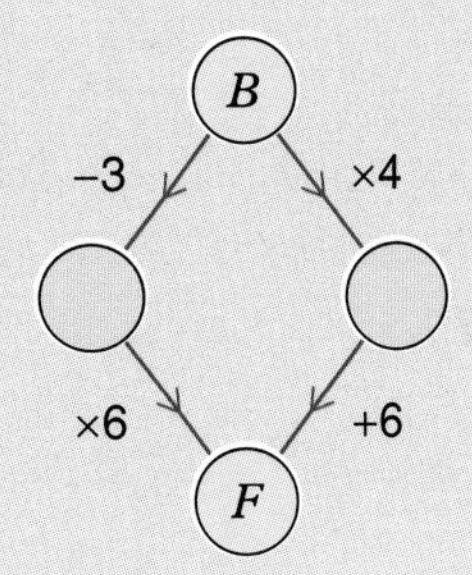

T **21** For one British pound you get about 1·5 euro.

a Copy and fill in this table.

British pounds (p)	10	20	30	40	50
Euro (€)	15				

b Is the number of euro directly proportional to the number of British pounds? Explain.

c Work out the ratio *euro : British pounds* for each pair of values on the table. What do you notice?

d Use a copy of this grid. Plot the graph of euro versus British pounds. Is it possible to draw a straight line through the points?

Euro and British pounds

y-axis: Euro (€), 0 to 80; x-axis: British pounds (£), 0 to 50

e Write a formula for the relationship between euro (e) and British pounds (p)

$e =$ ______

f Find the number of euro you would get for £80.

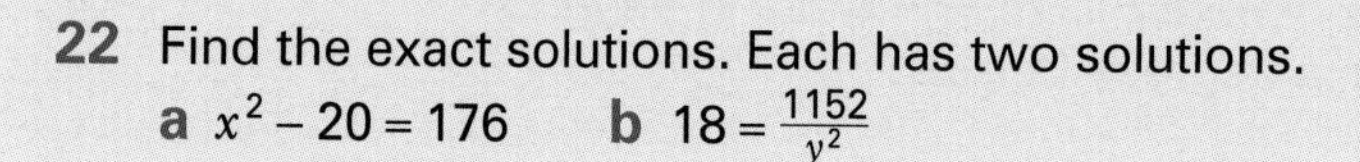

22 Find the exact solutions. Each has two solutions.

a $x^2 - 20 = 176$ **b** $18 = \frac{1152}{y^2}$

23 Use a calculator to find the solution to these to 2 d.p.

a $a^3 = 30$ **b** $b^3 - 2b = 70$ **c** $m^3 + m = 105$ **d** $4{\cdot}6n^3 = 7{\cdot}94$

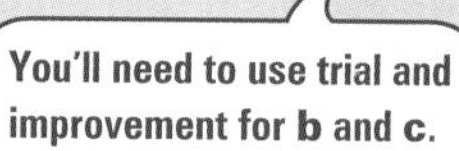

24 The product of three consecutive numbers is 195 054. Find the numbers.

T ***25** Use a copy of this table.

x	2	3	4
y			

Fill in the table for the values of x for $y = x^3 - 2x - 12$.
Continue to find the value of x to 2 d.p. that gives the value of y closest to 0.

9 Sequences and Functions

You need to know

Key vocabulary

arithmetic sequence, consecutive, continue, difference pattern, flow chart, function, general term $T(n)$, input, linear sequence, nth term $T(n)$, output, predict, rule, sequence, term

Reductions

To get a **reduced number** add the digits of the number together

Examples

		reduced number
16	$1 + 6 =$	7
32	$3 + 2 =$	5

If you get a two-digit number when you add, repeat until you get a single digit.

Examples

			reduced number
59	$5 + 9 = 14$	$1 + 4 =$	5
78	$7 + 8 = 15$	$1 + 5 =$	6

1 Try to find a pattern.

a

	reduced number
$8 \times 1 = 8$	8
$8 \times 2 = 16$	7
$8 \times 3 = 24$	6
$8 \times 4 = 32$	5
$8 \times 5 = 40$	4
⋮	⋮

b

	reduced number
2	2
$2 \times 2 = 4$	4
$2 \times 2 \times 2 = 8$	8
$2 \times 2 \times 2 \times 2 = 16$	7
$2 \times 2 \times 2 \times 2 \times 2 = 32$	5
⋮	⋮

c

	reduced number
$12 \times 1 = 12$	3
$12 \times 2 = 24$	6
$12 \times 3 = 36$	9
$12 \times 4 = 48$	3
⋮	⋮

d

	reduced number
$18 \times 1 = 18$	9
$18 \times 2 = 36$	9
$18 \times 3 = 54$	9
⋮	

2 Try some other tables or multiplications to see if the reduced numbers give a pattern.

Writing sequences from flow charts

Worked Example

Write down the sequence given by this flow chart.

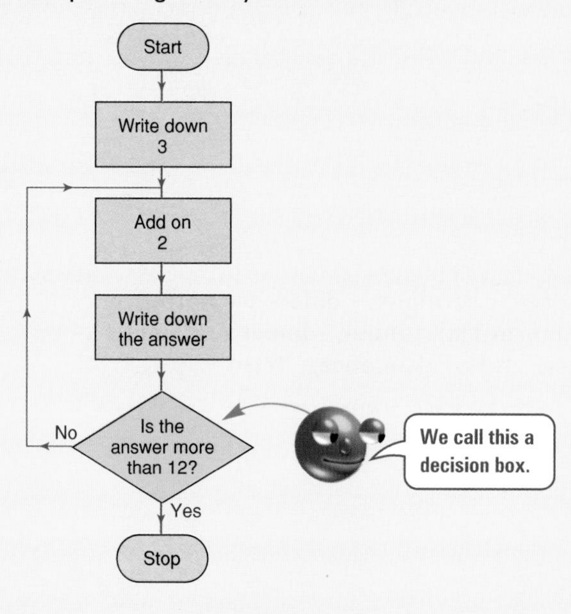

Answer

This flow chart gives the sequence **3, 5, 7, 9, 11, 13**.

Exercise 1

1 Write down the sequences given by these flow charts.

a

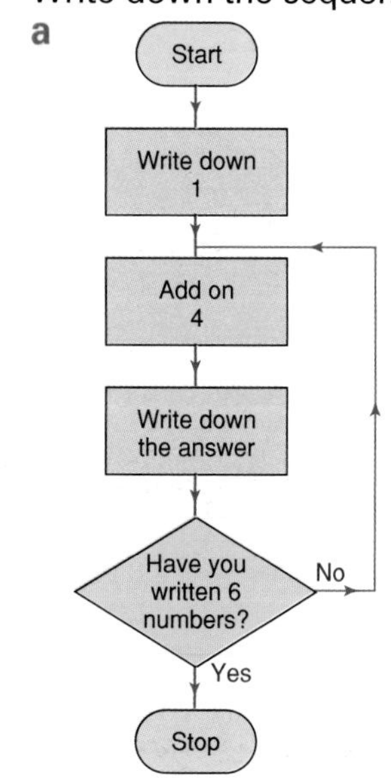

b

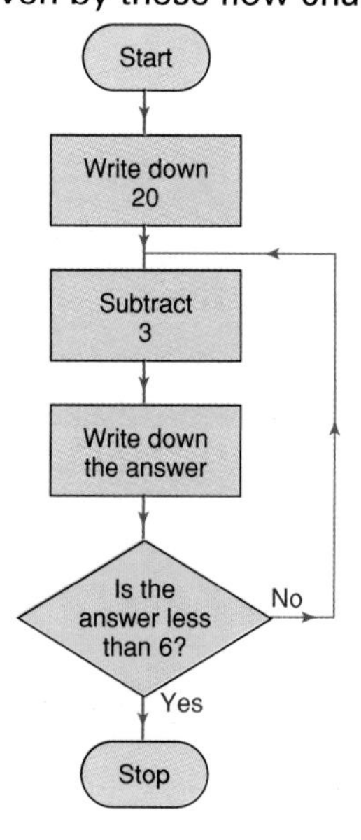

c

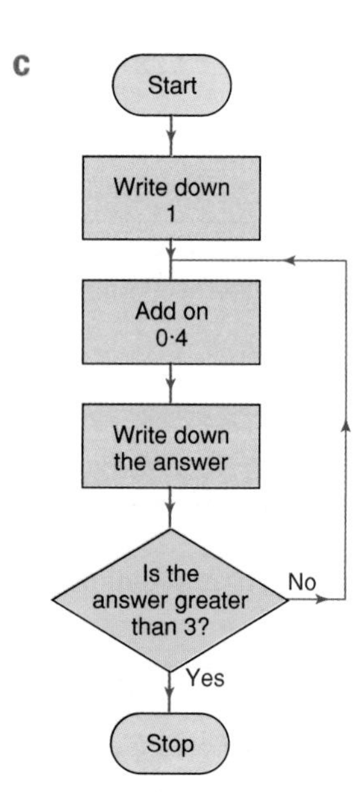

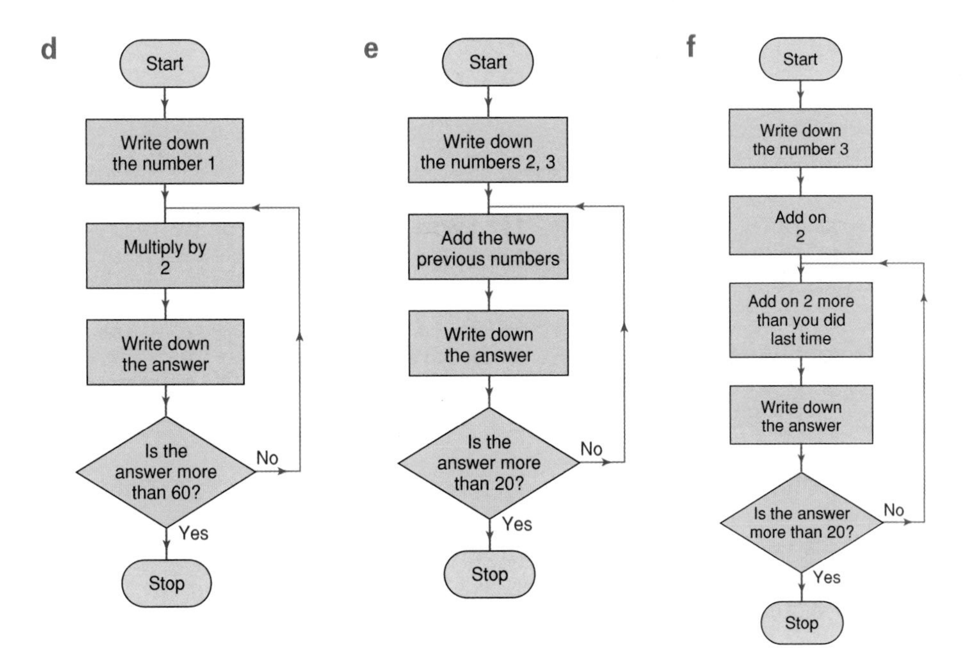

Review Write down the sequences given by these flow charts.

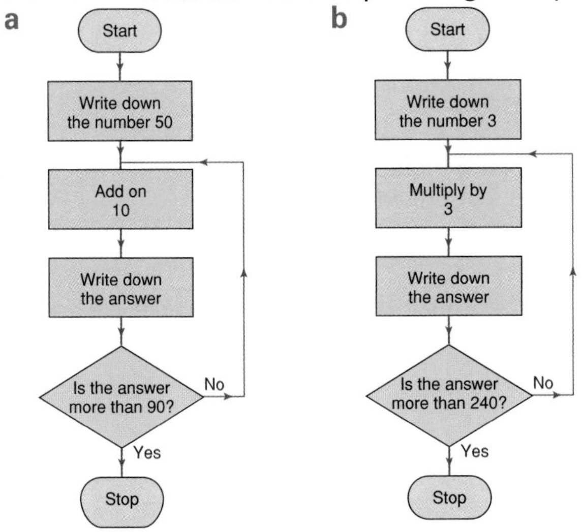

Generating sequences by multiplying and dividing

Maria started swimming every morning. On her first morning she swam one length. Each morning she swam twice as many lengths as the morning before.
She wrote down how many lengths she swam each day.

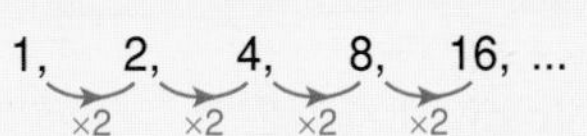

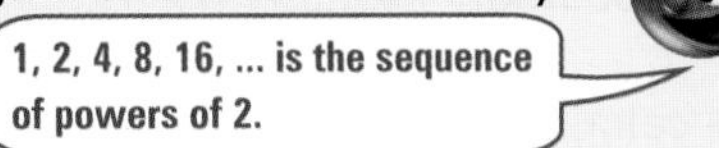

This sequence has been made by multiplying by 2 to get the next term.

Other examples of sequences made by **multiplying or dividing by a constant number** are:

$2, {}^-4, 8, {}^-16, 32, {}^-64, \ldots$ multiplying by $^-2$
$4, 2, 1, 0{\cdot}5, 0{\cdot}25, \ldots$ dividing by 2 (or multiplying by $\frac{1}{2}$)
$1, \frac{1}{3}, \frac{1}{9}, \frac{1}{27}, \frac{1}{81}, \ldots$ dividing by 3 (or multiplying by $\frac{1}{3}$).

A sequence made by multiplying or dividing by a constant number ascends or descends in unequal steps.

Exercise 2

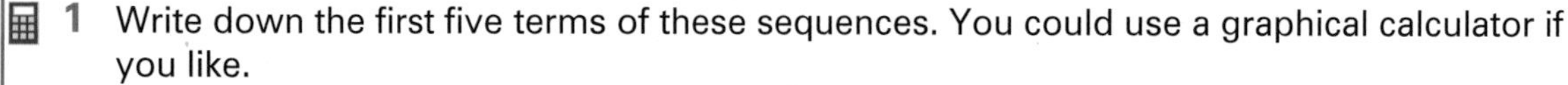

1 Write down the first five terms of these sequences. You could use a graphical calculator if you like.

- **a** **1st term** 2, **rule** multiply by 3
- **b** **1st term** 1000, **rule** divide by 10
- **c** **1st term** 1, **rule** divide by 2
- **d** **1st term** 1, **rule** multiply by $^{-}3$
- **e** **1st term** $^{-}2$, **rule** divide by $^{-}2$
- ***f** **1st term** $\frac{1}{32}$, **rule** multiply by 2

2 These sequences continue in the same way. Write down the next two terms.

- **a** 1, 2, 4, 8, ...
- **b** 1, 3, 9, 27, ...
- **c** 3, 6, 12, 24, ...
- **d** 64, 32, 16, ...
- **e** 100, 10, 1, ...
- **f** 50, 10, 2, 0·4, ...
- **g** 1, $\frac{1}{2}$, $\frac{1}{4}$, $\frac{1}{8}$, $\frac{1}{16}$, ...

3 Sam bought two goldfish in June. Each following month he bought twice as many goldfish as the month before.

- **a** How many goldfish did he buy in September that year?
- ***b** His goal was to have over 125 goldfish in total.
  In which month would he reach his goal?

***4** Rosie started with a 64 m long piece of rope. She cut it in half. She then cut *one* of these pieces in half. She repeated this over and over.
How many cuts would she need to make to get a piece 125 mm long?

Review Write down the first six terms of these sequences.
You could use a graphical calculator if you like.

- **a** **1st term** 3, **rule** multiply by 2
- **b** **1st term** 5, **rule** multiply by $^{-}5$
- **c** **1st term** 100, **rule** divide by 5
- **d** **1st term** $^{-}3$, **rule** divide by $^{-}1$
- **e** **1st term** $\frac{1}{2}$, **rule** divide by 2

Practical

You will need a graphical calculator.

1 Jeff used a graphical calculator to repeatedly multiply by 0·2. This is what was displayed.

- **a** Predict the next three terms.
- **b** Are the numbers getting bigger or smaller?
- **c** Write the terms as fractions.
- ***d** Write the terms using powers.

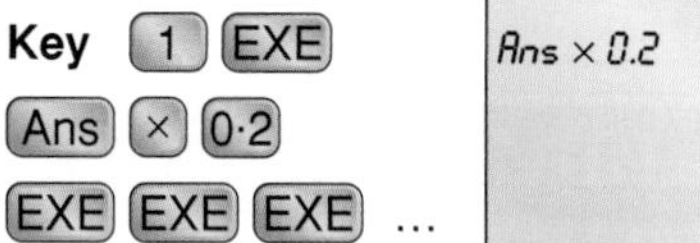

1
1
Ans × 0.2
0.2
0.04
0.008
0.0016

2 Use a graphical calculator to repeat question **1** but

- **i** start with 1 and repeatedly multiply by 0·3
- **ii** start with 1 and repeatedly multiply by 0·4.

3 **a** Use a graphical calculator to start with 1 and repeatedly divide by 2.
b Will the terms ever become zero or negative?

Discussion

- What do you think the answer to 33 333 × 11 will be?
  What do you think the answer to 33 333 333 × 11 will be? **Discuss.**
  Check using your graphical calculator.

  3 × 11 = 33
  33 × 11 = 363
  333 × 11 = 3663
  3333 × 11 = 36663

  How could you explain a rule for finding the answers? **Discuss.**

  Look at the number patterns below.
  Predict what the next few lines will be. Check using a calculator. **Discuss.**
  Discuss what the 'rule' for continuing each might be.

a	b	c	d
2 × 4 =	4 × 11 =	7 × 6 =	7 × 9 =
22 × 4 =	44 × 11 =	67 × 66 =	7 × 99 =
222 × 4 =	444 × 11 =	667 × 666 =	7 × 999 =
2222 × 4 =	4444 × 11 =	6667 × 6666 =	7 × 9999 =

e	f	g
9 × 9 =	9 × 6 = 54	37 × 99 = 3663
9 × 99 =	99 × 66 = 6534	37 × 999 = 36963
9 × 999 =	999 × 666 = 665334	37 × 9999 = 369963
9 × 9999 =	9999 × 6666 = 66653334	37 × 99999 = 3699963

Arithmetic sequences

3, 7, 11, 15, 19, ... We *add 4* to each term to get the next term.
⁻1, ⁻3, ⁻5, ⁻7, ⁻9, ... We *add ⁻2* to each term to get the next term.

Sequences made by adding a constant number to get the next term are called **arithmetic sequences**.
The first term in an arithmetic sequence is called a, and the constant number we add is called d.

A constant number means the same number each time. Arithmetic sequences are linear.

Worked Example
Write down the first five terms of the arithmetic sequence with
a $a = 3, d = 4$ **b** $a = 5, d = {}^-3$.

Answer
a The first term is 3.
We add 4 to get the next term.
The sequence is 3, 7, 11, 15, 19, ...

b The first term is 5.
We add ⁻3 to get the next term.
The sequence is 5, 2, ⁻1, ⁻4, ⁻7, ...

Exercise 3

1 Which of these are arithmetic sequences?
a 16, 20, 24, 28, 32, ... **b** 101, 97, 93, 89, ... **c** 2, 4, 8, 16, 32, ... **d** $\frac{1}{2}, \frac{3}{4}, 1, 1\frac{1}{4}, 1\frac{1}{2}, \ldots$

2 What is a in these arithmetic sequences?
a 6, 11, 16, 21, ... **b** 2, 5, 8, 11, 14, ... **c** 72, 63, 54, 45, ...
d ⁻12, ⁻9, ⁻6, ⁻3, 0, ... **e** 4, 2, 0, ⁻2, ⁻4, ... **f** 0·3, 0·6, 0·9, 1·2, ...
***g** $1\frac{3}{5}, 1\frac{1}{5}, \frac{4}{5}, \frac{2}{5}, \ldots$ ***h** 1·25, 1, 0·75, 0·5, 0·25, ...

3 What is d in each of the arithmetic sequences in question **2**?

4 Write down the first six terms of these arithmetic sequences.
a $a = 1, d = 3$ b $a = 10, d = 5$ c $a = 7, d = 7$ d $a = 100, d = {}^-5$ e $a = 63, d = {}^-9$
f $a = 4, d = {}^-1$ g $a = 0, d = {}^-3$ h $a = {}^-20, d = 3$ i $a = 0{\cdot}5, d = 0{\cdot}25$

5 A sequence is arithmetic.
a If d is a positive number, will the sequence be an ascending or descending sequence?
b What about if d is a negative number?

6 These are arithmetic sequences. Find the missing terms.
a 7, ■, 11, ■, ■ b 8, ■, 16, ■, ■ c 24, ■, ■, 18, ■ d 94, ■, ■, ■, 66, ■

7 The fifth term of an arithmetic sequence is 24. If $d = {}^-3$, what is a?

*8 Find the answers to these mentally.
a The start number is 3. The sequence is arithmetic. One of the terms is 33. Find all the possible values d could be.
b The start number is 64. To find the next term, subtract y. One of the terms is 16. Find all the possible values y could be.

*9 Copy this diagram.
Fill it in so that the sequences are arithmetic.

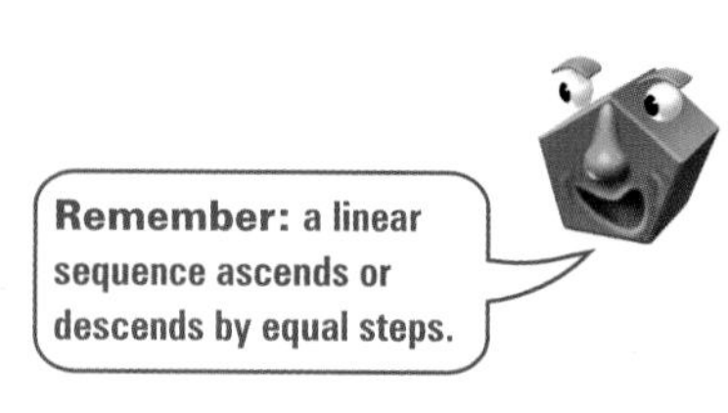

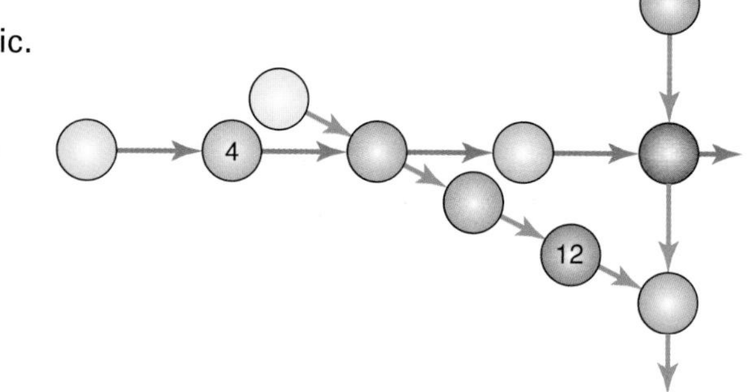

*10 Find the 10th term of an arithmetic sequence if the first term is $a + b$ and $d = 2a - b$.

Review 1 Write down the first five terms of these arithmetic sequences.
a $a = 7, d = 10$ b $a = 3, d = 4$ c $a = 3, d = {}^-5$ d $a = {}^-10, d = 3$ e $a = {}^-11, d = {}^-3$

Review 2 Write down the values of a and d for these arithmetic sequences.
a 11, 17, 23, 29, ... b 14, 9, 4, ⁻1, ⁻6, ... c ⁻6, ⁻3, 0, 3, 6, ... d 0·4, 0, ⁻0·4, ⁻0·8, ...

Predicting the next terms of a sequence

Discussion

Mrs James put this sequence on the board.
She asked her class to show how it might continue.

Penny wrote 1, 2, 3, 4, 5, 6, ...
Blair wrote 1, 2, 3, 4, 5, 8, 13, ...

Who is correct? **Discuss.**

How might the sequence 1, 2, 4, ... continue?
What about 3, 8, 13, ...?

Sometimes we can **predict** how a sequence continues by looking for a pattern.

Example 1, 4, 9, 16, ...
The first four terms are the first four square numbers.
We predict that the sequence continues 25, 36, 49, ...

Example 130, 122, 114, 106, ...
Each term is eight less than the term before.
We predict that the sequence continues 106 98 90 82 74

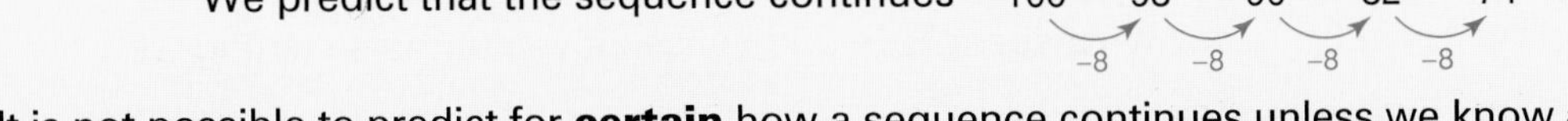

Deciding if a sequence ascends or descends in equal or unequal steps can help you predict the next term.

It is not possible to predict for **certain** how a sequence continues unless we know a rule for the sequence.

Examples 3, 6, 12, ... might continue as 3 6 12 24 48 96 192 (×2 ×2 ×2 ×2 ×2 ×2)

or as 3 6 12 21 33 48 66 (+3 +6 +9 +12 +15 +18)

Exercise 4

1 Predict the next three terms of these sequences. Say if the sequence ascends or descends and by equal or unequal steps.

a 12, 24, 36, 48, ... **b** 5, 8, 11, 14, 17, ... **c** 1, 2, 4, 7, 11, ...
d 400, 200, 100, 50, ... **e** 1, 3, 6, 10, ... **f** 10, 100, 1000, ...
g 30, 29, 27, 24, ... **h** 128, 64, 32, ... **i** 116, 111, 106, 101, ...
j 50, 48, 44, 38, 30, ... **k** 144, 121, 100, 81, ...

2 How might these sequences continue? Give more than one way.
a 2, 4, 6, ... **b** 0, 1, 3, ... **c** 2, 5, 9, ...

Review 1 Predict the next three terms of these sequences. For each sequence decide if it
A ascends by equal steps **B** ascends by unequal steps
C descends by equal steps **D** descends by unequal steps.
a 3, 6, 9, 12, ... **b** 120, 114, 108, 102, ... **c** 36, 49, 64, 81, ...
d 50 000, 5000, 500, 50 ... **e** 2, 4, 7, 11, ... **f** 2, 4, 8, 14, 22, 32, ...

Review 2 Renée was asked how the sequence 2, 4, 8, ... might continue. Give two possible answers Renée could give. Explain your answers.

Discussion

Which three numbers could complete this number square?

Discuss.

2	4			
2	3	4	3	2
3	1	3	1	2
3	4	4	3	2
4	3	2	1	1

Writing sequences from rules

To be certain how a sequence continues we need to know the **rule for the sequence**. Rules can be given in different ways.

1 Term-to-term rule

A term-to-term rule gives the rule for finding the next term.

Example **1st term** 8, **term-to-term rule** add consecutive numbers starting with 1

This rule gives the sequence

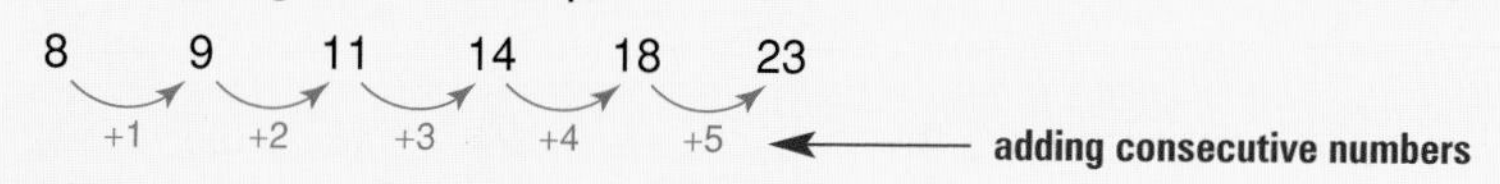

Example $T(2) = T(1) + 6$, $T(3) = T(2) + 6, \ldots$

The term-to-term rule is 'add 6 to the previous term.'

2 Rule for the nth term or general term

If we are given the rule for the nth term, we can find the terms of the sequence by substituting 1, 2, 3, 4, ... for n.

Example $T(n) = n^2 + 3$

$T(\mathbf{1}) = \mathbf{1}^2 + 3 = 4$
$T(\mathbf{2}) = \mathbf{2}^2 + 3 = 7$
$T(\mathbf{3}) = \mathbf{3}^2 + 3 = 12$
$T(\mathbf{4}) = \mathbf{4}^2 + 3 = 19$

The first four terms are 4, 7, 12, 19.

$T(20)$ is found by substituting 20 for n.

$T(\mathbf{20}) = \mathbf{20}^2 + 3$
$= 403$

Exercise 5

1 Write down the first ten terms of these sequences.
- **a** **first term** 224, **term-to-term rule** divide by 2
- **b** **first terms** 1, $^-1$, **term-to-term rule** add the two previous terms

2 The nth term of a sequence is given. Write the first ten terms.
- **a** $T(n) = 50 - 2n$ **b** $T(n) = 0{\cdot}5n + 1$ **c** $T(n) = 66 - 6n$ **d** $T(n) = n + \frac{1}{2}$ **e** $T(n) = 0{\cdot}2n$
- **f** $T(n) = n^2$ **g** $T(n) = n^2 + 3$ **h** $T(n) = 2n^2 + 2$ **i** $T(n) = n^2 - 3$

3 For each of the sequences in question **2** find
i $T(20)$ **ii** $T(25)$ **iii** $T(17)$ **iv** $T(38)$

4 Describe each of the sequences in question **2** using a term-to-term rule.
The answer to **a** could be **first term** 48, **rule** subtract 2.

5 $T(2) = a + T(1)$, $T(3) = a + T(2), \ldots$
Write down the first six terms of the sequence if
a $T(1) = 4, a = {}^-2$ **b** $T(1) = {}^-6, a = 1{\cdot}5$ **c** $T(1) = 4{\cdot}2, a = {}^-0{\cdot}2$ **d** $T(1) = \frac{3}{4}, a = \frac{1}{2}$.

6 $T(3) = T(1) + T(2)$, $T(4) = T(2) + (T3), \ldots$
Write down the first eight terms of the sequence if the first two terms are
a 1, 1 **b** 2, 1 **c** $^-2, {}^-1$ ***d** $\frac{1}{2}, \frac{1}{4}$ ***e** $a, b + 1$.

*7 **a** The nth term of a sequence is $\frac{n}{n^2 + 2}$.
The first term of the sequence is $\frac{1}{3}$.
Write down the next three terms of the sequence.
b Write down the first four terms of the sequence $T(n) = \frac{n + 1}{n^2 + 1}$.

*8 The next number in a sequence is the sum of the two previous numbers.
What are the missing numbers?
□, □, □, 1, 0, 1, 1, 2, 3, 5, 8

*9 Mrs Anderson wrote this on the board.
To find the next term of the sequence, add □.

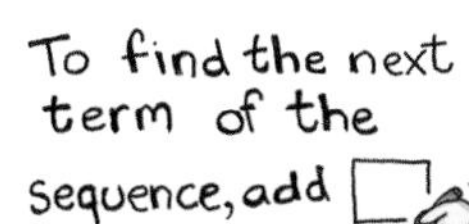

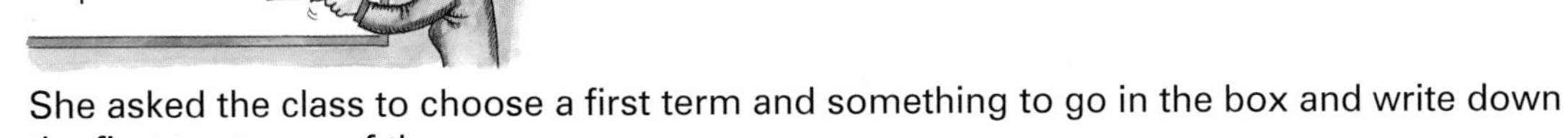

She asked the class to choose a first term and something to go in the box and write down the first ten terms of the sequence.
What first term and number for the box might these people have chosen? You could use a graphical calculator to help.
a Dorothy's sequence has every second number an integer.
b Zoe's sequence has every fourth number an integer.
c Casey's sequence has exactly eight two-digit numbers.
d Harim's sequence has every fourth number a multiple of 8.

Review 1 The nth term of a sequence is given. Write the first eight terms.
a $T(n) = 5n + 4$ **b** $T(n) = 99 - 9n$ **c** $T(n) = 2n - 0{\cdot}2$ **d** $T(n) = 110 - 5n$
e $T(n) = n + 0{\cdot}5$ **f** $T(n) = n^2 - 1$ **g** $T(n) = 2n^2$

Review 2 For each of the sequences in **Review 1**, find $T(21)$.

Review 3 Describe each of the sequences in **Review 1** using a term-to-term rule.

Review 4 $T(3) = T(1) + T(2)$, $T(4) = T(2) + T(3)$, ...
Write down the first six terms of the sequence if $T(1)$ and $T(2)$ are
a $^-1$, 2 **b** $\frac{3}{4}$, $\frac{1}{2}$ **c** a, b.

Review 5 A sequence has the term-to-term rule 'subtract □'.
Choose a first term and a number for the box that will give a sequence with every fourth term a multiple of 4.

Investigation

Fibonacci sequence

The **Fibonacci** sequence is made by starting with 1, 1 and adding the two previous terms together.
The first few terms are 1, 1, 2, 3, 5, 8, 13, 21, 34, 55, 89, ...
Fibonacci numbers occur naturally in nature and other places.

Example The spirals of sunflowers and pineapples.

1 Add the first six terms of the Fibonacci sequence. How much less than the eighth term is this?
Add the first seven terms. How much less than the ninth term is this?
What do you think the sum of the first nine terms would be? Do not add them.

2 Richie thinks that for any three consecutive Fibonacci numbers, the difference between the middle number squared and the product of the other two numbers is always 1.

Example 2, 3, 5 $3^2 = 9$ $2 \times 5 = 10$ $10 - 9 = 1$

Investigate to see if Richie is right.

3 **Investigate** the sequence for the number of rabbits at the beginning of each month.

Beginning of month 1 | Beginning of month 2 | Beginning of month 3 | Beginning of month 4 | Beginning of month 5

4 **Investigate** the sequence given by the number of cows in this family at the beginning of each of the first six years.
A cow has her first calf at the beginning of her third year. She continues to calve each year for the next 6 years.
Assume that this cow has only female calves.
Assume that her offspring also begin calving at the beginning of their third year and have only female calves.

5 Explore the sum of the squares of any two consecutive Fibonacci numbers.
You could fill in a table like this one.

Fibonacci numbers	Sum of squares	Differences
1, 1	$1^2+1^2=2$	
1, 2	$1^2+2^2=5$	3
2, 3	$2^2+3^2=13$	8

6 At Mr Bradshaw's front door there are four steps.
There are five different ways he can climb them if he can only climb them one or two steps at a time.

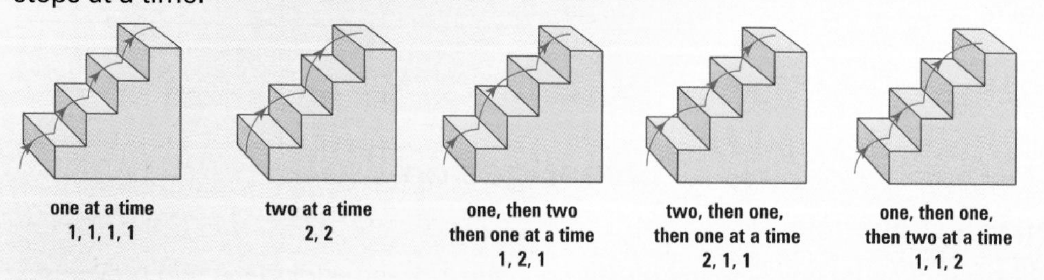

Investigate the number of ways to climb a staircase for different numbers of steps.
You could use a table like this to help.

Number of steps	Number of ways to climb
1	1
2	
3	

7 Find out more about Fibonacci sequences and where they occur in nature.

Practical

A **You will need** a graphical calculator.

Key this into your graphical calculator.

Ans ÷ 2 + 3 EXE

EXE EXE EXE EXE EXE …

25
25
Ans÷2+3
15.5
10.75
8.375
7.1875
6.59375

What happens if you keep pressing EXE ?
What value does the sequence get closer and closer to?

What if you start with a different number?
What if you divide by 4 instead of 2?
What if you divide by 5 instead of 2?
What if you divide by 2 and add 5 instead?
What if ...

B **You will need** a spreadsheet package.
Use a spreadsheet to generate the sequence given by these.

$T(n) = \frac{n}{n^2}$ $\qquad$ $T(n) = \frac{n}{n^2 + 1}$

Ask your teacher for the Generating Sequences ICT worksheet.

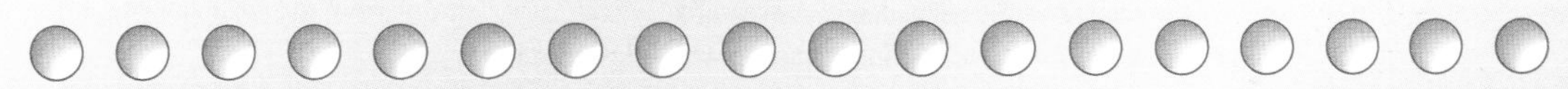

Investigation

Tens and units

Bjorn started with a two-digit number, TU.
He found $2U + T$.
He then found $2U + T$ for the new number.
He kept repeating this to get a number chain.

25 → (12) → 5 → 10 → 1 → 2 → 4 → 8 → 16 → 13 → 7 → 14 → 9 → 18 → 17 → 15 → 11 → 3 → 6 → (12)

Example
25
$2 \times 5 + 2 = 12$
$2 \times 2 + 1 = 5$
$2 \times 5 + 0 = 10$
$2 \times 0 + 1 = 1$
and so on.

He found that the number chain then repeated.

Investigate which number gives the shortest chain before it repeats.
What about the longest chain?

What if you found $3U + T$?
What if ...

Describing linear sequences

Discussion

2, 4, 6, 8, 10, ...
2×1 2×2 2×3 2×4 2×5

- The sequence $T(n) = 2n$ gives the multiples of 2.

  The sequence $T(n) = 2n + 1$ generates the odd numbers starting at 3.

  The sequence $T(n) = 3n + 1$ generates numbers that are all one more than a multiple of 3.

  How might you describe these sequences? **Discuss.**
  $T(n) = 3n - 1$ $T(n) = 2n - 1$ $T(n) = 2n + 2$ $T(n) = 2n - 6$
  $T(n) = 5n + 1$ $T(n) = 10n + 3$ $T(n) = 6n - 3$ $T(n) = 6n + 6$
- The sequence $T(n) = 20 - 2n$ generates the two times table backwards, starting at 18.
  How might you describe these sequences? **Discuss.**
  $T(n) = 6 - n$ $T(n) = 100 - 5n$ $T(n) = 110 - 10n$
- $T(n) =$ ____. What expression goes in the gap to generate these sequences? **Discuss.**
  — the ascending multiples of 4, starting at 4
  — numbers that are all one less than a multiple of 5, starting at 4
  — the descending multiples of 6, starting at 72
  — the ascending multiples of 7, starting at 14

$T(n) = 3n + 4$ $T(n) = 5n - 2$ $T(n) = 6n + 8$ $T(n) = 10n - 3$
All of these are called **linear sequences**.
We can describe a sequence by looking at the rule for the nth term.

A linear sequence has the rule $T(n) = an + b$. a and b are any numbers.

Examples $T(n) = 3n$ generates the ascending multiples of 3.

$T(n) = 3n + 1$ generates an ascending sequence with a difference of 3 between consecutive terms. It starts at 4.

$T(n) = 3n + b$ also generates an ascending sequence with a difference of 3 between consecutive terms. The starting number depends on the value of b. If b is a multiple of 3, the terms will also be multiples of 3.

The number multiplying n gives the difference between consecutive terms.

$T(n) = b - 3n$ generates a descending sequence with a difference of 3 between terms. If b is a multiple of 3, all the terms will be multiples of 3 also.

Exercise 6

1 a $T(n) = 5n$ generates the multiples of 5 starting at 5.
Describe the sequences these will generate.
i $T(n) = 5n + 5$ **ii** $T(n) = 5n + 20$ **iii** $T(n) = 5n - 5$

b $T(n) = 5n - 1$ generates numbers one less than multiples of 5.
Describe the sequences these will generate.
i $T(n) = 5n + 1$ **ii** $T(n) = 5n + 2$ **iii** $T(n) = 5n - 2$

2 What will the difference between consecutive terms be in the sequences generated by
a $T(n) = 6n + 2$ **b** $T(n) = 3n - 7$ **c** $T(n) = 7n - 3$ **d** $T(n) = 11n + 4$
e $T(n) = 15n - 3$ **f** $T(n) = an + 3$ **g** $T(n) = an + b$?

3 For what values of b will $T(n) = 3n + b$ generate multiples of 3?

4 $T(n) = 8n + b$
Describe the sequence generated when
a $b = 8$ **b** $b = 2$ **c** $b = {}^{-}8$.

5 **a** Describe the sequence generated by $T(n) = 88 - 8n$.
b Write a rule for the nth term of a sequence that descends by sevens.

6 $T(n) = 10n + b$
For what values of b will this generate a sequence of numbers where the units digits of all terms is b?

7 Describe the sequence generated by these. Say whether each is ascending or descending. Give the difference between terms, the starting number and any other relevant information about multiples.
a $T(n) = 5n$ **b** $T(n) = 5n - 2$ **c** $T(n) = 5n + 5$ **d** $T(n) = 5n - 5$
e $T(n) = 10n + 2$ **f** $T(n) = 7n - 3$ **g** $T(n) = 6 - 2n$ **h** $T(n) = 99 - 9n$

***8** $T(n) =$ ____
Write an expression to fill the gap that will generate these sequences.
a the ascending multiples of 9 starting at 9
b ascending numbers with a difference of 5, starting at 2
c descending multiples of 4 starting at 40

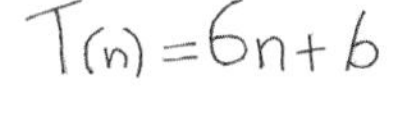

Review 1
a For what values of b is Anna correct?
b Describe the sequence generated by $T(n) = 6n - 1$.

Review 2 $T(n) = 4n + b$.
Describe the sequence that is generated when
a $b = 0$ **b** $b = 4$ **c** $b = 16$ **d** $b = {}^{-}1$ **e** $b = 2$.

Review 3
a Describe the sequence generated by $T(n) = 66 - 6n$.
b Write a rule for the nth term of a sequence that
i ascends by sixes **ii** descends by sevens.

Sequences in practical situations

Discussion

Pierre made photo frames from oval pieces of wood.

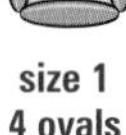

size 1
4 ovals

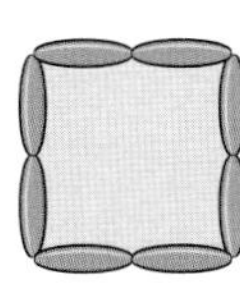

size 2
8 ovals

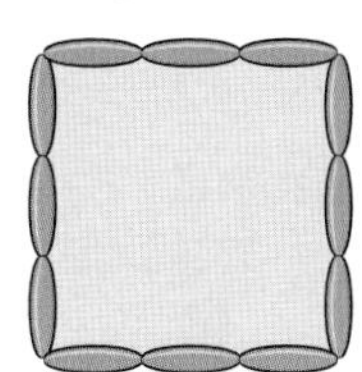

size 3
12 ovals

What sequence for ovals is generated? Predict the next few terms.
How did you make these predictions? **Discuss.**
What would the expression for the number of ovals needed for size n be?
How did you work it out? **Discuss.**

Worked Example

Jolene made wall hangings.

Size 1

Size 2

Size 3

a Draw the next two diagrams.

b Fill in this table.

Size	1	2	3	4	5	...
Number of coloured triangles						...

c Predict how the sequence for the coloured triangles might continue.

d Explain the sequence by referring to the diagrams.

Answer

a

b

Size	1	2	3	4	5	...
Number of coloured triangles	3	6	10	15	21	...

c We can use the differences to help predict the next few terms

Size	1	2	3	4	5	6	7	8	9
Number of coloured triangles	3	6	10	15	21	28	36	45	55
Increase		3	4	5	6	7	8	9	10

The number of triangles increases by one more each row.
The sequence continues as shown in red.

3, 6, 10, 15, 21, **28**, **36**, **45**, **55**, ...

You could check by drawing some more diagrams.

d A possible explanation might be:

Each time a new size is drawn, it has one more row of triangles.
This row has one extra triangle.
So the *increase* is one more each time.

Exercise 7

1 Abigail owned a gift shop called 'A for Awesome'.
She designed a logo for her sign.

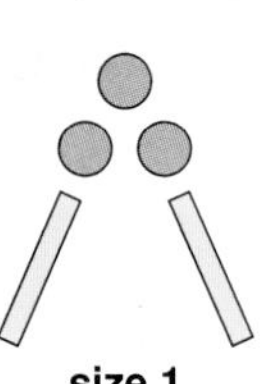

size 1

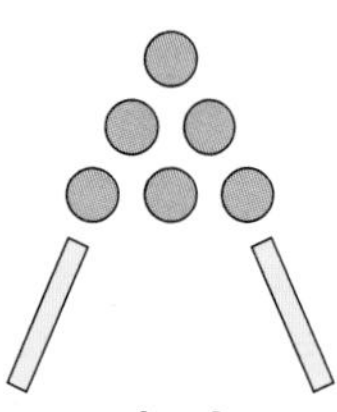

size 2

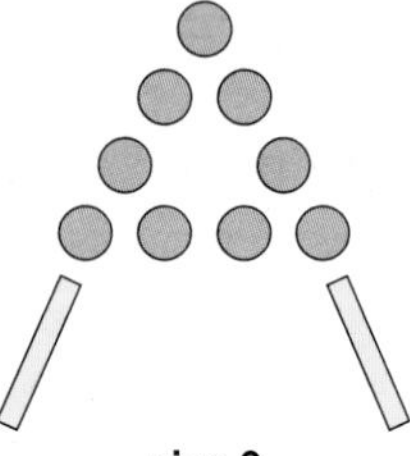

size 3

- **a** Draw a Size 4 logo.
- **b** What sequence is generated by the number of circles for Size 1, Size 2, Size 3 ...? Predict the next few terms.
  Explain how you got this sequence referring to the diagrams.
- **c** Write an expression for the number of circles in a Size n logo.
  Explain how you found this expression by referring to the diagrams.
  Check the terms you predicted in **b** to see if they are correct.

2 Boxes of soap are stacked to make a pattern for a display.

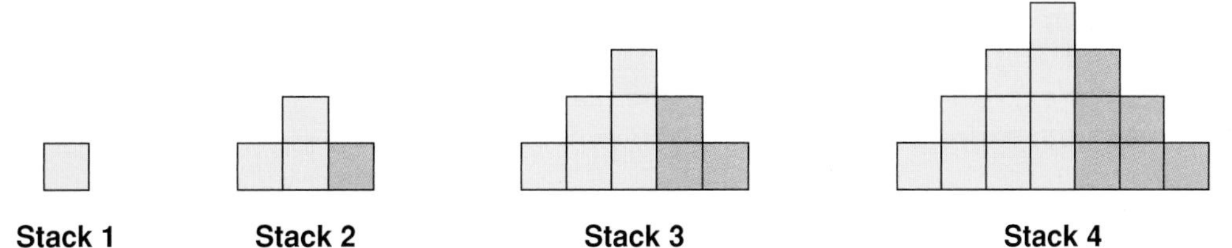

- **a** What sequence is generated by the number of boxes in Stacks 1, 2, 3, 4, ...
  Explain how you work out the next term of the sequence.
- **b** Write an expression for the number of boxes in the nth stack.
  Justify your expression by referring to the diagrams.
  Hint: Can you move the purple boxes to make a square stack?

3

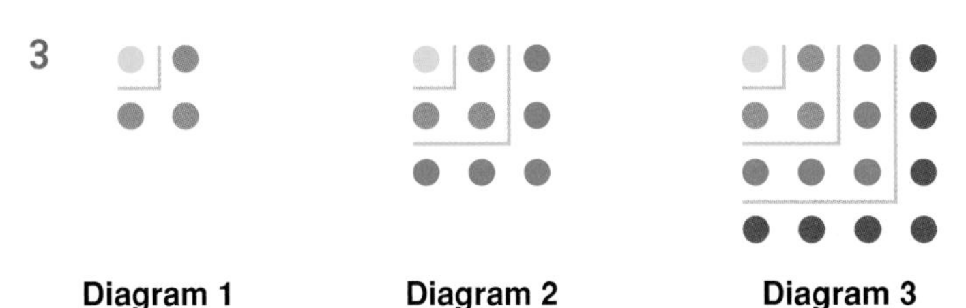

- **a** Draw Diagram 4.
- **b** Find the value of **i** $1 + 3$ **ii** $1 + 3 + 5$ **iii** $1 + 3 + 5 + 7$ **iv** $1 + 3 + 5 + 7 + 9$
- **c** What is the sum of
  - **i** the first six odd numbers
  - **ii** the first twenty odd numbers
  - **iii** the first n odd numbers?
- **d** Look at the diagonals in Diagram 3.
  In the first diagonal there is 1 dot, in the second 2 dots, in the third 3 dots and so on. How does this continue?
  - **i** Show from the diagonals that
    $1 + 2 + 3 + 4 + 3 + 2 + 1 = 4^2$.
  - **ii** Show from the diagonals in Diagram 4 that
    $1 + 2 + 3 + 4 + 5 + 4 + 3 + 2 + 1 = 5^2$
  - ***iii** Write a general rule for the sum of the dots in the diagonals of Diagram n.

4

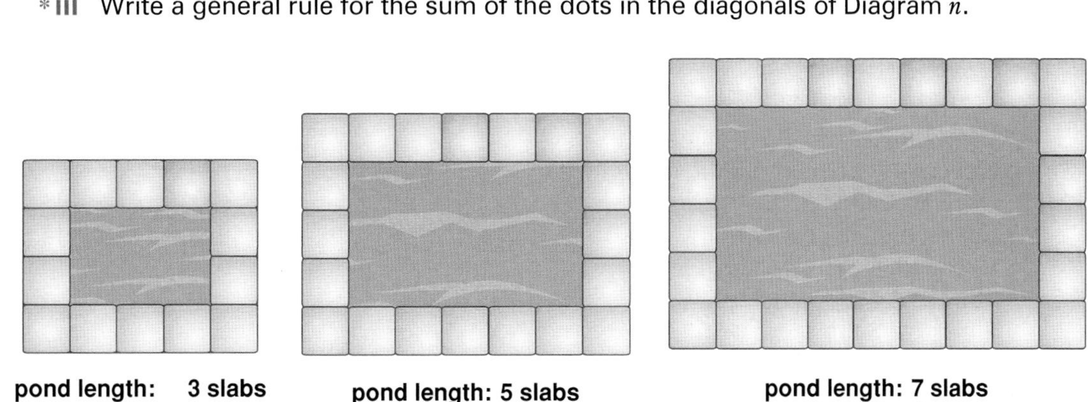

a Draw some more ponds.
Find the total number of concrete slabs needed to surround each pond.
Fill in the results on a table like this.

Length of pond (in slabs)	**Width of pond** (in slabs)	**No. of slabs to surround**
3	2	14
5	3	20

b Try to find an expression for the number of concrete slabs needed for a pond of length l and width w.
Justify your expression by referring to the diagram.
Compare your expression with those of others in your class.

5 Raphael wanted to know the maximum number of times lines crossed. He drew these diagrams.

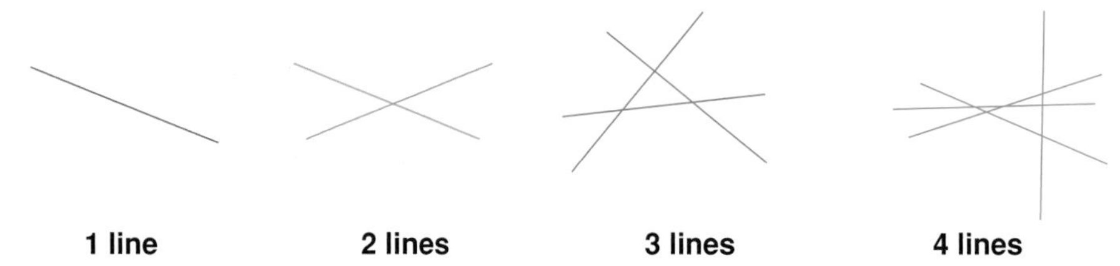

1 line **2 lines** **3 lines** **4 lines**

a Copy and fill in this table.

Number of lines	1	2	3	4	5	...
Maximum crossings						...
Increase						

b Predict how the sequence for maximum crossings might continue.
c Raphael explained the sequence by referring to the diagrams.

When we add a new line, it crosses each of the existing lines. This increases the maximum crossings by the number of existing lines.

Is Raphael correct?
Explain the sequence in your own words.

6

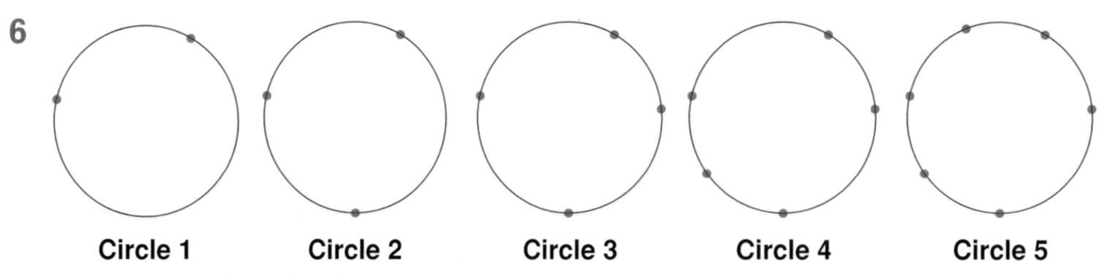

Circle 1 **Circle 2** **Circle 3** **Circle 4** **Circle 5**

a Draw circles like those shown.
b How many dots are on the next circle?
Draw this circle.
c On each circle, every point is to be joined to every other point with a straight line.
On Circle 1 there will be 1 line.
On Circle 2 there will be 3 lines.
On Circle 3 there will be 6 lines.
Join every point to every other point on each of your 6 circles.
d Copy and complete this sequence for the number of lines on the 6 circles.
1, 3, 6, ..., ..., ...

e Predict the next 3 terms of this sequence. Explain why the sequence continues like this by referring to the diagrams.

You could use a table showing the increase each time to help.

*__f__ In a room there are 4 people. Each person shakes hands with every other person. How many handshakes will there be?

*__g__ What is the connection between the number of handshakes between any number of people and the number of lines joining the dots on a circle?

*__h__ In a county cricket competition there are 14 teams. Each team plays every other team. How many games will there be?

7

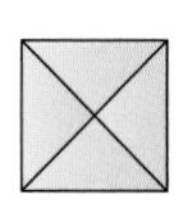

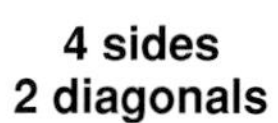

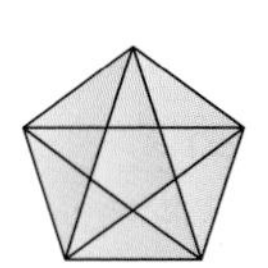

5 sides
5 diagonals

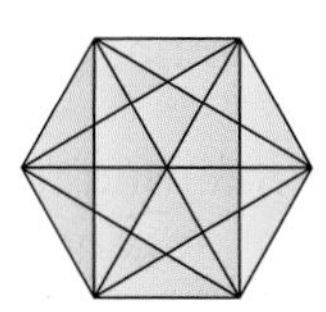

6 sides
9 diagonals

Use the explanation below to complete this formula for the number of diagonals, d, in an n-sided polygon. $d =$ ________

Each vertex is joined to all the other vertices, except the two next to it.
This gives $n - 3$ diagonals.
For n vertices this gives $n \times (n - 3)$ diagonals.
But each diagonal is only drawn from one end and $n \times (n - 3)$ counts each diagonal twice.

Solve the problems in questions 8 to 11 using sequences.

8 At 1 °C the speed of sound in air is approximately 332·1 metres per second and increases about 0·6 metres per second for each one degree increase in temperature.

a Write down the first few terms of the sequence that represents the speed of sound at temperatures of 1 °C, 2 °C, 3 °C, ...

b What is the approximate speed of sound at a temperature of 19 °C?

***9**

Steve has had 50 fence posts delivered to his farm gate. He carries these, one at a time, and positions them along a straight fence line in the holes he has already dug. The first hole is 60 metres from the gate and the rest of the holes are 5 metres apart.

a Write down the first three terms of the sequence that gives the distance Steve walks to position each post and go back to collect the next one.

b How far does Steve walk to position the 33rd post and walk back?

***10** When Sarah was ten months old, she had a vocabulary of 8 words. At twelve months her vocabulary was 10 words, at fourteen months it was 12 words and at sixteen months it was 14 words. Her parents decided that at this rate of increase, Sarah would be an old lady before she had a vocabulary of 1000 words. Were Sarah's parents right? How old would she be?

*11 A business lost £2500 during its first year of operation, £2100 during its second year and £1700 during its third year. If this improvement continues, what profit or loss will the business make in its 10th year of operation?

Review 1 Andrew was making Christmas decorations.

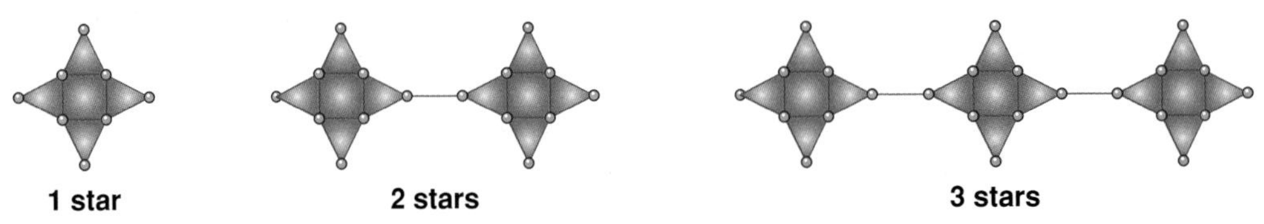

a What sequence is generated by the number of dots in 1 star, 2 stars, 3 stars, 4 stars, ... ? Predict the next few terms.
Explain how you got this sequence, referring to the diagrams.

b Write an expression for the number of dots needed for n stars.
Justify your expression by referring to the diagrams.

Review 2 Celine made these patterns with square and circular pieces of felt.

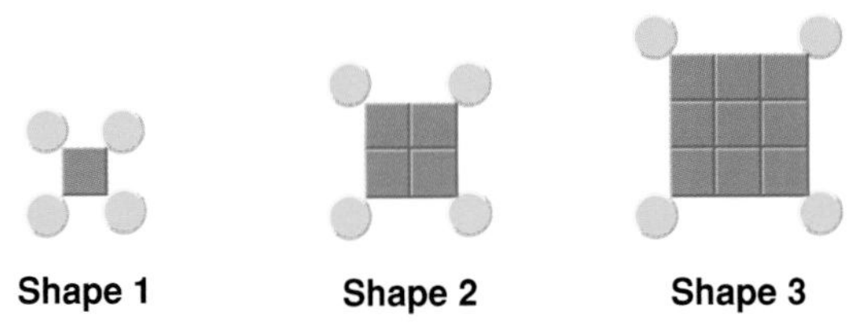

a What sequence is generated by the number of green squares in Shape 1, Shape 2, Shape 3, ...? Predict the next few numbers.

b Predict how many green squares there will be in Shape 8.
Explain how you made this prediction.

c Write an expression for the number of green squares in Shape n.

d Write an expression for the total number of green *and* circular pieces of felt in Shape n.
Justify your expression by referring to the diagram.

Review 3 Trent makes miniature fountains.

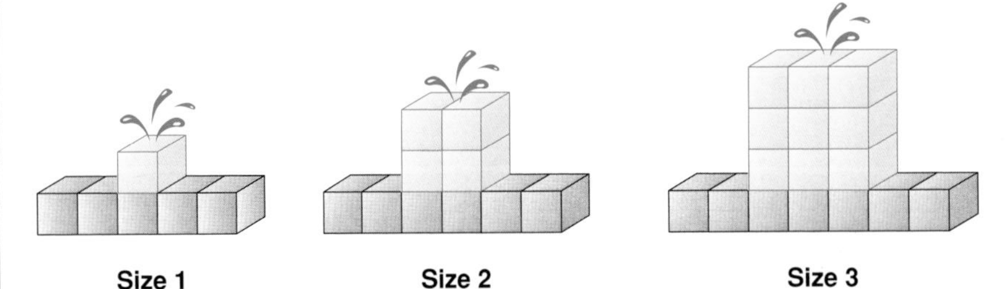

a Draw Size 4 and Size 5 fountains.

b Copy and fill in this table.

Size	1	2	3	4	5	...
Number of bricks	6	10				...
Increase	4					

c Predict the number of bricks needed for Sizes 6 and 7.
*Explain why the sequence continues like this by referring to the diagrams.

***Review 4** In Gaylene's maths group there are 6 students. One evening each student has one phone conversation with every other student. How many phone calls do these 6 students make?

Puzzle

1 Jake has a 6 metre long stick which he cuts into 25 pieces. Each piece that he cuts off is 1 cm longer than the preceding piece. What was the length of the first piece that Jake cut off?

2 This diagram shows the top 3 rows of a 'tower' that has been built from building blocks. Each row has one block less than the row below. If a total of 276 blocks was used, find the number of rows in this tower.

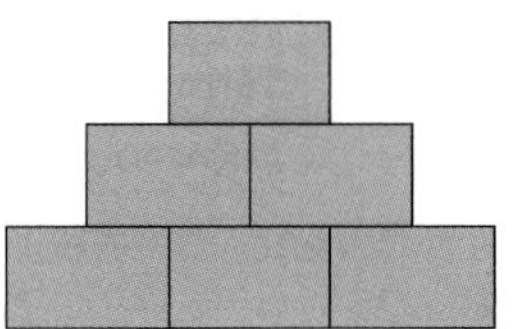

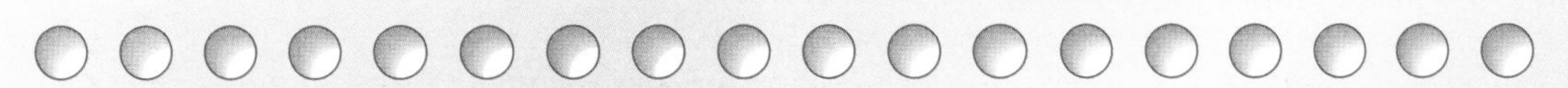

Investigation

The Log Cutter

You will need a graphical calculator.

A wood cutter has 20 logs in his shed.
Each morning he takes half to the market to sell and each evening he cuts 15 more logs.
How much money will he eventually make each day if he sells the logs at 67p each?

What if he originally had more logs?
Investigate further.

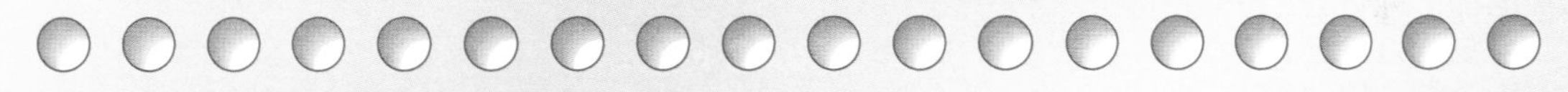

Investigation

Letters of the Alphabet

The letters H are as high as they are wide. Each is made from 1 cm squares. **Investigate** to find a relationship between the areas and perimeters. As part of your investigation make and test statements. Use a table to record your results.

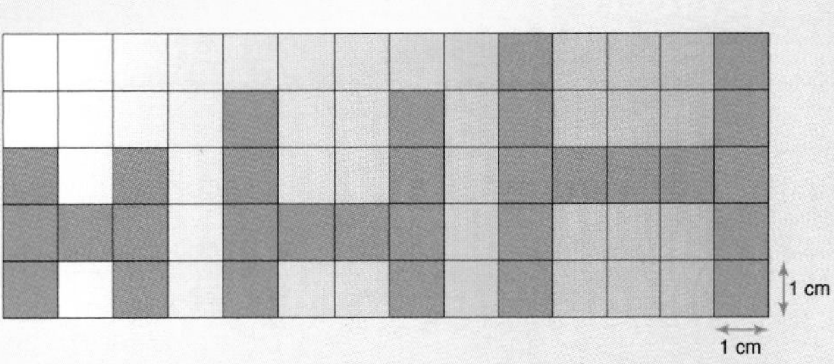

Predict the perimeter of an H that is 7 cm high.
Predict the area of an H that is 10 cm wide.
Predict the area of an H that has a perimeter of 34 cm.
Predict the perimeter of an H that has an area of 34 cm^2.

What if the letter H was replaced with the letter F?
What if the letter H was replaced with the letter E?
What if ...

Finding the rule for the nth term

To find the rule for the nth term, find the **difference** between consecutive terms.

Worked Example

Find the nth term of 25, 32, 39, 46 ...

Answer

Term	25, 32, 39, 46
Difference	7 7 7

The difference between consecutive terms is **7**, so the nth term is of the form $T(n) = 7n + b$

$T(\mathbf{1}) = 25$ $\quad 7 \times \mathbf{1} + b = 25$

so $\quad b = 18$

$T(n) = 7n + 18$

Check by testing a few more terms

$n = 2 \quad 7n + 18 = 7 \times 2 + 18 = 32$ ✓ $\quad n = 3 \quad 7n + 18 = 7 \times 3 + 18 = 39$ ✓ $\quad n = 4 \quad 7n + 18 = 7 \times 4 + 18 = 46$ ✓

Exercise 8

1 Find the nth term of these sequences.

a 30, 34, 38, 42, ... **b** 28, 37, 46, 55, ... **c** 58, 66, 74, 82, ...
d 80, 72, 64, 56, ... **e** 66, 59, 52, 45, 38, ... **f** 1·1, 1·2, 1·3, 1·4, 1·5, ...
g 3·3, 3·5, 3·7, 3·9, 4·1, ... **h** $^-7$, $^-17$, $^-27$, $^-37$, ... **i** $^-3$, $^-12$, $^-21$, $^-30$, $^-39$, ...

*2 Write a rule for each of the following sequences as $T(n) =$ ____. Then find the required term.

a Which term of 4, 7, 10, 13, ... is equal to 100?
b Which term of 2, 7, 12, 17, ... is equal to 197?
c Which term of 7, 5, 3, 1, ... is equal to $^-15$?
d Which term of 60, 56, 52, 48, ... is equal to $^-20$?

Review Find the nth term.

a 44, 52, 60, 68, 76, ... **b** 88, 81, 74, 67, 60, ... **c** 2·2, 2·5, 2·8, 3·1, 3·4, ...
d $^-7$, $^-13$, $^-19$, $^-25$, $^-31$, ...

Functions

Remember

We can find the **output** of a function machine if we are given the input.

Example 3, 8, 0·9, $^-7$ → [multiply by 3] → [subtract 2] → 7, 22, 0·7, $^-23$

$y = 3x - 2$ or $x \rightarrow 3x - 2$

Input	Output
3	7
8	22
0·9	0·7
$^-7$	$^-23$

We can show the inputs and outputs on a table.

We can also show a function using a mapping diagram.

Example $x \rightarrow \frac{x}{2} + 1$

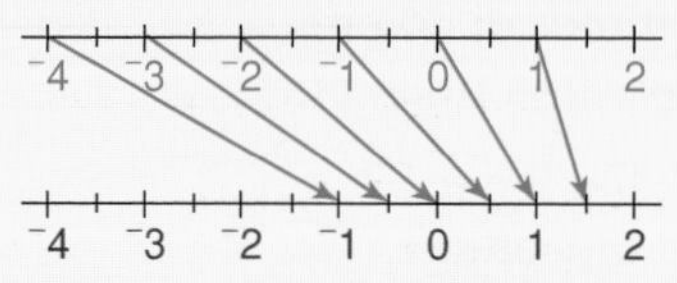

If $x \rightarrow x$ the function is called the **identity function**, because it maps every number onto itself. The number is unchanged.

Exercise 9

T

1 Use a copy of this.
Fill in the input/output table for each function machine.

a x → multiply by 3 → add 2 → y

Input	Output
5	
2	
0	
12	

b x → divide by 2 → subtract 1 → y

Input	Output
4	
10	
24	
68	

c x → add 3 → multiply by 3 → y

Input	Output
11	
21	
$^-2$	
$^-5$	

d x → multiply by 4 → subtract 1 → y

Input	Output
7	
1·5	
0·5	
$^-1$	

2 Draw input/output tables like the ones in question **1** for these function machines. The input values are given.
Find the output.

a 4, 7, 0, 3·5 → subtract 2 → multiply by 3 →

b 25, 0·5, $^-5$, $^-15$ → divide by 5 → add 2 →

c $x \rightarrow 2(x + 4)$ for inputs 0, $\frac{1}{2}$, 6·5, $^-3$

d $x \rightarrow \frac{x-3}{2}$ for inputs 7, $5\frac{1}{2}$, 9·6, $^-5$

T

3 Use a copy of these mapping diagrams.
Fill them in for the functions and input given.

a $y = 2x - 1$ for $x = 2, \frac{1}{2}, 1\frac{1}{2}, 1\frac{1}{4}$

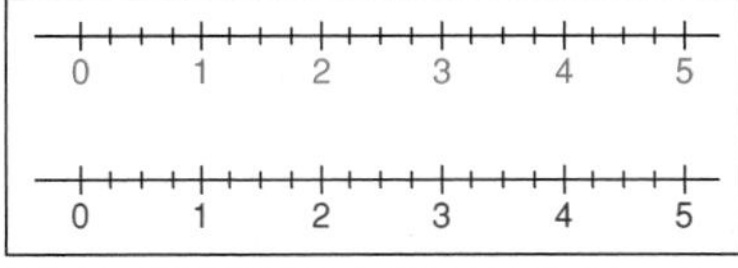

b $x \rightarrow 5x - 1$ for $x = \frac{1}{5}$, 1·2, $\frac{4}{5}$, 0·6

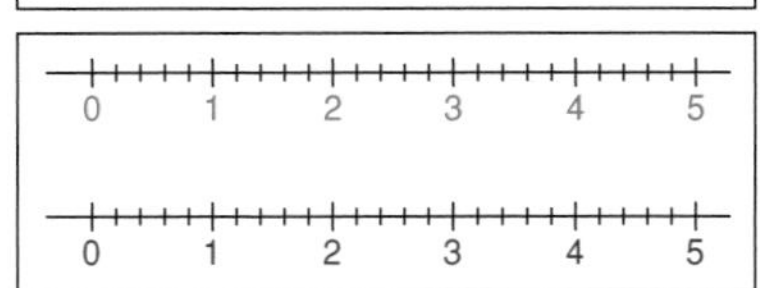

c $x \rightarrow \frac{x}{2} + 2$ for $x = {}^-2$, $^-1$·6, 0·8

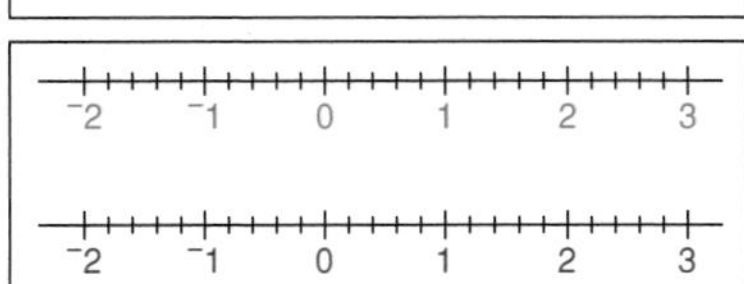

4 a Draw a mapping diagram for $x \rightarrow x$ for $x = {}^-8, {}^-3{\cdot}5, 0, 2{\cdot}5, 3\frac{1}{2}$.
b What special name does this function have?
c What would ${}^-37{\cdot}6$ map onto?

5 Draw a mapping diagram for these functions. The input values are given.
a $x \rightarrow 3x - 2$ for $x = 0, 1, 2, 3, 4$
b $x \rightarrow \frac{x}{2} + 2$ for $x = 2, 6, 1, {}^-2$
c $y = 4x - 3$ for $x = 2, 1{\cdot}5, 2{\cdot}5, {}^-1$
d $x \rightarrow 3(x-2) + 1$ for $x = 3, 1, 0{\cdot}5, 2\frac{1}{2}$
***e** $x \rightarrow \frac{x}{3} - 1$ for $x = {}^-9, 4{\cdot}5, 1\frac{1}{2}$

T **6 a** Use a copy of these mapping diagrams.
Fill them in for the function given. Use input values of x = 1, 2, 3, 4, 5 and 6.
i $x \rightarrow x + 2$ **ii** $x \rightarrow x + 3$ **iii** $x \rightarrow x - 1$

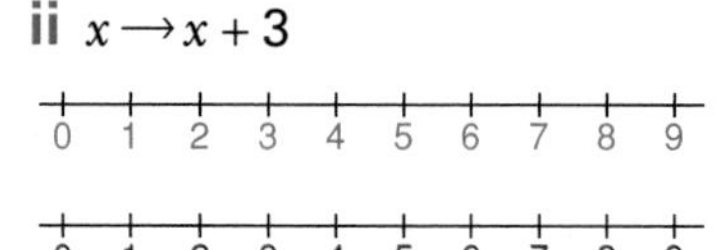

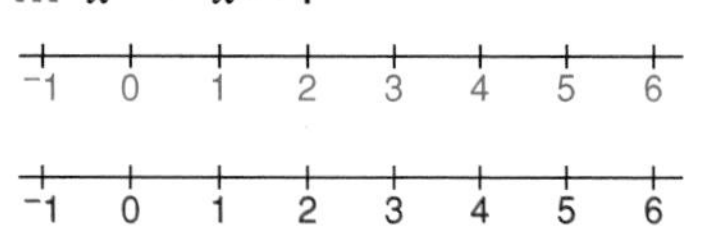

b What do you notice about the lines in each of the mapping diagrams in **a**?
c What can you say about the lines on a mapping diagram for a function of the form $x \rightarrow x + c$?

T **7 a** Use a copy of this mapping diagram.
Fill it in for the function $x \rightarrow 2x$ for x-values from 1 to 5.

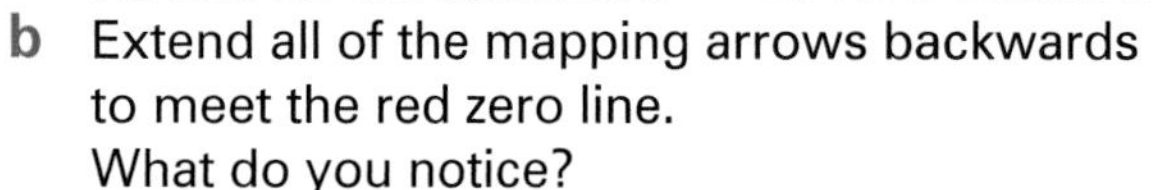

b Extend all of the mapping arrows backwards to meet the red zero line.
What do you notice?
c $x \rightarrow 2x$ is the function for the multiples of 2.
Repeat **a** and **b** for these.
i $x \rightarrow 3x$ **ii** $x \rightarrow 4x$
What do you notice?
d What can you say about the mapping arrows for multiples if they are extended backwards to the zero line?

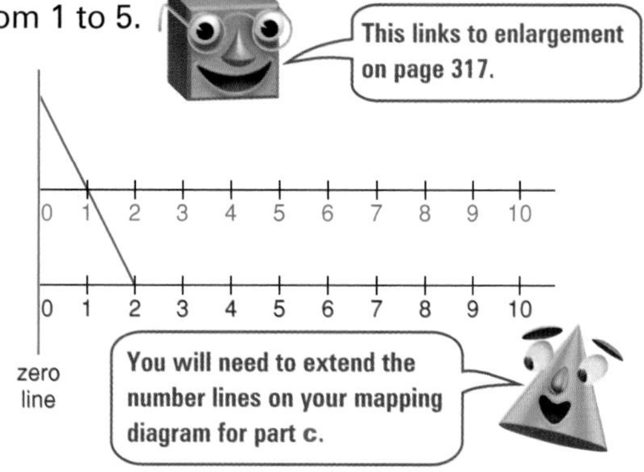

Review 1 Use a copy of this.
Fill in the input/output table for each function machine.

a $x \rightarrow$ multiply by 4 $\rightarrow$ subtract 2 $\rightarrow y$

Input	Output
2	
15	
2·5	
$\frac{1}{2}$	

b $x \rightarrow$ subtract 1 $\rightarrow$ divide by 2 $\rightarrow y$

Input	Output
3	
8	
2·4	
$2\frac{1}{2}$	

Review 2 Draw and fill in an input/output table for this function machine and given input.
$x \rightarrow \frac{x+1}{3}$ for the inputs ${}^-2, 3{\cdot}5, \frac{{}^-3}{4}, 16{\cdot}1$.

Review 3 Draw a mapping diagram for the function $y = \frac{x + \frac{1}{2}}{2}$ for $x = 2\frac{1}{2}, \frac{1}{2}, {}^-\frac{1}{2}, 5$.

Review 4 Match these functions with either **A** or **B**.

a $x \to 4x$ **b** $x \to x + 4$ **c** $x \to 5x$
d $x \to x + 6$ **e** $x \to x - 4$ ***f** $x = \frac{1}{2}x$

A The lines on a mapping diagram of this function will be parallel.
B If we extend the lines on a mapping diagram of this function backwards, they will meet at a point on the zero line.

Review 5 Explain using a diagram what the identity function is.

Practical

You will need a spreadsheet software package.

Ask your teacher for the Functions ICT worksheet.

Finding the function given the input and output

Discussion

Matthew and Millie were playing a game of 'Guess my Rule'.
Millie secretly filled in the rule on a function machine.
Matthew gave Millie some input numbers and she told him what the output would be for her function machine.
Matthew then drew this.

3, 5, 8, 2 → ? → 15, 25, 40, 10

How could Matthew work out Millie's operation? **Discuss**.

How could you work out the missing operation for this function machine?

4, 6, 1·5, $\frac{1}{2}$ → ? → ⁻2, 0, ⁻4·5, ⁻5$\frac{1}{2}$

They played the game with a two-step function machine. One operation was filled in and the other wasn't.
How could they find the missing operation in this? **Discuss**.

4, 0, 7, 3 → multiply by 2 → ? → 11, 3, 17, 9

What if neither operation was filled in and this was the result? **Discuss**.

7, 9, 5, 3, 11 → ? → ? → 13, 17, 9, 8, 21

Worked Example

Find the rule for this function machine.

9, 7, 13, 5, 17 → ? → ? → 15, 11, 23, 7, 31

Answer
Put the input and output in order on a table.
When the input increases in steps of 2, the difference is 4.
When the input increases in steps of 4, the difference is 8.
So if the input increased in steps of 1, the difference would be **2**.
The rule must be $x \rightarrow \mathbf{2}x + c$.
From the first input/output we see $c = {}^-\mathbf{3}$.

	+2	+2	+4	+4	
Input (x)	5	7	9	13	17
Output (y)	7	11	15	23	31
Difference		4	4	8	8

5 → [multiply by 2] → 10 → [subtract 3] → 7

When we check c for the other values, it is correct.

The rule is $x \rightarrow \mathbf{2x - 3}$ or $y = \mathbf{2x - 3}$

Exercise 10

1 Find the functions for these.

a 3, 1, 2, 4, 5 → [?] → [?] → 11, 5, 8, 14, 17

b 5, 1, 3, 2, 4 → [?] → [?] → 12, 0, 6, 3, 9

c 1, 4, 5, 3, 2 → [?] → [?] → $1\frac{1}{2}$, 3, $3\frac{1}{2}$, $2\frac{1}{2}$, 2

***d** 4, 3, 5, 1, 2 → [?] → [?] → $\frac{2}{3}$, $\frac{1}{3}$, 1, $\frac{-1}{3}$, 0

2 Find the functions for these.

a 9, 3, 1, 5, 7 → [?] → [?] → 17, 5, 1, 9, 13

b 5, 11, 13, 9, 7 → [?] → [?] → 15, 27, 31, 23, 19

c 4, 14, 6, 12, 8 → [?] → [?] → 10, 40, 16, 34, 22

d 1, 13, 3, 9, 5 → [?] → [?] → $^-2$, 22, 2, 14, 6

e 2, 10, 4, 12, 6 → [?] → [?] → 10, 34, 16, 40, 22

***f** 6, 10, 4, 8, 12 → [?] → [?] → $\frac{3}{4}$, $1\frac{3}{4}$, $\frac{1}{4}$, $1\frac{1}{4}$, $2\frac{1}{4}$

Review Find the functions for these.

a 1, 5, 3, 2, 4 → [?] → [?] → 9, 21, 15, 12, 18

b 4, 3, 5, 1, 2 → [?] → [?] → 3, 1, 5, $^-3$, $^-1$

c 4, 14, 6, 12, 8 → [?] → [?] → 19, 59, 27, 51, 35

d 7, 13, 3, 9, 5 → [?] → [?] → 17, 35, 5, 23, 11

Properties of functions when combining number operations

Discussion

- Find the outputs for each of these function machines.

  3, 5, 9, 15 → [add 4] → [add 3] → __, __, __, __

  3, 5, 9, 15 → [add 7] → __, __, __, __

  What do you notice about the outputs? **Discuss.**

  Can two additions, two subtractions or an addition and a subtraction always be replaced with a single addition or subtraction? **Discuss.**
  Try some examples.

  Can two multiplications, two divisions or a multiplication and a division be replaced with a single multiplication or division? **Discuss.**

- Rosalie thinks that some rules for function machines can be expressed in more than one way.

  2, 0, 4, 8 → [?] → [?] → 9, 3, 15, 27

  She worked out that the missing operations are → [multiply by 3] → [add 3] →
  She wrote the rule as $x \rightarrow 3x + 3$.
  What else could the missing operations be?
  How else could this rule be expressed? **Discuss.**

- Machine 2 is Machine 1 with the order of the operations reversed.
  Find the outputs for both.

  2, 8, 6, 4 → [add 3] → [multiply by 4] → **Machine 1**

  2, 8, 6, 4 → [multiply by 4] → [add 3] → **Machine 2**

  Why is the output not the same? **Discuss.**
  Is the output always different if the order of the operations is reversed? **Discuss.**

You need to know these **properties of functions when combining operations.**

1 Two additions, two subtractions or an addition and a subtraction will simplify to a single addition or subtraction.

Example → [add 3] → [subtract 2] will simplify to → [add 1] →

2 Two multiplications, two divisions or a multiplication and a division will simplify to a single multiplication or division.

Example → [multiply by 4] → [divide by 2] will simplify to → [multiply by 2] →

3 A function can sometimes be expressed in more than one way.

Example $x \rightarrow 4x - 4$ is equivalent to $x \rightarrow 4(x - 1)$

4 Changing the order of the operations will usually change the function.

Example x → [multiply by 3] → [add 2] → y is different from x → [add 2] → [multiply by 3] → y

$y = 3x + 2$ $\qquad$ $y = 3(x + 2)$

Exercise 11

1 What single operation could replace the two given?

a $x \rightarrow$ add 3 $\rightarrow$ add 5 $\rightarrow y$

b $x \rightarrow$ subtract 2 $\rightarrow$ subtract 4 $\rightarrow y$

c $x \rightarrow$ add 3 $\rightarrow$ subtract 7 $\rightarrow y$

d $x \rightarrow$ multiply by 2 $\rightarrow$ multiply by 3 $\rightarrow y$

e $x \rightarrow$ divide by 3 $\rightarrow$ divide by 2 $\rightarrow y$

f $x \rightarrow$ multiply by 15 $\rightarrow$ divide by 3 $\rightarrow y$

2 Find a function from the box that would give the same output as each of these.

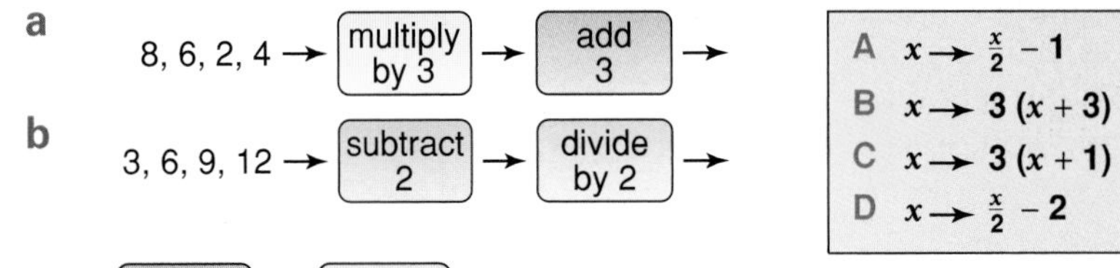

3

$x \rightarrow$ add 3 $\rightarrow$ multiply by 4 $\rightarrow y$

$x \rightarrow$ multiply by 4 $\rightarrow$ add 3 $\rightarrow y$

These function machines have the same two operations, add 3 and multiply by 4. Explain why you don't get the same output from both if you input the same numbers.

4 Find two different ways to write the rule for this function machine.

3, 1, 7, 5 $\rightarrow$? $\rightarrow$? $\rightarrow$ 12, 6, 24, 18

Review 1 What single operation could replace these?

a $\rightarrow$ add 3 $\rightarrow$ add 8 $\rightarrow$

b $\rightarrow$ multiply by 4 $\rightarrow$ multiply by 6 $\rightarrow$

c $\rightarrow$ subtract 4 $\rightarrow$ add 6 $\rightarrow$

d $\rightarrow$ multiply by 12 $\rightarrow$ divide by 4 $\rightarrow$

Review 2 Find two different ways of writing the function for this.

1, 2, 3, 4 $\rightarrow$? $\rightarrow$? $\rightarrow$ 4, 6, 8, 10

Inverse of a function

We can find the **inverse of a function** by doing the inverse operations in the reverse order. On a mapping diagram the inverse function maps the function back to itself.

Example

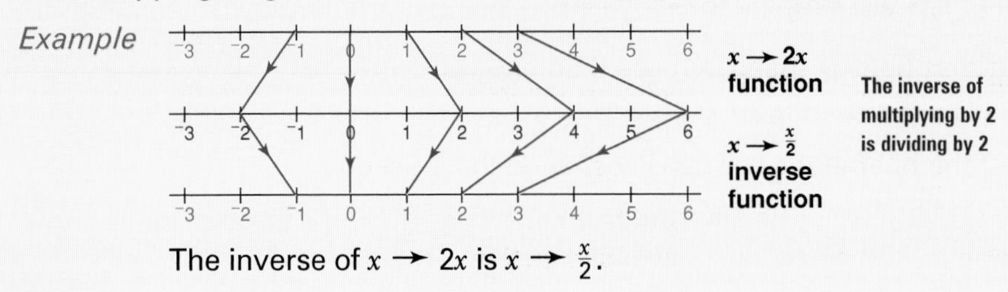

The inverse of $x \rightarrow 2x$ is $x \rightarrow \frac{x}{2}$.

Example
We can use an inverse function machine to find the inverse of a function.

Example

$x \rightarrow$ [subtract 2] $\rightarrow$ [multiply by 3] $\rightarrow 3(x-2)$ **function machine**

$\frac{x}{3}+2 \leftarrow$ [add 2] $\leftarrow$ [divide by 3] $\leftarrow x$ **inverse function machine**

The inverse of $x \rightarrow 3(x-2)$ is $x \rightarrow \frac{x}{3}+2$.

Start with x and work backwards doing the inverse operations.

Link to changing the subject of a formula see page 172.

We can use an inverse function machine to find the input if we are given the output.

Example

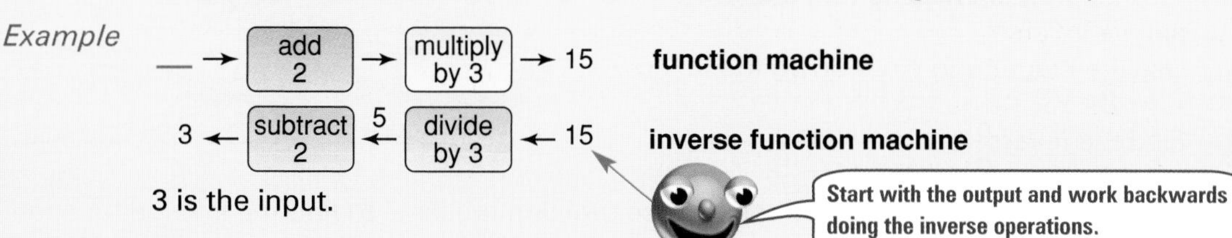

3 is the input.

Exercise 12

T **1** Use a copy of these mapping diagrams.
Fill each in to show the function and its inverse.
Write down the inverse function.

a $x \rightarrow x+3$

b $x \rightarrow 3x$

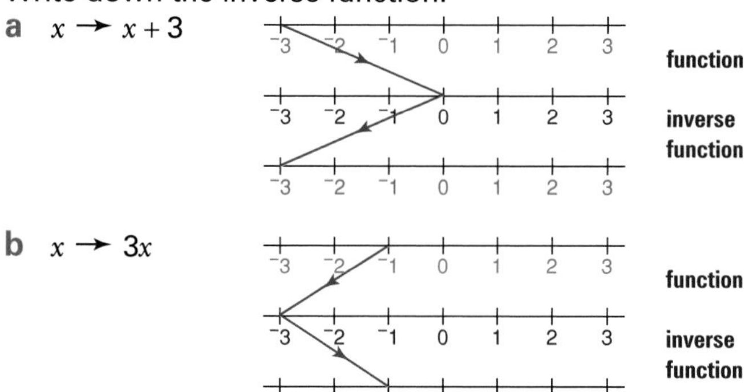

T **2** Use a copy of this.
Fill in the inverse function machine to find the inverse function.

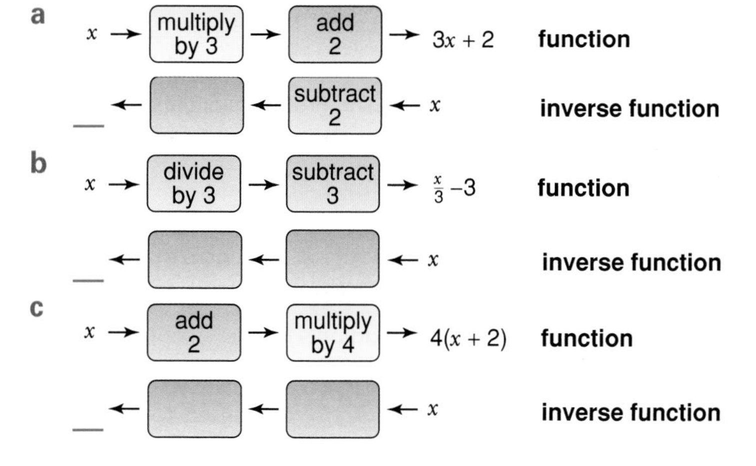

3 Find the inputs for these.

a __, __, __, __ → add 3 → multiply by 4 → 24, 48, 8, 14

b $x \rightarrow 3(x + 2)$ for outputs 21, 13·2, $^{-}15$, $8\frac{1}{4}$

c $x \rightarrow \frac{x}{4} + 3$ for outputs 5, $2\frac{3}{4}$, 3·7, 3·3

4 Find the inverse function for these.

a $x \rightarrow 2x + 1$ **b** $x \rightarrow 3x - 2$ **c** $x \rightarrow 2(x - 5)$ **d** $x \rightarrow \frac{x + 4}{8}$

e $x \rightarrow \frac{x}{2} + 4$ **f** $x \rightarrow \frac{1}{3}x - 1$ **g** $x \rightarrow \frac{1}{2}x + 15$

5 a This diagram shows the function $x \rightarrow 8 - x$ and its inverse.
Use the diagram to write down the inverse of $x \rightarrow 8 - x$.

***b** Find the inverse of $x \rightarrow c - x$.

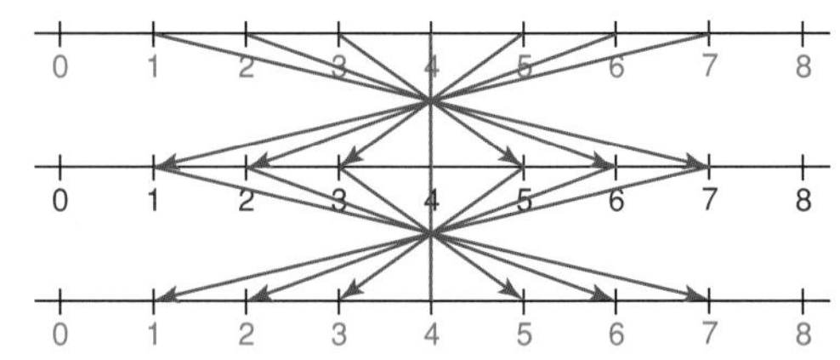

T

Review 1 Use a copy of this. Fill in the inverse function machine to find the inverse function.

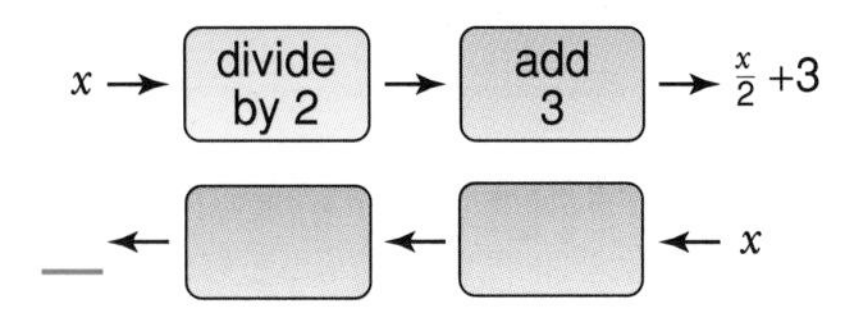

Review 2 Find the inverse functions for these.

a $x \rightarrow 4x - 3$ **b** $x \rightarrow 3(x + 2)$ **c** $x \rightarrow \frac{1}{4}x + 10$

Review 3 What is the inverse function of $x \rightarrow 20 - x$?

Summary of key points

A We can write a sequence from a **flow chart**.
See page 192 for an example.

B We can generate a sequence by **multiplying or dividing by a constant number**.

Examples 1, 2, 4, 8, 16, 32, ... multiplying by 2
729, $^{-}243$, 81, $^{-}27$, 9, $^{-}3$, ... dividing by $^{-}3$

C We can generate a sequence by **adding a constant term**. The sequence will always be linear (go up in equal steps).
These sequences are called **arithmetic sequences**. The first term is called a and the constant number added is called d.

Example If $a = 4$ and $d = {}^{-}2$ the sequence is 4, 2, 0, $^{-}2$, $^{-}4$, ...

D We can **predict** the next few terms of a sequence by looking for a **pattern**. To predict with certainty, we must know the **rule for the sequence**.

Example 1, 2, 3 could continue as 1, 2, 3, 4, 5, 6, ... (adding 1) or as 1, 2, 3, 5, 8, 13, ... (adding the two previous terms).

We can write a sequence if we know the **term-to-term rule** or the **rule for the *n*th term**. ***T*(*n*)** is the **notation** for the *n*th term (general term).

Examples **1st terms** 2, 3 **term-to-term rule** add the two previous terms generates the sequence 2, 3, 5, 8, 13, ...

The sequence with general term or *n*th term $T(n) = 50 - 5n$ is 45, 40, 35, 30, 25, ...

We can use the term-to-term rule to help us **describe a sequence**.

Example $T(n) = \mathbf{3}n + 1$ generates numbers with a difference of **3**, starting at 4. All terms are one more than a multiple of 3.

We can find the rule for the *n*th term in a **practical situation**.

Example

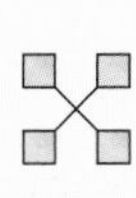

Shape 1

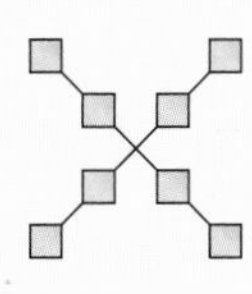

Shape 2

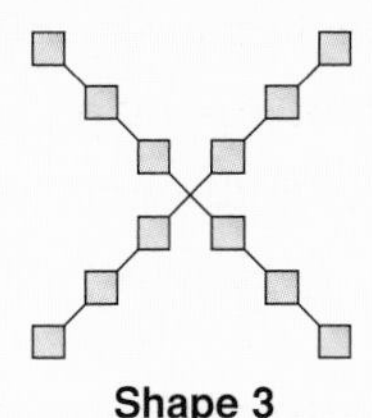

Shape 3

The expression for the number of squares in the *n*th shape is $4n$.
Each time a new shape is drawn 4 new squares are added.
There are *n* lots of 4 where *n* is the shape number.

We can find a **rule for the *n*th** term by finding the difference between consecutive terms.

Example

Term	18 29 40 51 62
Difference	11 11 11 11

The difference between consecutive terms is **11**, so the *n*th term is of the form $T(n) = \mathbf{11}n + b$

$T(1) = 18 \qquad 11 \times 1 + \mathbf{7} = 18$

$\boldsymbol{T(n) = 11n + 7}$

Check by testing some more terms.

The input and output of a function can be shown as a **table** or as a **mapping diagram**.

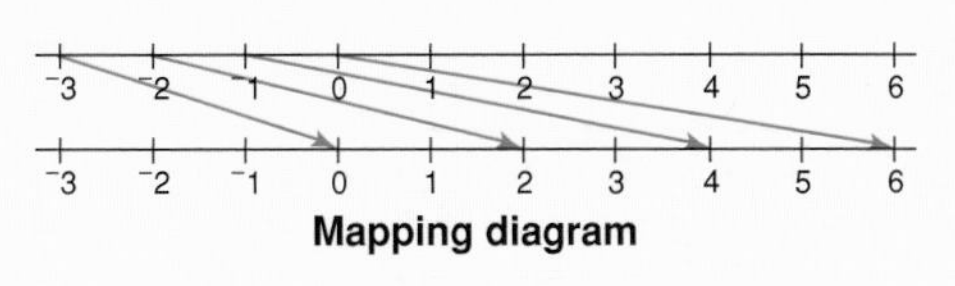

Mapping diagram

Input	Output
⁻3	0
⁻2	2
⁻1	4
0	6

table for $x \rightarrow 2(x + 3)$

You should know these **properties of a mapping diagram**.

The mapping arrows for functions of the form $x \rightarrow x + c$ produce parallel lines.

Example $x \rightarrow x + 2$

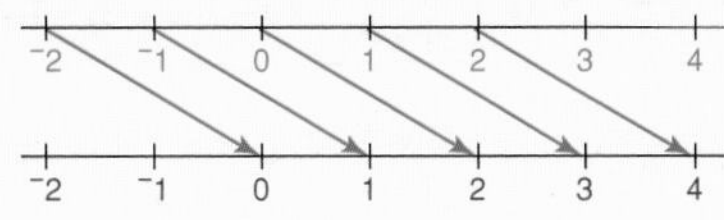

Mapping arrows for multiples, of the form $y = mx$, if extended backwards to meet the zero line, meet at a point.

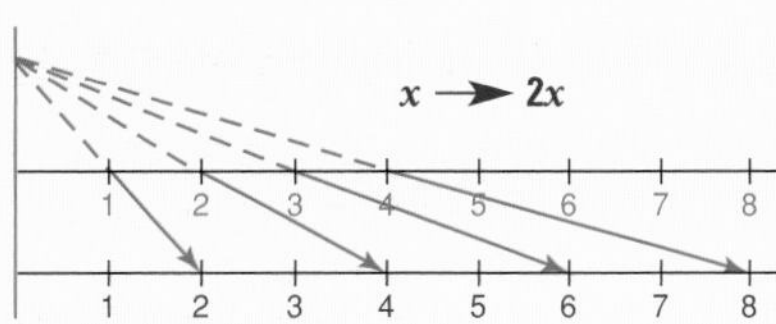

Example $x \rightarrow 2x$

$x \rightarrow x$ is called the **identity function**.

It maps any number onto itself.

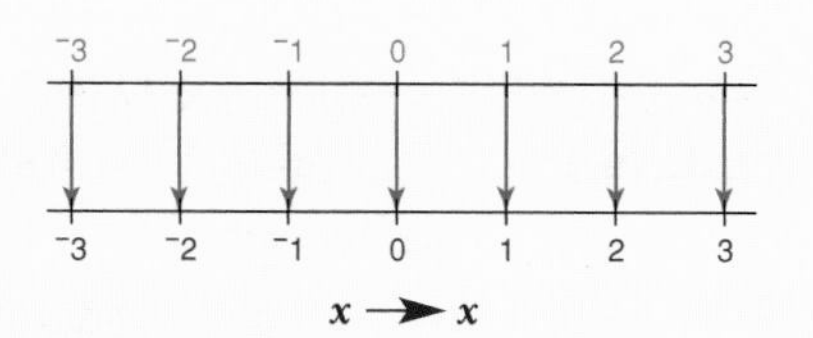

If we are given the input and output we can **find the rule** for the machine.

Example

2, 6, 4, 8 → [?] → [?] → 4, 16, 10, 22

We can find the rule using a difference table.

Each time the input increases by 2, the output increases by 6.

Try 'multiplying by 3' for the first operation.

From the first input and output, we find that the second operation is 'subtract 2'.

The rule for this machine is $\boldsymbol{y = 3x - 2}$.

This is correct when checked for the other values of input and output.

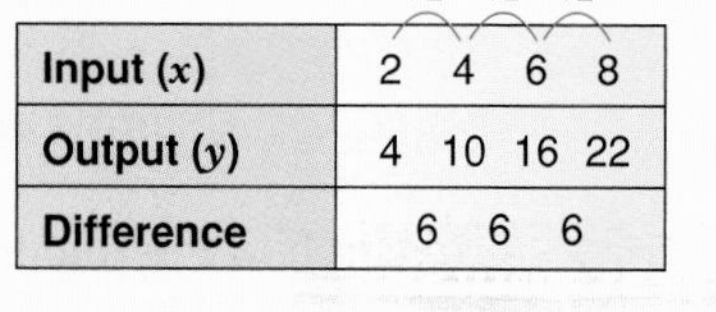

+2 +2 +2

Input (x)	2	4	6	8
Output (y)	4	10	16	22
Difference	6	6	6	

K

Know these **properties of functions** when combining operations.

1. Two additions, two subtractions or an addition and a subtraction will simplify to a single addition or subtraction.
2. Two multiplications, two divisions or a multiplication and a division will simplify to a single multiplication or division.
3. A function can sometimes be expressed in more than one way.
   Example $x \rightarrow 3x - 3$ is equivalent to $x \rightarrow 3(x - 1)$.
4. Changing the order of the operations will usually change the function.

L

The **inverse** of two combined operations is found by doing the inverse operations in the reverse order.

The **inverse** of a function maps the output back to the input.

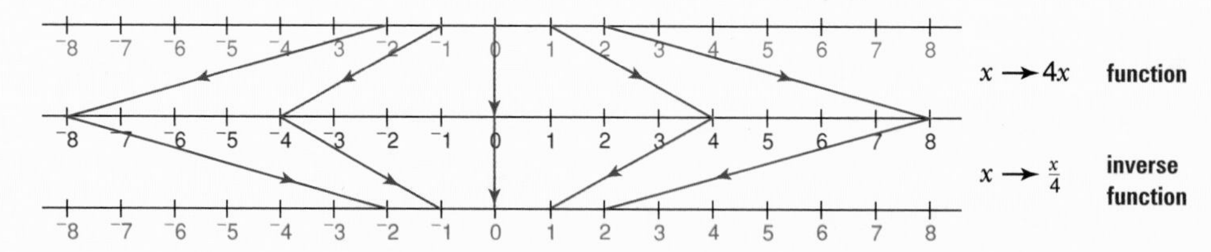

We use an inverse function machine to find the inverse of a function.

Example

x → [add 3] → $x + 3$ → [multiply by 2] → $2(x + 3)$ **function machine**

$\frac{x}{2} - 3$ ← [subtract 3] ← $\frac{x}{2}$ ← [divide by 2] ← x **inverse function machine**

The inverse of $x \rightarrow 2(x + 3)$ is $x \rightarrow \frac{x}{2} - 3$

Start with x and work backwards doing the inverse operations.

We can find the **input given the output** using an inverse function machine.

Example

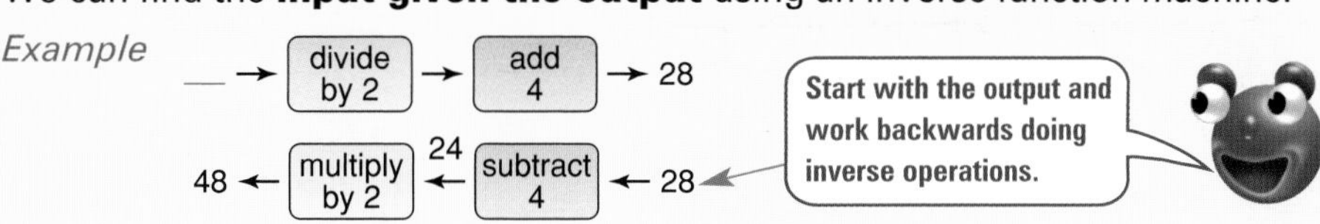

The input is 48.

Test yourself

1 Write down the sequence given by this flow chart.

2 Write down the first five terms of these sequences.
a **1st term** 3, **rule** multiply by 2
b **1st term** ⁻400, **rule** divide by ⁻2

3 Write down the first six terms of these arithmetic sequences.
a $a = 3, d = 2$ **b** $a = {}^-5, d = 3$ **c** $a = 5, d = {}^-2$

4 Predict the next three terms of these sequences.
a 87, 95, 103, 111, ... **b** 196, 169, 144, 121, ...
c ⁻3, ⁻17, ⁻31, ⁻45, ... ***d** $\frac{1}{2}, \frac{-3}{4}, {}^-2, {}^-3\frac{1}{4}, \ldots$

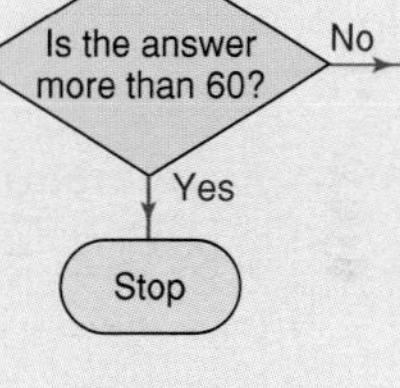

5 **a** How might the sequence 4, 6, 10, ... continue? Give two possible answers.
b What do you need to know to be able to write down how a sequence continues with certainty?

6 The nth term of a sequence is given. Write the first ten terms.
a $T(n) = 4n - 3$ **b** $T(n) = 77 - 7n$ **c** $T(n) = 3n - 0{\cdot}5$ **d** $T(n) = 0{\cdot}3n$
e $T(n) = n^2 + 1$ **f** $T(n) = 2n^2 - 1$ **g** $T(n) = n^2 - 3$

7 Describe the sequences in question **6** using a term-to-term rule.

8 Write down the first five terms of a sequence if $T(2) = T(1) \times b$ and
a $T(1) = 0{\cdot}1, b = 2$ **b** $T(1) = \frac{1}{2}, b = 4$ **c** $T(1) = 5000, b = 0{\cdot}2$.

9 Write down the first three terms of the sequence $T(n) = \frac{n}{n^2 + 1}$.

10 Find a sequence with the rule 'to find the next term subtract □' that has every fifth number an integer. Write down the first term and a value that could go in the box.

11 $T(n) = 6n - 1$ generates ascending numbers with a difference of 6 which are all one less than a multiple of 6. It starts at 5.
Describe these sequences in a similar way.
a $T(n) = 6n + 1$ **b** $T(n) = 66 - 6n$ **c** $T(n) = 6n + 12$

12 G

Diagram 1 Diagram 2 Diagram 3 Diagram 4

a 1, 5, 13, 25 are the first four terms of the sequence of the number of dots in the diagrams shown. Draw the next diagram in this sequence to find the fifth term.

b Copy and fill in this table.

c Predict the number of dots needed for diagram 7. Explain why the sequence continues like this, referring to the diagrams.

Diagram number	1	2	3	4	5
Number of dots	1	5	13	25	
Increase	4				

13 Find the nth term of these sequences. H

a 48, 56, 64, 72, ... **b** 75, 68, 61, 54, 47, ...

c 2·1, 2·3, 2·5, 2·7, 2·9, ... **d** $^{-}2$, $^{-}11$, $^{-}20$, $^{-}29$, ...

14 Which term of, 24, 30, 36, 42, ... is equal to 162? H

15 **a** Copy the input/output table. Fill in for the function $x \rightarrow 3(x - 2)$. I

b Draw a mapping diagram for $x \rightarrow 3(x - 2)$ for the input values $\frac{1}{2}$, 1·5, $\frac{-1}{2}$.

Input	Output
7	
$^{-}3$	
1·4	
$\frac{1}{2}$	

16 **a** If you drew a mapping diagram for $x \rightarrow x + 3$ what would you notice about the mapping arrows? I

b If you drew a mapping diagram for $y = 2x$ and extended the mapping arrows back to the zero line, what would you notice?

17 J

8, 6, 2, 4, 10 → [?] → [?] → 3, 2, 0, 1, 4

Find the rule for this function machine.
Use a difference table to help.

18 What single operation could replace the two given? K

a x → [add 5] → [subtract 6] → y **b** x → [divide by 2] → [multiply by 6] → y

19 Clark drew this function machine. K

x → [add 4] → [multiply by 3] → y

He then drew this function machine.

x → [multiply by 3] → [add 4] → y

Clark thinks that because the operations are the same as in the first function machine, that he will get 27 again if the input is 5. Explain why Clark is wrong.

20 Find the input values. M

a __, __, __ → [multiply by 4] → [subtract 2] → 62, 12, $^{-}34$

b __, __, __ → [add 6] → [divide by 2] → 21, 7·1, 12·6

21 Find the inverse functions. L

a $x \rightarrow 5x - 7$ **b** $x \rightarrow \frac{x + 3}{5}$ **c** $x \rightarrow \frac{1}{4}x - 3$

22 Each rung of a ladder is 0·75 cm shorter than the rung below it. If the bottom rung is 30 cm long, how long is the 12th rung? G

10 Graphs

You need to know

Key vocabulary

coordinate pair, equation (of a graph), gradient, intercept, linear function, linear relationship, notation, slope, steepness, straight-line graph

Its a Puzzle

You will need a piece of graph paper with 1 cm squares.

Draw a set of axes with both x- and y-values from $^-6$ to 6.

Mark these points: A(6, 6) B(6, $^-6$) C($^-6$, $^-6$) D($^-6$, 6) E($^-6$, 0) F(0, 6) G($^-3$, 3) H(3, 3) I($^-3$, $^-3$)

Join A to B, B to C, C to D, E to F, A to C, B to G, G to H, D to A, E to I. Cut out the shapes you have drawn.

What is the connection between these shapes and the picture, shown below, of someone running? **Discuss**.

Equations of straight-line graphs

Remember

An equation of the form $y = mx + c$ is the equation of a straight line. m and c stand for numbers.

$2y = 5x - 4$, $x + y = 3$, $2x - 3y = 12$, $y = \frac{1}{2}x + 3$ are all equations of straight lines. The last equation, $y = \frac{1}{2}x + 3$, is in the form $y = mx + c$. The other three equations may all be rearranged into the form $y = mx + c$ using inverse operations.

Worked Example

Rearrange each of the following into the form $y = mx + c$.

a $x + y - 3 = 0$ **b** $2y = 5x - 4$ **c** $\frac{y}{3} - 2x = 0$ ***d** $x - y = {}^-3$

Answer

We need to make y the subject of each. One method of doing this is to use inverse operations.

a

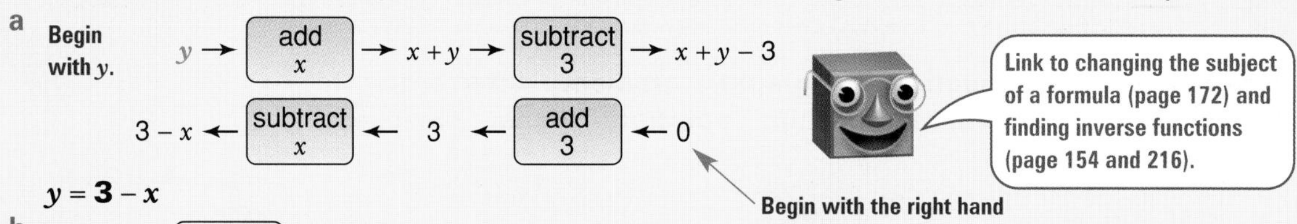

$y = \mathbf{3 - x}$

b

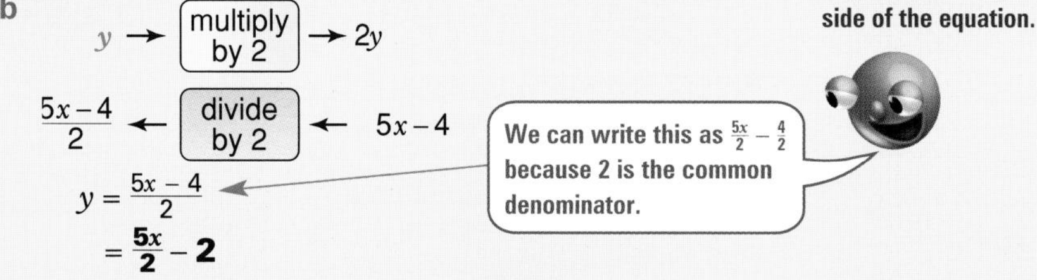

$y = \frac{5x - 4}{2}$

$= \mathbf{\frac{5x}{2} - 2}$

c

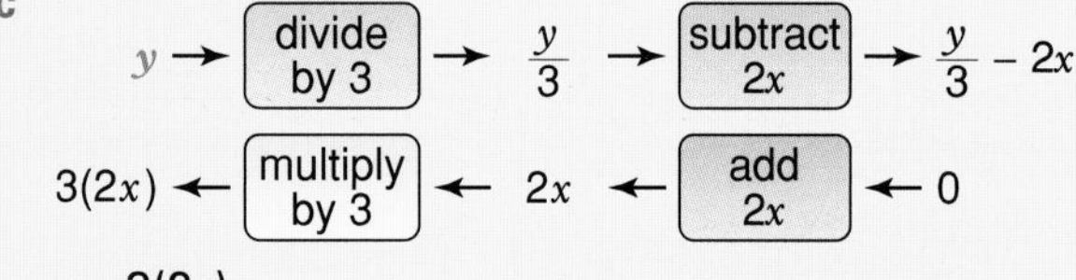

$y = 3(2x)$

$y = \mathbf{6x}$

***d**

y → multiply by $^-1$ → $-y$ → add x → $x - y$

$\frac{^-3 - x}{^-1}$ ← divide by $^-1$ ← $^-3 - x$ ← subtract x ← $^-3$

$y = \frac{^-3 - x}{^-1}$

$y = 3 + x$

Once the equation is in the form $y = mx + c$, we can construct a table of values and plot the graph.

Worked Example

Draw the graph of $x + y - 3 = 0$.

Answer

From **a** in the above worked example, the equation when rearranged is $y = 3 - x$.

We choose 3 or 4 values for x and construct a table.

To find the y-value, substitute the x-value into $y = 3 - x$.

x	$^-1$	0	1	2
y	4	3	2	1

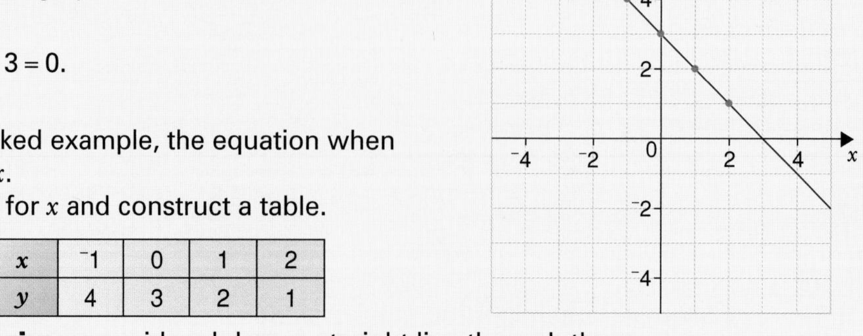

Plot these **coordinate pairs** on a grid and draw a straight line through them.

Discussion

Tracy used 'transforming both sides' rather than inverse operations to rearrange $2y = 5x - 4$ into the form $y = mx + c$.

$$2y = 5x - 4$$
$$\frac{2y}{2} = \frac{5x - 4}{2} \quad \text{dividing both sides by 2}$$
$$y = \frac{5x - 4}{2}$$

Compare Tracy's method with the method using inverse operations.

Use Tracy's method to rewrite $x + y - 3 = 0$, $\frac{y}{3} - 2x = 0$ and $x - y = ^-3$ in the form $y = mx + c$.

Discuss the advantages and disadvantages of each method.

Exercise 1

1 Write each of the following line equations in the form $y = mx + c$.

a $x + y = 3$	**b** $x + y = 6$	**c** $x + y = ^-7$	**d** $3y = 2x + 6$
e $2y = x - 4$	**f** $4y = 6x + 8$	**g** $x + 2y = 2$	**h** $x + 3y = ^-3$
i $2x + 2y = 1$	**j** $3x + 2y = ^-4$	**k** $x + 2y - 1 = 0$	**l** $2x + y + 1 = 0$

2 Rearrange these line equations in the form $y = mx + c$.

a $x - y = 4$	**b** $x - y = ^-1$	**c** $3x - y = 6$	**d** $2x - y = ^-2$
e $2x - 3y = 12$	**f** $3x - 2y = ^-6$	**g** $x - 4y = 4$	**h** $2x - 5y = 10$
i $x - y - 6 = 0$	**j** $4x - 2y + 1 = 0$	**k** $3x - 4y - 12 = 0$	

3 Draw axes with both x- and y-values from $^-6$ to 6. On these axes draw and clearly label the following lines.

$2x + y = 5$ $\quad 2y = 3x$ $\quad y = 3x - 2$ $\quad x + 2y = 8$

4 Draw a set of axes. Number both the x- and y-axes from $^-5$ to 8. On this set of axes, draw and label the following lines.

$2x - y = 0$ $\quad x - y = 4$ $\quad 2x - 5y = 10$ $\quad 3x - 4y + 8 = 0$

5 **a** Does the point $(^-2, ^-5)$ lie on the line $2x - 4y = 16$? How can you tell?

b Does the line $2x + y = 1$ go through the point $(2, ^-1)$?

6 Use a copy of this. [SATS 2002 paper 2]

a Each point on the straight line $x + y = 12$ has an x-coordinate and a y-coordinate that **add together** to make 12.
Draw the straight line $x + y = 12$.

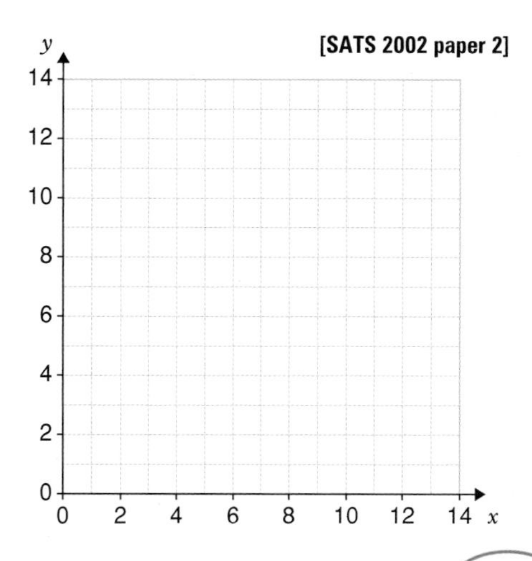

*b On your grid plot at least 6 points whose x-coordinate and y-coordinate multiply together to make 12.
Then draw the part of the curve $xy = 12$ that you would see on your grid.

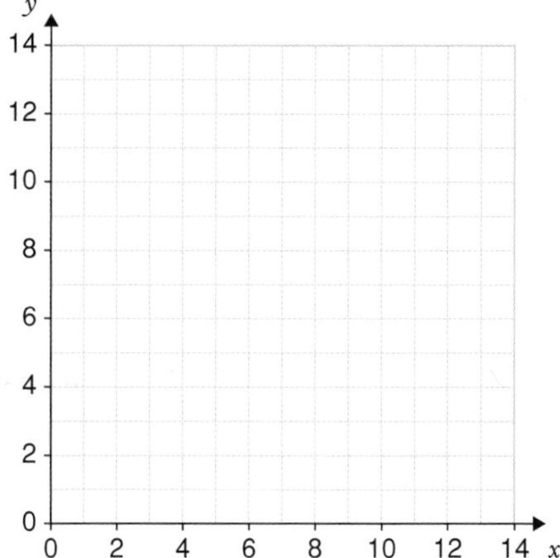

*7 Explain why the graph you drew in **6b**, $xy = 12$, is not a straight line.

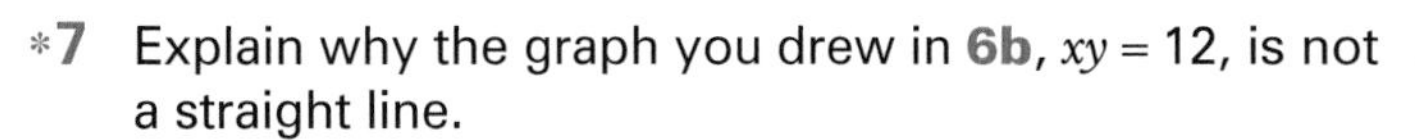

*8 Find the coordinates of the points of intersection of the following pairs of lines, by drawing both on the same set of axes.

a $y = x + 5$; $x + y = {}^{-}1$ **b** $x + y = 1$; $x - y = 5$
c $x + 2y = 3$; $x - y = 0$ **d** $2x + y = 1$; $x + y + 2 = 0$
e $2x + y + 3 = 0$; $x + 2y = 0$

Review 1 Write these line equations in the form $y = mx + c$.

a $x + y = {}^{-}2$ **b** $3y = x - 6$ **c** $x + 2y = 8$ **d** $6x + 3y = 2$
e $x - y = 7$ **f** $2x - 3y = {}^{-}18$ **g** $x + y - 4 = 0$

Review 2

a Draw a set of axes. Number both the x- and y-axes from $^{-}6$ to 8. On these axes, draw and label the following lines.

$2x + y = 4$ $x - 2y = 6$ $2x + 5y = 10$ $2x - y = 0$

b What are the coordinates of the point where the lines $x - 2y = 6$ and $2x - y = 0$ meet?

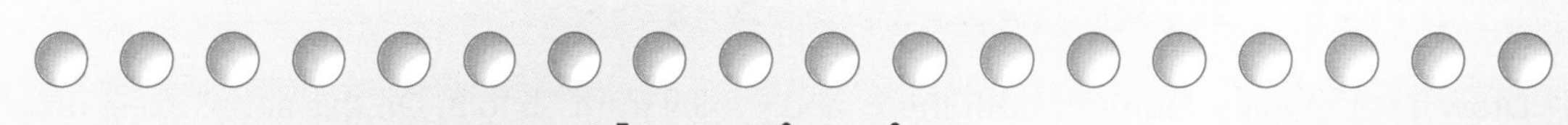

Investigation

Similarities and Differences

You will need graph paper, a graph plotting software package or a graphical calculator

- Graph these lines.
$y = x + 2$
$y = 2x + 2$
$y = 3x + 2$
$y = {}^{-}x + 2$
$y = {}^{-}2x + 2$
$y = {}^{-}3x + 2$

Try to draw some graphs on paper, some using a software package and some using a graphical calculator.

Describe the similarities and differences.

Make and test a statement about the position of the line $y = 4x + 2$.

What if the lines were
$y = 3x - 3$
$y = 2x - 3$
$y = x - 3$
$y = {}^{-}x - 3$
$y = {}^{-}2x - 3$?

What if the lines were $y = 3x - 1$
$y = 2x - 1$
$y = x - 1$
$y = {}^{-}x - 1$
$y = {}^{-}2x - 1$?

What if the lines were $y = 3x$
$y = 2x$
$y = x$
$y = {}^{-}x$
$y = {}^{-}2x$
$y = {}^{-}3x$?

- Graph these lines. $y = 2x + 2$
$y = 2x + 1$
$y = 2x$
$y = 2x - 1$
$y = 2x - 2$

Describe the similarities and differences.

Make and test a statement about the position of the lines $y = 2x + 3$ and $y = 2x - 3$.
What if the lines were $y = 3x + 2$, $y = 3x + 1$, $y = 3x$, $y = 3x - 1$, $y = 3x - 2$?
What if the lines were $y = x + 2$, $y = x + 1$, $y = x$, $y = x - 1$, $y = x - 2$?
What if the lines were $y = {}^{-}2x + 2$, $y = {}^{-}2x + 1$, $y = {}^{-}2x$, $y = {}^{-}2x - 1$, $y = {}^{-}2x - 2$?
What if ...

Remember

In $y = mx + c$ m represents the **gradient** or slope.
c tells us the **y-intercept**.

The gradient is the steepness of the line. The y-intercept is where it cuts the y-axis.

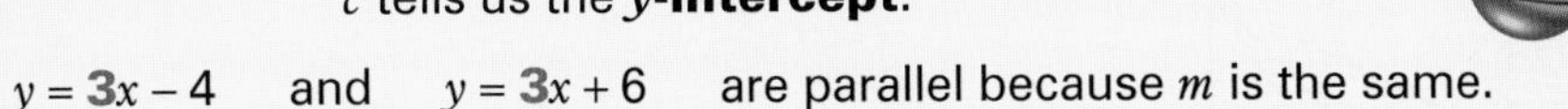

$y = 3x - 4$ and $y = 3x + 6$ are parallel because m is the same.

$y = {}^{-}3x + 2$ and $y = 5x + 2$ cross the y-axis at the same place because c is the same.

$y = 4x + 2$ has a steeper positive gradient than $y = 2x + 3$ because 4 is greater than 2.

Exercise 2

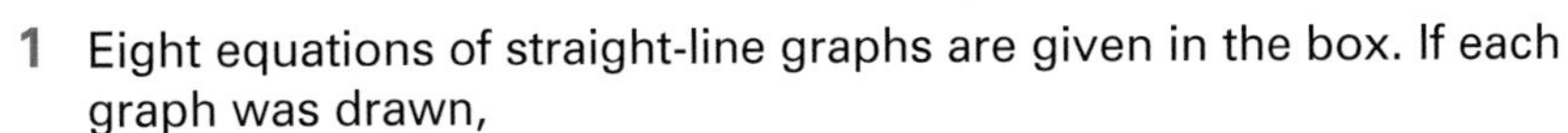

1 Eight equations of straight-line graphs are given in the box. If each graph was drawn,

$y = 2x + 4$
$y = 3x - 5$
$y = \frac{1}{2}x - 2$
$y = 8 - x$
$y = 5 - 2x$
$y = x + 2$
$y = x - 5$
$y = {}^{-}x$

a which would have the steepest slope
b which would have a positive slope
c which would cut the y-axis at (0, 2)
d which would cut the y-axis at (0, $^{-}5$)
e which two would be parallel? Is there more than one answer?

2 Write down the two lines in each list that are parallel.
a $y = 3 - 2x$, $y = 2x + 3$, $y = 3x + 2$, $y = 2x - 4$
b $y = x + 4$, $y = 4 - x$, $y = 2x + 4$, $y = 3 - x$
*__c__ $2y = 3x - 4$, $2y + 3x = 7$, $3y = 2x + 1$, $y = \frac{3}{2}x + 1$

3 Write down the gradient of each of these lines.

a $y = 2x + 5$ b $y = \frac{1}{3}x$ c $y = x - 2$ d $y = {}^{-}3x + 2$

e $y = {}^{-}\frac{1}{2}x + 6$ f $y = 2 + 3x$ g $y = 4 - x$ h $y = 3 + 5x$

i $y = 3 - 5x$ j $y = \frac{2}{3}x - 7$ k $y = {}^{-}\frac{3}{5}x$

4 Where do the lines given in question **3** cross the y-axis?

5 Write down the equation of a line parallel to

a $y = 2x - 1$ b $y = 7 - 3x$.

6 Write down the equation of a line which crosses the y-axis at

a (0, 4) b (0, $^{-}2$).

7 Match these equations with the lines in the diagram.

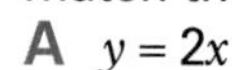

A $y = 2x$

B $y = x + 2$

C $y = 2 - x$

D $y = 3x - 9$

E $y = {}^{-}4$

F $x = {}^{-}4$

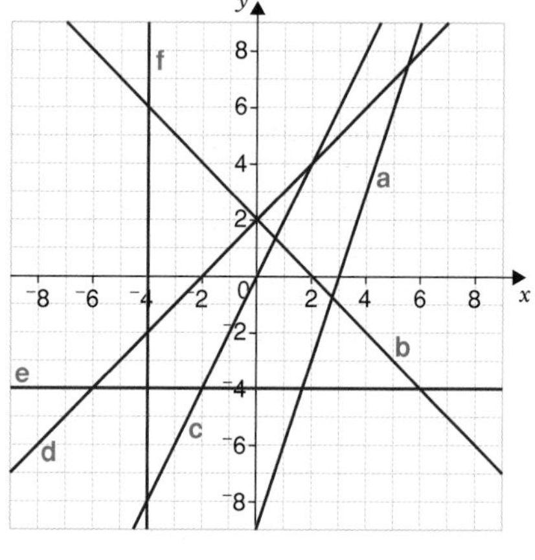

8 These straight-line graphs all pass through the point (10, 10). **[SATS 2000 paper 1]**

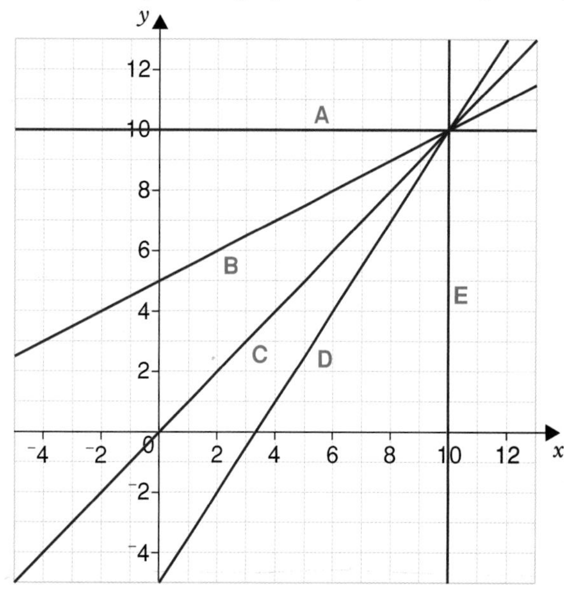

a What goes in the gaps to show which line has which equation?

i line ___ has equation $x = 10$.

ii line ___ has equation $y = 10$.

iii line ___ has equation $y = x$.

iv line ___ has equation $y = \frac{3}{2}x - 5$.

v line ___ has equation $y = \frac{1}{2}x + 5$.

b Does the line that has equation $y = 2x - 5$ pass through the point (10, 10)? Explain how you know.

c I want a line with equation $y = mx + 9$ to pass through the point (10, 10). What is the value of m?

9 Ella drew this diagram.
The line through the points A and D has the equation $y = 3$.
What is the equation of the line through the points
- **a** C and E
- **b** A and B
- **c** C and D
- **d** F and E?

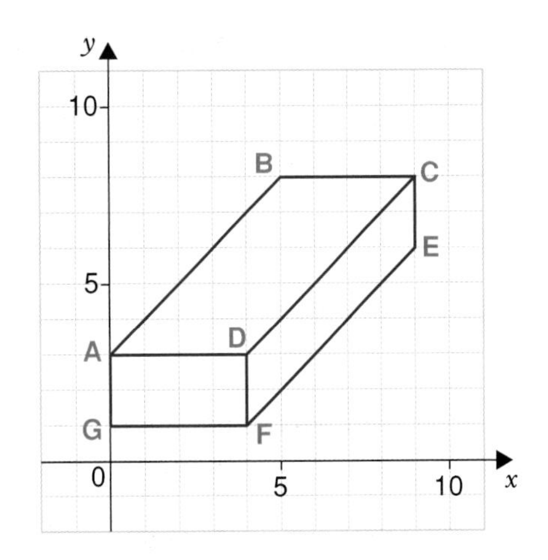

*** 10 a** A line with equation $y = mx + 7$ passes through the point (1, 9).
What is the value of m?

b A line with equation $y = mx - 2$ passes through the point (2, 4).
What is the value of m?

c A line with the equation $y = 2x + c$ passes through the point ($^{-}1$, $^{-}3$).
What is the value of c?

You might have to solve an equation to find the answer to these.

Review 1 Ella has drawn the graph of $y = 3x - 4$.
- **a** She wants to draw a line parallel to this.
  Which of the equations in the box should she choose?
- **b** She wants to draw a line with a steeper slope than $y = 3x - 4$.
  Which equation from the box should she choose?

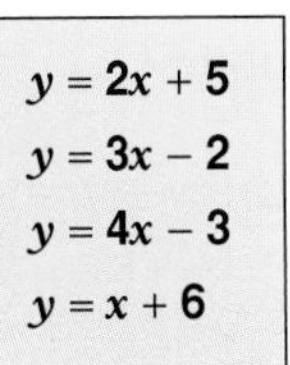
$y = 2x + 5$
$y = 3x - 2$
$y = 4x - 3$
$y = x + 6$

Review 2 Match these equations with the lines.
- **A** $y = x$
- **B** $y = {^{-}2}x$
- **C** $y = {^{-}2}x + 4$
- **D** $y = x + 4$
- **E** $y = {^{-}3}$
- **F** $x = {^{-}3}$

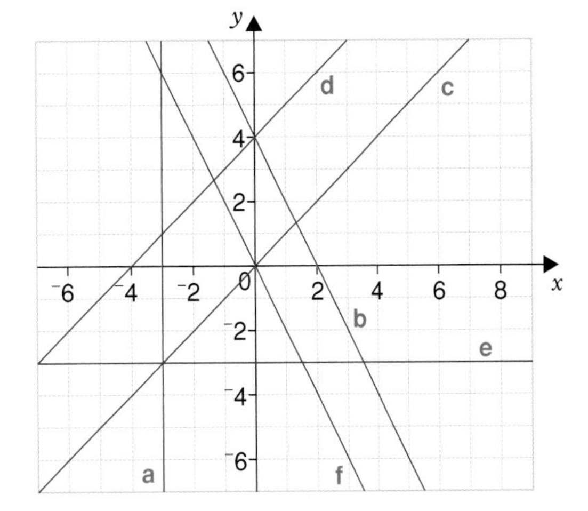

Review 3 A line with equation $y = mx - 3$ goes through the point (2, 7). What is the value of m?

Practical

You will need a graphical calculator.

1 This shows Hayden's and Emma's graphical calculator screens.
Suggest equations for these straight lines.
Check using your graphical calculator.

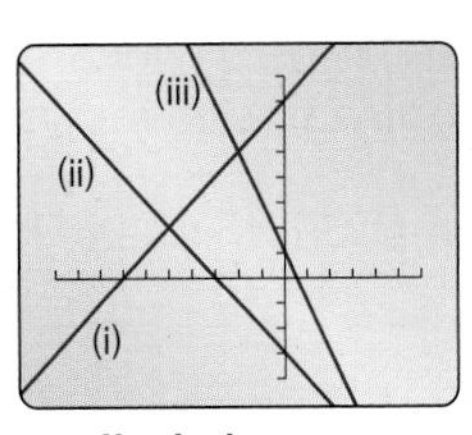

Hayden's screen

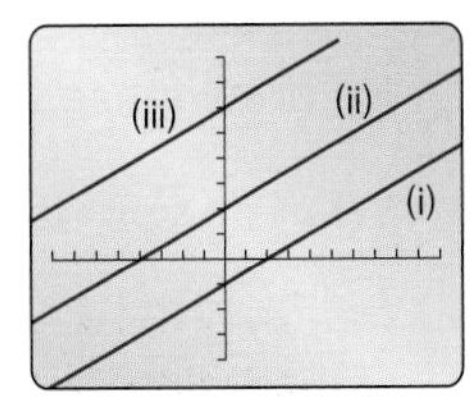

Emma's screen

2 Zoe drew this line on her graphical calculator.
It went through the point (2, 3).
Find some more straight lines that pass through this point.

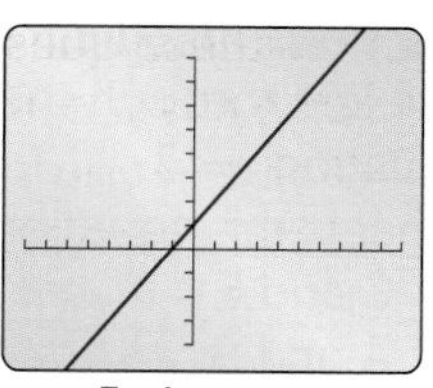
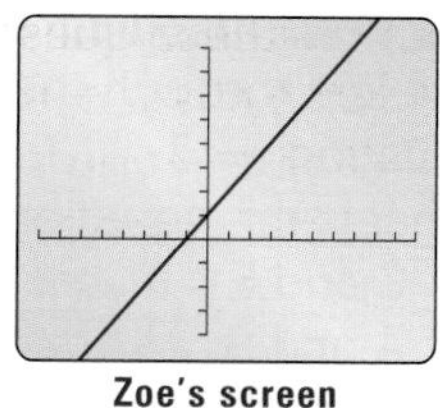

Zoe's screen

3 Create these displays using your graphical calculator.

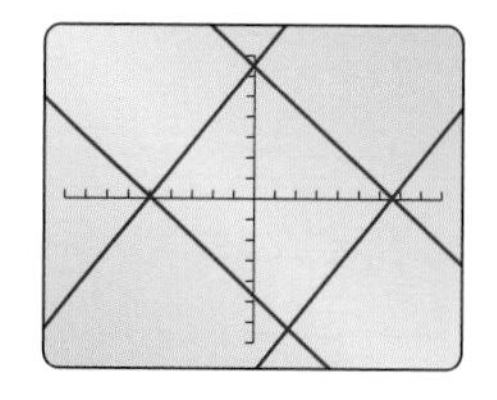
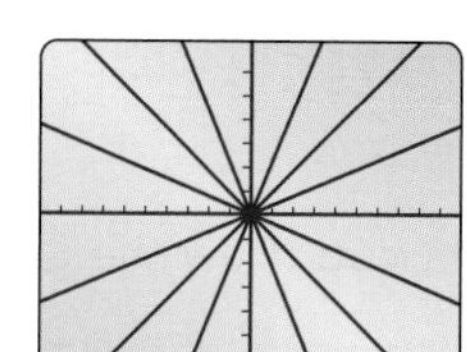
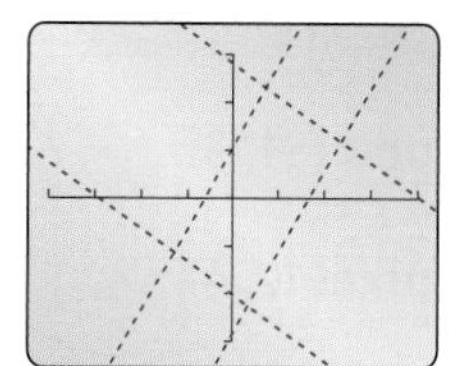
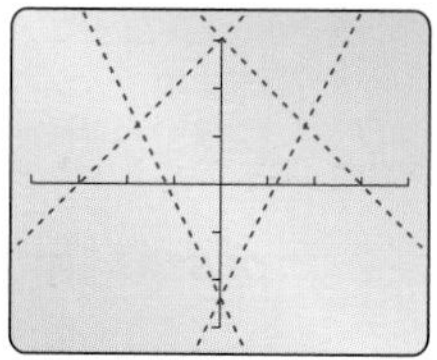

4 The lines $y = 3x - 2$ and $y = 4$ go through the point (2, 4).
Draw some more straight lines that pass through the point (2, 4).

5 Draw some straight lines that pass through each of these points.
a (2, 2) b ($^-4$, $^-4$) c (2, $^-3$) d (4, $^-6$)

6 Draw a line on your calculator screen that goes through these points.
a ($^-4$, 0) and (0, 4) b ($^-2$, 0) and (0, 4) c (0, $^-10$) and (10, 0)

Gradient of a straight line

Discussion

- James drew this table for $y = 2x + 1$.

x	0	1	2	3	4	5	6
y	1	3	5	7	9	11	13
Difference in y values		2	2	2	2	2	2

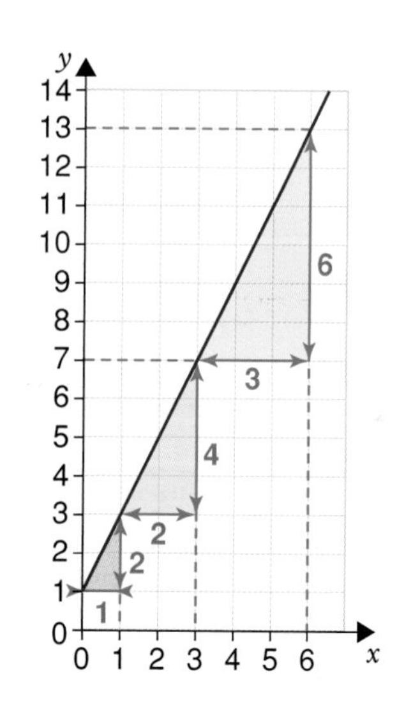

He drew the graph of $y = 2x + 1$.
He noticed that when x changed by 1, y changed by 2; when x changed by 2, y changed by 4; when x changed by 3, y changed by 6.

He drew these triangles to show this.

What do you notice about this ratio? **Discuss.**

$$\frac{\text{change in } y}{\text{change in } x} = \frac{3-1}{1-0} = \frac{7-3}{3-1} = \frac{13-7}{6-3} = \ldots$$

Is the ratio the same for all corresponding changes in y and changes in x? **Discuss.**

- **Discuss** these questions.
  Is the change in y proportional to the change in x?
  What can you say about the triangles James drew on his graph?
  Hint: think about enlargement.
  How are $y = 2x + 1$ and the ratio $\frac{\text{change in } y}{\text{change in } x}$ related?
- Draw the graph of $y = 3x + 1$ and find $\frac{\text{change in } y}{\text{change in } x}$.
  What do you notice?
- **What if** you drew the graph of $y = {}^-2x + 1$ or $y = {}^-3x + 1$?

$y = mx + c$ is a **linear function**.
The **change in y is proportional to the change in x.**

For any two points (x_1, y_1) and (x_2, y_2) that lie on the line,

$$\frac{y_2 - y_1}{x_2 - x_1} = \frac{y_3 - y_2}{x_3 - x_2} = m$$

$$m = \frac{\text{change in } y}{\text{change in } x} = \frac{y_2 - y_1}{x_2 - x_1}$$

gradient (pointing to m)

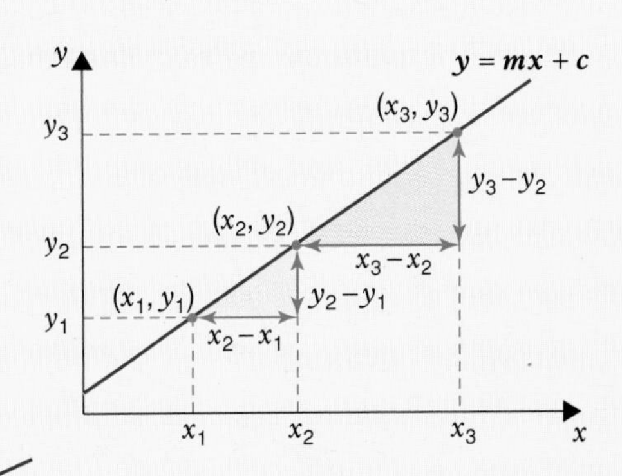

For a positive gradient, y increases as x increases.
For a negative gradient, y decreases as x increases.

Worked Example

Find the gradient of each of these lines.

a

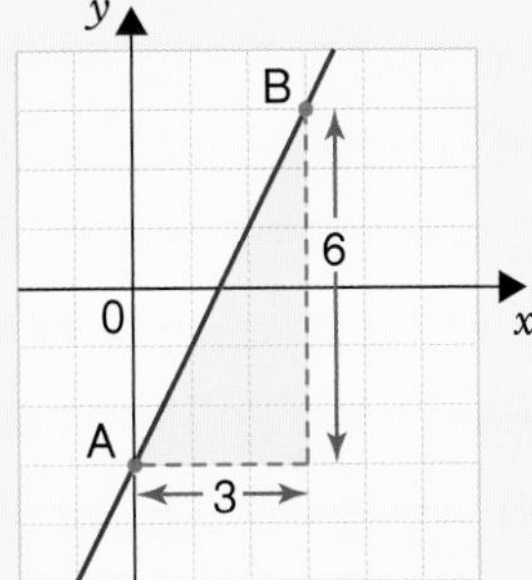

b

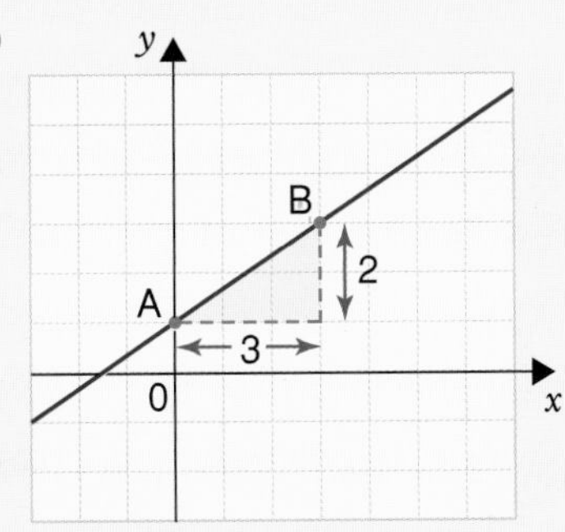

c

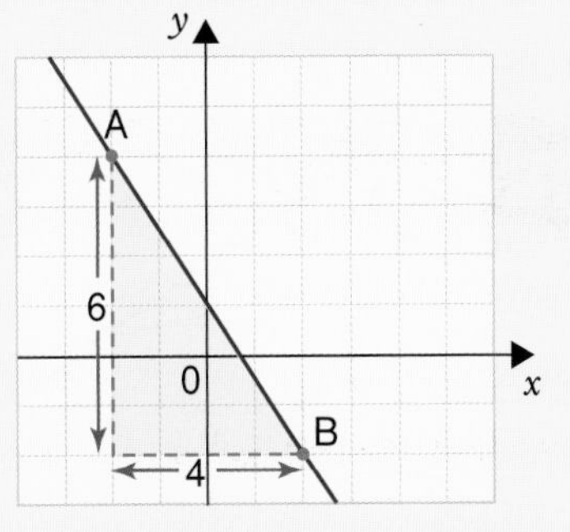

Answer

Let the points be $A(x_1, y_1)$ and $B(x_2, y_2)$

a gradient $= \frac{\text{change in } y}{\text{change in } x}$
$= \frac{y_2 - y_1}{x_2 - x_1}$
$= \frac{3 - {}^-3}{3 - 0}$ ← difference in y-coordinates / difference in x-coordinates
$= \frac{6}{3}$
$= \mathbf{2}$

b gradient $= \frac{\text{change in } y}{\text{change in } x}$
$= \frac{y_2 - y_1}{x_2 - x_1}$
$= \frac{3 - 1}{3 - 0}$
$= \mathbf{\frac{2}{3}}$

c gradient $= \frac{\text{change in } y}{\text{change in } x}$
$= \frac{y_2 - y_1}{x_2 - x_1}$
$= \frac{4 - {}^-2}{{}^-2 - 2}$
$= \frac{6}{{}^-4}$
$= \mathbf{\frac{{}^-3}{2}}$

Always check that a positive slope has a positive value and a negative slope has a negative value for the gradient.

Exercise 3

1

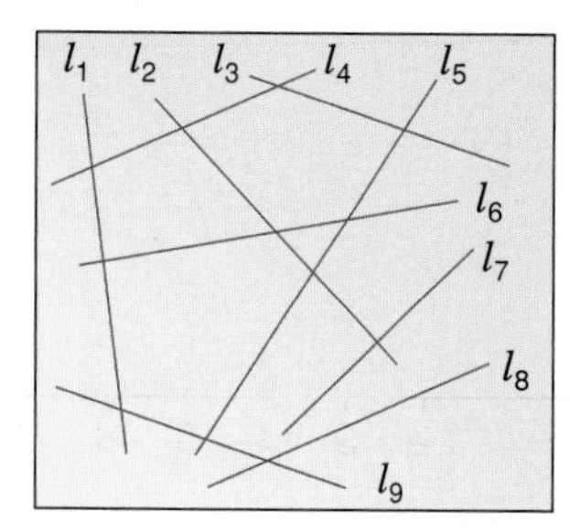

a Name the lines that have a positive gradient.
b Name the lines that have a negative gradient.
c There are two pairs of lines that have the same gradient. Name these pairs of lines.

2

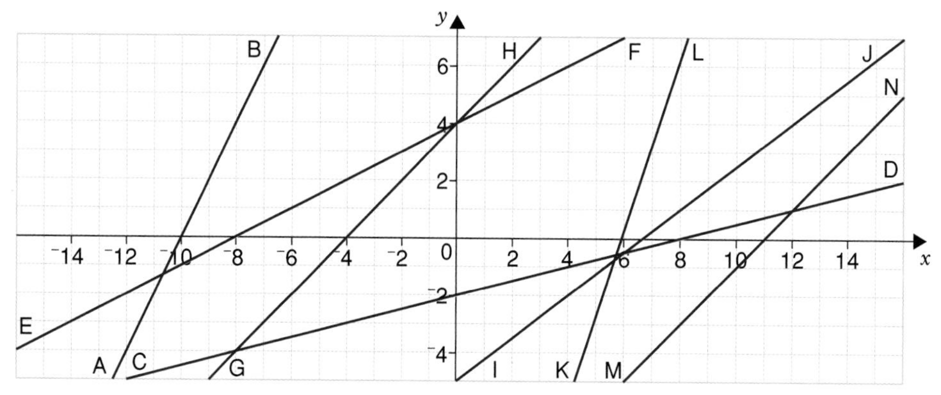

Find the gradients of these lines.

3

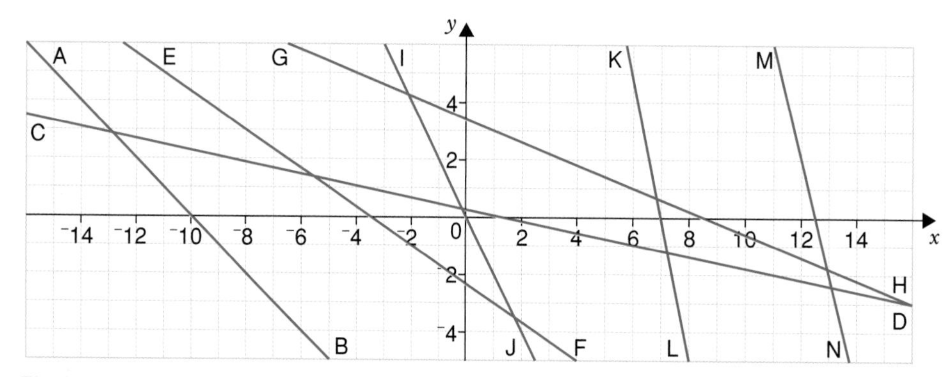

Find the gradients of these lines.

4 Find the gradients of these lines.

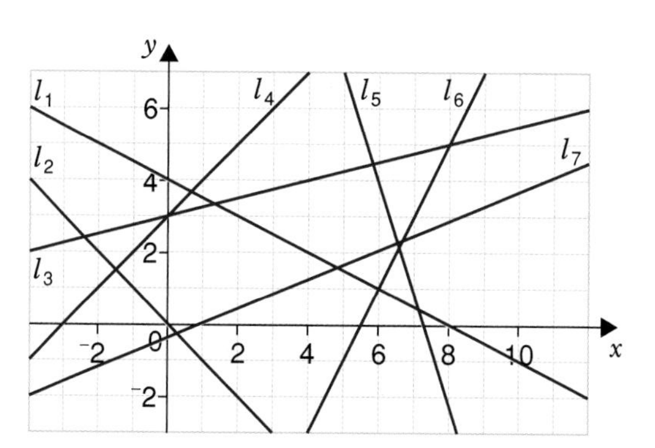

5 a Jack wanted to know the gradient of the line $y = 3$.
He wrote

$m = \frac{y_2 - y_1}{x_2 - x_1}$

$= \_\_\_\_$

$= \_\_\_\_$

Finish Jack's working to find the gradient of $y = 3$.

***b** Jack tried to find the gradient of $x = 4$.
He decided it was not possible to give a gradient for $x = 4$.
He is right. Explain why you think this is.

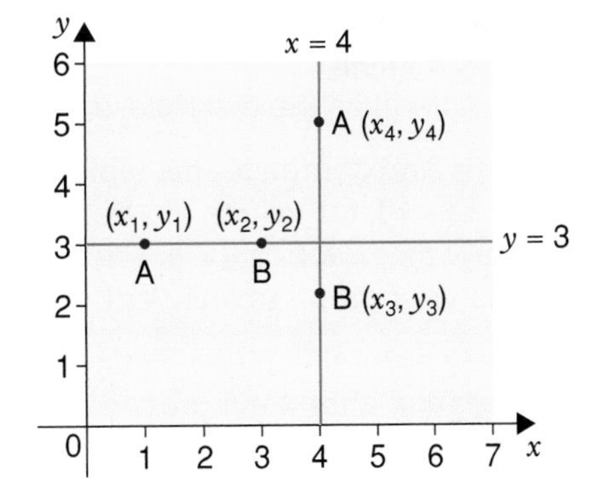

***6** The three vertices of a triangle are P(2, 1), Q(4, ⁻3), R(7, 0). Find the gradient of each side of this triangle.

Review 1 Find the gradient of each line on this graph.

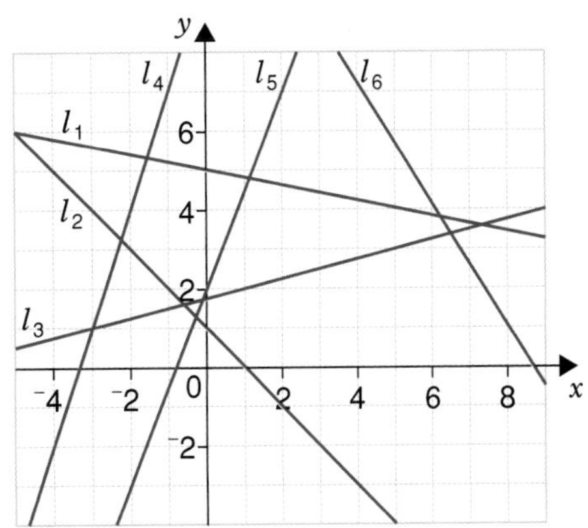

Review 2 The vertices of a quadrilateral are
A(⁻3, 2), B(1, 5), C(7, 4), D(3, ⁻1).

Find the gradient of the diagonals of this quadrilateral.

Reading and plotting real-life graphs

Reading graphs

We can **estimate** values from a graph.

Worked Example

A large vase holds about 16 pints.
Using the graph, estimate how many litres this is.

Answer

Each small square on the horizontal axis represents 4 pints.
Each small square on the vertical axis represents 4 litres.
16 pints is **about 10 litres**.

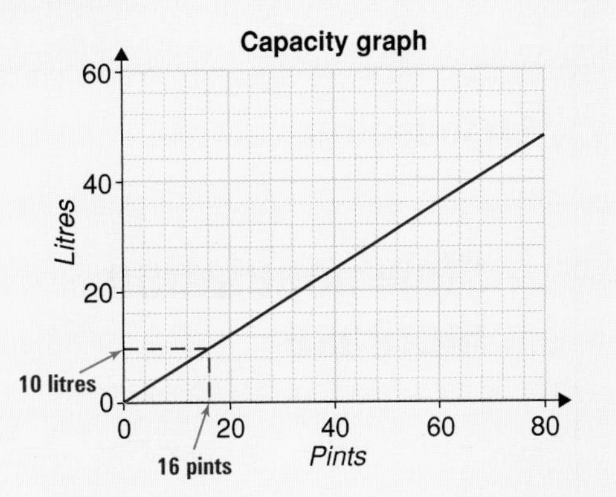

Plotting graphs

To plot a real-life graph

- decide how many points to plot
- construct a table of values using a formula
- choose suitable scales for the axes
- plot the points accurately
- draw a line through the points if it is sensible to do this
- give the graph a title and label the axes.

Worked Example

$d = 5t + 10$ gives the distance, d, in kilometres, of a plane from Hedgend after t minutes.

a Copy and complete this table of values for $d = 5t + 10$, for values of t from 0 to 5.

t	0		
d			

b Draw the graph of $d = 5t + 10$.

c Use your graph to answer these questions.

Darryl is on this plane. Estimate how far Darryl is from Hedgend after 4 minutes.

Katherine lives 17·5 km from Hedgend. Estimate how long it takes before the plane is flying over Katherine's house.

Answer

a

t	0	2	4
d	10	20	30

b

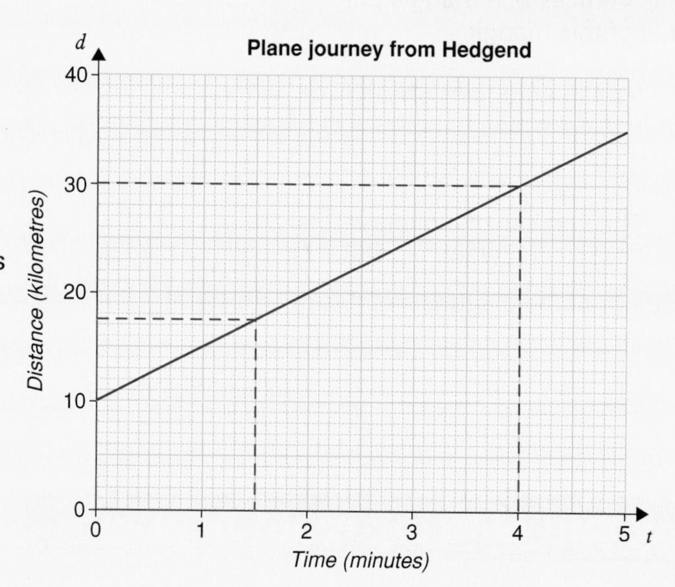

c Darryl is about 30 kilometres from Hedgend.
It takes about 1·5 minutes to be over Katherine's house.

Discussion

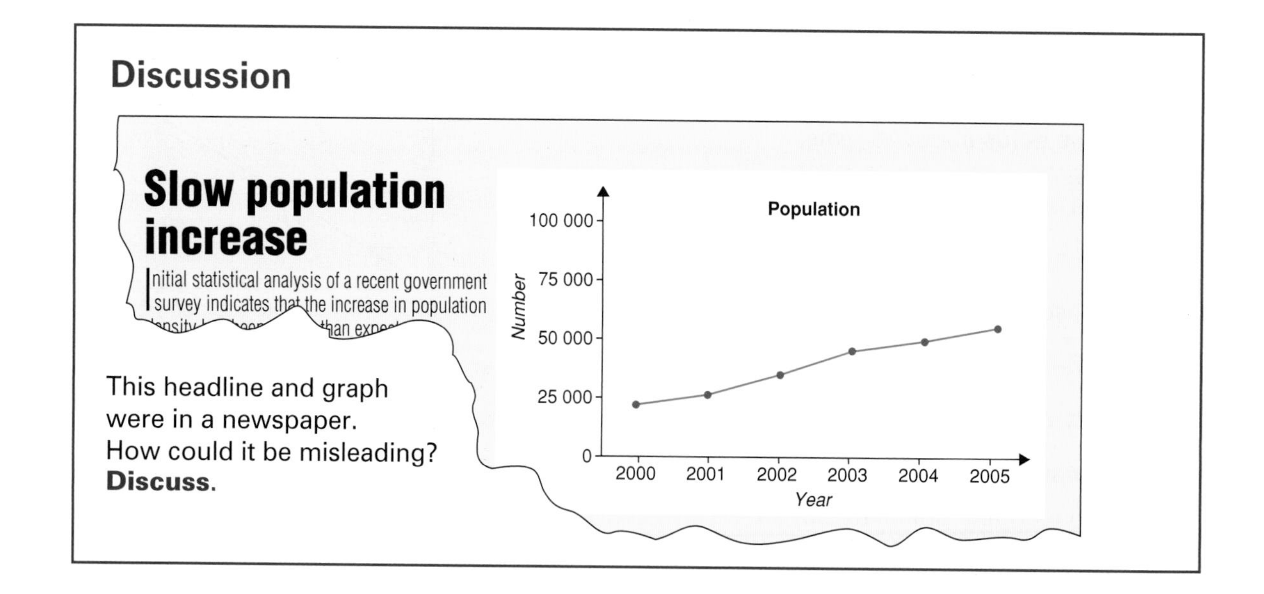

Slow population increase

Initial statistical analysis of a recent government survey indicates that the increase in population

This headline and graph were in a newspaper.
How could it be misleading? **Discuss.**

Exercise 4

T

1 The graph shows the average heights of 1 to 4 year olds.

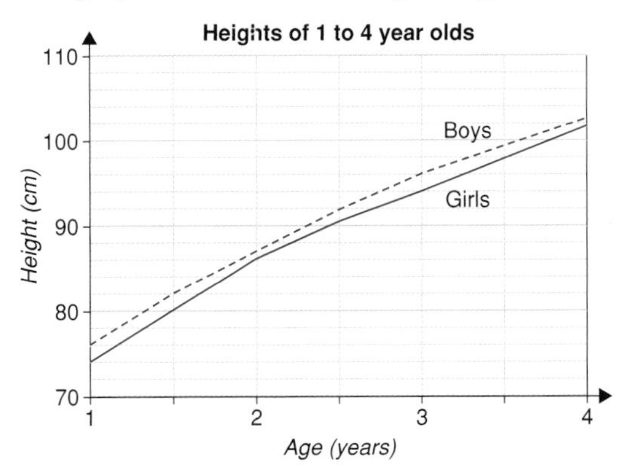

a Estimate the average height of $2\frac{1}{2}$-year-old girls.
b Estimate the average height of 3-year-old boys.
c Olivia is of average height. Her height is 98 cm.
Use the graph to find out how old Olivia is.
d Use a copy of this table.
Use the graph to fill it in.

Age of girl (in years)	Height in cm at start of year (approximate)	Height in cm at end of year (approximate)	Approximate growth in cm
1 to 2	74	86	12
2 to 3	86		
3 to 4			

e About how much taller, on average, are $2\frac{1}{2}$-year-old boys than $2\frac{1}{2}$-year-old girls?
f What does the shape of this graph tell you about the growth of 1 to 4 year olds?

T

2 You pay £2·60 each time you go to an aerobics class.
a Copy and complete this table.

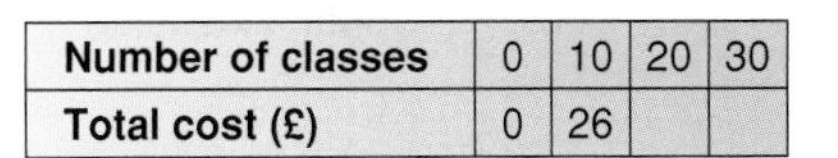

Number of classes	0	10	20	30
Total cost (£)	0	26		

b Use a copy of this grid.
Show the information in the table in **a** on the grid.
Join the points with a straight line.
c A different way of paying is to pay a yearly fee of £24. Then you pay £1·60 for each class.
Copy and complete this table.

Number of classes	0	10	20	30
Total cost (£)	24	40		

d Show the information in the table in **c** on the same graph.
Join these points with a straight line.
e For how many classes does the graph show that the cost is the same for both ways of paying?
f Janita wants to go to aerobics classes once every fortnight for a year.
Which way is cheaper for her to pay? Use the graph to estimate by how much.
Show how you did this.

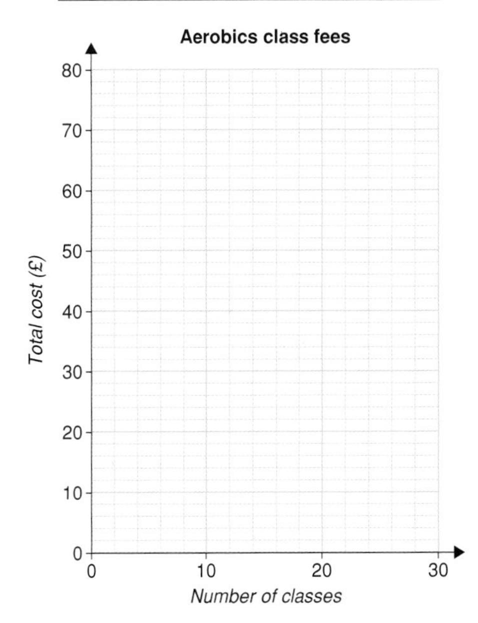

3 You can buy Australian \$2·50 with one British pound.

a Write a formula for changing pounds, p, to Australian dollars, d.

b Draw a graph of this relationship for p between 0 and 10. Have pounds on the horizontal axis.

c Use your graph to estimate the number of dollars that can be bought for £6.

d Dan bought 22 dollars. Estimate how many pounds Dan paid.

e Angela bought £60 worth of dollars. Use your answer for **c** to find how many dollars Angela got.

You will need to decide how many points to plot and choose a suitable scale for the axes.

4 Temperatures given in °C can be changed to °F by using the relationship $F = 1{\cdot}8C + 32$.

a Debbie drew the graph of this relationship. She began by plotting the points for which $C = 10$, $C = 40$ and $C = 100$.
Copy and complete these coordinates. (10, 50), (40, ___), (100, ___).

b Plot Debbie's 3 points. Draw the line that goes through these.

c Use your graph to change 75 °C to °F.

d Use your graph to change 75 °F to °C.

T

5 Joel was doing an experiment in science.
He measured the amount a spring extended when various masses were hung from it.
This table gives his results.

Mass (kg)	1	2	3	4	5	6
Extension (mm)	10	10·8	11·6	12·4	13·2	14

a Use a copy of the three sets of axes below.
Show the information in the table on all three.

b How does the choice of scale for the y-axis affect what the graph looks like?

c Joel was asked to write a conclusion for his experiment.
He wrote 'The spring extends lots more each time another kilogram is added to it'.
Which set of axes do you think Joel used for his graph?

d Do you think Joel's conclusion is correct? Explain why or why not.

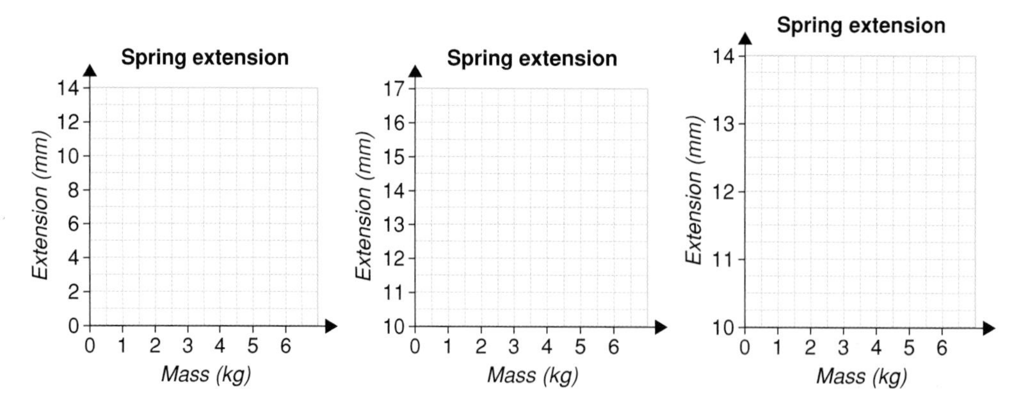

6 Explain why this graph could be misleading.

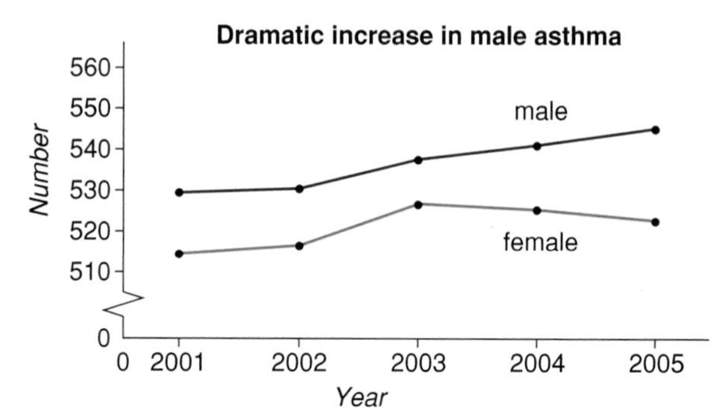

Review 1 When Trish mows the lawns she gets paid £1·50.

a Copy and complete this table.

No. of times lawns mown	1	2	3	4	5
Money earned (£)	1·50	3·00			

b Use a copy of this grid.
Show the relationship, money earned versus number of times lawn mown.

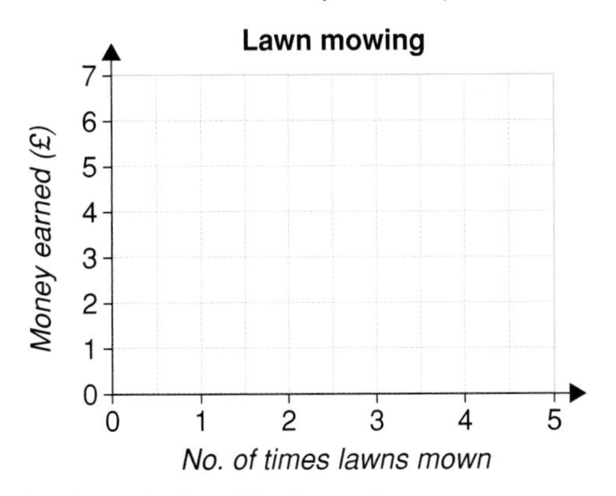

c Is the relationship linear?

d Should the points be joined with a straight line? Explain.

Review 2 Explain why this graph could be misleading.

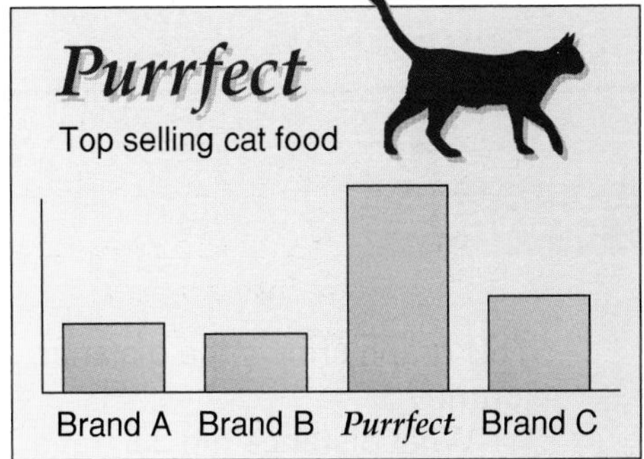

     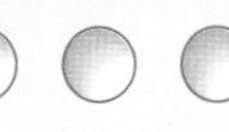 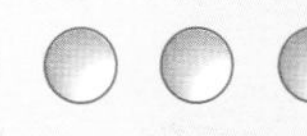

Investigation

Collision course

On a foggy English night, two boats are on the Thames.
One is travelling in the direction given by $y = 2x + 3$.
The other is travelling in the direction given by $y = \frac{x}{2} - 3$.

Draw a graph to show their paths?
Is it possible for the boats to collide? **Investigate**.
If so, give the coordinates of where they would collide.

Practical

You will need a graphical calculator.

Heat a container of water to over 60 °C. (Hot water from a tap will be adequate.)
Use a temperature probe attached to a graphical calculator to plot a cooling curve.

Distance–time graphs

We can tell some things about how an object is moving by looking at its **distance–time graph**.

Example Adriana drew this distance–time graph. She drove out of the drive and down the road. After 200 m she realised that she had forgotten something and so she stopped and then drove back home.

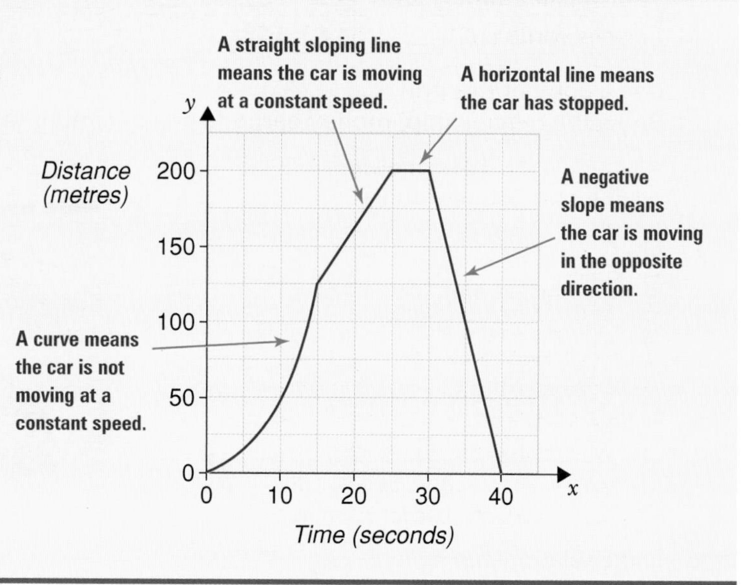

Exercise 5

1 This graph shows the distance Jill travelled on her cycle journey.

a What distance did Jill travel?

b How long did Jill's journey take?

c For how long did Jill stop during the journey?

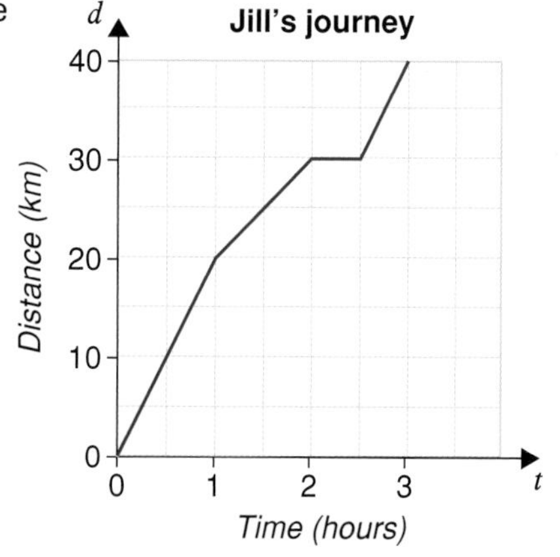

2 This graph shows Arshard's car journey from home to his aunt's, and back home again.

a How far away does his aunt live?

b For how long did he stop during the journey to his aunt's?

c How long did he stay at his aunt's?

d How far did he travel in the first hour?

e How long did it take him to travel home?

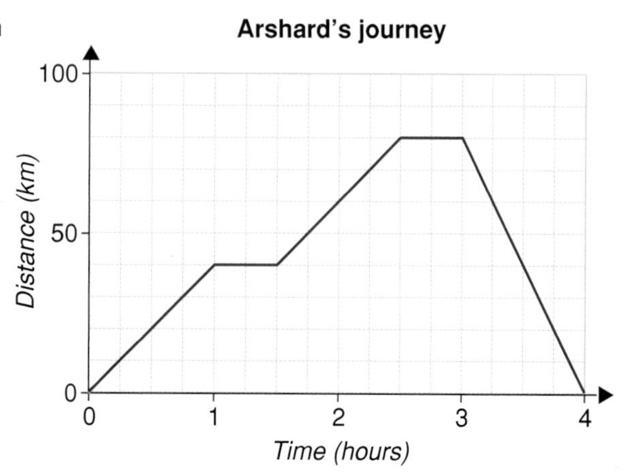

3 Robbie and Julie both threw a cricket ball up into the air.
This graph shows how high the two balls went.
Julie's ball went higher than Robbie's.

a Estimate how much higher Julie's ball went than Robbie's.

b Estimate the time after which their balls were the same height.

c Estimate the number of seconds that Julie's ball spent more than 8 m above the ground.

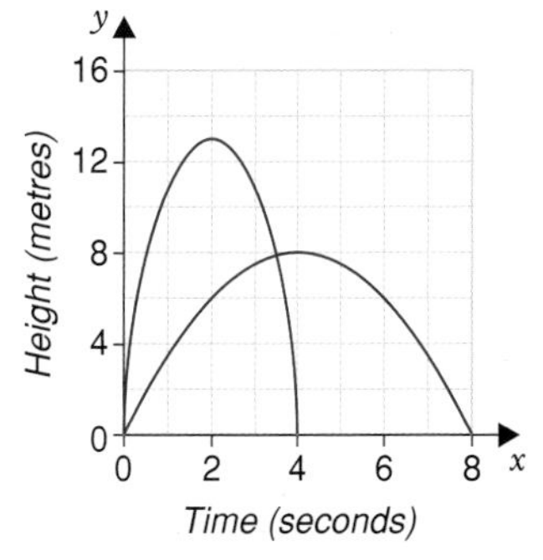

[SATS 2002 paper 2]

4 I went for a walk.
The distance–time graph shows information about my walk.
Which of these best describes my walk?

A I was walking faster and faster.

B I was walking slower and slower.

C I was walking north-east.

D I was walking at a steady speed.

E I was walking uphill.

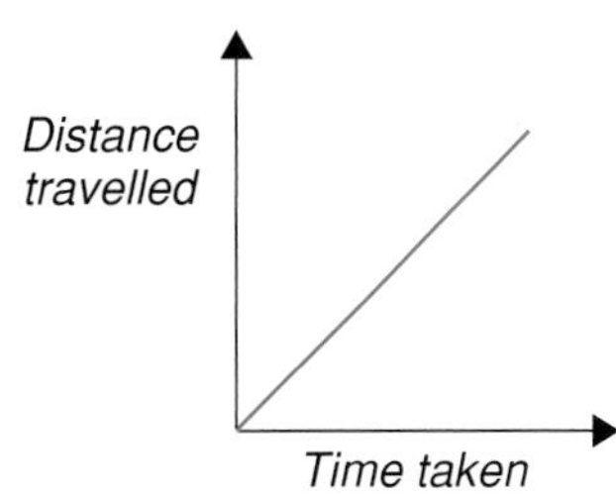

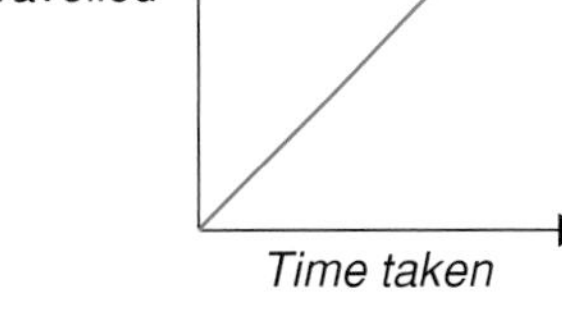

5

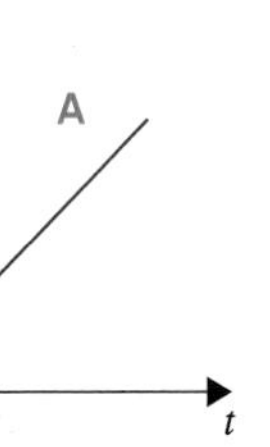

d B t

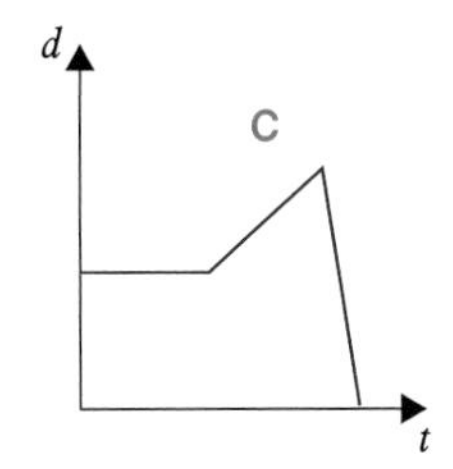

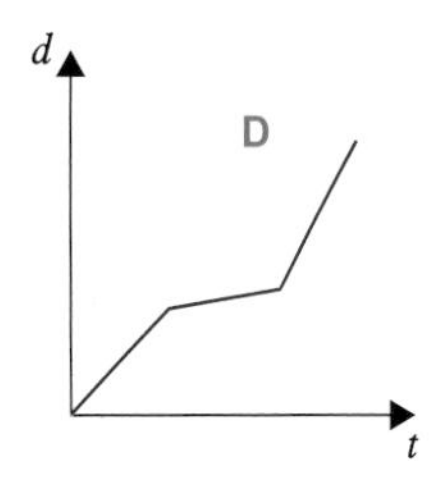

Jake cycles along a flat road, then up a hill, then down the other side. Which graph best describes Jake's cycle journey?

[SATS 2000 paper 2]

6 The graph shows my journey in a lift.
I got into the lift at floor number 10.

a The lift stopped at two different floors before I got to floor number 22.
What floors were they?

b For how long was I in the lift while it was moving?

c After I got out of the lift at floor number 22, the lift went directly to the ground floor.
It took 45 seconds.
Use a copy of the graph.
Show the journey of the lift from floor 22 to the ground floor.

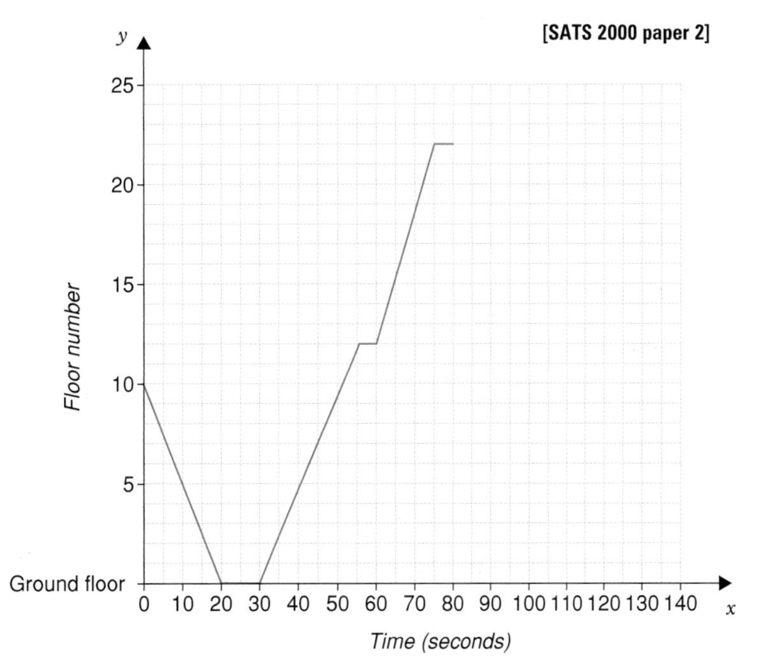

7 This graph shows the distance–time graph for a ride at a fun park.

a In which section of the graph, AB, BC, CD or DE, is the carriage
  i travelling at constant speed
  ii stopped
  ***iii** accelerating?

b After how many seconds did the ride stop?

c Estimate the maximum distance the ride reached from the start.

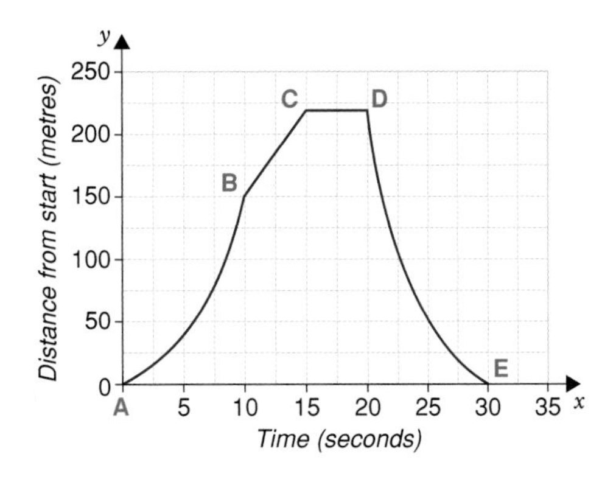

T

Review Use a copy of this.

The graph represents Guy's journey from home to London.

Guy stopped for lunch on the way.

a Write down the time at which he stopped.

b For how long did he stop?

c At 12 p.m. Guy had to slow down because of traffic.
For how many miles did he travel at this slower speed?

d Guy spent an hour visiting friends in London. He then returned home, travelling at a steady speed. It took him $2\frac{1}{2}$ hours.
Use this information to complete the graph of his journey.

***e** Between which times did Guy travel fastest?

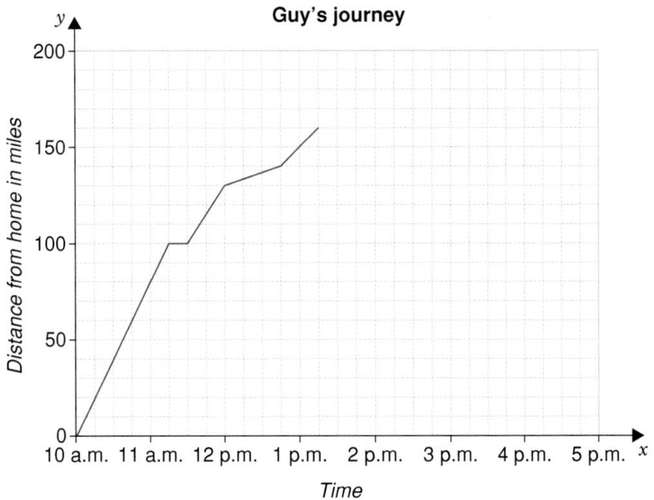

Interpreting and sketching real-life graphs

A graph shows the **relationship between variables.**

Examples

When x is large y is large.
As x increases in equal steps y increases by increasing amounts.

When x is large y is moving to 0.
As x increases in equal steps y decreases by decreasing amounts.

When x is large y becomes 0.
As x increases in equal steps y decreases by increasing amounts.

We can sketch a graph to show a relationship between two variables.

Example When the temperature of a closed can is increased the pressure inside the can increases also.
This sketch shows the relationship.

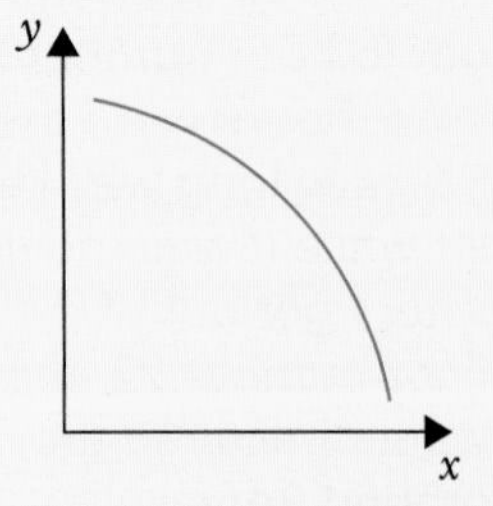

Exercise 6

1 Sketch a line graph to show these.

a The depth of water against time when each of these is filled with a steady flow of water. Sketch both on the same axes.

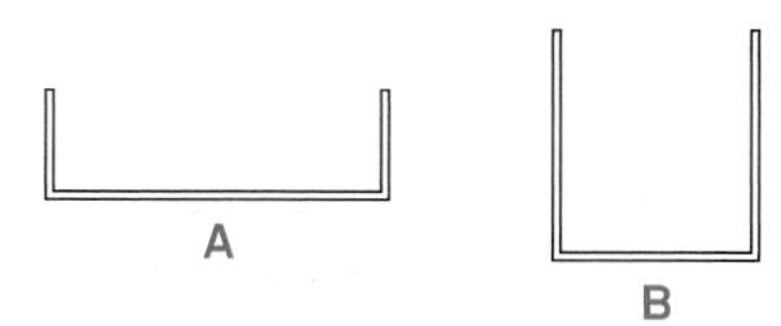

b The depth of sand against time when each of these is filled with a steady flow of sand. Sketch both on the same axes.

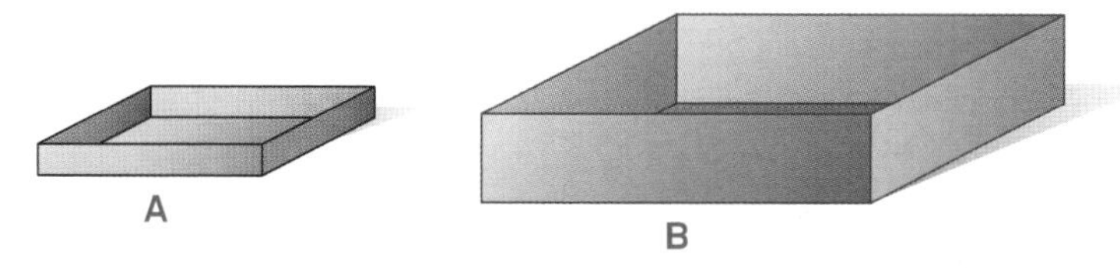

2 Three boys, Tom, Omar and Logan were in a race.
This graph shows their speed each second.

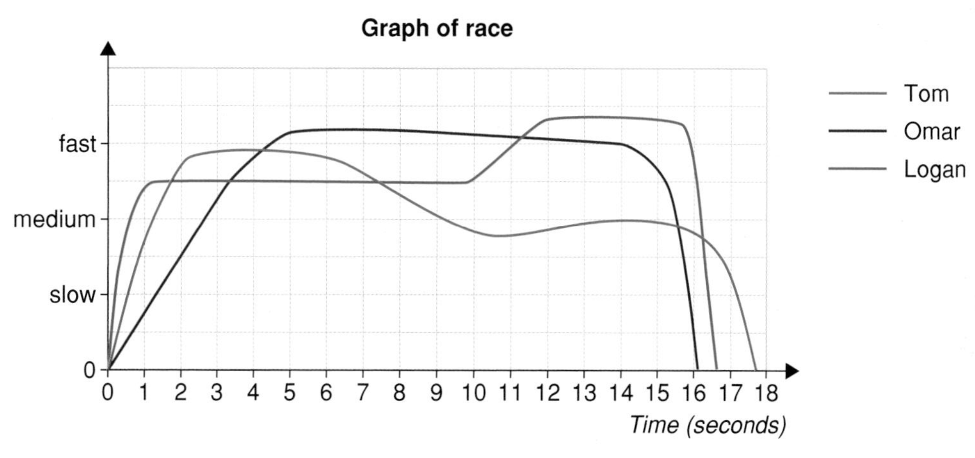

a Who won?

b Use the shape of the graph to tell a story about how each boy ran the race.

3 Three small plants are planted.
The first, p_1, is given plenty of water and plenty of sunlight.
The second, p_2, is given plenty of water and kept in a dark room.
The third, p_3, is not given much water and is kept in a dark room.
None of the plants die.
Which of these graphs do you think is the graph for each?

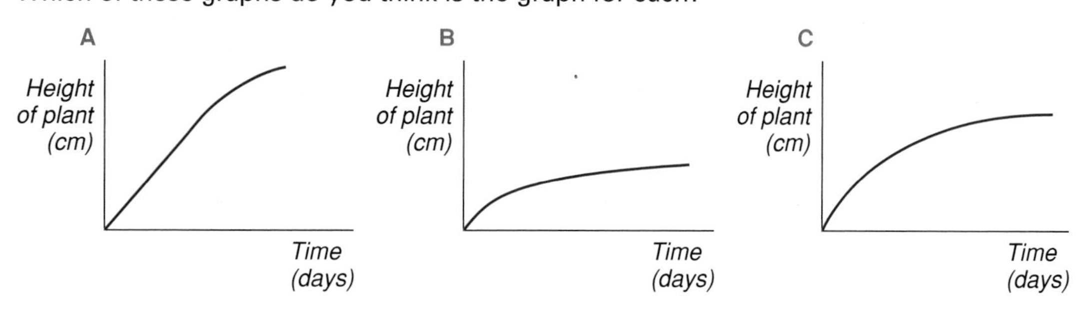

T

4 Three beakers are filled with boiling water.
The first beaker, A, is left in a room.
The second beaker, B, is put in the fridge.
The third beaker, C, is covered and wrapped in a fibre glass wrapping.
This shows the graph of temperature against time for beaker A.
Use a copy of this graph.
Sketch the lines for the other two beakers on the same grid.

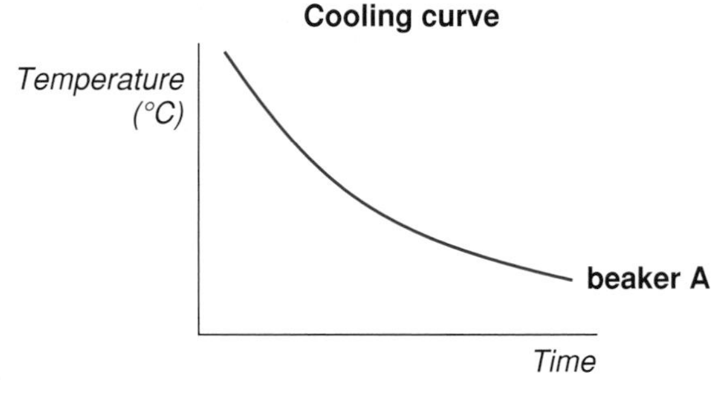

5 **a** A new manager at 'Gifts for Everyone' took over in August this year. Sales immediately dropped. He looks at this graph which shows **last year's** sales. Should the manager be worried? Explain by referring to the shape of this graph and the trends shown by it.

b Give a reason why sales might increase in December.

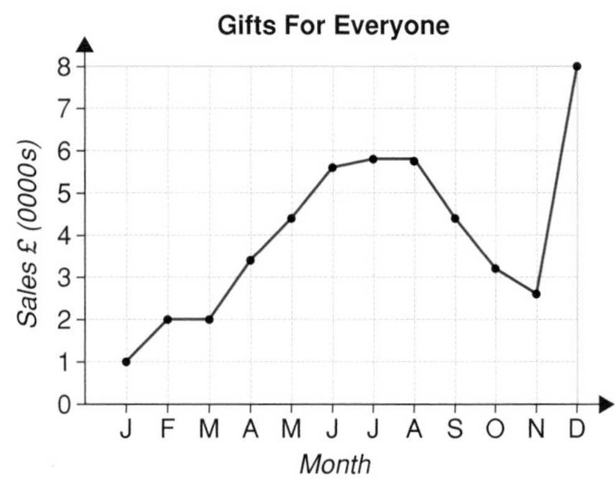

6 Match these graphs with the statements below.

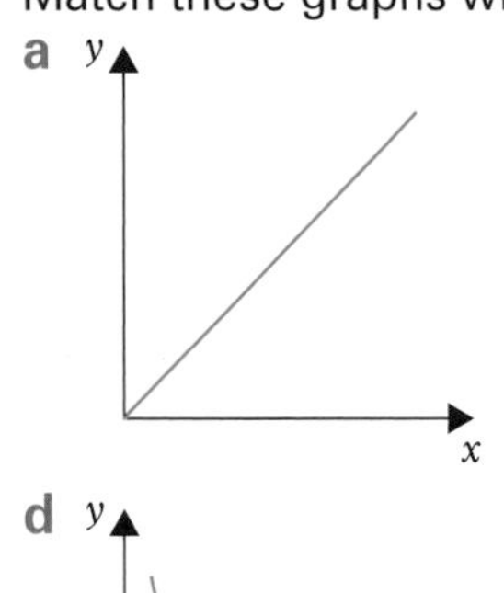

b

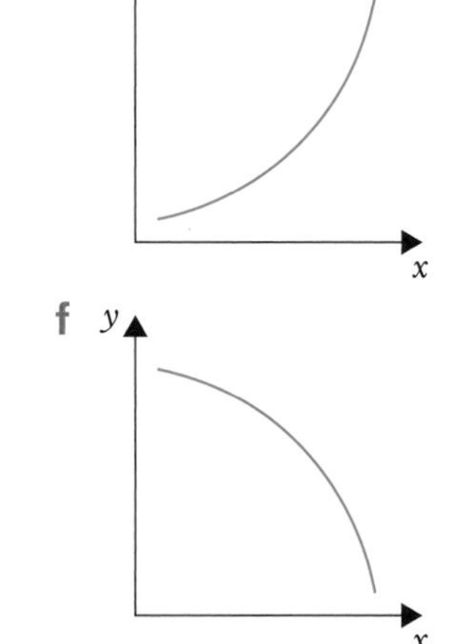

d **e** **f**

A the distance (y) plotted against time (x) travelled by a train moving at constant speed

B the number (y) of flies left in a room as fly spray begins to work slowly at first and then more rapidly, plotted against time (x)

C the temperature (y) of a hot water bottle left on the floor to cool, plotted against time (x)

D the volume of water (y) left in a bath being emptied at a constant rate, plotted against time (x)

E the distance (y) plotted against time (x), of a car which accelerates away from the lights then gradually slows down and stops

F the number of bacteria cells (y) in a piece of rotting meat, plotted against time (x)

7 For each of the graphs given in question **6**, choose one of these statements to describe the relationship between x and y.
As x increases by equal amounts
A y increases by equal amounts
B y increases by increasing amounts
C y increases by decreasing amounts
D y decreases by equal amounts
E y decreases by increasing amounts
F y decreases by decreasing amounts.

*__8__ The graph of depth of water against time when a steady flow of water pours into this container is shown.

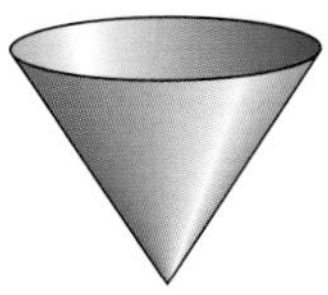

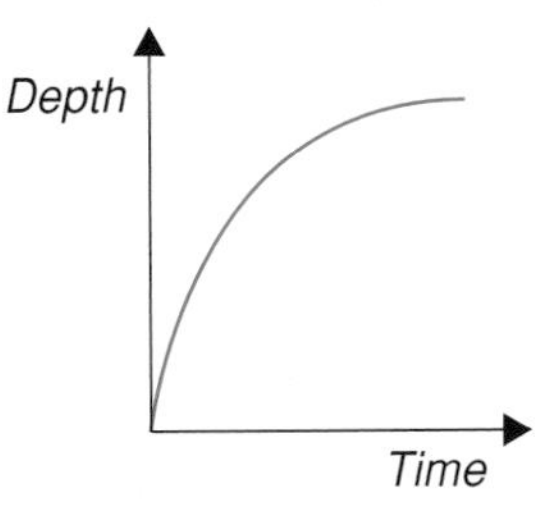

Sketch the graph of depth against time for water poured at a steady rate into these containers.

a

b

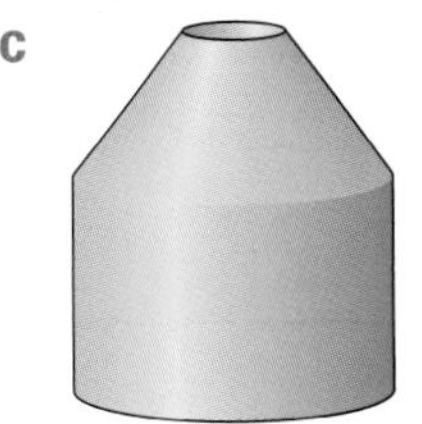

c

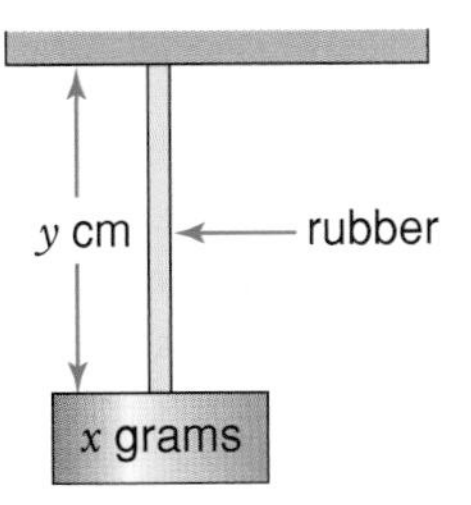

*__9__ Rosalie did four experiments as part of her science project. Sketch a graph of y against x for each of these situations.

a She suspended a mass of x grams from a piece of rubber which stretches to a length of y cm.

b She put two equal masses on a see-saw. One mass is x cm above the ground and the other is y cm above the ground.

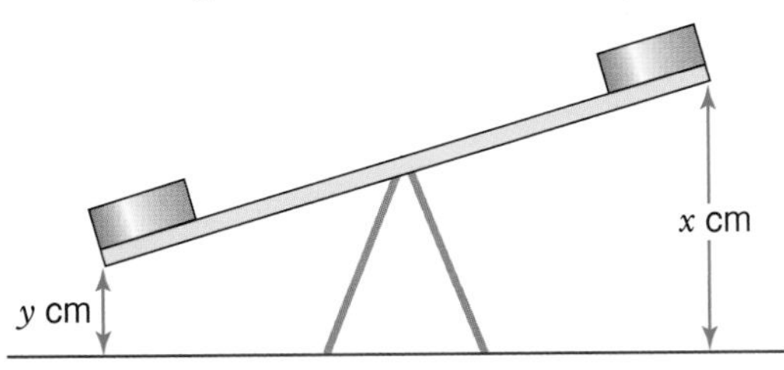

c She starts a chemical reaction which gives off bubbles. She measures the number of bubbles, y, given off per second, x.

d She put a 3 kg mass 8 cm from the pivot point of a see-saw. She balanced this exactly by putting a mass of y kg a distance x cm from the pivot point.

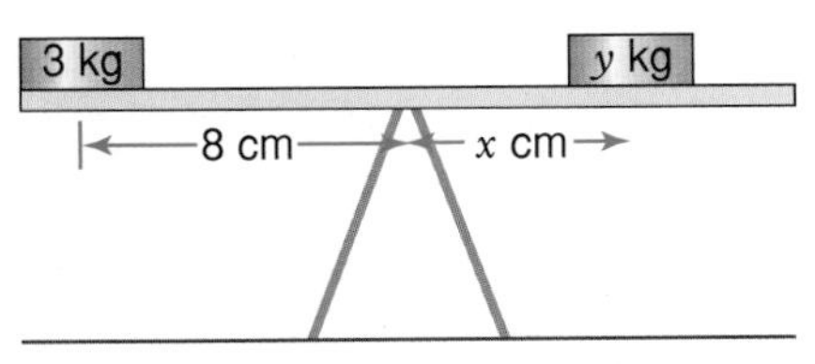

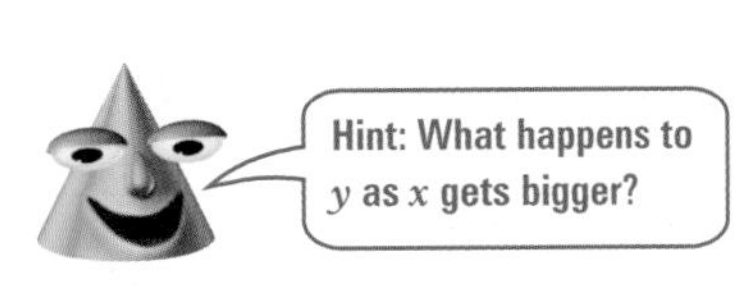

y cm
rubber
x grams

T

Review 1 Three beakers are filled with cold water.
The first beaker, A, is heated using a gas flame.
The second beaker, B, is left in the sun on a hot day to heat up.
The third beaker, C, has ice added to it.
This graph shows the relationship between temperature (°C) and time (minutes) for beaker B.

a Use a copy of this grid. On it sketch the graph of temperature versus time for beakers A and C.

b What does the shape of this graph for beaker B tell you about the temperature of the water over time?

c Which of these is true about the relationship between temperature and time for beaker B?
As time increases in equal amounts
A temperature increases in equal amounts
B temperature increases in increasing amounts
C temperature increases in decreasing amounts.

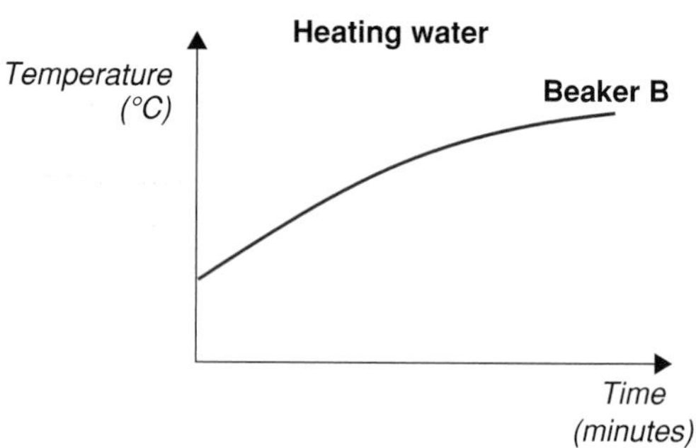

* **Review 2** Water drips from a tap into this container.

Sketch the graph of depth against time.

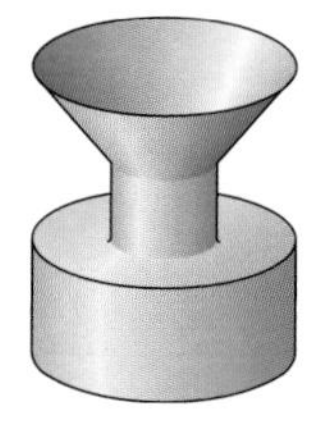

Discussion

- This graph shows the cost to laminate posters.
  How much does it cost to laminate a 12 cm^2 poster?
  What about a 10 cm^2 poster?
  What about a 60 cm^2 poster?
  What about ...
  Discuss.
  This is called a **step graph**. Why do you think this is?

Laminating costs

Cost (£): 0, 1, 2, 3, 4, 5

Size of poster (cm^2): 0, 20, 40, 60, 80, 100, 120, 140, 160

- How would you draw a step graph for these postal charges? **Discuss**.
  Draw the graph.

Weight up to	First Class	Second Class	Weight up to	First Class	Second Class
60 g	27p	19p	450 g	£1·48	£1·19
100 g	41p	33p	500 g	£1·66	£1·35
150 g	57p	44p	600 g	£2·00	£1·60
200 g	72p	54p	700 g	£2·51	£1·83
250 g	84p	66p	750 g	£2·69	£1·94*
300 g	96p	76p	800 g	£2·91	
350 g	£1·09	87p	900 g	£3·20	
400 g	£1·30	£1·05	1 kg	£3·49	

Costs for First Class items over 1 kg are £3·49 and then 85p for each extra 250 g.
* Items over 750 g cannot be sent Second Class.

- * **Discuss** how to draw the graph of y, the greatest integer less than or equal to x, versus x.
- What other examples of step graphs can you think of? **Discuss**.

Summary of key points

$y = mx + c$ is the equation of a straight line.

We can rearrange equations into the form $y = mx + c$.

Example $2y - 2x - 4 = 0$

start with y → y → [multiply by 2] → $2y$ → [subtract $2x$] → $2y-2x$ → [subtract 4] → $2y-2x-4$

$\frac{4+2x}{2}$ ← [divide by 2] ← $4+2x$ ← [add $2x$] ← 4 ← [add 4] ← 0 ← start with the right hand side

$y = \frac{4 + 2x}{2} = \frac{4}{2} + \frac{2x}{2} = \mathbf{2 + x}$ or $\mathbf{y = x + 2}$.

Once the equation is in the form $y = mx + c$ we can plot the graph by constructing a table of values.

Example $y = x + 2$

x	$^-2$	0	1	2
y	0	2	3	4

In $y = mx + c$, m represents **the gradient** or steepness of the line.

c represents the **y-intercept**.

If two equations have the same value of m, the lines are parallel.

If m is negative the graph has a negative slope.

If m is positive the graph has a positive slope.

For a straight line, the change in y is **proportional** to the change in x.

$\frac{\text{change in } y}{\text{change in } x} = m$, the gradient of the line

$m = \frac{y_2 - y_1}{x_2 - x_1}$ for any two points on the line.

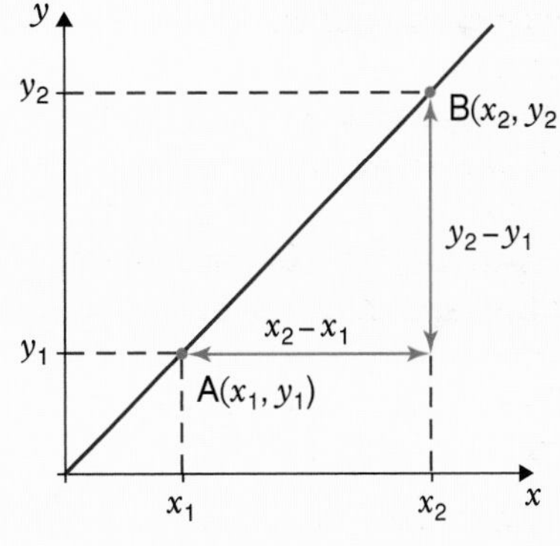

Sometimes we **plot a real-life graph** using a **formula**. To do this we

- decide how many points to plot
- construct a table of values using the formula
- choose suitable scales for the axes
- plot the points accurately
- give the graph a title and label the axes.

We can use the graph to estimate values.

Example This shows the graph of ferry charges for different lengths of cars. We estimate the charge for a 420 cm car as £84. We estimate the length of a car charged £115 as 575 cm.

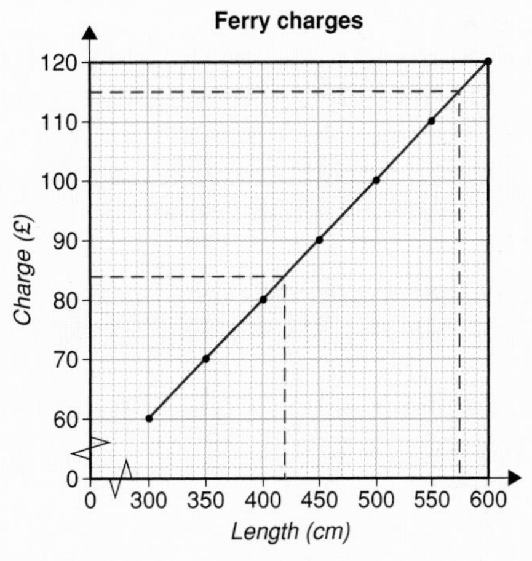

It is important to choose a suitable scale for the vertical axis or the graph can be misleading.

This shows a **distance–time graph**.
Lucy rowed a dinghy at a steady pace. She rowed 6 km in 3 hours. She rested for an hour then rowed steadily back to the start in 2 hours. Lucy drew this graph to show the relationship between distance and time.

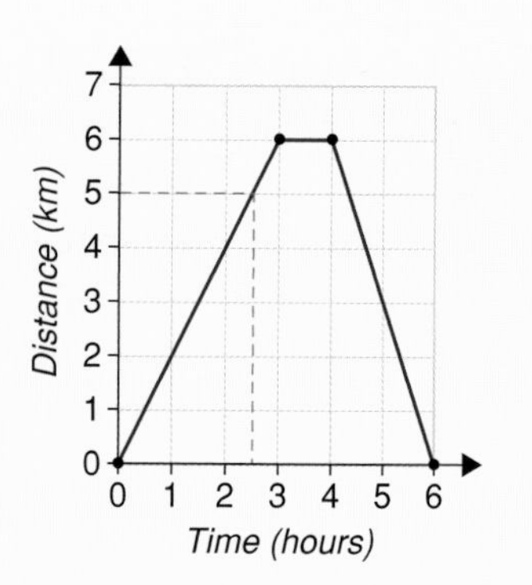

We can use the graph to estimate the distance she had travelled after $2\frac{1}{2}$ hours.
She had travelled about 5 km.

Sometimes we **sketch** a graph for a real-life situation or interpret a sketch.

Example This graph shows the depth of water against time for water pouring steadily into these containers.

A B

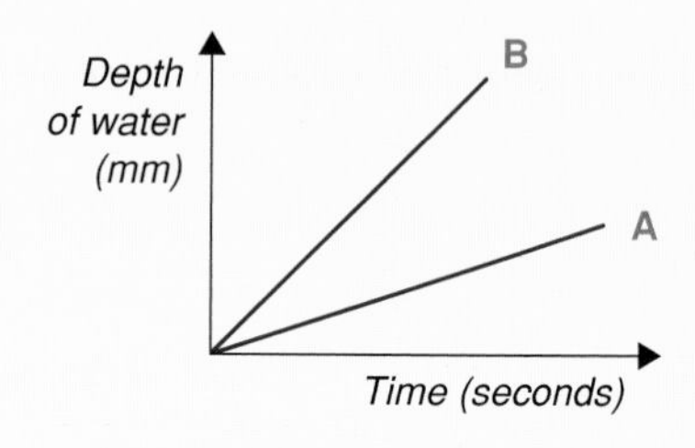

Test yourself

1 Rearrange these line equations into the form $y = mx + c$.
a $2x + y = 3$ **b** $x + 2y - 6 = 0$ **c** $3x - y = 1$

2 Draw a set of axes with x-values from $^-6$ to 6 and y-values from $^-6$ to 6.
Draw and label the lines given in question **1**.

3 Does the point $(1, ^-4)$ lie on the line $3x - 2y = 11$?
How can you tell?

4

A $y = ^-3x + 2$ B $y = 2x - 3$ C $y = 3x - 2$

Choose an equation from the box to match each of these.
You may use some equations more than once.
a crosses the y-axis at $^-3$
b has a positive gradient steeper than $y = 2x + 4$
c has a negative gradient
d is parallel to $y = 12 - 3x$
e has a gradient of 2
f cuts the y-axis at (0, 2)

5 Match these equations with the lines.
A $y = ^-x + 3$
B $y = ^-\frac{1}{2}x + 3$
C $y = x$
D $y = \frac{1}{3}x - 1$
E $y = ^-2x + 6$

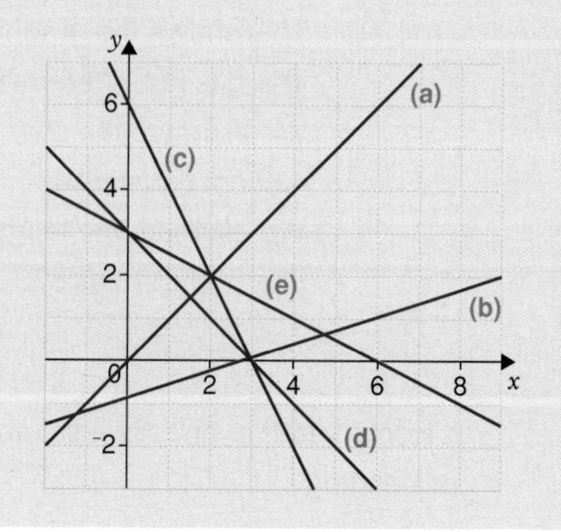

6 Find the gradient of each of these lines.

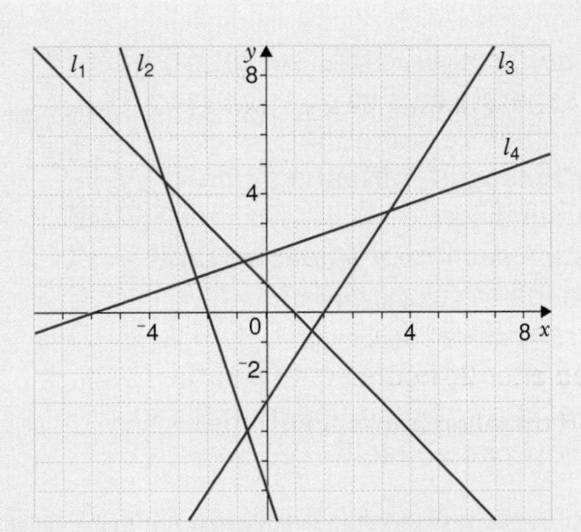

7 This graph gives the height of two boys.

a How tall was Nick when he was $2\frac{1}{2}$ years old?

b How tall was Adam when he was 3 years old?

c How tall was Adam when he was born?

d At what age were the boys the same height?

e About how much taller was Adam than Nick when they were both $2\frac{1}{2}$ years old?

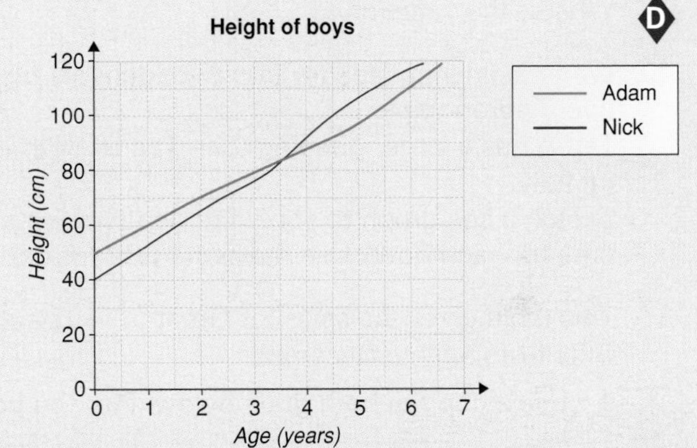

8 One litre is about 0·22 gallons.

a Write a formula to show the relationship between gallons, g, and litres, ℓ.

b Copy and complete this table.

ℓ	0	10	20
g			

c Draw a graph of gallons versus litres. Put litres on the horizontal axis.

d Use your graph to find the approximate number of gallons in 16 litres.

e A factory manufactures tanks. They label these tanks with both litres and gallons. These labels give the capacity to the nearest 10 litres and the nearest 10 gallons. Copy and complete the labels.

9 The crew of a movie set were testing fake bombs. At explosion time the fake bombs emit fireworks. This graph shows how high two fake bombs went during a test. Bomb A didn't reach as high as bomb B. The curves end when the bomb pretends to explode.

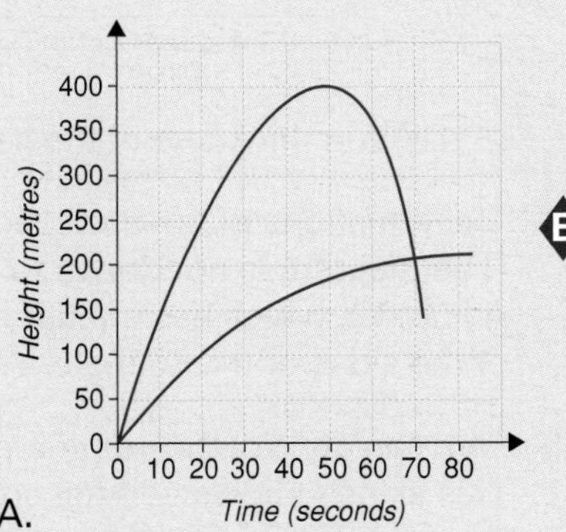

a Estimate how much higher bomb B went than bomb A.

b Estimate the time at which the bombs were at the same height.

c Estimate the time at which bomb A 'exploded'.

d Estimate the time that bomb B was more than 250 m above the ground.

e In the first 50 seconds, as time increases in equal steps, is the height of bomb B

A increasing in equal steps

B increasing in increasing amounts

C increasing in decreasing amounts?

10 Adrienne and her brother Afraaz set off from their home in separate cars to go to a disco. Afraaz left home before Adrienne.

a How many times did Adrienne stop on her way to the disco?

b What does the section of Afraaz's graph that has a negative slope tell us?

c Who got to the disco first?

d After Afraaz left home the second time, did he travel at a constant speed until he stopped? How can you tell?

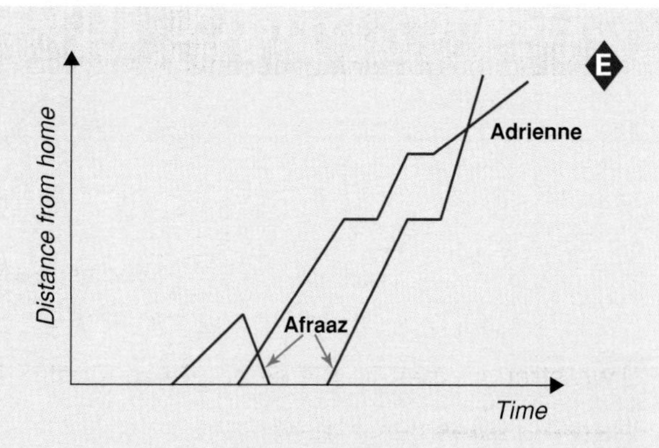

E

11 Jarod wants to draw a line with equation $y = mx + 4$ to pass through the point (5, 19). What is the value of m?

12 Water is flowing steadily into two identical bowls from two different taps.
Tap A has a larger opening than Tap B. Both are turned on fully.
Sketch a line graph to show the depth of water against time for each bowl. Use the same grid for both.

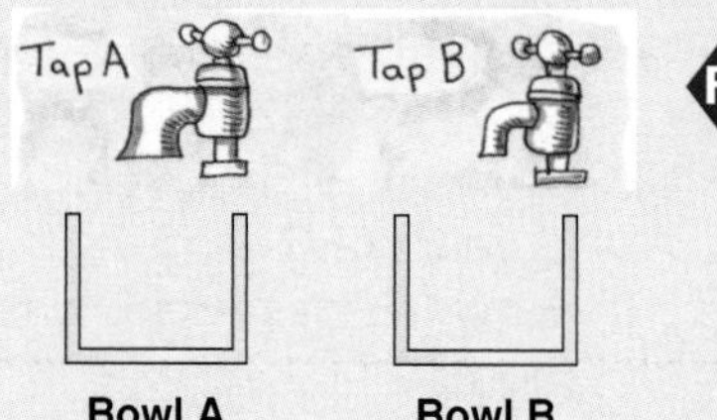

F

T

13 This table gives the sales for Great Gifts from July to December.

Month	Jul	Aug	Sep	Oct	Nov	Dec
Sales (000s)	11	12	10	13	16	18

D

a Use a copy of both grids below. Plot the points on each of them.

i

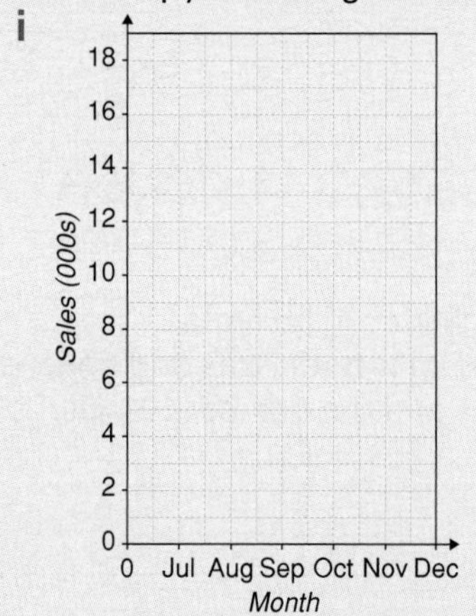

ii

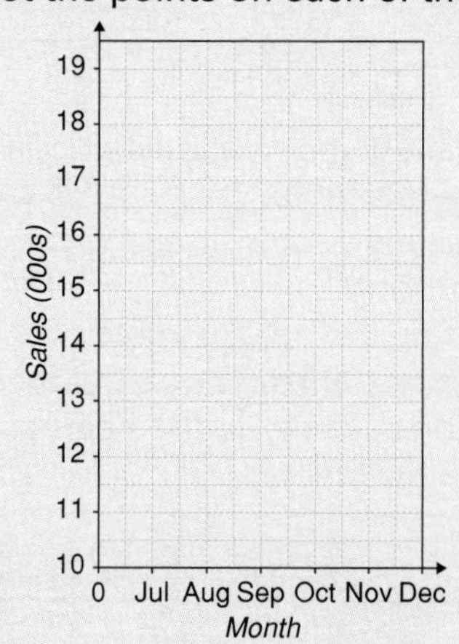

b Do you think one of the sets of axes gives a misleading graph? Explain.

14 The equations of the three sides of a triangle are shown. (The diagram is not drawn to scale.)
A line PQ, which has equation $y = 2x - 4$, is parallel to one of the sides of this triangle. Which one?

A, B, C; $2x - y + 1 = 0$; $x + 2y = 12$; $2y = x - 4$

B

15 Paul can control the volume of the CDs he plays at a disco. This sketch shows volume against the time during one song.
Describe how the volume changed between A and B on the graph.

Volume; O, A, B, R, S; Time

F

***16** Logan planted a baby plant. He measured the height of the plant (y) each week.
Sketch a graph of height (y) versus time (x).

F

Shape, Space and Measures Support

Lines and angles

line
infinite length

P ——— Q
line segment
finite length

Two straight lines on the same **plane** must either **intersect** or be **parallel**.

A **plane** is a flat surface.

Parallel lines never meet.
We show parallel lines with arrows.
SR is parallel to UT.

We show **perpendicular** lines using the symbol ⊥

AB is perpendicular to CD.

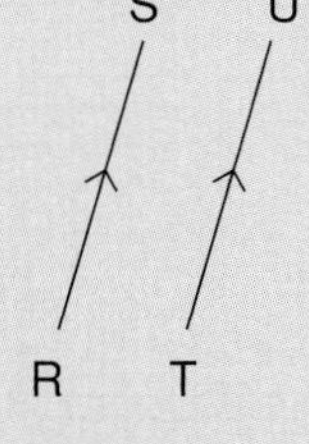

parallel lines

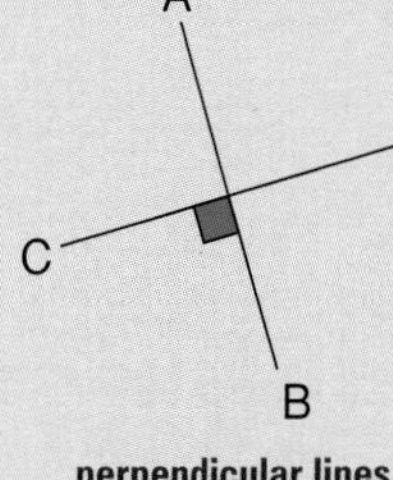

perpendicular lines

We name a line segment using two upper-case letters.

Parallel and perpendicular lines can be drawn using a **ruler and set square**.

We name this angle

1 using the letter at the vertex, ∠Q

or **2** using three letters, the middle letter being the vertex, ∠PQR or ∠RQP or $P\hat{Q}R$ or $R\hat{Q}P$.

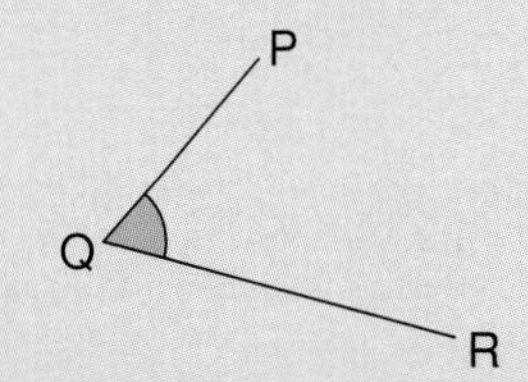

If there is more than one angle at the vertex always use three letters to name an angle.

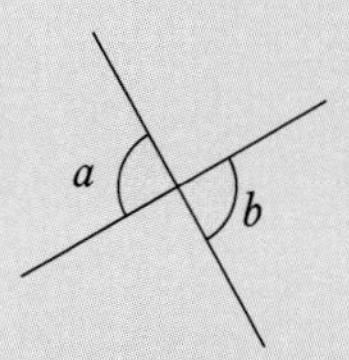

$a = b$
Vertically opposite angles are equal.

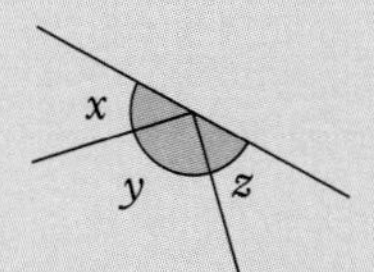

$x + y + z = 180°$
Angles on a straight line add to 180°.

x, y and z are adjacent angles on a straight line.

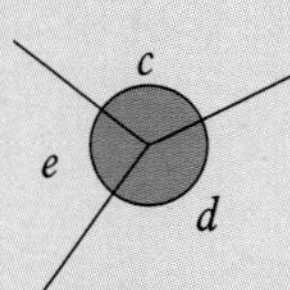

$c + d + e = 360°$
Angles at a point add to 360°.

Angles made with parallel lines

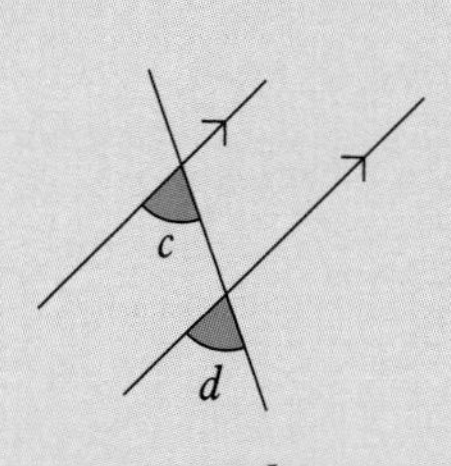

$c = d$
Corresponding angles are equal.

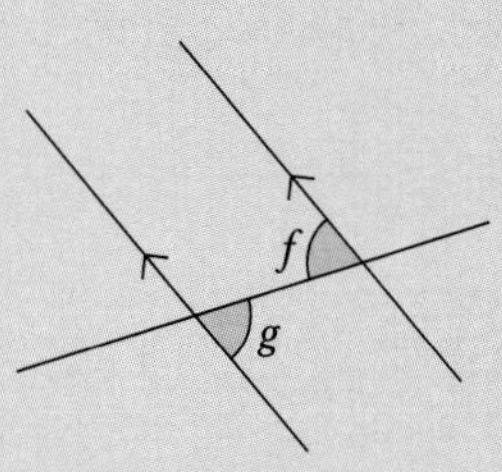

$f = g$
Alternate angles are equal.

Complementary angles add to 90°.
Supplementary angles add to 180°.

The interior angles of a triangle add to 180°.

Example $m + 57° + 64° = 180°$
$m = 180° - 57° - 64°$
$= 59°$

The exterior angle of a triangle is equal to the sum of the two opposite interior angles.

Example $f = 47° + 58°$
$= 105°$

Practice Questions 4, 27, 34, 37

2-D shapes

Naming triangles

A **triangle is named** using the capital letters at the vertices.
This triangle could be named as △STR or △SRT or △RST or ...
The side opposite each vertex is named with the lower-case letter of the vertex.

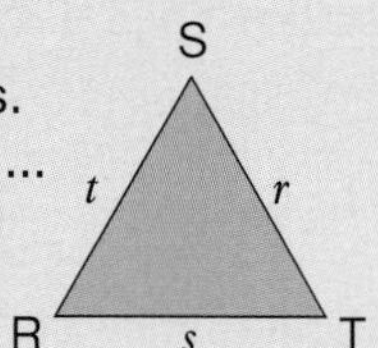

Properties of triangles A triangle is a 3-sided polygon.

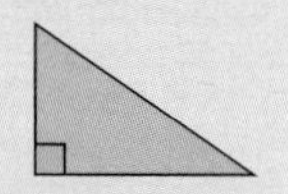
right-angled
one angle is a right angle

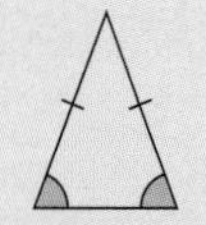
isosceles
2 equal sides
2 base angles equal

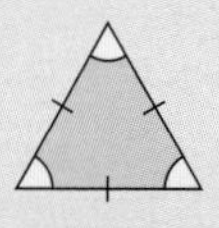
equilateral
3 equal sides
3 equal angles

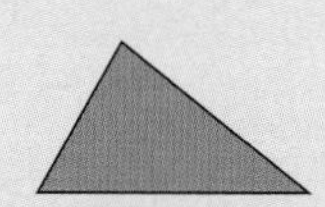
scalene
no 2 sides are equal
no 2 angles are equal

Properties of quadrilaterals A quadrilateral is a 4-sided polygon.

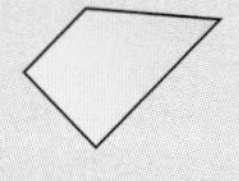

These are the **special quadrilaterals**.

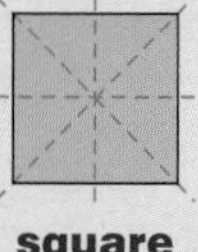
square

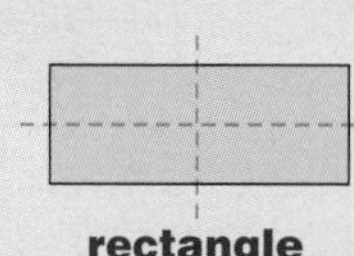
rectangle

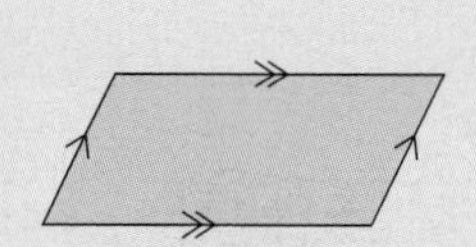
parallelogram

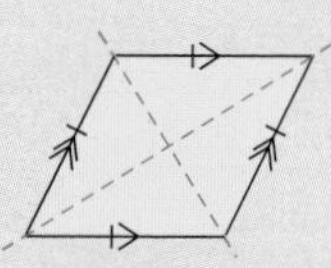
rhombus

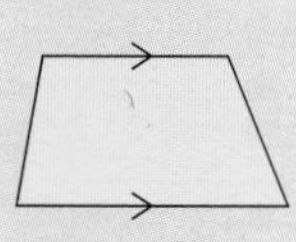
trapezium

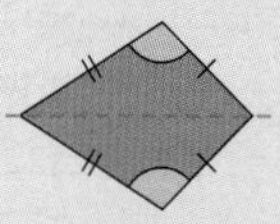
kite

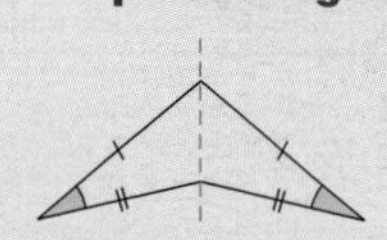
arrowhead or **delta**

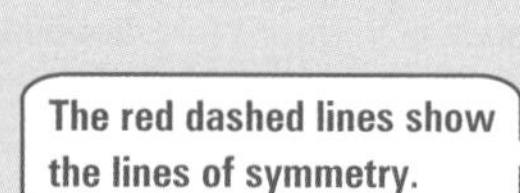

Some of the properties of these special quadrilaterals are shown in the following table.

	Square	Rhombus	Rectangle	Parallelogram	Kite	Trapezium	Arrowhead
one pair of opposite sides parallel	✓	✓	✓	✓		✓	
two pairs of opposite sides parallel	✓	✓	✓	✓			
all sides equal	✓	✓					
opposite sides equal	✓	✓	✓	✓			
all angles equal	✓		✓				
opposite angles equal	✓	✓	✓	✓	1 pair		
diagonals equal	✓		✓				
diagonals bisect each other	✓	✓	✓	✓			
diagonals perpendicular	✓	✓	✓		✓		✓
diagonals bisect the angles	✓	✓					

Polygons

A **polygon** is a 2-D shape made from line segments enclosing a region.
2-D is short for two-dimensional.

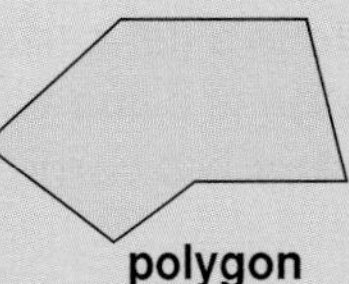
polygon

A 3-sided polygon is a triangle. A 4-sided polygon is a quadrilateral.
A 5-sided polygon is a pentagon. A 6-sided polygon is a hexagon.
A 7-sided polygon is a heptagon. An 8-sided polygon is an octagon.
A **regular polygon** has all its sides equal and all its angles equal.

A **concave** shape has at least one reflex angle.

A **convex** shape has no reflex angles.

convex

concave

Practice Questions 5, 6, 9, 14, 28, 30, 48

Constructions

We use compasses and a ruler to construct the **perpendicular bisector** of a line segment, BC.

Open the compasses to a little more than half the length of BC. With compass point first on B and then on C, draw arcs to meet at P and Q.

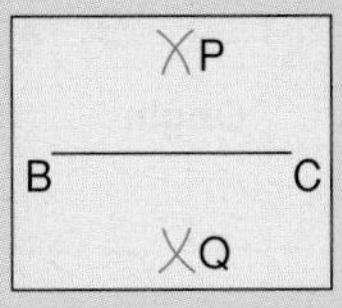

Draw the line through P and Q. R is the point which bisects BC.

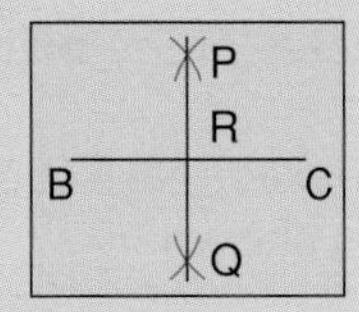

We use compasses and a ruler to construct the **bisector of an angle P**.

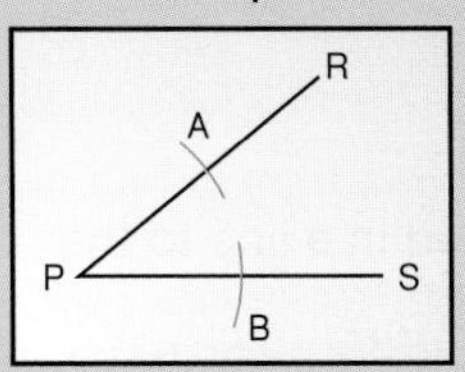

Open out the compasses to a length less than PR or PS. With compass point on P, draw arcs as shown.

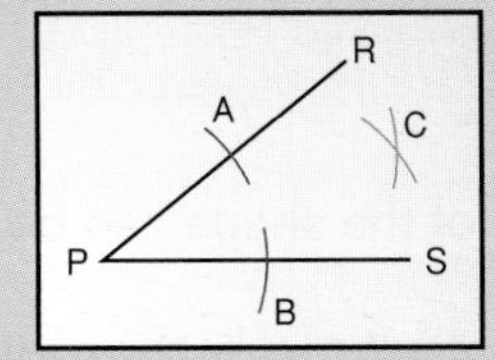

With compass point on first A, then B, draw arcs to meet at C.

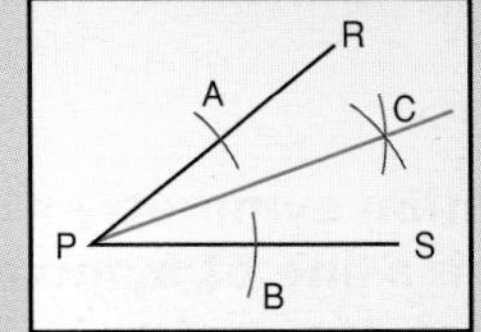
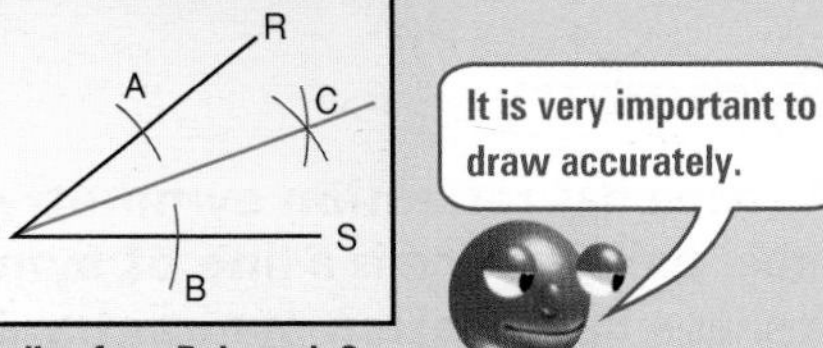

Draw the line from P through C. This line, PC, is the bisector of angle P.

We can **construct triangles and quadrilaterals** using a set square and ruler or compasses and ruler.

Examples

To construct this triangle:
1 Draw PR 2·6 cm long.
2 Draw an angle of 85° at R.
3 Draw RQ 2·8 cm long.
4 Join P to Q.

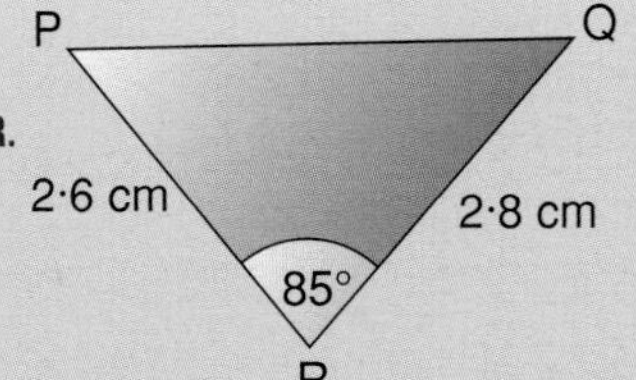

To construct this triangle:
1 Draw AB 2·8 cm long.
2 Open compasses to 2·5 cm and with point on A draw an arc.
3 Draw an arc from B, 2 cm long.
4 Complete the triangle.

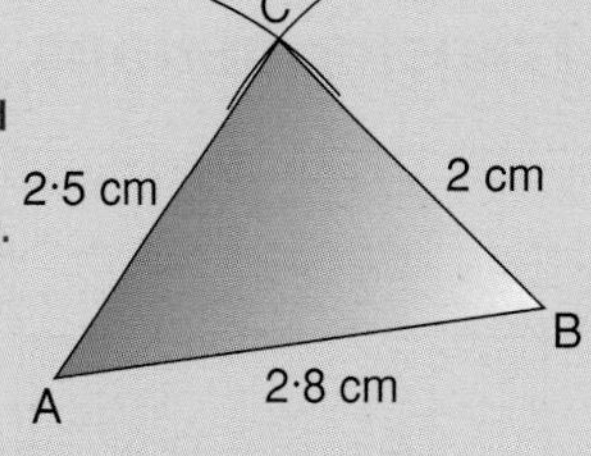

Practice Questions 36, 38, 40

3-D shapes

3-D stands for three-dimensional. 3-D shapes have length, width and height.

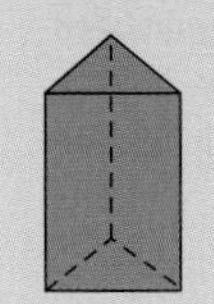
Triangular prism

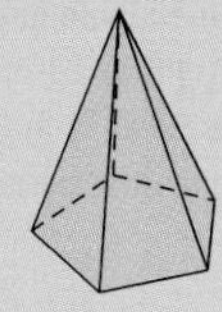
Pyramid (pentagonal base)

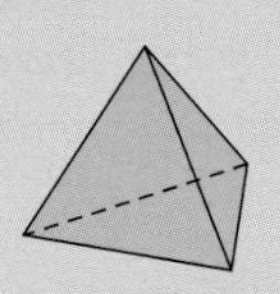
Tetrahedron (triangular-based pyramid)

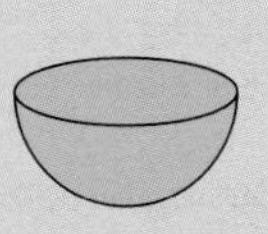
Hemisphere

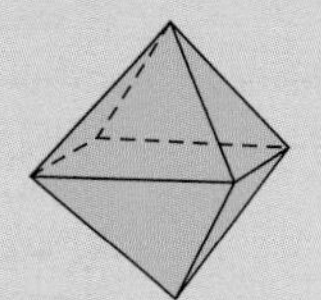
Octahedron

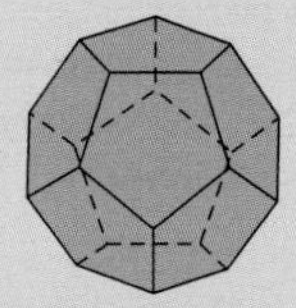
Dodecahedron

A **face** is a flat surface.
An **edge** is a line where two faces meet.
A **vertex** is a corner where edges meet.

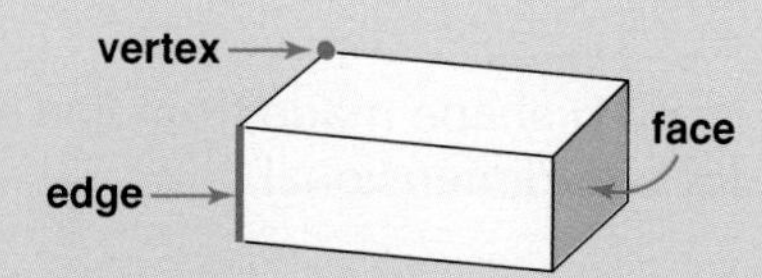

A 2-D shape that can be folded to make a 3-D shape is called a **net**.

Example This net folds to make a tetrahedron.

Practice Questions 15, 16, 19, 24, 31, 39, 42

Coordinates

We use **coordinates** to give the position of a point on a grid.
The coordinates of A are ($^-2$, 1).
We always give the x-coordinate first.
The coordinates of the **origin** are (0, 0).
The x- and y-axes make four quadrants as shown.

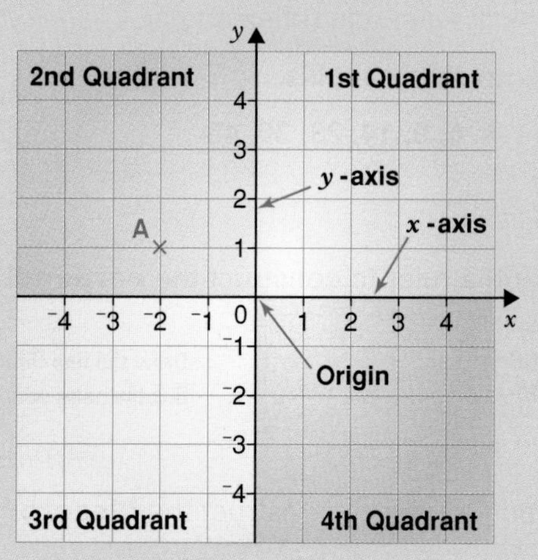

Practice Question 32

Symmetry

A shape has **reflection symmetry** if one half of the shape can be reflected in a line to the other half. The line is a **line of symmetry**.
A shape has **rotation symmetry** if it fits onto itself **more than once** during a complete turn.
The **order of rotation symmetry** is the number of times a shape fits exactly onto itself during one complete turn.
If a shape has rotation symmetry of order 1, we say it does not have rotation symmetry.

Practice Question 11

Transformations

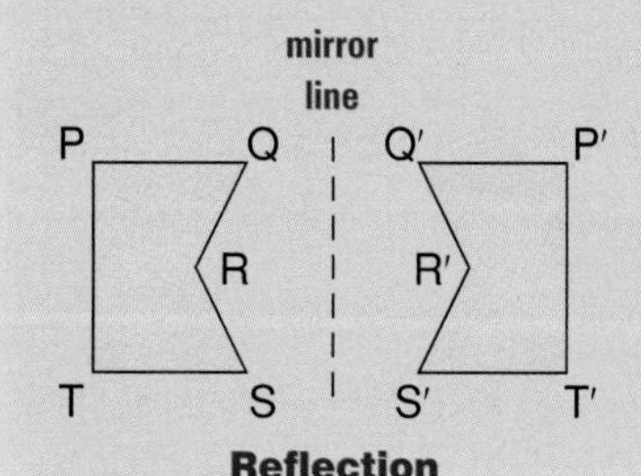

Reflection

Corresponding points are equidistant from the mirror line.
If you reflect the image in the mirror line you get back to the original shape (**self-inverse**).

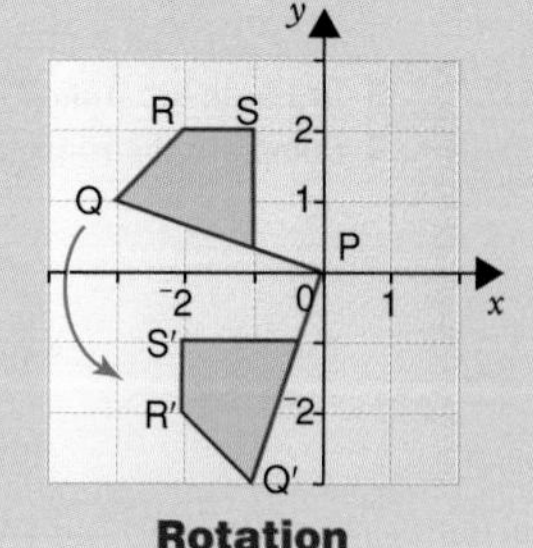

Rotation

To rotate we need

- the angle of rotation
- the centre of rotation.

PQRS has been rotated 90° about the origin.
90° means 90° **anticlockwise**.
The **inverse** rotation is 90° clockwise about the origin or 270° anticlockwise.

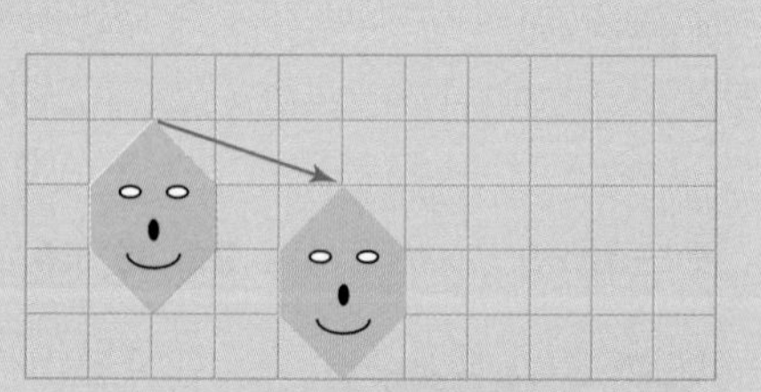

Translation

To translate we slide the shape without turning. The blue shape has been translated 3 units to the right and 1 unit down. The **inverse** translation is the same number of units in the opposite directions.

To **enlarge** a shape we need to know
- the scale factor
- the centre of enlargement.

PQR has been enlarged to P′Q′R′ by a scale factor 2, centre of enlargement, O.
Each point on P′Q′R′ is two times as far from O as the corresponding point on PQR.

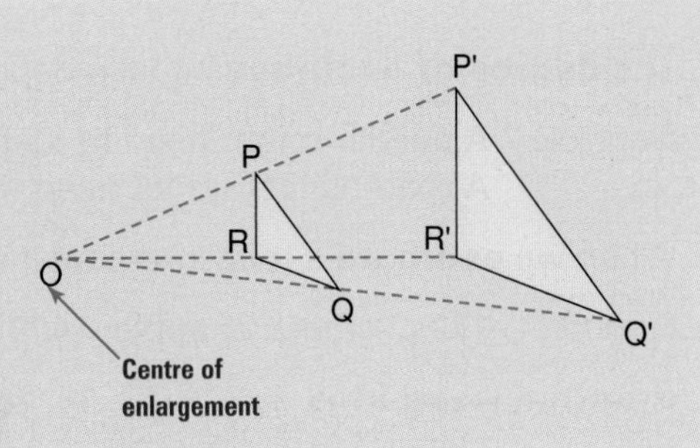

Practice Questions 2, 12, 13, 22, 29, 33, 43, 49, 50, 51

Measures

Metric conversions

You need to know these **metric conversions**.

length	mass	capacity (volume)	area	time
1 km = 1000 m	1 kg = 1000 g	1 ℓ = 1000 mℓ	1 ha = 10 000 m^2 (hectare)	1 minute = 60 seconds
1 m = 100 cm	1 tonne = 1000 kg	1 ℓ = 100 cℓ		1 hour = 60 minutes
1 m = 1000 mm		1 cℓ = 10 mℓ		1 day = 24 hours
1 cm = 10 mm		1 ℓ = 1000 cm^3		1 year = 12 months or 52 weeks and 1 day
		1 mℓ = 1 cm^3		1 year = 365 days or 366 in a leap year
		1 m^3 = 1000 ℓ		1 decade = 10 years
				1 century = 100 years
				1 millennium = 1000 years

Examples

0·58 ℓ = (0·58 × 1000) mℓ
= **580 mℓ**

520 cm^3 = (520 ÷ 1000) ℓ
= **0·52 ℓ**

67 cm = (67 ÷ 100) m
= **0·67 m**

53 900 m^2 = (53 900 ÷ 10 000) ha
= **5·39 ha**

8000 cm^3 = **8 ℓ** or **8000 mℓ**

7240 ℓ = (7240 ÷ 1000) m^3
= **7·24 m^3**

Metric and imperial equivalents.

These are **rough metric equivalents** for some **imperial units**.

length	mass	capacity
1 mile ≃ 1·6 km	1 pound (lb) is a bit less than $\frac{1}{2}$ kg	1 pint ≃ 600 mℓ
1 yard ≃ 1 m	1 oz ≃ 30 g	1 gallon ≃ 4·5 ℓ
1 inch ≃ 2·5 cm		

These are **rough imperial equivalents** for some **metric units**.

length	mass	capacity
1 km ≃ 0·625 miles	1 kg ≃ 2·2 lb	1 ℓ ≃ 1·75 pints
(8 km ≃ 5 miles)		
1 m ≃ 1 yard or 3 feet		

Remember: These are estimates.

When **reading scales** you need to work out the value of each small division.
When measuring we must choose the **degree of accuracy**, the **unit** and a suitable **measuring instrument**.

Example When measuring the length of a book we could measure it to the nearest centimetre using a ruler.

The **degree of accuracy** chosen depends on the situation.

Example A builder might need to measure to the nearest millimetre.
A road builder might need to measure to the nearest metre.

When we **estimate** a measurement it is a good idea to give **a range** for the estimate.

Example 150 g < mass of apple < 400 g

Practice Questions 1, 3, 7, 17, 18, 20, 21, 23, 35

Perimeter, area and volume

Perimeter is the distance around the outside of a shape.

Area is the amount of space covered by a shape.
Area is measured in km^2, m^2, cm^2, mm^2 or hectares.

The height and base must be perpendicular.

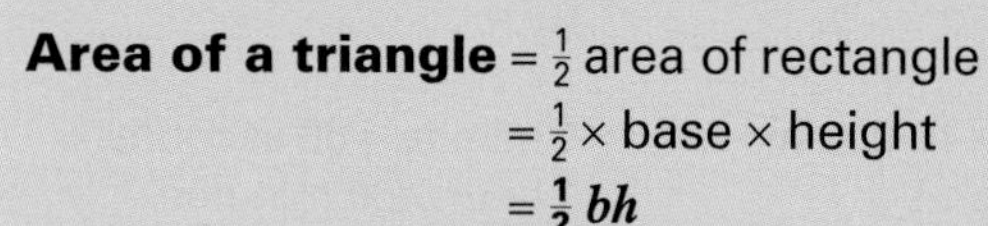

Area of a triangle = $\frac{1}{2}$ area of rectangle
= $\frac{1}{2}$ × base × height
= $\frac{1}{2}bh$

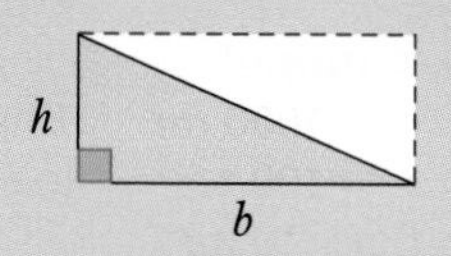

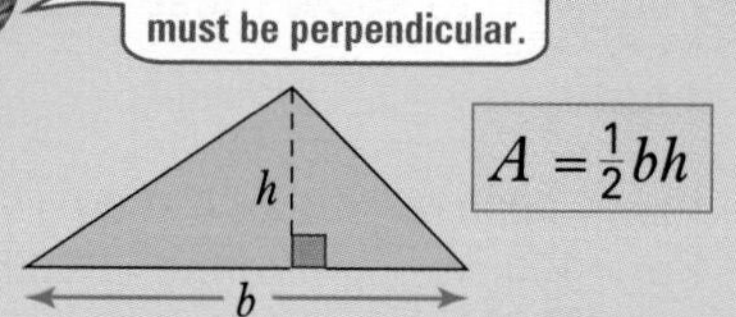

Area of a parallelogram = bh

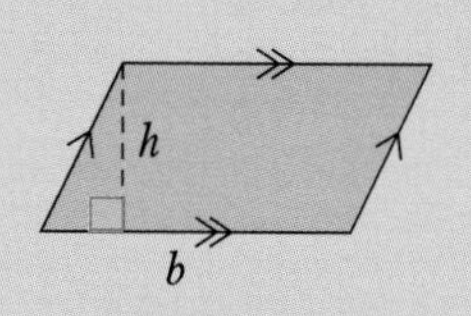

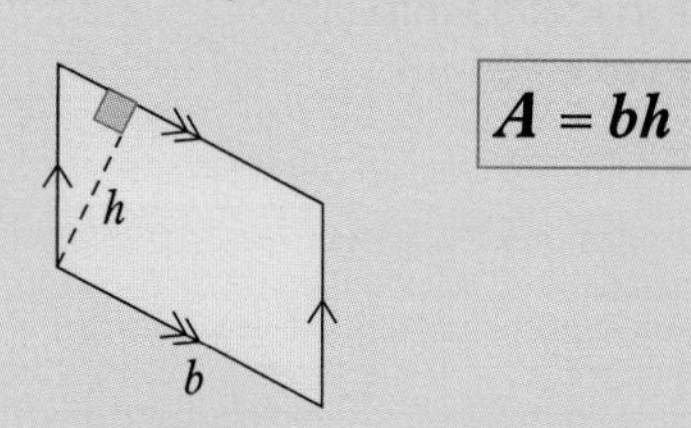

Area of a trapezium = $\frac{1}{2}(a + b)h$.

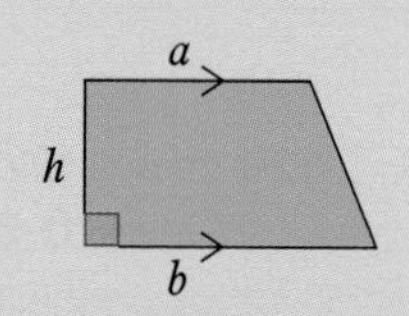

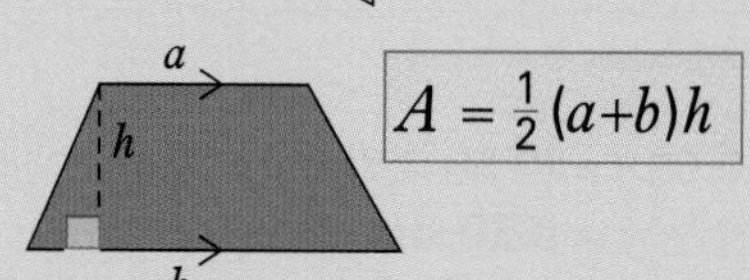

We sometimes find the area of a shape by dividing it into rectangles.
Area of shape = area of A + area of B.
Area = 5 × 8 + 4 × 3
= 40 + 12
= **52 cm²**

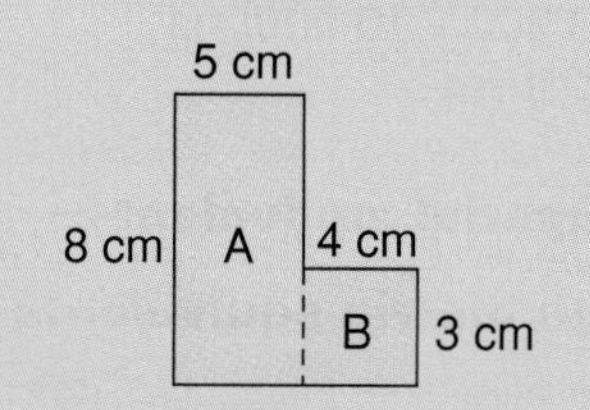

The **surface area of a cuboid** = 2(length × width) + 2(length × height) + 2(height × width)
= $2lw + 2lh + 2hw$

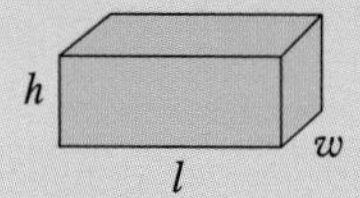

Example Surface area = $2lw + 2lh + 2hw$
= 2 × 12 × 6 + 2 × 12 × 5 + 2 × 5 × 6
= 144 + 120 + 60
= **324 cm²**

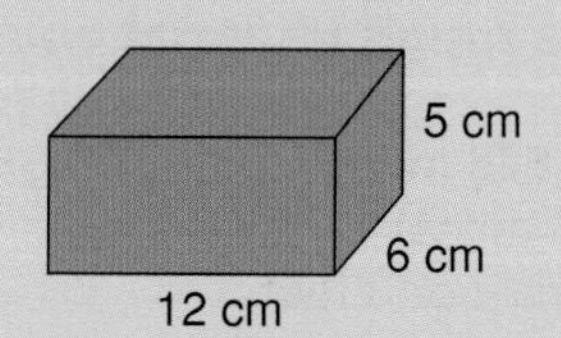

Volume is the amount of space taken up by a solid.
Volume is measured in mm^3, cm^3, m^3 or ℓ, pints, gallons.

Volume of a cuboid = length × width × height
= lwh

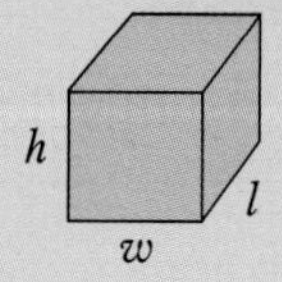

Practice Questions 10, 25, 26, 41, 44, 45, 46, 47

Compass directions

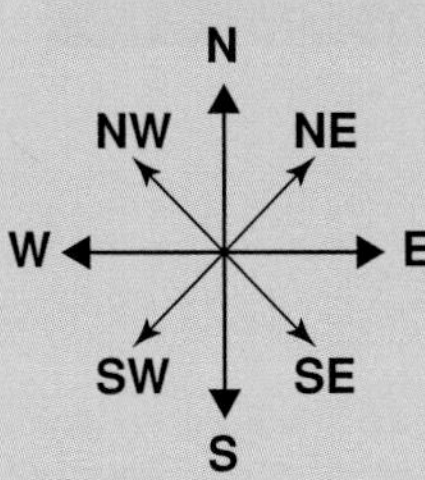

A Northerly wind blows from the North.
A South-Westerly wind blows from the South-West.

Practice Question 8

Practice Questions

1

km	m	mm	tonne	kg	g	cm	ℓ	mℓ

State which unit of measurement the following are most likely to be measured in. (Choose from the units in the box.)

a the depth of water in a bath
b the length of a sports ground
c the amount of water in a bath
d the mass of a ship
e the mass of a mouse
f the distance between two villages

2 Which grey shape is
a a translation of the red shape
b a rotation of the red shape
c a reflection of the red shape?

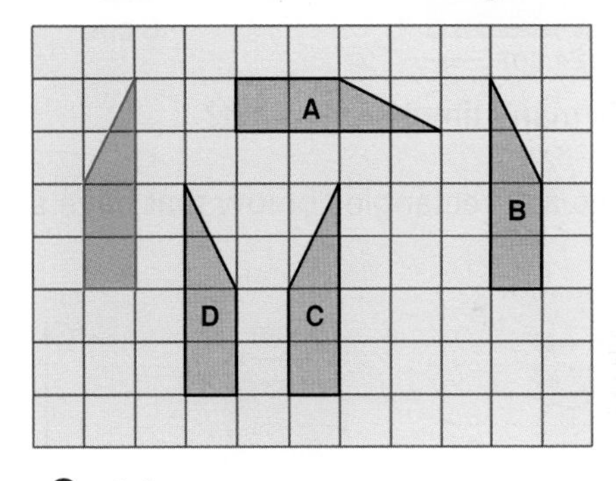

3 **a** The thickness of a ruler could be
A 1 cm **B** 1 mm **C** 0·1 mm.
b The mass of a train could be
A 1 tonne **B** 100 tonne **C** 100 kg.
c The capacity of a tank could be
A 100 tonne **B** 100 mℓ **C** 100 ℓ.

4 Name the shaded angles.

a

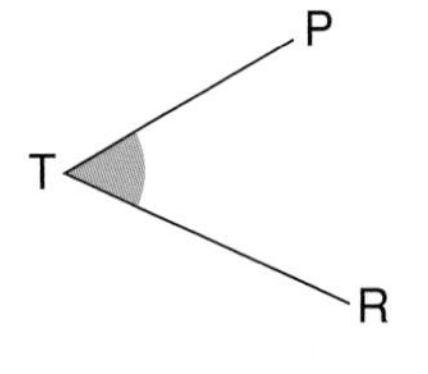

b

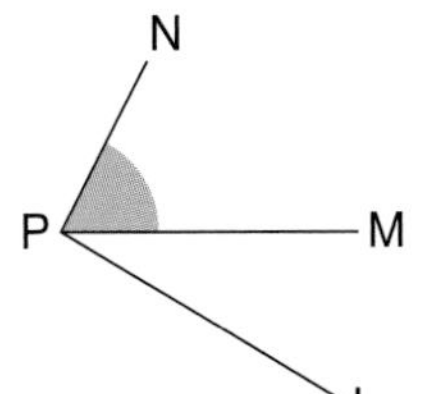

c

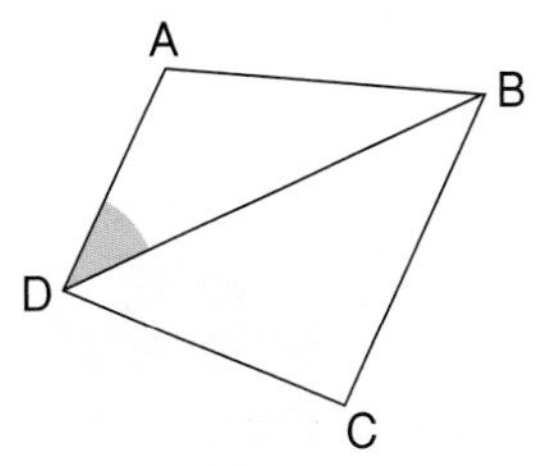

5 Name these triangles.

a

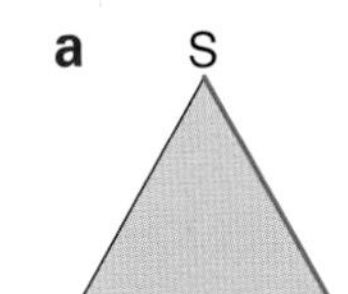

b

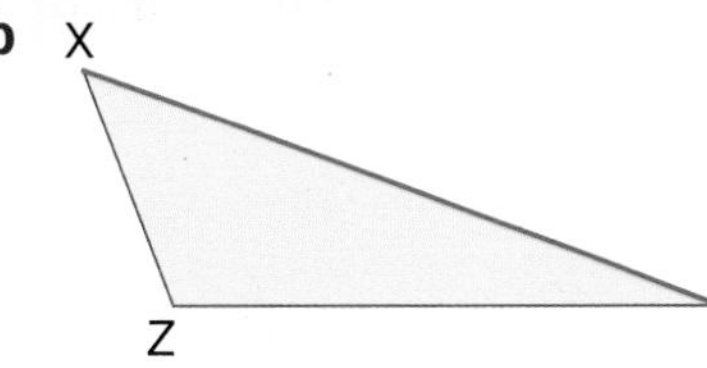

6 Using a lower-case letter, name the red side of each triangle in question **5**.

7 A video recorder is set to begin and finish recording as shown. If a 180 minute tape is used, would there be enough room on this tape to also record another programme which runs for 55 minutes?

Start	22 : 40
Finish	01 : 15

8 Rebecca was facing North. She turned

$\frac{1}{4}$ turn clockwise then

$\frac{1}{2}$ turn anticlockwise then

$\frac{1}{4}$ turn anticlockwise.

What direction is she facing now?

9 a I have a rectangle made out of paper.
The rectangle measures 10 cm by 14 cm.
I want to **fold** the rectangle **in half** to make a smaller rectangle.
I can do this in two different ways.
What size could the smaller rectangle be? Write both ways.

10 cm

14 cm

b I have a square made out of paper. The square measures 24 cm by 24 cm.
I keep folding it in half until I have a rectangle that is 6 cm by 12 cm.

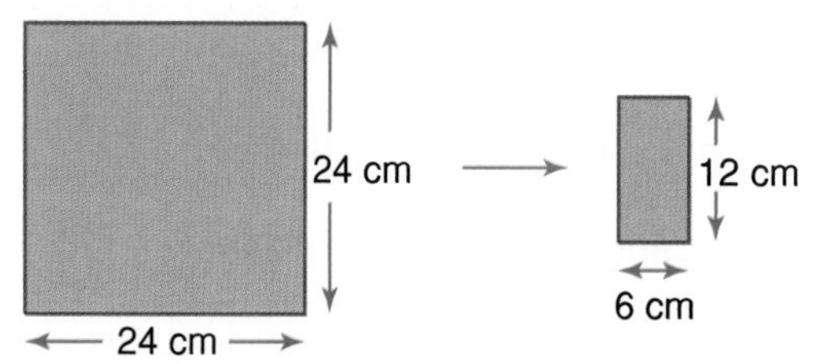

How many times do I fold it?

10 a Name any rectangles below that have an area of 12 cm^2. **[SATs 2001 paper 2]**

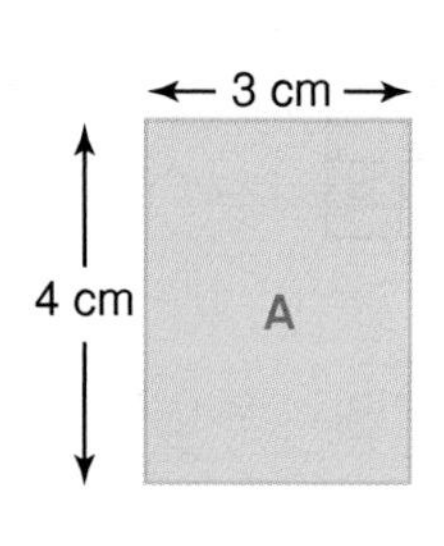

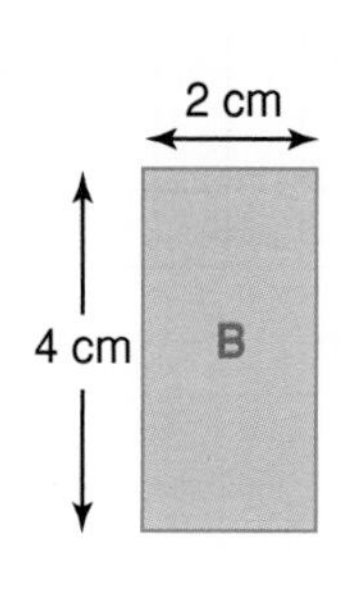

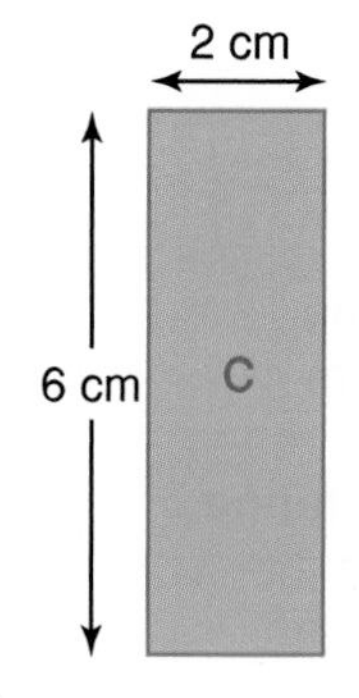

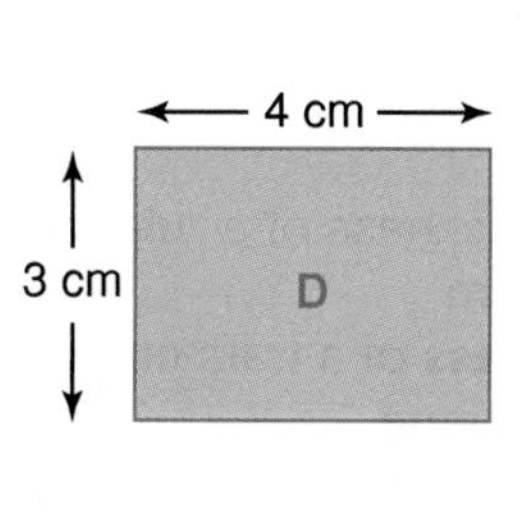

b A **square** has an area of **100 cm^2**.
What is its **perimeter**?
Show your working.

11 Describe the symmetry of these shapes fully.

a

b 

c 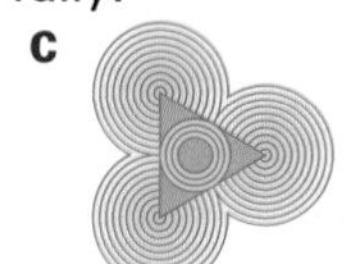

d 

12 a Use a copy of this.
Draw the image of the green shape after the translation given.

b What is the name given to this polygon?

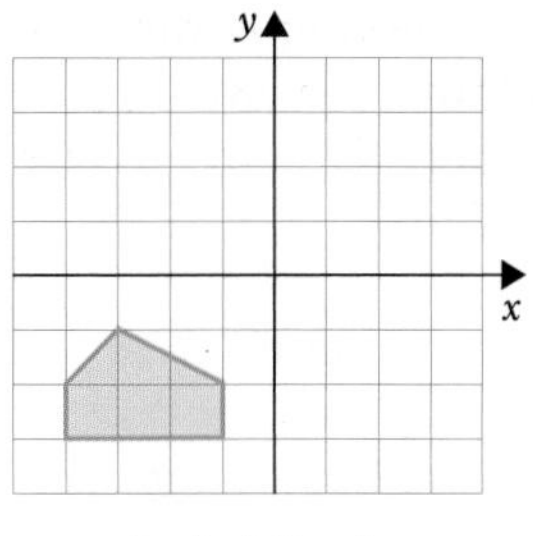

5 units right and 3 units up

13 What translation is the inverse of the translation given in question **12**?

14 Write true or false for these.

a An arrowhead and an isosceles triangle both have one line of symmetry and two equal angles.

b A parallelogram has diagonals that bisect the angles and has two lines of symmetry.

c A rhombus has two lines of symmetry and diagonals that bisect each other.

d An equilateral triangle has three lines of symmetry and three angles of 60° each.

e A right-angled triangle never has line symmetry.

15 Imagine three cubes. Two cubes are blue and one cube is red. Put them together in a line so that the red cube is in the middle.

a How many blue faces are showing?

b How many red faces are showing?

16 Imagine you are flying above a cuboid-shaped building.
What is the maximum number of faces you can see?

17 Find the measurements given by pointers A, B and C. In part **d** you will need to estimate.

a

C A B
cm 2 3

b

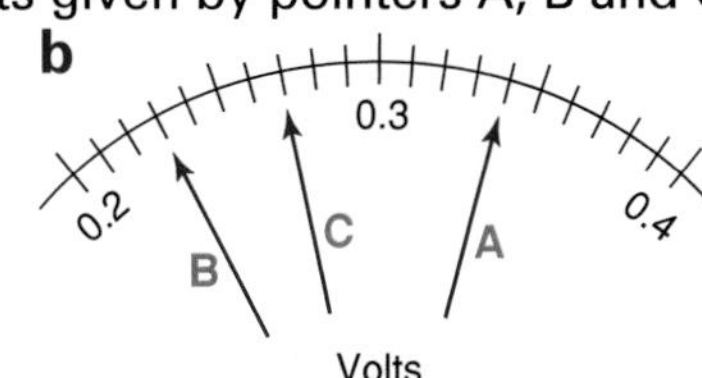

c

450
300
600
A
150
750
B
C
0
900
grams

d

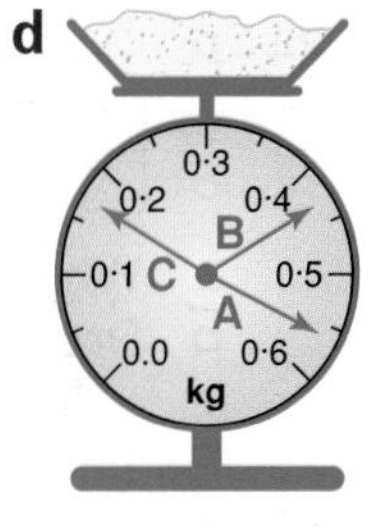

18 a Winstone travelled 25 miles from Andover to Newbury. How many kilometres is this?

b Winstone had a suitcase with him. It was 65 cm long and weighed 14 lb. About how long was this case, in inches? About how heavy was it, in kilograms?

c In the suitcase, Winstone had a 3-litre bottle of juice. About how many pints is this?

19 a I have a paper circle.
Then I cut a sector from the circle. It makes this net.
Which 3-D shape below could I make with my net?

[SATs 2002 paper 2]

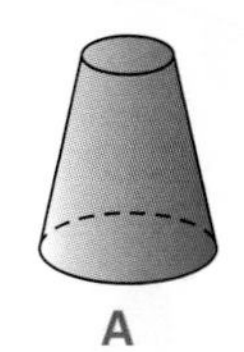
A

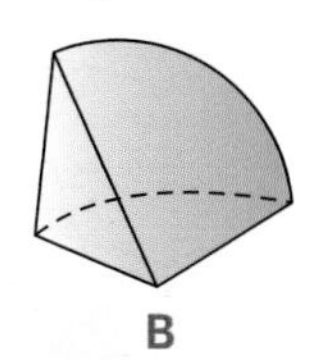
B

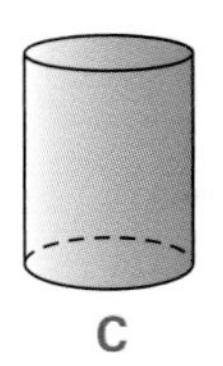
C

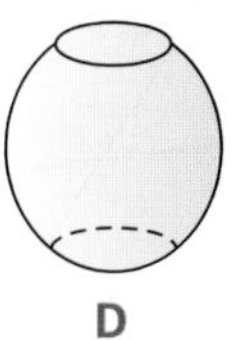
D

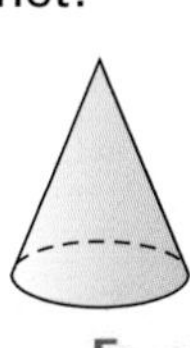
E

b Here is a sketch of my net.
Make an **accurate drawing** of my net.

Not drawn accurately

8·5 cm
110°
8·5 cm

20 Find the missing numbers.

a 39 mm = ___ cm　**b** 2800 m = ___ km　**c** 340 cm = ___ m
d 580 mm = ___ m　**e** 960 mℓ = ___ ℓ　**f** 57 g = ___ kg
g 6152 mℓ = ___ ℓ　**h** 4900 kg = ___ tonne　**i** 0·07 kg = ___ g
j 5·26 cℓ = ___ ℓ　**k** 72 kg = ___ tonne　**l** 90 mℓ = ___ cm^3
m 3·6 ℓ = ___ cm^3　**n** 5720 ℓ = ___ m^3　**o** 43 000 m^2 = ___ ha

21 Choose the best range.

a **A** 50 g ⩽ mass of a calculator ⩽ 200 g
B 1 kg ⩽ mass of a calculator ⩽ 1·5 kg
C 500 g ⩽ mass of a calculator ⩽ 1 kg

b **A** 0·5 m ⩽ length of a bike ⩽ 1 m
B 1 m ⩽ length of a bike ⩽ 2 m
C 1·5 m ⩽ length of a bike ⩽ 2·5 m

T **22** Use a copy of this diagram.

a Reflect the shape in the y-axis.
b Reflect the image you got in **a** in the x-axis.
c What single transformation is equivalent to **a** and **b**?
d Would you get the same result if you had reflected the shape in the x-axis and then reflected the image in the y-axis?

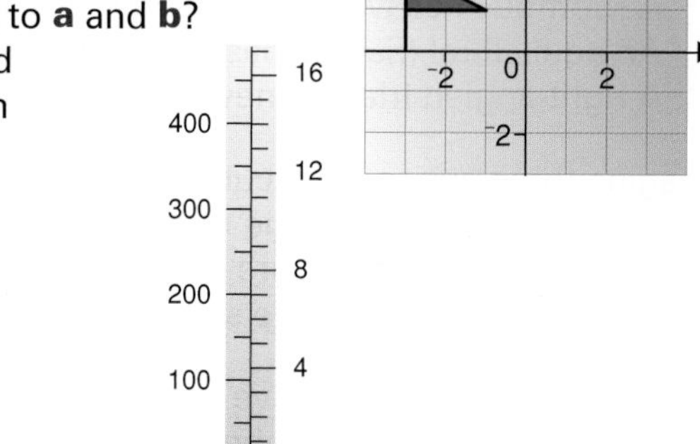

23 This scale measures in grams and ounces. Use the scale to answer these questions.

a About how many ounces is 200 grams?
b About how many grams is 15 ounces?
c About how many ounces is 1 kilogram? Explain your answer.

24  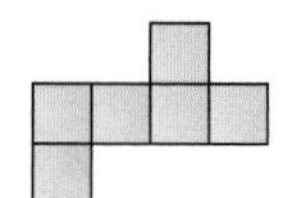 B 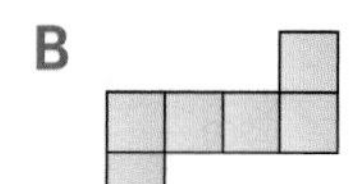 C 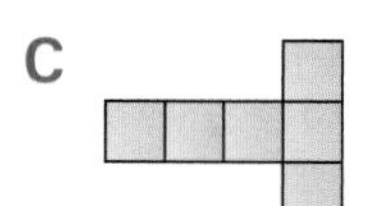

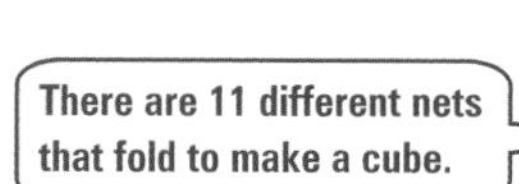

a Which of these nets will fold to make a cube?
b Draw another two nets that will fold to make a cube.

25 Find the perimeter of these.

a 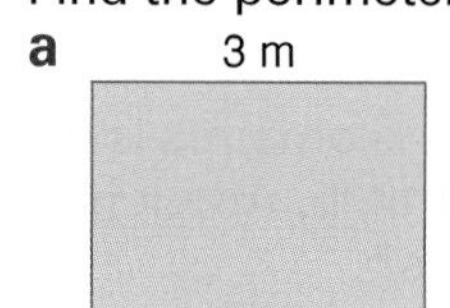

b 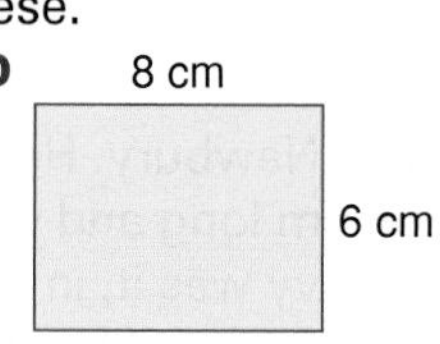

c 5 mm, 13 mm, 12 mm

d 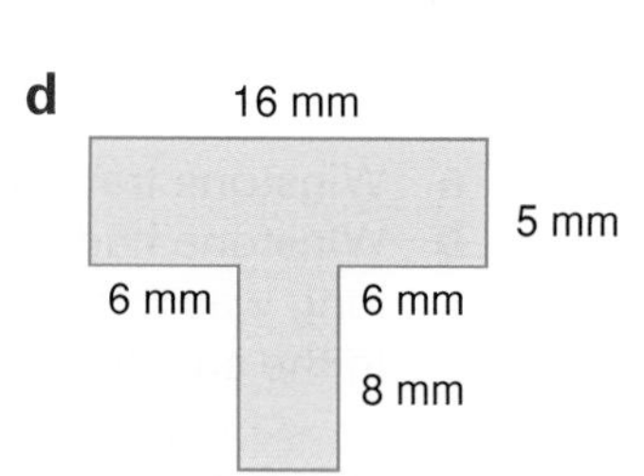

26 Find the area of each of the shapes in question **25**.

27 Find the angles marked with letters.

a 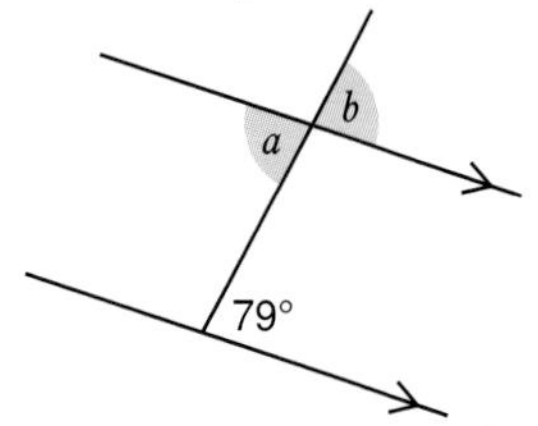

b 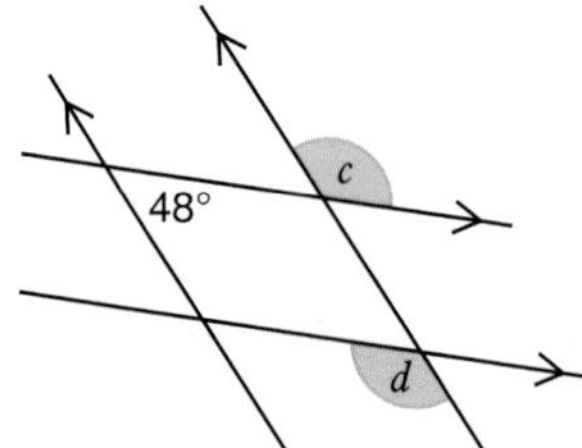

c 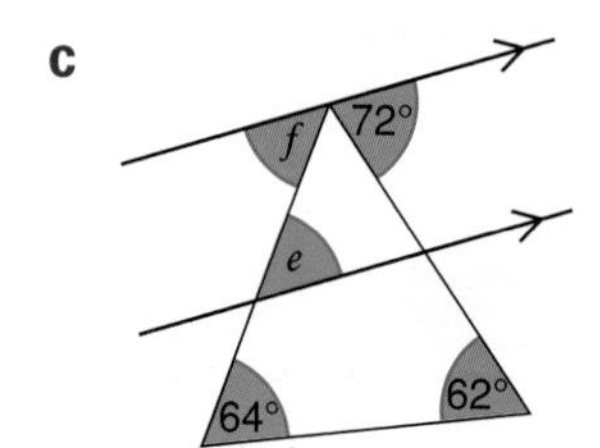

28 **a** A quadrilateral with two lines of symmetry is a
A parallelogram　**B** square　**C** rectangle　**D** kite.

b A quadrilateral with two equal diagonals is a
A parallelogram　**B** kite　**C** rhombus　**D** none of these.

29 The shape A has been enlarged to the shape A′
What is the scale factor for this enlargement?

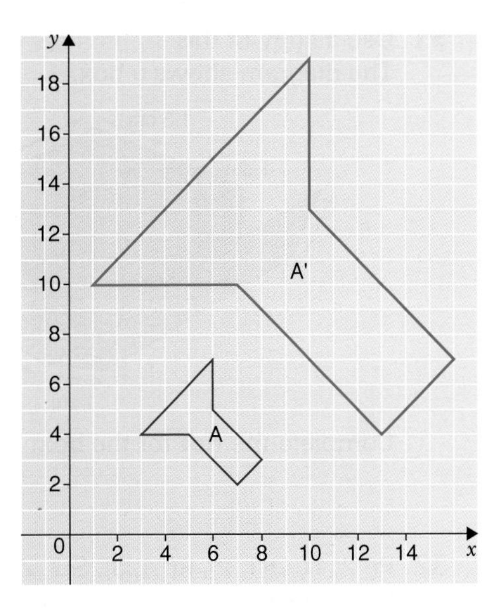

T

30 Use a copy of this.
Four squares join together to make a bigger square.

[SATs 2002 paper 1]

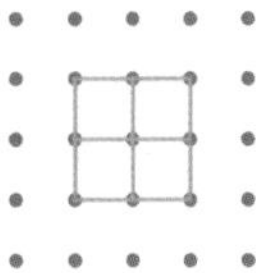

a Four congruent triangles join together to make a bigger triangle.

Draw **two more** triangles to complete the drawing of the bigger triangle.

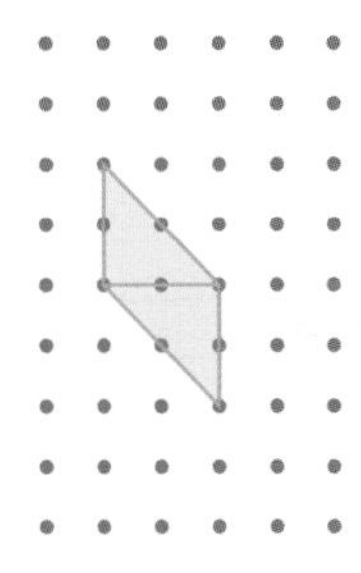

b Four congruent trapeziums join to make a bigger trapezium.

Draw **two more** trapeziums to complete the drawing of the bigger trapezium.

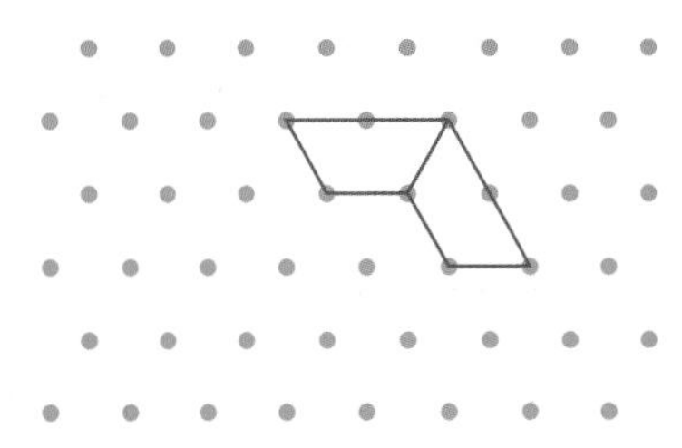

c Four congruent trapeziums join to make a **parallelogram**.

Draw **two more** trapeziums to complete the drawing of the parallelogram.

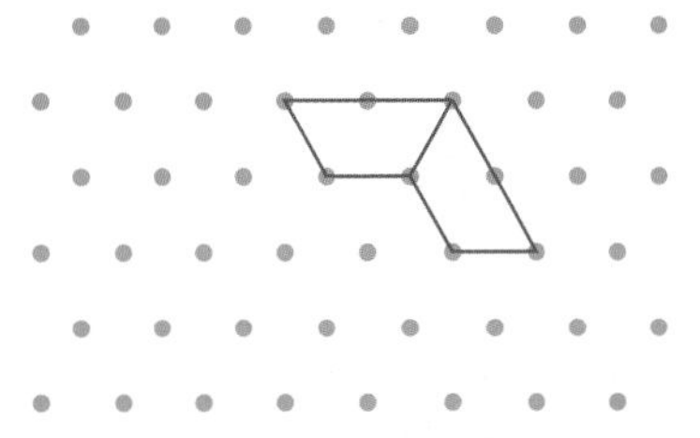

T

31 Use a copy of this. **[SATs 2001 paper 1]**
The diagram shows a box.

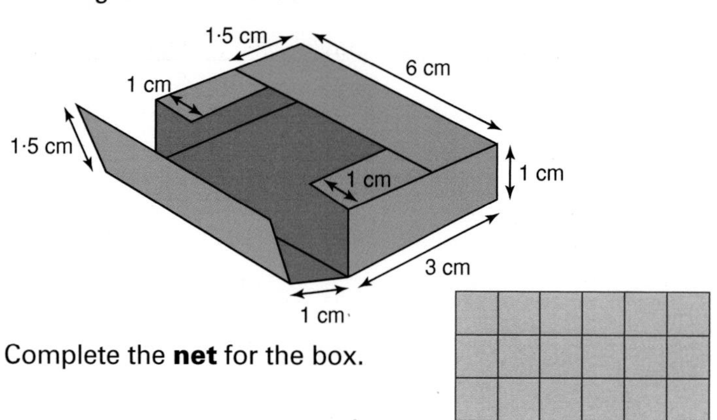

Complete the **net** for the box.

32 P($^-$2, 1), Q(1, 2), R($^-$1, 3), S(**?**, **?**)
Plot P, Q and R.
Find the coordinates of S so that PQRS is
a a parallelogram **b** an arrowhead **c** a kite.

T

33 Use a copy of these shapes.
The centre of rotation and angle of rotation are given.
Draw the image shapes.
a (0, 0), 90° **b** (1, 1), 180°

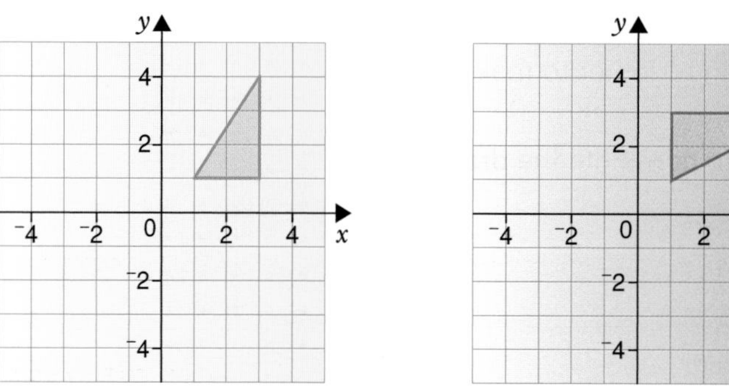

34 Find the angles marked with letters.

a
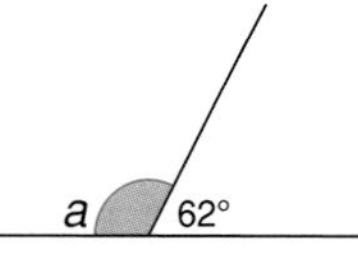

b
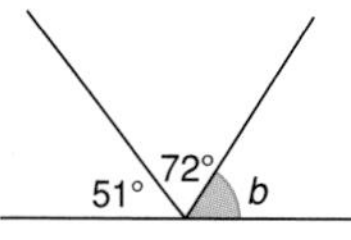

c
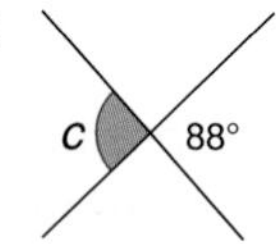

d
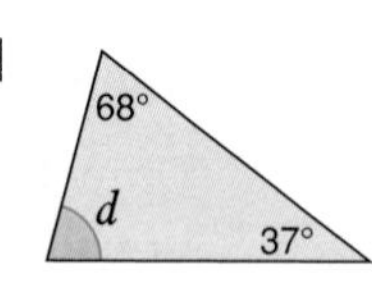

e
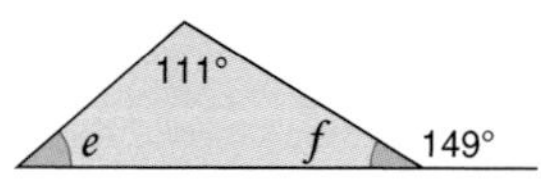

f
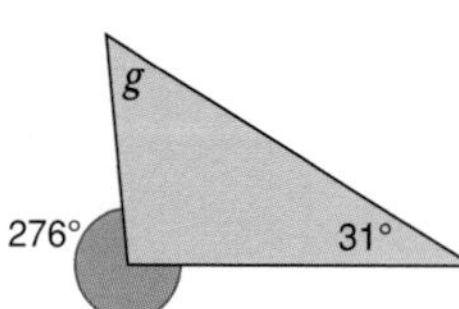

g
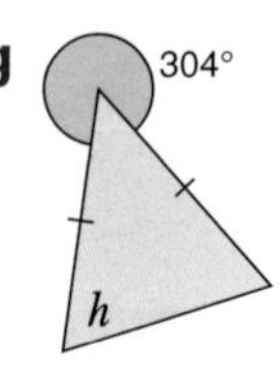

h
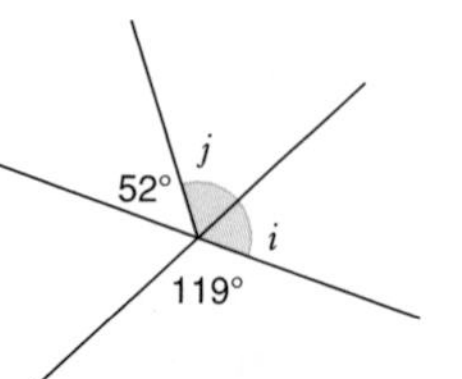

i
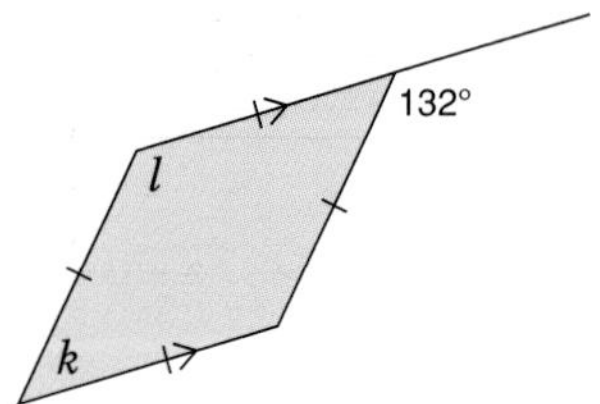

j
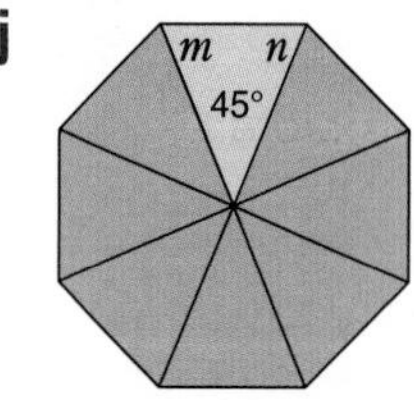

regular octagon

35 Michael is having a party.

a He needs 30 m of red paper to decorate the room. Each roll is 750 cm long. How many rolls will he need?

b He decides he needs 50 ℓ of fizzy drink. Each bottle contains 125 cℓ. How many bottles will he need?

c His best friend travels 50 miles to come to the party. About how many kilometres is this?

d He orders 10 kg of sausage rolls and savouries. About how many pounds is this?

36 Use a compass and ruler to accurately construct this triangle. On your drawing, measure the size of ∠RPQ. Give your answer to the nearest degree.

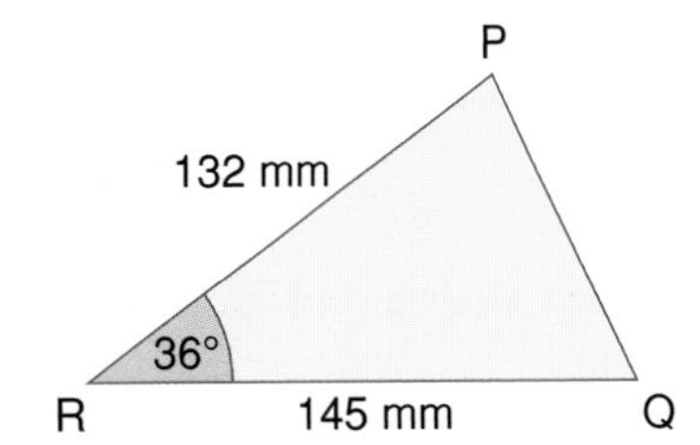

37 Use a copy of this.

[SATs 2001 paper 1]

The diagram shows two isosceles triangles inside a parallelogram.

a On the diagram, mark another angle that is 75°. Label it 75°.

b Calculate the size of the angle marked k. Show your working.

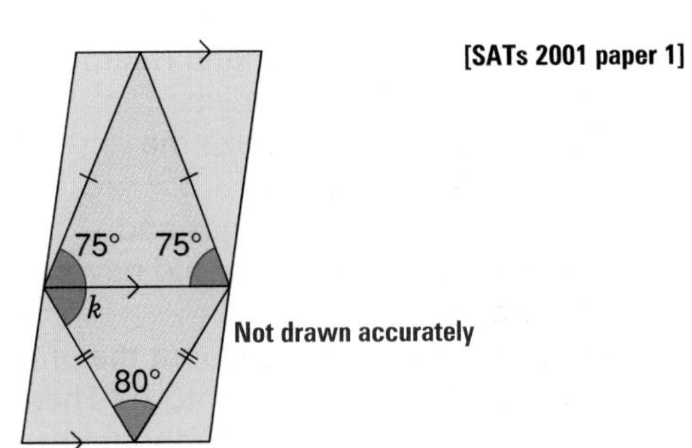

38 Use a copy of this diagram.

a Construct the perpendicular bisector of AB.

b Label the point where it crosses AC as L.

c On your diagram, measure AL.

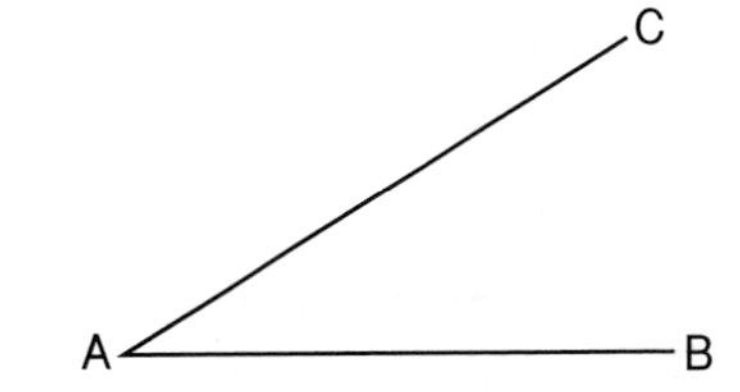

39 Wiri folded this net of centimetre squares to make a box 3 cm high and 2 cm wide.

a How long is the box?

b What is the volume of the box?

c Jayne made a box which had the same volume as Wiri's. Her box was 5 cm wide and 1 cm high. How long was Jayne's box?

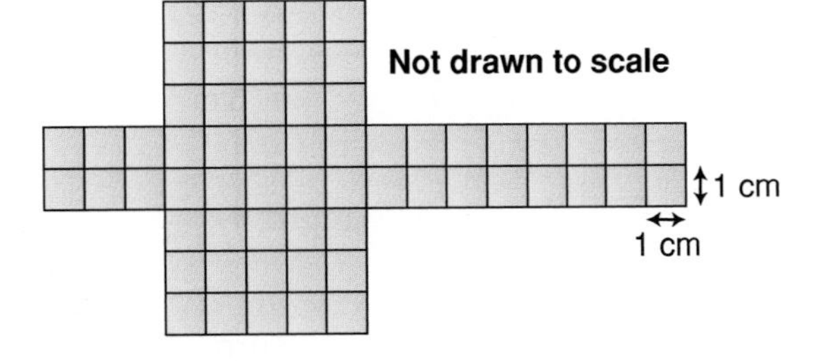

40 This sketch shows a square, a rectangle and a triangle. Use your drawing instruments to construct this diagram.

a On *your* diagram, measure the lengths of AB and CD. Give your measurements to the nearest millimetre.

b *Estimate* the size of each angle in your triangle. Check your estimates by measuring the size of the angles with a protractor.

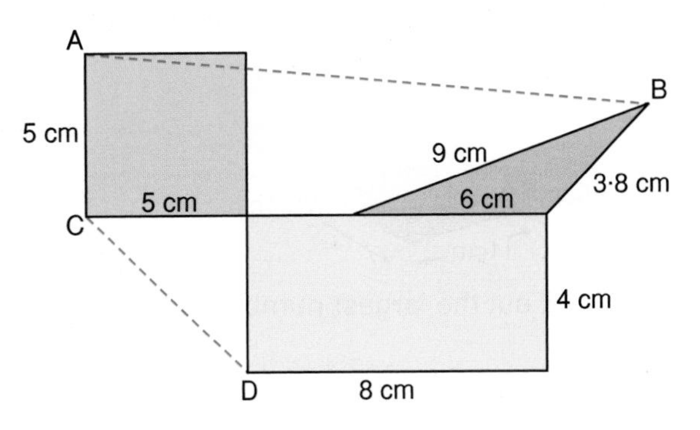

41 Lindy is painting a wall beside some stairs.
It consists of three shapes, one of which is a rectangle.
Find the area of each of these shapes.
Give the answers in square metres.

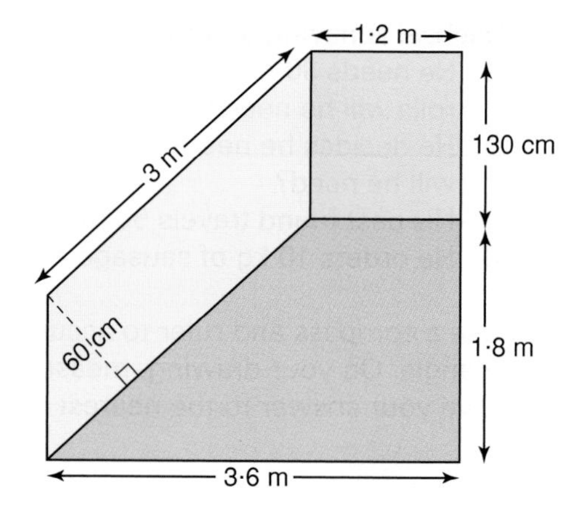

42 **a** Use your compasses and ruler to draw a net for the pyramid.
b Use your protractor and ruler to draw a net for the prism.
c Fold your nets to make the pyramid and prism.
d How many faces, edges and vertices does the pyramid have?

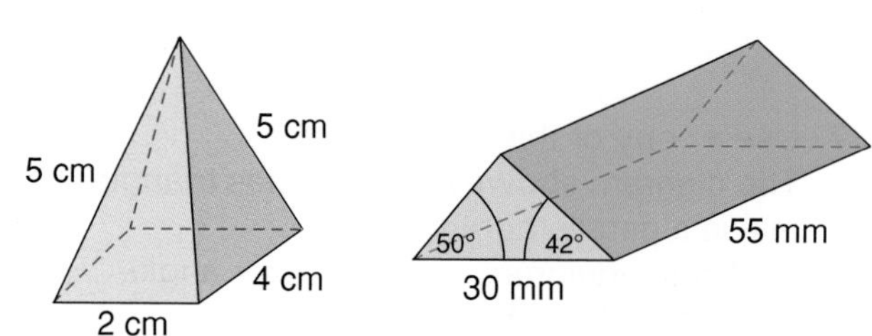

43 A rotation through 270° about the origin maps A onto A′.
Which of these maps A′ onto A? There are two correct answers.
A a rotation through 270° about (0, 0)
B a rotation through 90° clockwise about (0, 0)
C a rotation through 90° about (0, 0)
D a rotation 270° clockwise about (0, 0)

44 Better Gardens design paved courtyards with a garden in the middle.
Find the area of paving in these courtyards.

a

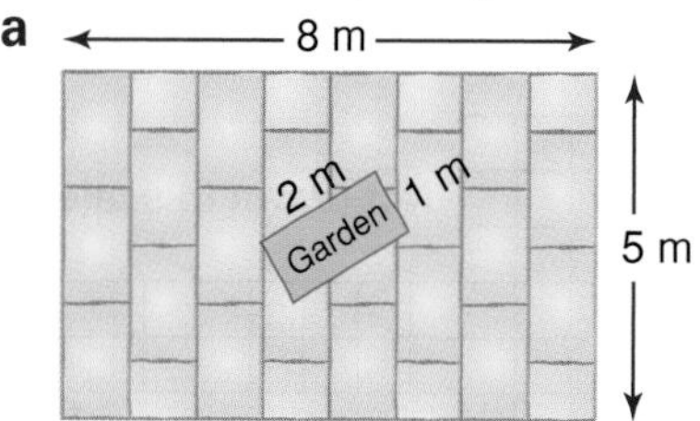

b

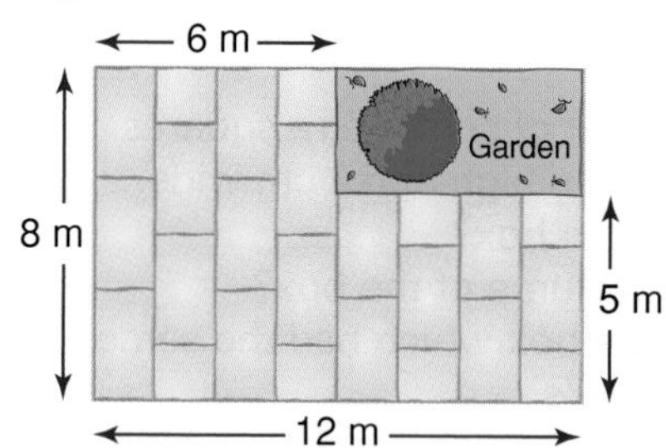

45 Boxes of mushrooms measure 3·5 cm by 5·5 cm by 7·1 cm.
A store puts them in a display rack like this one.

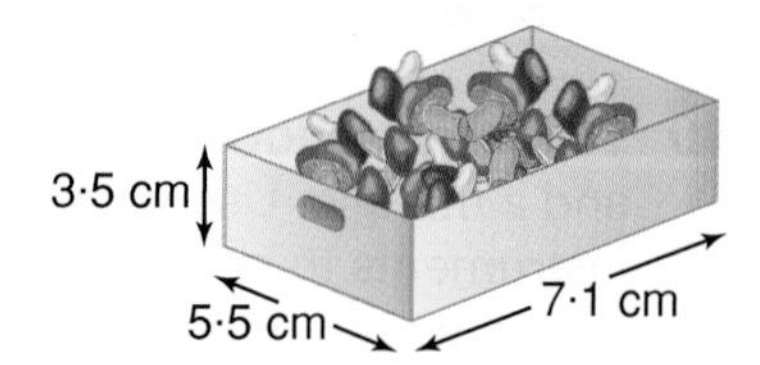

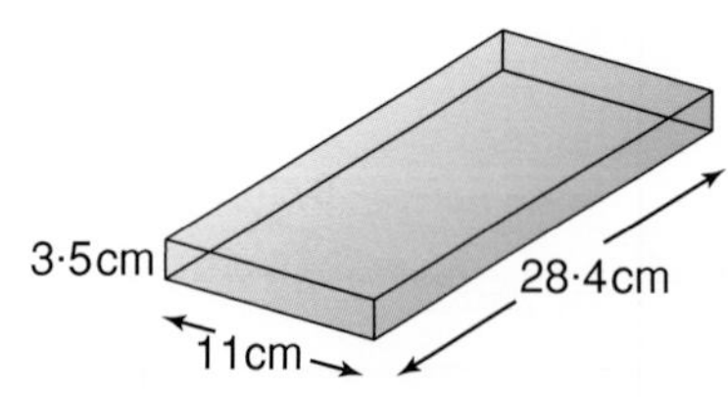

Work out the largest number of boxes of mushrooms that can lie flat in the tray.

46 Find the surface area of this wooden T.

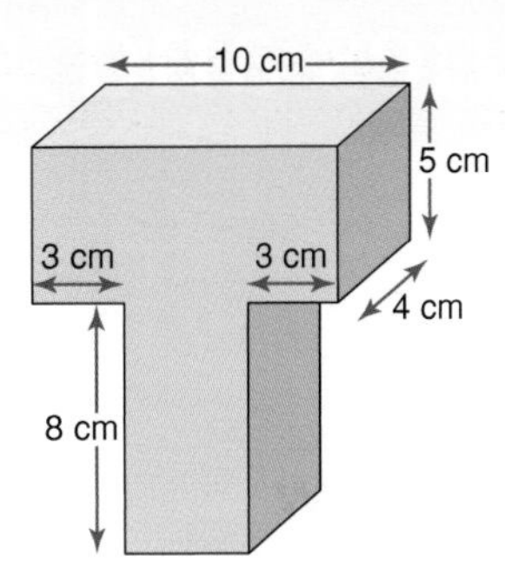

47 **a** What is the total surface area of this cuboid?
b Jasmine wants to cover the sides and bottom with silver glitter and the rest with red glitter to make a Christmas decoration.
What area of the decoration will be
i silver glitter **ii** red glitter?

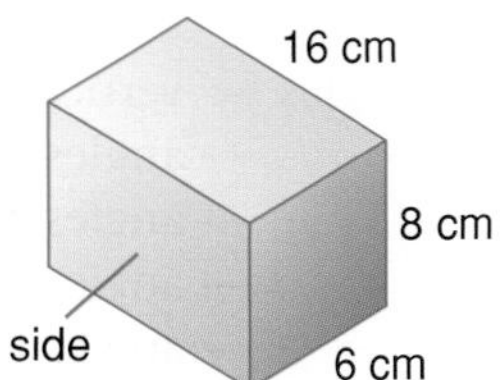

48 **a** Finish these Logo instructions to draw a regular hexagon.
repeat ___ [fd 100 rt ___]
b Finish these Logo instructions to draw a parallelogram.
repeat 2 (fd 20, rt 50, ___ 30, rt ___)

49 Explain why this shape will tessellate.

50 Triangles A, B, C and D are drawn on a grid.

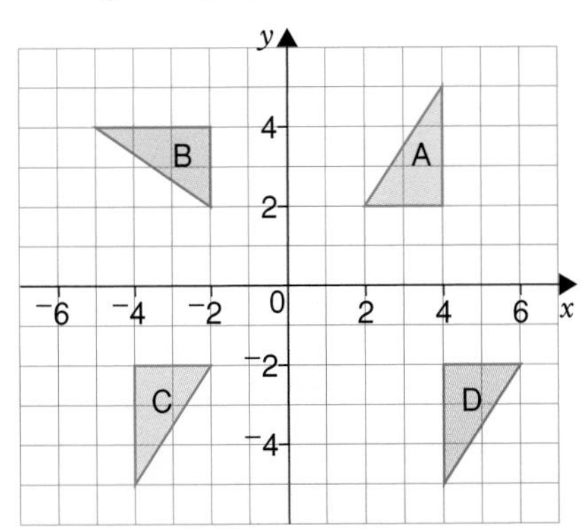

a Find a single transformation that will map
i A on to C **ii** C on to D.
b Find a combination of two transformations that will map
i B on to C **ii** C on to D.

T

51 Use a copy of this.
a Enlarge this 'mouse' by a scale factor of 2, centre of enlargement (0, 0).
Write down the coordinates of A′.
b Use another copy of the 'mouse'.
Enlarge it by a scale factor of 2, centre of enlargement (0, 1).
Write down the coordinates of A′.

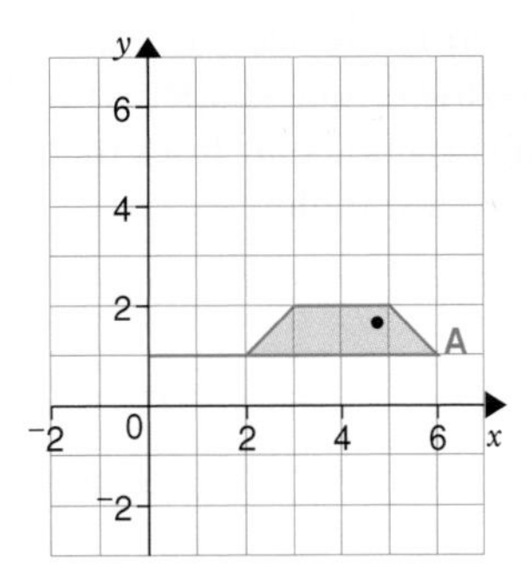

11 Lines and Angles

You need to know

✓ lines and angles page 249

✓ calculating angles — vertically opposite angles page 249
— angles on a straight line
— angles at a point
— angles made with parallel lines
— angles in a triangle

Key vocabulary

adjacent, alternate angles, complementary angles, corresponding angles, exterior angle, interior angle, prove, supplementary angles

Picture Perfect

1. This 'picture' has been constructed using a ruler and set square.
   It is made up of parallel and perpendicular lines.
   Construct your own 'picture' using parallel and perpendicular lines. Use a set square and ruler.
2. Have a competition to see who can find a picture from a magazine or newspaper with the most parallel and perpendicular lines **or** the most acute or obtuse or reflex angles.

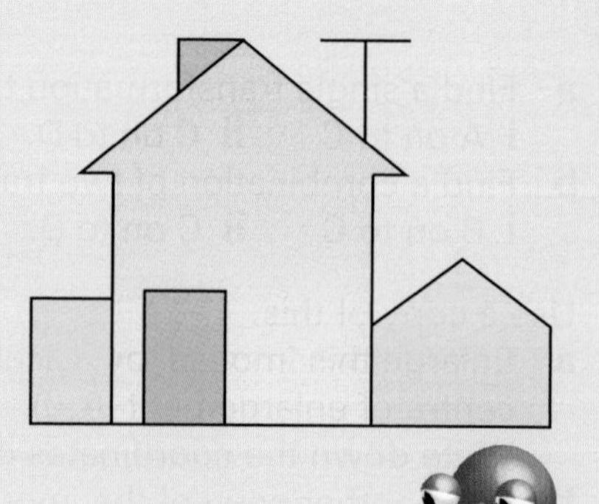

Make your picture as interesting as possible.

Using geometrical reasoning to find angles

Sometimes it is not obvious how to find the size of an angle. Often you can use **geometrical reasoning**. It helps to

This is called geometric reasoning.

1 write down what you know
2 work out any angles that it looks like you might need
3 use what you know from **1** and **2** to write down the steps needed to find the angle or prove its value. Always give reasons.

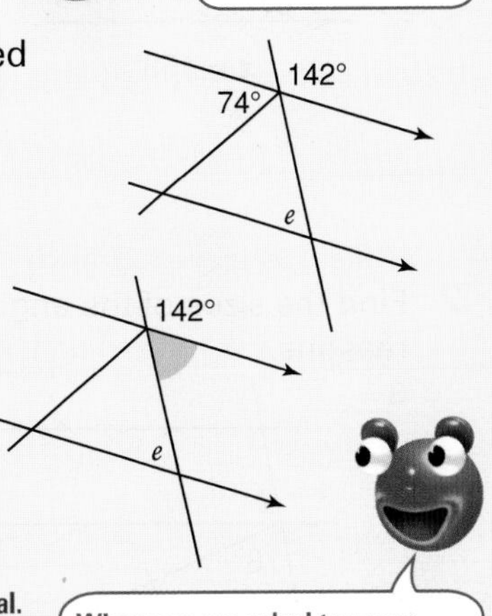

Worked Example
Prove that $e = 38°$.
Show your working clearly and give reasons.

Answer
e is an alternate angle to the shaded angle.
We can find the shaded angle using angles on a straight line.

shaded angle + 142° = 180°	**Angles on a straight line add to 180°.**
shaded angle = 180° − 142°	
= 38°	
e = **38°**	**Alternate angles on parallel lines are equal.**

When you are asked to prove something, you must show each step one by one and give reasons.

Note: when you prove something, you need an equation and a reason for each fact you write down.

There is often more than one way to find the answer.
Prove $e = 38°$ in the above worked example in a different way.

Sometimes you are asked to write and solve equations.

Worked Example
Write and solve equations to find the value of a and x.
Show your working clearly, giving reasons.

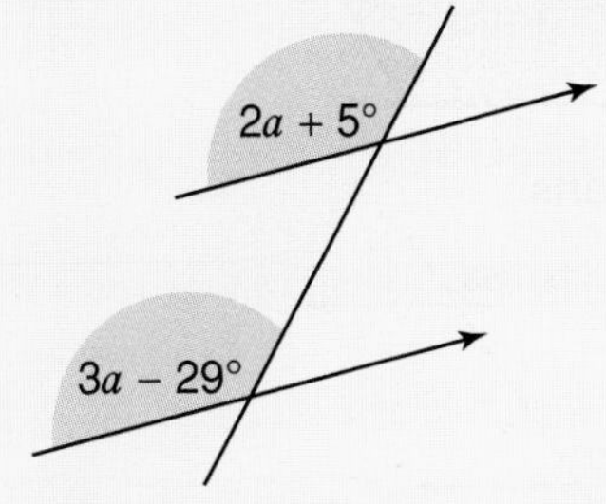

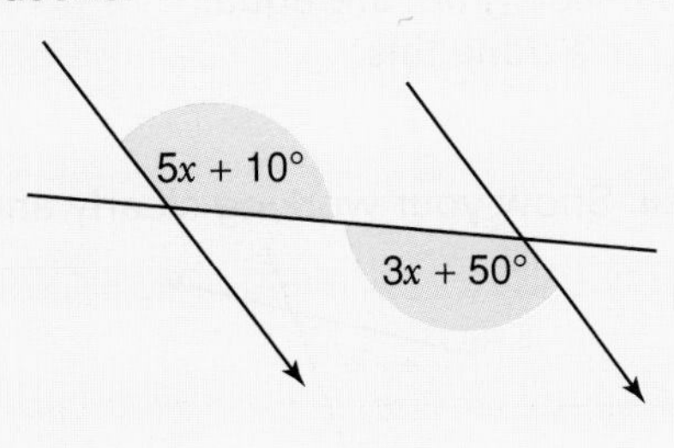

Answer

$3a - 29° = 2a + 5°$	**corresponding angles on parallel lines are equal**
$a - 29° = 5°$	**subtracting $2a$ from both sides**
$a = 34°$	**adding 29° to each side**
$5x + 10° = 3x + 50°$	**alternate angles on parallel lines are equal**
$2x + 10° = 50°$	**subtracting $3x$ from both sides**
$2x = 40°$	**subtracting 10° from both sides**
$x = 20°$	**dividing both sides by 2**

Exercise 1

1 For each of these, prove that x has the value shown. Show all your working clearly and give reasons.

a

$x = 153°$

b

$x = 53°$

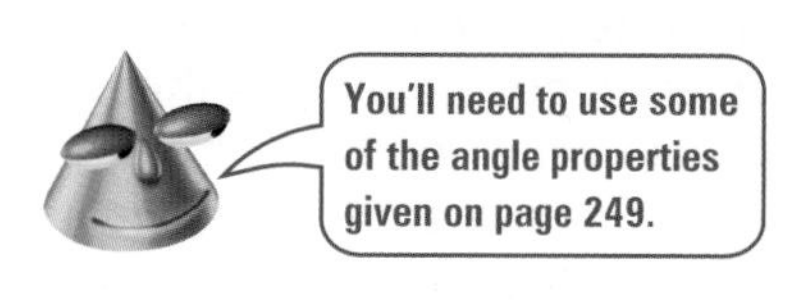

2 Find the sizes of the angles marked with letters. Show your working clearly and give reasons.

a

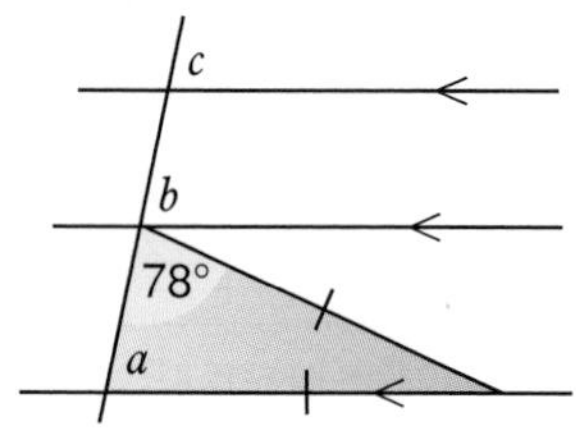

b

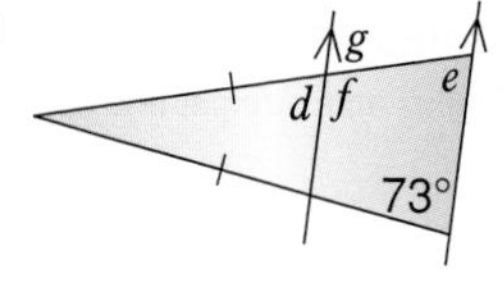

c

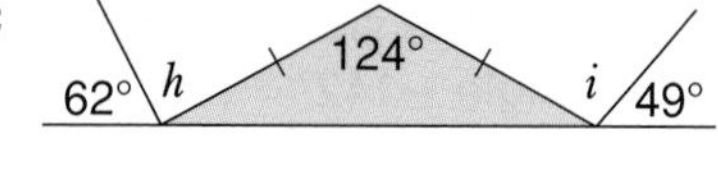

3 Naseem drew these triangles.

AB is parallel to CD.
AB = AC
BC = BD

Find the size of all the other interior angles.
Show your working clearly and give reasons.

4 Maria used alternate angles to prove that the opposite angles of a parallelogram are equal. Show how she might have done this.

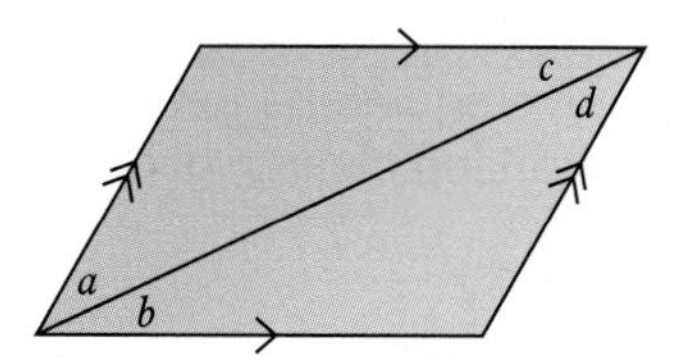

5 Calculate the value of n. Show your working clearly and give reasons.

a

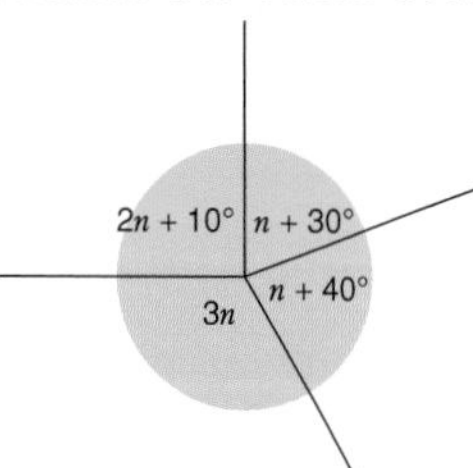

b

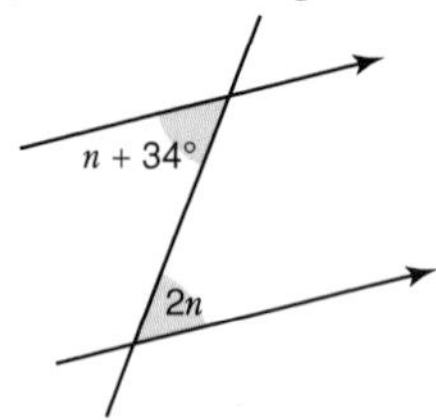

c

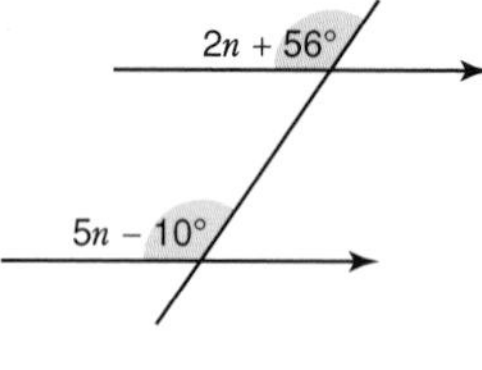

d

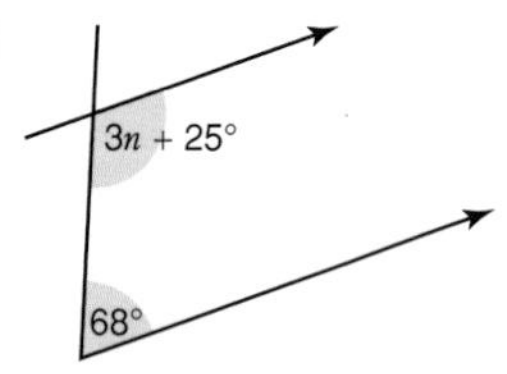

e

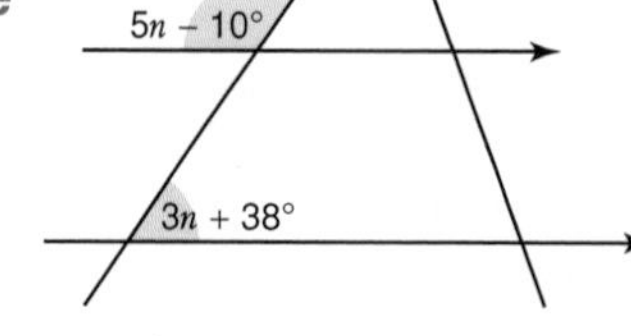

f

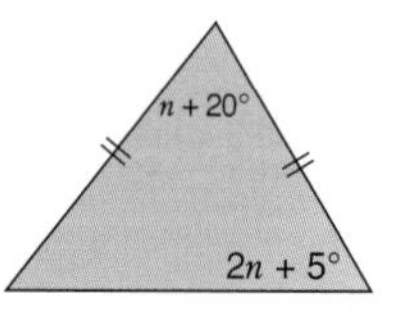

g

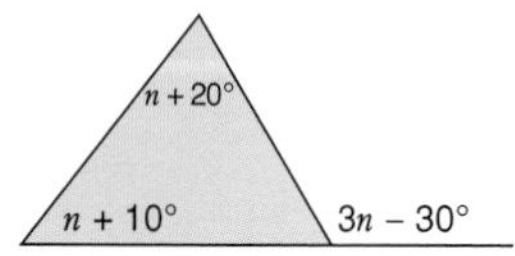

h

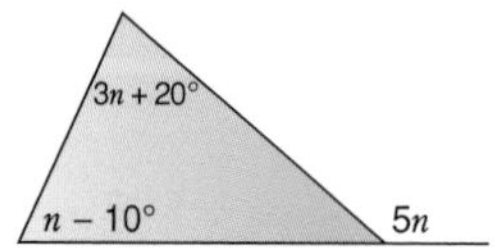

6 Prove that

a $a = 60°$

b $b = 25°$.

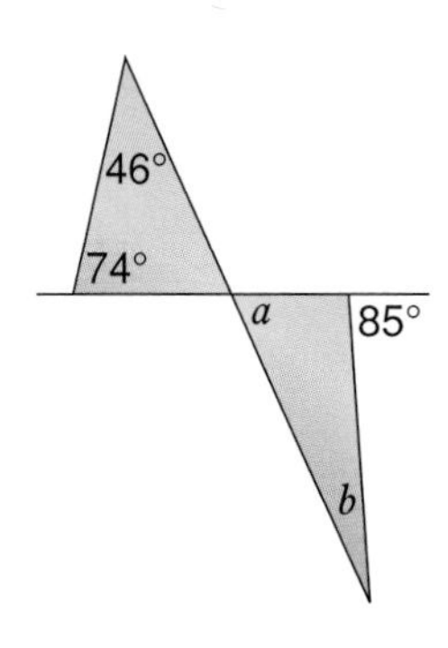

7 **a** Explain why the lines AB and CD are not parallel.

b Name a pair of parallel lines. Give a reason for your answer.

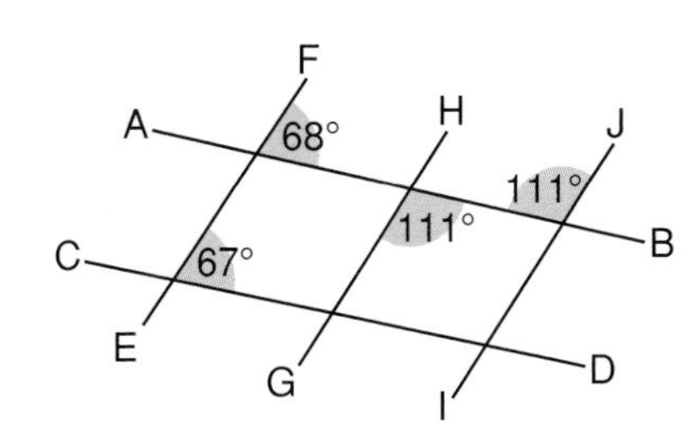

8 The diagram shows the positions of three points, P, Q and R. The distances PQ and PR are equal. Find the sizes of angles a, b and c. Write down all your working and give reasons.

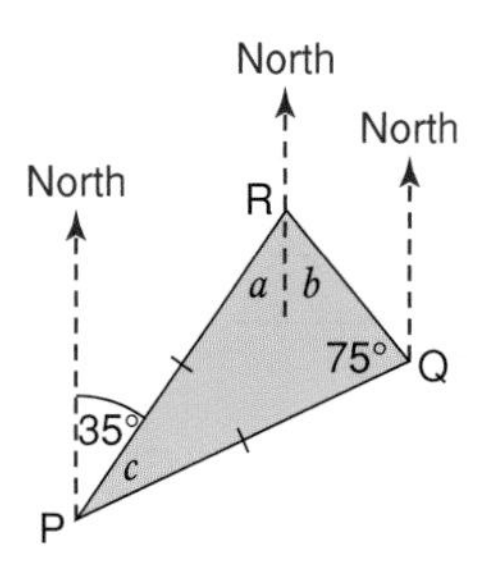

9 The angle at the vertex of a regular pentagon is 108°. Two diagonals are drawn as shown to make three triangles.

a Calculate the size of the angles in each triangle.

b The purple triangle and one of the pink triangles are placed together like this. Explain why the triangles fit together to make a new triangle. What are its angles?

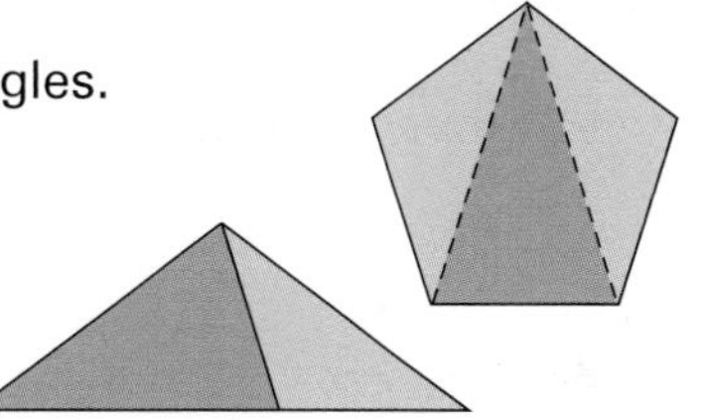

*__10__ Prove that $x = 73°$.

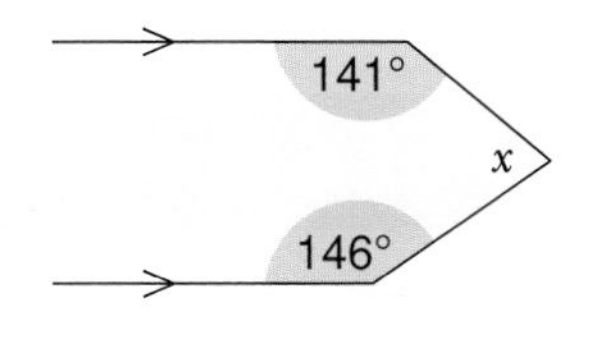

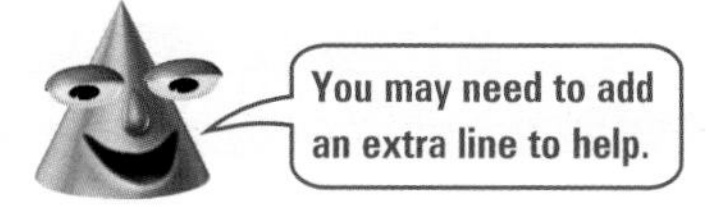

*__11__ Prove that $y = 19°$

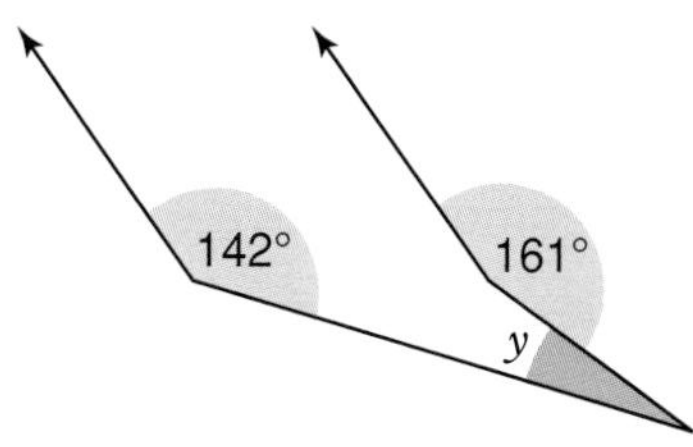

*__12__ Prove that $y = 360 - x - z$.

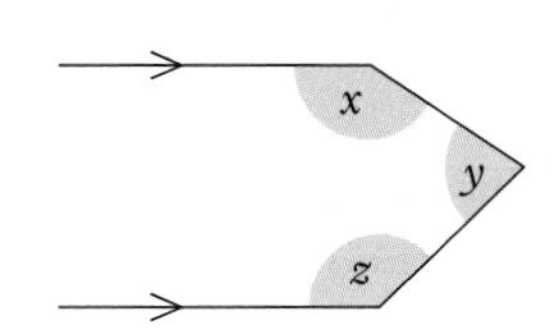

Review 1 Find the size of the angles marked with letters. Write down all your working and reasons.

a

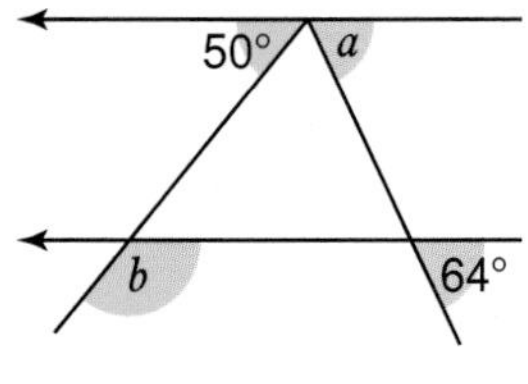

b

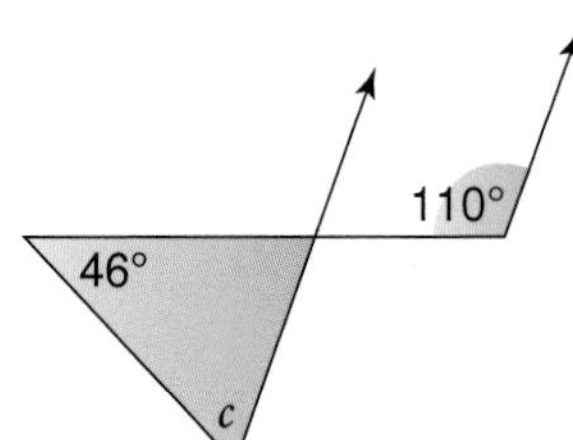

c

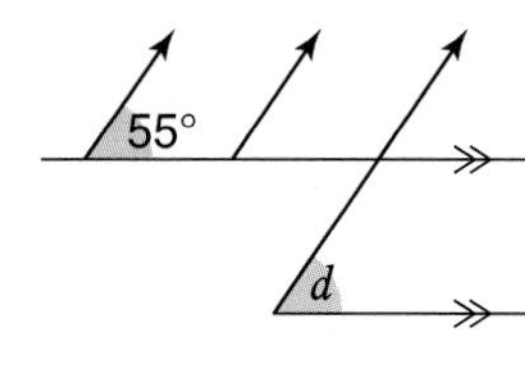

Review 2 Write and solve equations to find the value of x.

a

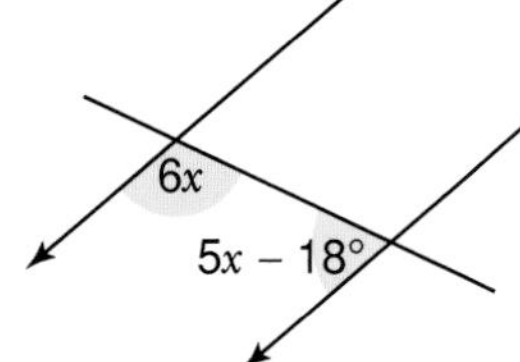

b

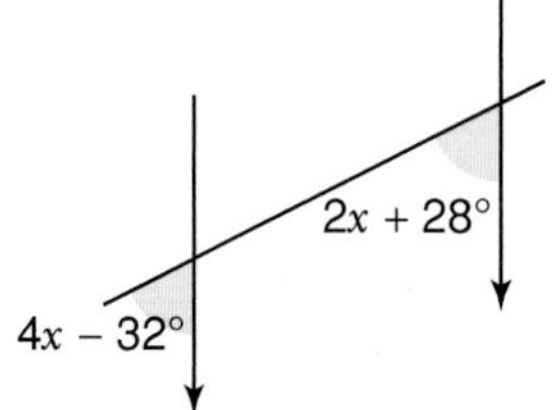

Review 3 Explain why the lines HG and FE are parallel.

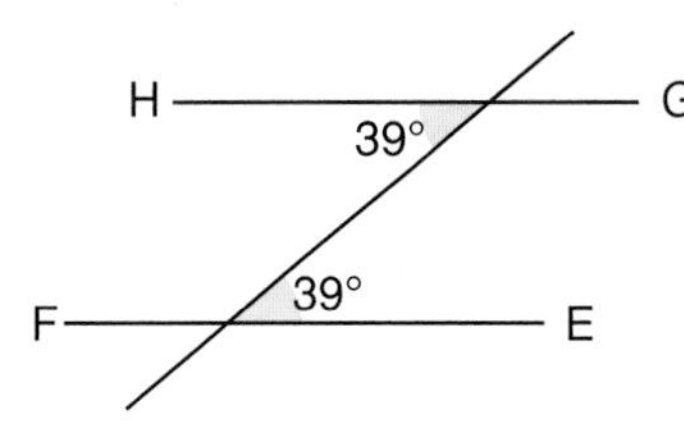

* **Review 4** Prove that $x = 57{\cdot}5°$.

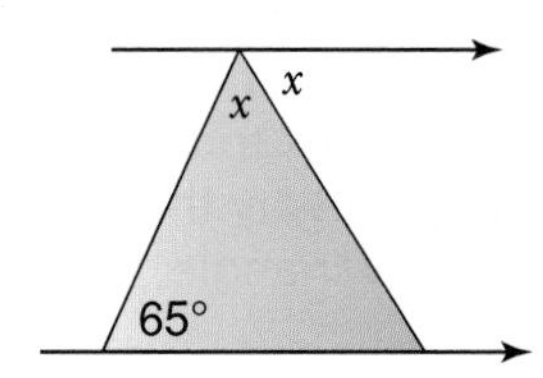

Interior and exterior angles of a polygon

Practical

You will need a dynamic geometry software package or acetate sheets.

Draw two pairs of parallel lines like these.

What shape is the green-shaded shape?

Explore the relationships between angles a, b, c and d.

Do these relationships stay the same if you rotate **one** pair of parallel lines? Explain why or why not.

If you use acetate draw each pair of lines on a separate sheet.

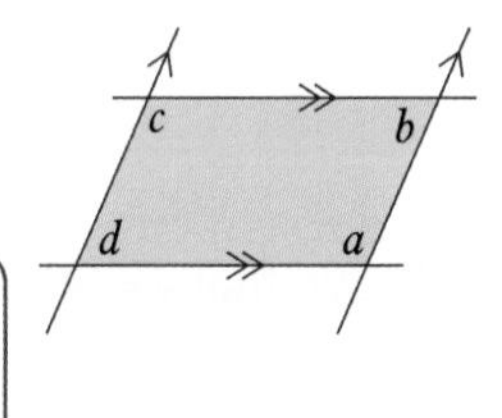

Ask your teacher for A Quadrilateral between Parallel Lines ICT worksheet.

Discussion

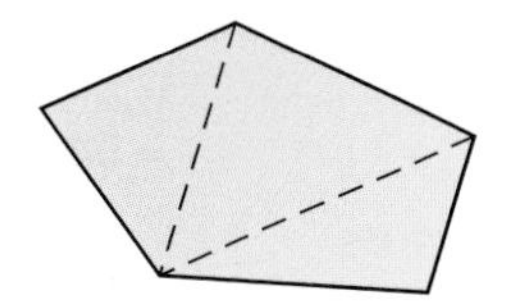

- This pentagon can be divided into three triangles by drawing all the diagonals from one vertex.

  How many triangles can a hexagon be split into by drawing all the diagonals from one vertex? **Discuss**.

  What if it was a quadrilateral?
  What if it was an octagon?
  What if it was a heptagon?
  What if ...

  What goes in the gap? **Discuss**.
  A polygon with n sides can be split into ________ triangles.

- The pentagon above is split into three triangles.
  What then is the sum of the interior angles of a pentagon? **Discuss**.

  What is the sum of the interior angles of a quadrilateral, hexagon, ... ?

  What is the sum of the interior angles of an n-sided polygon?

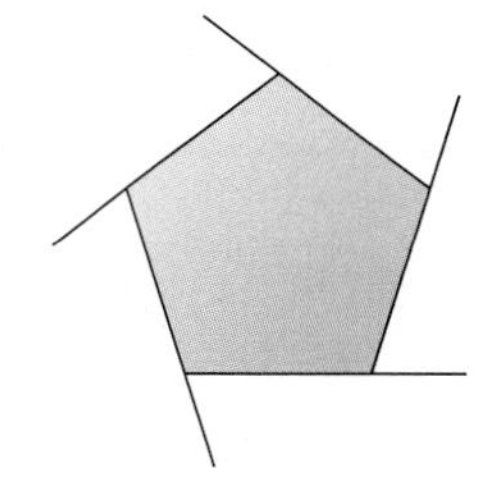

- At *each* vertex of this pentagon, what is the sum of the interior and the exterior angle? **Discuss**.
  Use this answer and the answer to the sum of the interior angles of a pentagon to find the sum of the exterior angles of a pentagon.

  What is the sum of the exterior angles of a quadrilateral, hexagon, ... ? **Discuss**.

Practical

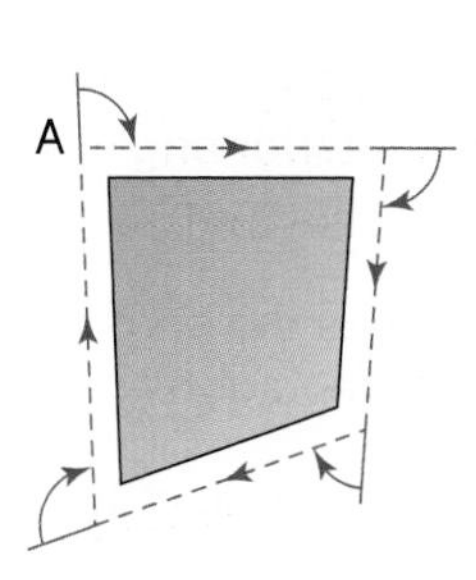

A Mark a polygon on the floor or pavement.
You could use chalk or string.
Start at one vertex (A).
Walk along one side.
When you get to the next vertex, turn and walk along the next side.
Continue until you are back facing the way you were when you started.
Through what angle have you turned altogether?

What is the sum of the exterior angles of this quadrilateral?

What if you walked around a pentagon instead?
What if you walked around a hexagon instead?
What if ...

B **You will need** a dynamic geometry software package.
Ask your teacher for the Exterior Angles of Polygons ICT worksheet.

The sum of the interior angles of an n-sided polygon is $(n - 2) \times 180$.

Example $p + q + r + s + t = (5 - 2) \times 180$
$= 3 \times 180°$
$= 540°$

Worked Example
Find the value of n.

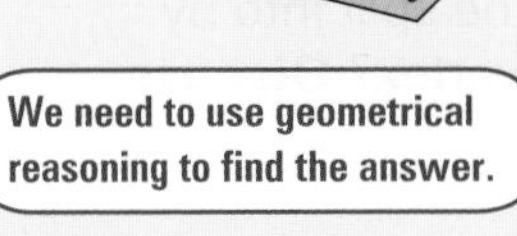

Answer
We label the unknown angles we need to find.

$k = 52{\cdot}5°$ — **base angles of isosceles △ are equal**

$l + 52{\cdot}5° + 52{\cdot}5° = 180°$ — **angles of a △ add to 180°**

$l = 180° - 52{\cdot}5° - 52{\cdot}5°$
$= 75°$

$m = 75°$ — **vertically opposite angles are equal**

$n + 88° + 52° + 75° + 158° = 540°$ — **interior angles of a pentagon**

$n = 540° - 88° - 52° - 75° - 158°$

$n =$ **167°**

The sum of the exterior angles of a polygon is 360°.

Example $n + 72° + 38° + 54° + 61° + 31° = 360°$
$n = 104°$

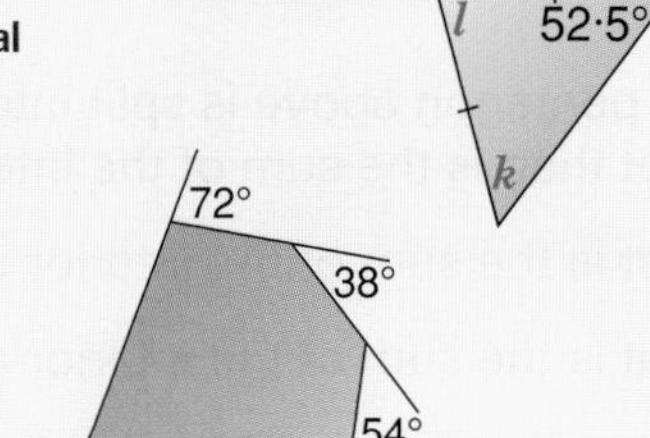

Exercise 2

Only use a calculator if you need to.

1 What goes in the gap?
- **a** A quadrilateral can be divided into ___ triangles.
- **b** A pentagon can be divided into ___ triangles.
- **c** A hexagon can be divided into ___ triangles.
- **d** An octagon can be divided into ___ triangles.
- **e** An n-sided polygon can be divided into ___ triangles.

2 Using the formula for the sum of the interior angles of an n-sided polygon, find the sum of the interior angles of
a a heptagon **b** a decagon **c** a 12-sided polygon.

3 What is the sum of the exterior angles of
a a heptagon **b** a decagon **c** a 12-sided polygon **d** an equilateral triangle?

4 Find the size of the angle marked as x.

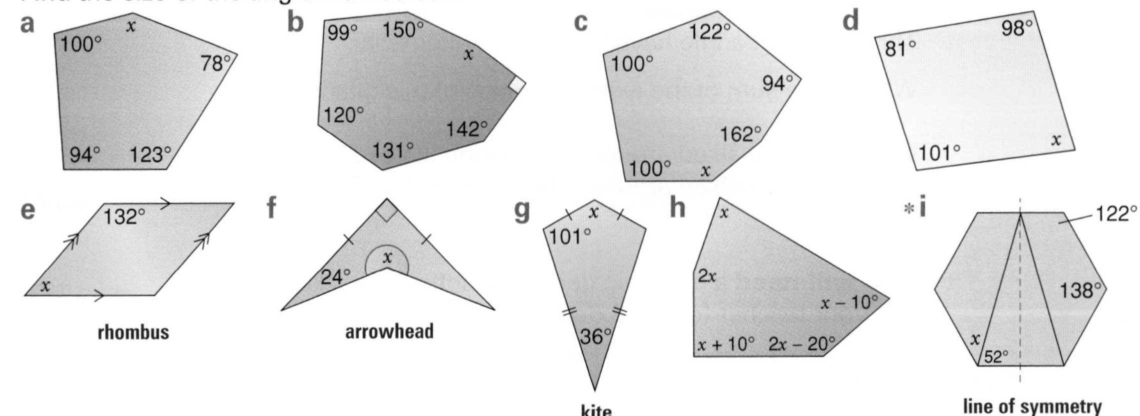

5 Calculate the value of a.

a

b

c

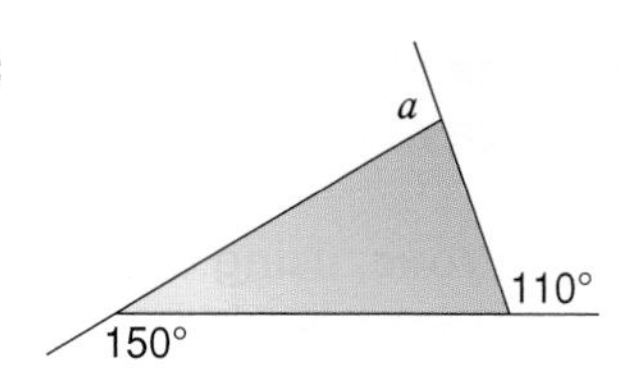

d

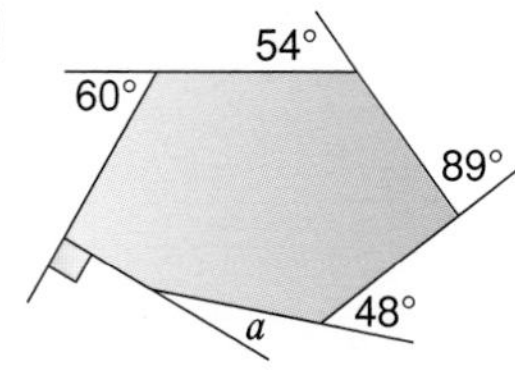

e

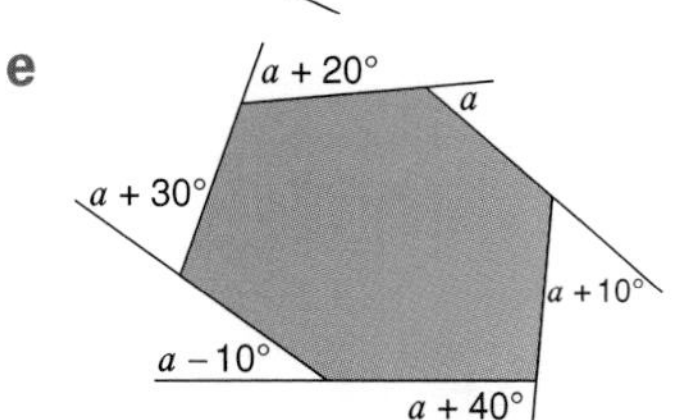

6 Find the sizes of the angles marked with letters. Show your working and give reasons.

a

b

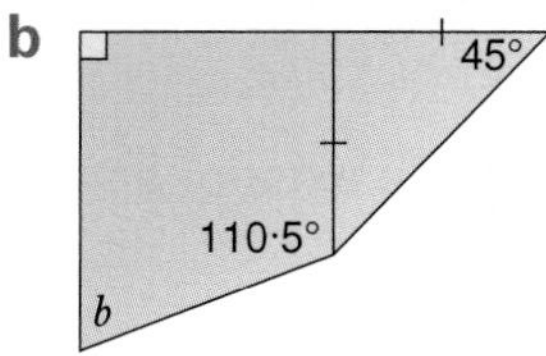

c

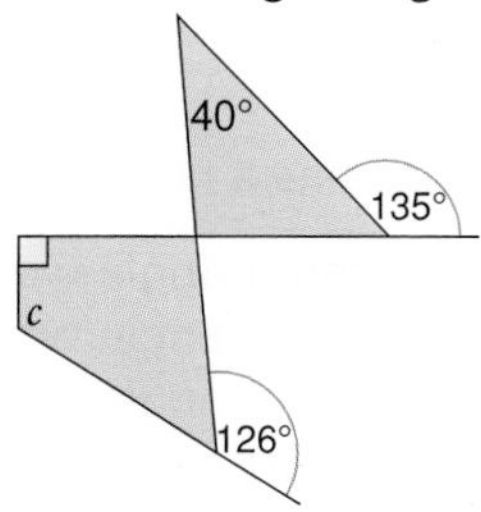

7 **Prove**, using diagrams and giving reasons, that

a the sum of the interior angles of a quadrilateral is 360°

b the sum of the interior angles of a pentagon is 540°

c the sum of the interior angles of a hexagon is 720°.

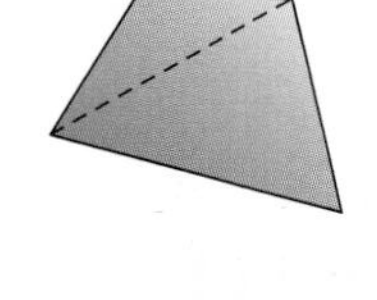

8 Find the size of each interior angle of a

a regular octagon **b** regular pentagon **c** regular hexagon

d regular 12-sided polygon.

Remember: A regular shape has all its sides and angles equal.

9 Find the size of each exterior angle of a

a regular pentagon **b** regular 20-sided polygon **c** equilateral triangle.

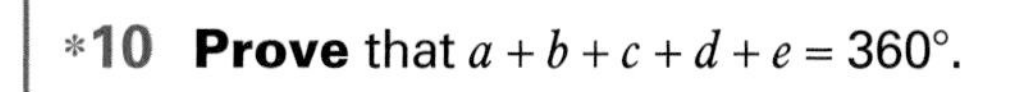

*__10__ **Prove** that $a + b + c + d + e = 360°$.

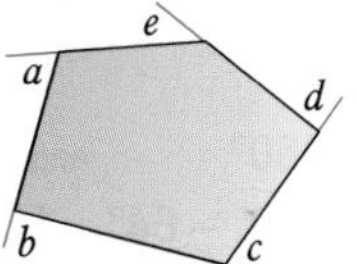

*__11__ The ratio of the size of angles a, b and c is $a : b : c = 2 : 3 : 4$.

a Find the size of each angle a, b and c.

b Find the ratio of $d : e : f$.

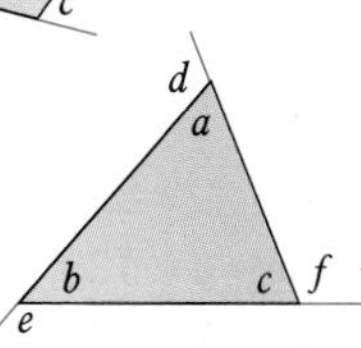

Review 1 Find the value of y.

a

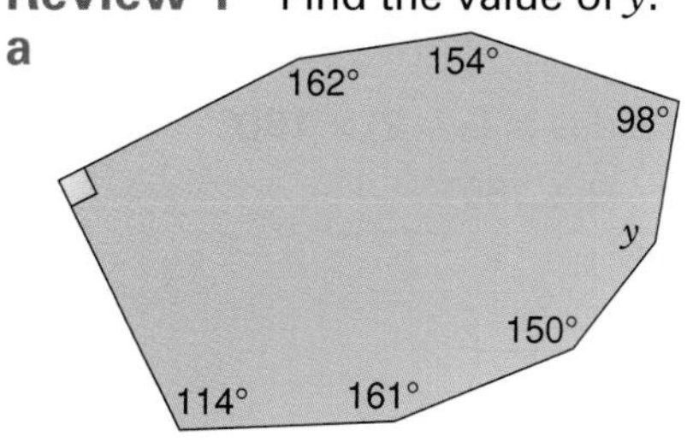

b

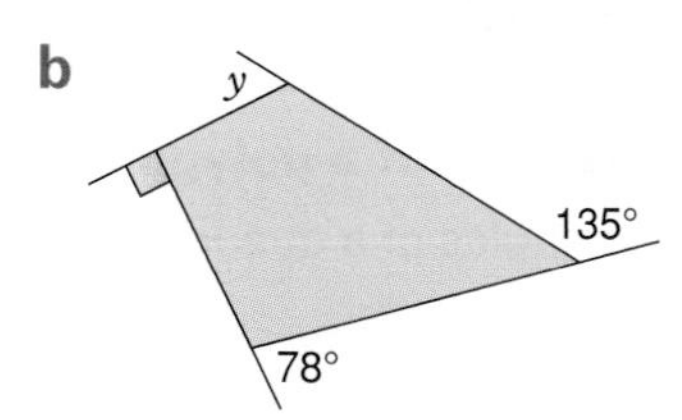

c

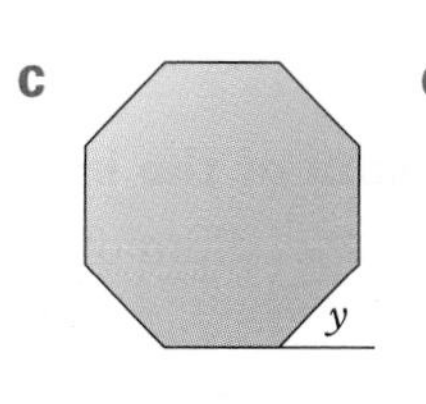

d

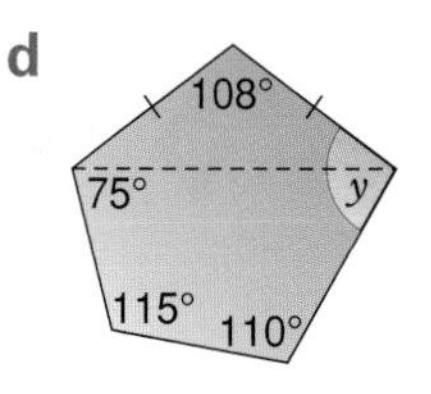

Review 2
a Find the size of each interior angle of a regular decagon.
b Find the size of each exterior angle of a regular 12-sided polygon.

Review 3 Prove, using diagrams and geometric reasoning, that each of the interior angles of a regular octagon is 135°.

Review 4 Find the values of x and y.

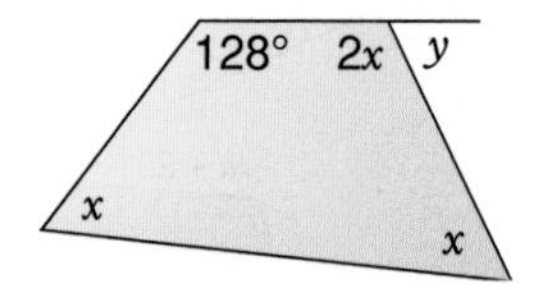

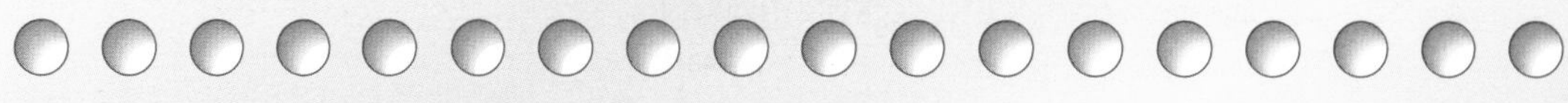

Investigation

Polygons

How many sides has a regular polygon with an exterior angle of 40°?

What if each exterior angle was 72°?
What if each exterior angle was 15°?
What if each exterior angle was 80°?
What if each exterior angle was 50°?
What if ...
Investigate.

Summary of key points

We often use **geometrical reasoning** to find an unknown angle or prove something. Write down the steps clearly, one by one, and give reasons.

Example Prove that $x = 96°$.

A kite has one pair of equal angles.
So $a = b$.

$a + b + 64° + 128° = 360°$ — angles of a quadrilateral add to 360°
$a + b = 360° - 64° - 128°$
$= 168°$
$b = \frac{168°}{2}$
$= 84°$
$b + x = 180°$ — angles on a straight line add to 180°
$84° + x = 180°$ — substitute $b = 84°$
$x = \mathbf{96°}$

64°
a
b
x
128°
kite

B The **sum of the interior angles of a polygon** with n sides is **$(n - 2) \times 180°$**.

Examples Sum of interior angles of a quadrilateral $= (4 - 2) \times 180°$
$= 360°$

Sum of interior angles of hexagon $= (6 - 2) \times 180°$
$= 720°$

The sum of the exterior angles of any polygon is 360°.

You should be able to explain, using geometrical reasoning, why B and C are true.

Example A pentagon can be divided into 3 triangles.
The sum of the interior angles is $3 \times 180° = 540°$,
because the angle sum of each triangle is 180°.

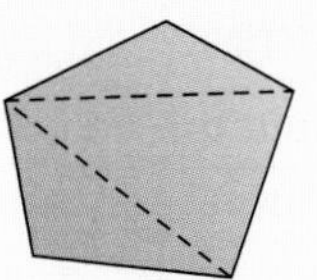

A pentagon has 5 vertices.
At each vertex the sum of the interior and exterior angle = 180°.
The total sum of the 5 interior and 5 exterior angles = $5 \times 180°$
= 900°.

Sum of exterior angles = Total sum – Sum of interior angles
= 900° – 540°
= **360°**

Test yourself

1 Find the sizes of the angles marked with letters. Show your working clearly and give reasons.

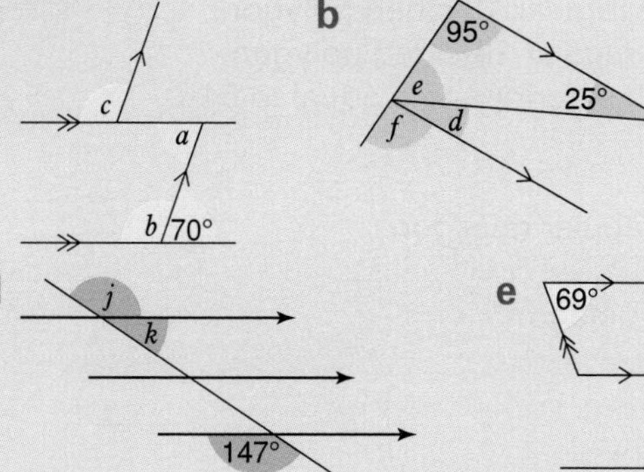

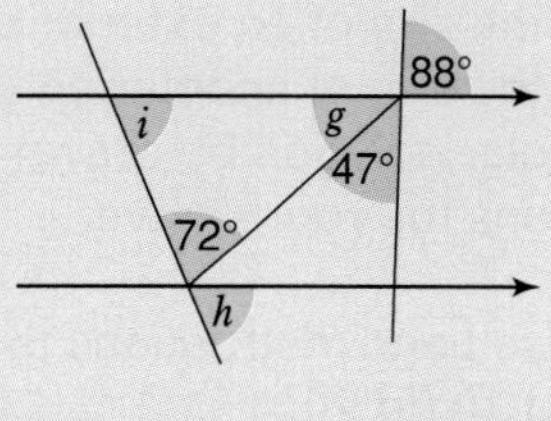

2 Calculate the value of x. Show your working clearly and give reasons.

A

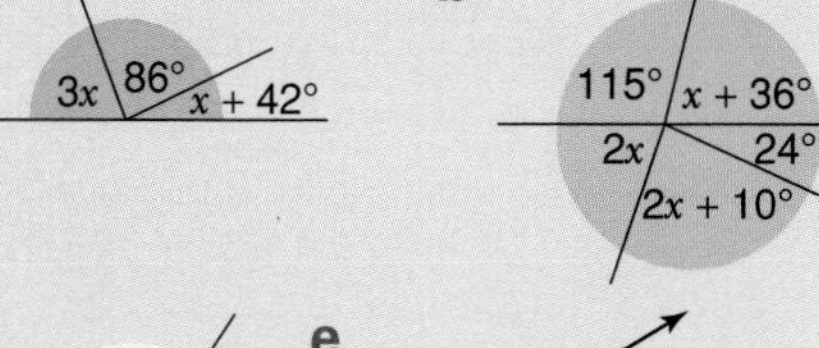

c $x - 30°$, $x - 25°$, $2x + 15°$

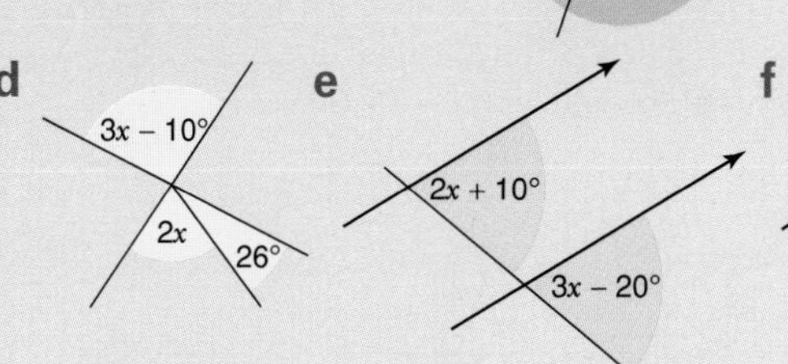

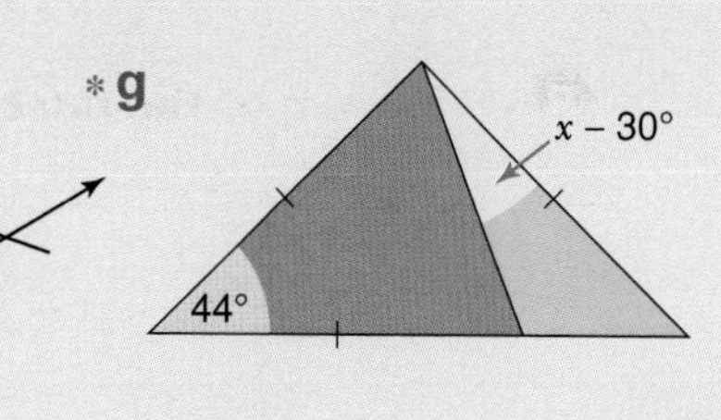

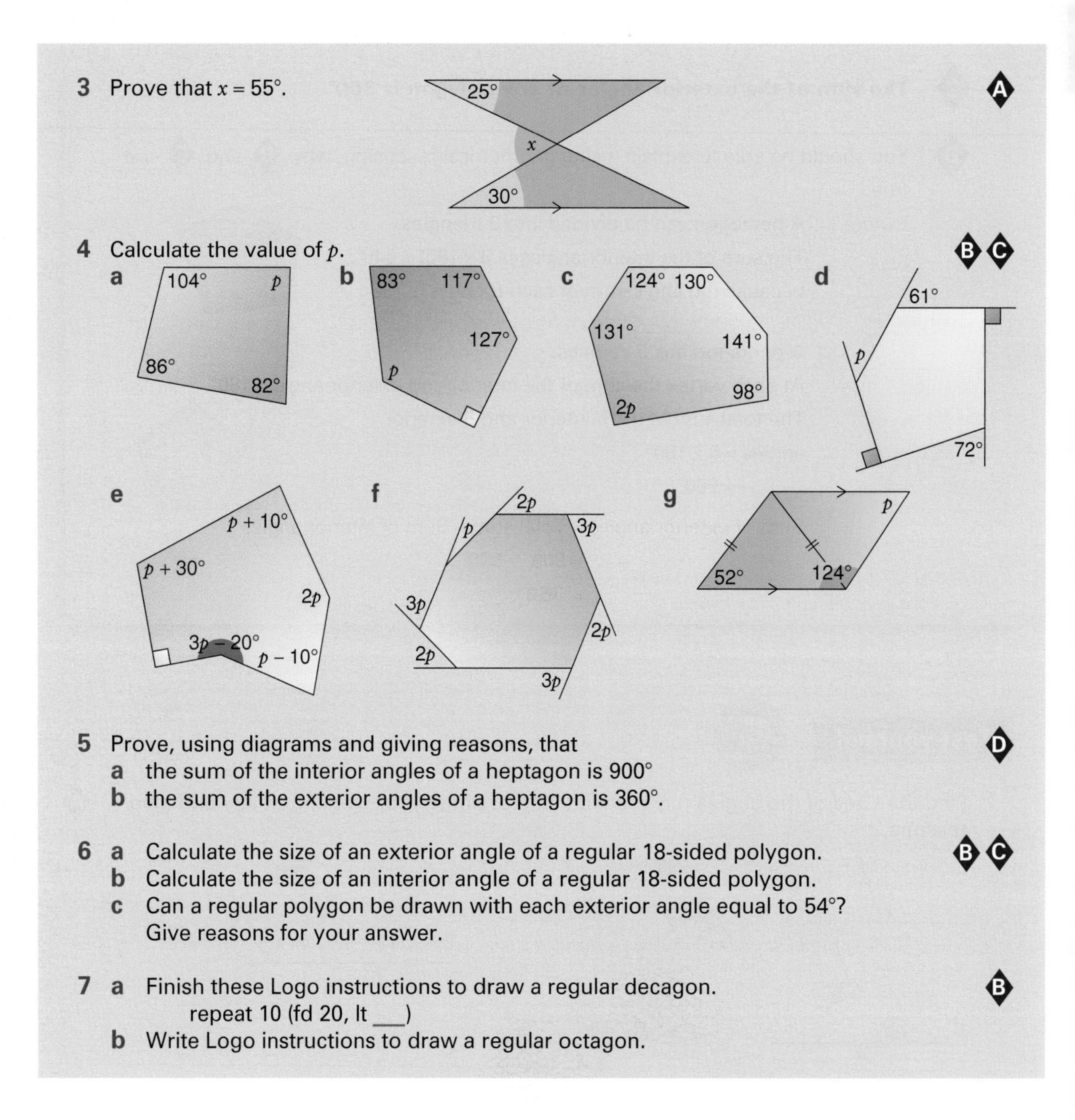

3 Prove that $x = 55°$.

4 Calculate the value of p.

a **b** **c** **d**

e **f** **g**

5 Prove, using diagrams and giving reasons, that

a the sum of the interior angles of a heptagon is 900°

b the sum of the exterior angles of a heptagon is 360°.

6 **a** Calculate the size of an exterior angle of a regular 18-sided polygon.

b Calculate the size of an interior angle of a regular 18-sided polygon.

c Can a regular polygon be drawn with each exterior angle equal to 54°? Give reasons for your answer.

7 **a** Finish these Logo instructions to draw a regular decagon.

repeat 10 (fd 20, lt ___)

b Write Logo instructions to draw a regular octagon.

12 Shape, Construction and Loci

You need to know

✓ 2-D shapes	— naming triangles — properties of triangles — properties of quadrilaterals — polygons	page 250
✓ constructing triangles and quadrilaterals		page 251
✓ 3-D shapes		page 251

Key vocabulary

arc, bisect, bisector, chord, compasses, elevation, equidistant, heptagon, isometric, loci, locus, midpoint, perpendicular bisector, plan view, plane, regular, tessellate, tessellation, triangular prism, view

T

Piecing it Together

Ask your teacher for a set of tangram pieces or a large copy of this diagram.

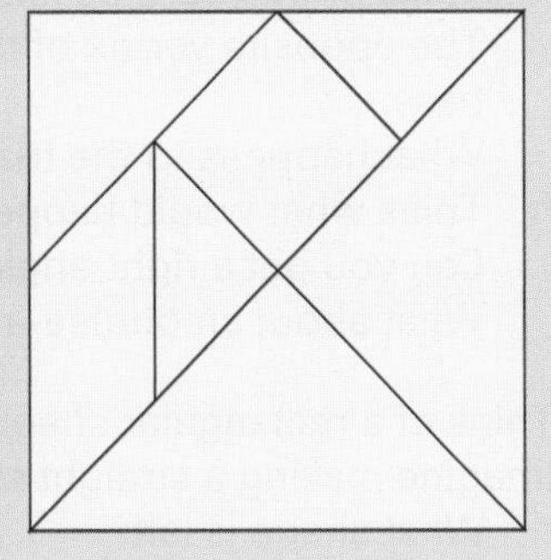

Cut your diagram into 7 pieces.

Use all of your 7 tangram pieces to make these shapes.

Visualising and sketching 2-D shapes

Discussion

What might Dylan's answer be?

Exercise 1

1 **a** Imagine a quadrilateral with one line of symmetry.
What could it be?
b What if it also has diagonals that meet at right angles?
What could it be now?

2 Imagine a quarter of a shape is
a a right-angled isosceles triangle **b** a parallelogram.
What is the shape? Is this the only possible shape?

3 **a** Think of an equilateral triangle with one side horizontal. Call this side the base.
Think of this base as fixed.
The opposite vertex of the triangle moves slowly in a straight line perpendicular to the base.
What happens to the triangle?
b Think what would happen if the opposite vertex moves parallel to the base.
Can you get a right-angled triangle?
What about an obtuse-angled triangle?

4 Think of a rectangular sheet of paper.
Imagine making a straight symmetrical cut across one corner.
a What shape is left?
b What if you made a series of cuts, always parallel to the first cut.
What shapes is it possible to make?

5 **a** Suppose you were to draw a set of parallel lines 4 cm apart on each of two sheets of acetate.
Imagine placing one set on top of the other so that the two sets are perpendicular.
What shapes would you see?
b What happens as you rotate one of the sheets about a fixed point (intersection of two lines).
c Repeat **a** and **b** for one set of lines 4 cm apart and the other set 2 cm apart.

Review Repeat question **4** of the exercise for a square sheet of paper.

Practical

1 **You will need** some pictures or posters with geometrical patterns.
Examples: tiling patterns, Escher drawings, wallpaper, posters with patterns, ...

Describe the patterns. Give as much detail as possible.
You could work individually, in pairs or in groups.
Find an interesting way to present your work. You could use a computer to make a poster or booklet.

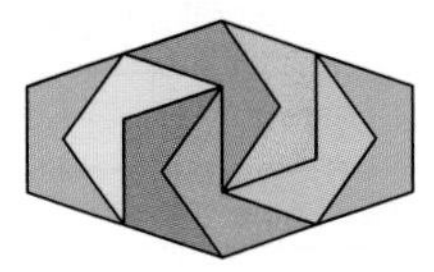

2 **You will need** to choose one group of objects such as bridges or quilts or fences or seeds or cones or buildings or windows or churches ...

Describe the shapes you see. Suggest reasons why these objects are made by people or by nature to be particular shapes.

Example Roofs are often a triangular shape because a triangle is a very strong shape.

Investigation

Rectangles and Squares

1 Joshua drew a 4 by 2 rectangle on squared paper.
He found he could **cut** it into squares in three different ways.

He was only allowed to cut along grid lines.

or

or

How many ways can you *cut* a 6 by 3 rectangle into squares? **Investigate**.
What about a 7 by 3, 8 by 3, ... rectangle?

2 Can a 5 by 3 rectangle be cut into 7 squares? What about 8 squares? 9 squares?
Give mathematical reasons for your answers. **Investigate** other rectangles.

Using properties of shapes

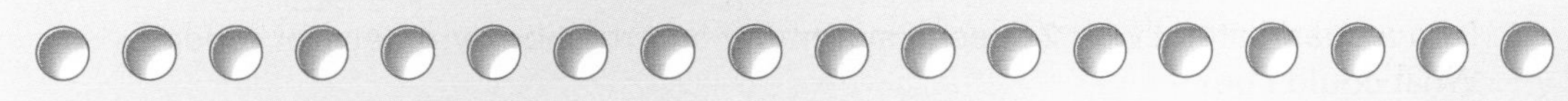

T

Investigation

Properties

You will need a dynamic geometry software package.
Ask your teacher for a copy of the Different Parallelograms ICT worksheet.

Exercise 2

T

1 Use a copy of this table. Fill it in.

Quadrilateral	Diagonals equal	Diagonals cross at right angles	Diagonals bisect each other	Diagonals bisect the angles	Sides	Angles
Square						4 right angles
Rectangle						
Parallelogram	✗	✗	✓	✗	opposite sides equal	opposite angles equal
Rhombus	✗	✓	✓	✓	4 equal	
Kite						
Isosceles trapezium						
Trapezium						
Arrowhead						

2 Write true or false for these. Explain your answer.

a A kite and a rhombus have diagonals that bisect the angles.

b A rhombus can never be split into two equilateral triangles.

c In any triangle the largest angle is opposite the longest side and the smallest angle is opposite the shortest side.

3 Use the properties of the shape to find angle A and the length of AB in each of the following. Show your reasoning clearly.

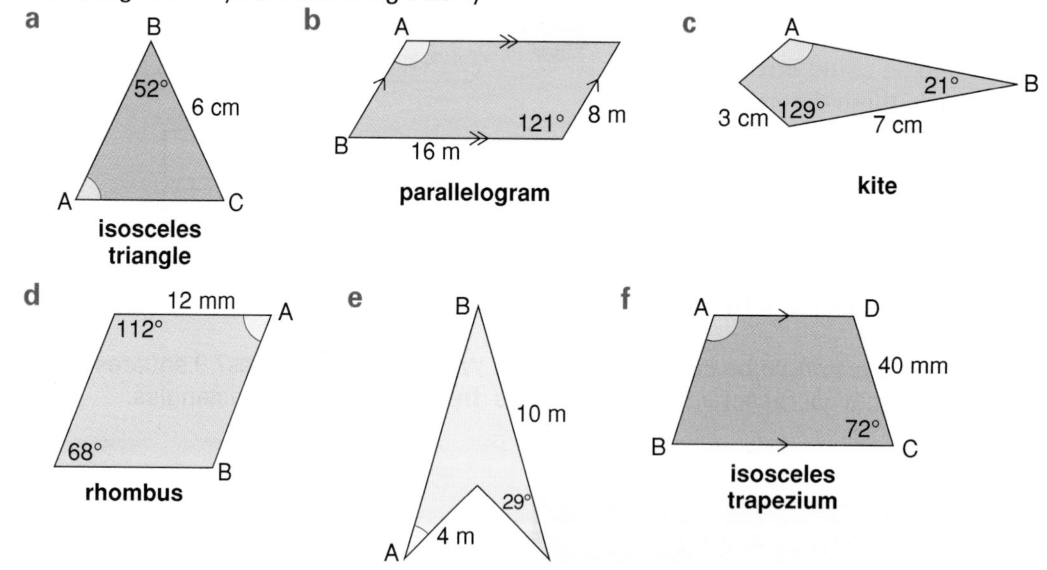

4 a I am a quadrilateral with 2 lines of symmetry and rotation symmetry of order 2. What could I be?

b I am a quadrilateral with no lines of symmetry and rotation symmetry of order 2. What could I be?

5 Which of these statements are true? Explain why or why not.

a All equilateral triangles are isosceles triangles.

b All rectangles are squares.

c All squares are rectangles.

d All squares are rhombuses.
e All kites are quadrilaterals.
f All quadrilaterals are parallelograms.
g Some rectangles are squares.
h Some parallelograms are kites.
i All rhombuses are parallelograms but a parallelogram is not necessarily a rhombus.

6 Marcy bought tiles which were the shape of a rhombus.
She put three tiles together to make this shape.
a Find the value of angles a, b, c and d.
b Write down all the things you know or can deduce about the shape formed by the three tiles.

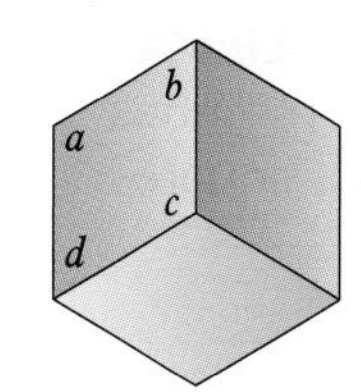

7 PQRS is a rhombus.
An equilateral triangle, PTU, is drawn inside the rhombus.
PS = PU.
a Show that a, b, c and d are all equal.
***b** Find the size of the angles a, b, c and d.

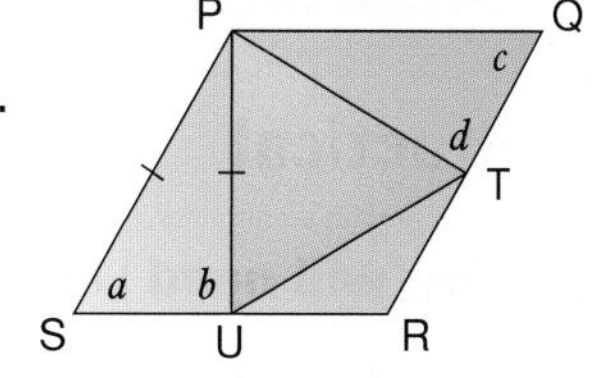

8 **a** Kylie wanted to sort these quadrilaterals.

rhombus **parallelogram** **trapezium**
kite **arrowhead (delta)** **rectangle** **square**

She drew a tree sorting diagram.
What questions should she put in A, B, C and D?

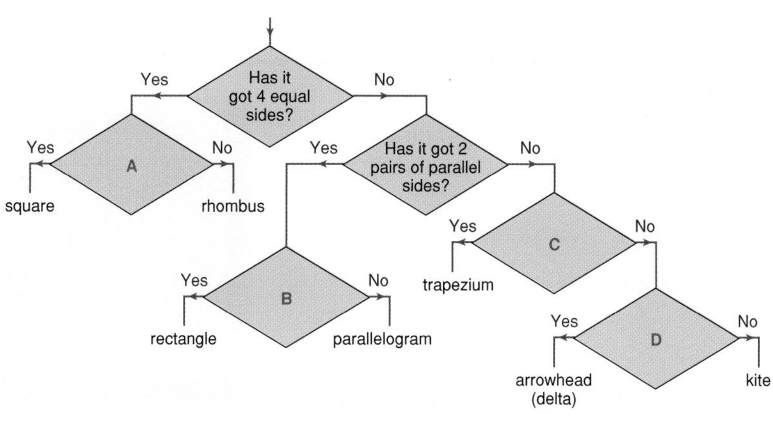

b Draw a different tree sorting diagram to sort Kylie's shapes.
***c** Draw a tree sorting diagram that sorts triangles and quadrilaterals according to their symmetry properties.
Compare your diagram with those of others.

***9** How many isosceles triangles can be drawn inside $A\hat{B}C$ as shown if $A\hat{B}C = 18°$?
Explain.

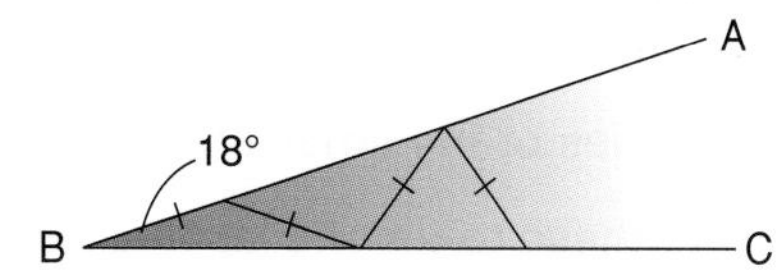

Review 1 Can a parallelogram be split into two isosceles triangles?
What about two scalene triangles?

Review 2 ABC is an equilateral triangle.
BCDE is a rhombus.
a Find the length of ED.
b Find the size of angle BCD.
***c** Find the size of angle CBE.

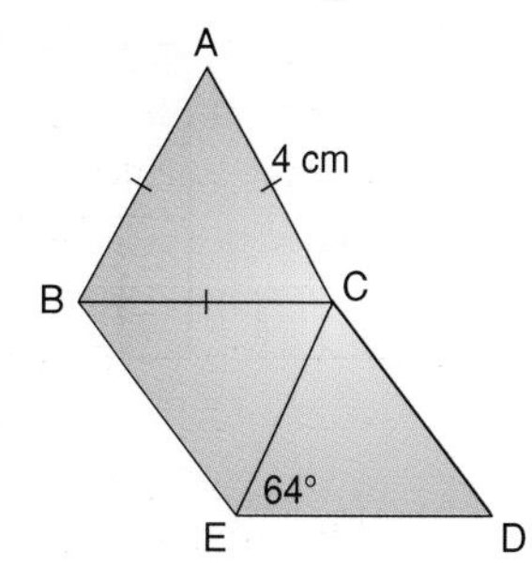

Review 3 Are these statements true or false? Explain your answers.

a Some quadrilaterals are trapeziums.
b Some rhombuses are squares.
c Some isosceles triangles are equilateral triangles.

Review 4 What do the quadrilaterals in each of these lists have in common?

List A	List B	List C	List D
rhombus	rectangle	rhombus	rectangle
square	square	square	rhombus
		kite	square
			parallelogram

Practical

You will need Logo.

Ask your teacher for the Making Shapes with Logo ICT worksheet.

Tessellations

Remember
A shape will **tessellate** if it can be used to completely fill a space with no overlapping and no gaps.

Example

A **tessellation** is made by reflecting, rotating or translating a shape.

Exercise 3

You will need some squared paper and thin card for this exercise.

1 Make 4 copies of the small shape.

Tessellate these to tile the large area.
Which of reflection, rotation and translation did you use?

a

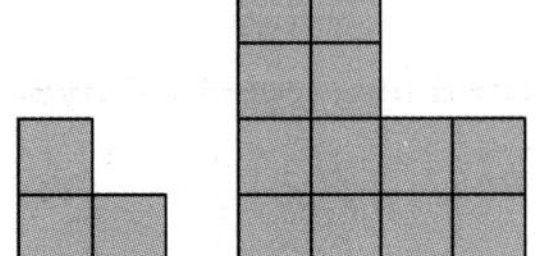

b

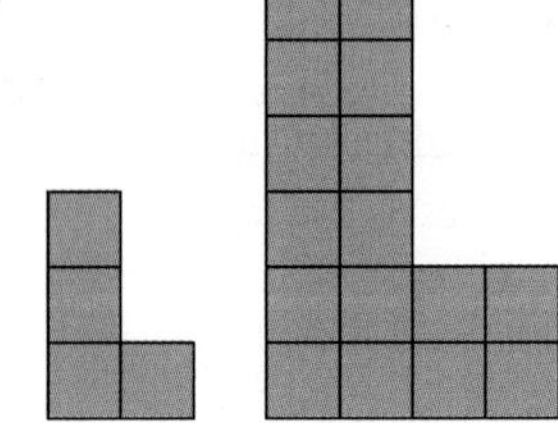

2 Tessellate one of the quadrilaterals shown below, or a quadrilateral of your choice, to tile an area. Which of reflection, rotation and translation did you use to make the tessellation? Colour your design using not more than 4 colours.

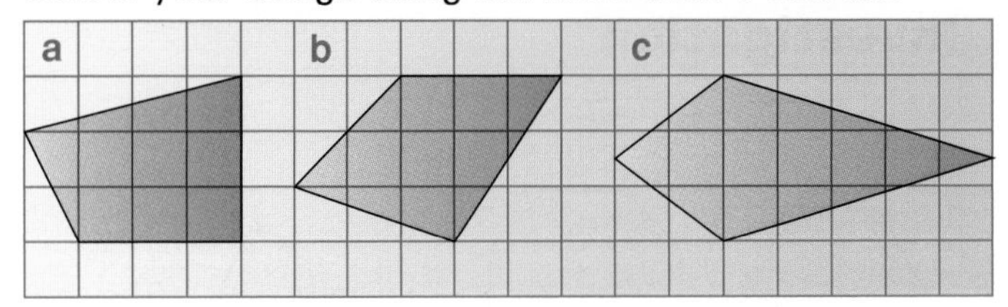

3 On thin card, draw and cut out a rhombus.
Use this rhombus to create the shape shown.
Tessellate this shape.
Explain how you did this.

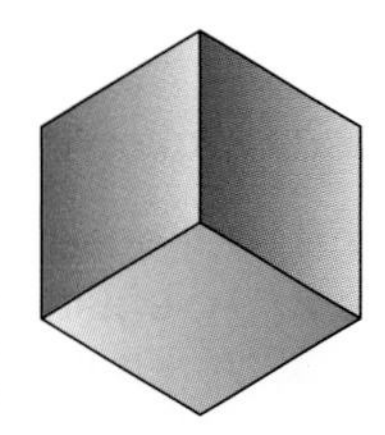

4 Make a cardboard template as follows.

Draw and cut out a square.

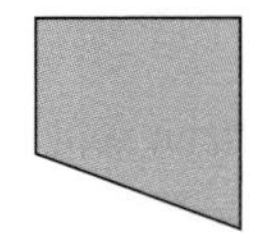
Cut off part of the square.

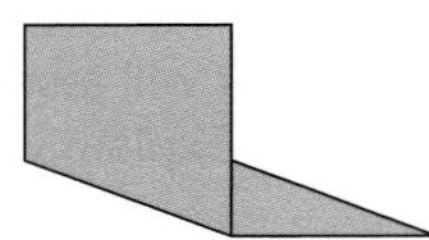
Put this part somewhere else.

You can cut off more than one part as long as you put them somewhere else.
Use your template to make lots of these shapes.
Tessellate your shape and then colour the design.
Which of reflection, rotation and translation did you use?

You can cut off curved shapes if you like.

Review Use some copies of these shapes.
Which ones will tessellate?

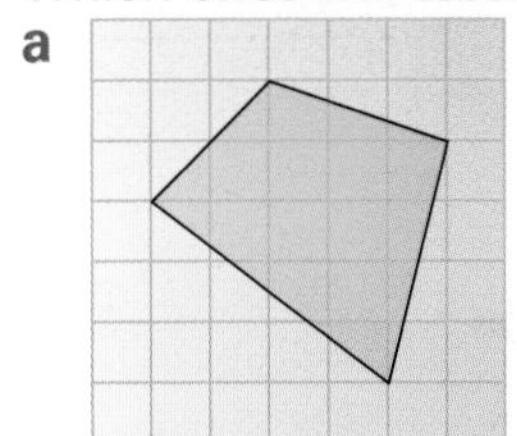

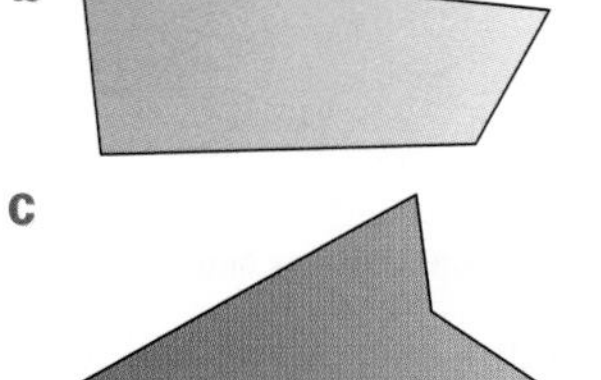

Practical

You will need card.

Design a paving stone.
Make a cardboard template of your paving stone.

Go into the school grounds.
Experiment with your paving stone template to tile an area in different ways.

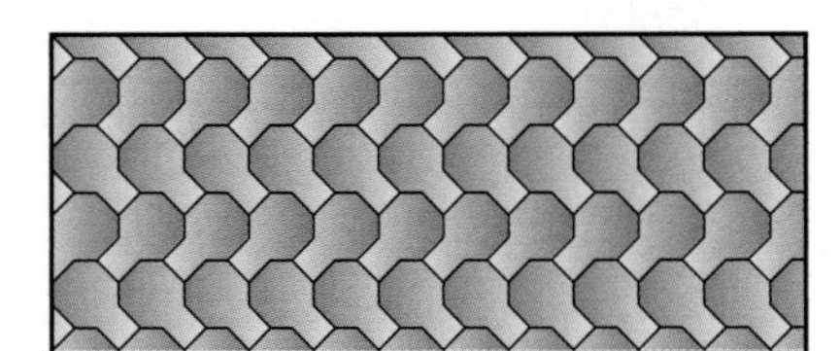

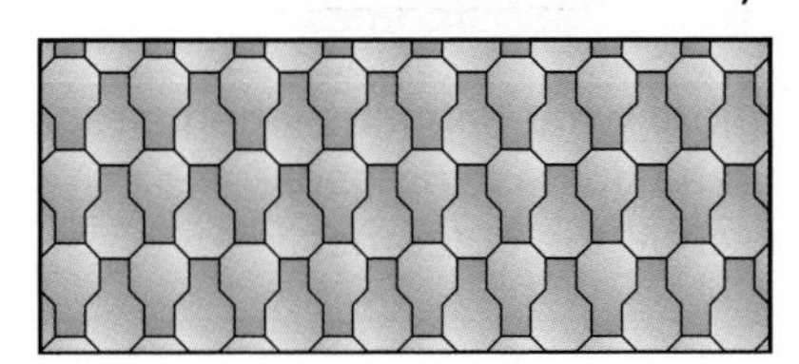

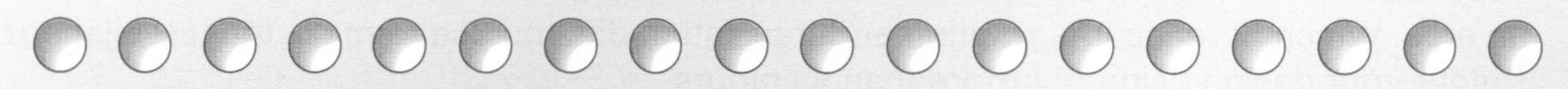

Investigation

Tessellating Polygons

1

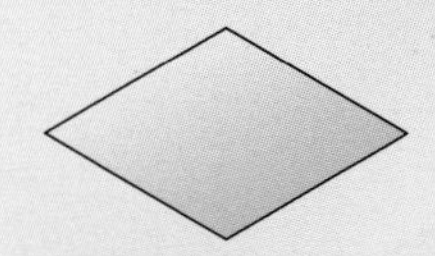 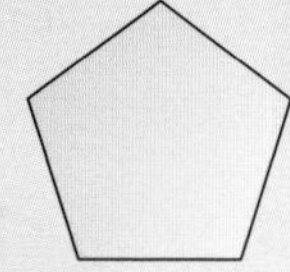 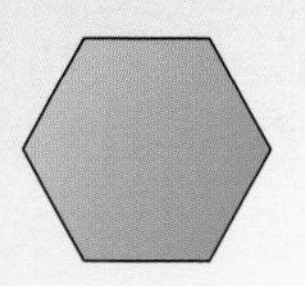 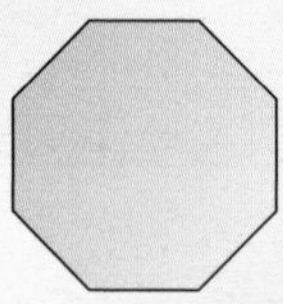 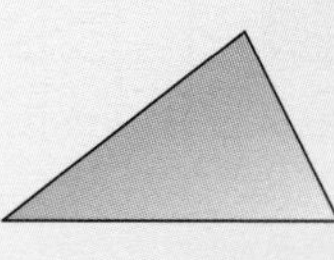

Investigate the following statements, which may be true or false. Explain.

Statement 1: All quadrilaterals tessellate.
Statement 2: A regular pentagon will not tessellate.
Statement 3: A regular hexagon will tessellate.

Make and test a statement about a regular octagon or about an irregular hexagon or about any other polygon you choose.

2 There are three regular polygons that will tessellate.
Which three are they? **Investigate**. Explain why these three will tessellate but the other regular polygons will not.

*__3__ There are eight tessellations that can be made using a combination of two or three regular polygons.
Two of these are shown.

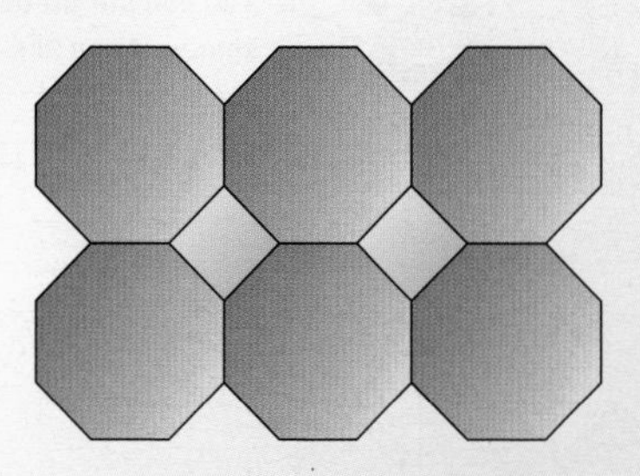

regular octagons and squares

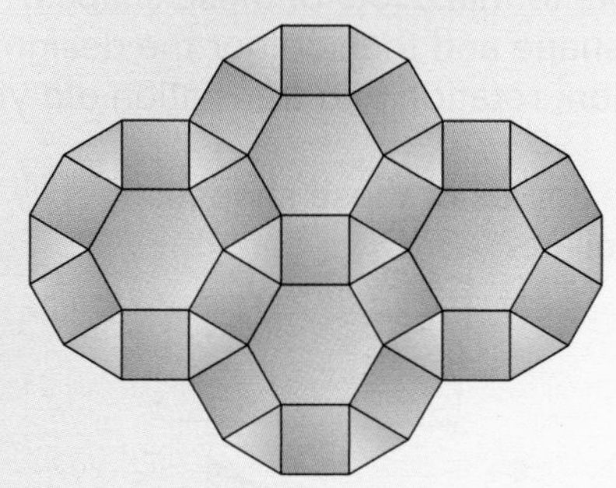

regular hexagons squares and equilateral triangles

What are the other six? **Investigate**. Explain why each will tessellate.

*__4__ A tessellation can be made by overlaying octagons and squares on octagons and squares.

Examples

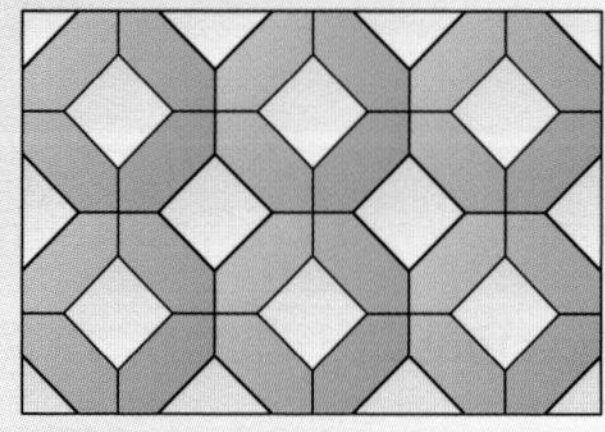

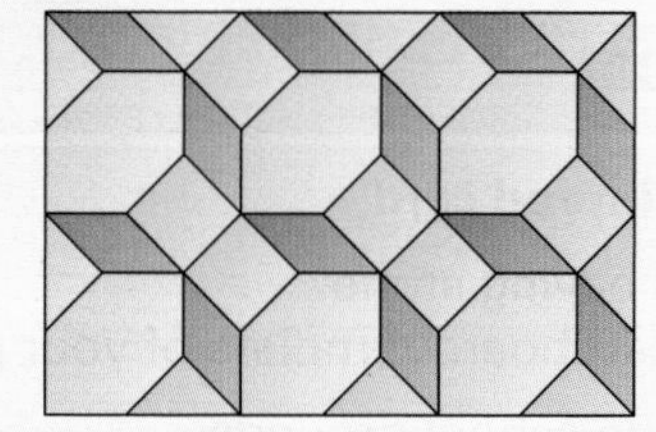

Explore other tessellations that can be made this way.
Describe the outcomes.

*__5__ Explore what happens when a regular polygon that will *not* tessellate is used to cover an area but leaving gaps.
Describe the outcomes.

Congruence

Remember
Congruent shapes are exactly the same shape and size. If two shapes are congruent, a tracing of one can fit exactly on top of the other.

Practical

1 **You will need** a copy of these shapes and some tracing paper.

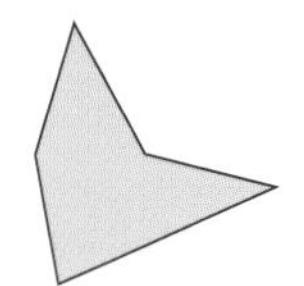 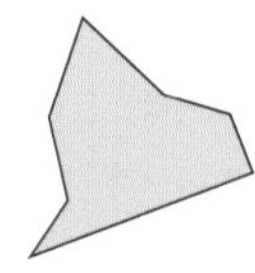 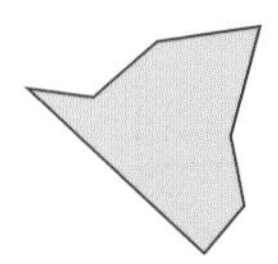 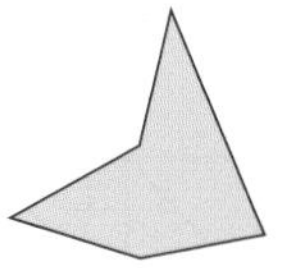 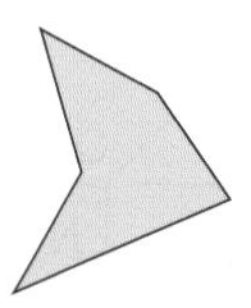 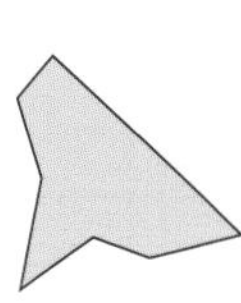

Use tracing paper to find out which of these shapes are congruent.
Explain how you mapped one congruent shape onto the other.

2 These two shapes are **congruent**.
Use tracing paper to work out which
 a angles are equal
 b lengths are equal.

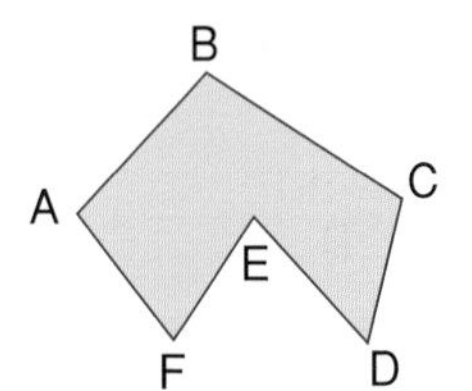

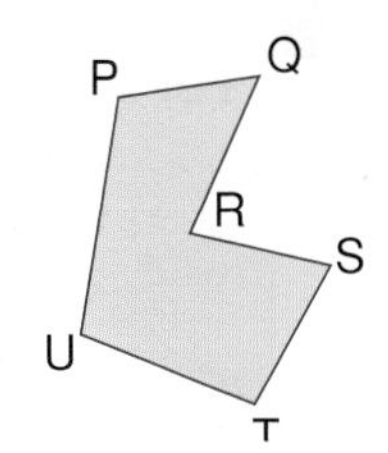

In congruent shapes
corresponding sides are equal
corresponding angles are equal.

Example

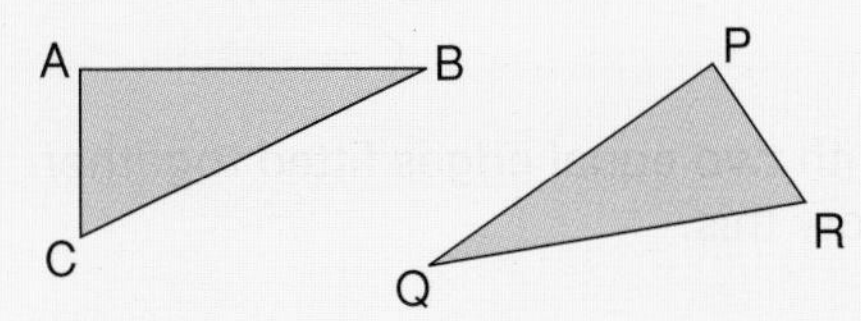

$\angle A = \angle P$, $\angle C = \angle R$, $\angle B = \angle Q$
AB = PQ, AC = PR, BC = QR

One congruent shape is mapped onto another by a translation, reflection or rotation or some combination of these.

Example In the example above, to map ABC onto PQR, the triangle ABC is reflected and rotated.

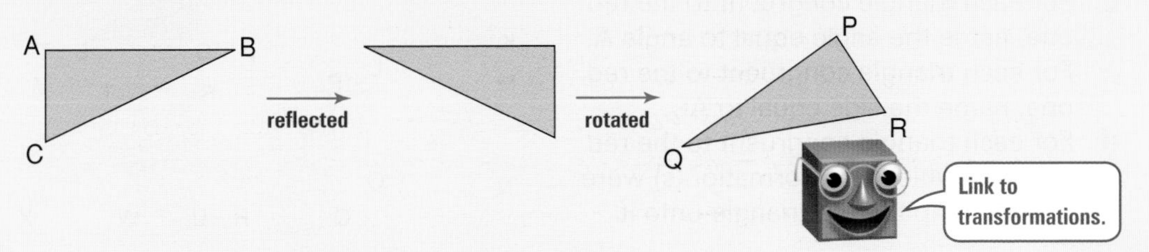

Exercise 4

1 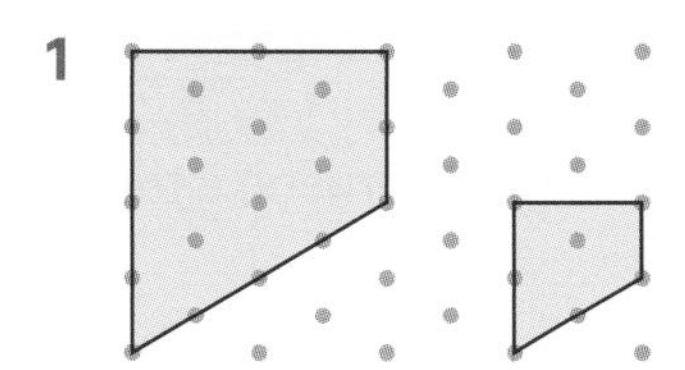

Explain why these two shapes are not congruent.

2 **a** Name all the congruent shapes in this diagram.

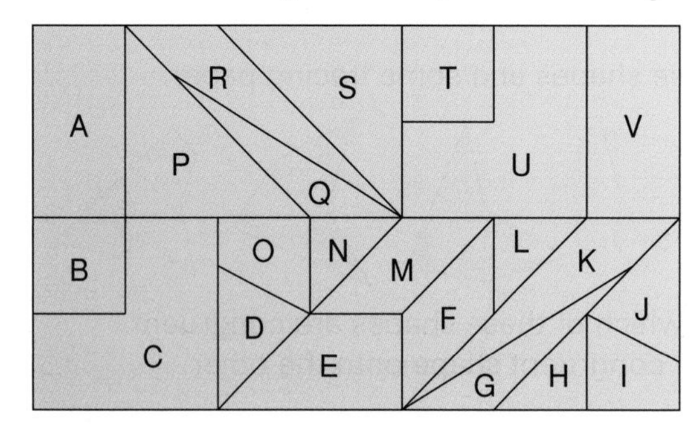

b For each pair, say which transformation(s) map one onto the other.

3 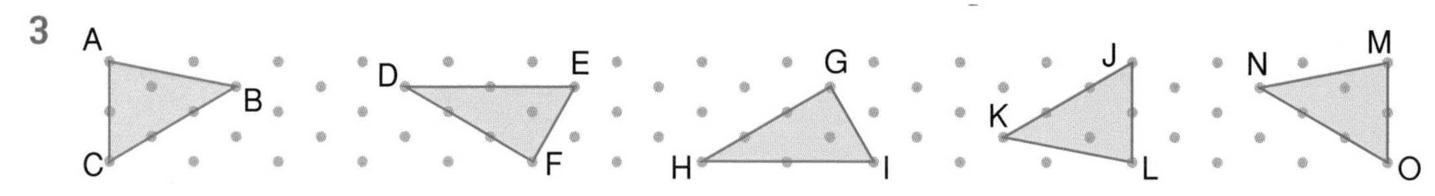

a Which shapes are congruent to the red shape?
b For each of the congruent shapes, name the angle that is equal to angle
i A **ii** B **iii** C.
c For each of the congruent shapes name the side that is equal to
i AB **ii** BC **iii** AC.
d For each congruent shape, say which transformation(s) were used to map the red shape onto it.

*__4__ Two congruent scalene triangles are joined with two equal edges fitted together.
a What shape is made? Explain how you know this.
Is this the only possible shape?
b What if the triangles have a right angle?
c What if the triangles are isosceles?
d What if the triangles are equilateral?

Review

a Name all the shapes that are congruent to the red shape.
b For each triangle congruent to the red one, name the angle equal to angle A.
c For each triangle congruent to the red one, name the side equal to AC.
d For each triangle congruent to the red one, say which transformation(s) were used to map the red triangle onto it.

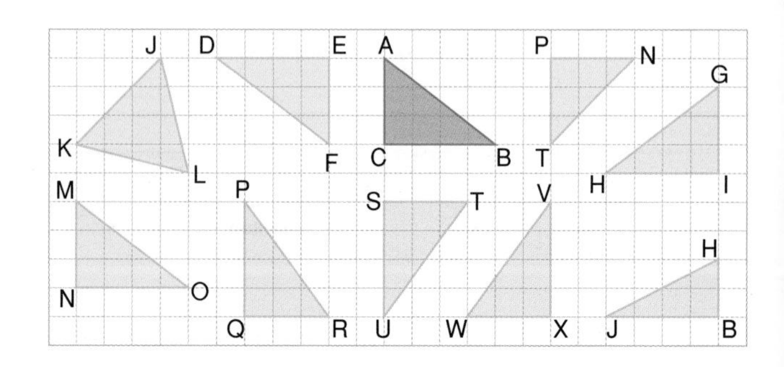

Practical

You will need a 3 × 3 pin board or 3 × 3 dotty grid
a 4 × 4 pin board or 4 × 4 dotty grid
a 5 × 5 pin board or 5 × 5 dotty grid.

Helena divided this pinboard into two congruent halves.
Find as many ways as possible of doing this.
Divide a 4 × 4 pinboard or 4 × 4 grid into two non-congruent halves.
Do this in as many ways as possible.
What about a 5 × 5 pinboard or grid?

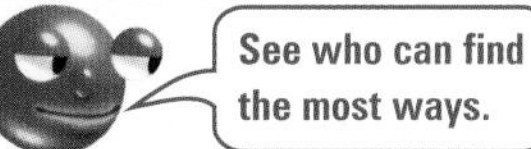

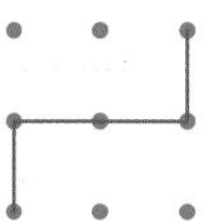

Circles

A **circle** is a set of points equidistant from its centre.
The **circumference** is the distance round the outside of the circle (shown in red).
The **radius** is the distance from the centre to the circumference.
An **arc** is part of the circumference.
A **sector** is the area made by an arc and two radii.

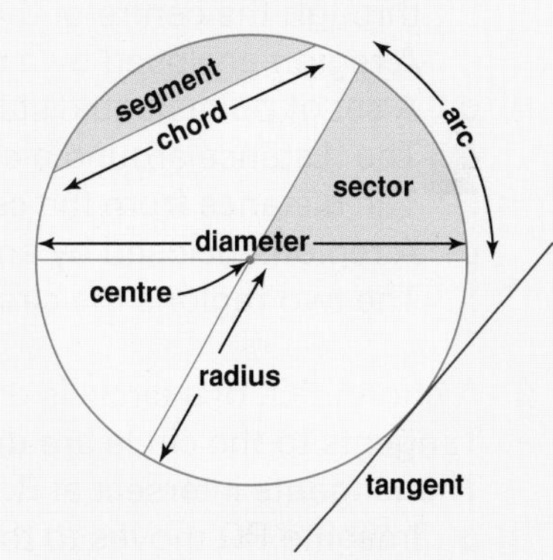

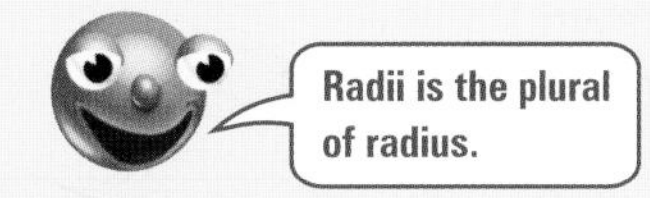

Practical

You will need a dynamic geometry software package.

Ask your teacher for the Squares and Circles ICT worksheet.

Imagine a line moves towards a circle.

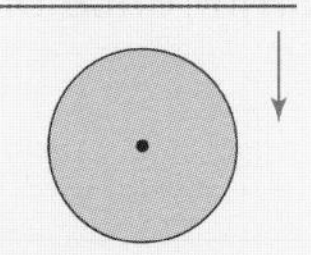

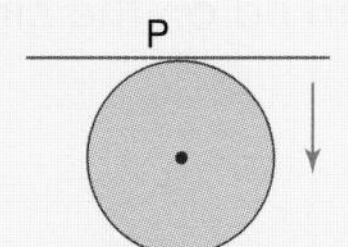

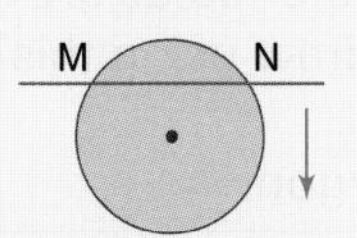

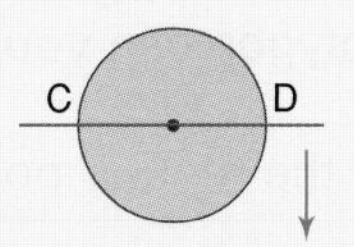

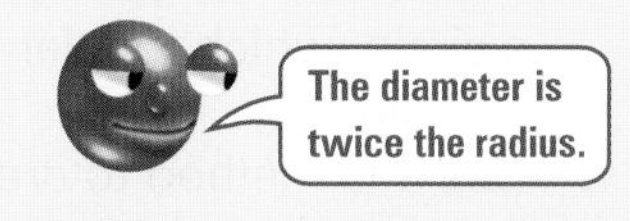

When the line

- touches the circle at P, it is called a **tangent** to the circle at that point
- intersects the circle at two points M and N, the line segment MN is called a **chord** of the circle, and this chord divides the area enclosed by the circle into two regions called **segments**
- passes through the centre, the line segment CD becomes the **diameter** and divides the area enclosed by the circle into two **semicircles**.

Exercise 5

1 Name the parts of the circle shown in red. The • shows the centre of the circle.

a **b** **c** **d** **e** **f**

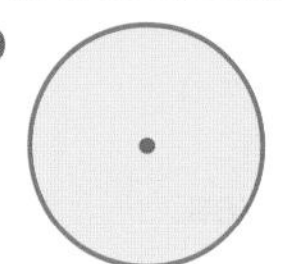
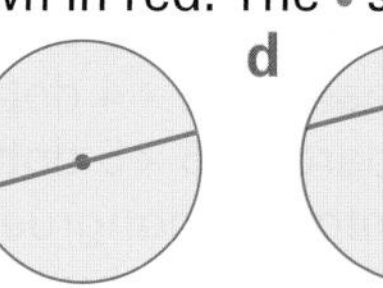
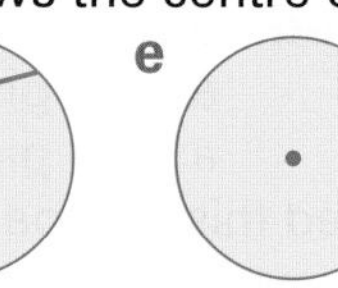
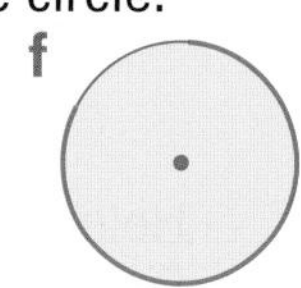

2 Name the regions shaded in green.

a **b** **c** **d** **e**

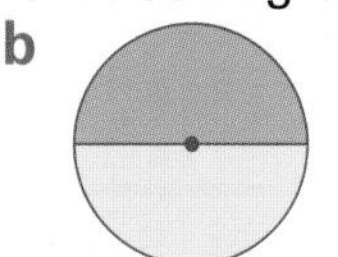
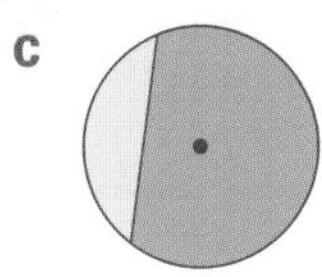
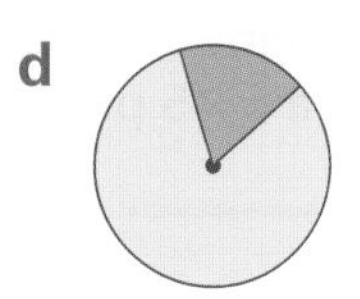
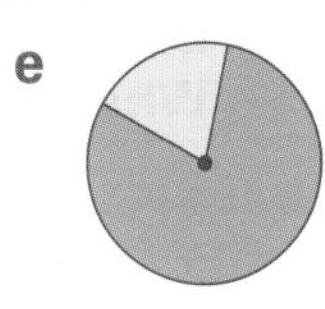

3 Match these definitions with the correct name from the box.

- **a** A line that touches the circle at just one point, P.
- **b** A line that joins two points on the circumference and passes through the centre of the circle.
- **c** A region enclosed by a chord and an arc.
- **d** A set of points equidistant from another fixed point.
- **e** The distance around a circle.
- **f** The distance from the centre to the circumference.
- **g** A region enclosed by an arc and two radii.
- **h** The two regions the circle is divided into by the diameter.

A. circle
B. circumference
C. radius
D. arc
E. diameter
F. tangent
G. chord
H. segment
I. semicircle
J. sector

4 A chord, PQ, of a circle is drawn.
Tangents to the circle are drawn at P and Q.
The tangents intersect at R.

- **a** Imagine PQ moves to the left.
  What happens to R?
- **b** Imagine PQ moves to the right.
  What happens to R?

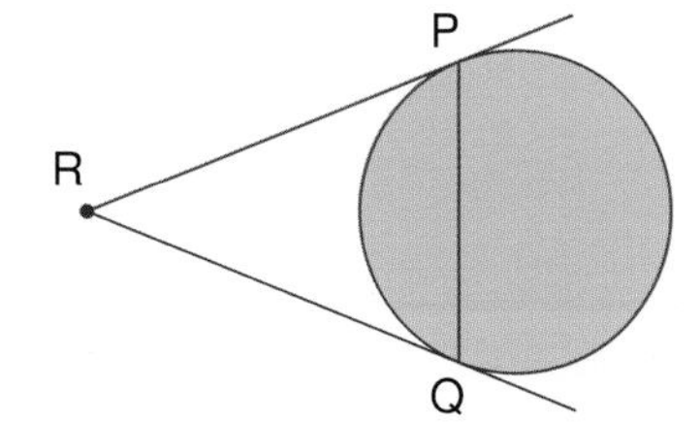

5 Hazel wanted to draw a regular pentagon.
She did this by drawing a circle and dividing the circumference into five equal sections.

- **a** Explain how she might have done this.
- **b** How could she continue to draw the pentagon?
- **c** How could she draw a regular hexagon in a similar way?
- ***d** Explain why any regular polygon can be constructed this way.

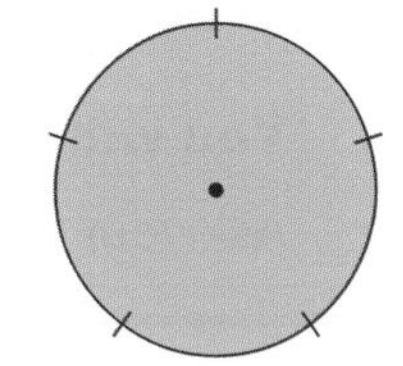

***6** **a** If chords of length equal to the length of the radius are marked on the circumference of a circle, which regular polygon can you draw using this?
Explain why.
b Use this method to draw the regular polygon.

Review 1 Draw a circle.
On your circle label these parts.
a radius **b** circumference **c** arc **d** segment **e** sector **f** chord

Review 2 Draw a regular octagon by dividing the circumference of a circle into equal arcs.
Explain why an octagon can be constructed in this way.

Practical

1

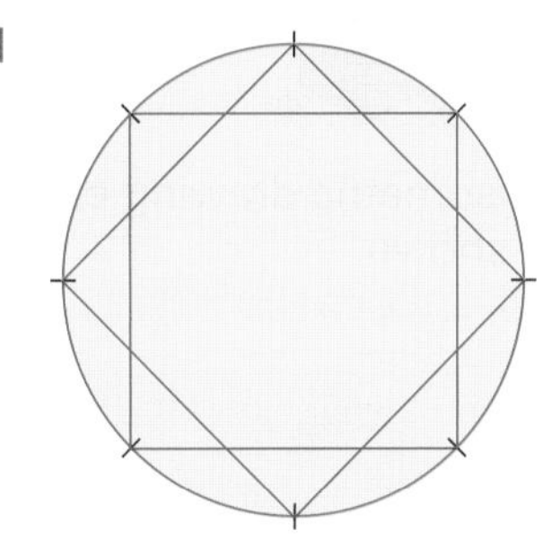

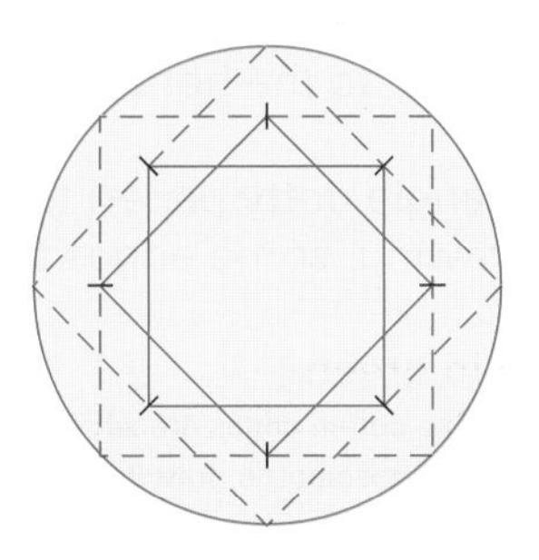

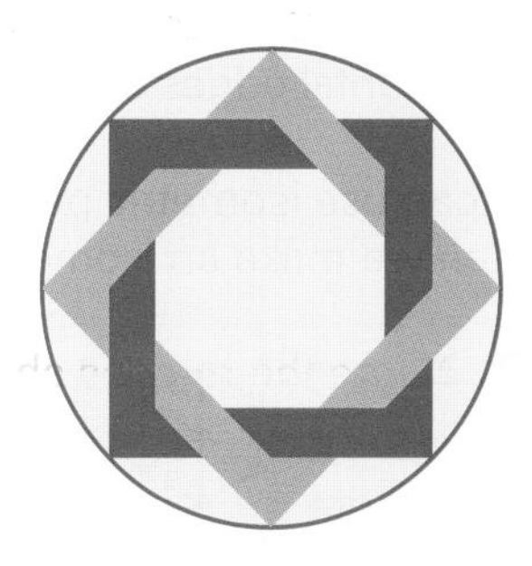

Use these diagrams to help you draw the interlocking square design.

2 Draw an interlocking pentagonal or interlocking hexagonal design.

3-D shapes

Practical

You will need a photograph or poster and a partner.

1 Look at a photograph or poster. Describe to your partner, the 3-D shapes you can visualise from the picture or poster.
Give as much detail as possible. Use words like face, edge, vertex, intersect, point, parallel, perpendicular.

T 2 **You will need** a copy of the nets below.

Rosalee drew this net.
She imagined how it would fold to make a cube.
She then coloured one set of
- edges that would meet at a point in blue
- parallel edges in red
- parallel faces in green
- perpendicular edges in purple
- perpendicular faces in pink.

Use a copy of these nets.
Mark a set of each of the above on each net.

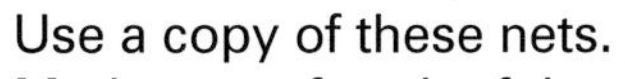

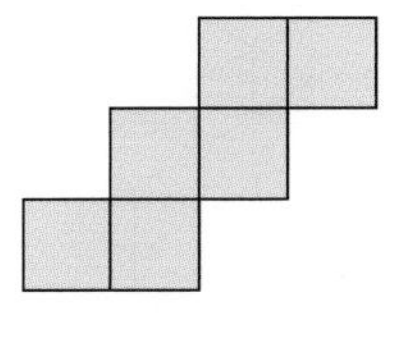

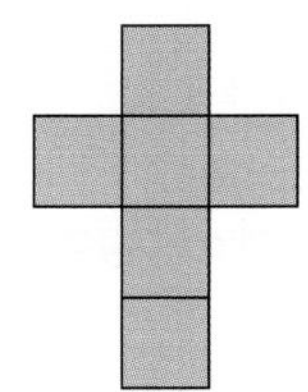

You could cut the net out and fold it to check **after** you have put the colours on.

Draw some more nets for a cube. Do the same for these.

This is an **isometric drawing of 3 cubes**.
On an isometric drawing

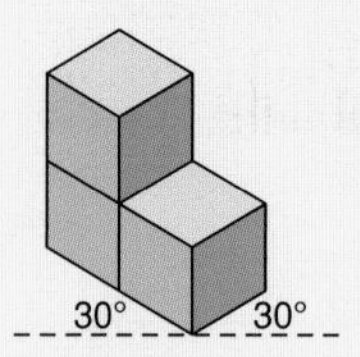

- vertical edges are drawn as vertical lines
- horizontal edges are drawn at 30° to the horizontal.

We can use isometric paper (triangle dotty paper) to make isometric drawing easier.
When we make an isometric drawing, some edges are not shown.

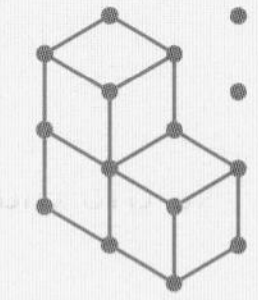

Example

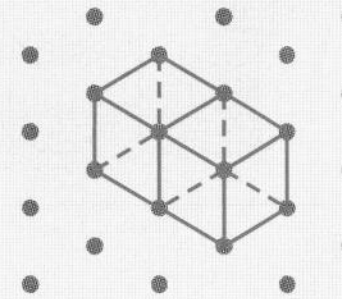

The dotted edges are not shown on an isometric drawing.

Exercise 6

1 **You will need** isometric paper for this question.
Draw these shapes on isometric dot paper.

a

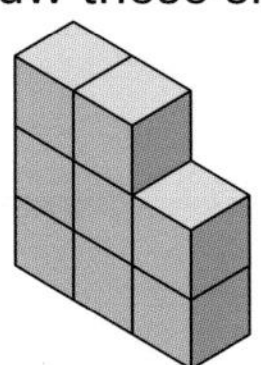

b

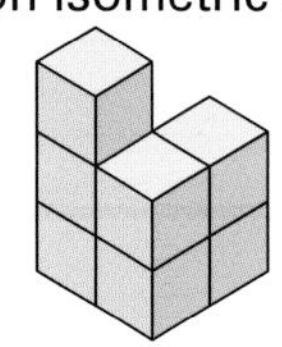

c

2 On the drawings you made in question **1**, use dashed lines to show the position of the hidden edges.

Review You will need isometric paper.
Draw these shapes on your paper.

a

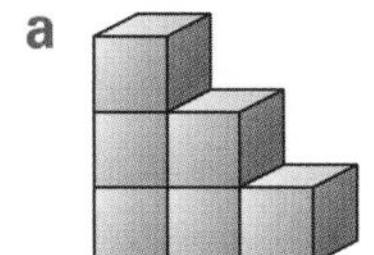

b

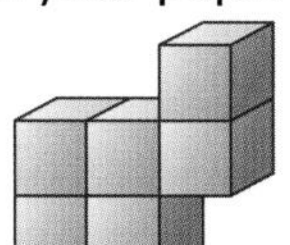

c

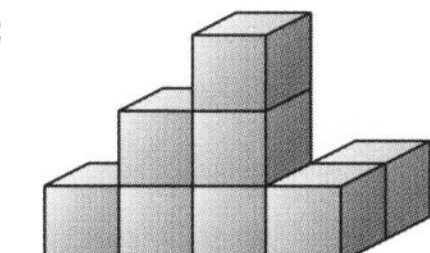

On your drawings, use dashed lines to show the hidden edges.

Investigation

Smallest and Largest Surface Area

Jaffar made a shape from 8 cubes.
He drew it on isometric paper.
He worked out the surface area was 28 units^2.
What other shapes could he make with 8 cubes?
Which one has the greatest surface area? **Investigate**.
Which one has the smallest surface area?
Draw these on isometric paper.

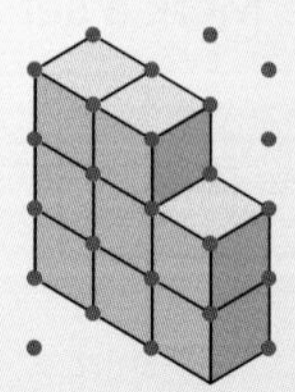

What if he used 9 cubes?
What if he used 10 cubes, 11 cubes, 12 cubes?
What if ...

Plans and elevations

Discussion

Charles flew over the pyramids of Egypt and looked down from above. He drew this diagram to show what he saw from directly above.

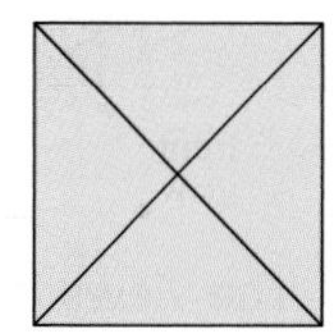

Is the diagram correct? **Discuss.**

These diagrams are views of solids when seen from directly above. What might these solids be? Explain why. **Discuss.**

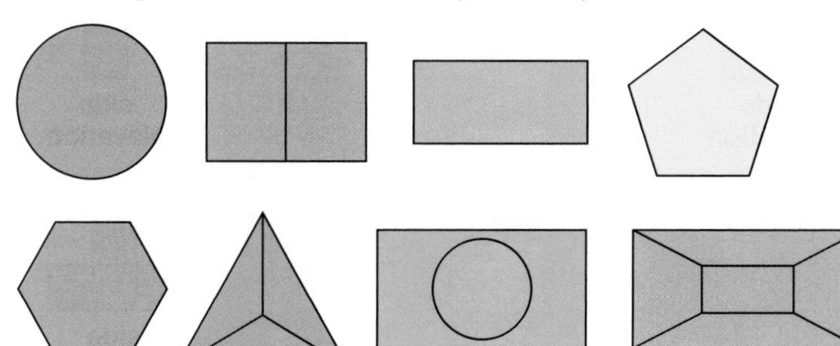

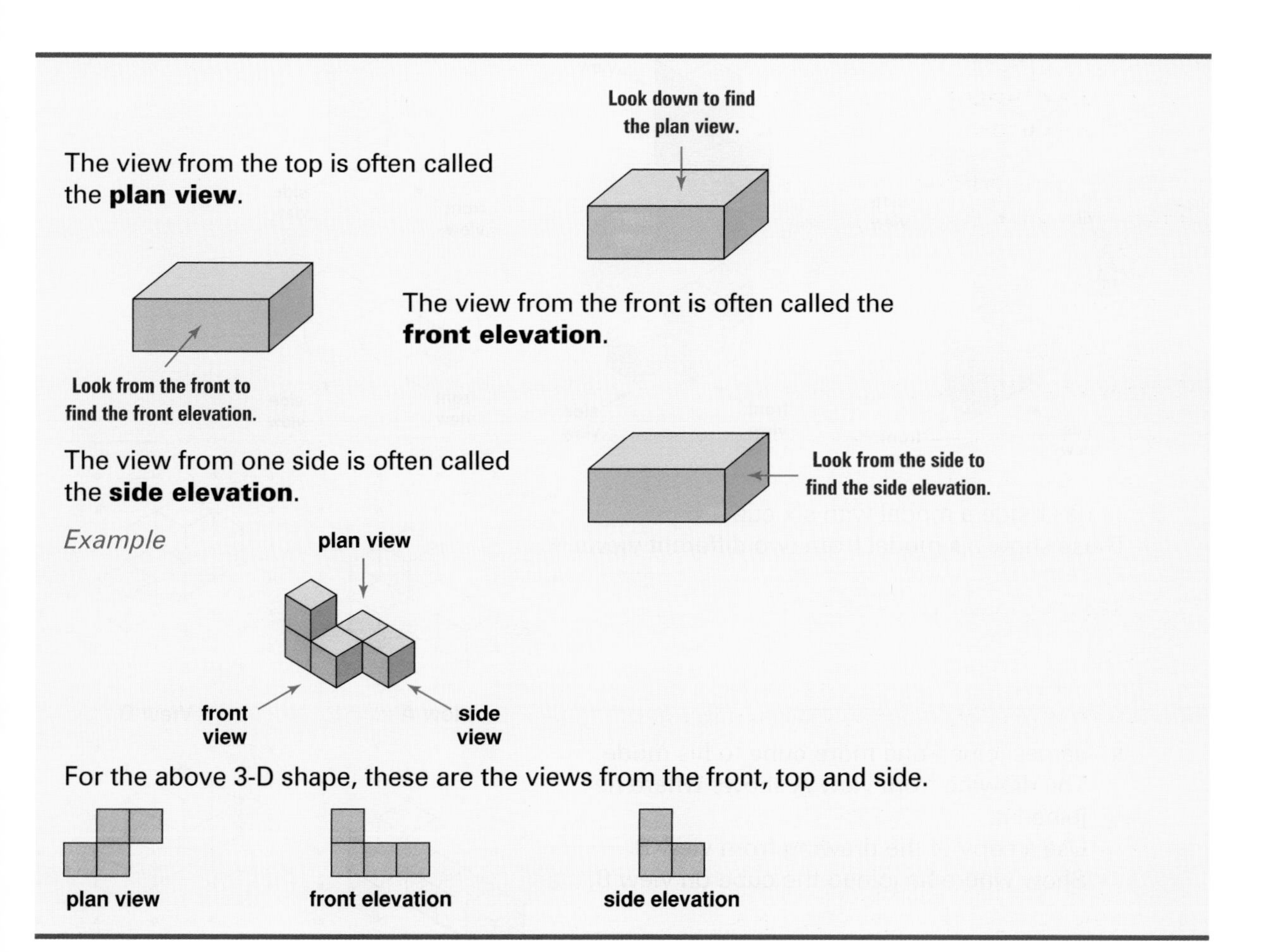

The view from the top is often called the **plan view**.

The view from the front is often called the **front elevation**.

The view from one side is often called the **side elevation**.

Example

For the above 3-D shape, these are the views from the front, top and side.

Exercise 7

1 A B C D

front view side view

The views from the front, top and side of these 3-D shapes are shown below.
Which views belong to which shape?

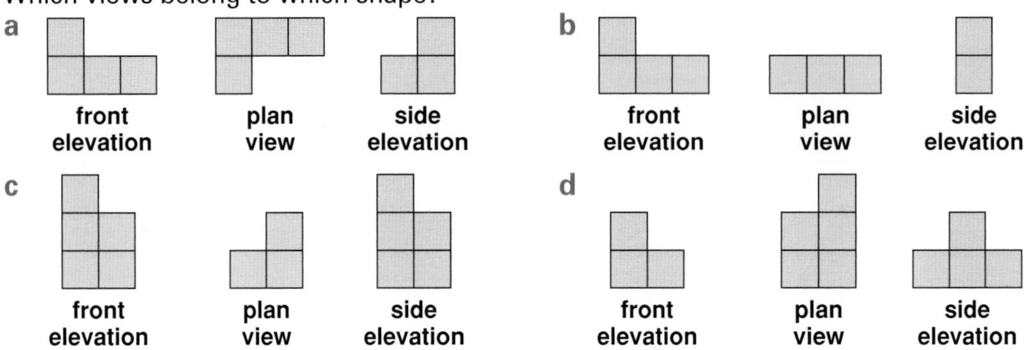

2 Draw the front elevation, side elevation and plan view of these shapes.

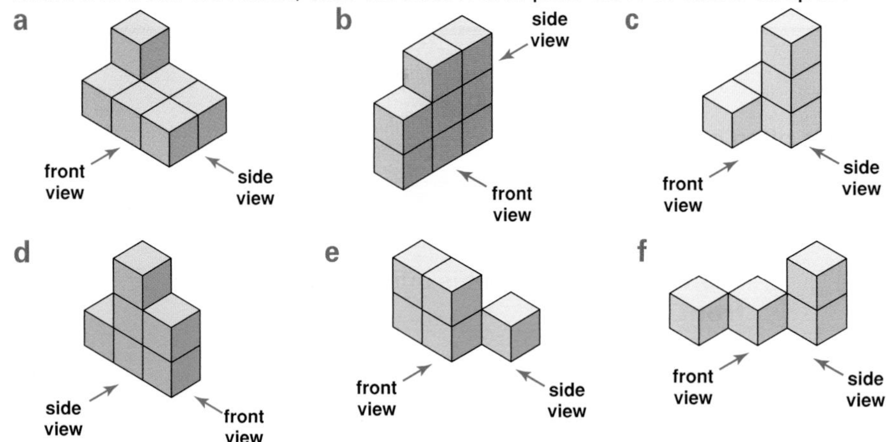

T

3 James made a model with six cubes.
These show his model from two different views.

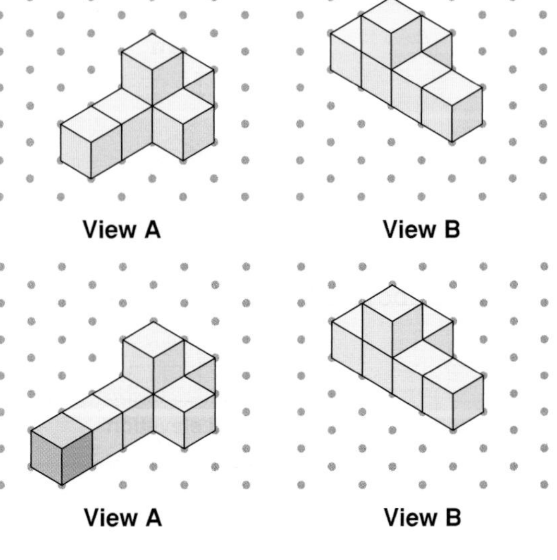

a James joined one more cube to his model.
The drawing from view A shows where he joined it.
Use a copy of the drawing from view B.
Show where he joined the cube on view B.

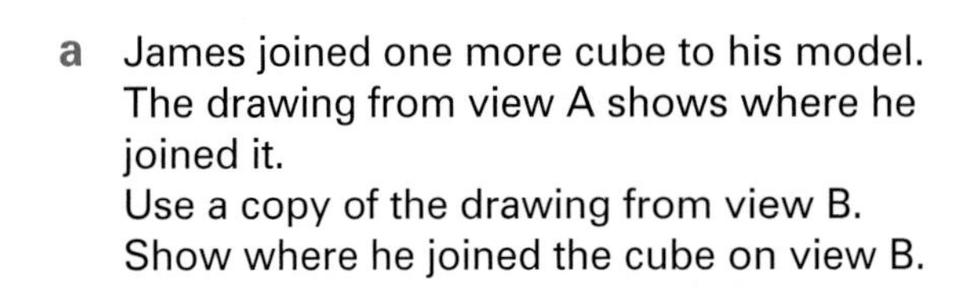

b James moved the cube to a different position.
Use a copy of the drawing from view B.
Complete it.

c This time James uses two more cubes to make a different shape.
Use a copy of view B.
Complete it.

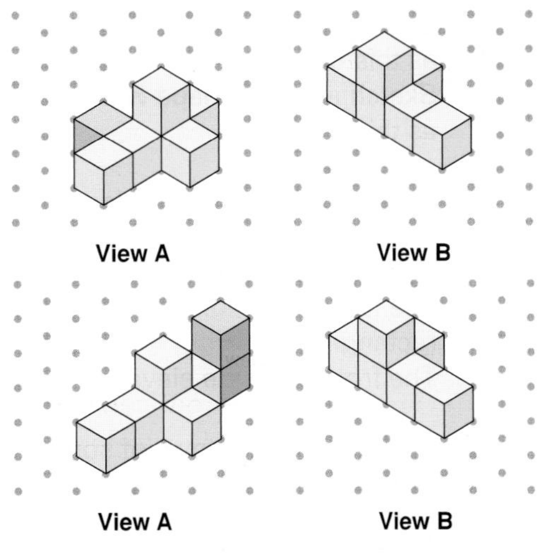

4 The diagram on the left represents a plan view of the solid on the right.
The number in each square tells us how many cubes are on that base.

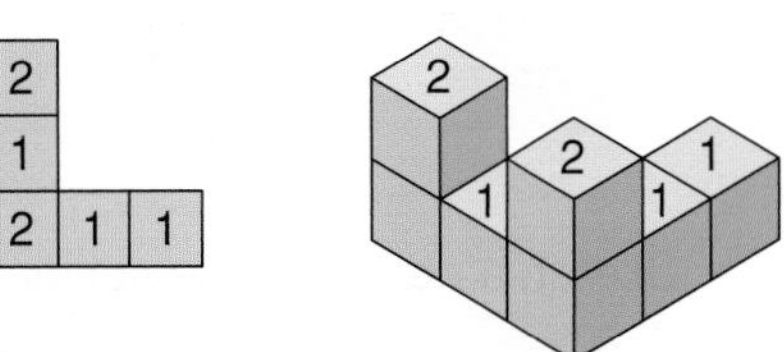

On isometric paper, draw the shapes that these represent.
Draw each from the view given by the red arrow.

a 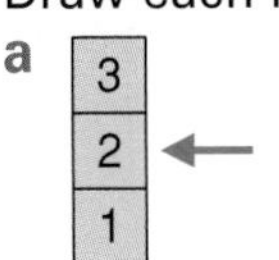

b 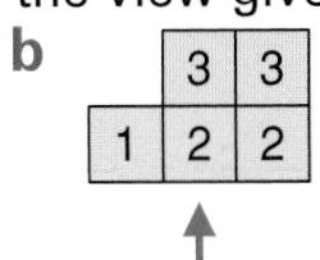

c 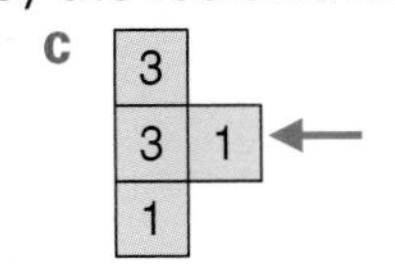

d 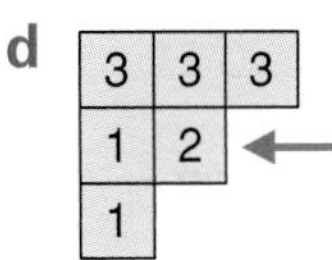

e 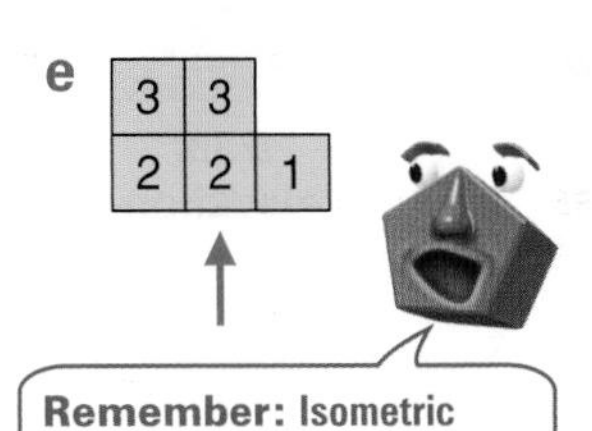

Remember: Isometric paper is triangle dotty paper.

5 **a** Which of these shapes have the same plan view?
b Which have the same front elevation?
c Which have the same side elevation?

A 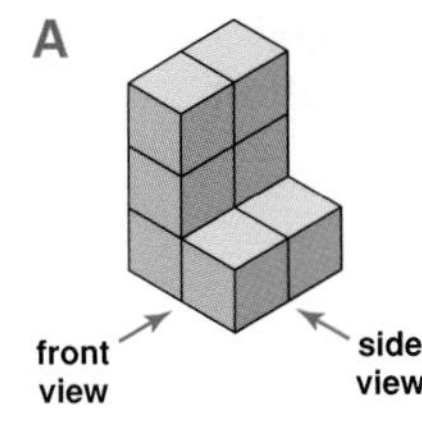

B 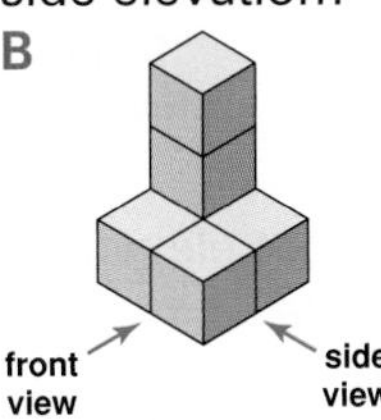

C 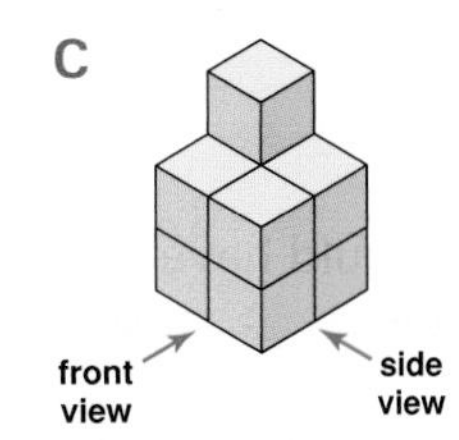

D 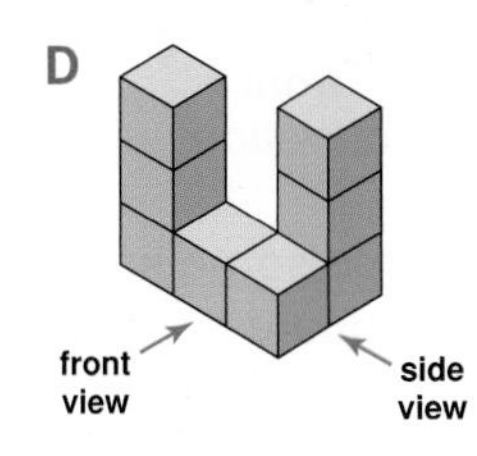

Note: You could build these shapes using multilink or centimetre cubes.

6 **a** Sketch or make other shapes with the same side elevation as this one.
b Sketch or make other shapes with the same front elevation as this one.

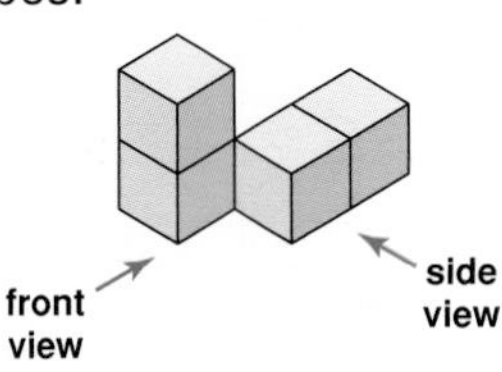

7

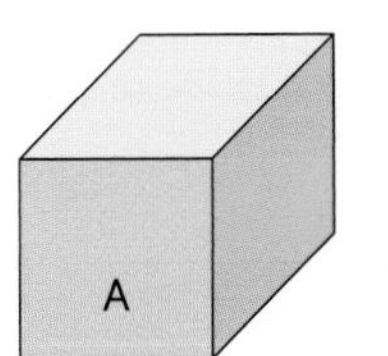

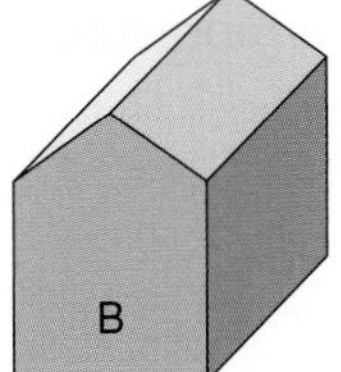

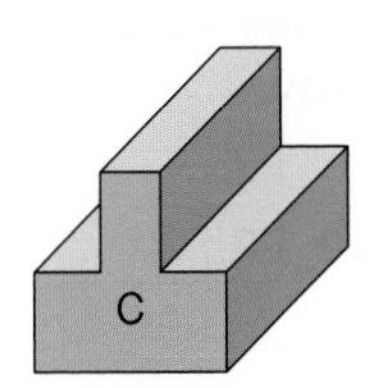

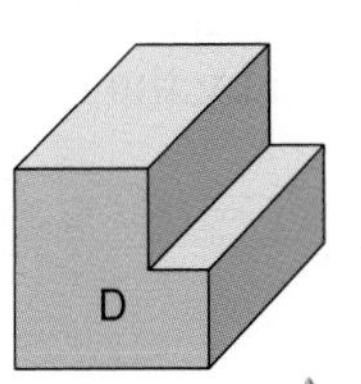

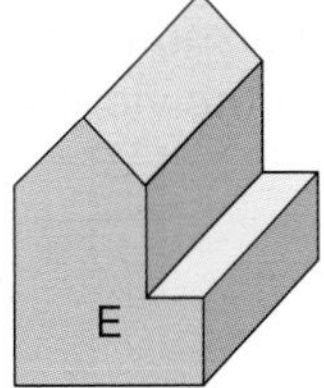

a Which of these shapes have the same plan view?
b Which have the same front elevation?
c Which have the same side elevation?

8 **a** Sketch another shape with the same side elevation as this one.
b Sketch another shape with the same front elevation as this one.

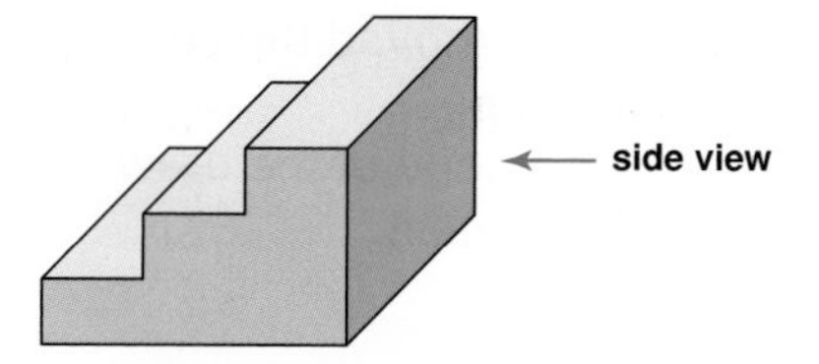

9 Sketch the solid that these describe.
a The front and side elevations are both triangles and the plan view is a square.
b The front and side elevations are both rectangles and the plan view is a circle.
c The front elevation is a rectangle, the side elevation is a triangle, and the plan view is a rectangle.
d The front and side elevations and the plan view are all circles.

*__10__ These show the shadows of some solids.
Describe a possible solid for each shadow.
Is there more than one answer?

a  **b**  **c** 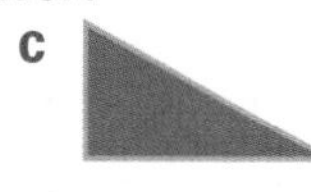 **d** 

*__11__ Sketch a net for these solids.

a

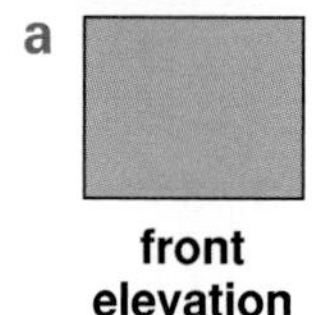
front elevation

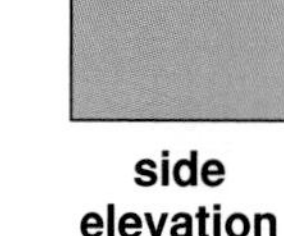
side elevation

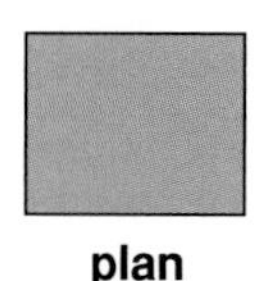
plan view

b

front elevation

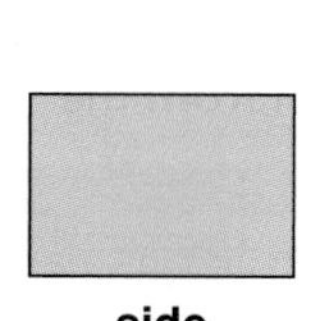
side elevation

plan view

c

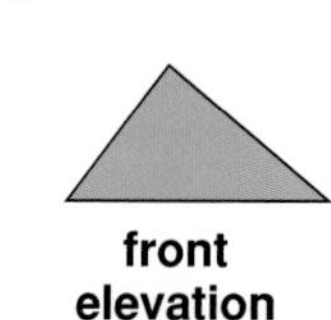
front elevation

side elevation

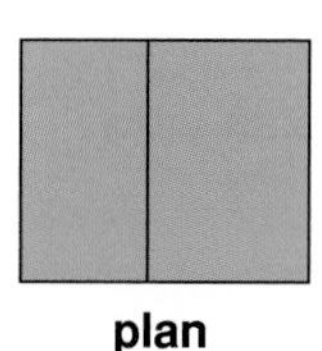
plan view

d

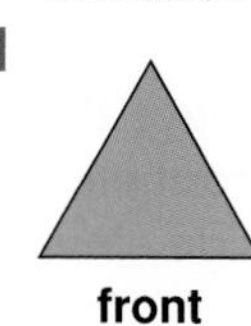
front elevation

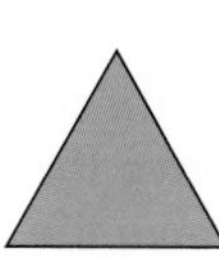
side elevation

plan view

*__12__ Name some solids that could have these as a front elevation.
a square **b** rectangle **c** isosceles triangle

Review 1 Draw the front elevation, side elevation and plan view of these shapes.

a

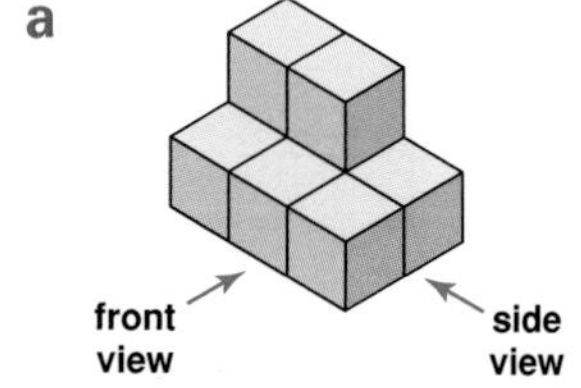

b

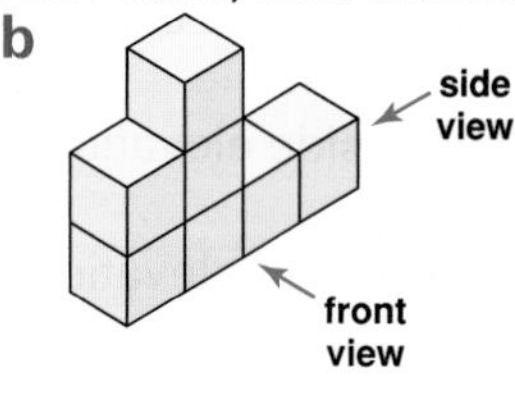

c

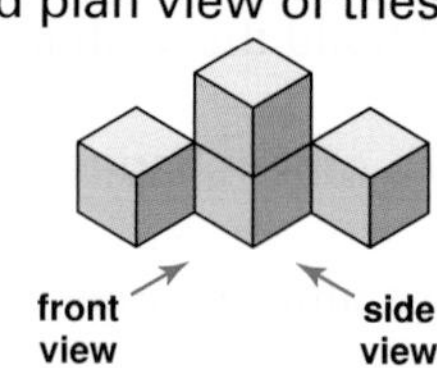

Review 2 On isometric paper, draw the shapes which these represent.
Draw each from the view shown by the red arrow.

a

2	3	
2	1	1

b

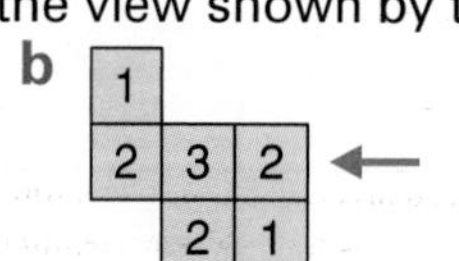

c

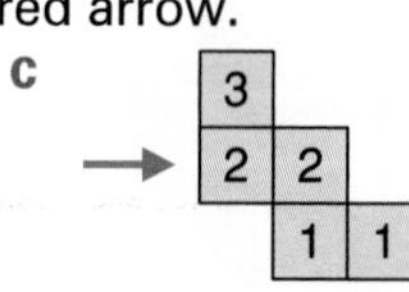

Review 3 Sketch the solid with side elevation a triangle, and front elevation and plan view both rectangles.

Review 4 These show the shadows of two solids. What could each be? Is there more than one answer?

a 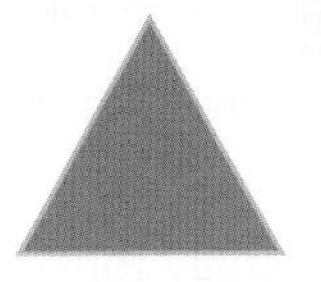 b

Review 5 Sketch a net for this solid.

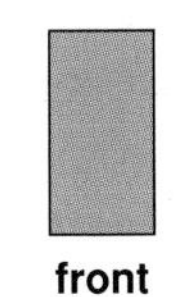

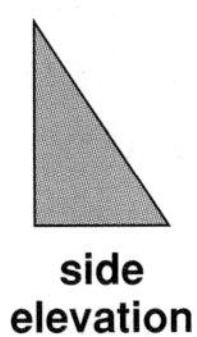

Practical

You will need multilink or centimetre cubes and some isometric paper.

1 Build some 3-D shapes using cubes. Draw the front elevation, side elevation and plan view of each.

2 Which of these is it possible to build? Is there more than one way to build some of them?

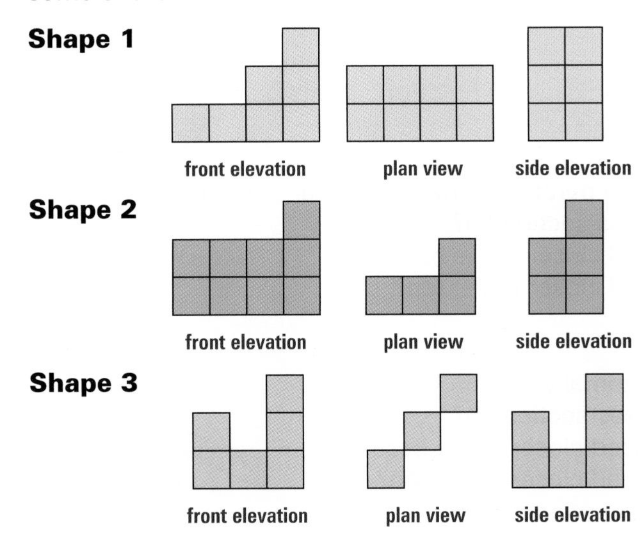

3 **Work in pairs.**
Use cubes to build a 3-D shape.
Draw the front elevation, side elevation and plan view.
Ask your partner to build the shape from your drawings.

4 Build and then sketch the shapes with plans and elevations as shown below.

a

plan view front elevation side elevation

b

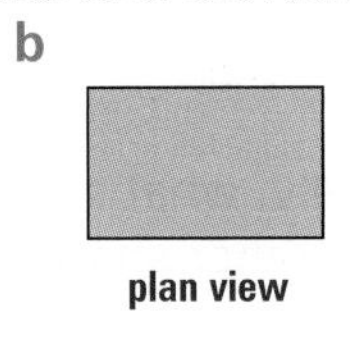

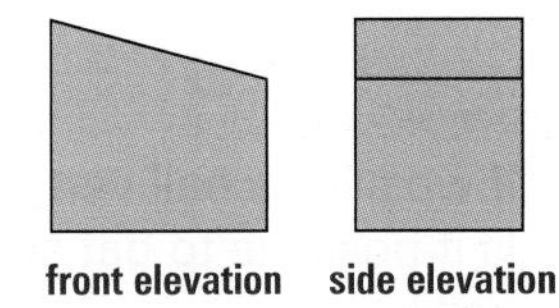

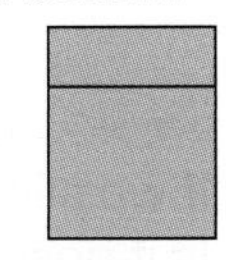

c

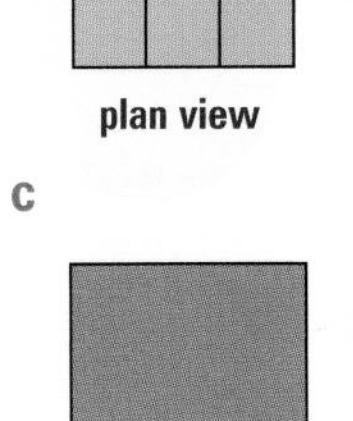

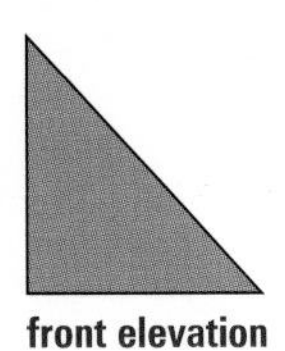

d

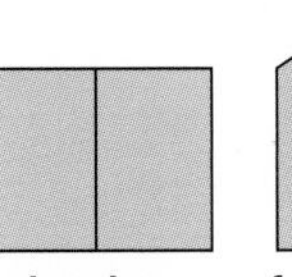

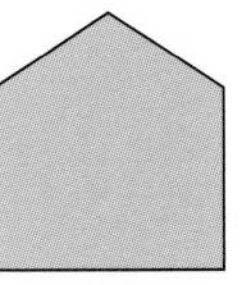

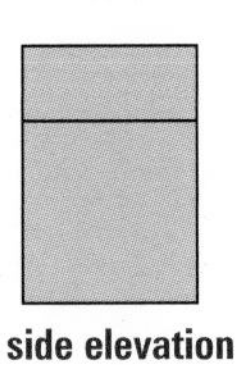

5 Choose a building in the school grounds or a piece of equipment in the gymnasium or workshop.
Draw the plan view, front elevation and side elevation.
See if another student can identify the building or object from your drawing.

6 Use cardboard to make a model of one of the school buildings.
Draw the plan and front and side elevations on a sheet of paper. Display these along with the model.

Cross-sections

When we slice a shape, the face that is made is called a **cross-section**.

Example If we slice a cuboid as shown the cross-section is a rectangle.

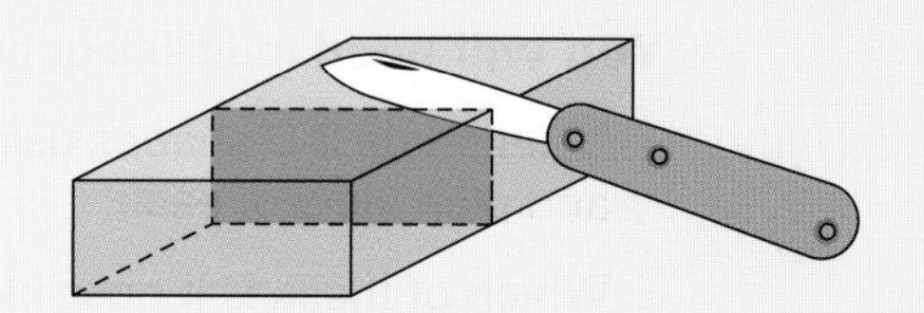

Exercise 8

1 Tonya sliced a square-based pyramid horizontally near the top.
 a What shape will the cross-section be?
 b What if Tonya had sliced closer to the base?
 c What if Tonya had sliced closer to the top?

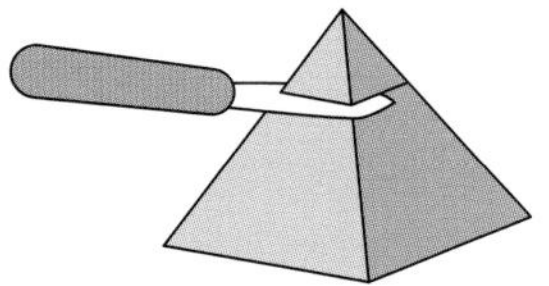

2 Brittany sliced a cone horizontally.
 a What shape will the cross-section be?
 b What if Brittany had sliced closer to the base?
 c What if Brittany had sliced closer to the top?

3 **a** If we slice a cube vertically as shown, what shape is the cross-section?
 b Is it possible to slice a cube to get
 i a rectangle **ii** a triangle **iii** a pentagon **iv** a hexagon?
 If so, explain how.

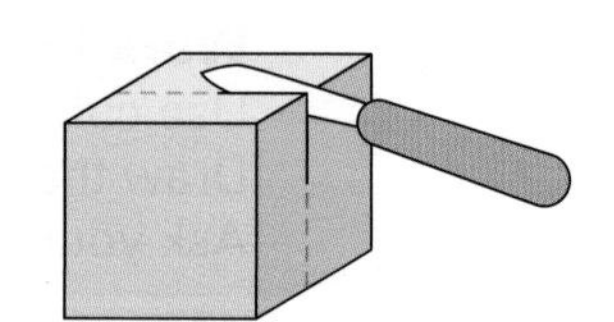

Review
a If we slice a ball vertically, what shape will the cross-section be?
b Is it possible to get any other shaped cross-section by slicing a ball?

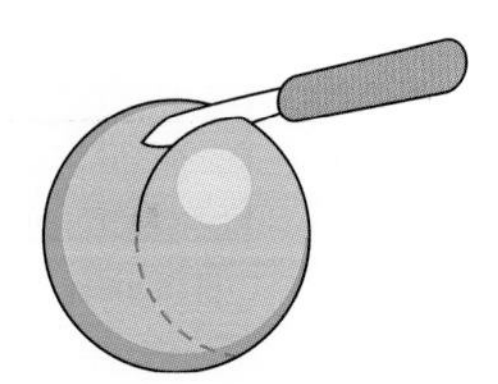

Construction

See page 25 for how to construct
- the mid-point and perpendicular bisector of a line segment
- the bisector of an angle.

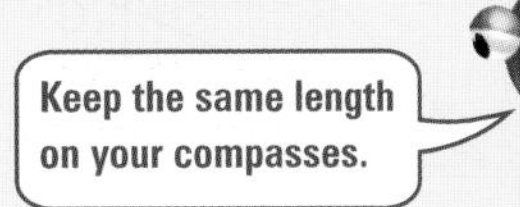

This shows **the construction of the perpendicular from A to the line segment BC**.

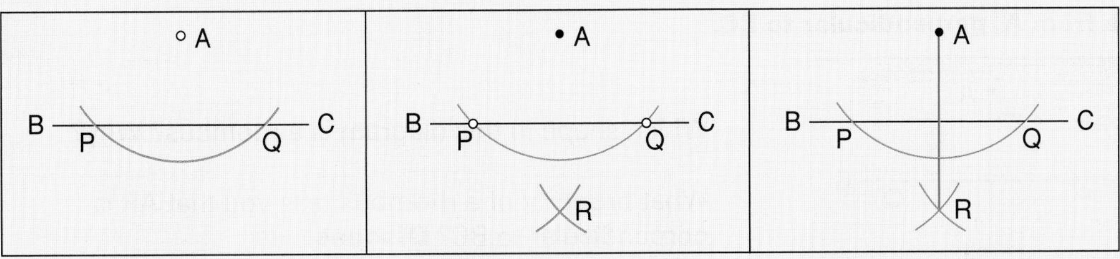

Open out the compasses. With the point on A, draw an arc to cross BC at P and Q.

With the point firstly on P, then on Q, draw two arcs to meet at R.

Join A and R. AR is the perpendicular from A to the line segment BC.

This shows the construction of the **perpendicular from a point P on a line segment BC**.

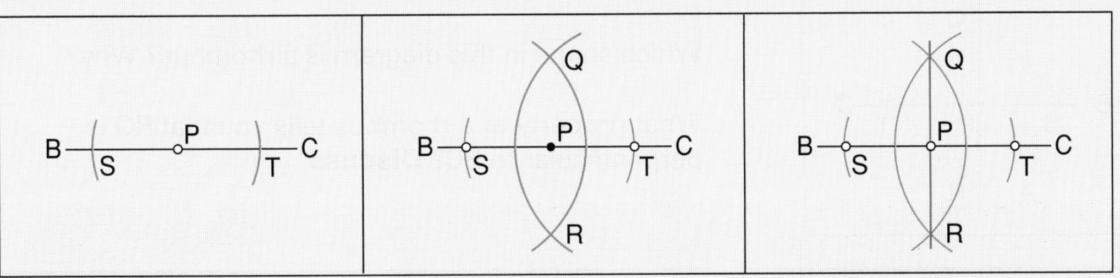

Open out the compasses to less than half the length of BC. With the point on P, draw arcs, one on each side of P. Label where they cross BC as S and T.

Open out the compasses a little more. With compass point first on S and then on T, draw arcs so they cut at Q and R.

Draw the line through Q and R. QR is the perpendicular from P on the line segment BC.

Discussion

The diagrams below are the final diagrams from each of four constructions.
Some dotted lines have been added.

Perpendicular bisector of the line BC

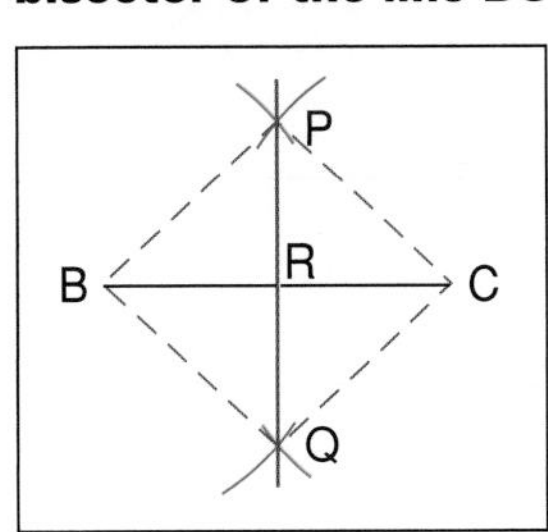

Which shape in this diagram is a rhombus? Why?

What property of a rhombus tells you that R is the mid-point of BC? **Discuss**.

What property of a rhombus tells you that PQ is the perpendicular bisector of BC? **Discuss**.

Bisector of the angle P

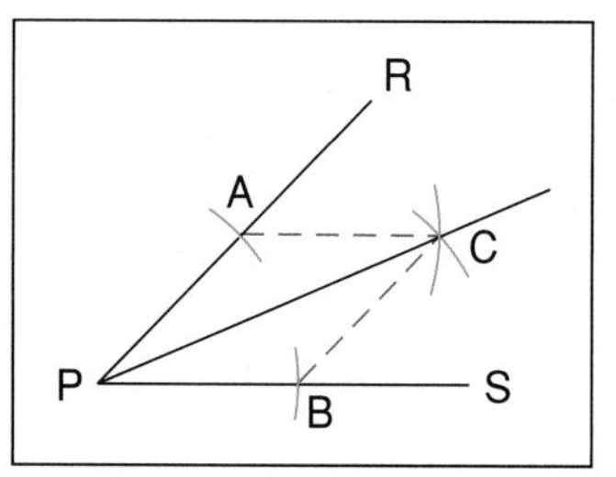

Which shape in this diagram is a rhombus? Why?

What property of a rhombus tells you that angle RPC = angle SPC? **Discuss.**

Line from A, perpendicular to BC.

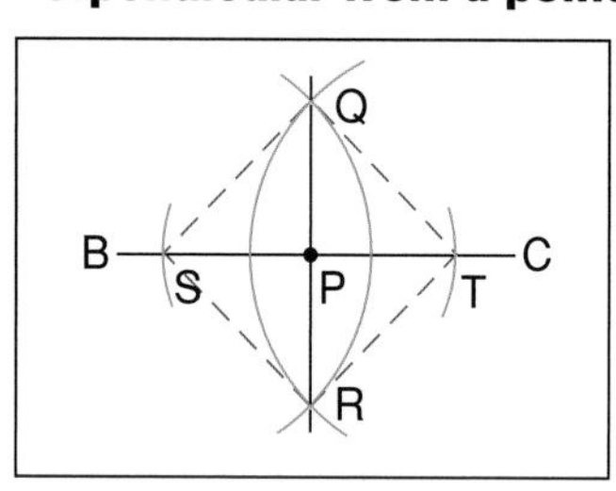

Which shape in this diagram is a rhombus? Why?

What property of a rhombus tells you that AR is perpendicular to BC? **Discuss.**

Perpendicular from a point P on a line segment BC

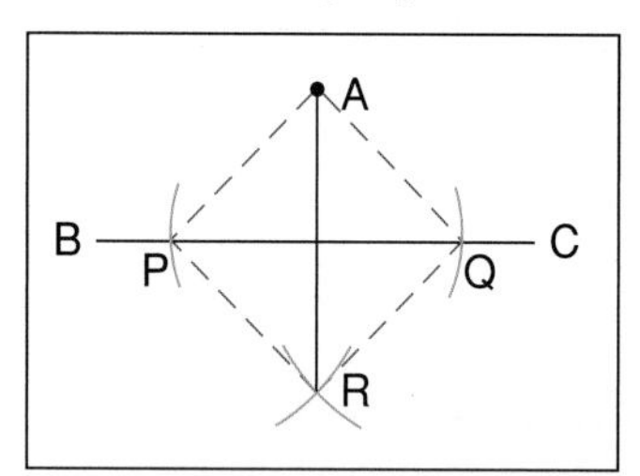

Which shape in this diagram is a rhombus? Why?

What property of a rhombus tells you that RQ is perpendicular to BC? **Discuss.**

Exercise 9

Draw and measure carefully in this exercise.

1 Use a copy of this diagram.
- **a** Construct the line through P which is at right angles to the line segment BC.
- **b** Label as X, the point where this line meets AB.
- **c** Measure the length of BX.

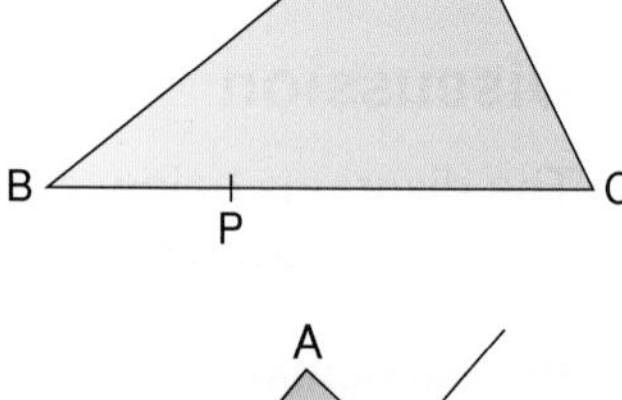

2 Draw a large triangle, ABC, in your book.
Measure to find three points one third of the way from A to C, C to B and B to A.
- **a** Construct a line from each of these points, perpendicular to the side.
- **b** What shape do the three lines make?

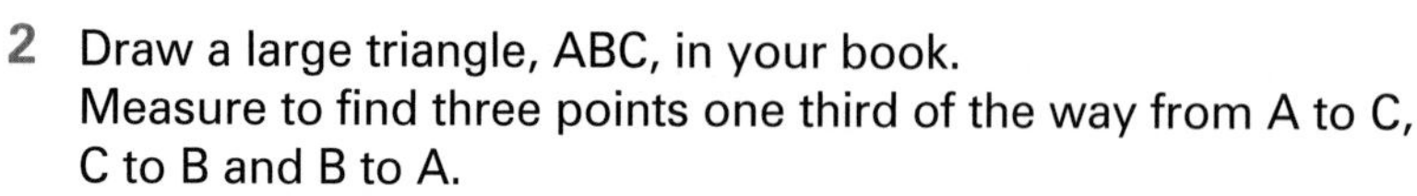

3 Draw a large triangle PQR. Bisect each of the angles.
What do you notice?

4 Draw a large triangle STR. Construct the perpendicular bisector of each side.
What do you notice?

5 Draw a large triangle, PQR, in the middle of a sheet of squared paper.

a Construct the line perpendicular to QR that passes through P.
b Construct the line perpendicular to PR that passes through Q.
c Construct the line perpendicular to PQ that passes through R.

What do you notice?

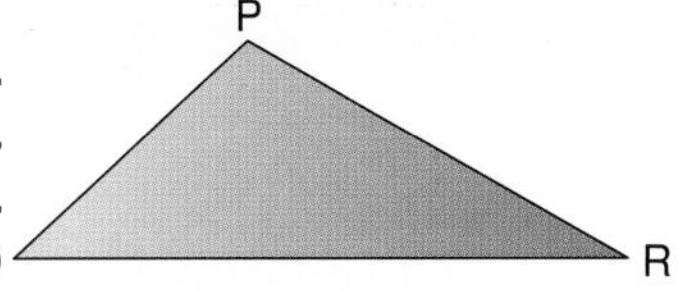

6 Use a ruler and compasses to construct a regular hexagon. Draw two diagonals of the hexagon to form a right-angled triangle.
Explain why it is a right-angled triangle with angles of 30° and 60°.

***7** Use a ruler and compasses to construct a rhombus and a square with the same area.

Review 1 Three cycle tracks are shown, AD, DC and AC.
The council wants to put a new cycle track perpendicular to AC, from B to DC.
Trace the diagram.
Use your compasses to construct the line of the new cycle track.

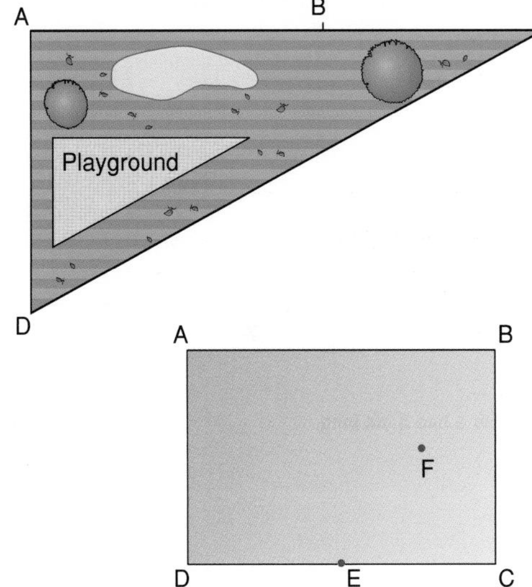

Review 2 Use a copy of this diagram.

a Construct the line from E, perpendicular to DC.
b Construct the line from F, perpendicular to BC.

Puzzle

The 10 pieces of a circle puzzle may be constructed as shown.

1. On thin card, draw a circle, centre O and radius 6 cm.
2. Draw in the diameter AB.
3. Construct the diameter CD so that CD is perpendicular to AB.
4. Bisect OC. Label the mid-point P.
   Bisect OD. Label the mid-point Q.
5. Through Q, construct the line RS so that RS is parallel to AB.
6. Join AP, AQ, BP, BQ.
7. Carefully cut out the 10 pieces.

Rearrange the 10 puzzle pieces to make interesting shapes such as those shown below.

Constructing triangles

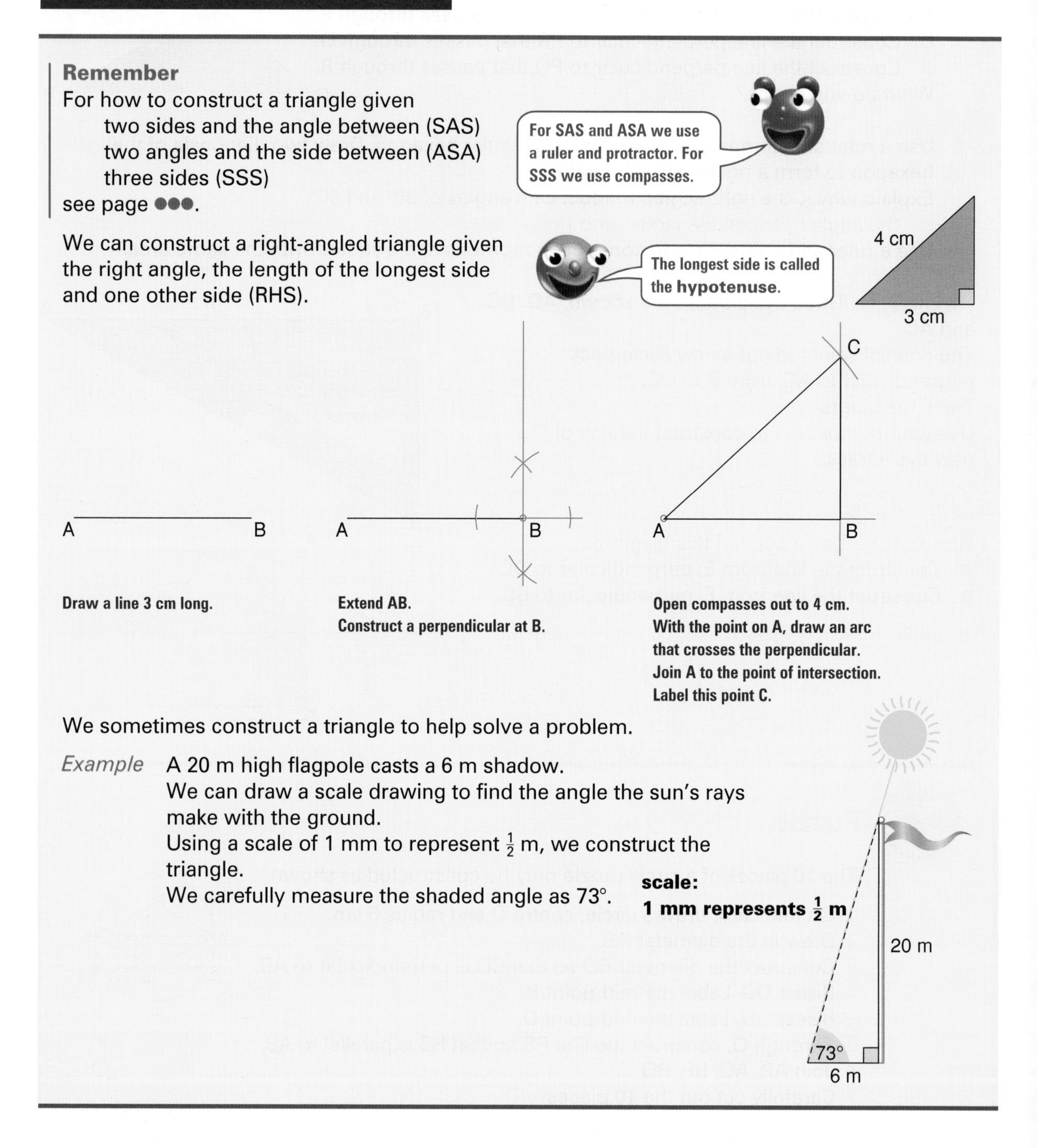

Remember
For how to construct a triangle given
two sides and the angle between (SAS)
two angles and the side between (ASA)
three sides (SSS)
see page ●●●.

For SAS and ASA we use a ruler and protractor. For SSS we use compasses.

We can construct a right-angled triangle given the right angle, the length of the longest side and one other side (RHS).

The longest side is called the **hypotenuse**.

Draw a line 3 cm long.

Extend AB.
Construct a perpendicular at B.

Open compasses out to 4 cm.
With the point on A, draw an arc that crosses the perpendicular.
Join A to the point of intersection.
Label this point C.

We sometimes construct a triangle to help solve a problem.

Example A 20 m high flagpole casts a 6 m shadow.
We can draw a scale drawing to find the angle the sun's rays make with the ground.
Using a scale of 1 mm to represent $\frac{1}{2}$ m, we construct the triangle.
We carefully measure the shaded angle as 73°.

scale:
1 mm represents $\frac{1}{2}$ m

Exercise 10

1 Is it possible to construct triangle PQR such that

a P = 60°, Q = 30°, R = 90°
b P = 60°, Q = 60°, R = 60°
c PQ = 5 cm, QR = 3 cm, PR = 2 cm
d PQ = 4 cm, QR = 4 cm, PR = 3 cm
e PQ = 7 cm, QR = 3 cm, PR = 3 cm
f QR = 8 cm, PR = 6 cm, R = 50°
****g** P = 30°, Q = 45°, PR = 6 cm
****h** QR = 7 cm, PR = 4·95 cm, Q = 45°?

2 In a gale, a tree falls against a house. The top of the tree is 5 m up the wall of the house.
The foot of the tree is 7 m from the wall of the house.
Make a scale drawing. Use a ruler and protractor to find
a the height of the tree
b the angle between the tree, after it has fallen, and the ground.

3 A 12-foot ladder is leaned against a wall so that its base is 3·5 feet from the wall.
Construct a triangle to scale. Use a ruler and protractor to find
a how far up the wall the ladder reaches
b the angle between the ladder and the ground.

*__4__ How many different triangles could be drawn using any three of the following lengths for the sides?
8 cm   7 cm   6 cm   5 cm   4 cm   3 cm

Review 1 Is it possible to construct triangle EFG such that
a E = 40°, F = 60°, EF = 5 cm   **b** FG = 7 cm, EG = 4 cm, F = 45°?

Review 2 At an adventure playground a 12 m slide ends 5 m from the bottom of a rock wall.
Make a scale drawing. Use a ruler, and protractor to find
a the height of the wall
b the angle the slide makes with the ground.

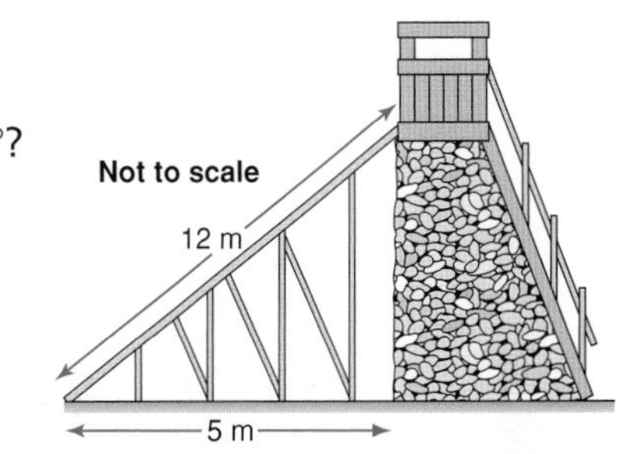

Practical

You will need a dynamic geometry software package.

Ask your teacher for the Constructions on a Line and Constructing a Triangle with Three Known Lengths ICT worksheet.

Loci

Discussion

- The arrow on the end of the minute hand of a clock moves so that it is always the same distance from a fixed point.
What path does the arrow make?

- The path of a bouncing ball might look like this.

What might the paths of these look like? **Discuss.**
Sketch the paths.
the head of a boy jumping from a tree
a man's head on an aeroplane taking off
a child's head when the child is swinging on a swing
a woman's hand as she chops a piece of wood
a point at the middle of a car wheel as the car goes along the road
a girl's head on a roller coaster

A **locus** is a set of points that satisfy a rule or set of rules.

Example The locus of a horse on a rein, moving so that it is **always the same distance from a person** standing in the middle of a field, is **a circle**.

Practical

You will need a group of about 10 to 20 people and two objects such as two rubbish tins (A and B).

1 Put one of the objects in the middle of a space.
Ask everyone to stand so that they are all the same distance from the object.
What is the locus of the heads?

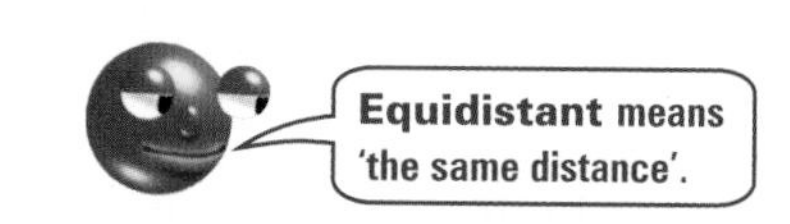

2 Put two objects, A and B, about 3 m apart.
Ask everyone to stand so that each person is **equidistant** from A and B.
Which of these gives the locus of the heads?
A. the line joining A and B
B. a circle.
C. the perpendicular bisector of the line joining A and B.

Joel is standing so he is an equal distance from A and B.

3 Ask 12 pupils to stand in two rows, A and B, that meet at one end to form an angle as shown.
Ask everyone else to stand so that each person is equidistant from row A and row B. What is the locus of the heads of everyone not standing in rows A or B?

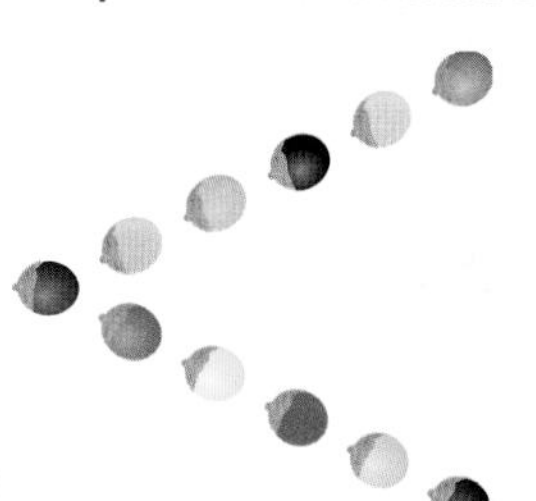

4 Ask 6 pupils to stand in a row.
Ask everyone else to stand so that they are always the same distance from the people in the row.
What is the locus of the heads of these people?
(Not the six standing in the row.)

Note: You could use counters instead of people.

*5

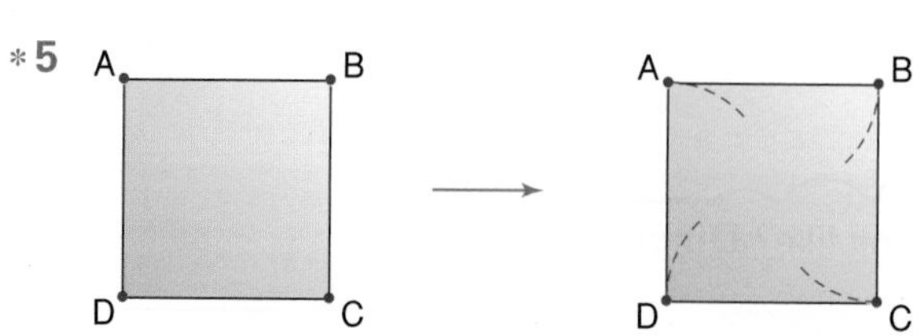

Four students stand at the corners A, B, C and D of a square. At the same instant, each student begins walking clockwise, at the same speed, towards the next student; the first part of each path is shown. Find the paths followed by each of the students. Try this yourselves.

Exercise 11

1 Jenni was going skiing. She stuck her ski pole in the ground. Before starting skiing she walked so that she was always the same distance from her ski pole.
Describe Jenni's locus in words.

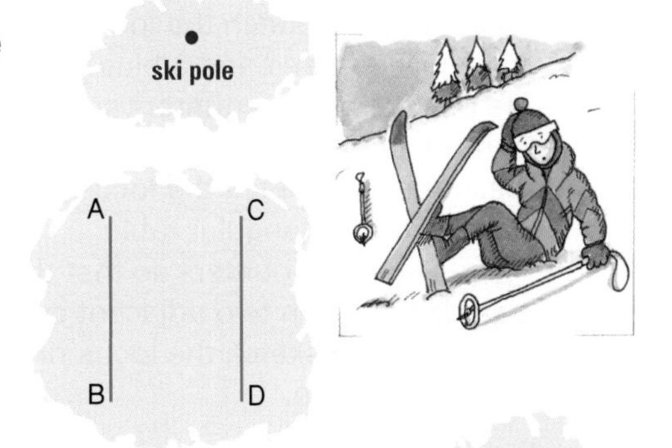

2 Jenni skis downhill between two rows of trees (shown by AB and CD). She skis so that she is always the same distance from the trees on either side of her.
Sketch Jenni's locus.
Describe Jenni's locus in words.

3 On the downhill run there are two marker posts (shown by R and Q). Jenni skis so that she is never closer to one than the other.
Sketch Jenni's locus.
Describe Jenni's locus in words.

4 **a** As Jenni skis past the learner's tow-rope (shown by ST) she stays exactly 5m from it. Assume Jenni can only ski on one side of the rope.
Sketch and describe this locus.
b What if Jenni can ski on either side of the rope?

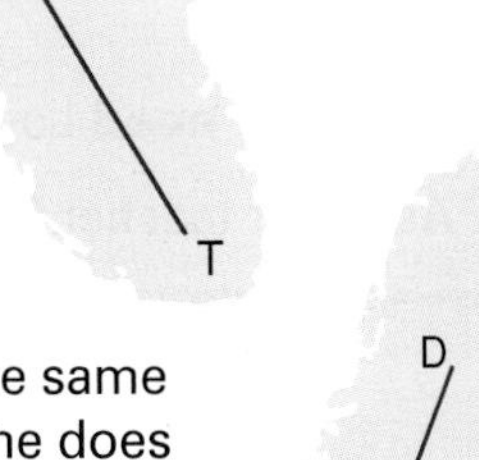

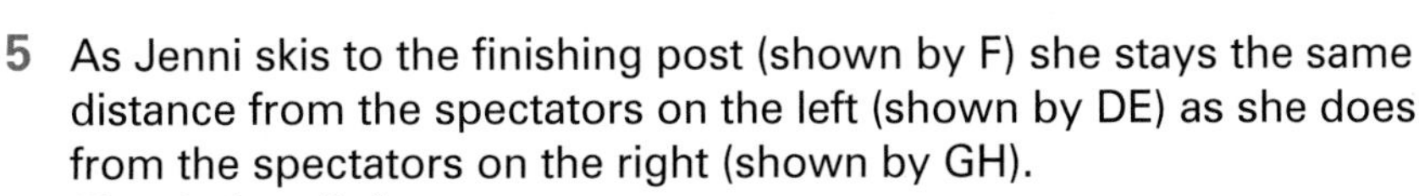
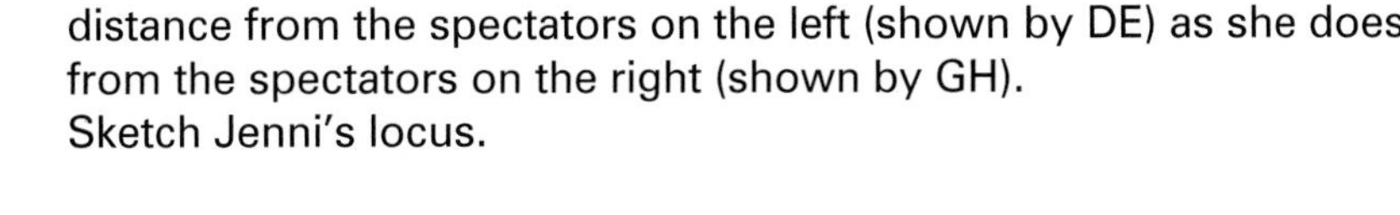

5 As Jenni skis to the finishing post (shown by F) she stays the same distance from the spectators on the left (shown by DE) as she does from the spectators on the right (shown by GH).
Sketch Jenni's locus.

6 At the skifield hire shed, a box of ski boots is rolled along the floor.
What path would the point P move along?

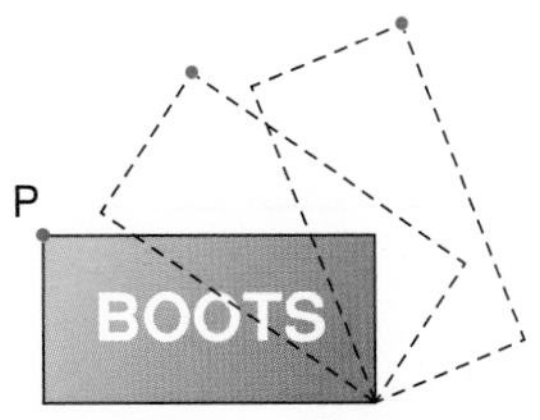

***7** Jenni cycled to catch the bus to go skiing. She has three reflectors on her wheel, at A, B and C. As her wheel travels along the road, what path would each of the reflectors travel?
Sketch the path for each.

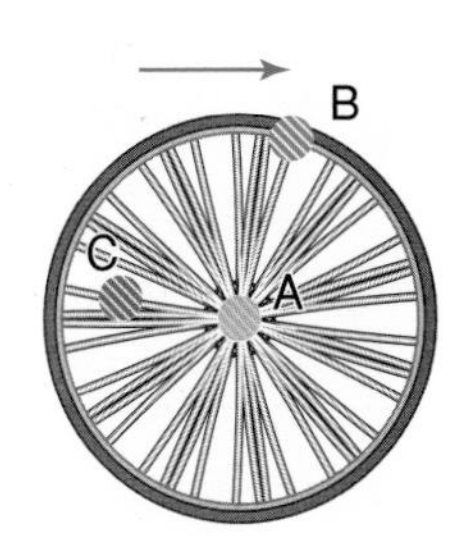

***8** **a** The locus of all points that are the same distance from (3, 3) as from ($^-1$, 3) is a straight line.
Make a copy of this grid.
Draw this straight line.
b The locus of all points that are the same distance from the x-axis as they are from the y-axis is **two** straight lines.
Draw both straight lines on a set of axes.

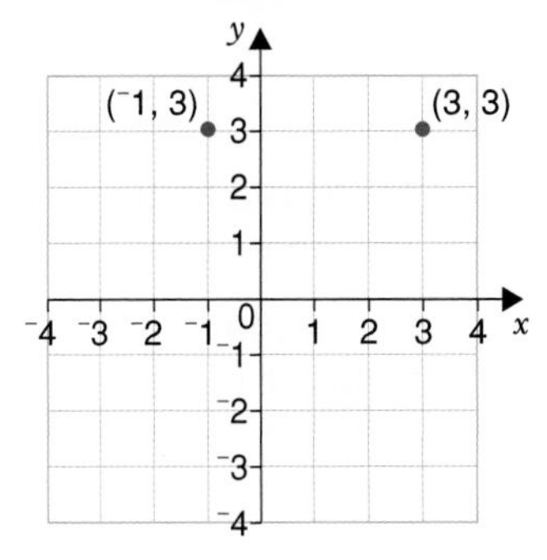

Review 1 Jake had some blue, red and yellow counters.

a Jake placed a blue counter in the middle of a table.
He placed yellow counters so that their centres were all the same distance from the centre of the blue counter.
Describe and sketch the locus on which the centres of the yellow counters lie.

b Jake placed a red and a blue counter on the table, 30 cm apart.
He placed yellow counters so that their centres were always equal distances from the centres of the red and blue counters.
Describe and sketch the locus on which the centre of the yellow counters lie.

c Jake had a rectangular table.
He placed red counters so that their centres were equidistant from two adjacent edges of the table.
Describe and sketch the locus on which the centre of the red counters lie.

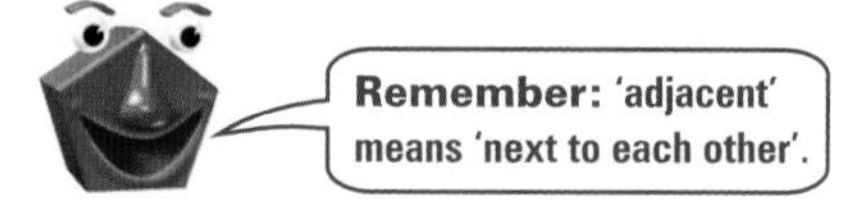

Review 2 Plot the points (2, 1) and (2, $^{-}3$) on a grid.
Draw the locus of all points that are the same distance from (2, 1) and (2, $^{-}3$).

T

Practical

You will need Logo.

Ask your teacher for the Investigating Loci ICT worksheet.

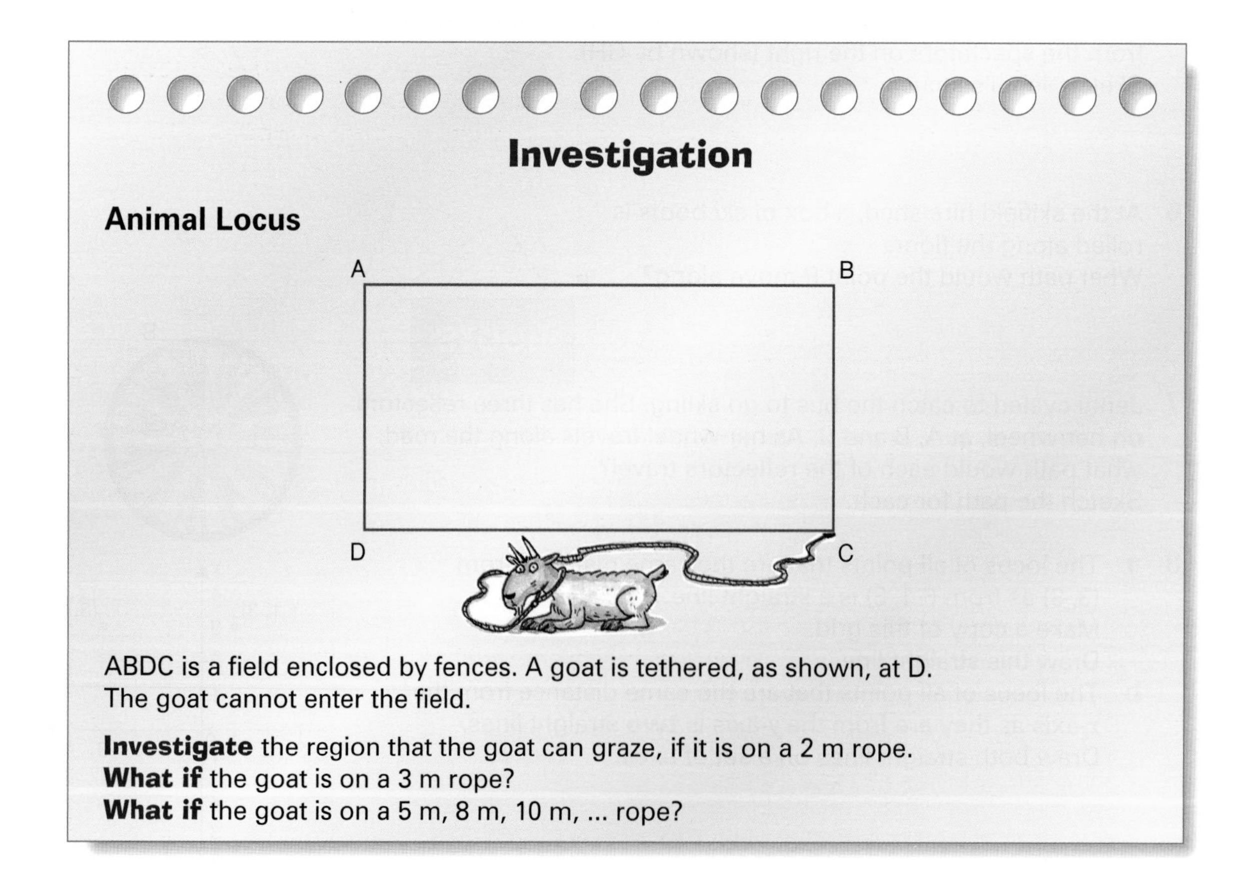

Investigation

Animal Locus

ABDC is a field enclosed by fences. A goat is tethered, as shown, at D.
The goat cannot enter the field.

Investigate the region that the goat can graze, if it is on a 2 m rope.
What if the goat is on a 3 m rope?
What if the goat is on a 5 m, 8 m, 10 m, ... rope?

Game

Spy catcher — a game for two players.

You will need a larger copy of this board and a different coloured small counter for each player.

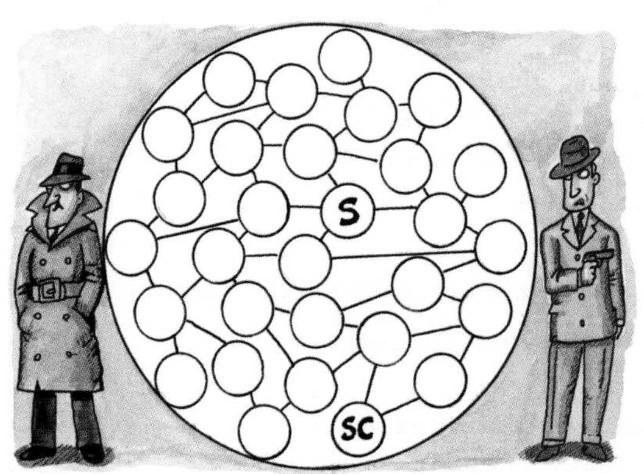

The circles represent towns. The lines connecting the circles represent roads.

To play
- Choose one player to be the spy (S), the other is the spy catcher (SC).
- Beginning with the spy catcher, take turns to move along a road into the next town.
  Only one road may be moved along each turn.
- The spy catcher tries to move into a town where the spy is before the spy can move out.

Note The spy catcher can always catch the spy by moving to a particular town along a particular road. The spy catcher must prevent the spy from moving into this town along this road. Locate this town and this road.

Summary of key points

We can **visualise or sketch a 2-D shape** from a description.

We use the **properties of 2-D shapes** to solve problems.

Example

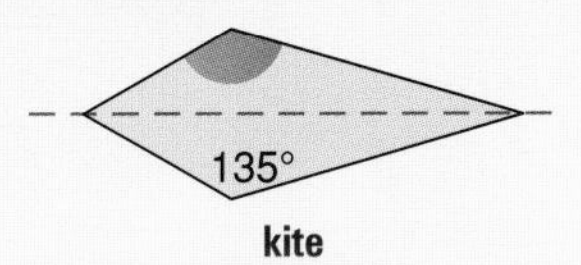

The shaded angle is 135° because a kite is symmetrical.

A **tessellation** is made by reflecting, rotating or translating a shape.

Example

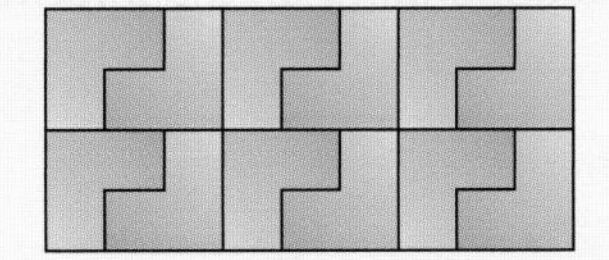

This tessellation could be made by rotating and translating the shape.

Congruent shapes are exactly the same shape and size.
In congruent shapes, corresponding sides and angles are equal.

Example

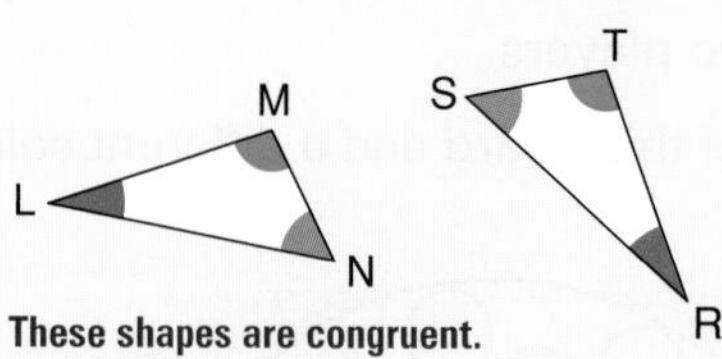

These shapes are congruent.

LM = RT
MN = TS
LN = RS

One congruent shape is mapped onto another by a translation, reflection or rotation or some combination of these.

The **parts of a circle** are shown in this diagram.
The definitions are given on page 285.

segment
chord
arc
sector
diameter
centre
radius
tangent

We often **draw 3-D shapes** using **isometric** (triangle dotty) paper.
An isometric drawing has

vertical edges drawn as vertical lines
horizontal edges drawn at 30° to the horizontal.

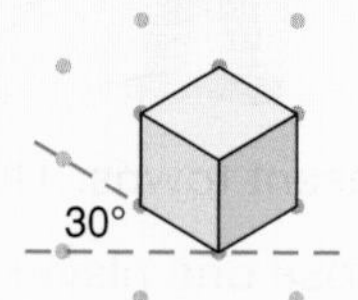

G

The view from the top of a shape is called the **plan view**.
The view from the front is called the **front elevation**.
The view from the side is called the **side elevation**.

Example

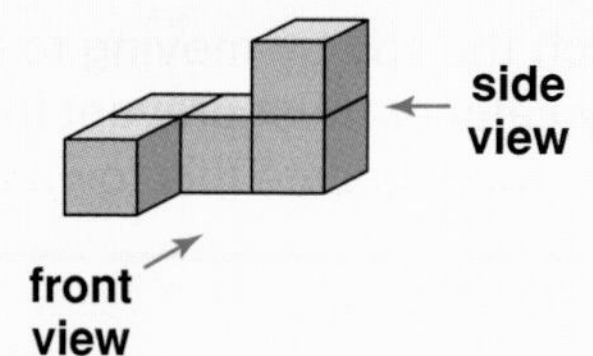

plan view

front elevation

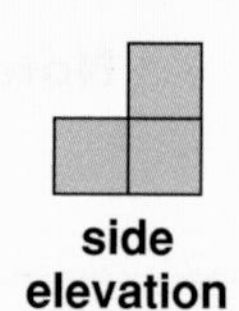

side elevation

H

When we slice a shape, the face that is made is called a **cross-section**.

Example If we slice a square-based pyramid horizontally as shown, the cross-section is a square.

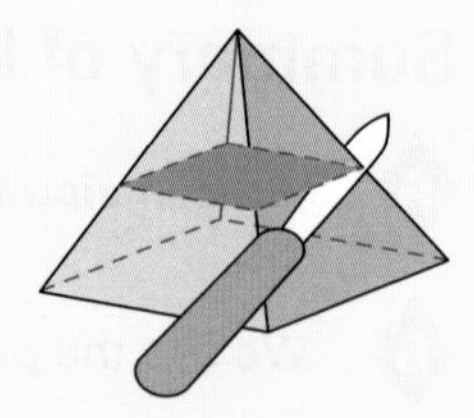

Constructions

see page 295 for these constructions.

- perpendicular from a point to a line
- perpendicular from a point on a line

We **construct triangles** in different ways depending on the information we are given.
The methods are shown on page 298.

A **locus** is a set of points that satisfy a rule or a set of rules.

Examples The locus of the head of a boy jumping from a tree might look like this.

The locus of a robot moving so that it is always the same distance from a fixed point is a circle.

Your locus if you walked so that you were always the same distance from two trees would be the perpendicular bisector of the line joining the two trees.

Test yourself

1 a Imagine a special quadrilateral with two lines of symmetry. What could it be?

b Suppose it also has rotation symmetry of order 2. What could it be now?

2 ABDC is an arrowhead. Show that $A\hat{B}D = 22°$.

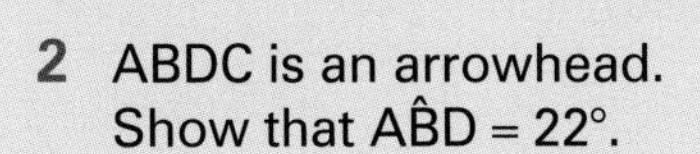

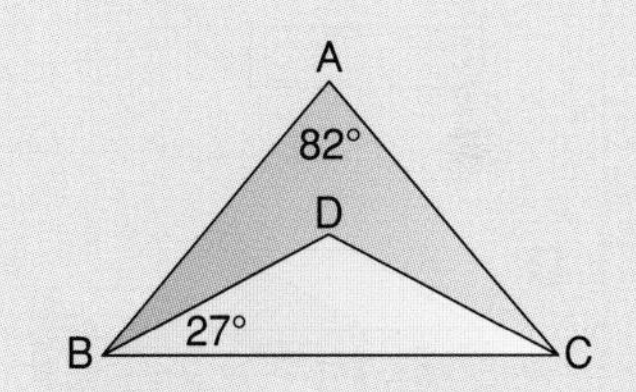

3 Use isometric (triangle dotty) paper to tessellate this shape. Describe how you did this using the words reflection, rotation or translation.

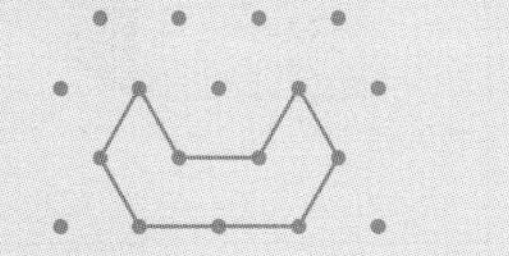

4 ABCD and EFGH are congruent.

a What is the size of angle G?

b What is the length of HG?

c Which of these would map ABCD onto FGHE?

A reflection **B** rotation **C** reflection and translation

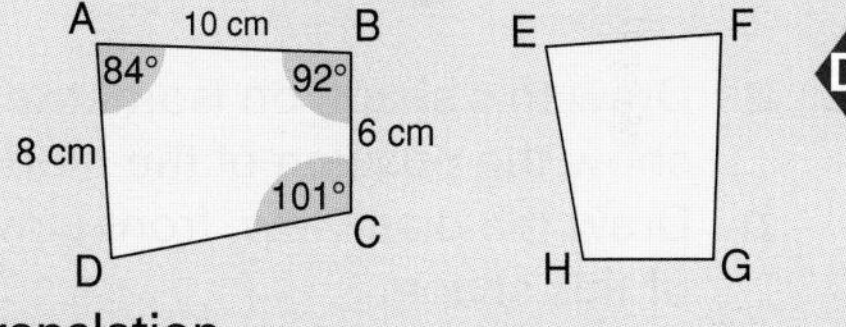

5 A line and a circle move towards each other. Match the description below with a diagram from the box. The line

a makes a tangent to the circle

b intersects the circle at two points, making a chord that is not a diameter

c passes through the centre of the circle, making a diameter

d divides the area enclosed by the circle into two regions called segments

e divides the area enclosed by the circle into two regions called semicircles.

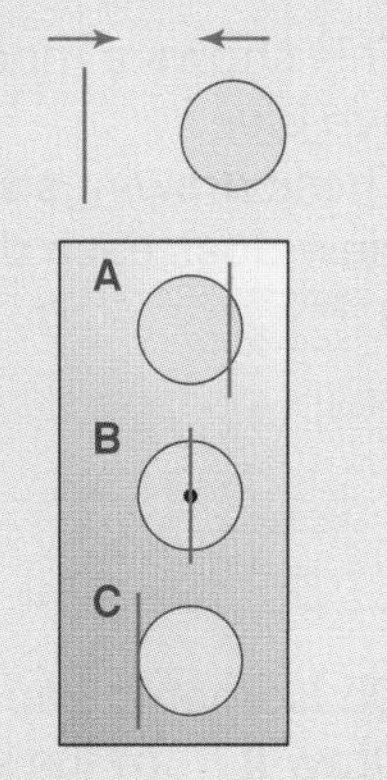

6 Draw a circle.
Label these parts on your circle.

radius sector arc circumference

7 Match the shapes with the sketches of their plans and elevations. G

A	i plan view, front elevation, side elevation
B	ii plan view, front elevation, side elevation
C	iii plan view, front elevation, side elevation
D	iv plan view, front elevation, side elevation

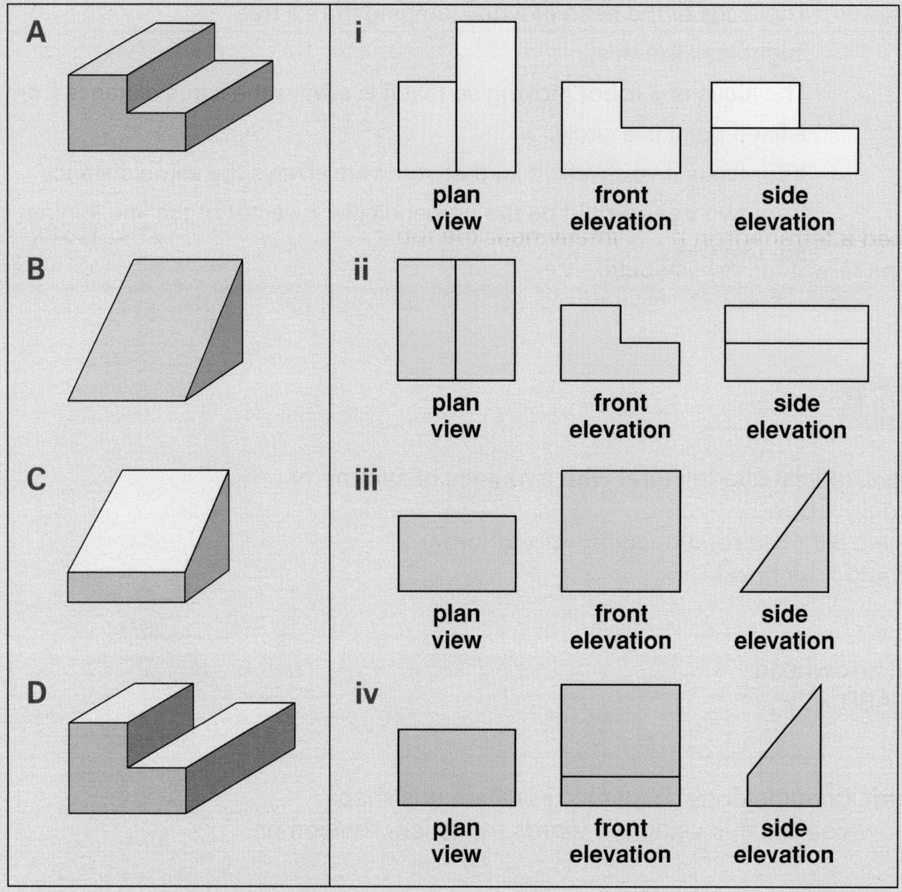

8 **a** Draw this shape on isometric paper. Using dashed lines, show the position of the hidden edges. F G

b Draw the plan view, front elevation and side elevation of this shape.

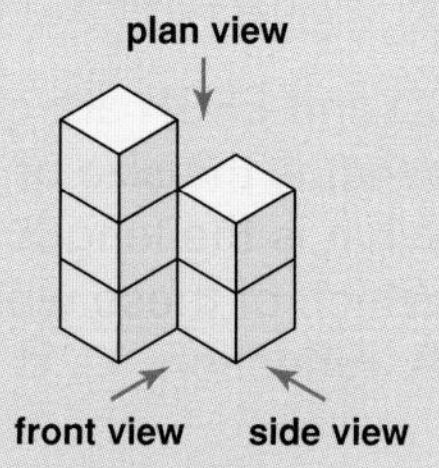

plan view
front view side view

9 This shows a model with nine cubes, four red and five blue. F G
These drawings show the four side views. Which view does each drawing show?

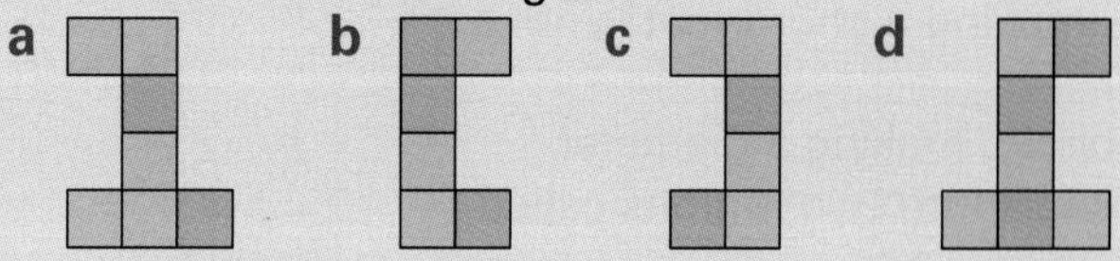

a **b** **c** **d**

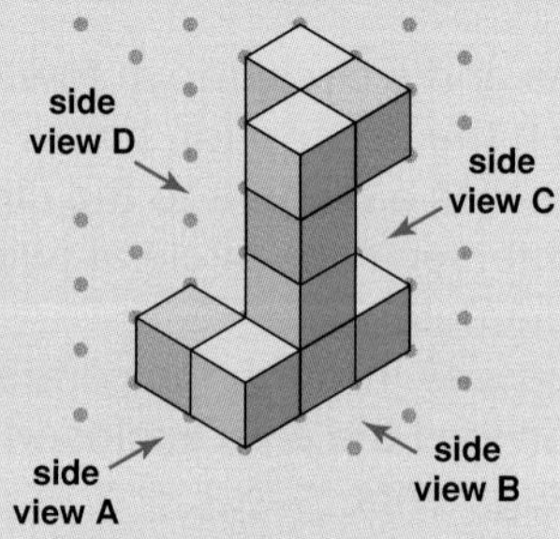

side view D
side view C
side view A
side view B

10 On isometric paper, draw the shape that this represents. Draw it from the view given by the red arrow. F

3	2	2
	1	1

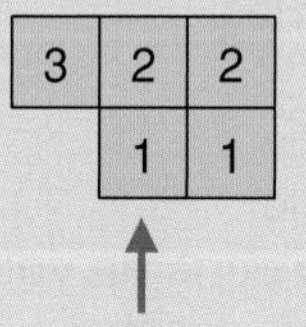

11 Sketch a net for this solid.

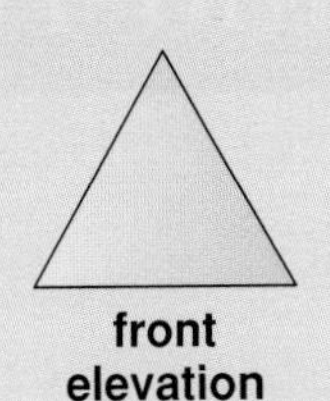

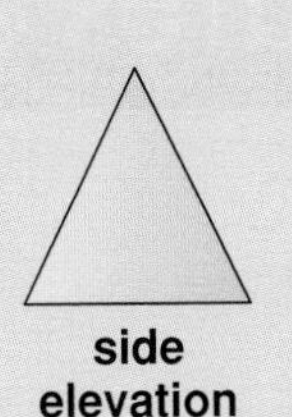

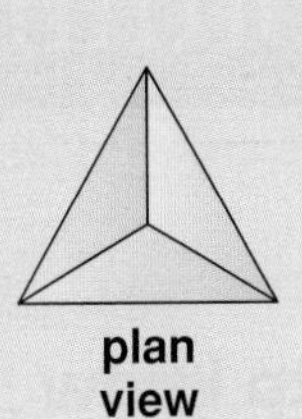

12 Marlon sliced a tetrahedron horizontally near the top.
- **a** What shape will the cross-section be?
- **b** What if Marlon sliced the tetrahedron vertically through the top?

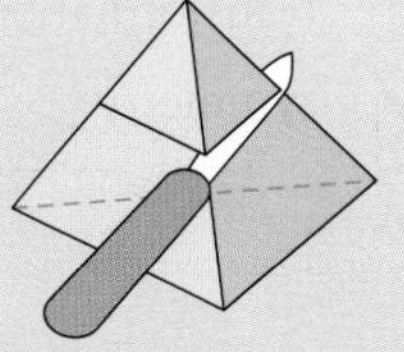

13 Use a copy of this diagram.
- **a** Use compasses to construct the line through A that is at right angles to XY.
- **b** Name the point where this line meets YZ as B.
- **c** Bisect angle AYZ.
- **d** Name the point where this bisector meets AB as C.
- **e** Measure the length YC to the nearest millimetre.

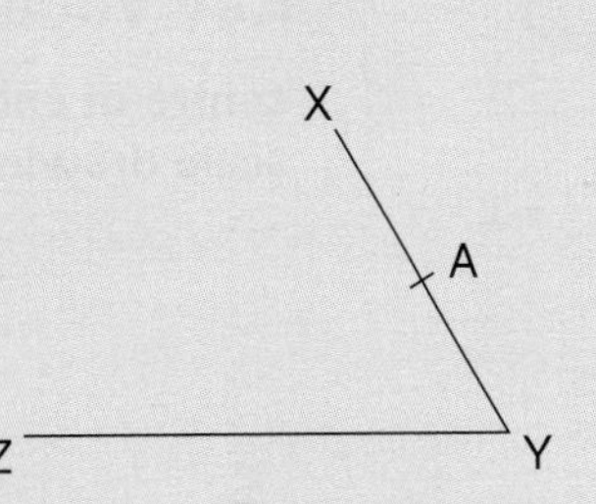

14 The foot of a 3.5 m long staircase is 3 m away from a wall.
Use compasses and a ruler to construct a scale drawing.
On your scale drawing measure carefully to find
- **a** the angle the slope of the staircase makes with the floor (shown shaded in purple)
- **b** the height up the wall the staircase reaches.

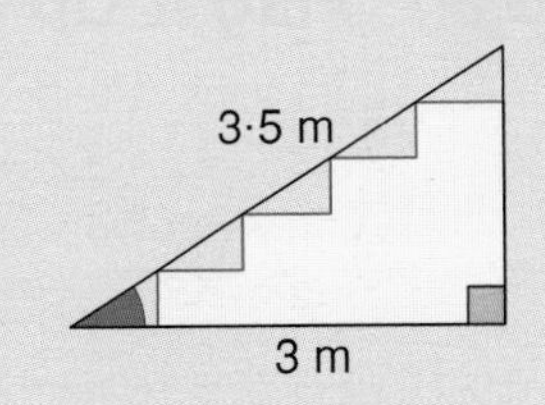

15 Sketch and describe the locus of these.
- **a** the head of a child on a merry-go-round
- **b** a woman running so that she is always the same distance from two trees
- **c** a man walking so that he is always the same distance from two fences which are at right angles to each other

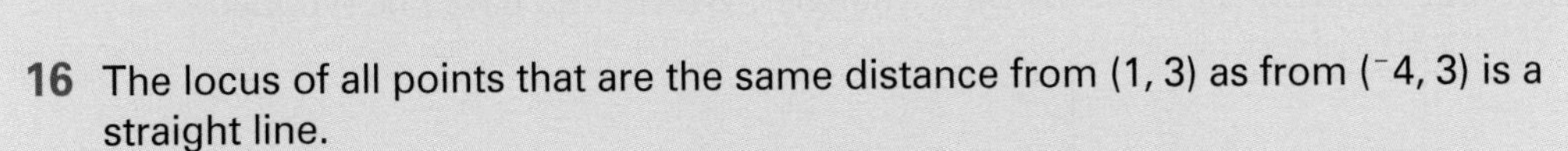

16 The locus of all points that are the same distance from (1, 3) as from ($^{-}4$, 3) is a straight line.
Plot the points (1, 3) and ($^{-}4$, 3) on a set of axes and draw the straight line.

17 This shape is made from five cubes. Two pairs are joined only by a common edge.
Use isometric paper to draw a different shape made from five cubes in which two pairs are joined by just a common edge.

13 Coordinates and Transformations

You need to know

✓ transformations page 252

✓ symmetry page 252

Key vocabulary

centre of enlargement, enlarge, enlargement, map, plan, scale, scale drawing, scale factor

Flag It Up

Each of these shapes is made from the same two identical triangles.

Make some different shapes using these triangles. Which ones have line symmetry?

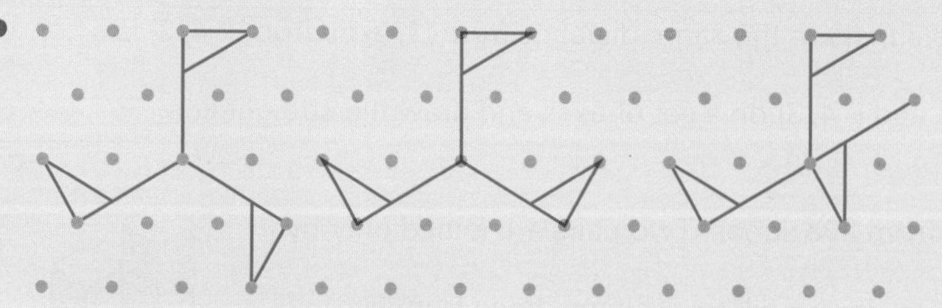

Three 'flags' are used to make patterns.

Make some other patterns using these flags. Which ones have been made by rotating one of the flags? Do any of the diagrams have line symmetry?

- Make a pattern of your own which has line symmetry. Choose a shape other than a triangle or a flag.
- Make a pattern of your own by rotating a shape.

Coordinates and transformations

Worked Example

The points $(^-2, 4)$, $(3, 2)$, $(^-1, 1)$ and $(^-3, 2)$ are the vertices of a quadrilateral.
Write down the coordinates of the vertices of the quadrilateral after

a reflection in the x-axis
b rotation 270° about the origin
c translation of 1 unit to the right and 5 units down.

To remind yourself about reflection, rotation and translation see page 252.

Answer

a

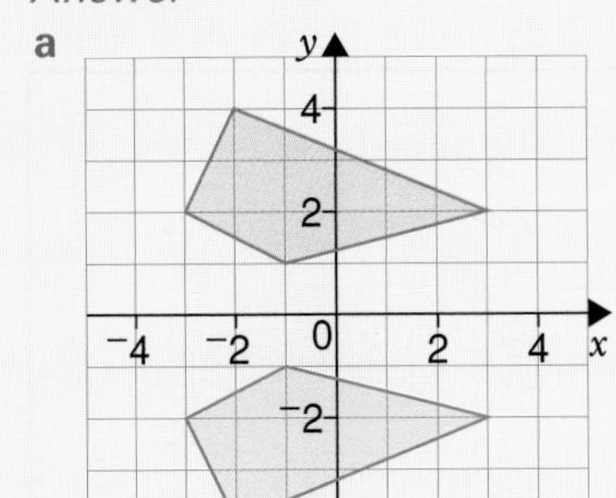

The red shape has been reflected to give the blue shape.
Coordinates of image are
$(^-2, ^-4)$, $(3, ^-2)$, $(^-1, ^-1)$ and $(^-3, ^-2)$

b

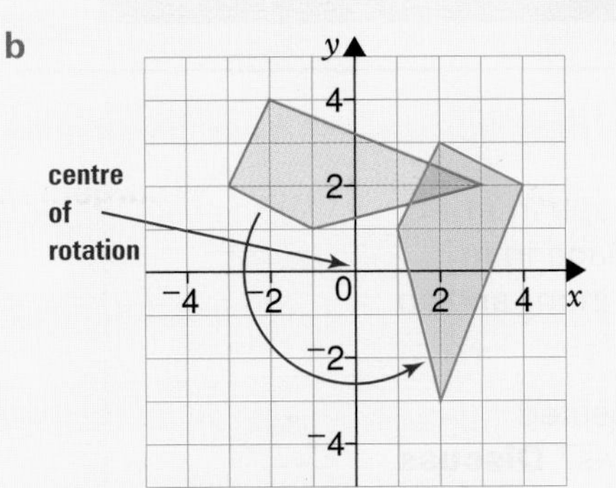

Use tracing paper and a pin to rotate the red shape 270° about (0, 0) to give the blue shape.
Coordinates of image are
$(4, 2)$, $(2, ^-3)$, $(1, 1)$ and $(2, 3)$.

c

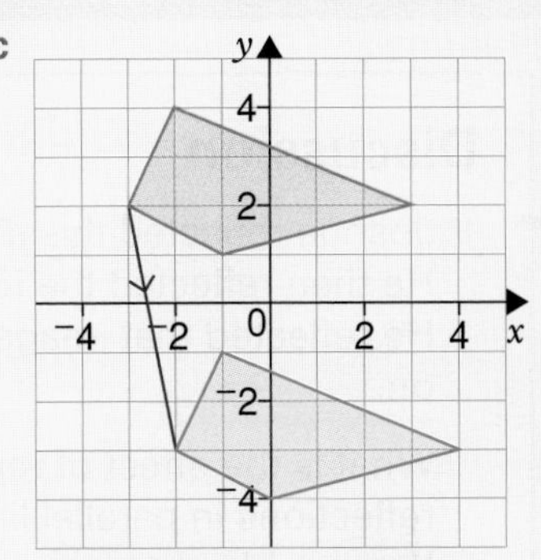

Each vertex on the red shape has been moved 1 unit right and 5 units down to give the blue shape.
Coordinates of image are.
$(^-1, ^-1)$, $(4, ^-3)$, $(0, ^-4)$ and $(^-2, ^-3)$.

Exercise 1

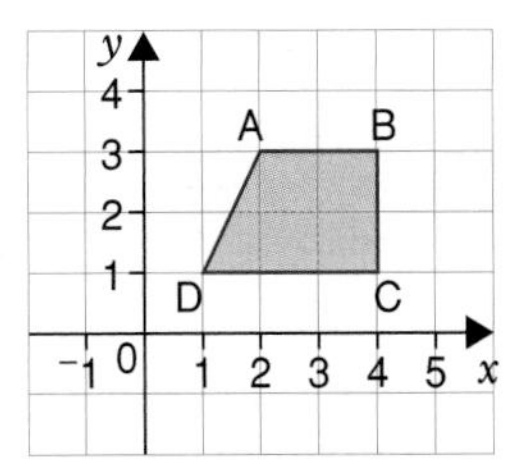

1 Draw a set of axes with x-values and y-values from $^-8$ to 8.
Copy this shape onto your axes.

a Write down the coordinates of A, B, C and D.
b Write down the coordinates of A′, B′, C′ and D′ after
  i reflection in the x-axis
  ii reflection in the y-axis
  iii rotation 180° about (1, 1).

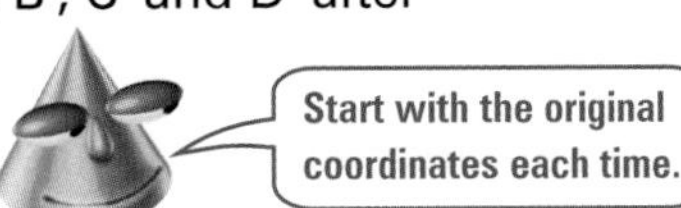

2 The points $(^-4, ^-2)$, $(^-1, 2)$ and $(2, ^-1)$ are the vertices of a triangle. Plot these points on a grid.
Write down the coordinates of the vertices after

a translation 2 units to the left and 4 units down
b rotation 90° about the origin
c reflection in the line $y = {^-1}$.

3 Three of the vertices of a square are $(2, 3)$, $(4, ^-1)$ and $(0, ^-3)$. Plot these points on a grid.

a Write down the coordinates of the 4th vertex.
b Write down the coordinates of the 4 vertices after a
  i translation 3 units to the right and 1 unit up
  ii rotation 270° about the origin
  iii reflection in the line $y = x$.

Start with the original coordinates each time.

Review Copy this shape onto your axes.

a Write down the coordinates of the vertices of the shape.

b Write down the coordinates of the vertices after

i translation 5 units to the left and 3 units down

ii reflection in the x-axis

iii reflection in the line $x = 1$

iv rotation 180° about the origin.

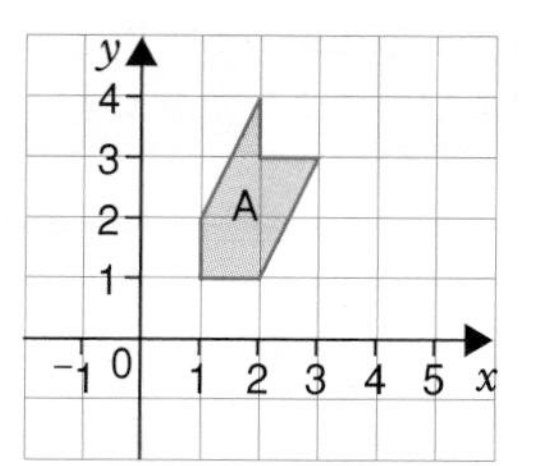

Start with the original shape each time.

Combinations of transformations

Discussion

- Joshua reflected this shape in m_1. He then reflected the image in m_2. He reflected that image in m_3 and so on.

  What is the effect of repeated reflections in parallel lines? **Discuss.** You could try some more reflections in parallel lines.

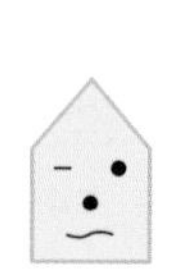

m_1 m_2 m_3 m_4 m_5 m_6

- Sajid reflected this shape in the x-axis and then he reflected the image in the y-axis.
  What single transformation would give the same result? **Discuss.**
  If Sajid did these reflections in the opposite order, what would happen? **Discuss.**
  What single transformation would give the same result? **Discuss.**

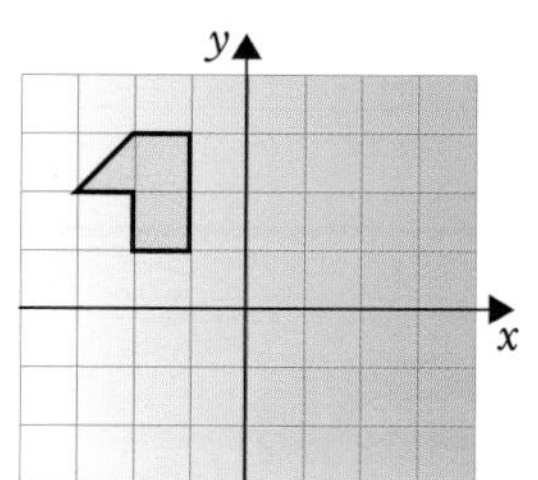

Exercise 2

T **1** Use one copy of this diagram for each part of the question.

a Reflect the shape in the line $x = {}^-2$. Reflect the image in the line $x = 2$.
What single transformation is equivalent to these two reflections?
A translation **B** rotation **C** reflection

b Translate the shape 2 units right and 4 units up and **then** translate the image 1 unit left and 2 units down.
What single translation is equivalent to these two?

c Rotate the shape 90° about the origin. Then rotate the image 180° about the origin.
What single rotation is equivalent to these two?

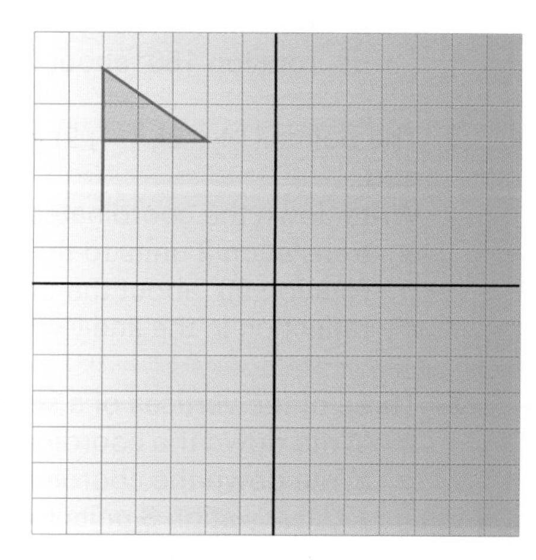

2 What goes in the gaps to make these true?
Choose from the box.

a Reflection in two parallel lines is equivalent to a ______.
b Reflection in two perpendicular lines is equivalent to a ______.
c Two rotations about the same centre are equivalent to a single ______ about the same centre.
d Two translations are equivalent to a single ______.

half-turn rotation
rotation
reflection
translation

3 Copy this diagram.

a Rotate ΔABC 90° about C.
b Reflect ΔABC in the mirror line m.
c What name is given to the quadrilateral formed by the three triangles?

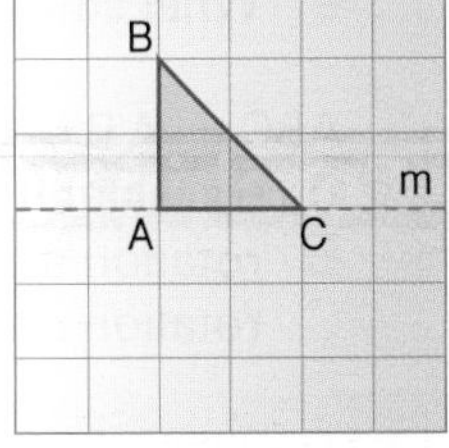

4 Copy this diagram.

a Rotate ΔABC 90° about A to A′B′C′.
b Reflect ΔABC in the mirror line m.
c Reflect $\Delta A'B'C'$ in the mirror line m.
d What shape is formed by the four triangles?

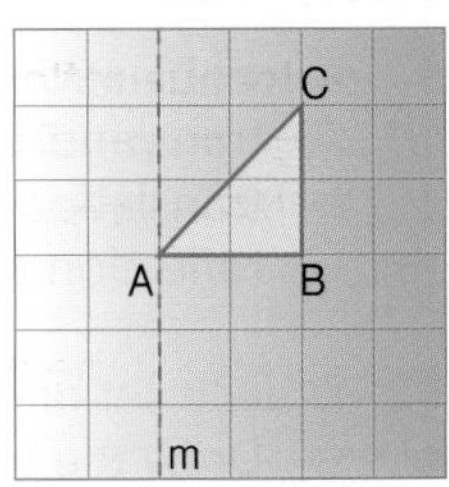

5 **a** When a shape is translated, which of these stay the same?
lengths **angles**
b What about when a shape is rotated?
c What about when a shape is reflected?
d When a shape is translated, rotated or reflected is the image always congruent to the original shape?
e Will the image remain congruent if a shape is transformed by a combination of translation, rotation and reflection?

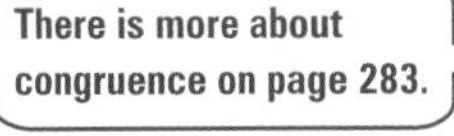
There is more about congruence on page 283.

T

6 Use two copies of this.

a Reflect ABC in the x-axis to give A′B′C′.
Reflect A′B′C′ in the y-axis to give A″B″C″.
Identify the equal lengths and equal angles in the original and A″B″C″.
b Rotate ABC a three-quarter turn about the origin to give A′B′C′.
Translate A′B′C′ 1 unit left and 2 units down to give A″B″C″.
Identify the equal lengths and equal angles in the original and A″B″C″.

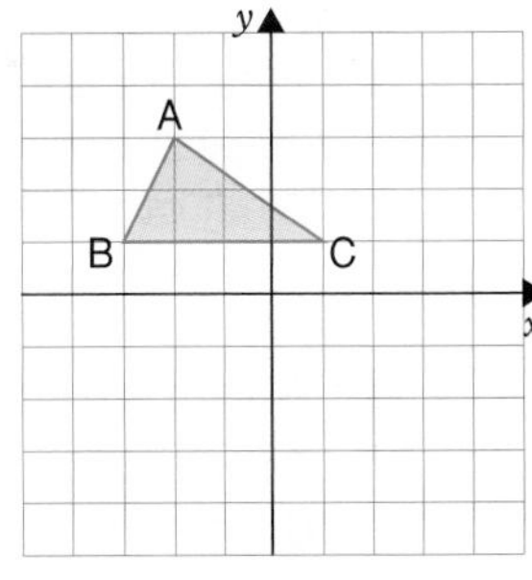

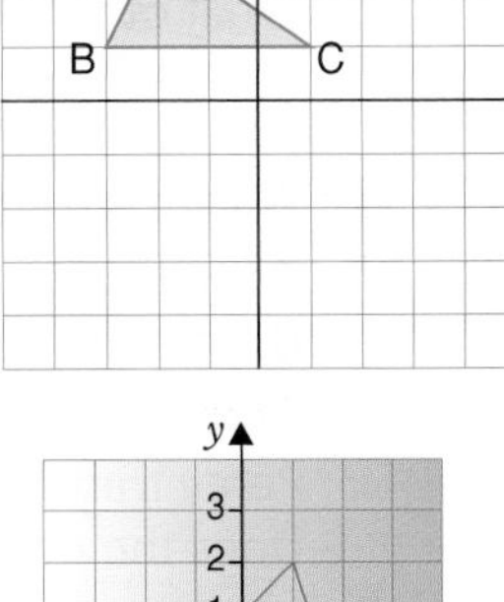

T

7 Use a copy of this.
Reflect this quadrilateral in the y-axis.
Then reflect both shapes (object and image) in the x-axis.
In the pattern made by all of the shapes combined, identify
a equal angles **b** equal lengths.

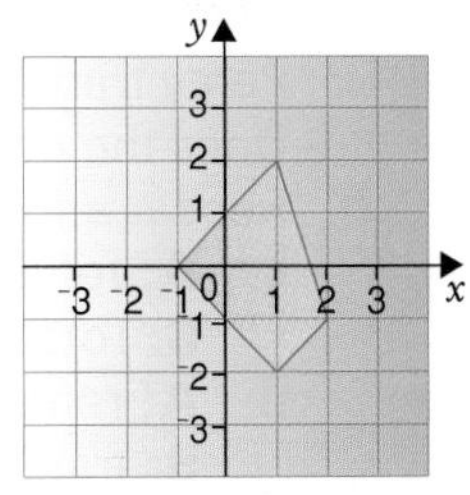

8 *Imagine* these transformations or combinations of transformations.
What shape will the **combined** object and image form?

a A square is reflected along one of its sides.
b A right-angled triangle is reflected along its longest side.
c A square is rotated 90° about any one of its corners.
d A square is rotated twice by 90° about any one of its corners.
e A square is rotated three times by 90° about any one of its corners.
f An isosceles triangle is rotated by 180° about the mid-point of its base.
g A scalene triangle is rotated by 180° about the mid-point of one of its sides.

In **d** and **e** the corner could be different each time.

9 **a** Draw a set of axes with x from $^-4$ to 5 and y from $^-3$ to 4.
b Draw the triangle with vertices at P(3, 2), Q(3, 0) and R(1, 0).
c $\Delta PQR \rightarrow \Delta P'Q'R'$ under reflection in the y-axis.
Draw $\Delta P'Q'R'$.
Give the coordinates of P′, Q′, R′.
d $\Delta P'Q'R'$ is rotated about (0, 0) through a half turn to $\Delta P''Q''R''$.
Give the coordinates of P″, Q″, R″.
e What single transformation would map ΔPQR onto $\Delta P''Q''R''$?

$\Delta PQR \rightarrow \Delta P'Q'R'$ means ΔPQR maps onto $\Delta P'Q'R'$.

10 ABC → A′B′C′ under a combination of
A translation and reflection
B rotation and translation
C rotation and reflection.

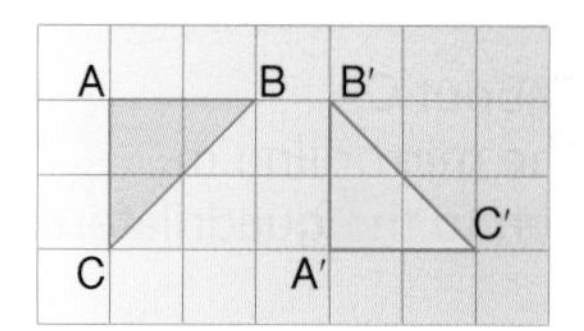

11 The diagram on the left maps onto the diagram on the right under a combination of
A translation and enlargement
B translation and rotation
C reflection and rotation.

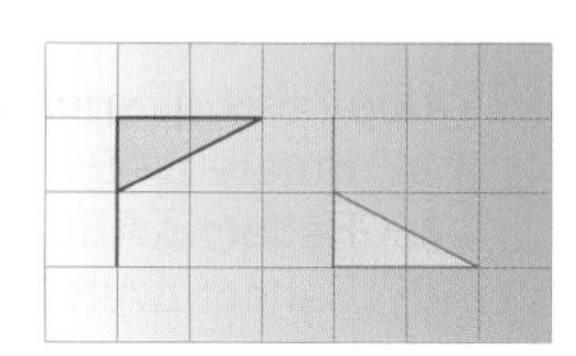

12 Draw a set of axes with x from $^-3$ to 7 and y from $^-5$ to 5.
a Draw the triangle ABC with vertices A(1, 3), B(6, 3), C(1, 1).
b ΔABC is reflected in the x-axis to $\Delta A'B'C'$. Draw $\Delta A'B'C'$. Write down the coordinates of A′, B′ and C′.
c $\Delta A'B'C'$ is rotated about (2, 0), through 270° anticlockwise, onto $\Delta A''B''C''$.
Draw $\Delta A''B''C''$. Give the coordinates of A″, B″ and C″.
d Which single transformation maps ΔABC onto $\Delta A''B''C''$?
A reflection **B** rotation **C** translation **D** enlargement

13 A is reflected in the dashed line ($y = {}^-x$) to give A′.
A′ is then rotated through 270° about (0, $^-2$) to give A″.
a What combination of reflection and translation maps A onto A″?
b What other combinations of transformations map A onto A″?

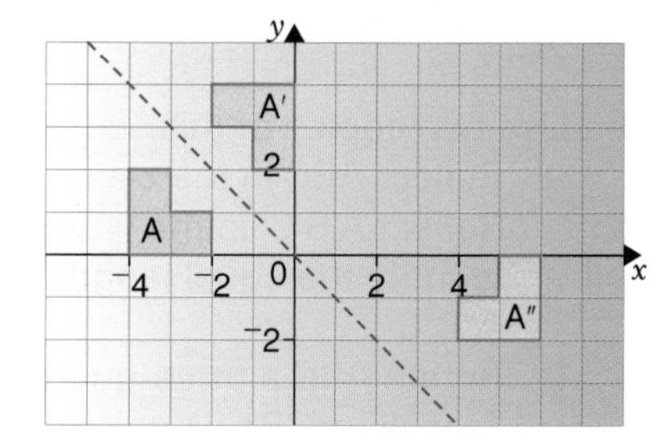

14 **a** Wasim wrote down two transformations.

A: reflection in the y-axis
B: rotation 270° about the origin

Wasim transformed a shape first by A then by B.
Percy transformed the same shape first by B then by A.
Does the order in which these transformations, A and B, are carried out matter?

b Repeat **a** for these transformations.
i **A:** reflection in the x-axis
B: rotation 90° about the origin
ii **A:** reflection in the line $y = x$
B: translation 2 units right then 1 unit down
iii **A:** translation 3 units right and 2 units up
B: rotation 270° about the origin.

15 Some congruent shapes are shown on this grid.
Describe a combination of two transformations that will map A onto each of the other shapes.

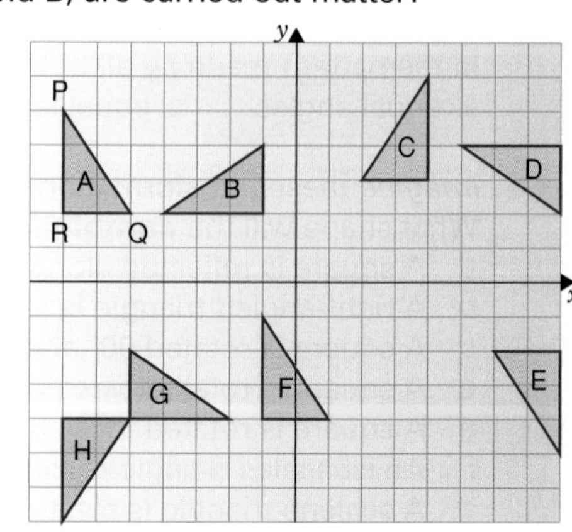

Review 1 ABC is a right-angled triangle.

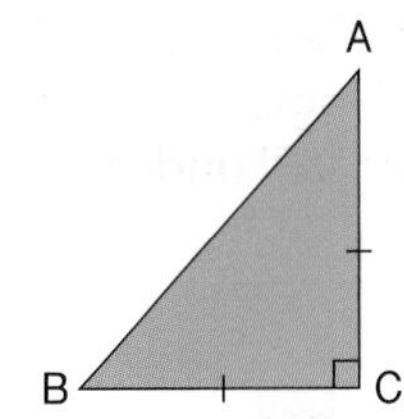

a ABC is reflected in the line BC.
What shape do the object and image together make?

b This new shape is reflected in the line AC extended.
What shape do all the shapes combined make?

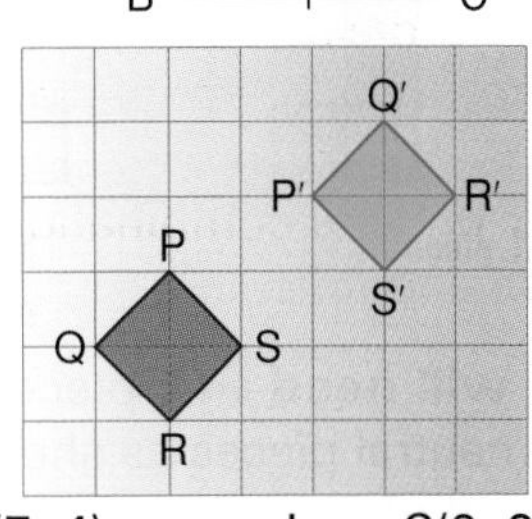

Review 2 PQRS → P′Q′R′S′ under a combination of

A translation and reflection
B translation and rotation
C rotation and reflection.

Review 3 Triangle ABC has vertices A(2, 4), B(7, 4) and C(2, 2).

a Draw ABC on squared paper.

b ABC is reflected in the x-axis to A′B′C′. Draw A′B′C′.
Write down the coordinates A′, B′ and C′.

c A′B′C′ is reflected in the y-axis to A″B″C″. Draw A″B″C″.
What are its coordinates?

d There is a single transformation that maps ABC onto A″B″C″.
Describe this transformation as fully as you can.

Review 4

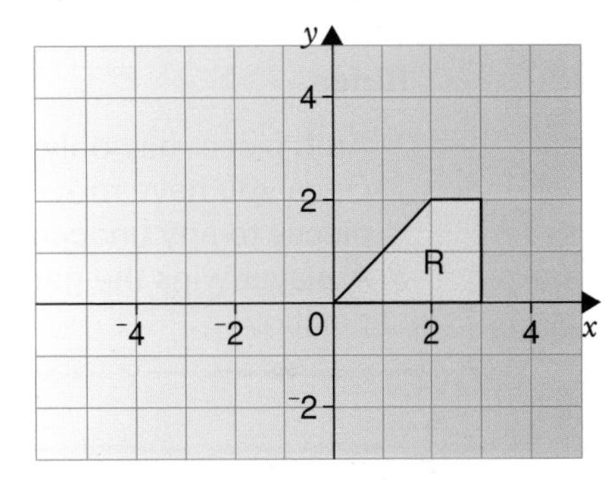

a Copy this diagram.

b R is translated 2 units left and 2 units up to T.
Draw T.
Write down the coordinates of the vertices of T.

c T is rotated 90° clockwise about the point (⁻2, 2) to V.
Draw V.
Write down the coordinates of the vertices of V.

d If these transformations were done in the other order, so that R was rotated first then translated, would you expect the final position of R to be V?
Do the transformations in this order. Was V the final position of R?

Practical

How have transformations been used in this sequence of diagrams?

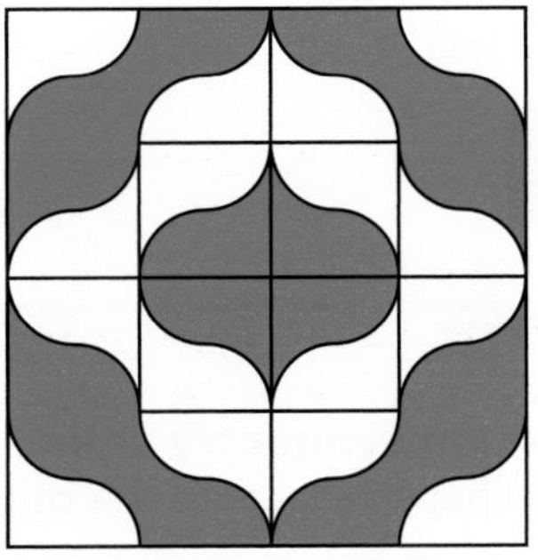

Design a cover for one of your school books or for the school magazine. You could begin with a basic shape, such as the first shape in the diagrams above, and use reflections, rotations and translations of this shape.

The L Game – a game for 2 players

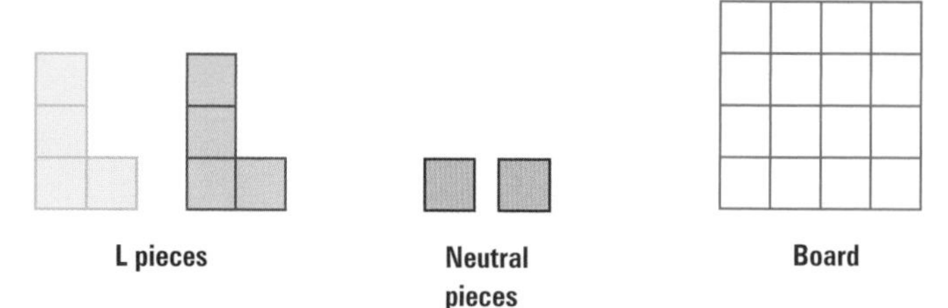

You will need an **L** piece for each player as shown.
Two neutral pieces as shown.
A board of 16 squares.

The **L** pieces and the neutral pieces could be made from thin card.

To play

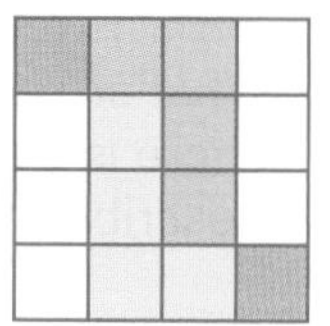

- Place the **L** pieces and the neutral pieces on the board as shown.
- Take it in turns to move your **L** piece to a new position. A new position is one in which at least one of the squares covered was not covered in the previous position. You may reflect your **L** piece, as well as translate or rotate it.

Rules

1 An **L** piece may only be moved into an empty square.
2 Once you have moved your **L** piece, you may also move one of the neutral pieces to any unoccupied square.
3 A player wins the game when the other player is unable to move to a new position.

Practical

You will need a dynamic geometry software package.

Ask your teacher for the Transformations ICT worksheet.

Symmetry

Remember
A shape has **reflection symmetry** or **line symmetry** if one or more lines can be found so that one half of the shape reflects to the other half.

A shape has rotation symmetry if it fits onto itself **more than once** during a full turn.
The **order of rotation symmetry** is the number of times a shape fits onto itself in a 360° turn.

Practical

A **You will need** a 3 × 3 pinboard or square dotty paper.

Marina made this polygon on her 3 × 3 pinboard.
She wrote this

It has 7 sides.
It is not regular.
It has one line of symmetry and no rotation symmetry.

Explore what other polygons you can make on a 3 × 3 pinboard.
Decide if each of your polygons is regular and write down the symmetry properties of each.
What is the maximum number of sides a polygon on this pinboard can have.

B **You will need** a 5 × 5 pinboard or square dotty paper.

Heiata divided her 5 × 5 pinboard into two congruent halves.
She wrote, 'The pattern has rotation symmetry of order 2'.

Find as many ways as possible to divide a 5 × 5 pinboard into congruent halves.
Describe the rotation symmetry of each.

Note: You could have a competition to see who can find the most ways.

Discussion

- 

  Is this shape symmetrical? How do you know? **Discuss**.

- 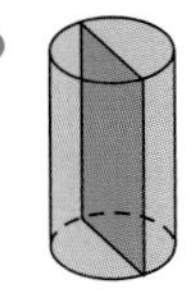 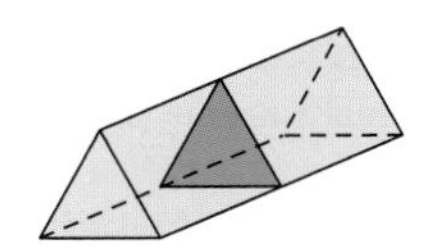

  What is the shaded part called? **Discuss**.

- 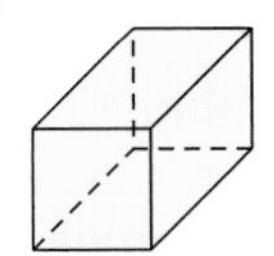 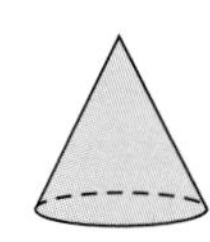 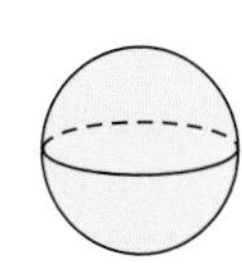 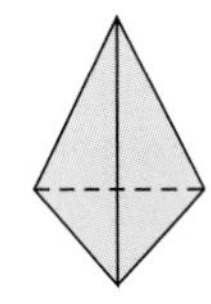

  How can we tell if these shapes are symmetrical? **Discuss** how to draw a shaded part that cuts the shape in half.

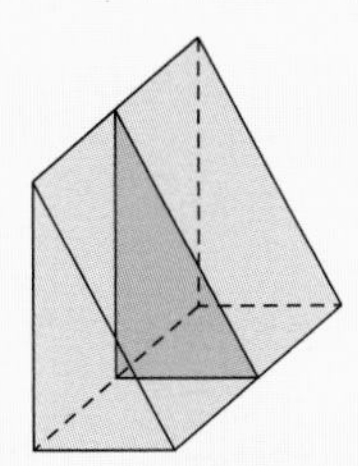

This solid shape is symmetrical.
The shaded shape is called a **plane of symmetry**.
The plane of symmetry divides a solid shape into two equal parts.

Exercise 3

1 Describe the reflection and rotation symmetries of

a an equilateral triangle
b a regular hexagon
c a square
d an isosceles triangle
e a parallelogram
f a rhombus
g a rectangle
h a trapezium (non isosceles)
i an isosceles trapezium
j a kite.

There is more about symmetry and 2-D shapes on page 000.

2 Each of these statements is true.
How could you show that each is true using the symmetry properties of the shape?
The answer to **a** might be, '**If we fold an isosceles triangle along its line of symmetry, the base angles fit exactly on top of one another**.'

a The base angles of an isosceles triangle are equal.
b The angles of an equilateral triangle are all equal.
c The line of symmetry of an isosceles triangle passes through a vertex and is the perpendicular bisector of the third side.
d The opposite angles and opposite sides of a parallelogram are equal.
e The diagonals of a rhombus bisect each other.
f The diagonals of a parallelogram bisect each other.
g The angles between a pair of opposite sides of a parallelogram and a diagonal are equal.

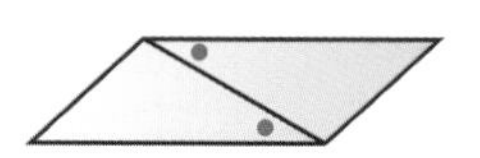

3 Which of these shapes are symmetrical?

a 
b 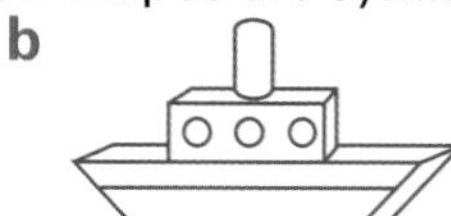
c 
d 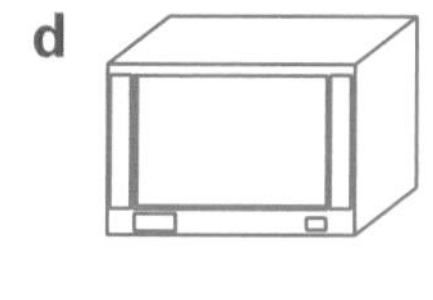
e 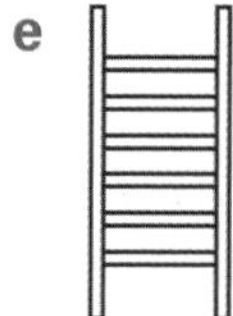

4 Which shapes have

a just one plane of symmetry
b more than two planes of symmetry?

A 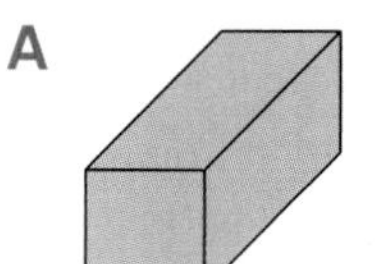
B 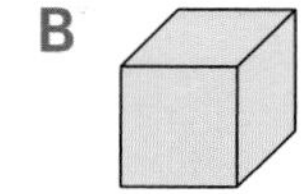
C 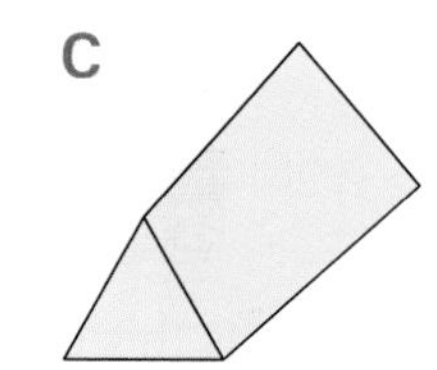

T

5 Use some copies of the shapes in question **4**.
Draw all the planes of symmetry for each shape.
Use a separate copy for each plane of symmetry.

You could draw the shapes on isometric dot paper first and then draw the planes of symmetry.

Review 1 How could you use the symmetry properties of a rhombus to show that

a opposite angles and sides are equal
b the diagonals bisect the angles?

Review 2 Is this shape symmetrical if you ignore the label?

Review 3 Use some copies of this shape.
Draw all the planes of symmetry.
Use a separate copy for each plane of symmetry.

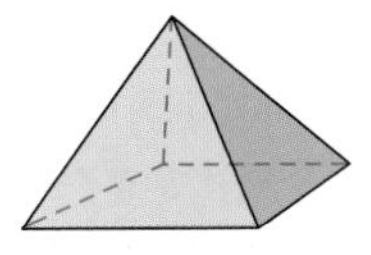

Square-based pyramid

Practical

You will need four multilink or centicubes and isometric paper.

Make all the shapes possible with the four cubes.
Draw each on isometric paper.
Prove that no other shapes are possible.

For each, identify any planes of symmetry.

Investigate the ones with no planes of symmetry to see if they will fit together to make a symmetrical pair.

Enlargement

Remember
To draw an enlargement you need to know the **scale factor** and the **centre of enlargement**.
When the scale factor is **2**, each distance from O to the image is **2** times the distance from O to the object.
So $OA' = 2 \times OA$ and $OB' = 2 \times OB$ and $OC' = 2 \times OC$.

Note: The centre of enlargement does not change its position after an enlargement.

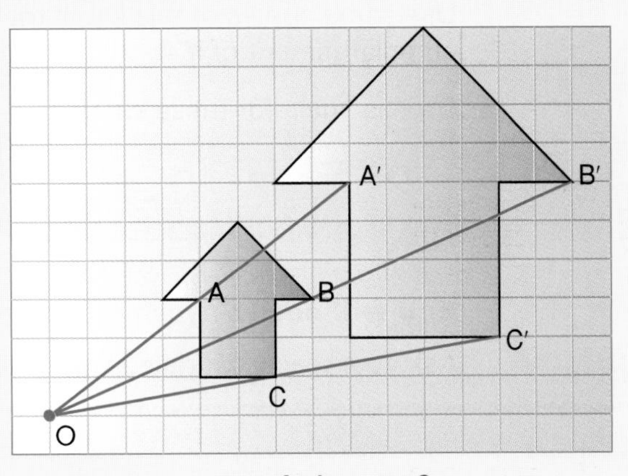

centre of enlargement O.
scale factor 2.

Practical

A **You will need** three A4 sheets of 1 cm square dotty paper and 3 copies of the table on the next page.

1. Draw this 'alien' in the middle of a sheet of square dotty paper.
2. Begin with the red dot as your centre of enlargement. Enlarge the alien by a scale factor of 2. Fill in your table for this enlargement.
3. Repeat **2** for scale factors of 3 and 4.

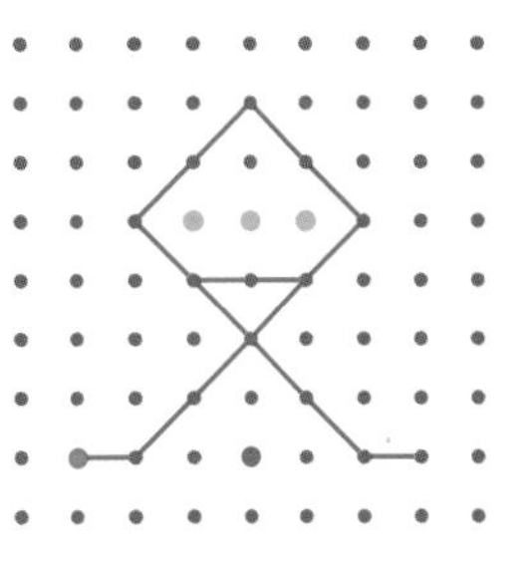

Repeat **1**, **2** and **3** using first the blue dot and then the green dot as your centre of enlargement. Use a new sheet of dotty paper and a new table for each new centre of enlargement.

Length (cm)	Length on original	Scale Factor		
		2	3	4
distance between eyes	2	4		
height of alien	6	12		
width of head	4			
length of foot	1			
length of leg	2·8			

Rosalyn noticed that when she enlarged by a scale factor of **2**, the ratio *height of alien on enlargement to height of alien on the original* is 12 : 6 = **2** : 1.
Is the ratio of corresponding lengths on the enlargement and the original always 2 : 1 when the scale factor is 2?
What is the ratio of the corresponding lengths when the scale factor is 3?
What about for scale factor 4?

Do the angles of a shape change when it is enlarged?
Use your enlargements of the alien to decide.

T

B **You will need** some copies of the shapes below.

Ashrad built this larger square from 16 smaller ones.

Use some copies of this small triangle to build an enlargement of it.

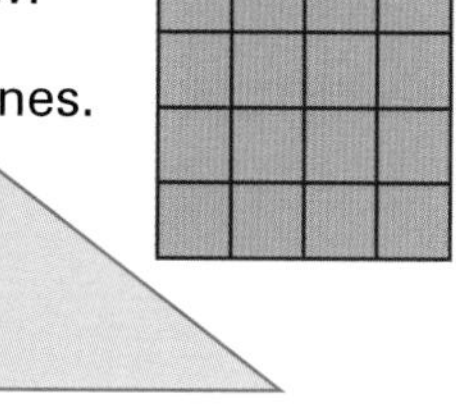

Do this again for these shapes.

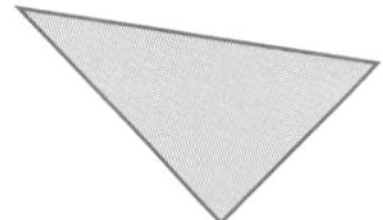

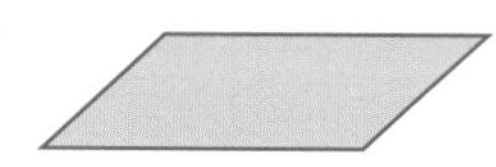

C **You will need** a dynamic geometry software package.

Ask your teacher for a copy of the Two Quadrilaterals ICT worksheet.

For **any enlargement** the ratio of two corresponding line segments is equal to the scale factor.

Example ABC has been enlarged to A′B′C′ by scale factor **3**.

So $\frac{A'C'}{AC} = \mathbf{3}$, $\frac{C'B'}{CB} = \mathbf{3}$ and $\frac{A'B'}{AB} = \mathbf{3}$.

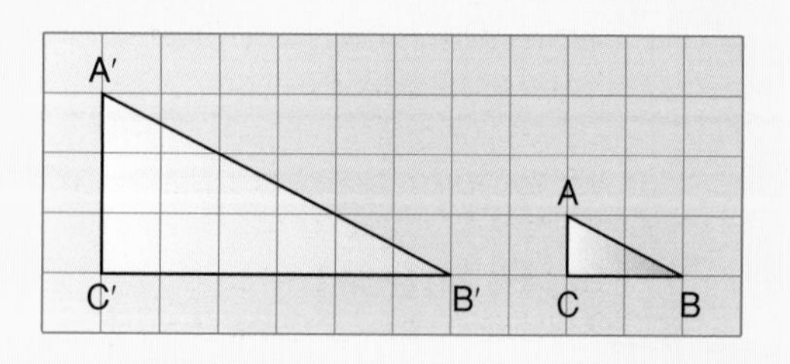

Exercise 4

1 When a shape is enlarged, which of these is **always** the same on the original and the image?

lengths **angles**

2 **a** **b** **c** **d** **e**

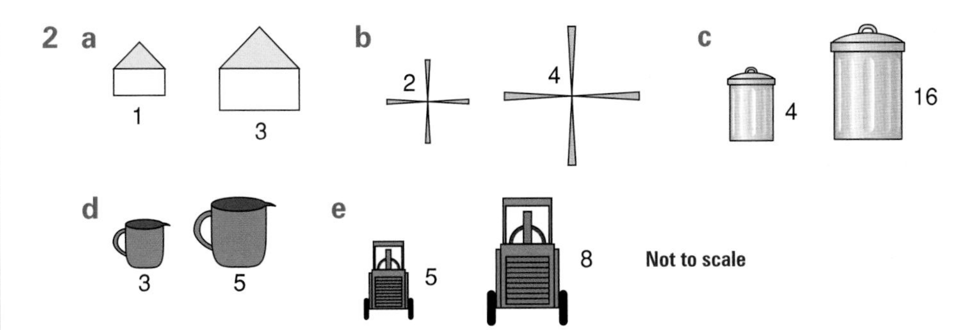

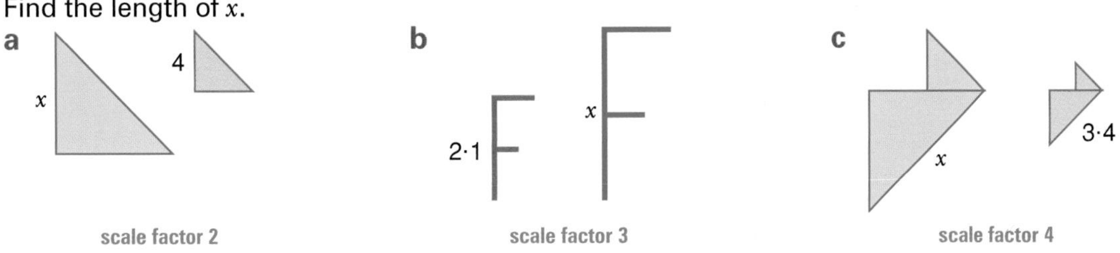

In each of these diagrams the smaller shape has been enlarged to the larger shape.
Use the dimensions given to find the scale factor of each of these enlargements.

3 The red shape has been enlarged to the green shape.
The scale factor for each enlargement is given in green.
Find the length of x.

a x 4 scale factor 2

b 2·1 x scale factor 3

c x 3·4 scale factor 4

4 The bigger star is an enlargement of scale factor 1·5 of the smaller star.

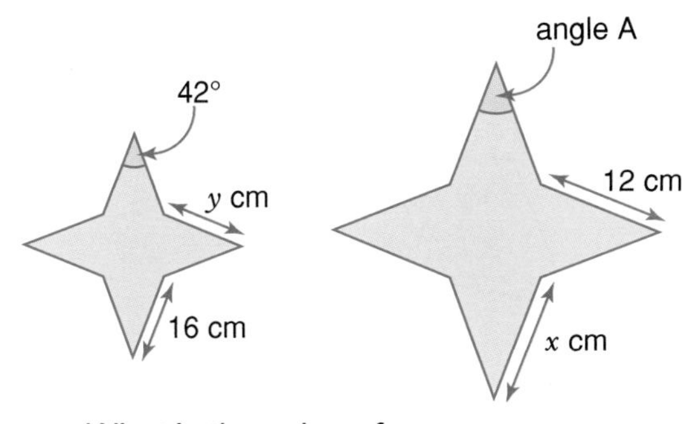

What is the value of
a x **b** y **c** angle A?

5 Use three copies of this.

a Write down the coordinates of A, B, C and D.

b Enlarge ABCD by a scale factor of 2 and centre of enlargement (0, 0).
Write down the coordinates of the image, A′B′C′D′.

c Repeat **b** for scale factor 3.

d Repeat **b** for scale factor 4.

e What do you notice about the coordinates of ABCD and the coordinates of the image in each of **b**, **c** and **d**?

***f** Is what you found out in **e** true if the centre of enlargement is *not* the origin? Check by enlarging the shape by scale factor 2, 3 and 4 using centre of enlargement (1, 0).

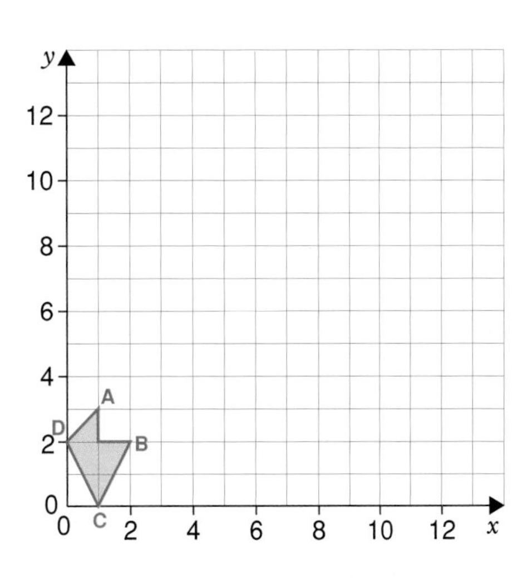

6 Copy these shapes onto squared paper.

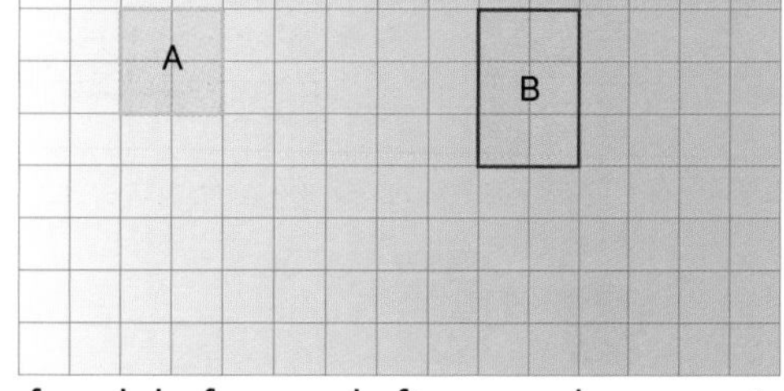

a Find the perimeter of each.

b Enlarge A by scale factor 2 and any centre of enlargement to get A′.
What is the perimeter of A′?

c Enlarge B by scale factor 3 and any centre of enlargement to get B′.
What is the perimeter of B′?

d Draw some other rectangles. Find the perimeter of each before and after an enlargement. For each enlargement, choose the scale factor and the centre of enlargement.

e What do you notice happens to the perimeter of a shape when it is enlarged by scale factor n?
Copy and finish this statement.
Perimeter of image = perimeter of object ______.

7 The perimeter of the shape is given. If the shape is then enlarged by the scale factor shown in green, what will the perimeter of the image shape be?

a

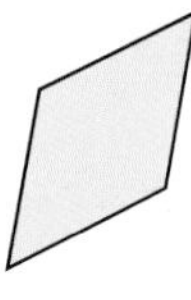

perimeter 10 m
scale factor 2

b

perimeter 13 cm
scale factor 3

c

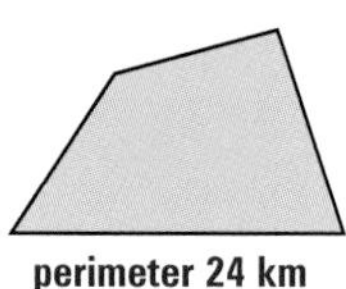

perimeter 24 km
scale factor 4

8 P($^-2$, 3), Q(1, 5), R($^-2$, 7) is enlarged by scale factor 3 and centre of enlargement the origin. Without drawing the enlargement, write down the coordinates of P′, Q′ and R′.

Review 1 ABCD has been enlarged by a scale factor of 2 to give A′B′C′D′.

a What is the size of ∠A′?

b What is the length of B′C′?

c The perimeter of ABCD is 14 m. What is the perimeter of A′B′C′D′?

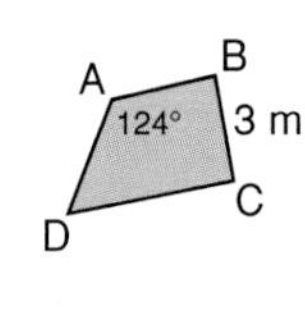

Review 2 The green shape has been enlarged to the purple shape. Find the scale factor of the enlargement.

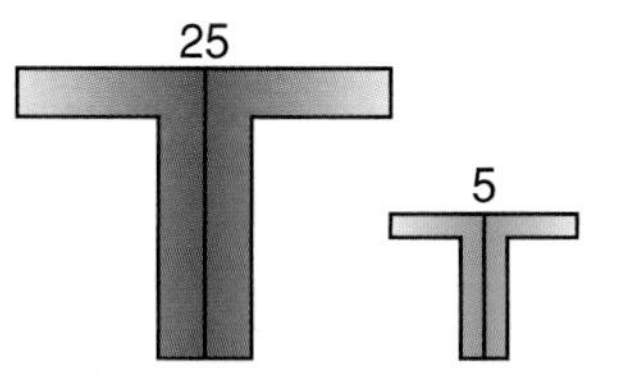

Review 3 A(1, 0), B($^-2$, 3), C(1, 4) is enlarged by a scale factor of 2 and centre of enlargement (0, 0). Without drawing the enlargement, write down the coordinates of the image A′B′C′.

Investigation

Paper Ratios

You will need a sheet of A3 paper.

A piece of A3 paper folded in half gives two sheets of A4 paper.
A piece of A4 can be folded to give two pieces of A5 paper.
A5 can be folded to give A6.
A6 can be folded to give A7.
Investigate the ratios of paper sizes A3 to A7.

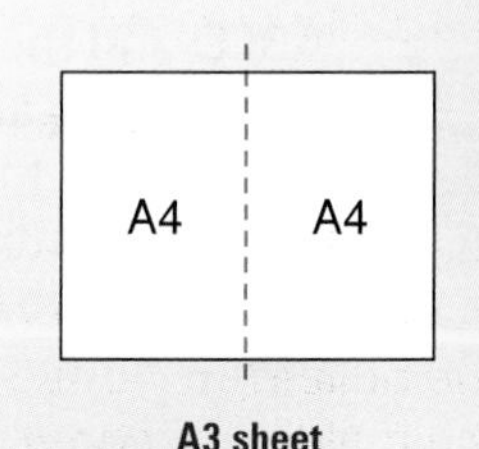

A3 sheet

Is each an enlargement of another?
If so, what is the scale factor of the enlargement?

Can the sheets be aligned corner to corner to give a centre of enlargement?
Investigate.

Practical

A **You will need** a shape drawn in the centre of a sheet of A4 paper.

Enlarge your shape using a photocopier.
Give both copies to someone else and ask them to guess the scale factor for the enlargement.
Repeat for other enlargements.

B A pantograph is an instrument which is used to enlarge a drawing.
Find out how a pantograph is constructed and how it is used to make an enlargement.
You could make a pantograph using cardboard joined together with split pins.
You could use your pantograph to make an enlargement of a drawing.

Scale drawing

A **scale drawing** is a drawing of something in real life. It has a scale so we can work out what each length in the drawing represents in real life.

A scale of 1 cm to 1 m means that 1 cm on the scale drawing represents 1 m, or 100 cm, in real life.

This can be written as 1 cm to 1 m
1 cm to 100 cm
1 cm represents 1 m
1 : 100.

Example Peter made a scale drawing of his bedroom. The scale he used was **1 : 50**.
On his drawing his room is 8 cm by 9 cm
In real life
the width of his room is 8×50 cm = 400 cm
the length of his room is 9×50 cm = 450 cm.
His room is 400 cm by 450 cm or 4 m by 4·5 m.

We can use comparison with heights and lengths we know to estimate other heights or lengths in a scale drawing.

Example In the scale drawing the man is 1 cm high and the building is 3 cm high.
The building is three times as tall as the man.
The height of an average man is about 1·8 m.
The building is about $3 \times 1{\cdot}8 = 5{\cdot}4$ m.
The building is about $\mathbf{5\frac{1}{2}}$ **m** tall.

Worked Example

Find the scale of this drawing if in real life the table is 2 m long.

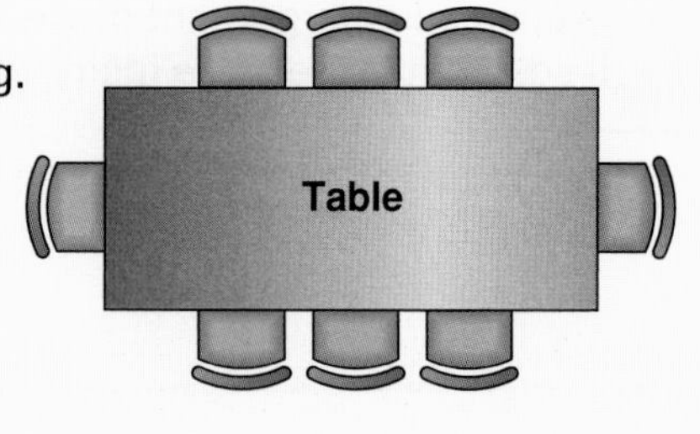

Answer

The table in the drawing is 4 cm long.
To find the scale we have to find how many times 4 cm has been scaled up to give 2 m.

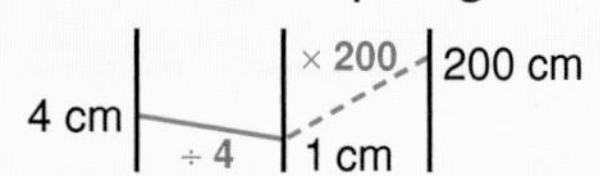

So the scale is 1 to 50 or **1 cm represents 50 cm.** $\div 4 \times 200 = \times 50$

Another way of finding the scale is

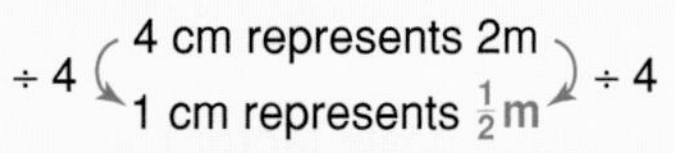

So the scale is **1 cm represents 50 cm**.

Exercise 5

1 Estimate the height of these.
   a the child
   b the garage
   c the tree
   d the light

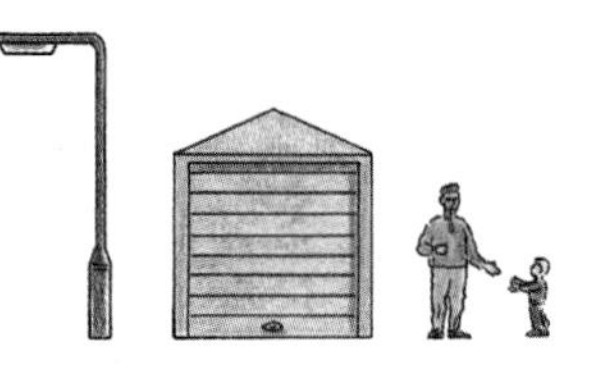

2 This picture shows an average adult woman.
Estimate the length of the fence.

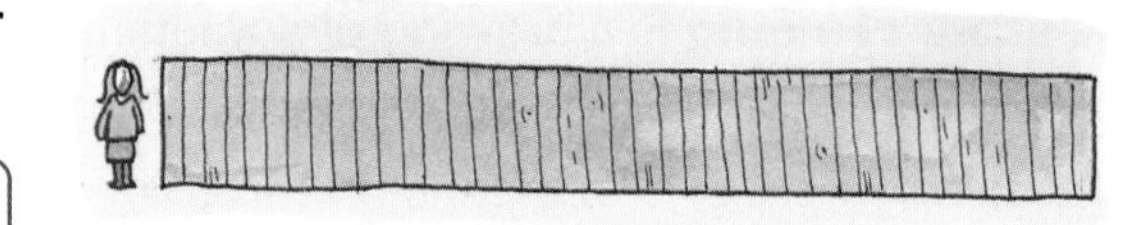

3 Jafar made an accurate floor plan of his house.
He used the scale 1 cm represents 400 cm.
On his plan the games room measured 2 cm × 3 cm.
What were the actual dimensions of this room?

4 At the scene of an accident the police made a scale drawing.
They used the scale 1 : 200.
On the drawing one of the skid marks was 84 mm long. How long were the actual skid marks in
   a mm   **b** m?

5 A scale of 1 : 50 000 was used to draw a plan of the proposed extension to the Eastern Motorway. On the plan this extension was 75 cm long.
What was the actual length, in km, of this proposed extension?

6 Carrie made a scale drawing of a yacht in design and technology.
She used the scale 1 inch represents 9 feet.
What is the real-life length of these? The length on the scale drawing is given.
   a mast 4 inches   **b** length of yacht 3·5 inches
   c width of yacht 1·2 inches   **d** width of sail at bottom 1·8 inches.

7 The scale of a map is 1 : 250 000
What actual distance (in km) do these measurements, taken from the map, represent?
a 5 cm ***b** 5 mm ***c** 150 mm

8 In science a scale drawing of a bone is 4 mm long. In real life the bone is 40 cm long.
What is the scale on this drawing?

9 In a scale drawing a bed is 4 cm long. In real life the bed is 2 m long.
What is the scale on this drawing?

10 In a scale drawing a field that is 50 yards long in real life is drawn as 1 inch long.
What is the scale on this drawing?

11 Find the scale of each of these scale drawings.

a

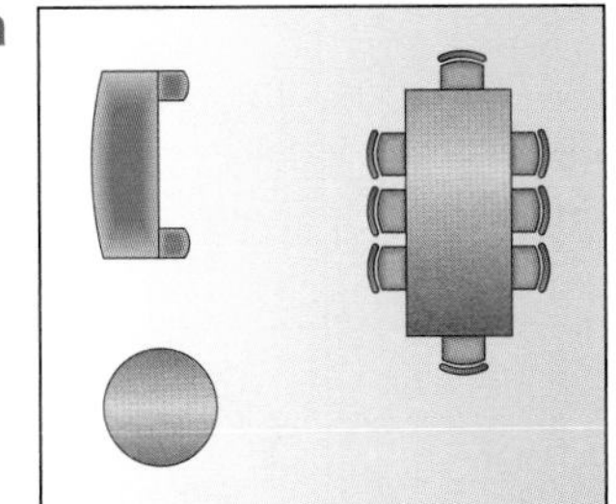

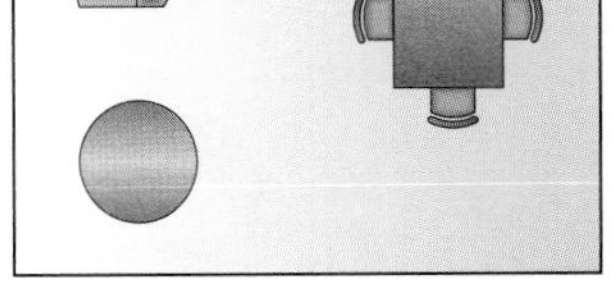

In real life this room is 4 m by 3·5 m.

b

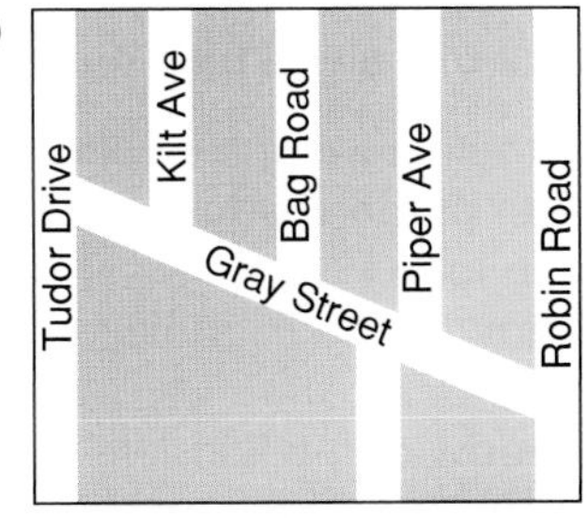

In real life Grey Street is 1·8 km long.

c

In real life the perimeter of the school is 1 km.

***12** Sylvia drew this scale drawing of part of her school grounds. The bike sheds are each 6 m wide and 6·75 m long. What scale did Sylvia use? Give your answer in the form 1 : n.

Review 1 This picture shows a whale.
A man of average height is standing by it.
Estimate the length of the whale.

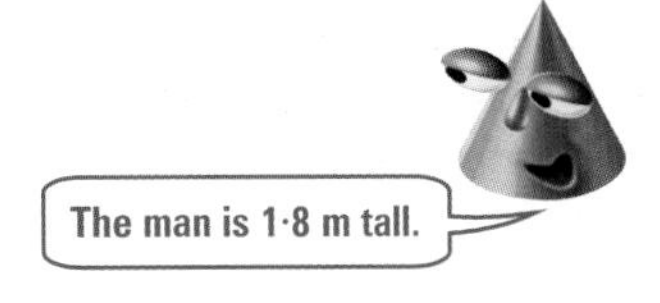

Review 2 The scale 1 mm represents 2 m is used on a scale drawing of the school grounds.

a On this scale drawing, the distance between the gymnasium and the tennis courts is 15 mm. How far is it from the gymnasium to the tennis courts?

b On the scale drawing, the distance from the main gate to the main entrance is 32 mm. How far is this in real life?

Review 3 On a map a road is 4 cm long. In real life the road is 1 km long. What is the scale on this map?

Review 4 Brad drew this scale drawing of his room. In real life the room is 4 m by 4·5 m. What is the scale of this drawing?
Give you answer in the form 1 : n.

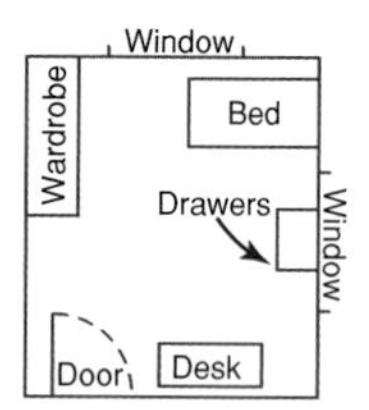

Practical

You will need a metre rule and a tape measure.

Bryn wanted to know the height of the school flagpole.
He measured the length of the shadow of a one metre rule.
The shadow was 64 cm long.
He measured the shadow of the flagpole.
It was 252 cm long.
He worked out the shadow of the flagpole was *about* four times longer than the shadow of the metre rule. **$252 \div 64 \approx 4$**

So the flagpole must be about 4 times as high as the metre rule.

The flagpole is about $4 \times 1 = 4$ m high.

Use Bryn's method to estimate the height of a tall tree, a lamp post, a flagpole, a building or some other tall structure.

Making a scale drawing

Worked Example
A decorator wants to make a scale drawing of Kate's living room.
She made this sketch.
She used a scale of 1 cm represents 2 m.
Make the scale drawing.

8 m
1·2 m
Door
5 m
3·6 m
1·6 m
Window

Answer
We must work out the lengths for the scale drawing.

8 m is drawn as 4 cm.
3·6 m is drawn as 1·8 cm.
1·2 m is drawn as 0·6 cm.
5 m is drawn as 2·5 cm.
1·6 m is drawn as 0·8 cm.

The scale drawing is shown.

To make a scale drawing.

1 make a rough sketch with the real-life measurements written on it
2 decide on the scale
3 work out what each measurement will be on the scale drawing.

Practical

You will need a ruler, tape measure and scissors.

1. Make a scale drawing of a room in your school or house.
   Make separate scale drawings of the furniture. Cut them out.
   Design a layout for this room that is different from the present layout.
   Good scales to choose might be 1 cm represents 1 m or 2 cm represents 1 m.
2. Make a scale drawing of the grounds of your school. Make scale drawings of each building at your school. Place these on the scale drawing of the grounds.
   A good scale to use might be 1 cm represents 10 m.
3. Make a scale drawing of some part of your community. You could choose a park or the streets around your school or your High Street or something else.
4. Make a scale drawing of a piece of clothing. You could choose a piece that you already own or you could design one.
5. Make a scale drawing of some aspect of life you have studied in another subject. It might be a farm, a town, a type of house or an article of clothing. These could be from another country or another period of history that you have studied. You will need to make a sketch first and estimate measurements.

Exercise 6

1 Make an accurate scale drawing from each of these sketches.
Use the scale given in green.

a

1·5 m
2 m
0·5 m
4 m

1 cm represents 0·5 m

b

12 km
3·2 km
Home
Bank
5·4 km
8 km
Supermarket
School
Pool
5 km

1 cm represents 2 km

2

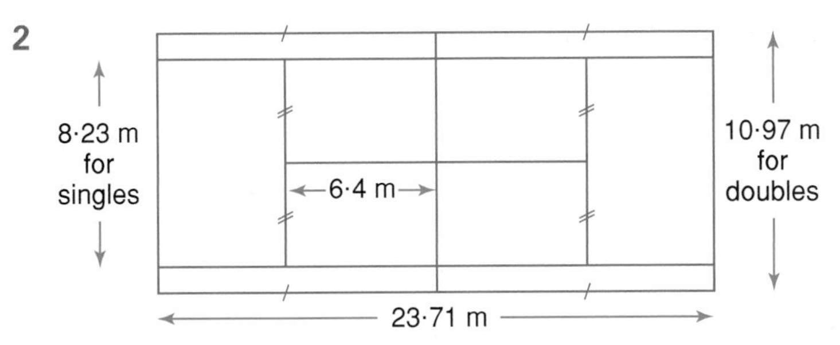

This is a sketch of a tennis court.
Choose a suitable scale and make an accurate scale drawing of the tennis court.

*3 **a** How far is the ship from the lighthouse?

b A yacht is 14 km due South of the ship. Make a scale drawing showing the position of the lighthouse, the ship and the yacht using the scale given. Use your scale drawing to find the distance from the lighthouse to the yacht.

Scale 1: 500 000

Review 1 Make an accurate scale drawing of this sketch. Use the scale given in green.

1 cm represents 5 m

60 m
20 m
15 m
House
30 m
20 m
3·5 m

Review 2 This sketch shows one half of a jumper. The figures on the sketch are in cm. The sketch is *not* drawn to scale.

Make a scale drawing of this sketch. Use the scale 1 : 10, or another scale of your choice.

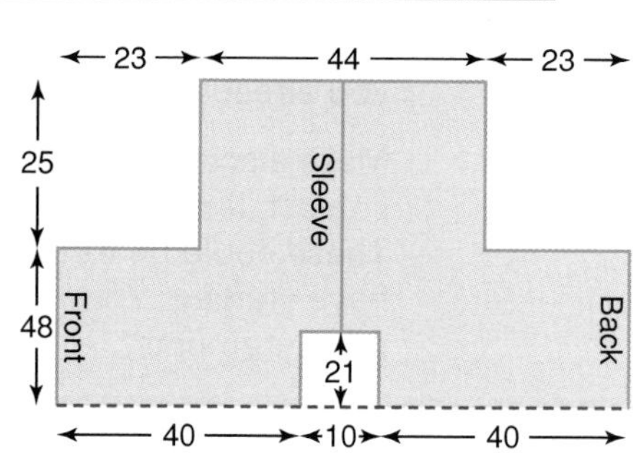

Reading scale drawings

Worked Example

This is the scale drawing Latif made of his family's car. Find the length of this car.

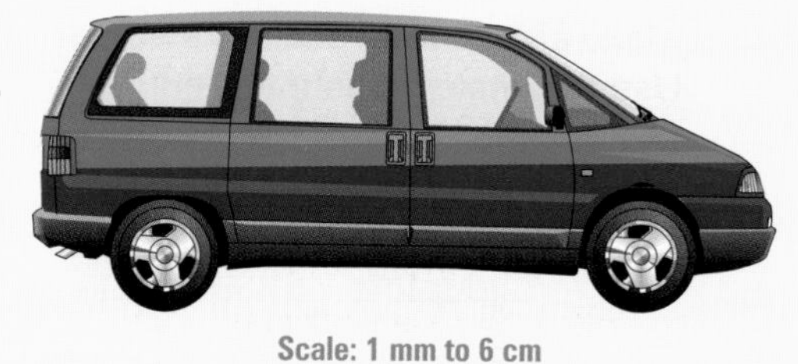

Scale: 1 mm to 6 cm

Answer

Measuring the length on the drawing we get 65 mm.
1 mm on the drawing represents 6 cm on the van.
65 mm on the drawing represents 65 × 6 cm = 390 cm
= 3·9 m

The length of the car is **390 cm** or **3·9 m**.

Exercise 7

1

LAKE DEEP WATER
Raft
Diving rock
Rubbish bin
PICNIC AREA
Table
Boat Ramp
Boat Hire
Shop
Scale 1 : 1000

The map on the previous page shows a picnic area near the Turner's home.

- **a** What is the map distance from the shop to the boat hire?
- **b** What is the actual distance from the shop to the boat hire?
- **c** Find the actual distance from the table to the rubbish bin.
- **d** Verity swam from the diving rock to the raft. How far did she swim?
- **e** Annabel and Anthony hired a canoe and paddled from the boat ramp to the raft and back again. How far did they paddle altogether?

2

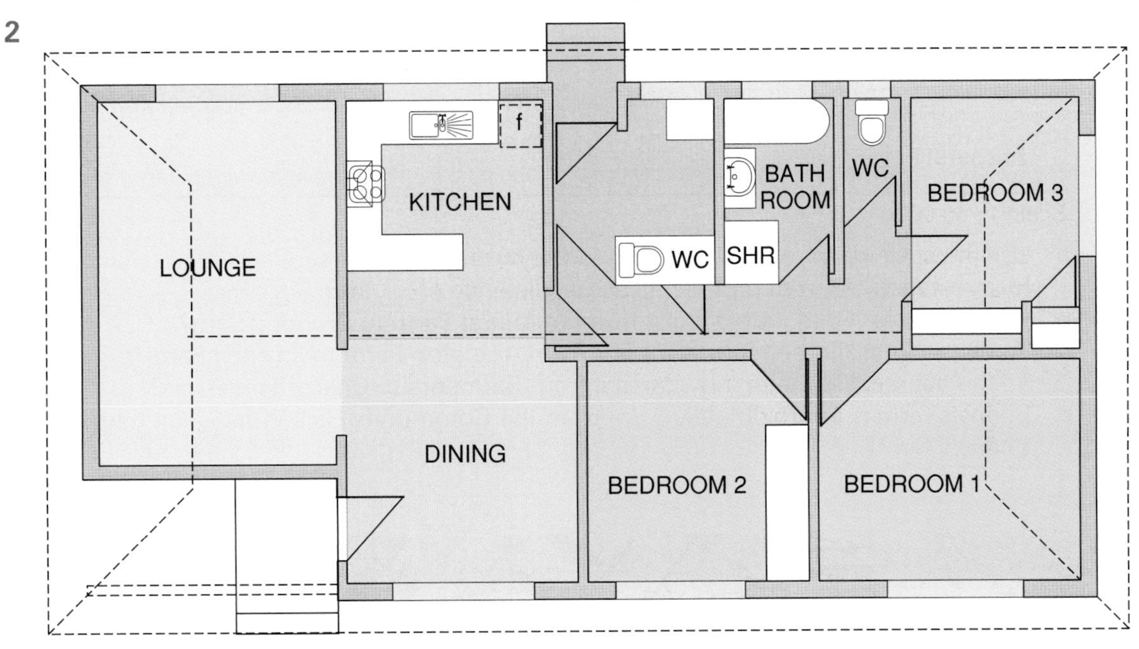

scale 1:100

This is a scale drawing of a new bungalow on the Fairview Housing Estate.

- **a** What are the inside dimensions of the lounge?
- **b** What are the inside dimensions of bedroom 1?
- **c** How long is the window in bedroom 3?
- **d** The outside dotted line is the edge of the roof. What are the dimensions of the roof?
- **e** How deep is the wardrobe in bedroom 2?
- **f** How long is the bath?

3 This shows the cutting layout for a shirt.
The pattern pieces are as follows:

1 – front
2 – back
3 – front band
8 – collar
9 – sleeve

- **a** What is the width of the collar at the fold line?
- **b** What is the length of the back at the fold line?
- **c** What are the dimensions of the front band?
- **d** What is the width of the sleeve at its widest point?
- **e** What is the length of the front at the selvage line?

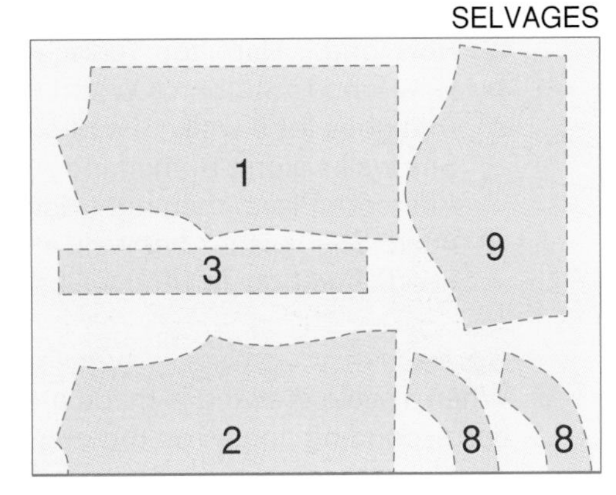

Scale: 1 mm to 2 cm

4

Community Hall

Appleton Fm

Saunders

Tall Trees Fm

Bromley Barn

Cherry Tree Fm

Wadd Fm

Upton Fm

Iborden Fm

Tolehurst Fm

Scale 1 : 50 000

a Use the scale given on the map to find the missing numbers.
Map distance of 1 mm represents actual distance of ___ mm = ___ m.

b To the nearest 50 m, how far is it from Tolehurst Farm to Bromley Barn?

c To the nearest kilometre, how far is it from Appleton Farm to Iborden Farm?

d To the nearest kilometre, how far is it from Saunders to Cherry Tree Farm?

e Shona's farm is approximately 5 km from the Community Hall. What is the name of Shona's farm?

5

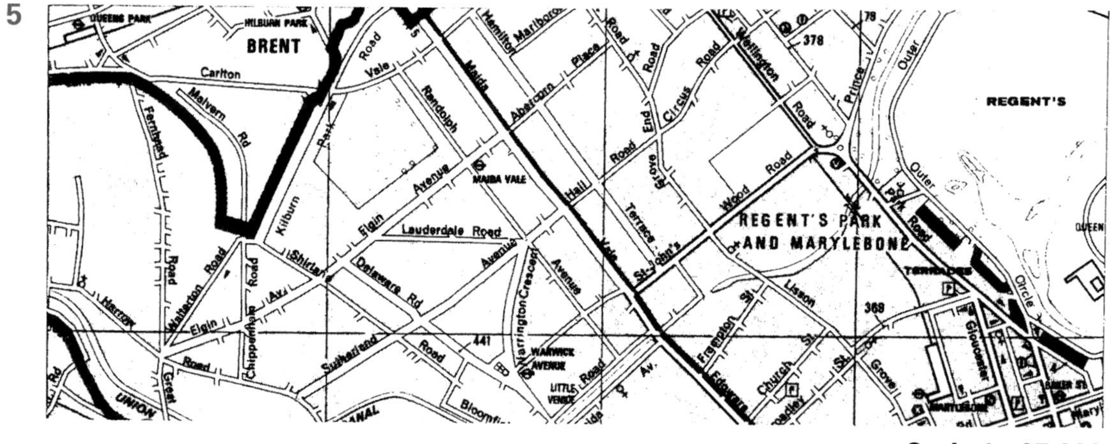

Scale 1 : 25 000

a How long is Hamilton Terrace? (Give the answer to the nearest 100 metres.)

b How long is St-John's Wood Road? (Give the answer to the nearest 100 metres.)

c Ann goes for a walk. She begins at the corner of Harrow Road and Sutherland Avenue. She walks along Sutherland Avenue, turns left into Randolph Avenue, then right down Abercorn Place, then right down Hamilton Terrace, then left down St-John's Wood Road. She finishes her walk at the corner of St-John's Wood Road and Wellington Road. Find how far Ann walked, to the nearest kilometre.

6 When a scale drawing is made are the angles on the scale drawing the same as the corresponding angles on the original?

Review

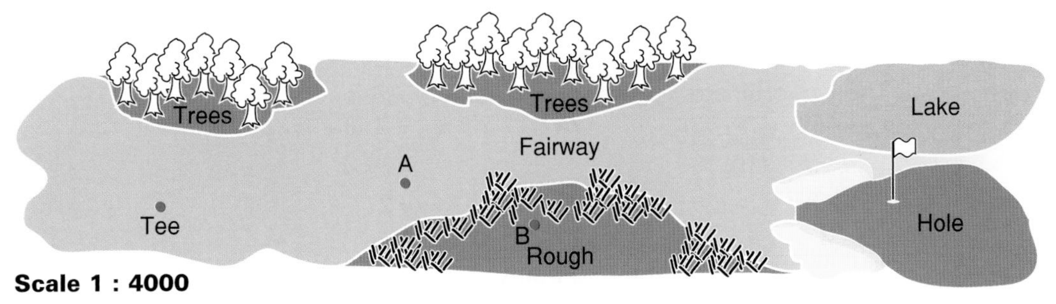

Scale 1 : 4000

This is a map of the No. 5 hole at the Summerhill Golf Course. Hetty's wood shot, from the tee, landed on the fairway at A. Keith's landed in the rough at B.

a On this map find the distance (in mm) from the tee to A and B.
b What distance (in m) did Hetty's ball travel?
c How far did Keith's ball travel?

Practical

1 Use a map of your town or district to find distances between landmarks. You could write some questions and give these to someone else to answer.

2 Use a road map of some part of Great Britain to find the distance, 'as the crow flies', between towns and cities.

3 Find a scale drawing and work out actual measurements. You could use a scale drawing of a boat, an aeroplane, a park or of something else.

Finding the mid-point of a line

Discussion

- Two points, A and B, have the same y-coordinate. The x-coordinate, x_m, of the mid-point of AB is found from

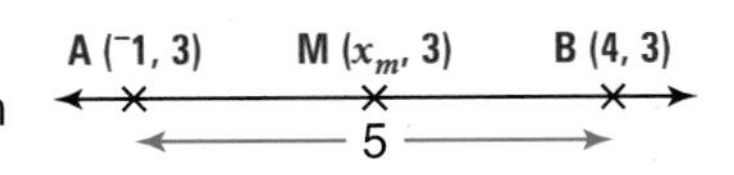

$$x_m = {}^{-}1 + \tfrac{1}{2} \times 5 = 1\tfrac{1}{2} \quad \textbf{or} \quad x_m = \tfrac{1}{2}({}^{-}1 + 4) = \tfrac{1}{2} \times 3 = 1\tfrac{1}{2}$$

x-coordinate of A **length of AB** **x-coordinate of A** **x-coordinate of B**

If A is the point $(x_1, 4)$ and B is the point $(x_2, 4)$, how could you write a general rule using x_1 and x_2 for finding the x-coordinate of the mid-point of AB? **Discuss.**

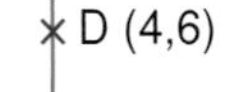

- How could you complete this to find the y-coordinate of the mid-point, y_m, of CD? **Discuss.**

$$y_m = {}^{-}2 + \tfrac{1}{2} \times \_\_\_ = \_\_\_ \quad \textbf{or} \quad y_m = \tfrac{1}{2}(\_\_\_ + \_\_\_) = \_\_\_$$

If C is the point $(4, y_1)$ and D is the point $(4, y_2)$, how could you write a general rule using y_1 and y_2 for finding the y-coordinate of the mid-point of CD? **Discuss.**

Example A($^-3$, 4) B(7, 4)

The x-coordinate, x_m, of the mid-point of AB is

$x_m = {}^-3 + \frac{1}{2}(7 - {}^-3)$ **or** $x_m = \frac{1}{2}({}^-3 + 7)$

$= {}^-3 + \frac{1}{2}(10)$ ← length of AB $\quad = \frac{1}{2} \times 4$

$= {}^-3 + 5$ $\quad = 2$

$= 2$

The coordinates of the mid-point of AB are **(2, 4)**.

Exercise 8

1 Find the coordinates of the mid-point of these horizontal lines, JK.

a J(2, 4), K(8, 4) **b** J(0, 8), K(10, 8) **c** J($^-2$, 3), K(2, 3) **d** J($^-5$, 1), K(6, 1)

2 Find the coordinates of the mid-points of these vertical lines, PQ.

a P(5, 4), Q(5, 10) **b** P($^-3$, 1), Q($^-3$, 9) **c** P(4, $^-3$), Q(4, 7) **d** P(2, $^-3$), Q(2, $^-4$)

Review Find the coordinates of the mid-points of these horizontal and vertical lines, CD.

a C(3, 4), D(11, 4) **b** C(6, $^-1$), D(6, 9) **c** C($^-5$, 3), D($^-5$, 10) **d** C($^-3$, $^-2$), D($^-6$, $^-2$)

Discussion

The x-coordinate of the mid-point of AB is the same as the x-coordinate of the mid-point of AC. How could you use the properties of a rectangle to explain why? **Discuss.**

The y-coordinate of the mid-point of AB is the same as the y-coordinate of the mid-point of BC. Why? **Discuss.**

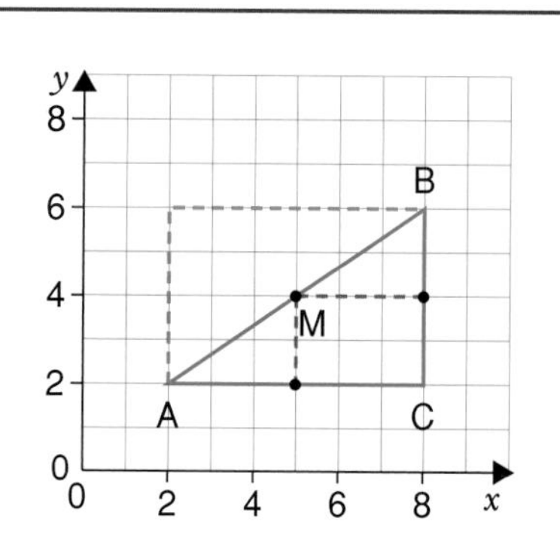

The **mid-point of a line segment joining A(x_1, y_1) to B(x_2, y_2)** is given by $\left(\frac{x_1+x_2}{2}, \frac{y_1+y_2}{2}\right)$.

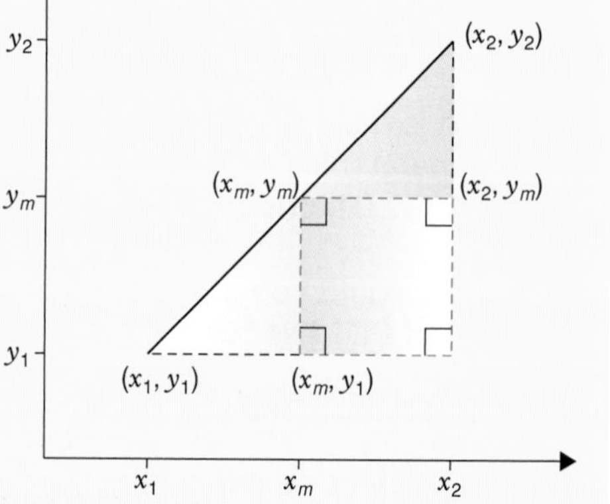

Worked Example

Find the mid-point of the line joining P($^-3$, 4) and Q(5, $^-10$).

Answer

The coordinates of the mid-point are given by $\left(\frac{x_1+x_2}{2}, \frac{y_1+y_2}{2}\right)$.

Let P be the point (x_1, y_1) and Q be the point (x_2, y_2).

Then $x_1 = {}^-3$, $y_1 = 4$, $x_2 = 5$, $y_2 = {}^-10$.

Coordinates of mid-point $= \left(\frac{{}^-3+5}{2}, \frac{4+{}^-10}{2}\right)$

$= \left(\frac{2}{2}, \frac{{}^-6}{2}\right)$

$=$ **(1, $^-3$)**

$\frac{1}{2}(x_1 + x_2)$ is the same as $\frac{x_1 + x_2}{2}$.

$\frac{x_1 + x_2}{2}$ is the **mean of the x-coordinates.**

Exercise 9

1 Plot the points A and B on a grid.

Find the coordinates of the mid-point of the line joining A and B using $\left(\frac{x_1+x_2}{2}, \frac{y_1+y_2}{2}\right)$.

Check on your grid that this gives the correct answer.

a A(2, 6), B(4, 8) **b** A(0, 3), B(4, 7) **c** A(1, 4), B(4, 8) **d** A(3, 2), B(8, 7)
e A($^-1$, 3), B(5, $^-2$) **f** A($^-3$, $^-5$), B(3, 2) **g** A($^-3$, $^-6$), B($^-4$, $^-7$)

2 Find the mid-point of the line AB.

a A(1, 2), B(5, 6) **b** A(2, 5), B(6, 11) **c** A($^-3$, 2), B($^-6$, 4)
d A(5, $^-2$), B(3, $^-6$) **e** A($^-4$, 2), B(5, $^-3$) **f** A(3, $^-1$), B($^-8$, $^-7$)
g A($^-6$, $^-4$), B($^-8$, $^-7$) **h** A($^-2$, $^-11$), B($^-5$, 7) **i** A(12, $^-13$), B($^-8$, 3)
j A(24, $^-18$), B($^-6$, 7)

Review Find the mid-point of the line joining CD.
a C(5, 4), D(3, 8) **b** C($^-3$, 4), D($^-9$, 6) **c** C($^-3$, $^-4$), D($^-13$, $^-9$) **d** C($^-3$, 6), D(4, $^-2$)

Summary of key points

Coordinates and transformations

Example The red shape has been reflected in the y-axis, to give the green shape.
The coordinates of the vertices of the original shape are ($^-3$, 0), ($^-3$, 3), ($^-1$, 3) and ($^-3$, 1).
The coordinates of the vertices of the image are (3, 0), (3, 3), (1, 3) and (3, 1).

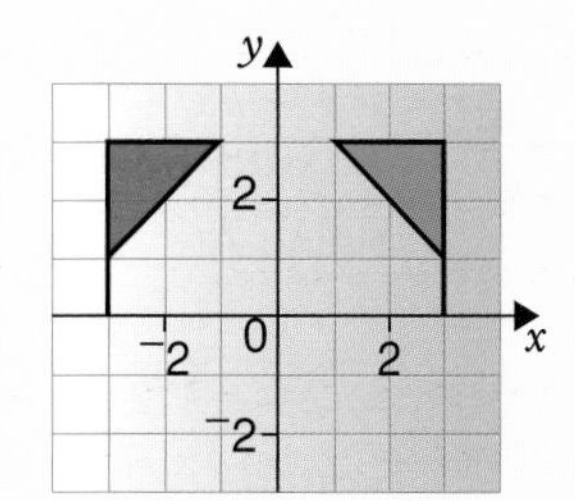

We can transform shapes using a **combination of transformations**.

Example A has been reflected in the y-axis and then translated 3 units down.

Some combinations have a predictable result.

A reflection in two parallel lines is equivalent to a translation.
A reflection in two perpendicular lines is equivalent to a half-turn rotation.
Two rotations about the same centre are equivalent to a single rotation.
Two translations are equivalent to a single translation.

Symmetry

Some 3-D shapes are **symmetrical**.

Example The shaded shape is called a plane of **symmetry**.
It divides the shape into two congruent pieces.

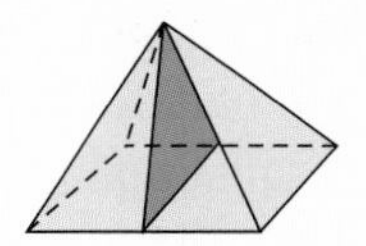

When a shape is **enlarged** the ratio of any two corresponding line segments is equal to the **scale factor**.
Under enlargement, angles always stay the same but lengths do not.
If a shape with perimeter, n is enlarged by a scale factor of **2** the perimeter of the image is $\mathbf{2} \times n$.

A **scale drawing** represents something in real life.

Scale: 1 mm represents 6 cm

Example This is a scale drawing of a car.
Each millimetre on the drawing represents 6 cm in real life.
So 1 mm on the drawing represents 60 mm in real life.
The car is 51 mm on the drawing.
In real life it is $51 \times 60 = 3060$ mm
$= 306$ cm
$= 3{\cdot}06$ m.

To make a scale drawing you need to:
1 make a rough sketch with the real-life measurements written on it,
2 choose a scale,
3 work out what each measurement on the scale drawing will be.

The **mid-point** of the line segment joining $A(x_1, y_1)$ to $B(x_2, y_2)$ is given by
$\left(\frac{x_1 + x_2}{2}, \frac{y_1 + y_2}{2}\right)$.

Example The mid-point, M, of the line joining AB is
$\left(\frac{3 + 13}{2}, \frac{4 + 10}{2}\right) = (8, 7)$.

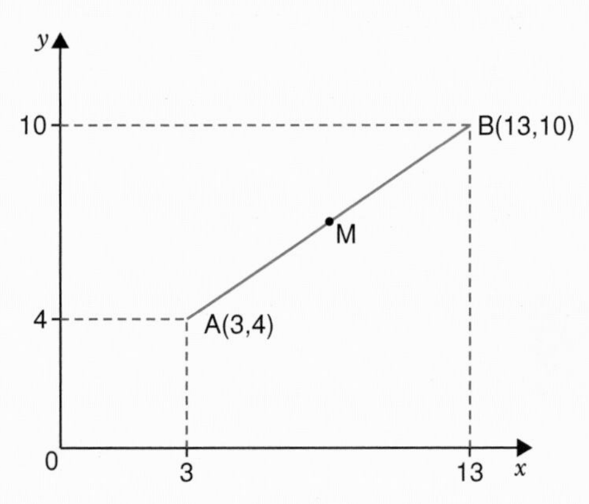

Test yourself

1 A($^-$2, 3), B(1, 3), C($^-$3, $^-$1)
Plot A, B and C to make a triangle ABC.
Write down the coordinates of A, B and C after these transformations.
a translation 4 units to the right **b** rotation 90° about (0, 0) **c** reflection in the x-axis.

2 Which of these are true?
a A reflection in two parallel lines is equivalent to a translation.
b A reflection in two perpendicular lines is equivalent to a half-turn rotation.
c Two rotations about the same centre are equivalent to a single rotation.
d Two translations are equivalent to a single translation.

3 Write true or false for these.

- **a** If ABC → A′B′C′ by rotation ∠A = ∠A′.
- **b** If ABC → A′B′C′ by translation AB = A′B′.
- **c** If ABC → A′B′C′ by reflection ∠B = ∠B′ and AC = A′C′.
- **d** If ABC → A′B′C′ by enlargement AB = A′B′.
- **e** If ABC → A′B′C′ by enlargement ∠C = ∠C′.

4 The red triangle is reflected in the y-axis to the green triangle.

- **a** Describe the transformation which maps the red triangle onto the blue triangle.
- **b** Describe a combination of two transformations that will map the green triangle onto the blue triangle.

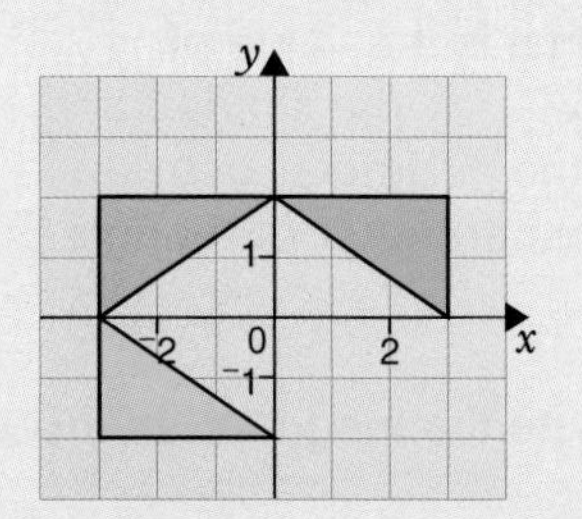

5 A right-angled isosceles triangle is rotated 90° about its right angle. What shape will the combined object and image form?

6 P maps onto P″ by first reflecting P in the line $y = x$ and then rotating the image, P′, by 270° about the origin.

- **a** Show that P could also be transformed to P″ by a combination of a reflection and a translation.
- **b** Show that P could also be transformed to P″ by another combination of two transformations.
- **c** Describe a combination of two transformations that could map Q onto Q″.
- **d** Give another answer for **c**.

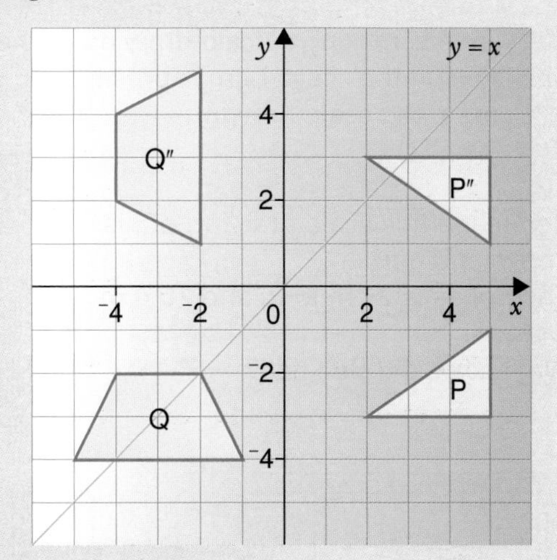

7 Some transformations are given:

A: reflection in y-axis — **B**: reflection in the line $x = 1$
C: translation 2 units right — **D**: translation 3 units left and 1 unit up
E: rotation 90° about (0, 0) — **F**: rotation 180° about (0, 0)

- **a** If ABC → A′B′C′ under any of the transformations **A** to **F** what can you say about
  - **i** the length of AB and the length of A′B′
  - **ii** ∠B and ∠B′?
- **b** Gendi transformed a shape first by **A** and then by **B**. Which of these is the same as these two transformations?
  A translation **B** rotation **C** reflection
- **c** Matthew transformed a shape first by **C** and then by **D**. Which of these is the same as these two transformations?
  A translation **B** rotation **C** reflection
- **d** Write down the rotation that is the same as doing **E** and then **F**.
- **e** Will doing **A** then **E** give the same result as doing **E** then **A**?

8 Use two copies of this shape.
Sketch a different plane of symmetry on each.
Shade your planes.

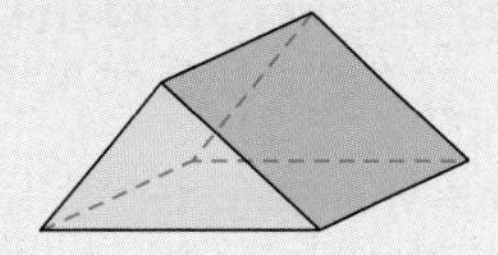

9 The red shape has been enlarged to the purple shape.
Find the scale factor for the enlargement.

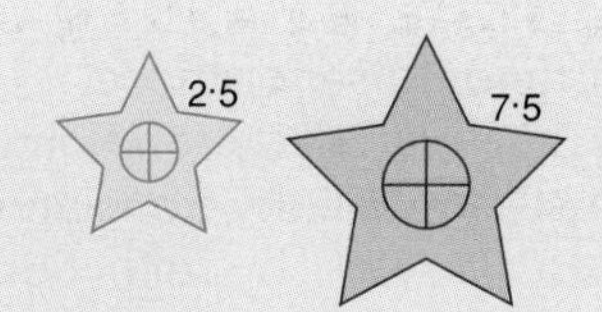

10 **a** The large triangle, B, is an enlargement of the small triangle, A.
What is the value of
i x **ii** the shaded angle?

b If the perimeter of triangle A is 12 cm, what is the perimeter of triangle B?

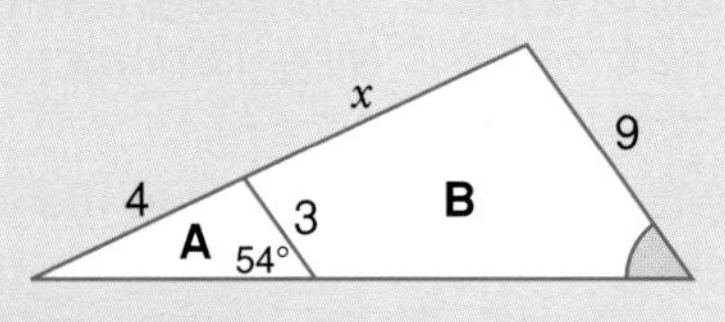

11 Estimate the length of the house in this picture.
The man is about 1·8m tall.

12 Maryanne is making a scale drawing of the school grounds.
She is using the scale 1 cm represents 10 m. How long in real life is a length that is
a 4 cm on the scale drawing **b** 15·5 cm on the scale drawing?

13 Ben made this scale drawing of his house and garden.
The house is 24 m long and 10·4 m wide.
What scale did Ben use?

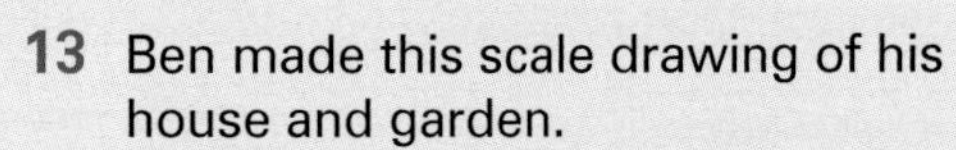

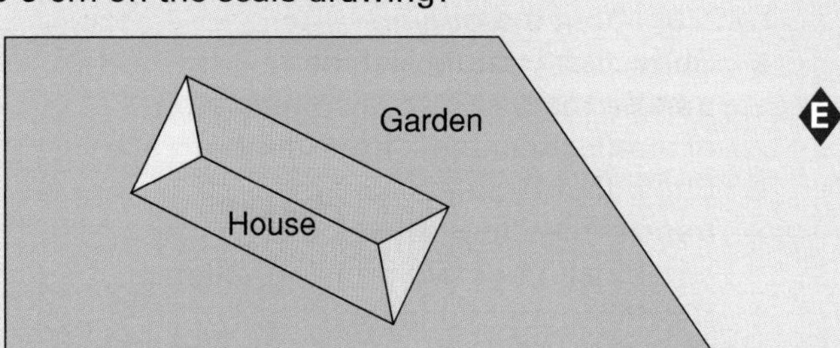

14

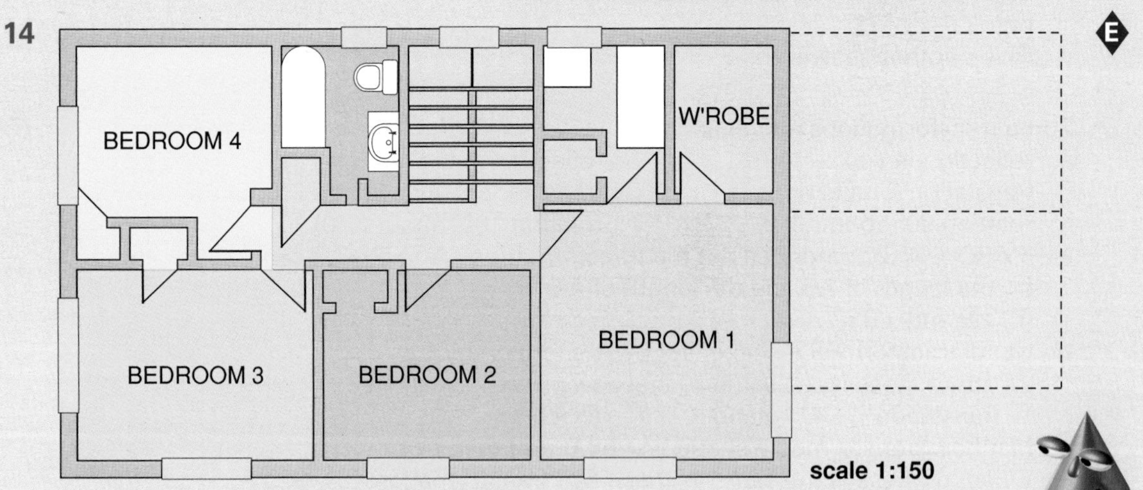

scale 1:150

This plan is of the top floor of a house.

a How long is the outside wall that runs alongside Bedrooms 3 and 4?
b What are the inside dimensions of Bedroom 1?
c What are the inside dimensions of Bedroom 3?
d The dotted lines show the ceiling of the sitting room which is at ground level.
What are the dimensions of this sitting room?

Give your answers to the nearest 10 cm.

15 Write down the coordinates of the mid-points of AB.
a A(3, 9) B(4, 17) **b** A($^{-}3$, 4) B($^{-}7$, $^{-}2$)

14 Measures, Perimeter, Area and Volume

You need to know

Key vocabulary

bearing, three-figure bearing, circumference, chord, displacement, foot, yard, hectare, pi, sector, segment, tonne, volume: cubic millimetre (mm^3), cubic centimetre (cm^3), cubic metre (m^3)

Makeover

Paint For two coats of paint, allow 1 litre of paint for every 8 square metres to be painted.

Curtain material width needed: twice the width of the windows
length needed: length of windows + 20 cm extra

Wallpaper

	WALLPAPER CHART														
height of room in metres	distance around the room in metres (m)														
	9	10	11	13	14	15	16	18	19	20	21	23	24	25	26
2·00 m – 2.30 m	4	5	5	6	6	7	7	8	8	9	9	10	10	11	12
2·31 m – 2.50 m	5	5	6	6	7	7	8	8	9	9	10	11	11	12	13
2·51 m – 2.70 m	5	6	6	8	8	8	9	10	11	11	12	13	13	13	14
2·71 m – 3.00 m	6	6	7	8	9	9	10	11	12	12	13	14	14	15	16
3·01 m – 3.20 m	6	7	8	8	10	10	10	12	12	13	13	14	15	16	16
	Number of rolls needed														

Use this information to do a project on decorating a room in your house, or school, in the colour scheme of your choice.

As part of your project, take measurements, draw plans, work out how much paint, wallpaper, curtain material etc. you will need and find the total cost.

Metric conversions, including area, volume and capacity

Remember

Length 1 km = 1000 m 1 m = 100 cm 1 m = 1000 mm 1 cm = 10 mm

Mass 1 tonne = 1000 kg 1 kg = 1000 g

Capacity 1 ℓ = 1000 mℓ 1 ℓ = 100 cℓ 1 cℓ = 10 mℓ 1 ℓ = 1000 cm^3 1 mℓ = 1 cm^3
1000 ℓ = 1 m^3

Area 1 hectare = 10 000 m^2

Time 1 hour = 60 minutes
1 minute = 60 seconds
1 day = 24 hours
1 week = 7 days
1 year = 365 days (366 in a leap year)
1 decade = 10 years

See page 253 for more on metric conversions.

Discussion

- How many squares of side 1 cm will fit on a square of side 1 m? **Discuss.**

  How many squares of side 1 mm will fit on a square of side 1 cm? **Discuss.**

- How many cubes of side 1 mm will fit inside a cube of side 1 cm? **Discuss.**

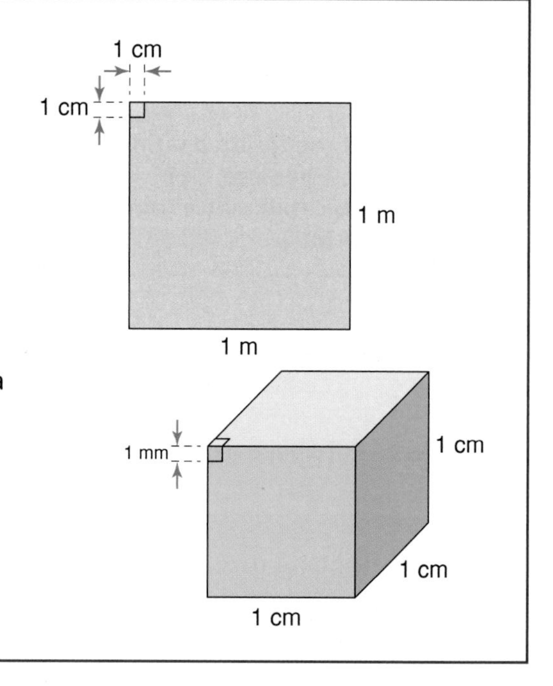

Area	**Capacity**
1 m^2 = 10 000 cm^2	**1 cm^3 = 1000 mm^3**
1 cm^2 = 100 mm^2	**1 m^3 = 1 000 000 cm^3**

Worked Example

What is the volume of this tank in
a m^3 **b** cm^3 **c** mℓ **d** ℓ?

There is more about the volume of a cuboid on page 254.

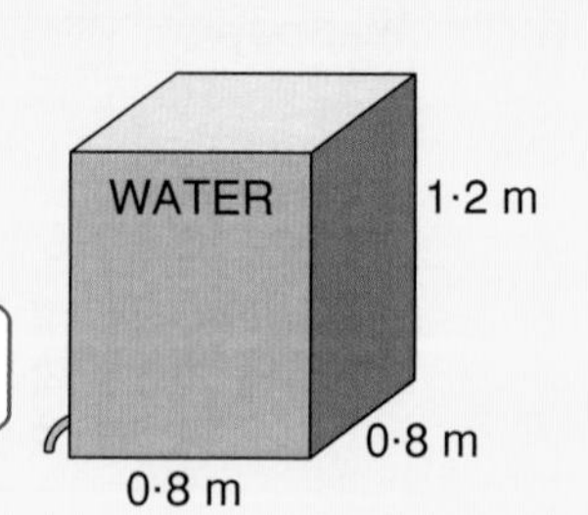

Answer

a Volume of a cuboid = $w \times l \times h$
= 0·8 × 0·8 × 1·2
= **0·768 m^3**

b 0·768 m^3 = (0·768 × 1 000 000) cm^3
= **768 000 cm^3**

c 1 cm^3 = 1 mℓ
So volume of tank in mℓ = **768 000 mℓ**

d 1 ℓ = 1000 cm^3
So 768 000 cm^3 = (768 000 ÷ 1000) ℓ
= **768 ℓ**

Exercise 1

1 Change

a 700 mm to m **b** 37 mm to cm **c** 280 cm to m **d** 1450 cm to m
e 0·8 ℓ to mℓ **f** 1564 g to kg **g** 360 mm to m **h** 400 cℓ to ℓ
i 5680 g to kg **j** 36 mm to m **k** 0·86 m to cm **l** 50 mℓ to ℓ
m 49 g to kg **n** 3 mm to m **o** 9 g to kg **p** 7600 kg to tonnes
q 420 kg to tonnes **r** 8·2 tonnes to kg **s** 0·3 tonnes to kg **t** 85 mℓ to cℓ.

2 Change

a 3·6 ha to m^2 **b** 83 ha to m^2 **c** 86 423 m^2 to ha **d** 80 m^2 to ha
e 0·62 ha to m^2 **f** 4·6 m^3 to ℓ **g** 43 cm^3 to mℓ **h** 8420 cm^3 to ℓ
i 3·5 ℓ to cm^3 **j** 5682 ℓ to m^3 **k** 0·82 ℓ to cm^3 **l** 0·28 cm^3 to mℓ
m 32 cm^3 to ℓ **n** 864 cm^3 to cℓ.

3 How many minutes in
a 4 hours **b** 5·5 hours **c** 3·25 hours **d** 6 hours 35 minutes?

4 How many hours and minutes in
a 180 minutes **b** 95 minutes **c** 500 minutes **d** 426 minutes?

5 How many hours in
a 4 days **b** $6\frac{3}{4}$ days **c** 1 week 3 days **d** 3 weeks 1 day 15 hours **e** 462 minutes?

6 Change
a 1324 months to years and months **b** 587 months to decades, years and months
c 862 weeks to years and weeks *__d__ 4000 minutes to days and hours and minutes.

7 Change

a 5 m^2 to cm^2 **b** 9·6 m^2 to cm^2 **c** 8 cm^3 to mm^3
d 5000 mm^2 to cm^2 **e** 9·25 m^2 to cm^2 **f** 0·042 m^3 to cm^3
g 987 200 cm^2 to m^2 **h** 92 700 000 cm^3 to m^3 **i** 780 000 mm^3 to cm^3
j 8·4 cm^2 to mm^2 **k** 330 000 mm^2 to cm^2 **l** 0·421 m^3 to cm^3
m 48 000 square centimetres to square metres
n 7·5 cubic centimetres into cubic millimetres
o 89·62 square centimetres to square millimetres.

8 What is the area of this poster in
a m^2 **b** cm^2 **c** mm^2?

THE Moon People

1·2 m

0·75 m

9 What is the volume of this dish in
a cm^3 **b** mm^3 **c** mℓ **d** ℓ?

8 cm

14 cm

28 cm

T

Review 1

4·23 5600 84 000 42·3 84 000   0·423 5·6

U (above 840)

84 000 0·056 56 000 **840** 4·275 5600   0·0056 42 700 42·3 56 56 000 0·056

U (above 840)

0·423 0·056   4·23 5600 84 000   5600 **840** 423 42 700 0·056

56 56 000 427·5 8400   4·23 56 000   0·84 0·423 0·084 0·084

U (above 840)

0·056 0·423 0·056 84 000   4·23 5600 56 000 **840** 5·6 42 700 0·056 427·5

42·75 84 000 0·056 0·0056 0·423 0·084 5·6

Use a copy of this box.
What goes in the gap? Write the letter beside each question above its answer in the box.

U 0·84 tonnes = **840** kg
I 4230 m^2 = ___ ha
G 4275 ℓ = ___ m^3
E 8·4 m^2 = ___ cm^2
A 0·0427 m^3 = ___ cm^3
C 5600 cm^3 = ___ m^3
L 840 cm^2 = ___ m^2
T 42·3 mℓ = ___ cℓ
H 5·6 m^3 = ___ ℓ
Y 8·4 ℓ = ___ cm^3
M 4·23 cm^2 = ___ mm^2
F 840 000 cm^3 = ___ m^3
N 560 cm^2 = ___ m^2
O 5·6 ha = ___ m^2
R 42·3 cm^3 = ___ mℓ
D 427 500 cm^3 = ___ ℓ
B 560 000 cm^2 = ___ m^2
S 560 mm^2 = ___ cm^2
P 42 750 mm^3 = ___ cm^3

Review 2 How many

a minutes in 5·4 hours
b hours and minutes in 572 minutes
c hours in 2 weeks 4 days
d years and months in 486 months?

Working with measures

The following exercise gives you practice at **estimating measures** and **solving measures problems**.
You often need to know metric conversions or the rough metric and imperial equivalents to solve problems.
You also sometimes need to know the formulae for the perimeter and area of shapes.

See page 253 for the metric and imperial equivalents.

When estimating, it is useful to compare to something you know well.

Example
A door is about 2 m high so the height of a window is about 1 m.

Exercise 2

1 Write down an approximate measurement for each of these.
Give a range for each.

a the depth, in cm, of a step in a staircase
b the amount of water, in ℓ, an electric jug kettle holds
c the thickness, in mm, of your front door key
d the area, in cm^2, of your school desk
e the volume, in mℓ, of a mug of coffee
f the height, in m, of your bedroom window
g the temperature, in °C, today
h the area, in cm^2, of a slice of toast
i the area, in cm^2, of a computer screen
j the width, in cm, of your face at the cheekbones
k the mass, in kg, of a bike
l the mass, in kg, of your desk

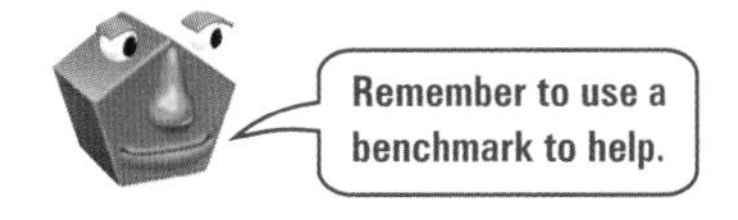

2

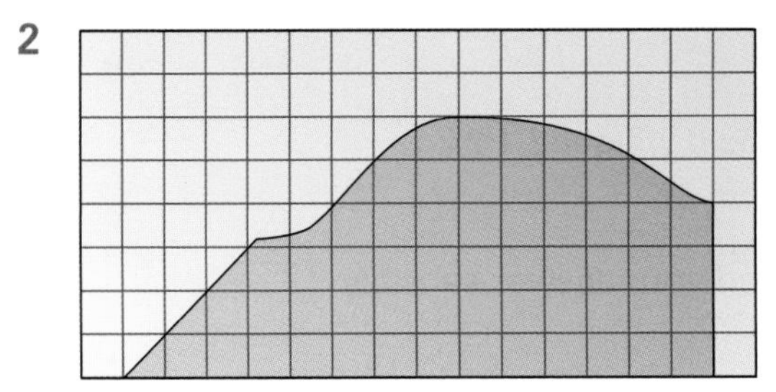

Each square has a length of 50 m.

a What is the area of each square?
b How many squares make 1 hectare?
c By counting squares, find the area of land sketched.
(Give your answer to the nearest hectare.)

3 This is a sketch of a rectangular park.
How many metres wide is this park?

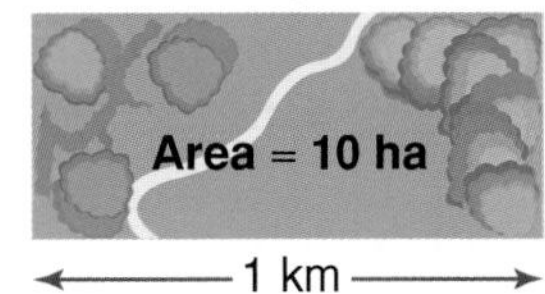

4 a Bill bought a 1 ℓ pack of ice cream. He used 55 cℓ of this to make a frozen pudding. Was there enough ice cream left to make another of these frozen puddings?
b How much ice cream, in cm^3, was left?

5 Use this chart to work out how much

a 85p is in rand
b £55 is in ringgit
c £1050 is in euro
d 1000 yuan is in pounds
e 5000 Jamaican dollars is in pounds
f 20 000 rupees is in pounds.

TODAY'S CONVERSION RATES

for **£1** you get

10·8	rand	**13·16**	yuan (China)
79·32	Jamaican dollars	**6·04**	ringgit (Malaysia)
1·57	euro	**92·24**	rupees

6 a What length of this wallboard, in cm, is needed to replace an 8 foot length?
b How much will it cost?

Wallboard
£16.50 per 30 cm

7 The diameter of a red blood cell is 0·000714 cm.
The diameter of a white blood cell is 0·001243 cm.

a Work out the difference between these two diameters.
Give your answer in millimetres.

b How many white cells would fit across a needle point which has a diameter of 0·12 mm?

8 An old knitting pattern uses twenty 1 oz balls of wool.
How many 25 g balls are needed to make this pattern?

9 This diagram shows a square inside a square.
Find the area of the red square.

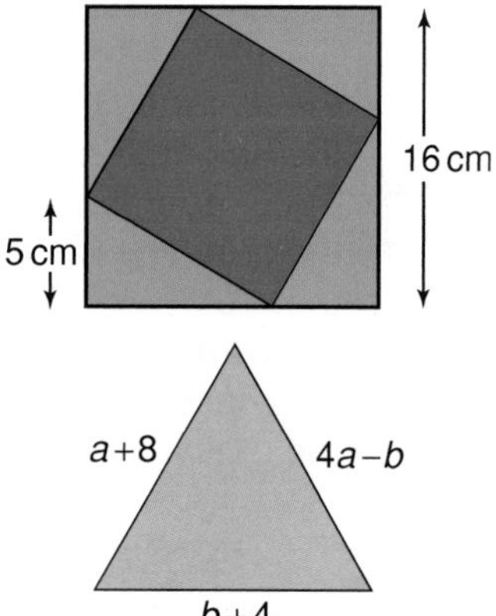

10 The length of a rectangle is 6 cm more than its width.
Its area is 112 cm^2.
What is its perimeter?

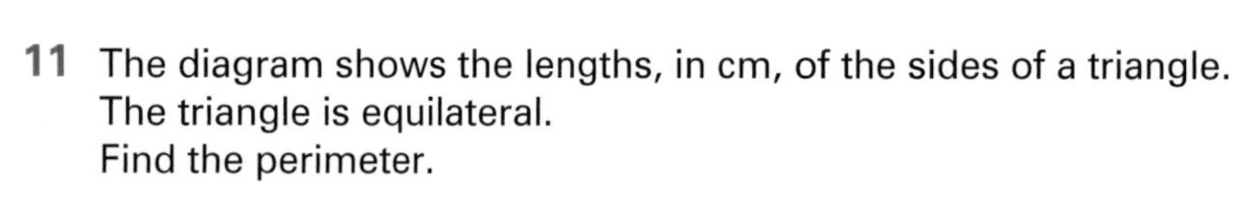

11 The diagram shows the lengths, in cm, of the sides of a triangle.
The triangle is equilateral.
Find the perimeter.

12 A uniform block of cheese in the shape of a cuboid weighed 1·2 kg.
5 cm of the cheese was cut off its length.
The remaining block weighed 0·75 kg.
What was the length of the original block? Round your answer sensibly.

13 **a** A 4 ℓ can of paint can cover 54 m^2.
How many cans are needed to paint the walls and door of a room 6·2 m by 4·8 m and 3·2 m high? The room has two windows 2·4 m wide and 1·6 m high.

***b** About how many litres of paint will be left over?

14 A dripping tap fills a 1 pint container in 4 minutes.
How much water will drip from the tap in 24 hours?
Give your answer in **a** pints **b** mℓ **c** ℓ.

***15** **a** Express 40 km/h in mph.
b Express 20 km/h in metres per second.

Review 1 Write down an approximate measurement for each of these.
Give a range. Use metric units.

a the area, in cm^2, of the top of a filing cabinet
b the volume, in ℓ, of a laundry basket
c the mass, in kg, of a telephone

Review 2 A bottle of juice holds 1500 mℓ.

a Joel drank 35 cℓ. How many cℓ are left?
b Four friends share the rest equally. How many cm^3 does each get?

Review 3 A rectangular park is 5·6 ha. One side is 160 m long.
How long is the other side?

Review 4

a The price of petrol was quoted in an old log book as £1·80 per gallon.
How much is this per litre?
b If it is now 68p per litre, how much is this per gallon?

* **Review 5** Pippa's grandmother used this recipe to make pavlova.
Change the amounts to metric measures.

Pavlova

$\frac{1}{4}$ pound of sugar
6 eggs
2 tsp vinegar
Beat egg white until stiff.
Add sugar and vinegar slowly.
Pour into a 10 inch mould.
Bake at 380° Fahrenheit for 1 hour.

* **Review 6** James changed £1000 into US dollars for a trip. He didn't go on the trip because of ill health. He changed the dollars back into pounds. If 1% commission is charged for **each** transaction, how much did he lose?
Note £1 = US\$1·58

Perimeter and area

Investigations

1 Moving vertices

You will need a pinboard or square dotty paper.

a ABCD is a rectangle.
A and B are translated 1 unit to the right to A′ and B′.
How does this affect the area?
What if A and B are translated 2 units to the right to A″ and B″?
What if A and B are translated 3 units to the right?
What if ...
Find the area of each parallelogram and explain what is happening.

A A′ A″ B B′ B″
D C
1 cm
1 cm

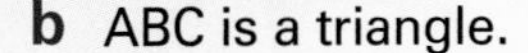

b ABC is a triangle.
A has been translated 1 unit right to A_1.
How does this affect the area?
What if A is translated 2 units right to A_2?
What if A is translated 3 units right?
What if ...
Find the area of each triangle and explain what is happening.

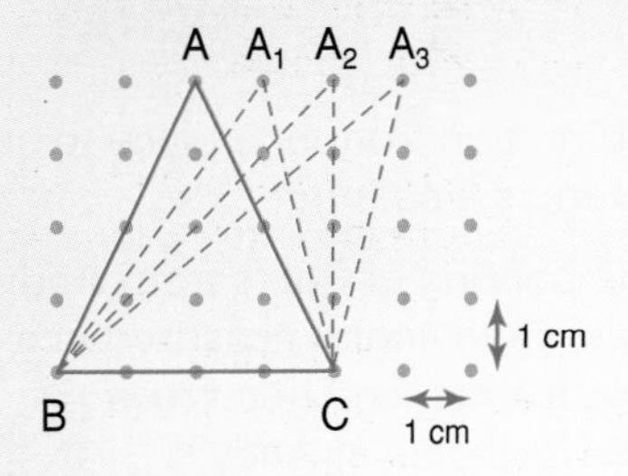

c Repeat **b** but translate A by 1, 2, 3, 4, ... units up.

2 Hexagons

Hetty made this shape with five regular hexagons of side n.
What is the perimeter of Hetty's shape?
What is the smallest perimeter for a shape made with five regular hexagons of side n?
What about six regular hexagons of side n?
What about seven regular hexagons of side n?
What about eight, nine, ... regular hexagons of side n?

3 Triangles in cubes

Bryan made a triangle by joining three of the vertices of a cube.
He said,
'The triangle I made is a right-angled scalene triangle.'
How many different shaped triangles can you make by joining three vertices of a cube?
How many are isosceles?
How many are equilateral?
Which one has the greatest area? Justify your choice.

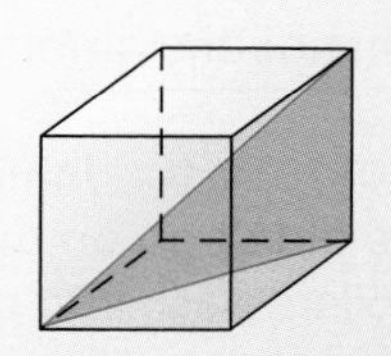

Bearings

Discussion

How might a pilot give the position of the aircraft to the control tower? **Discuss.**

How might a sailor give the position of the ship from a lighthouse? **Discuss.**

Who else might need to give their position? **Discuss.**

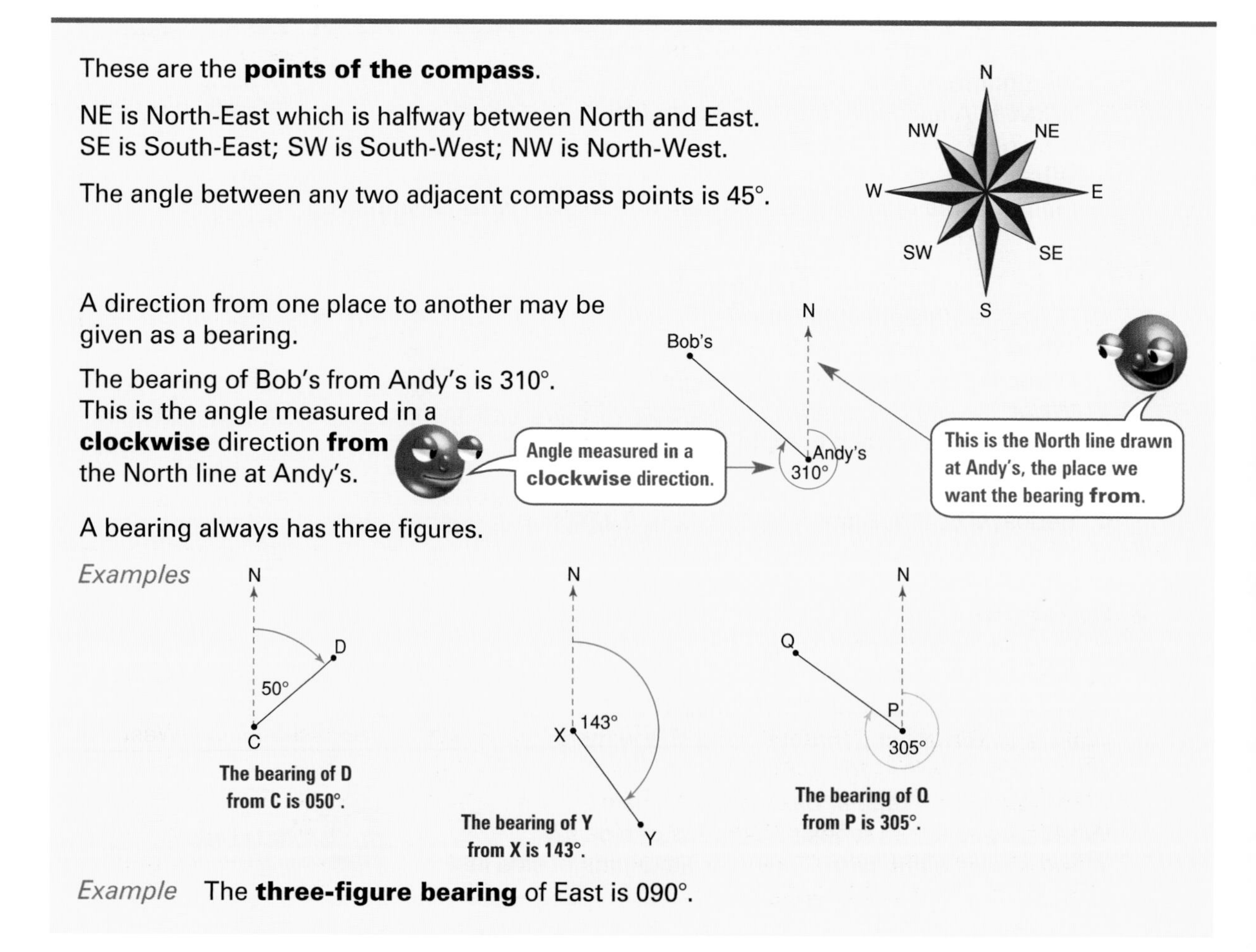

These are the **points of the compass**.

NE is North-East which is halfway between North and East.
SE is South-East; SW is South-West; NW is North-West.

The angle between any two adjacent compass points is 45°.

A direction from one place to another may be given as a bearing.

The bearing of Bob's from Andy's is 310°.
This is the angle measured in a **clockwise** direction **from** the North line at Andy's.

A bearing always has three figures.

Examples

The bearing of D from C is 050°.

The bearing of Y from X is 143°.

The bearing of Q from P is 305°.

Example The **three-figure bearing** of East is 090°.

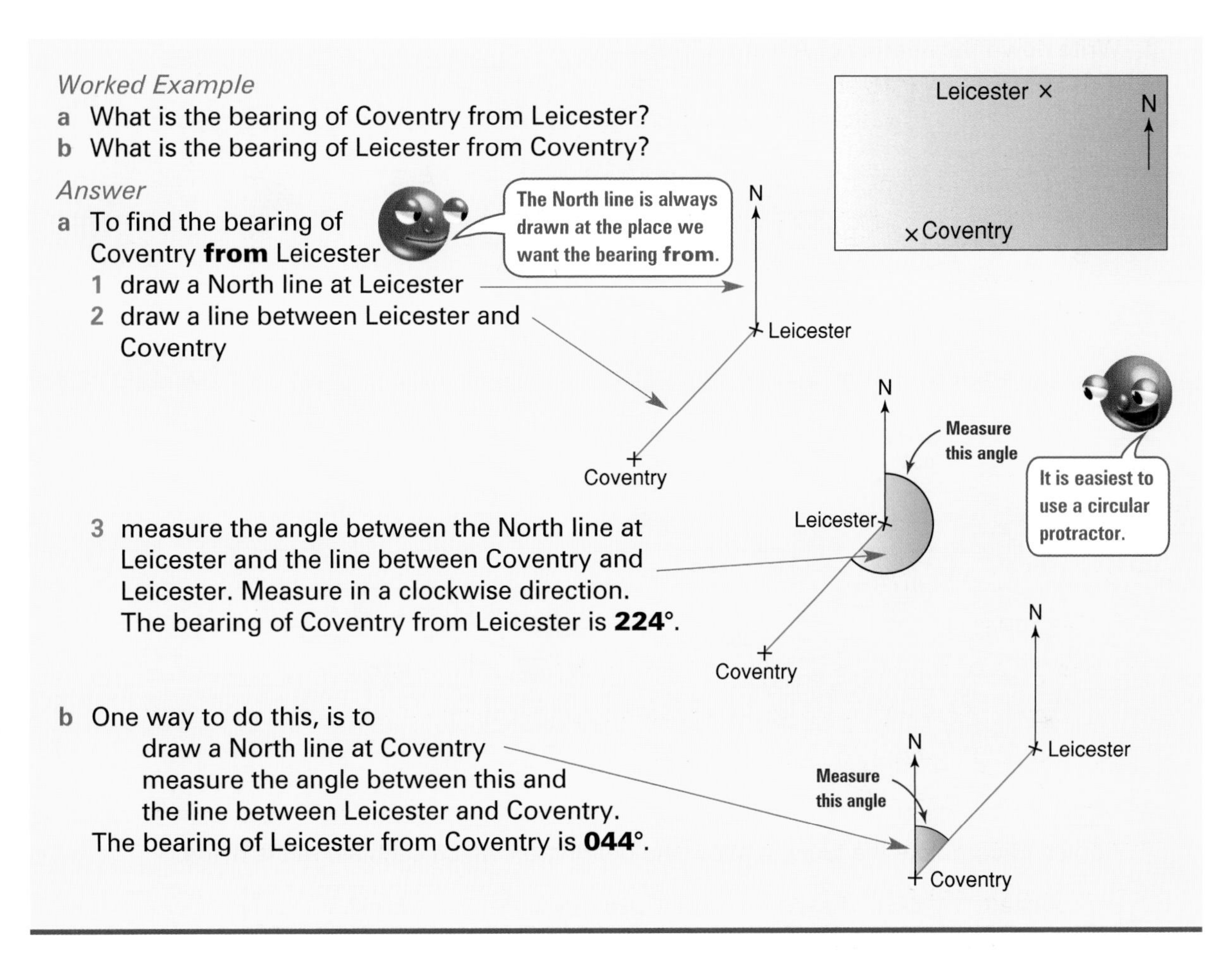

Worked Example

a What is the bearing of Coventry from Leicester?

b What is the bearing of Leicester from Coventry?

Answer

a To find the bearing of Coventry **from** Leicester

1. draw a North line at Leicester
2. draw a line between Leicester and Coventry
3. measure the angle between the North line at Leicester and the line between Coventry and Leicester. Measure in a clockwise direction. The bearing of Coventry from Leicester is **224°**.

b One way to do this, is to
draw a North line at Coventry
measure the angle between this and
the line between Leicester and Coventry.
The bearing of Leicester from Coventry is **044°**.

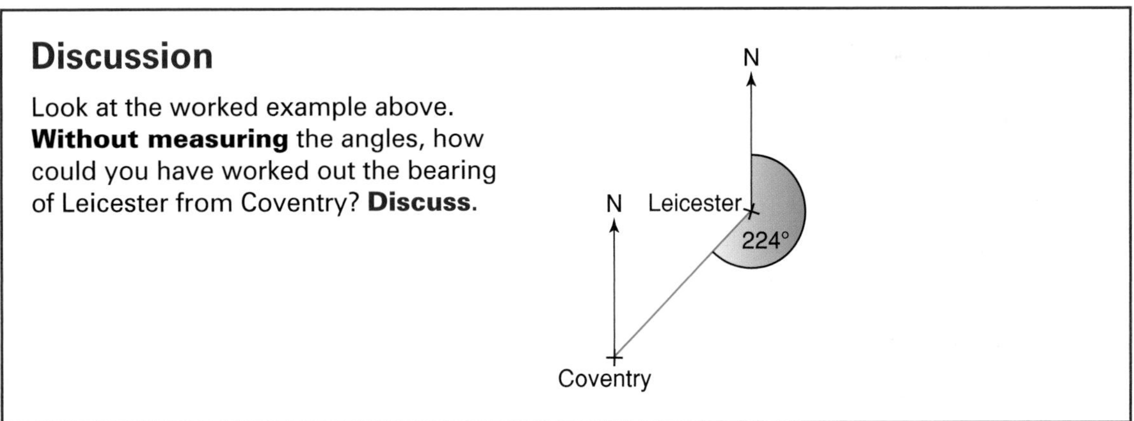

Discussion

Look at the worked example above. **Without measuring** the angles, how could you have worked out the bearing of Leicester from Coventry? **Discuss.**

Exercise 3

1 a	The bearing 090° is the same as	**A** North	**B** South	**C** East	**D** West.
b	The bearing 000° is the same as	**A** North	**B** South	**C** East	**D** West.
c	The bearing 270° is the same as	**A** North	**B** South	**C** East	**D** West.
d	NW is the same as a bearing of	**A** 045°	**B** 135°	**C** 225°	**D** 315°.
e	SE is the same as a bearing of	**A** 045°	**B** 135°	**C** 225°	**D** 315°.
f	SW is the same as a bearing of	**A** 045°	**B** 135°	**C** 225°	**D** 315°.

2 Write down the bearing of A from B.

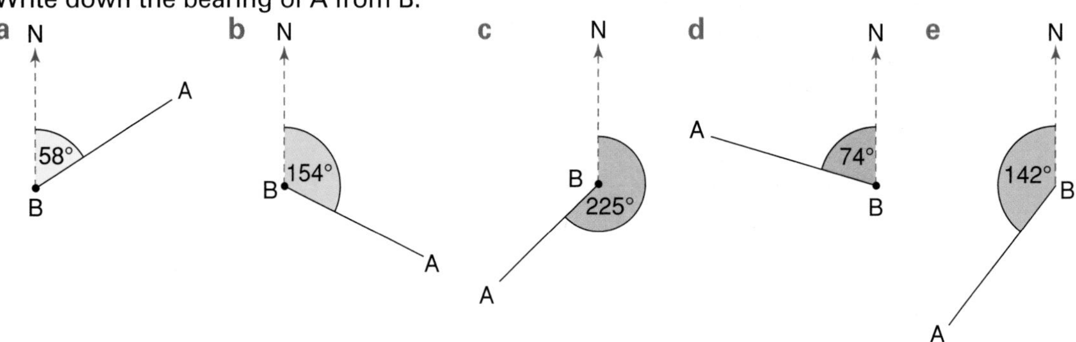

3

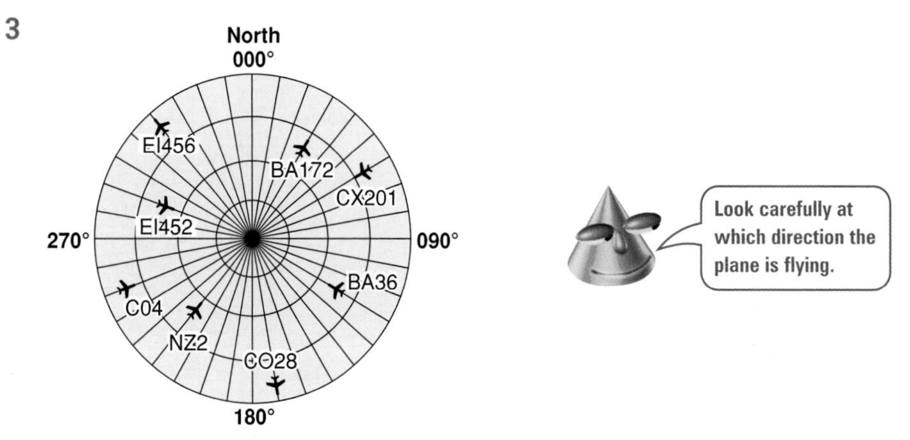

Copy and complete the table to show the bearing on which each aircraft is flying.

Aircraft	BA36	BA172		CO28	NZ2		EI452	
Bearing	300°		320°			070°		240°

4 Draw diagrams to show these bearings.
a 130° **b** 021° **c** 220° **d** 320°

5 Use your protractor to find the bearing of A from B.

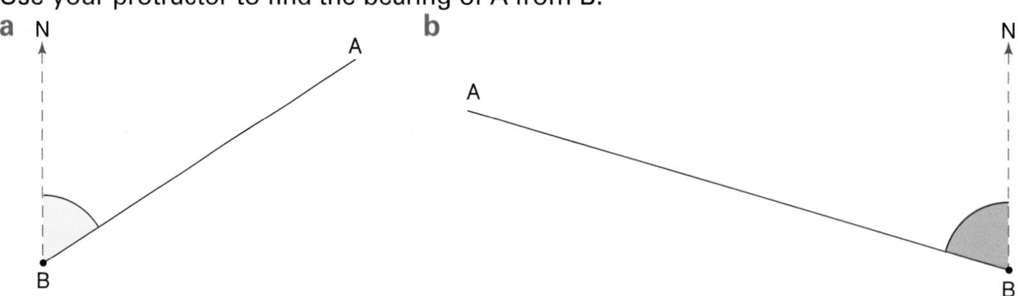

6 For the diagrams given in question **5**, find the bearing of B from A.

7 What is the bearing of **S from T**?

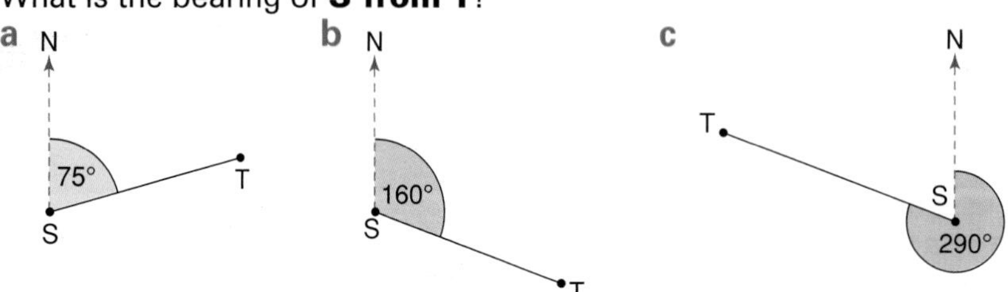

8 Use a larger copy of this diagram.
This diagram shows the positions of three towns.
Measure and write down the bearing of these.

a Waterford from Kilkenny
b Clonmel from Kilkenny
c Clonmel from Waterford
d Kilkenny from Waterford
e Kilkenny from Clonmel

N
N
Waterford
Kilkenny
N
N
Clonmel

9

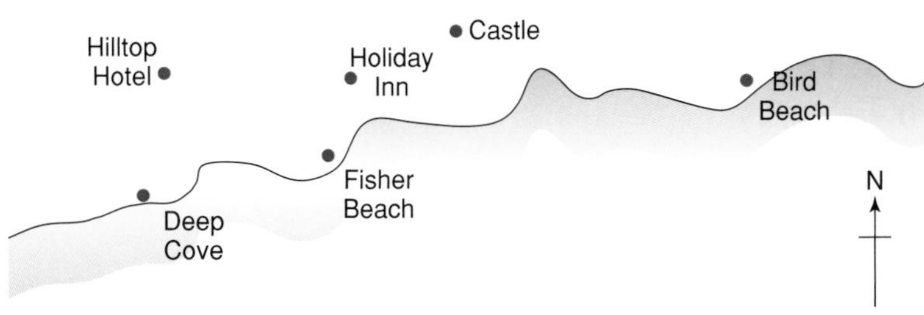

Measure and write down the bearing of these.

a Holiday Inn from Deep Cove
b Fisher Beach from Hilltop Hotel
c Fisher Beach from Bird Beach
d The Castle from Fisher Beach
e Hilltop Hotel from Fisher Beach
f Bird Beach from the Castle

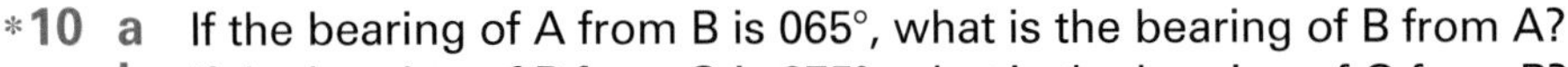

*** 10** **a** If the bearing of A from B is 065°, what is the bearing of B from A?
b If the bearing of P from Q is 275°, what is the bearing of Q from P?

*** 11** **a** A ship sails from Portsmouth on a course of 202°. It returns along the same shipping route. What is its course on the return journey?
b Write down a general rule for finding the bearing of a return journey.

Review 1

a West is the same as a bearing of **A** 000° **B** 090° **C** 180° **D** 270°.
b The bearing 135° is the same as **A** NE **B** SE **C** NW **D** SW.

Review 2 Three ships are near one another.
This diagram shows their positions.
Big Red is due North of Victoria.
Victoria is due West of Blue Star.
Blue Star is South-East of Big Red.

a What is the bearing of Blue Star from Victoria?
b What is the bearing of Blue Star from Big Red?
c What is the bearing of Big Red from Blue Star?

N
Big Red ×
Victoria ×
× Blue Star

Due North means exactly North. Due West means exactly West.

Review 3

a What bay is on a bearing of about 235° from the airport?
b What is the bearing of the Village from Cave Bay?
c What is the bearing of Sandy Bay from Two Tree Hill?
*** d** Without measuring, write down the bearing of Two Tree Hill from Sandy Bay.

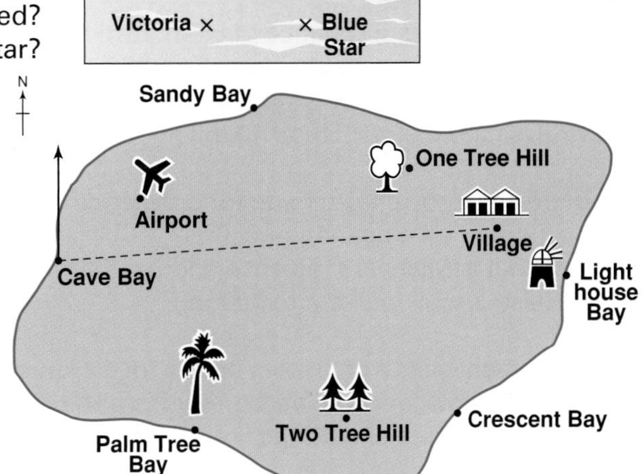

We sometimes make a **scale drawing** to answer questions about distances and bearings.

Worked Example

Bridget is a secret agent. She is picking up a message.
Bridget begins at the village hall. She walks due East for 200 m, then on a bearing of 300° for 100 m.
What is Bridget's distance and bearing from the village hall? Draw a scale diagram to find the answer.

Answer

A sensible scale to use would be 1 mm represents 2 m.
Then 200 m is represented by 100 mm.
100 m is represented by 50 mm.

Draw a scale drawing.

Mark H.
Draw a line, HA, 100 mm long on a bearing 090°.
From A, draw a line AB, 50 mm long, on a bearing of 300°.

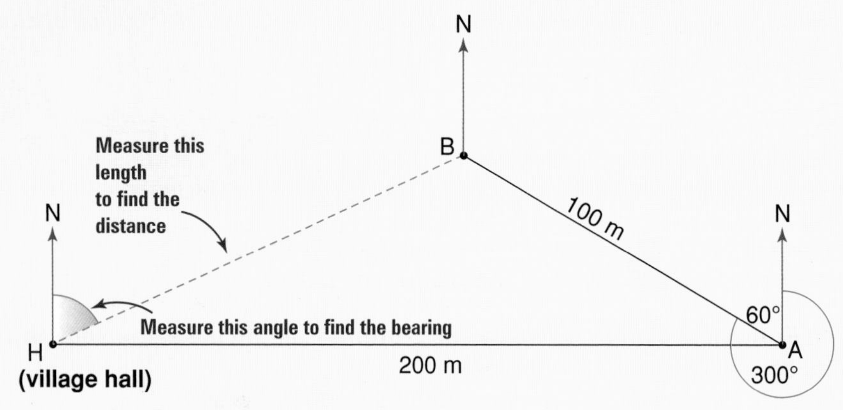

Now measure HB. HB is about 62 mm.
62×2 m = 124 m
Measure the angle between the North line at H and BH. It is about 65°.
Bridget is now **124 m from the village hall.**
Her **bearing from the village hall is 065°.**

Exercise 4

Draw scale drawings to answer these questions.

1 Marie cycles a distance of 8 km from F to G on a bearing of 090°. From G she cycles 6 km to H. The bearing of H from G is 180°.
What is Marie's bearing and distance from F?
(Use a scale of 1 cm to 1 km.)

2 Michael rows from Ryde for a distance of 800 m on a bearing of 300°, then he rows for 1200 m on a bearing of 060°.
What is Michael's distance and bearing from Ryde?
(Use a scale of 1 cm to 100 m.)

*3 A yacht sails from Hiton. It sails for 12 km on a course of 240°, then for 8 km on a course of 150°. How far is the yacht now from Hiton?
What course should the crew of the yacht set if they want to return to Hiton by the shortest route?

Review Susan walks 7 km from A to B on a bearing of 120°. She then walks 6 km from B to C on a bearing of 210°.
What is Susan's distance and bearing from A?
(Use a scale of 1 cm to 1 km.)

Practical

You will need a map showing the four cities closest to where you live.
Using the map, measure the bearings and distances 'as the crow flies' between the centres of the four cities closest to where you live.
Make a scale drawing of the positions of these four cities.
Use a scale of 1 : 500 000.

A scale of 1 : 500 000 means 1 unit on the map represents 500 000 units in real life.

Circumference and area of a circle

Practical

You will need some circular objects such as coins, saucers, ...

1 a The circumference of a circular object, such as a coin or saucer or the top face of a cylinder, can be found as follows.

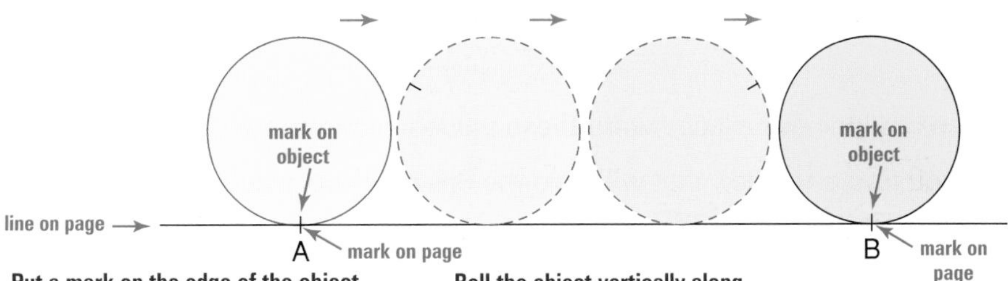

Put a mark on the edge of the object. Draw a line on your page. Put a mark, A, on the line. Line up the two marks.

Roll the object vertically along the line until the mark on the object touches the page again. Make a mark, B, on the line.

Measure the length of AB, in millimetres.

b Repeat **a** for five or six objects.

c Measure the diameter, in mm, of each object as accurately as possible.

d Fill in your results in a table like this one.

e Calculate $\frac{C}{d}$ for each object.

Give the answer to 1 d.p.
What do you notice?

Object	Circumference, C (in mm)	Diameter, d (in mm)	$\frac{C}{d}$

2 Another way of finding the circumference of a cylindrical object, such as a baked bean tin, is to measure with a tape measure.

Another way is to wind a length of cotton around the tin a number of times. Why do you think the cotton is wound around a number of times, rather than just once?

Use this method to find the ratio $\frac{C}{d}$ for a number of tins of different sizes.

What do you notice?

The meaning of π

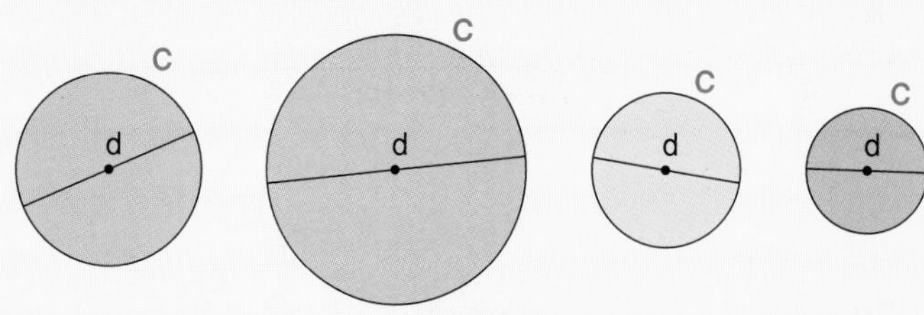

C is the circumference of a circle, d is the diameter. The ratio $\frac{C}{d}$ is the same for any circle, regardless of how big or small the circle is.

This ratio $\frac{C}{d}$ is called π (pronounced 'pi'). To 1 d.p. the value of π is 3·1.

Most calculators have a π key. This gives π to 9 decimal places as 3·141 592 654.
π has been worked out, accurate to many millions of decimal places, on the computer.

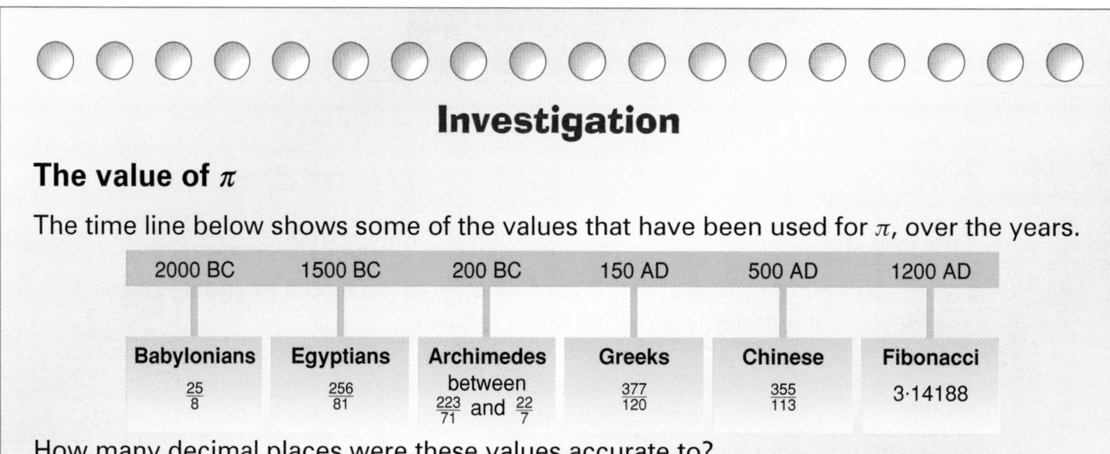

Investigation

The value of π

The time line below shows some of the values that have been used for π, over the years.

How many decimal places were these values accurate to?

Can you find a fraction that will give the 8th digit more accurately than the fraction the Chinese used? **Investigate**.

Circumference of a circle

For all circles, the ratio $\frac{C}{d}$ is equal to π.

$\therefore$ $\boldsymbol{C = \pi d}$ gives the **circumference, *C*, of a circle** of diameter, d.

Remember: The diameter of a circle, d, is twice the radius, r.
$\boldsymbol{C = \pi d}$ and $\boldsymbol{C = 2\pi r}$ both give the **circumference of a circle**.

When calculating with π we usually use the π key on the calculator or $\pi = 3{\cdot}14$ or $\pi = \frac{22}{7}$.
When estimating we use $\pi \approx 3$.

Worked Example
Find the circumference of these circles.
Use $\pi = 3{\cdot}14$.

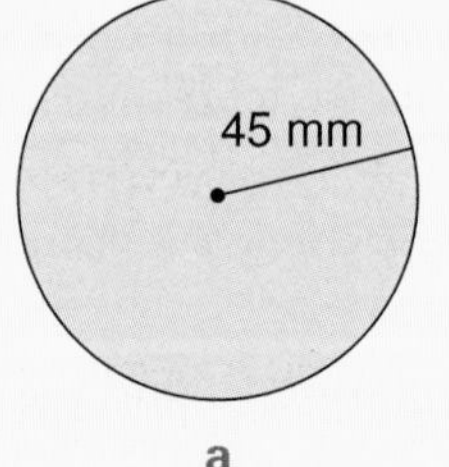

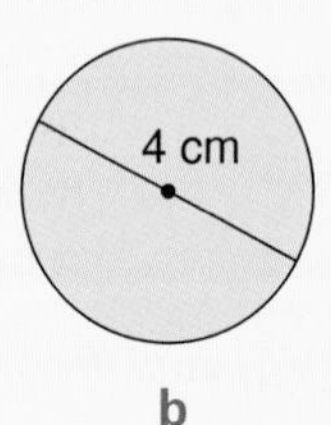

Answer

a $C = 2\pi r$
$C = 2 \times 3{\cdot}14 \times 45$
$=$ **282·6 mm**

b $C = \pi d$
$C = 3{\cdot}14 \times 4$
$=$ **12·56 cm**

Worked Example

Nathaniel's motorcycle has wheels of diameter of 40 cm.

a How far has Nathaniel travelled when the wheels have turned 300 times each?

b How many rotations does each wheel make for every 5 km travelled?

Answer

a $C = \pi d$

$= \pi \times 40$

$= 125{\cdot}6637061$ cm

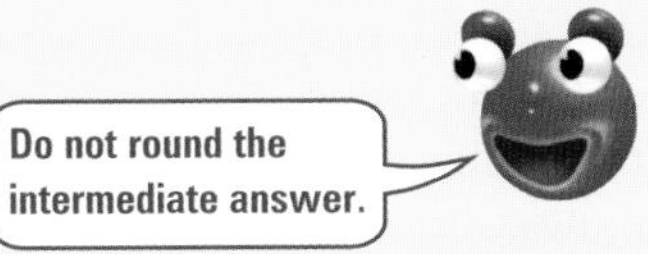

Distance travelled for 300 turns $= 300 \times 125{\cdot}6637061$

$= 37\,699$ cm to the nearest cm

$=$ **376·99 m**

Key Shift π × 40 × 300 =

b 5 km = 5000 m

5000 m $= x$ turns $\times \pi d$

5000 m $= x \times \pi \times 0{\cdot}4$ m

40 cm = 0·4 m.

$$x = \frac{5000}{\pi \times 0{\cdot}4}$$

$x = 3978{\cdot}9$ or about **3979** to the nearest turn

Exercise 5

Except for question 1 and Review 1.

1 Find the circumference of each of these circles. Use $\pi = 3{\cdot}14$.

a 8 mm

b

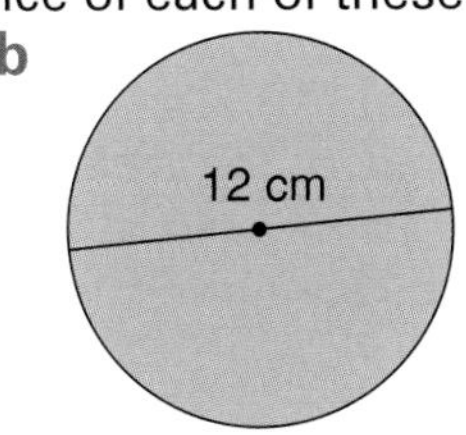

c

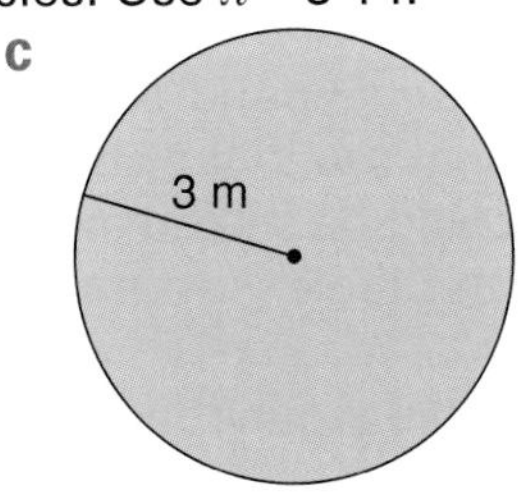

d

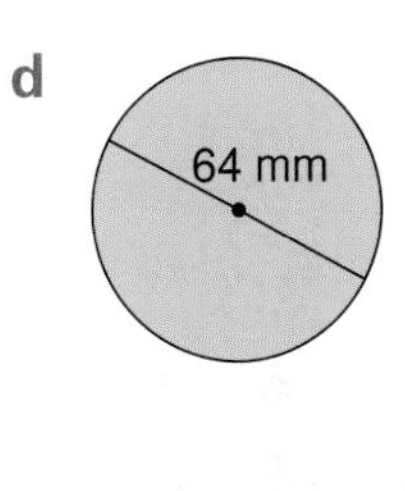

Use the π key on your calculator for questions 2–10 and Reviews 2 and 3. Round your answers sensibly.

2 An African Violet is in a pot. The top of the pot has a diameter of 9 cm. What is the distance around the edge of the top of this pot?

3 A ribbon is tied around a hat, as shown. Find the total length of the ribbon, if the bow needs 30 cm.

18 cm

4 A circular tablecloth overhangs a circular table, as shown. Mai-Lin buys 3·4 m of braid to sew around the edge. Is 3·4 m enough to sew around the outside edge?

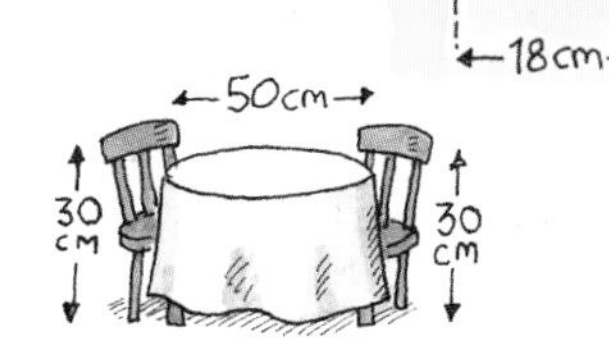

5 Samantha's bicycle has 0·6 m diameter wheels. On the way to school, the front wheel rotates 500 times. How far does Samantha live from school?

6 **a** The larger wheel on Bryce's wheelchair is 58 cm in diameter.
Bryce pushes it around exactly 20 times.
How far, in m, has Bryce moved?

b Bryce gets a new electronic wheelchair with a large wheel radius of 26 cm.
How many times must the wheel rotate for him to travel 860 cm?

7 A circular kart track has a diameter of 60 m.

a How far does a kart travel in four circuits of the track?

b How many circuits, to the nearest circuit, must be made to travel 1·5 km?

8 How many rotations must a cycle with a wheel of radius 28 cm make for every 5 km travelled?

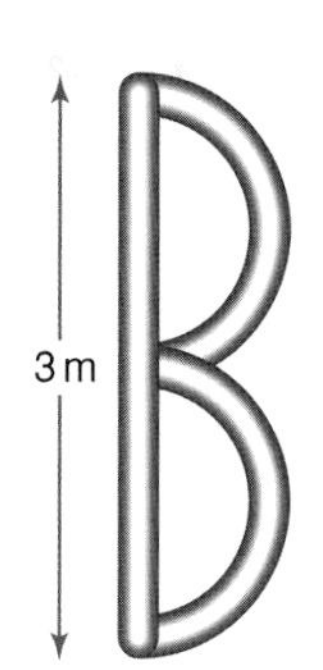

9 Barbara owns a gift shop which she calls 'B'.
She makes a sign with tube from a straight side and two equal semicircles.
What is the length of the tube?

*** 10** A circle has a circumference of 240 cm.
What is the radius of the circle?

Review 1 Find the circumference of these circles. Use $\pi = 3{\cdot}14$.

a **b**

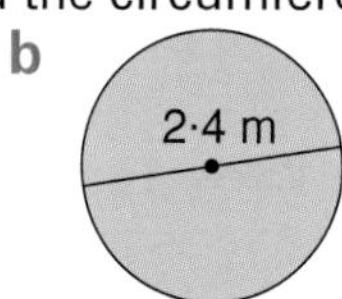

Review 2

a The diameter of King Arthur's Round Table is 5·5 m.
If each person needs 45 cm round the circumference of the table, what is the maximum number of people that could be seated around it?

b If 50 people sat around this table, how much room round the circumference would each have?
Is it possible to sit 50 people around the table?

Review 3 Roger's bicycle has wheels of diameter 68 cm.
How many times do the wheels on Roger's bicycle rotate in 4 km?

A about 200 times **B** about 2000 times **C** about 20 000 times

* Investigation

Race Track

This is a circular race track.

Two horses, A and B, train on the track.
The horses train so that B is always on the inside of A and 4 m from A.

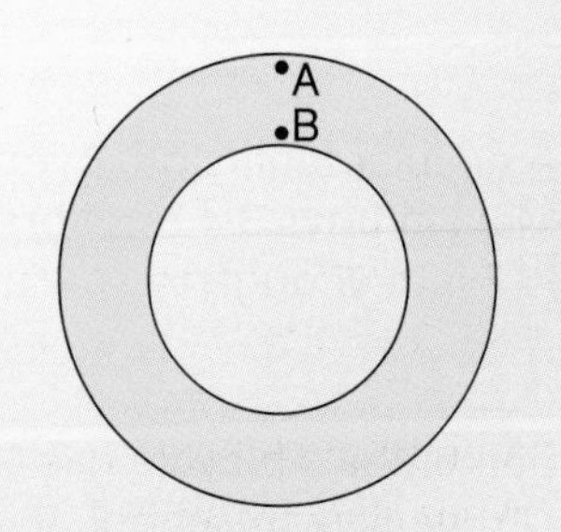

Investigate the difference in the distances run by A and B for varying diameters of track.

Area of a circle

The **area of a circle** can be calculated using the formula $A = \pi r^2$.

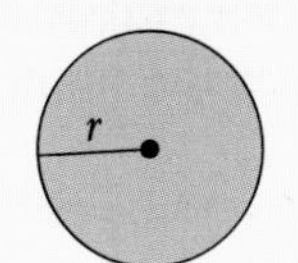

Worked Example

Find the area of these circles.
Use $\pi = 3{\cdot}14$.
Give the answers to 2 d.p.

a 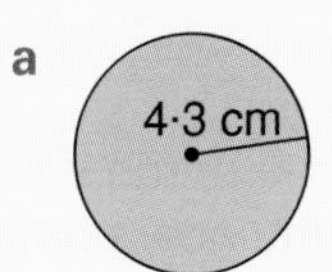

b 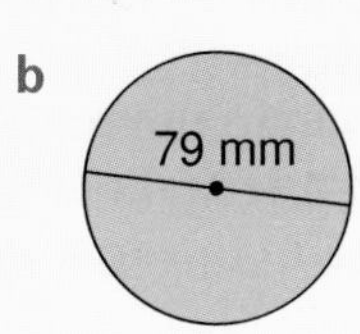

Answer

a $A = \pi r^2$

$A = 3{\cdot}14 \times 4{\cdot}3^2$ **Key** 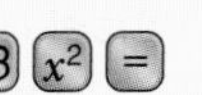

$= \mathbf{58{\cdot}06\ (2\ d.p.)}$

b $d = 79$ mm $r = 39{\cdot}5$ mm $A = \pi r^2$

$= 3{\cdot}14 \times 39{\cdot}5^2$

$= \mathbf{4899{\cdot}19\ mm^2\ (2\ d.p.)}$

Worked Example

An advertising board is in the shape of a rectangle with a semicircular end.
The rectangular part is 1·2 m long and 60 cm wide.
Find the area of the board.

Answer

Area of board = area of rectangle + area of semicircle

Area of rectangle $= lw$

$= 1{\cdot}2 \times 0{\cdot}6$

$= 0{\cdot}72\ \text{m}^2$

Area of semicircle = half of area of circle $= \frac{1}{2}$ of πr^2

$= \frac{1}{2}$ of $\pi \times 0{\cdot}3^2$ diameter of circle = 60 cm

$= 0{\cdot}5 \times \pi \times 0{\cdot}3^2$ radius = 30 cm = 0·3 m

$= 0{\cdot}14\ \text{m}^2$ (2 d.p.)

Area of board $= 0{\cdot}72\ \text{m}^2 + 0{\cdot}14\ \text{m}^2$

$= \mathbf{0{\cdot}86\ m^2\ (2\ d.p.)}$

Exercise 6

Use the π key on your calculator.

1 Find the area of these circles. Give the answers to 1 d.p.

a 9 m

b 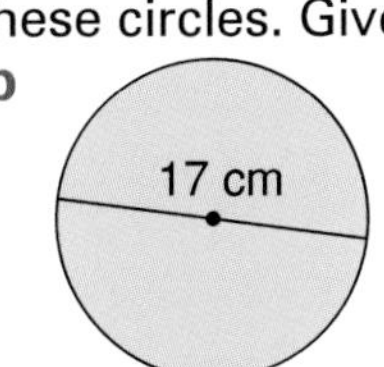

c 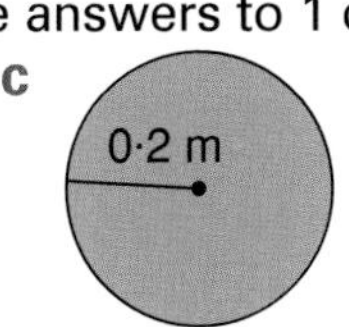

d 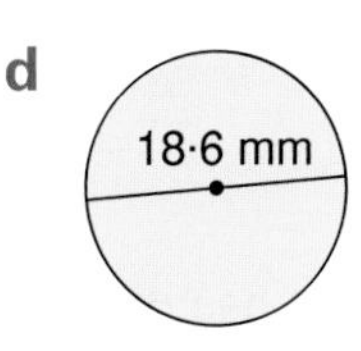

2 A canvas cover is made to fit exactly over a circular goldfish pond.
The diameter of this pond is 2·8 m.
What is the area of the canvas cover?

3 A circular picture, of diameter 29 cm, is covered with glass.
What is the area of the glass?

4 In the McMath solar telescope, in Arizona, there are two flat round mirrors. One has a radius of 1 m and the other a diameter of 122 cm. What is the difference in the areas of these mirrors? Give the answer in cm^2.

5 Here is a circle and a square.
The radius of the circle is 10 mm.

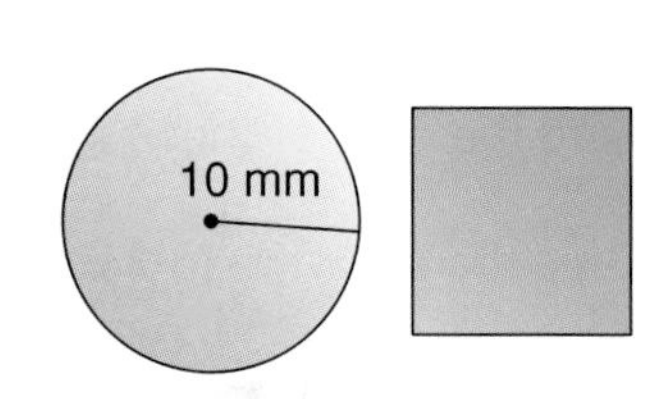

- **a** Find the area of the circle to the nearest mm^2.
- **b** The ratio of the area of the circle to the area of the square is 2 : 1. What is the area of the square to the nearest mm?
- **c** What is the length of the side of the square?

6 Calculate the area of the shaded shapes.

a

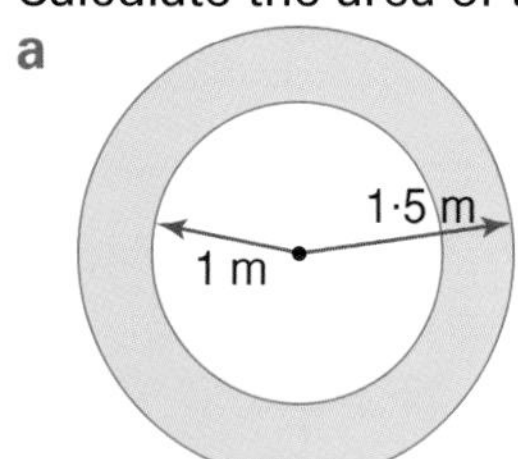

b

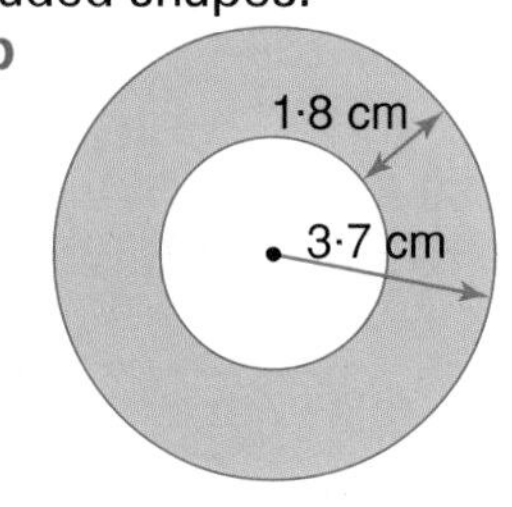

7 A gate from a courtyard to a garden is in the shape of a rectangle with a semicircular arch. The rectangular part is 1·6 m wide and 2·4 m high. What is the area of the gate?

2·4 m

1·6 m

*8 **a** 'Delicious Pastries' make small and large pies. The larger pies have twice the surface area of the smaller ones. If the smaller pies have a diameter of 12 cm, what is the diameter of the larger pies?

b Janna eats a 120° sector of a large pie. What area of pastry covers the top of this piece?

120°

*9 This diagram shows a circle and a square. The circle touches the sides of the square. What percentage of the diagram is shaded yellow?

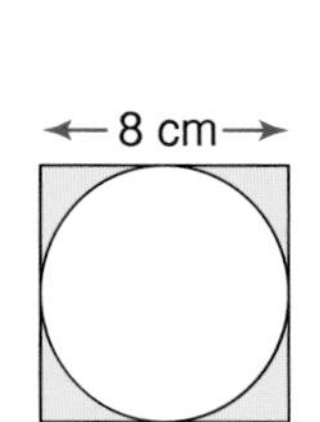

Review 1 Find the area of these to 1 d.p.

a

b

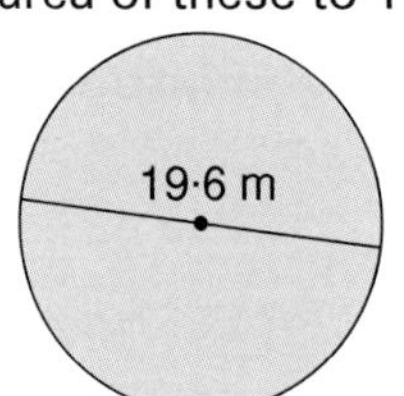

c

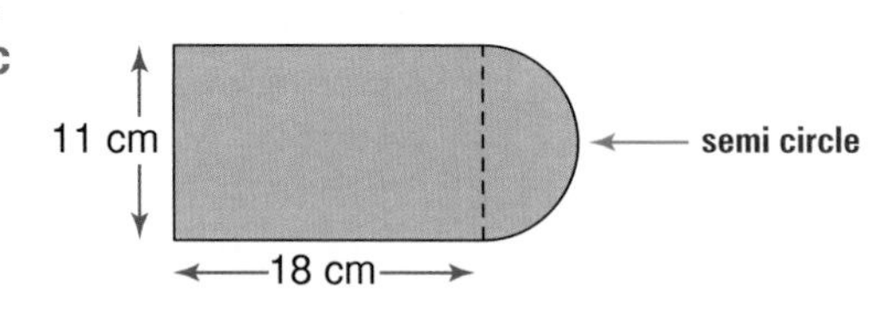

Review 2

- **a** A circular table mat has diameter of 15 cm. What is the area of this table mat?
- **b** A matching coaster covers 25% of the surface area of the placemat. What is the diameter of the coaster?

Review 3 A clockface of diameter 16 cm is mounted on a rectangular board of dimensions 24 cm × 22 cm. Find the area of board *not* covered by the clockface.

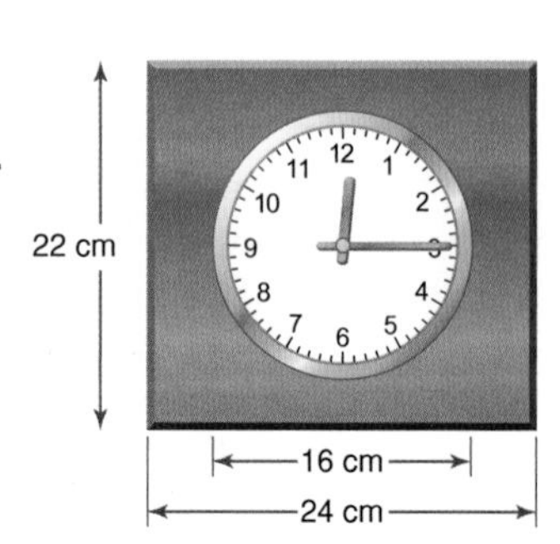

Puzzles

1

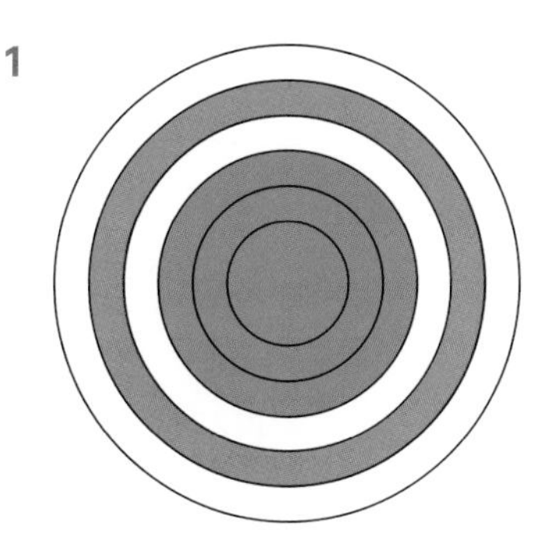

The six equally spaced circles all have the same centre.
Which of the two shaded regions has the greater area?

2 Tim and Neville walk at the same speed. They both begin at Q and finish at P. Tim walks along the large semicircle and Neville walks along the three small semicircles. Before they began their walk Neville claimed that he would get to P before Tim. Did he?

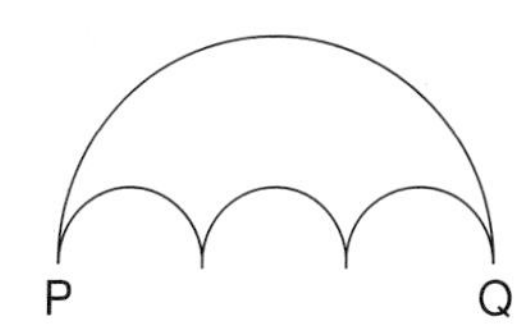

*3 Lightning Lady and Stirling Monarch set off from the same point on their circular training track, with their jockeys riding them in opposite directions. The first time they meet, Lightning Lady has run 500 m. The next time they meet Stirling Monarch still has 200 m to go to complete his first lap. How far is it around their training track?

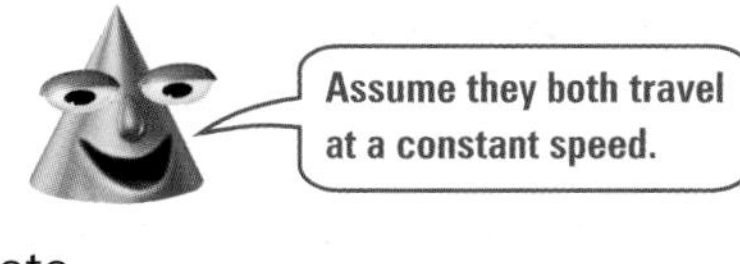

Practical

1 **You will need** a dynamic geometry software package.
 a Draw a circle inside a square.
 b Draw a square inside a circle with the same diameter as the circle in **a**.
 What is the ratio of areas of the squares?

2 Design a running track.

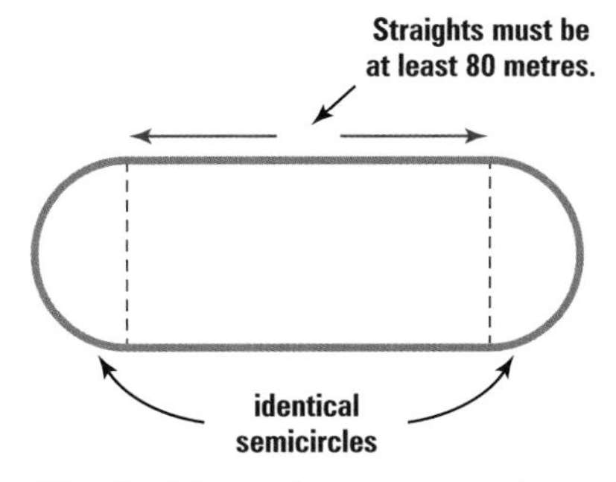

Total inside perimeter must be 400 m.

The inside perimeter must have this shape and features.
What is the greatest area the running track can enclose?

*If the track itself must be 10 m wide, what is the smallest rectangular field needed to contain it?

Surface area and volume of a prism

Discussion

The area of the cross-sections of these prisms are given.

How could you use these to find the volume of each prism? **Discuss**.

The volumes of both can be found from the same formula. **Discuss** a possible formula.

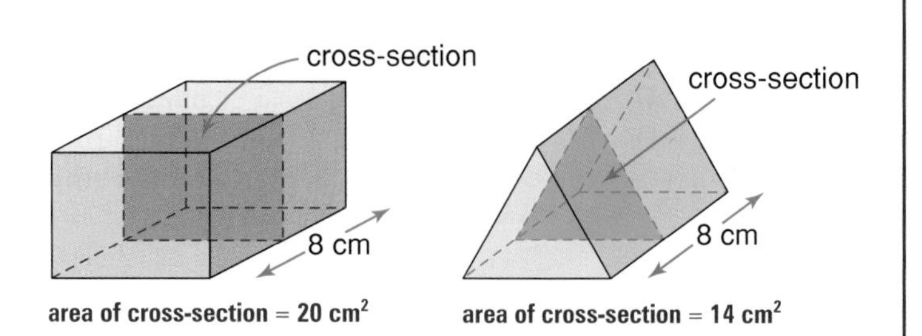

Could the formula you discussed be used to find the volume of any of these shapes? Which ones? **Discuss**.

A

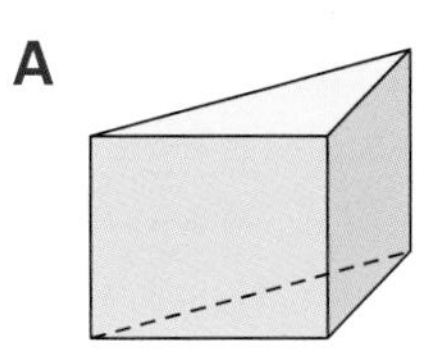

B

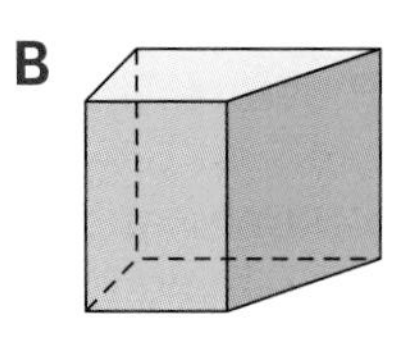

C

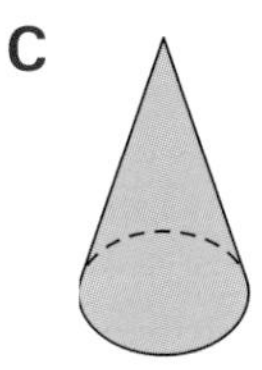

D

E

F

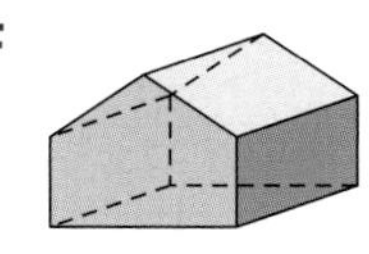

Remember

The surface area of a cuboid is given by the formula

surface area $= 2lw + 2lh + 2wh$

The surface area of other prisms is the sum of the areas of the faces of the prism.

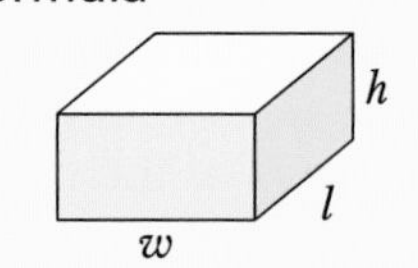

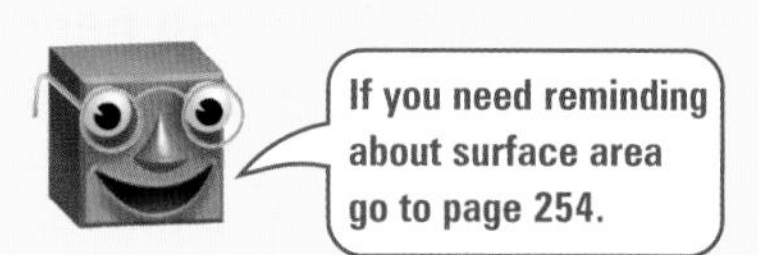

A **prism** has a **constant cross-section** throughout its length.
The cross-section is congruent and parallel to the two congruent end faces.

Examples

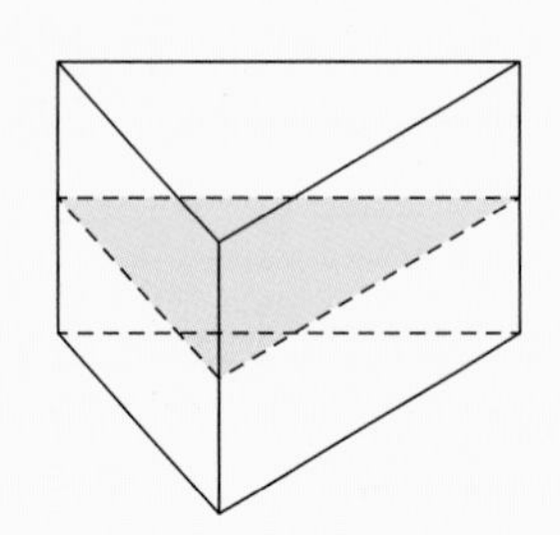

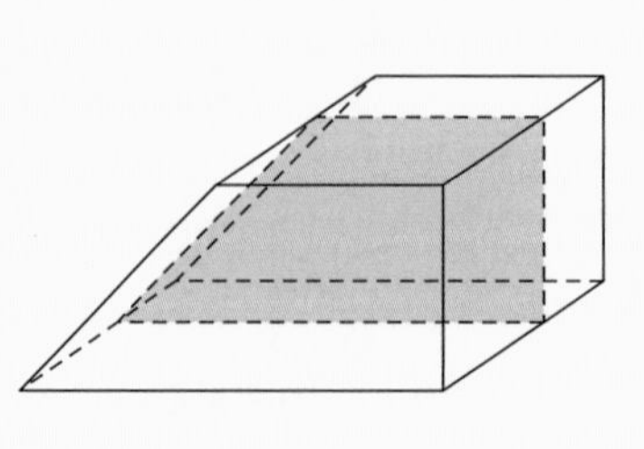

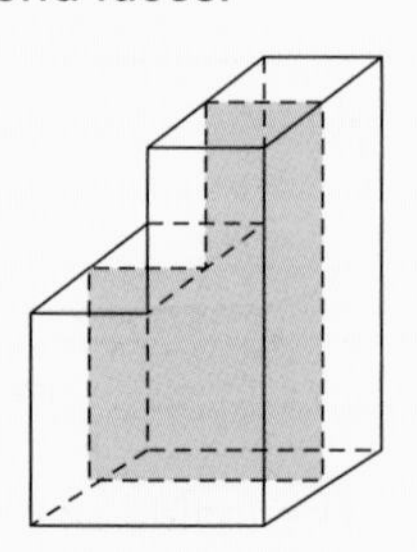

Volume of a prism = area of cross-section × length

Example Volume of prism = area of cross-section × length

Area of cross-section $= \frac{1}{2}bh$

$= \frac{1}{2} \times 8 \times 6$

$= 24 \text{ cm}^2$

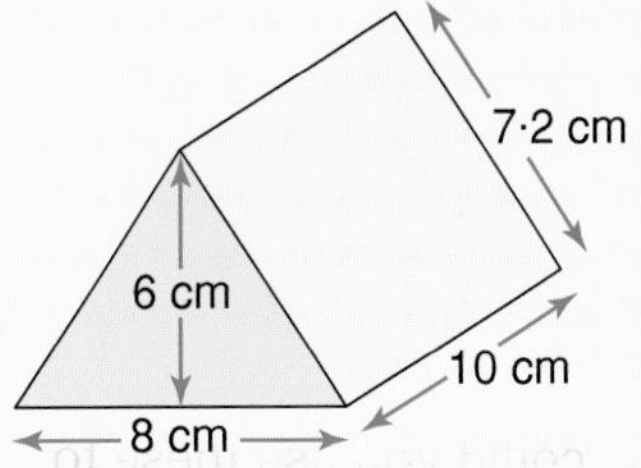

Volume $= 24 \times 10$

$= \mathbf{240\ cm^3}$

Surface area of prism = 2 × area of end triangle + 2 × area of side + area of base

$= 2 \times 24 + 2 \times 7{\cdot}2 \times 10 + 10 \times 8$

$= 48 + 144 + 80$

$= \mathbf{272\ cm^2}$

Exercise 7

Round your answers sensibly.

1 Find the volume of these shapes. The cross-section is shown shaded.

a

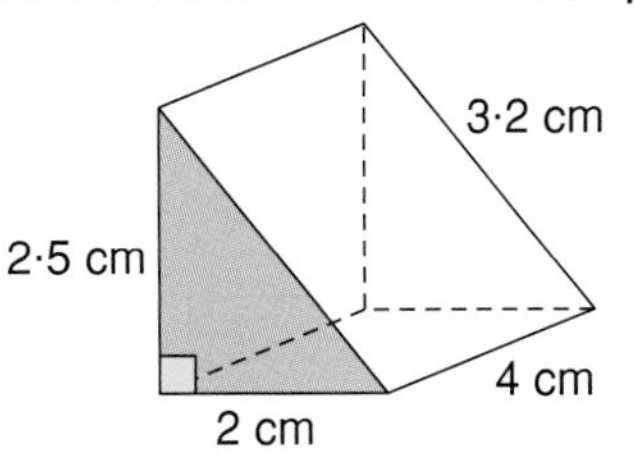

b

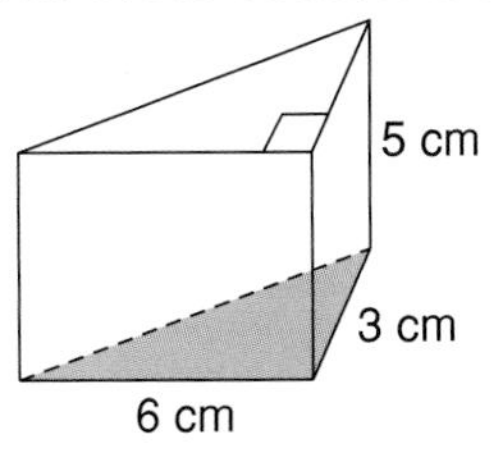

c

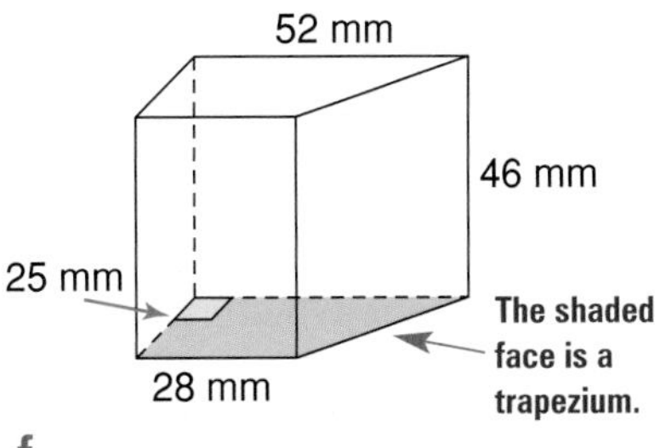

d

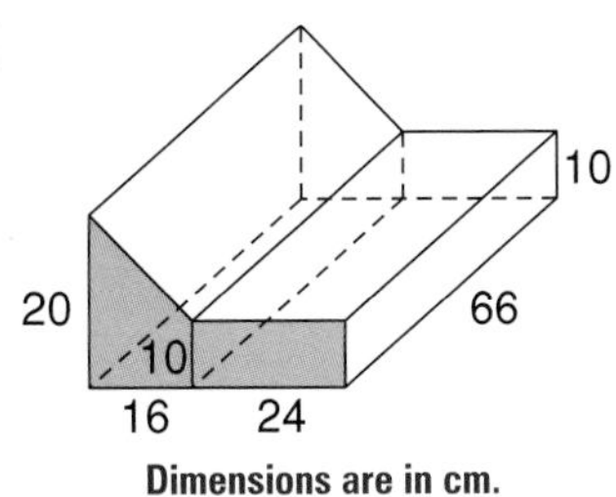

e

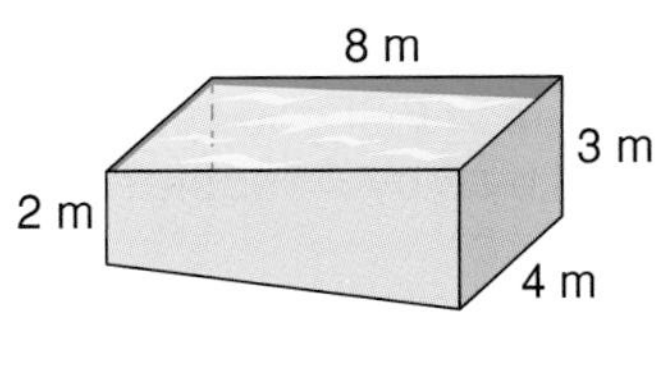

f

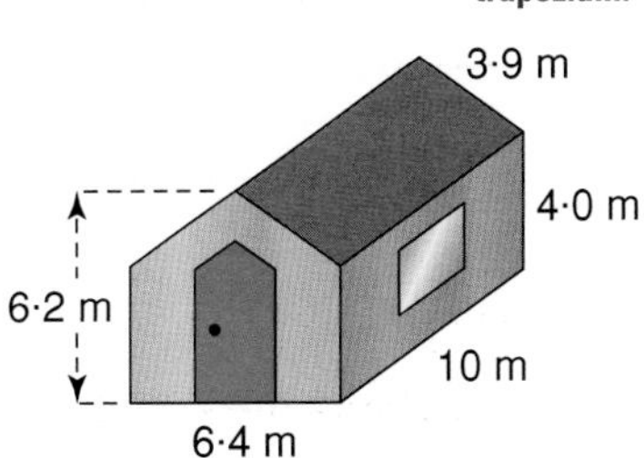

2 Find the surface area of the shape in question **1a**.

3 **a** This display stand has cross-sectional areas C_1, C_2 and C_3, all of length d.
Show that the volume of the stand is

$$V = (C_1 + C_2 + C_3)d.$$

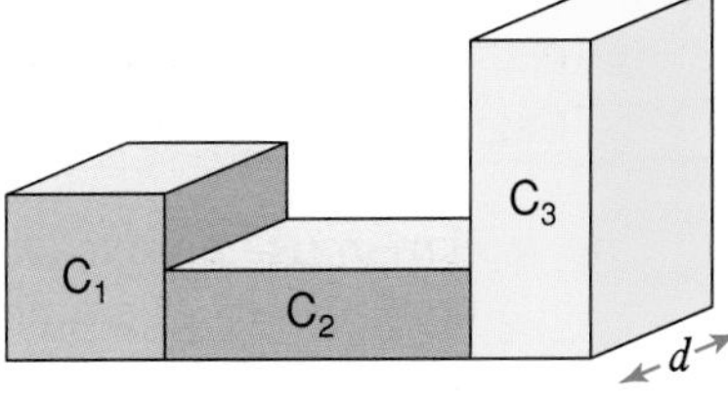

b Write an expression for the volume of this prism.
It has cross-sectional areas L_1, L_2, L_3 and L_4 all of length p.

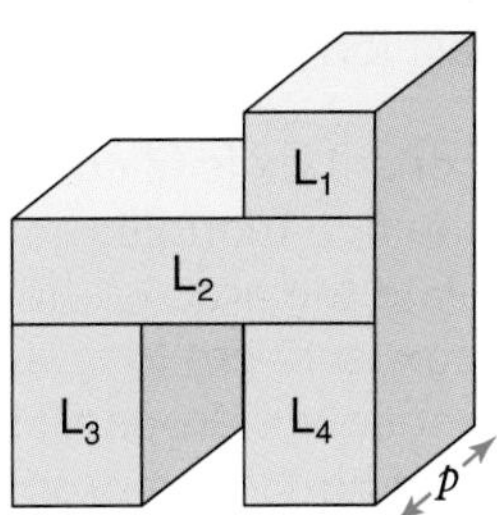

4 Find the volume of these steel beams.

a 24 cm, 10 cm, 12 cm, 28 cm, 1 m

b

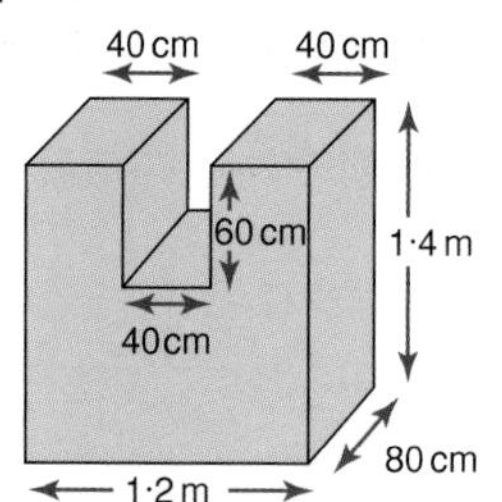

Not to scale

c

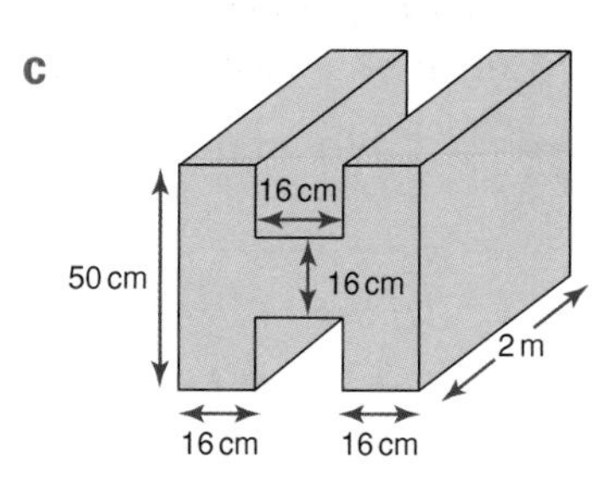

5 Find the surface area of the shape in question **4a**.

6 A wheelchair ramp is in the shape of a prism.
- **a** The purple shaded end is a trapezium. Calculate its area.
- **b** Calculate the volume of the ramp.
- **c** The **whole** ramp is to be covered in rubber. What area of rubber is needed?

(2·5 m, 3·8 m, 1·4 m, 2·0 m, 6·0 m)

7
- **a** What is the total surface area of this box?
- **b** Find the length of each side.

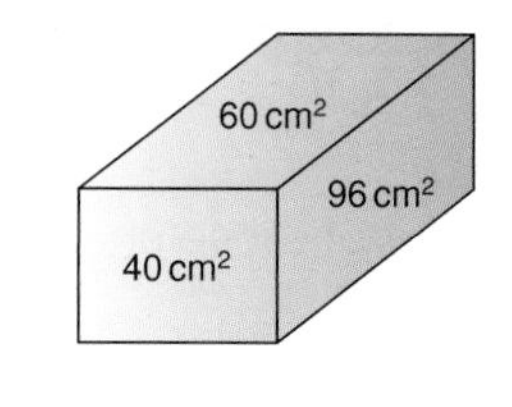

8 Suwani has some small cubes with each edge 1·5 cm. She makes a larger cube out of the small cubes. The volume of the larger cube is 216 cm^3. How many small cubes did she use?

9 A shape which is 10 cm long has ends shaped as shown in the diagram. Find its volume.

(24 cm, 16 cm, 14 cm)

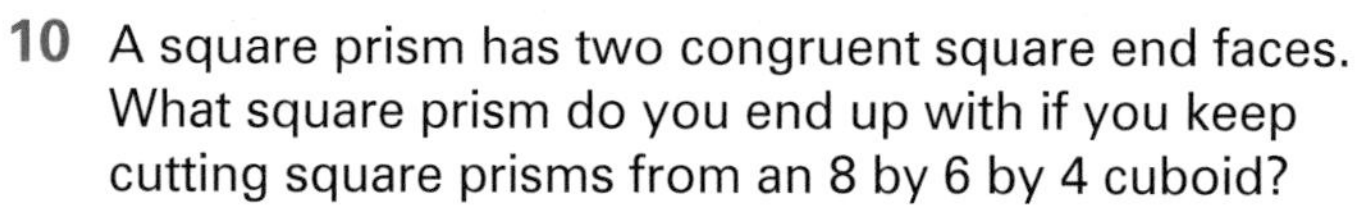

10 A square prism has two congruent square end faces. What square prism do you end up with if you keep cutting square prisms from an 8 by 6 by 4 cuboid?

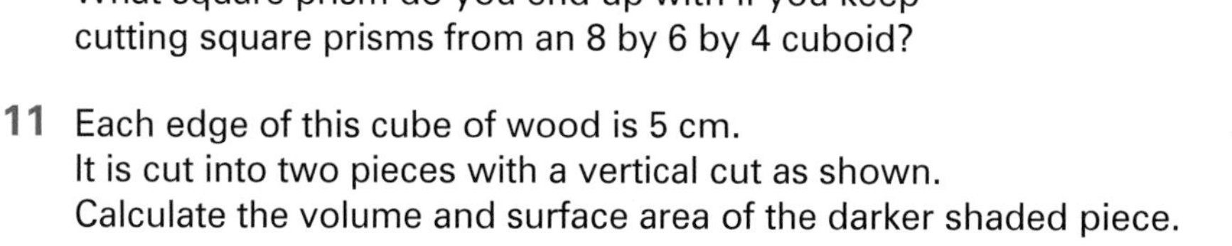

11 Each edge of this cube of wood is 5 cm. It is cut into two pieces with a vertical cut as shown. Calculate the volume and surface area of the darker shaded piece.

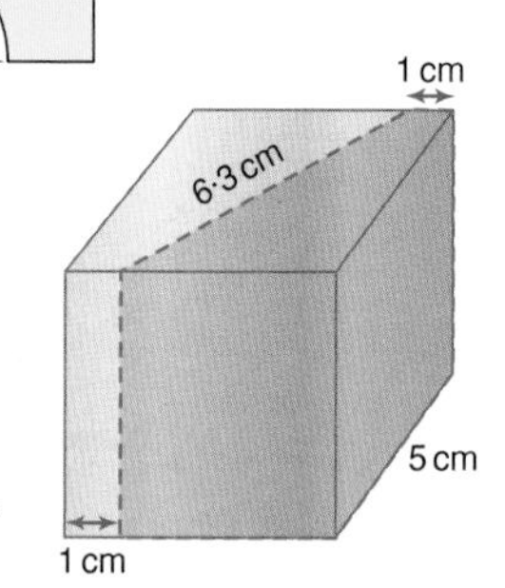

12 Workmen are digging a ditch which is to be 30 m long and 1·25 m deep. The cross-section is a symmetrical trapezium, as shown in the diagram. The ditch must be 1·5 m wide at the bottom and 2·5 m wide at the top.
- **a** What are the values of a, b and c on the cross-section?
- **b** Find the area of the cross-section.
- **c** Calculate how many m^3 of dirt are to be excavated from the ditch.

The workmen use an excavator and the dirt is taken away in lorries that can take 12 m^3 at a time.
- **d** How many full loads of dirt are taken away?
- **e** The last lorry is only partially filled up. What fraction of a full load does the last lorry take?

13 A pentagonal box is filled with chocolate drops. Each of the five triangles in the pentagon are identical.
- **a** Calculate the area of the pentagon.
- **b** The box is 12 cm long. Calculate its volume in cm^3.
- **c** The chocolate drops fill 90% of the box. 1 cm^3 of chocolate drops are 0·6 grams. What mass of chocolate drops are in the box?

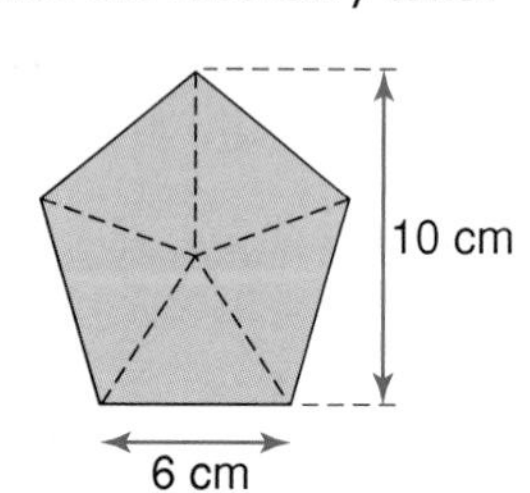

*14 This diagram represents a swimming pool. How much water is in the pool if the water comes to 20 cm below the top?

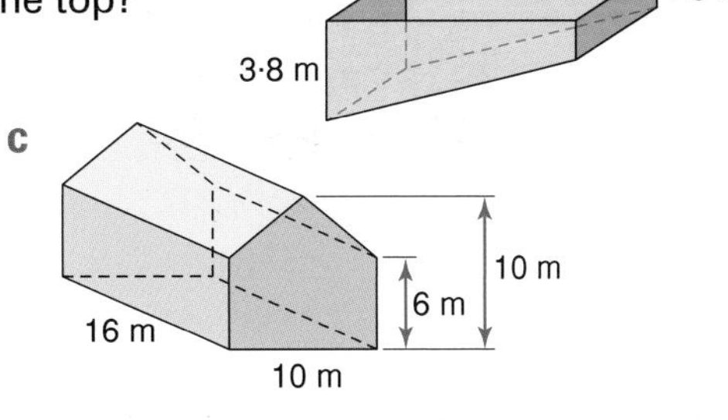

Review 1 Find the volumes of these shapes.

a

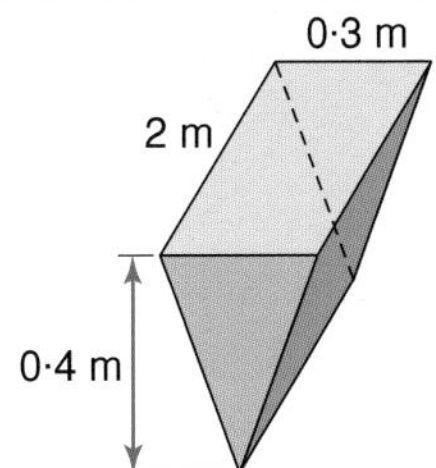

b

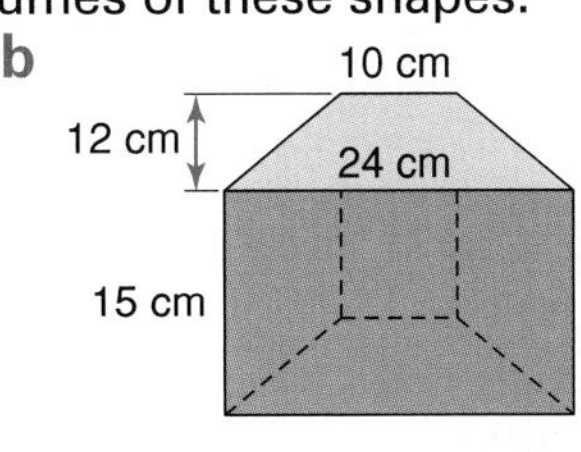

c

Review 2

a Find the volume, in m^3, of this T-shaped girder.

b Find the surface area of the girder.

Review 3 An airline has a rule that hand baggage must obey:

length + width + depth must be no more than 1 m

What dimensions for the hand baggage would give the greatest volume if only whole numbers of centimetres are allowed?

***Review 4** A tent is the shape of a triangular prism. It is 2·6 m long and the triangular end is 2·5 m wide and 1·9 m high. The sloping side is 2·3 m long.

a Find the volume of the tent.

b The tent has a ground sheet. What total area of material was used to make this tent?

Puzzle

How many different cuboids can be made using exactly one million cubes?

Summary of key points

You need to know these **area, volume and capacity conversions**.

area	**capacity**
1 m² = 10 000 cm²	**1 cm³ = 1000 mm³**
1 cm² = 100 mm²	**1 m³ = 1 000 000 cm³**

B **Solving measures problems**

You often need to convert between metric units or between metric and imperial units to solve problems.

C **Bearings** are always taken in a clockwise direction from North and always have 3 digits.

Examples

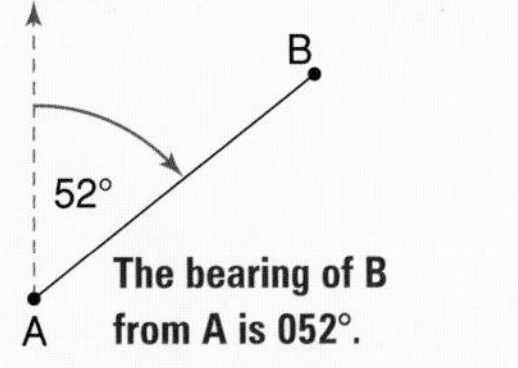

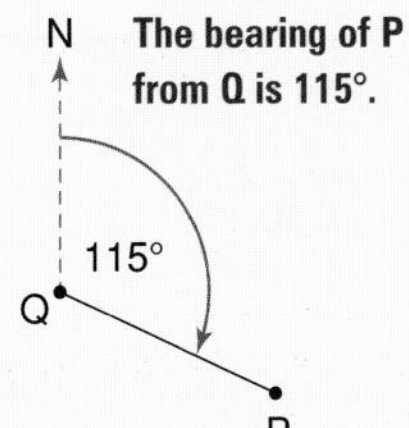

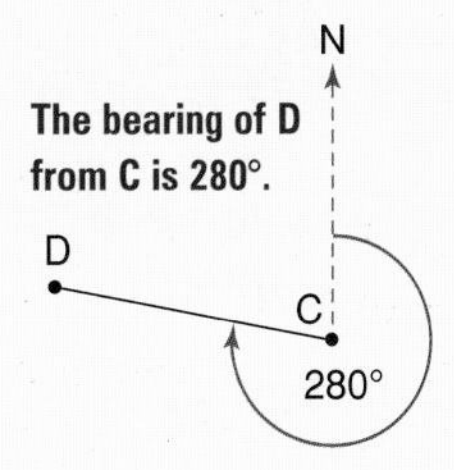

Sometimes we make a scale drawing to answer questions about bearings.

$C = \pi d$ or $C = 2\pi r$ give the circumference, C, of a circle of diameter, d and radius, r.

$\pi = 3{\cdot}1$ (1 d.p.) or $3{\cdot}14$ (2 d.p.) or you can use the π key on your calculator.

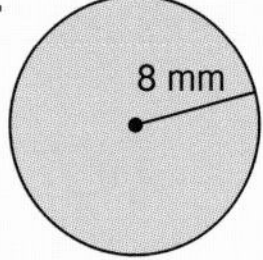

Example $C = 2\pi r$

$= 2 \times \pi \times 8$

$= 50{\cdot}3$ mm (1 d.p.) using the π key on a calculator

$A = \pi r^2$ gives the area, A, of a circle of radius, r.

Example The area of the circle in D is

$A = \pi r^2$

$= \pi \times 8^2$

$= 201{\cdot}0$ mm^2 (1 d.p.) using $\pi = 3{\cdot}14$

A **prism** has a constant cross-section throughout its length.

The cube is a common example of a prism.

Volume of a prism = area of cross-section × length

Example This is a symmetrical prism.

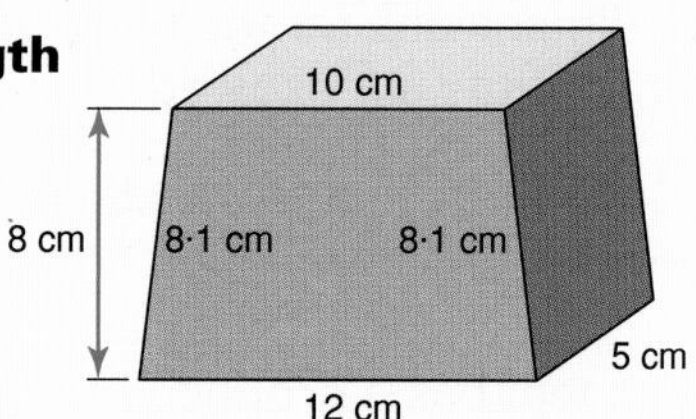

The cross-section is a trapezium.

Area of a trapezium $= \left(\frac{a+b}{2}\right)h$.

Volume of prism = area of cross-section × length

Area of cross-section $= \left(\frac{10+12}{2}\right) \times 8 = 88$ cm^2

Volume $= 88 \times 5$

$= \mathbf{440}$ **cm^2**

Surface area = sum of areas of all faces

$= 2 \times$ area of cross-section $+ 2 \times$ area of sides + area of top + area of base

$= 2 \times 88 + 2 \times (8{\cdot}1 \times 5) + 5 \times 10 + 12 \times 5$

$= 176 + 90 + 50 + 60$

$= \mathbf{367}$ **cm^2**

Test yourself

1 Change these. A

a	4·2 ha to m^2	**b**	800 m^2 to ha	**c**	5·1 m^3 to ℓ
d	52 cm^3 to mℓ	**e**	5860 cm^3 to ℓ	**f**	4936 ℓ to m^3
g	5 m^2 to cm^2	**h**	4 cm^3 to mm^3	**i**	853 200 mm^3 to cm^3
j	8·4 cm^2 to mm^2	**k**	0·62 m^3 to cm^3	**l**	52 800 mm^2 to cm^2

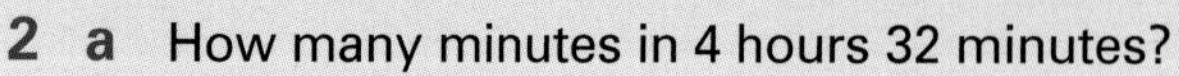

2 **a** How many minutes in 4 hours 32 minutes? A

b How many hours and minutes in 724 minutes?

c Change 472 weeks to years and weeks.

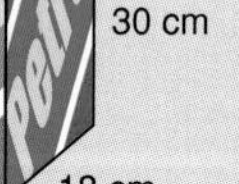

3 What is the volume of this petrol can in

a cm^3 **b** mm^3 **c** mℓ **d** ℓ?

4 What is the surface area of the petrol can in question **3** in

a cm^2 **b** mm^2?

5 A railway wagon, when not loaded, weighs 16·55 tonnes.
When it is loaded with 24 containers, it weighs 22 430 kg.
What is the mass, in kg, of each container?

6 A rectangular park has an area of 48 ha. One side is 400 m.
How long, in kilometres, is the other side?

7 Jake bought 3 ℓ of custard to make a custard pie.
He used 120 cℓ of the custard for his pie.
How many cm^3 of custard was left?

8 **a** Rewrite this sentence, replacing the imperial units with metric units.
Marcia cycled about **2 miles** to the local shop to buy **4 pints** of milk, **5 pounds** of apples, **3 oz** of chocolate chips, **3 yards** of tape and **4 inches** of ribbon.
b Rewrite this sentence, replacing the metric units with imperial units.
Robbie lives **10 km** from school. He has a dog that has a mass of **5 kg** and a rabbit that lives in a **2 metre** long hutch. Each day Robbie, the dog and the rabbit drink **4 litres** of milk between them.

9 Estimate these, using metric units. Give a range for each.
a the length of your longest school building
b the mass of ten of these books
c the capacity of an egg cup

10 This diagram shows the positions of several attractions at a theme park.
Measure and write down the **bearing** of these.
a Terror Cave from Peter Pan
b Roller Ride from Peter Pan
c The Steam Boat from Peter Pan
d The Pirate Ship from Peter Pan
e The Pirate Ship from Terror Cave

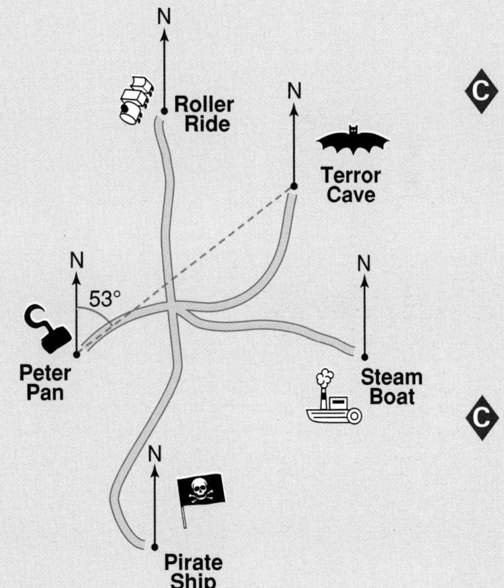

11 Use the diagram in question **10**.
What is on a bearing of
a 180° from Roller Ride
b 300° from Terror Cave?

12 Find the area and circumference of these circular placemats.

a

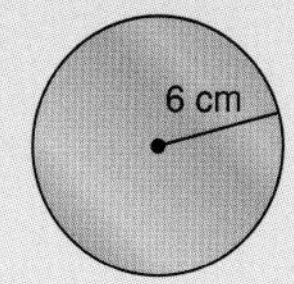

b

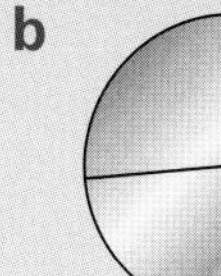

c

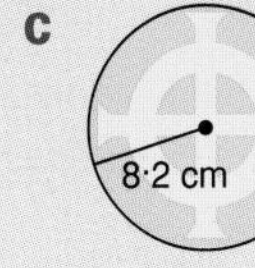

13 Kieren had the Irish crest printed on the back of a t-shirt. The diameter of the circle was 20 cm.
  a Find the area of the design.
  b Find the circumference.

E

14 The end of this garden is semicircular. Find
  a the area of the garden
  b the perimeter of the garden.

E

15 Find the volume of these solids.

F

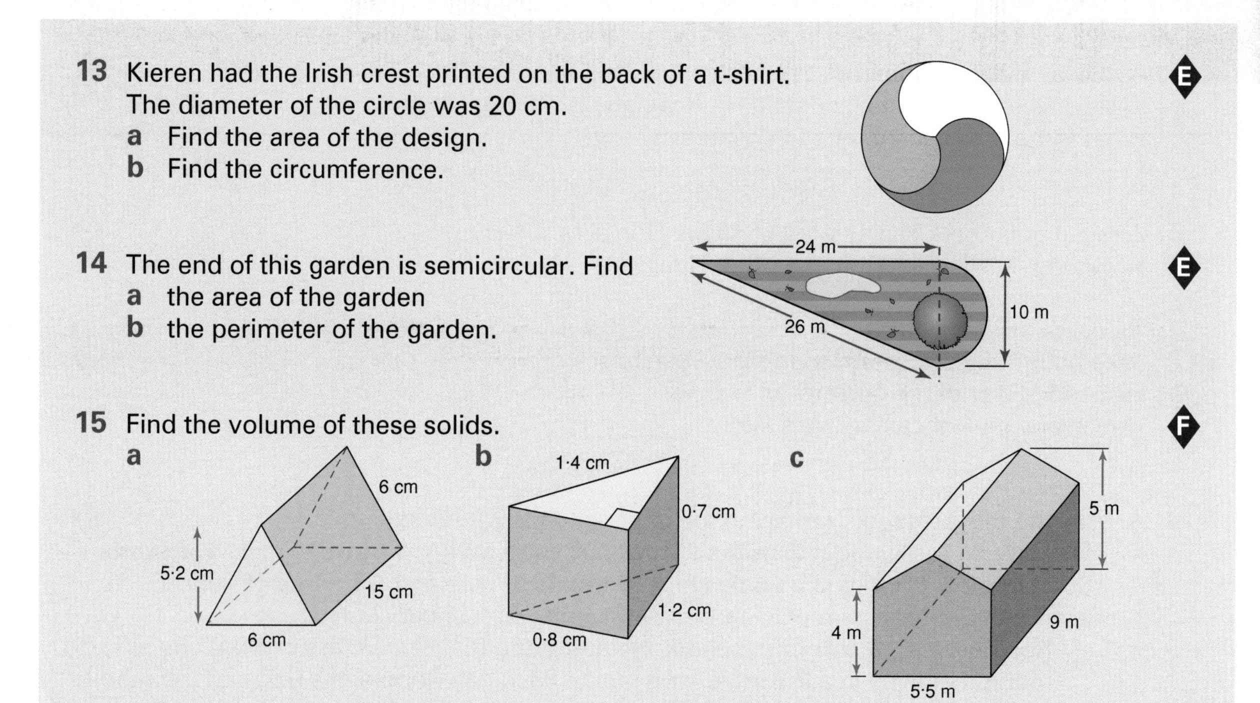

16 Find the surface area of the solid in question **15b**.

F

Handling Data Support

Planning and collecting data

Use these steps to plan a survey.

Step 1 **Decide on the purpose of the survey or specify the problem.**
Think of related questions you might want to explore.

Step 2 **Decide what data needs to be collected.**

Step 3 **Decide where to collect data from, and how much to collect.**

- You could use a primary source, for example
  a survey of a **sample** of people
  an experiment – observe, count or measure.
- You could use a secondary source, for example reference books, CD-ROMs, websites, books, newspapers, ...

The **sample** size should be as large as is sensible.

Once the survey has been planned, a **collection sheet** or a **questionnaire** often has to be designed.

Example Liz wrote down the colour of the chemicals in the science lab.

Colour	Tally	Frequency
Colourless	~~IIII~~ ~~IIII~~ IIII	14
Blue	~~IIII~~ I	6
Yellow	III	3

Frequency table

Example

How much time do you spend each week

a reading?
0 to 2 hours ☐ 2 to 4 hours ☐ 4 to 6 hours ☐ 6 to 8 hours ☐ ≥8 hours ☐

b watching TV?
0 to 3 hours ☐ 3 to 6 hours ☐ 6 to 9 hours ☐ 9 to 12 hours ☐ ≥12 hours ☐

Questionnaire

Practice Questions 15, 26

Discrete or continuous data

When collecting data you need to decide if it is **discrete** data or **continuous** data.
Discrete data can only have certain values. It is usually found by counting.

Example The number of people at a concert must be a whole number.

Continuous data can have any values within a certain range. It is usually found by measuring.

Example The temperature of humans usually lies between 35° and 40 °C.

Practice Question 13

Displaying discrete data

These diagrams show some ways of **displaying discrete data**.

Pictogram
Number of drinks sold

Each symbol represents 10 drinks	
Hot chocolate	
Coffee	
Juice	
Cola	

Note: A whole symbol represents 10.
So half a symbol represents 5.

Bar chart
Sometimes bar charts are drawn sideways.

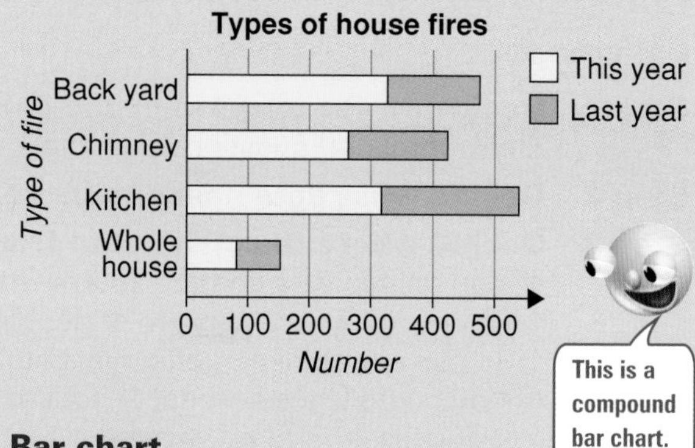

Bar-line graph

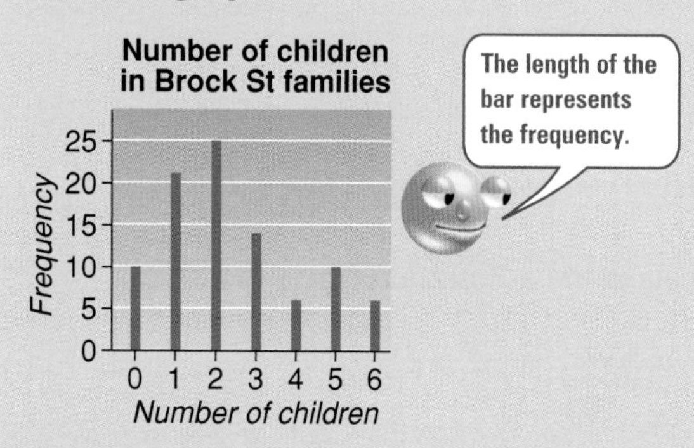

Bar chart
Sometimes two sets of data are shown on a bar chart.
Always give a key.

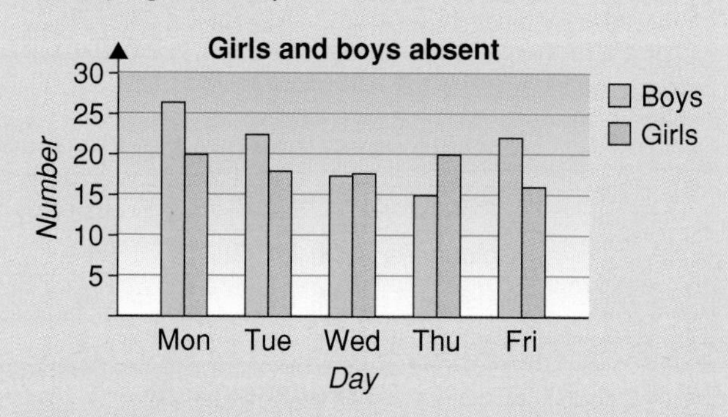

Line graph

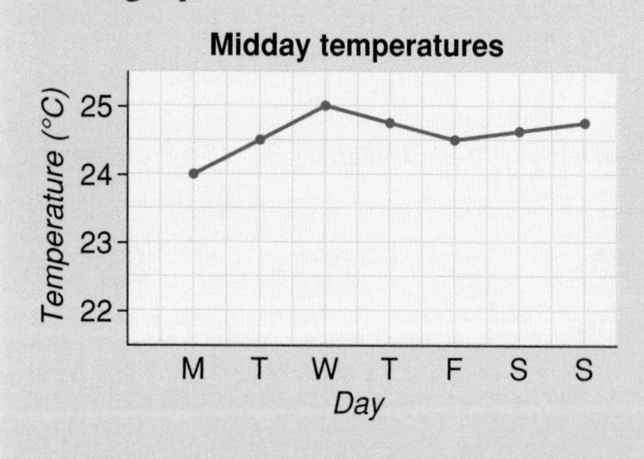

Pie chart
A pie chart shows the **proportion** in each category.

Transport to school

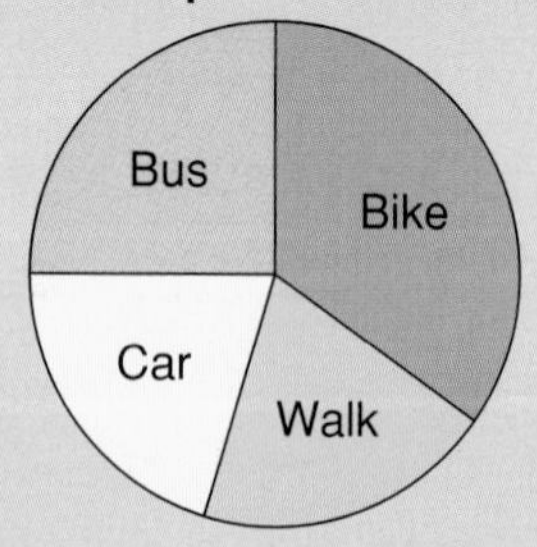

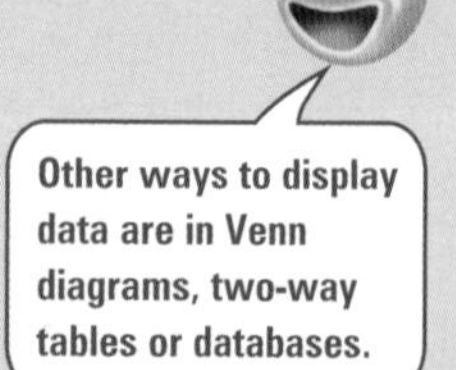

Pictograms, bar charts and pie charts are used to show **categorical** data.
Categorical data is non-numerical data.
Line graphs are usually used to show changes over time.

Always — give your graph a **title**
— label any axes
— have values at equal intervals on the axes.

Once we have displayed data on a graph, we can use the graph to help **interpret the data**.

Practice Questions 5, 10, 12, 21, 22, 27, 28, 29, 31, 32, 38

Grouped data

Sometimes **discrete data** is grouped into **equal class intervals** (groups of equal size).
Fran drew this frequency table to show the number of pupils at schools in her region.

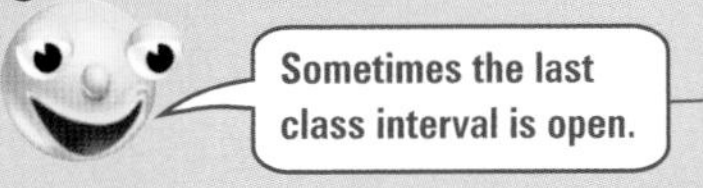

Number of pupils	Tally	Frequency
0–99	II	2
100–199	I	1
200–299	IIII	4
300–399	卌 I	6
>400	II	2

We can use a **bar chart** to show **grouped discrete data**.

Example The data in this frequency table is shown on the bar chart.

Number of pupils	Tally	Frequency
1–5	卌 I	6
6–10	卌 IIII	9
11–15	卌 卌	10
16–20	卌 II	7
21–25	卌 卌	10
26+	卌 I	6

equal class intervals

open class interval

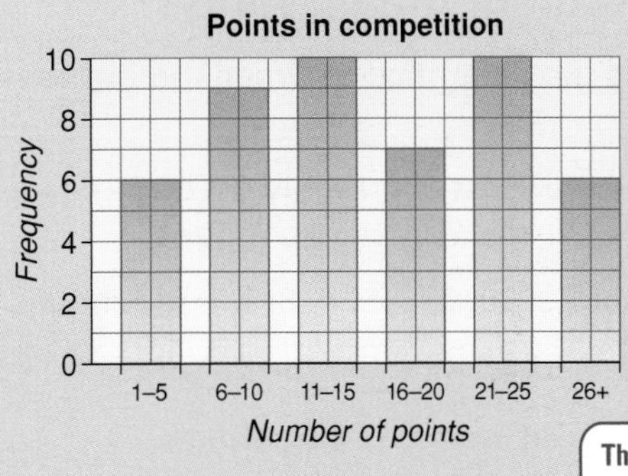

The class interval is written under each bar.

When we collect **continuous data** we usually group it into **equal class intervals** on a **frequency table**.

Example This frequency table shows the length of books in room 7.

Length in cms	Tally	Frequency
$10 < \ell \leq 15$	卌 卌 III	13
$15 < \ell \leq 20$	卌 卌 卌 II	17
$20 < \ell \leq 25$	卌 II	7
$25 < \ell \leq 30$	IIII	4
$30 < \ell \leq 35$	II	2

$10 < \ell \leq 15$ means a length greater than 10 but less than or equal to 15.

For continuous data we draw a **frequency diagram**.

Example This frequency diagram shows the data in the table above.

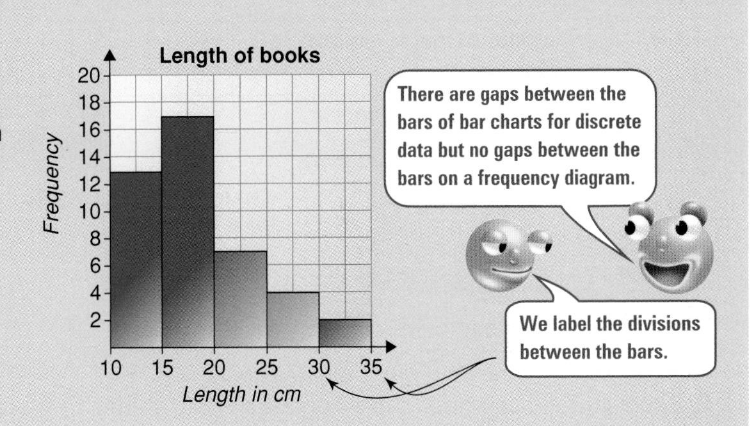

Practice Questions 4b, 24, 33

Drawing pie charts

To draw a pie chart
- find what fraction of the whole, each sector should represent
- multiply this fraction by 360° to find the angle at the centre of each sector.

Example **Hockey Matches**

Won	15
Lost	3
Drawn	6
	24

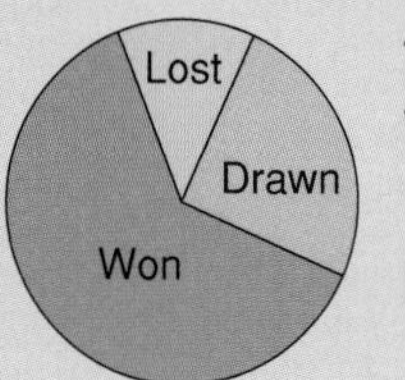

The angles are:

Won $\frac{15}{24} \times 360° = 225°$

Lost $\frac{3}{24} \times 360° = 45°$

Drawn $\frac{6}{24} \times 360° = 90°$

Practice Question 34

Mode, median, mean, range

The **range** is the difference between the **highest data value** and the **lowest data value**.

The **mode** is the most commonly occurring data value.
Sometimes a set of data has two modes.

Example For 8, 9, 9, 3, 6, 9, 8, 8, 7, 5 the modes are 8 and 9.

If data is grouped, we find the **modal class**. It is the class interval with the highest frequency.

Example **Age of people in a village**

Age (years)	0–9	10–19	20–29	30–39	40–49	50–59	60–69	70–79	80+
Frequency	3140	2987	2864	3346	3162	2834	2172	1832	1436

The modal class is 30–39.

$$\textbf{Mean} = \frac{\textbf{sum of data values}}{\textbf{number of data values}}$$

Example 4, 8, 6, 2, 3, 9, 3, 6

$$\text{Mean} = \frac{4+8+6+2+3+9+3+6}{8} = \frac{41}{8} = 5{\cdot}1 \text{ (1 d.p.)}$$

Sometimes the data is given in a **frequency table**.

Example Kelly collected this data on the number of major exports twenty countries have.

Number of major exports	0	1	2	3	4
Frequency	5	8	4	2	1
Total number of exports	0	8	8	6	4

Multiply the number of major exports by the frequency.

$$\text{Mean} = \frac{\text{sum of data values}}{\text{number of data values}}$$
$$= \frac{0+8+8+6+4}{5+8+4+2+1}$$

(numerator: sum of total number of exports; denominator: sum of frequencies)

$$= \frac{26}{20}$$
$$= \mathbf{1{\cdot}3}$$

We can find the mean using an **assumed mean**.

1. Assume the mean is a particular value.
2. Subtract the assumed mean from each data value.
3. Find the mean of the differences you found in **2**.
4. Add your answer to **3** to the assumed mean.

Example Joe's temperature over five hours was
37 °C 37·4 °C 38·1 °C 37·5 °C 37·8 °C.
1 Assume the mean is 37·5 °C.
2 The subtracted values are $^{-}$0·5 °C, $^{-}$0·1 °C, 0·6 °C, 0 °C and 0·3 °C.
3 The mean of the differences is $\frac{0·3}{5}$ = 0·06.
4 The mean is **37·56 °C**.

The **median** is the middle value when a set of data is arranged in order of size.
When there is an even number of values, the median is the mean of the two middle values.

Example 5, 6, 3, 9, 6, 4, 2, 1
In order these are 1, 2, 3, **4**, **5**, 6, 6, 9.
Median = $\frac{4+5}{2}$
= 4·5

Practice Questions 3, 4a, 6, 7, 11, 16, 19, 36, 37

Comparing data

To **compare data** we can use the **range** and one of the **mean**, **median** or **mode**.

Example Paula and Melanie both wanted to be chosen to represent the school in a 15 km run.
The mean and range of their last ten races is given in the table.

	Mean	Range
Paula	1 hour 48·62	19 minutes
Melanie	1 hour 49·25	3 minutes

We could choose Paula because her mean time is better or we could choose Melanie because her times are more consistent.

Practice Question 30

Probability

We can describe the probability of an event happening using one of these words.

certain **likely** **even chance** **unlikely** **impossible**

Examples It is **certain** that June follows May.
It is **likely** that you will eat breakfast tomorrow.
There is an **even chance** of getting a head when you toss a coin.
It is **unlikely** that you will see the Queen tomorrow.
It is **impossible** to draw a triangle with 4 sides.

Probability is a way of measuring the chance or likelihood of a particular outcome.
We can show probabilities on a probability scale.

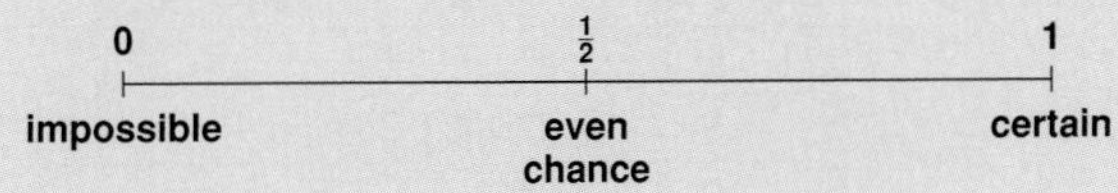

Example The probability of getting a 3 when you roll a dice is shown by the arrow.

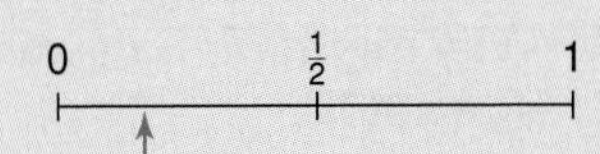

All probabilities lie from 0 to 1.

A counter is taken from this bag, without looking. This is called **at random**.
It could be red, blue or purple.
These are called the possible **outcomes**.

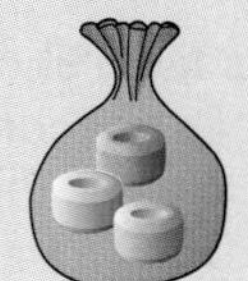

The possible outcomes of spinning this spinner twice are:
1, 1 or 1, 2 or 2, 1 or 2, 2

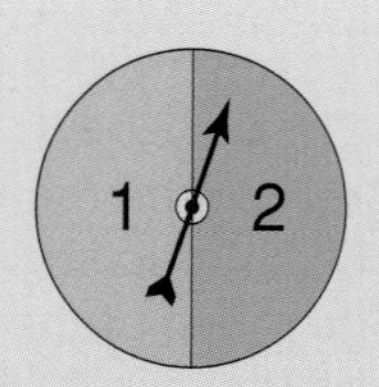

Equally likely outcomes have an equal chance of happening.

Example A counter is taken at random from the bag above.
Getting red, getting blue or getting purple are all equally likely outcomes, because there is the same number of each colour counter.

For equally likely outcomes

probability of an event = $\frac{\textbf{number of favourable outcomes}}{\textbf{number of possible outcomes}}$

Example The letters in the word HORSES are put in a tub.
A letter is taken at random.
The probability of getting an H is $\frac{1}{6}$ (one in six).

The probability of getting an S is $\frac{2}{6}$ or $\frac{1}{3}$ (two in six).

When a probability experiment is repeated the results are likely to be slightly different each time.

Example Junita tossed a coin 50 times. She got 28 heads.
She tossed it another 50 times. She got 22 heads.

Practice Questions 1, 2, 8, 9, 14, 17, 18, 20, 23, 25, 35

Practice Questions

1 Decide if each of these is **certain**, **very likely**, **likely**, **unlikely**, **very unlikely** or **impossible**.

a Someone will visit your house in the next week.
b A shopper at your local supermarket will buy chocolate today.
c You will read 50 books this week.
d A polar bear will give you a kiss tomorrow.
e It will get dark tonight.

2 Put the events in question **1** in order from most likely to happen to least likely to happen.

3 Mr Salt drew this table to show the sizes of women's trainers he sold last week.

a What is the modal shoe size?
b What is the range of shoe sizes?
c Why might Mr Salt want to know the modal shoe size sold?

Shoe size	Tally	Frequency
4	𝍸 III	8
5	𝍸 𝍸 I	11
6	𝍸 II	7
7	III	3
8	I	1

4 This table shows the number of questions pupils got correct in a multi-choice test.

a What is the modal class?
b Draw a bar chart to show the data.

Questions correct	Tally	Frequency
0–9	𝍸 II	7
10–19	𝍸 𝍸 III	13
20–29	𝍸 𝍸 𝍸 II	17
30–39	𝍸 II	7

5 These pictograms show the items sold at a café one Saturday and Sunday.

Each symbol represents 10	
Rolls	
Ice creams	
Coffee	
Juice	
	Saturday

Each symbol represents 10	
Rolls	
Ice creams	
Coffee	
Juice	
	Sunday

a How many coffees were sold on Sunday?
b How many more ice creams were sold on Saturday than Sunday?
c One of the days was hot and the other was cooler. Which day do you think was which? Explain.

6 Find the mode of each of these data sets.
a 5, 7, 11, 5, 8, 3, 5, 7, 5, 2
b 13, 19, 15, 19, 19, 19, 13, 13, 24, 16, 18, 13
c 2·6, 4·1, 8·3, 5·2, 7·1, 6·4, 3·9

7 Find the mean, median and range of each of these sets of data.
a 5, 8, 7, 12, 16, 24, 29, 30, 40.
b £5721, £8632, £5200, £5832, £6381, £5200, £3852, £4729
c 8·3 g, 9·6 g, 7·2 g, 15·2, g, 12·6 g, 10·4 g, 11·6 g, 13·9 g, 15·2 g, 9·6 g.

8 A door has a security lock. To open the door you must press the correct buttons. **[SATs 2002 paper 2]**
The code for the door is one letter followed by a single digit number.
For example: B6.

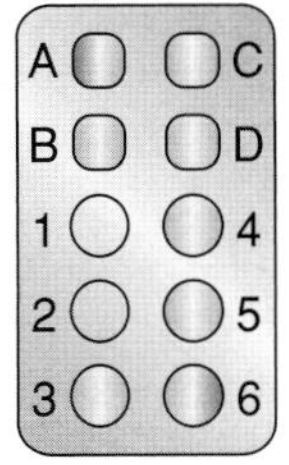

a How many **different** codes are there altogether? Show your working.
b I know that the correct code begins with D.
I press D, then I guess the single digit number.
What is the probability that I open the door?

9 A school has a new canteen. A special person will be chosen to perform the opening ceremony. **[SATs 2000 paper 2]**
The names of all the pupils, all the teachers and all the canteen staff are put into a box.
One name is taken out at random.
A pupil says:

> There are only three choices.
> It could be a pupil, a teacher or one of the canteen staff.
> The probability of it being a pupil is $\frac{1}{3}$.

The pupil is **wrong**. Explain why.

10 This table gives the population, to the nearest thousand, of a town and the projected population for future years.

Year	1985	1990	1995	2000	2005	2010	2015	2020
Population	28 000	30 000	31 000	29 000	28 000	26 000	26 000	28 000

a Use a copy of this grid.
Draw a line graph for the data.

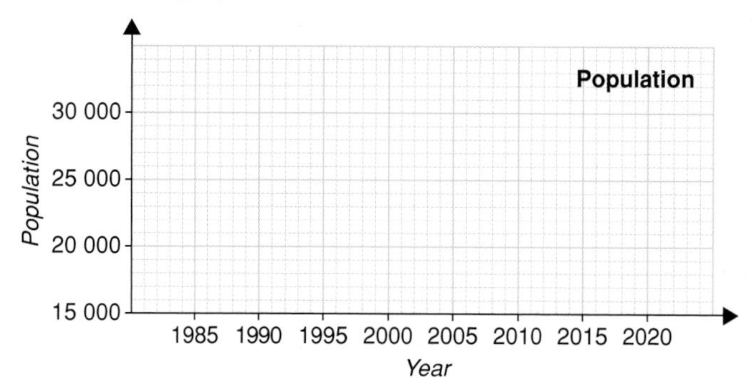

b What was the increase in population between 1985 and 1995?
c What is the expected increase in population between 2010 and 2020?
d It seems that the population in the middle of 1987 was 29 000. Is this necessarily so? Explain your answer.

11 The mass of iron filings picked up by a magnet on five different trials is given below.

85 g 96 g 110 g 98 g 97 g

a Find the mean using the assumed mean method.
b What is the range of the masses?
c What is the median mass?

12 Jayne gathered data on the cars that were for sale from all the car dealers in her district. She summarised the data in this table.

	British	Non British
New	124	74
Used	94	65

a How many cars were for sale altogether?
b How many of the cars were new?
c How many of the used cars were British?

13 Which of the following data is discrete and which is continuous?
Year of birth Waist measurements Time spent watching TV
Length of babies Size of families

14 The letters of the alphabet are put in a box.
One letter is chosen without looking.
a Copy and finish this list of outcomes.
a, b, c, ...
b Does each letter have an equal chance of being chosen?
c Is it equally likely that a vowel or a consonant will be chosen? Explain.

15 For each of these questions, how would you collect the data?
A Questionnaire or data collection sheet
B Experiment
C Secondary source, such as website, book, newspaper, CD-ROM, ...
a What percentage of countries in the world speak English?
b Do adult males, aged 25 to 60, watch more TV than teenage boys?
*** c** Do boys or girls know their times tables better?

16 Pete got a mean of 12 and a range of 5 for his six science experiments.

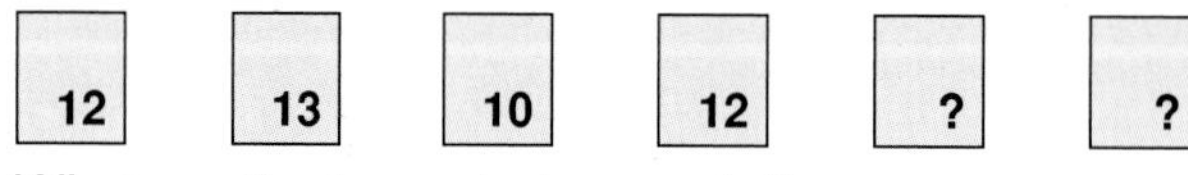

What are the two missing marks?

17 In each box of cereal there is a free gift of a card. **[SATs 2000 paper 2]**
You cannot tell which card will be in a box. Each card is equally likely.
There are four different cards A, B, C or D.

a Zoe needs card A.
Her brother Paul needs cards C and D.
They buy one box of cereal.
  i What is the probability that the card is one that Zoe needs?
  ii What is the probability that the card is one that Paul needs?

b Then their mother opens the box.
She tells them that the card **is not** card A.
  i Now what is the probability the card is one that Zoe needs?
  ii What is the probability that the card is one that Paul needs?

18 **a** A spinner has eight equal sections. **[SATs 2002 paper 1]**
  i What is the probability of scoring 4 on the spinner?
  ii What is the probability of scoring an **even** number on the spinner?

b A different spinner has six equal sections and **six numbers**.
On this spinner, the probability of scoring an **even** number is $\frac{2}{3}$.
The probability of scoring **4** is $\frac{1}{3}$.
Copy this spinner.
Write what numbers could be on this spinner.

19 This table shows the number of experiments done by the pupils in 8JT.
Find the mean number of experiments done.
Give your answer to 1 d.p.

Number of experiments	Frequency	Total
0	5	
1	3	
2	8	
3	12	
4	7	

20 This spinner is spun.

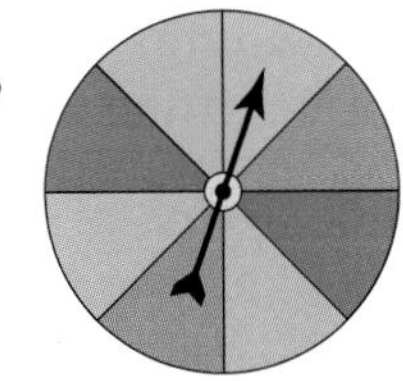

a Which colour is it least likely to stop on?
Give a reason for your answer.

b Copy this scale.

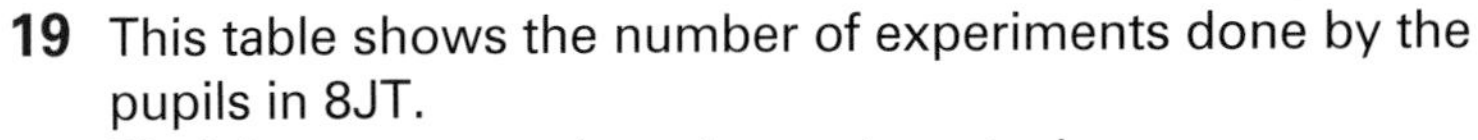

Mark with an X the probability that the colour will be blue.

c Write down the probability that the colour will be orange .

21 This table shows part of a database.

Name	Age	Male/female	Tutor group	Learns French	Learns German
Nick	14	M	104	Yes	No
Jackie	14	F	104	No	Yes
Rishi	15	M	104	Yes	Yes
Vera	14	F	105	Yes	No
Winstone	15	M	105	No	Yes
Caroline	14	F	104	Yes	Yes

a List the pupils who are 14 and in tutor group 104.
b List the pupils who learn French and German.

22 This graph shows the sales for 'Computers Today'.

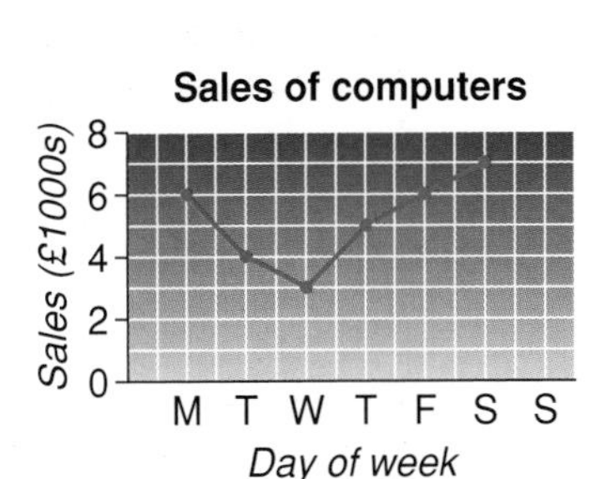

a On which day did they sell £5000 worth of computers?
b What were the sales, in pounds, on Saturday?
*__c__ Can you tell how much had been sold at midday on Wednesday?

23 Theo throws a dice. What is the probability he will get

- **a** an odd number
- **b** a 1 or a 2
- **c** a number bigger than 2
- **d** a number bigger than 6
- **e** a number less than 7?

24 The number of questions answered correctly by a class in a sports quiz were:

12	18	25	19	27	29	28	38	11	15
18	20	28	29	36	30	8	20	10	14
19	9	24	3	32	29	30	36	26	23

- **a** Put this data onto a frequency chart with class intervals 0–4, 5–9, 10–14, 15–19, 20–24, ...
- **b** Regroup the data using the intervals 0–9, 10–19, 20–29, 30–39.
- **c** Which class intervals do you think are more useful? Explain why.
- **d** Draw a bar chart using the intervals given in **b**.
- **e** Which of these do you think better explains the shape of the graph?
  - **A** All of the questions were very easy.
  - **B** Three pupils were away when the class was preparing for the test.

Remember to give your graph a title and label the axes.

25 There are 30 cubes in a bag. Tina takes one without looking.
She writes down its colour and then **puts the cube back** in the bag.
She does this 30 times.
Tina records her results in a chart.

blue	7
yellow	3
green	11
red	4
orange	5

- **a** Tina thinks there must be 7 blue cubes in the bag because there are 7 blues on the chart.
  Explain why Tina could be wrong.
- **b** Tina thinks there can't be any white cubes in the bag because there are none on the chart.
  Explain why Tina could be wrong.

26 Bene wanted to know if Year 8 girls could skip faster than Year 8 boys.

- **a** Write down two things Bene might find out from his survey.
- **b** What data does Bene need to collect?
- **c** Design a collection sheet for this data.
- **d** How could Bene collect the data?
- **e** What does Bene need to consider when choosing a sample size?

27 This bar chart shows the number of items the two families put in recycling bins one week.

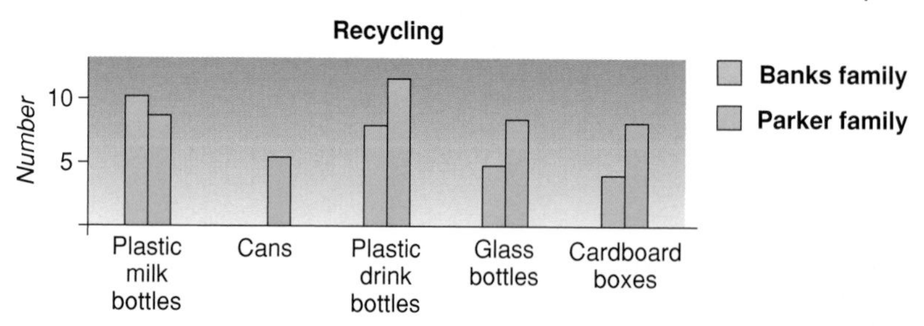

- **a** Which item did the Parker family recycle most that week?
- **b** The Banks family didn't recycle any cans that week.
  Which of these is the **least** likely reason?
  - **A** The Banks family forgot to put them out.
  - **B** The Banks family never ate food from cans.
  - **C** The Banks family didn't use cans that week.
- **c** Use the graph to compare what the two families recycled that week.

28 Some American and British people were surveyed about the sort of TV programme they liked best.
These pie charts show the results.
Jacinta looked at these charts and said 'More American people surveyed than British people surveyed like comedy'.
Explain why Jacinta is wrong.

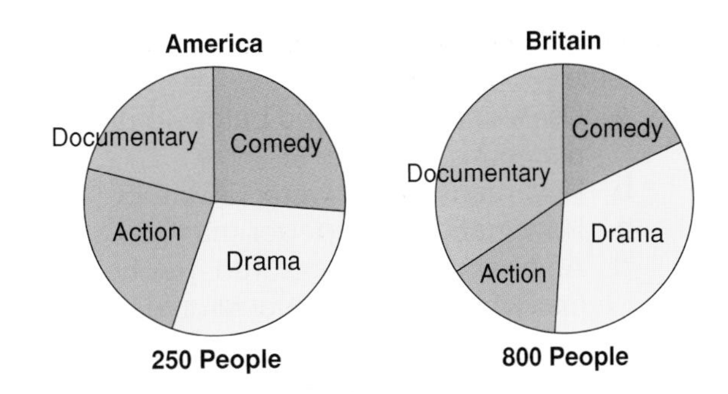

29 a This frequency chart shows the number of boys and girls from six classes that belonged to the computer club. Draw a bar chart to show the data. Put the data on the same bar chart.

Class	8N	8P	8A	8T	8Y
Boys	12	16	8	12	8
Girls	3	7	7	10	13

b Write a few sentences comparing the boys and girls.

30 This table gives the number of pupils away in two classes.

	Week 1	Week 2	Week 3	Week 4	Week 5	Week 6	Week 7	Week 8
Class 9L	0	0	0	10	14	16	0	0
Class 9T	4	5	5	7	5	4	6	4

The mean and range for each class are given.

	mean	range
9L	5	16
9T	5	3

Write a sentence comparing the number of pupils away in 9L and 9T.

31 a How many of each age group would there be if there are 200 members in the squash club?
b Why do you think there are fewer people older than 60 than any other age range?

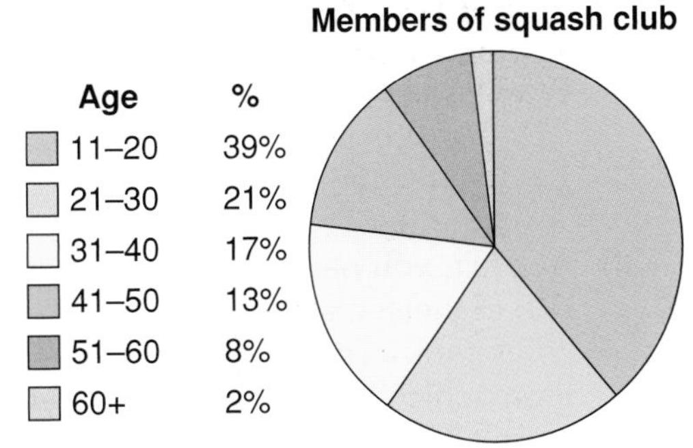

32 A survey was carried out on vehicle registration.
This table gives the number of cars, lorries and motorcycles without correct registration in three areas.

	Cars	Lorries	Motorcycles
Area 1	58	12	18
Area 2	14	8	15
Area 3	63	7	22

Draw a compound bar graph to show this data.
Have the vehicles on the horizontal axis.

T

33 Tara recorded the heights of some pupils in this table.

Height h (cm)	Frequency
$120 \leqslant h < 125$	2
$125 \leqslant h < 130$	5
$130 \leqslant h < 135$	8
$135 \leqslant h < 140$	14
$140 \leqslant h < 145$	11
$145 \leqslant h < 150$	9
$150 \leqslant h < 155$	3
$155 \leqslant h < 160$	1

a Use a copy of the grid below to draw a frequency diagram.

b How many pupils were 135 cm or taller?

c How many were shorter than 155 cm?

d Another pupil is measured. Her height is 140 cm. Into which class interval should her height be placed?

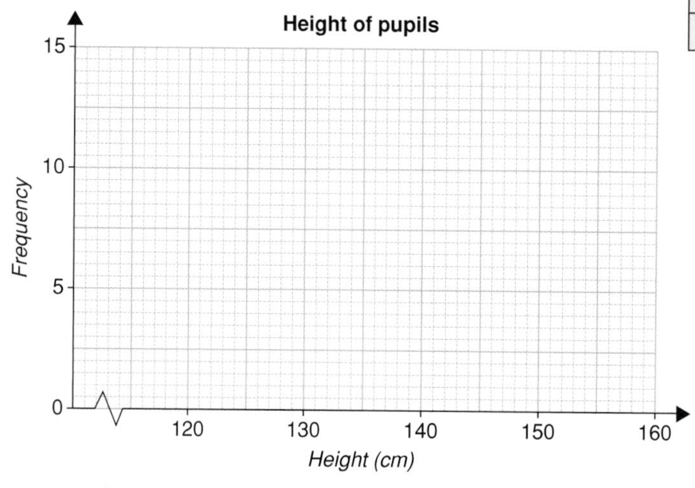

34 A vet is investigating to see if colour plays any part in cat health. This list gives the colour of the cats seen by this vet during one day of her investigation.

Ginger	38
Black	22
Black and white	21
Tortoiseshell	9

Draw a pie chart to illustrate this data.

35 At a fair, you paid £1 to take a marble at random from one of these bags.
If you took a red marble you won £1·50.
From which bag is it most likely you will get a red marble? Explain.

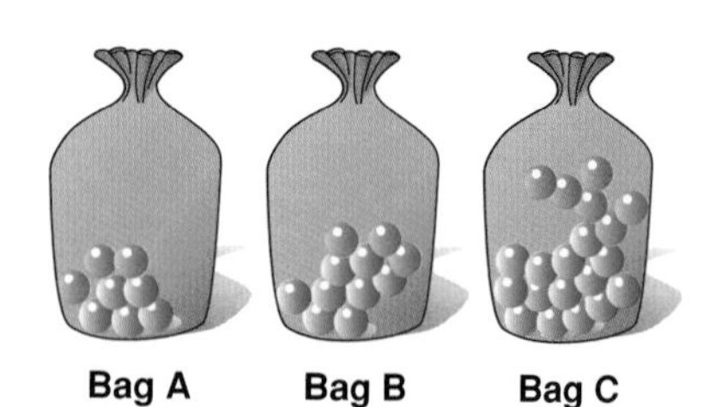

36 Pansy got these marks in three tests: 80, 86, 94.
She had one more test to sit.
She wanted a mean of 85 for the four tests.
What mark must she get in her fourth test?

37 Abbie and Brendon had a competition to see who could hold their breath underwater the longest. They decided to find the mean of three tries.

Abbie's results were	58 sec	65 sec	51 sec
Brendon's results were	●	57 sec	●

They both had the same mean. Brendon's range was half of Abbie's.
What are Brendon's two missing scores?

38 This graph shows the amount spent at a shop on cameras, video cameras and digital cameras.

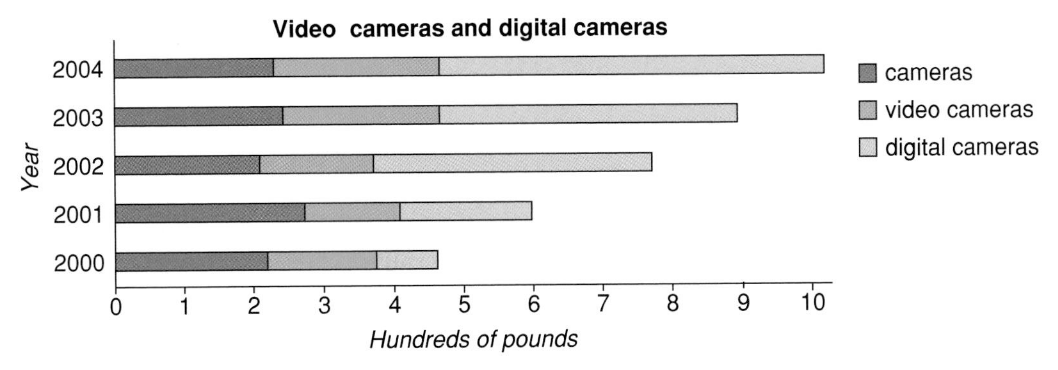

a How did spending change over five years?

b Using the data in the graph, predict what the results might look like this year and in ten years time.

15 Data Collection

You need to know

✓ planning and collecting data page 361

✓ discrete or continuous data page 361

Key vocabulary

continuous, data log, discrete, interrogate, primary source, sample, secondary source, two-way table

At the polls

- Coming up to elections, polls are often taken. These are surveys and the results are often used to sway voters.

  Think of some other examples where the results of surveys are used to change people's opinions.

- The election office of the Deseat Party rang twelve people and asked if they were going to vote for the Deseat Party or another party. Nine said they would vote for Deseat.

  In their advertisement in the paper the next day the headline said:

75% of people say they will vote for Deseat. Join the majority!

POLLING STATION

VOTE DESEAT

Is this a reasonable statement to make? Explain.

Two-way tables

Data collected from surveys can sometimes be summarised in a **two-way table**.
A **two-way table** can be used to display two sets of data.
We often use them to compare data.

Example This table shows the number of males and females who passed their driving test on the first and second attempts.

	1st attempt	2nd attempt
Male	27	13
Female	35	5

Worked Example
This table shows the number of accidents before and after lights were put at an intersection.

a How many accidents in total were there before the lights were installed?
b How many minor accidents were there in total?
c Compare the accidents before and after the lights were installed.

	Before lights	After lights
Minor	27	12
Serious	5	2
Resulting in death	2	0

Answer
a We add up the numbers in the 'before lights' column.

$27 + 5 + 2 = 34$

There were 34 accidents before the lights were installed.

b We add up the numbers in the minor accidents row.

$27 + 12 = 39$

There were 39 minor accidents.

c A possible answer is:

There were fewer accidents after the lights were installed and none of them resulted in death.

Exercise 1

1 This table shows the hair colour of the boys and girls in Deepak's class.
  a How many black-haired boys were in Deepak's class?
  b How many boys were in the class?
  c How many brown-haired pupils were in Deepak's class?

	Boy	Girl
Black	8	6
Blonde	2	3
Red	1	0
Brown	5	7

2 The drinks males and females had one day at a café are shown in this table.
  a How many males in total had drinks?
  b How many in total had tea?
  c How many females had tea or water?
  d How many had drinks in total?
  e Compare the choices of males and females.

	Male	Female
Tea	4	12
Coffee	14	6
Water	3	7
Fizzy	7	1
Milk	0	2

3 This table shows the main meal choices of people at a café and the time taken, to the nearest minute, for the meal to arrive at the table.

	Main meal				
Time taken (minutes)	**Beef**	**Lamb**	**Chicken**	**Vegetarian**	**Pasta**
1–5	0	0	2	5	1
6–10	4	3	1	2	6
11–15	3	1	4	0	3
15+	4	3	1	0	0

a How many of the beef meals took longer than 15 minutes?
b How many people ordered chicken?
c How many meals took 6–10 minutes to arrive?
d How many vegetarian meals took more than 10 minutes to arrive?
e If you were in a hurry, what would be the best meal to order?

4 This two-way table shows the number of students achieving grades A to E in examinations in maths and science.
a How many students achieved the same grade in both subjects?
b How many of the students who achieved grade B in maths achieved a different grade in science?
c What does the table suggest about the grades achieved by these students in maths and science?

		Maths grade				
		A	**B**	**C**	**D**	**E**
Science grade	**A**	5	3	2	0	0
	B	3	4	4	1	0
	C	1	5	7	0	1
	D	0	1	0	6	5
	E	0	0	0	4	6

5 A study on alcohol was done in the Netherlands.
These tables show the percentage of problem drinkers aged 16 years and above in three cities.

Utrecht

Age	Men	Women
16–24	23%	14%
25–34	17%	7%
35–44	14%	7%
45–54	15%	8%
55–69	6%	3%
Total	16%	8%

Rotterdam

Age	Men	Women
16–24	19%	9%
25–34	12%	5%
35–44	16%	5%
45–54	15%	4%
55–69	11%	4%
Total	14%	5%

Parkstad Limburg

Age	Men	Women
16–24	16%	4%
25–34	10%	2%
35–44	8%	3%
45–54	11%	2%
55–69	6%	3%
Total	10%	3%

a Which city has the greatest proportion of problem drinkers?
b Which city has the greatest proportion of problem female drinkers aged 55–69?
c Compare male and female problem drinkers.

6 These tables show the percentage of male and female smokers at a factory in 1983, 1993 and 2003.

2003

Age	Male %	Female %
16–19	24	30
20–24	40	38
25–34	36	31
35–49	26	26
50–59	30	31
60+	19	21

1993

Age	Male %	Female %
16–19	31	32
20–24	42	39
25–34	35	32
35–49	38	32
50–59	34	34
60+	27	21

1983

Age	Male %	Female %
16–19	38	36
20–24	49	45
25–34	48	41
35–49	49	44
50–59	50	46
60+	41	40

Write some sentences comparing
a the trends over time
b male and female smokers.

7 Use a copy of this.
A retirement home has 59 residents.
a Complete the two-way table.
b How many females are there in the home?
c How many females are 70 or over?

	M	F	Totals
Under 70	12		21
70 and over	16		
Totals			59

8 There are two local tourist attractions near Ronan's school.

Cathedral Castle

Visitors can get to each by car or bus.
Design a two-way table Ronan could use to collect information on how visitors get to each of the attractions.

9 Patty did a survey on how people got to school and how long it took them, to the nearest minute.
She collected the information in a two-way table.
What might Patty's table look like?

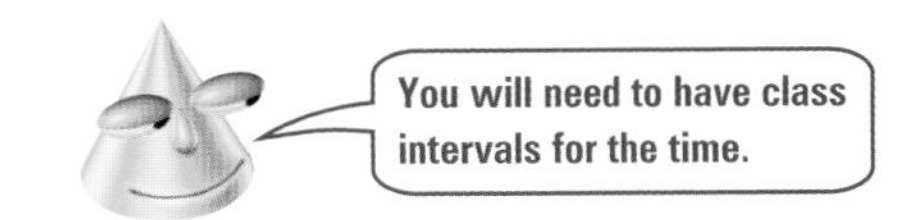

***10** The price of a ten-year-old second-hand car depends on its condition and mileage.
The table shows what percentage of the original price the second-hand value retains.
James thinks that the mileage being low becomes more important as the condition gets better.
Do you think he is right? Explain.

	Mileage		
Condition	**High**	**Medium**	**Low**
Excellent	25%	35%	55%
Good	20%	25%	30%
Poor	15%	18%	20%

***11** Rob was studying the differences between the learning ability of rats and mice. He tested them by trying to teach them to find their way out of a maze, roll a ball into a hole and retrieve food from a cage. He recorded the results and worked out what percentage of rats and mice achieved the tasks.
Design a two-way table to summarise Rob's results.

Review 1 This table shows the entries in a mixed team triathlon.
In each team there is a cyclist, a swimmer and a runner.
a How many male cyclists are there?
b How many females are entered altogether?
c How many teams are entered in the triathlon?
d Compare the male and female entrants in the triathlon.

	Cyclist	Swimmer	Runner
Male	12	5	15
Female	8	15	5

Review 2 Use a copy of this.
Students in a play either sang, had a main part or had a minor part.
65 students took part in the play altogether.
a Complete the two-way table.
b How many Year 8 students took part in the play?
c How many minor parts were there altogether?

	Singing	Main	Minor	Totals
Year 7	12	0		18
Year 8	17		8	
Year 9		4	12	
Totals		6		65

Review 3 Ali did a survey on male and female participation in a science fair. She recorded the percentage of males and females from her school who participated in these five sections:

'animal behaviour' 'plants' 'chemistry at home' 'how does it work?' and 'seeing and hearing'

Design a two-way table for her results.

Planning a survey

Discussion

- What do you already know about collecting data? **Discuss.**
- How do you decide what questions you want answered?

These are the headlines for some articles that appeared in a newspaper.

For which of these issues might a survey have been carried out?
What questions might the person who did the survey want answers to?
What related questions might the survey also explore?
Where might the data have been collected?
Who might have been surveyed?
How many might have been surveyed? What related questions might the survey also explore? **Discuss.**

For which of these surveys might a questionnaire have been used?
What questions should have been included?
What questions might have been included? **Discuss.**

- Do you think it is possible to draw incorrect conclusions from a survey? How could this happen? Is this likely? **Discuss.**
- Think of an issue that some people in your school, or your community, are concerned about.
  Think of the questions you want answered.
  Think of the sorts of answers you might anticipate.

  How could you find out how many people are concerned about this issue?
  How could you find out how strongly these people feel about this issue? **Discuss.**

Sources of data

Once you have decided what questions and related questions you want answered you need to **plan a survey**.
When planning a survey you need to consider a number of things.
Sources of data
Sample size
Design of questionnaire or collection sheets
Lets look at them in turn.

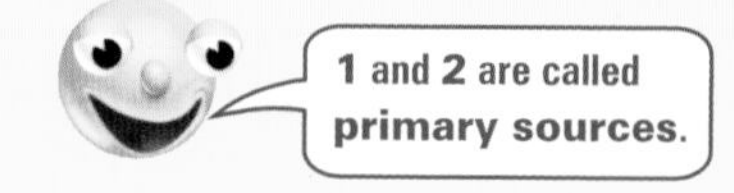

Remember
Data can be gathered from

1. a questionnaire or survey of a sample of people
2. an experiment which may use technology, such as a data logger, graphical calculator or computer
3. **secondary sources** such as reference books, websites, printed tables or lists, CD-ROMS, newspapers, historical records, **interrogating** a database.

When **planning a survey** it is helpful to identify all the possible sources of data. Then you must decide how you are going to collect the data.

Example If you are investigating factors that affect how well a vacuum cleaner sucks up dirt you will need to find published data on tests done or conduct your own tests.

Discussion

- Elijah wants to know which type of pie sold at the canteen is most popular. He could gather the data by doing a survey **or** he could use the canteen's sales record.

Discuss the advantages and disadvantages of each method of collection.
Think about which method will:

address the question better
take less time
use fewer resources
give the greatest insight to the question and related questions.

- For each of these, **discuss** the answers you might anticipate, any related questions and the possible sources of data.
  Discuss how you might collect the data.
  - What factors on the day affect athlete performance?
  - What effect does power, in watts, have on vacuum cleaner performance?
  - What factors affect weed growth in lawns?
  - Have rainfall patterns changed in Britain in the last 50 years?
  - What do people think of shops with a no return of goods policy?

Sample size

When we use a questionnaire or carry out a survey it is often not practical to ask everyone possible. We choose a **sample** to ask.

Example If you were doing a survey on whether people think the local taxi service is as good as it used to be, it is probably not possible to ask everyone who uses taxis in the local area.

A **sample** should be as large as it is sensible to make it.
The results that you get from the sample should be representative of everyone the survey relates to.

Example If you wanted to know if the age of a chicken affected what size egg it laid, you would need to survey chickens of all different ages.
It is important that the chickens are similar in all other ways, e.g. living conditions, breed, ...

Discussion

- **Discuss** the problems if a sample is
  a too small
  b too big.

- How many people do you think you should survey so that the data collected is representative? **Discuss**.
  Will it change depending on the survey?

Collection sheets and questionnaires

Remember
We group **continuous data** in equal class intervals.

To collect data, you sometimes need to design a questionnaire or a data collection sheet. You need to decide what units to use for any measurements and how accurate you want the data to be.

Example

Questionnaire on health

How many times have you been to the doctor in the last year?

0 □ 1 □ 2 □ 3 □ more than 3 □

How many days off work or school have you had in the last year?

0 □ 1–3 □ 4–6 □ 7–10 □ more than 10 □

Data collection sheet

Height of plant (mm)	Tally	Frequency
$0 < h \leqslant 10$		
$10 < h \leqslant 20$		
$20 < h \leqslant 30$		
$30 < h \leqslant 40$		
$40 < h \leqslant 50$		

Here are some **guidelines for writing a questionnaire**.

1 Allow for any possible answers.

Example
not at all □
up to 1 hour □
between 1 and 2 hours □
between 2 and 3 hours □
more than 3 hours □

rather than:
up to 1 hour □
between 1 and 2 hours □
between 2 and 3 hours □

2 Give instructions on how you want the questions answered.
Example: Tick one of these boxes.

3 Do not ask for information that is not needed. *Example*: Name of person.

4 Avoid questions that people may not be willing to answer.

5 If your questions are asking for opinions, word the questions so that *your* opinion is not evident.

6 Make the questions clear and concise.

7 Keep the questionnaire as short as possible.

Once you have written a questionnaire, it is a good idea to **trial** it. That is, have a few people answer the questionnaire. If any of these people have problems answering your questionnaire, you will need to refine it. Check the wording is clear and all possible responses are covered.

Exercise 2

1 Jonathon said 'most pupils in the school don't wear their school tie on most days'.
Some pupils decide to do a survey to see if Jonathon is right.

a Natalie says 'we can ask 10 people in our class if they wear their tie'.
Give two reasons why Natalie's method might not give very good data.

b Kylie says 'we could stand outside the gym at lunchtime and record how many people come out and how many don't have a tie'.
Give two reasons why Kylie's method might not give very good data.

2 Which of these would be a better question to put in a questionnaire to find out what people think of cats as a pet.

A Do you like cats? Yes ☐ No ☐

B I think a cat is a good pet.
Strongly agree ☐
Agree ☐
Not sure ☐
Disagree ☐
Strongly disagree ☐

3 Nia was doing a survey on how much Year 8 pupils helped at home. She wrote this question for her questionnaire.

I help at home enough. Yes ☐ No ☐

a Give a reason why Nia's question may not give very good data.
b Write a question for Nia's questionnaire that would give better data.

4 Anthony designed this data collection sheet to collect data on the length of run up and height jumped by people in a high jump competition.
Give two things about it that could be improved

Name	Length of run up to nearest 5 m	Height jumped to nearest 10 cm
Greg Bate		1·2 m

5 Which of the following statements are likely to get responses that would not be very useful? Explain your answer.
Rewrite these statements so the responses would give you more useful information.

a I get 8 hours sleep each night. Always ☐ Sometimes ☐ Never ☐
b I do my maths homework. Always ☐ Sometimes ☐ Never ☐
c I go out on a Friday night. Always ☐ Sometimes ☐ Never ☐
d I play sport. Always ☐ Sometimes ☐ Never ☐

6 Design a data collection sheet or questionnaire to collect data for each of these.
a Joanna wants to know if the number of hours people sleep depends on age.
b Pablo wants to know if the acceleration of a car depends on engine size.

7 Write a short questionnaire with three or four questions to find out attitudes on one of these.

the school report   fairly-traded goods

8 Choose a survey topic from these.

A What factors affect how often people eat take-aways?
B Are people happy with the local bus service?

a Write down what questions you want answered.
b Write down some possible results.
c Write down what data you need to collect and, if relevant, the accuracy needed.
d Design a collection sheet or questionnaire to collect the data.
e Suggest a suitable sample size if this is relevant to the question.

Review Tony's class is doing a survey on the factors that may affect the distance thrown by people in a discus competition between 500 competitors of all ages.
a Tony decided to ask 10 people what they thought affected the distance they threw.
Give two reasons why this method may not give very good data.

b Brady designed this collection sheet to gather data.

Name	Distance thrown	Run up	Feel well?

Give two ways Brady could improve this collection sheet.

c Mariana wrote this question for her questionnaire.

I think I threw well. Yes ☐ No ☐

Give a reason why Mariana's question may not give very good data.
Write a question for Mariana that would give better data.

Discussion

Julia's group designed and trialled the following questionnaire. After it was trialled they rewrote parts of it.
Which parts do you think they may have rewritten? What do you think the final questionnaire may have looked like? **Discuss**.
Julia's group wrote the questionnaire to test some predictions. What might these have been? **Discuss**.

Please complete this questionnaire on school hours.
Place a ✓ in the appropriate boxes.

1. Female ☐ Male ☐
2. Age: 11 12 13 14 15 16 17 18
3. I would like school to begin and finish
   - the same time as now ☐
   - 30 minutes later ☐
   - 1 hour later ☐
   - more than 1 hour later ☐
4. How long does it take you to get to school?
   - less than $\frac{1}{2}$ hour ☐
   - between $\frac{1}{2}$ hour and 1 hour ☐
   - more than 1 hour ☐
5. How do you travel to school?
   train ☐ bus ☐ cycle ☐ walk ☐ car ☐ other ☐
6. Do you work after school? Yes ☐ No ☐
   If your answer was yes, please answer 7 & 8.
7. What time do you begin work? ____________
8. How long does it take you to get to work?

Thank you for completing this questionnaire.

- How could the results from this questionnaire be organised? What sorts of tables and graphs could be used? Could a computer database be helpful? **Discuss**.

Remember

To **plan a survey**:

1. Decide on the question you want answered and any related questions.
2. Decide what data needs to be collected and how accurate it needs to be.
3. Decide where to collect the data from and how much to collect (sample size).
4. Design a questionnaire or data collection sheet.
5. Trial your questionnaire or data collection sheet on a few people first, then refine it if necessary.

Practical

Plan a survey.
You could choose one of the surveys already mentioned in this chapter **or** you could use one of the suggestions below **or** you could make up your own.
Check your choice with your teacher.

Follow the steps given above.
Design a collection sheet or questionnaire. Remember you may need to group the data. Decide first if it is continuous or discrete data.
Collect the data.

Suggestions
Acceleration of popular cars.
Jumping or throwing distances and the factors that affect these.
Distribution of grass/weeds in different parts of the school.
Attitudes to the legal driving age.
Attitudes to underage drinking.

Summary of key points

A **two-way table** displays two data sets in a table.

Example This two-way table shows the ages of students in school sports teams.

	Netball	Football	Hockey
Under 14	16	25	27
14 to 16	36	53	26
Over 16	29	34	28

Before collecting data you need to do these.

1 Identify possible **sources of data** — questionnaire
— survey a sample of people
— experimental data
— the Internet
— databases or printed data or lists.

2 Decide on the **sample size**. If it is too big this takes too much time and costs too much. If it is too small, the data will not be representative of all the data the survey relates to.

3 Design a **collection sheet** or write a **questionnaire**.
Decide on the units to use for any measurements and how accurate you want the data to be.

When **writing a questionnaire**, use the guidelines given on page ●●●.

To plan a survey follow these steps.

1 Decide on the question you want answered and any related questions.

2 Decide what data needs to be collected. Decide how accurate the data needs to be.

3 Decide how to collect the data and, if appropriate, the **sample** size.

4 Design a questionnaire or data collection sheet.

5 Trial your questionnaire or collection sheet on a few people and refine it if necessary.

Test yourself

1 This table shows the instruments played by the students in a music group.

	Violin	Piano	Guitar
Male	1	3	6
Female	4	5	1

a How many males played the guitar?
b How many males were in the group?
c How many guitar players were there in the group?
d Compare the instruments played by males and females in the group.

2 A charity is doing a survey on the age and gender of people in a retirement home. All of the residents are over 60.
Design a two-way table to collect the data.

3 Mia was doing a survey on attitudes to immediate family.
She wrote this question for her questionnaire.

I like all the people in my family. Yes ☐ No ☐

a Give a reason why Mia's question may not give very good data.
b Write a suitable question for Mia's questionnaire.
c Once Mia's questionnaire had been refined, she gave it to ten of her relatives to fill in. Give two reasons why Mia's sample might not give very good data.

4 For each of the following
a name some sources of data
b give a suitable sample size
c design a questionnaire or data collection sheet.

A How far can people jump from a standing start compared with if they have a run up? Does practice improve the standing-start jump?
B Does power affect the efficiency of hairdryers?

5 Write some questions suitable for a questionnaire to find out people's attitude to teenage smoking.

6 Choose a survey topic.
Plan your survey using the steps given in D
Design a data collection sheet or questionnaire to collect the data.
You could choose a topic already mentioned in this chapter or choose one of your own.

16 Analysing Data, Drawing and Interpreting Graphs

You need to know

Key vocabulary

correlation, distribution, line graph, population pyramid, scatter graph, stem-and-leaf diagram

Watch It!

Television viewing by gender and age
United Kingdom
Hours per person per week

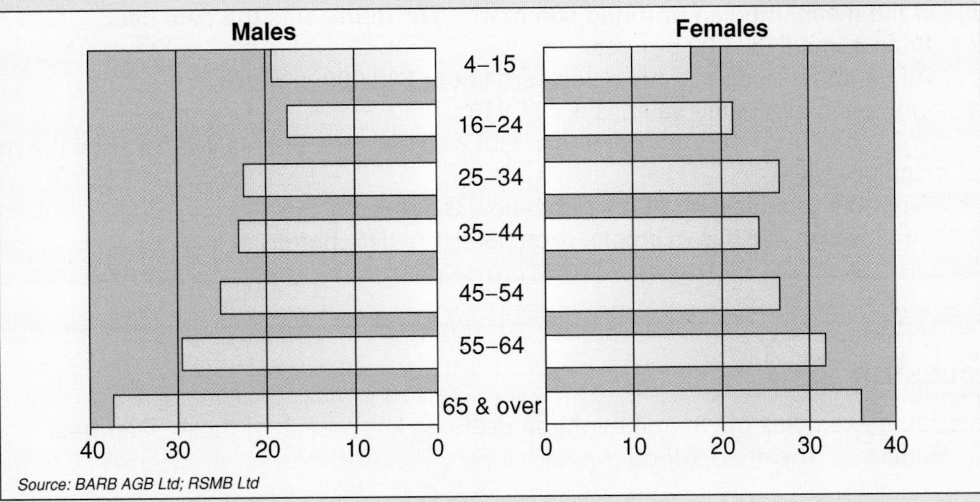

What does this graph tell you about

– the differences between male and female TV viewing
– the differences in TV viewing between age groups?

Mode, median, mean

Remember
range = highest data value–lowest data value

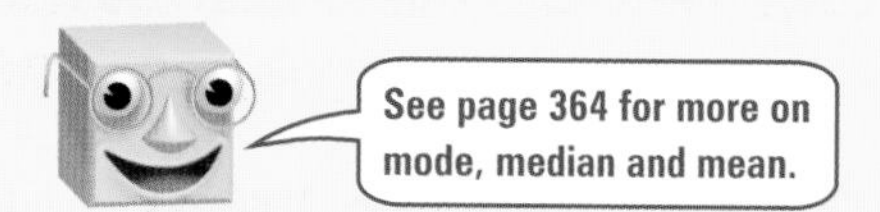

The **mode** is the most commonly occurring data value.

The **modal class** is the class interval with the highest frequency.

The **median** is the middle value of a set of ordered data.
If there is an even number of values, the median is the mean of the two middle values.

Mean = $\frac{\textbf{sum of data values}}{\textbf{number of values}}$

We sometimes find the mean using a calculator or spreadsheet.
Sometimes we find the mean using an assumed mean.

The median, mean and mode are all ways of summarising the data into a single number.

The median is useful for comparing a set of data values with a middle value.
Examples About half of the class got more than 72% in a test.
About half of the contestants in a javelin competition threw more than 50 m.

The mean gives us an idea of what would happen if there were 'equal' shares.
Example If three people got 78, 56 and 27, the mean tells us what each would get if the marks were shared between them equally.

The mode is useful for identifying the 'most popular' or in reporting opinion polls.
Example A clothing store might want to know the best-selling jeans.

Often one of the mean, median or mode summarises or represents the data best.
Example Brian earns £150 000 per year.
All the other people in his office earn about £20 000 per year.
The mean of all their salaries is £31 818.
The mean does not represent this data well because Brian's salary raises the mean significantly.
All of the salaries except one are below the mean.
The median, for this example, represents the data better.

Discussion

- Give some examples of when it might be useful to know each of these. **Discuss.**
  median mean mode
- In which situations is the mean not a useful statistic to represent the data? **Discuss.**
  What about the median?
  What about the mode?

Exercise 1 **Only use a calculator if you need to.**

1 These are the numbers of raffle tickets sold by the girls in a netball team.
10 10 10 10 8 10 0 2
a Find the mean, median and mode.
b Does the mode represent the data well?
Give a reason for your answer.
c Does the mean represent the data well?
Give a reason for your answer.

2 The cost of tickets to 9 shows at the town hall were
£25 £31 £20 £27 £30 £20 £22 £33 £26.
a Find the mean, median and mode.
b Does the mean represent the data well?
Give a reason for your answer.

3 These are the marks of ten people who entered a quiz.
100 120 140 142 143 145 145 145 482 498
a Find the mean and median.
b Does the median represent the data well?
Give a reason for your answer.
c Does the mean represent the data well?
Give a reason for your answer.

4 Which of the mean, median or mode would represent the data best in these cases? Explain.
a Mr Langham gave his class a test. All of the class except three people got 60% or more. The other three got less than 20%.
b Mrs Talley gave her class a test. The marks were all between 50% and 60%.
c The shoe size of Year 8 girls at Brooklands School was either 4, 5, 6 or 7. Most were size 5.
d The heights of Year 8 boys were between 1·2 m and 1·7 m. They were fairly evenly spread with a cluster between 1·4 m and 1·6 m.

***5** A small business has a manager and eight other staff. Their salaries in order are.
£70 000 £30 000 £19 000 £17 000 £15 000 £15 000 £12 500 £11 000 £10 000
a Find the mean, median and mode of these salaries.
b Which of these statistics would you be more likely to use if you were the manager and one of the other employees wanted a rise in salary?
c What if you were the employee who earned £11 000 negotiating for a rise in salary with the manager?

Review 1 This is how long Rachel had to wait for a bus each day this week.
1 min 3 min 3 min 13 min 14 min 4 min 11 min
a Calculate the mean time Rachel had to wait.
b Work out the median waiting time.
c What is the mode of the waiting times?
d Does the mean or median represent this data better?
Give a reason.
e Does the mode represent the data well?

Review 2 A farmer has thirty steers to sell. The weights in kg are

84	85	86	88	88	89	90	90	90	91
94	94	94	95	96	96	97	100	101	101
102	104	124	126	373	382	387	390	392	394

The masses are in order and the total mass is 4623.

a Find the mean, median and mode mass.

b If you were the farmer and wanted to quote an 'average mass' to a possible buyer, which of the mean, median or mode would you use? **Explain.** Would this 'average' be a fair one to quote?

Calculating the mean using a calculator or spreadsheet

Remember

$$\textbf{mean} = \frac{\textbf{sum of data values}}{\textbf{number of values}}$$

See page 364 to remind yourself about finding the mean from a frequency table.

When finding the mean of a large set of data, use a calculator or spreadsheet.

Example This table gives the scores of 285 entrants in a quiz competition.

Score	1	2	3	4	5	6	7	8	9	10
Number of people	8	16	23	26	31	87	41	32	17	4

Using a **calculator**, we can find the mean of the **distribution**.

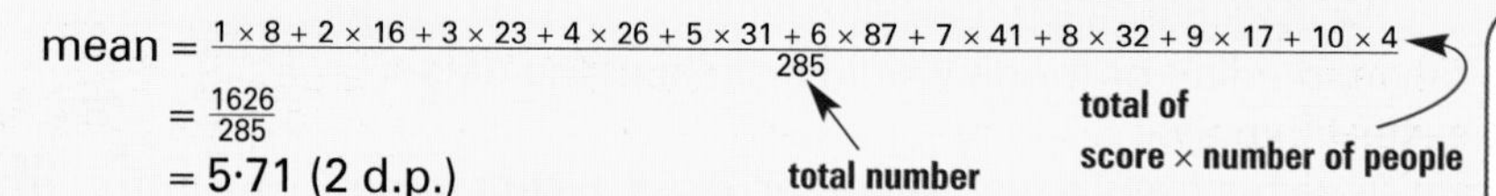

$$\text{mean} = \frac{1\times8+2\times16+3\times23+4\times26+5\times31+6\times87+7\times41+8\times32+9\times17+10\times4}{285}$$

$$= \frac{1626}{285}$$

$$= 5{\cdot}71 \text{ (2 d.p.)}$$

total of score × number of people

total number of people

A data set given in a frequency table is sometimes called a **distribution**.

Using the **spreadsheet**.

	A	B	C	D	E	F	G	H	I	J	K	L
1	Score	1	2	3	4	5	6	7	8	9	10	
2	No. of people	8	16	23	26	31	87	41	32	17	4	=SUM(B2:K2)
3	Total	=B1*B2	=C1*C2	=D1*D2	=E1*E2	=F1*F2	=G1*G2	=H1*H2	=I1*I2	=J1*J2	=K1*K2	=SUM(B3:K3)/L2

Exercise 2

Use a spreadsheet or a calculator to find the mean to 2 d.p., of these distributions. Do some first using a calculator then use a spreadsheet to check your answers.

1 Score rolled on a dice

Score	1	2	3	4	5	6
Number of throws	83	92	73	87	72	91

2 Visitors per patient at St George's Hospital

Visitors per patient	0	1	2	3	4	5	6	7	8
Frequency	6	25	14	9	23	14	17	3	2

3 Number of minutes taken by a person to solve a problem

Number of minutes	0	1	2	3	4	5	6	7	8	9
Frequency	29	38	43	37	28	43	38	41	40	39

Review Laura tossed a dice with eight triangular faces.
It had the numbers 1 to 8 on it.
This table gives her results.

Score	1	2	3	4	5	6	7	8
Number of throws	64	72	58	69	57	61	70	65

Use a spreadsheet or calculator to find the mean score.

The next exercise will give you practice at calculating and interpreting the mean, median, mode and range.

Exercise 3

1 This table shows the lengths of 100 paperback books.

a What is the modal class for length?

b If the shortest length was 22·6 cm and the longest length was 34·6 cm, what is the range?

Length (cm)	Frequency
$22 \leq l < 24$	5
$24 \leq l < 26$	27
$26 \leq l < 28$	26
$28 \leq l < 30$	31
$30 \leq l < 32$	8
$32 \leq l < 34$	2
$34 \leq l < 36$	1

2 Rajshree did an experiment in PE to see which was his better kicking foot.
This table shows the number of goals scored out of 5 kicks.

a What is the mean number of goals scored out of 5 with
i the left foot **ii** the right foot?

b What is the modal number of goals scored out of 5 with
i the left foot **ii** the right foot?

c Which foot do you think is Rajshree's better kicking foot? Explain.

Left foot		Right foot	
Goals out of 5	Frequency	Goals out of 5	Frequency
0	3	0	1
1	6	1	7
2	8	2	1
3	2	3	0
4	1	4	8
5	0	5	3

3 This table shows the humidity (percentage of moisture in the air) at three weather stations on Mount Snowdon between 01:00 and 05:45 one day.

	Humidity (%)		
Time (GMT)	Summit 1085 m	Clogwyn 780 m	Llanberis 105 m
05:45	64·0	94·0	87·0
05:30	76·0	91·0	86·0
05:15	65·0	86·0	86·0
05:00	59·0	82·0	85·0
04:45	63·0	84·0	85·0
04:30	57·0	88·0	84·0
04:15	48·0	86·0	83·0
04:00	46·0	81·0	82·0
03:45	47·0	83·0	81·0
03:30	43·0	88·0	80·0

	Humidity (%)		
Time (GMT)	Summit 1085 m	Clogwyn 780 m	Llanberis 105 m
03:15	66·0	90·0	78·0
03:00	19·0	90·0	76·0
02:45	58·0	90·0	77·0
02:30	57·0	91·0	80·0
02:15	59·0	88·0	80·0
02:00	67·0	86·0	80·0
01:45	57·0	88·0	80·0
01:30	55·0	97·0	81·0
01:15	55·0	98·0	85·0
01:00	65·0	99·0	87·0

a Find the mean, median and range of the humidity at each place.

b Which place has the least consistent humidity reading? Explain.

c Does the mean represent humidity well at
i the Summit **ii** Clogwyn **iii** Llanberis?

4 This is a population pyramid for Bangladesh in the year 2000.

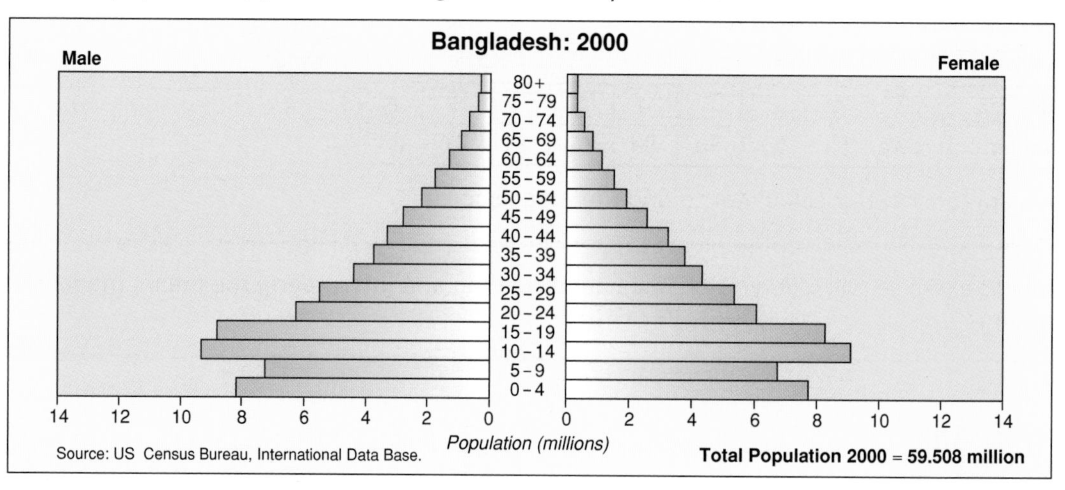

a What is the modal class?
b A newspaper headline said 'Explosion in teenage population in Bangladesh'. Comment on the accuracy of this headline.
c **Sketch** what the population pyramid would look like if the ages were grouped in 15-year intervals.
Which pyramid do you think shows the trends better?
Which pyramid is easier to read? Explain.

5 Adam wrote down these masses of salt used in an experiment to test solubility.

2·6 g 3·2 g 2·4 g 3·6 g 2·7 g 3·5 g

Use the assumed mean method to find the mean mass.

6 In a maths test the mean was exactly 72%. Four students were absent but sat the test later. When these marks were added in, the new class mean was exactly 73%.
a What can you say about the mean of the marks of the four students?
b What was the mean mark of the four students if the total number of students in the class is 32?

7 The median score of three people in a game is 40. The mean is 44 and the range is 20. What are the three scores?

8 Five friends got scores in the range 0 to 10 for a geography project.
One of them said, 'The range of our marks is 5, the mean is 6 and the mode is 7'.
Is this possible? If it is, what scores could they have got?

9 A group of five students got a mean of 15 in an English test.
The test was out of 20 and their range was 10 and their median was 16.
What possible marks did the five get?

10 a Miriam has these cards.
The mean value of the expressions is $5x$.
What expression is on the third card?

$5x - 10$ $5x$?

b Write a set of three expressions that have a mean value of $6x$.
c What is the mean of these three expressions?

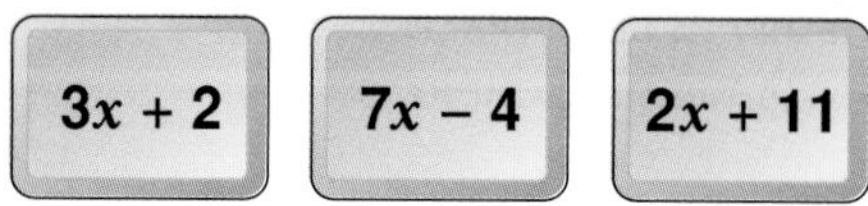

Write your expression as simply as possible.

*11 Ben, Cass and Maddy have a mean age of 11.
The range of their ages is 10.
Maddy is the youngest and Cass is the eldest.
What is the lowest possible age of
a Maddy **b** Cass?

*12 In science, Jasper worked out the mean temperature as 80 Fahrenheit. The teacher told him he wanted the mean in degrees Celsius.
To convert degrees Fahrenheit to degrees Celsius, the formula is:

$$C = \frac{5}{9}(F - 32)$$

Explain what Jasper needs to do to find the mean in degrees Celsius.

Review 1 This chart gives the distances, in metres, jumped from a standing start by the pupils of 8P.
a What is the modal class for distance?
b If the shortest distance jumped was 0·92 m and the longest was 1·73 m, what is the range?

Distance (d in m)	Frequency
$0.8 \leq d < 1.0$	2
$1.0 \leq d < 1.2$	10
$1.2 \leq d < 1.4$	10
$1.4 \leq d < 1.6$	5
$1.6 \leq d < 1.8$	4

Review 2 Owen has three number cards.
The mode of the three numbers is 6.
The mean of the three numbers is 7.
What are the three numbers?

Review 3 Jessie played three games in a competition.
Her mean score was 5 and her range was 8.
What did she score if her median was **a** 3 **b** 5?

* **Review 4** The mean age of the members of a computer club is 18 years 7 months.
The range is 9 years 5 months.
Robert joins the club. His age is 19 years and 7 months.

a Will the mean age of the members
- **A** increase by more than 1 year
- **B** increase by exactly 1 year
- **C** increase by less than 1 year
- **D** stay the same
- **E** change but we can't tell how?

b Will the range of the ages of the members
- **A** increase by more than 1 year
- **B** increase by exactly 1 year
- **C** increase by less than 1 year
- **D** stay the same
- **E** maybe change or maybe not, we can't tell?

Puzzle

1 Pablo has grown seven plants as part of an experiment.
The maximum height is 20 cm and the range is 10 cm.
What might the heights of the seven plants be if the
a mean is 15 cm b median is 14 cm c mode is 12 cm
d mode is 15 cm and the median is 14 cm?

2 In PE eight students were given marks out of 10 for their folders.
Seven of the marks are 7, 4, 8, 8, 7, 3, 5.
What is the eighth mark if
a the mean is 6·25 b the mode is 8
c the median and the mode are the same
*d the mean is the same as the median?

Finding the median, range and mode from a stem-and-leaf diagram

This is a **stem-and-leaf diagram**.
It shows the temperatures on one day at 21 cities in Europe.
The digit on the left of the line is the 'stem' and the digit on the right is the 'leaf'.
The temperatures on the bottom row of the diagram are 30 °C, 31 °C and 33 °C. In this example the stem is the tens digit and the leaf is the units digit.

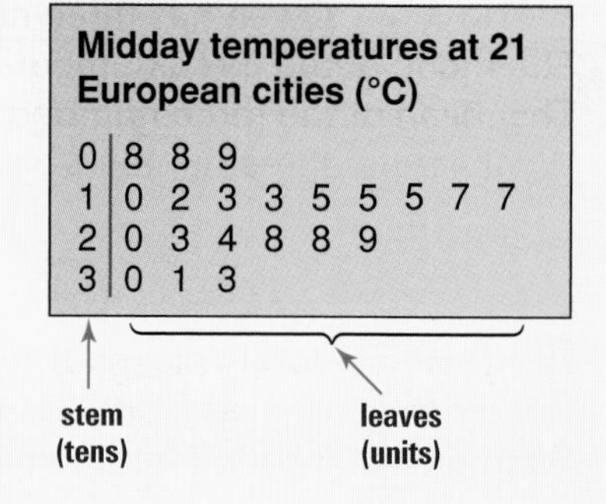

Midday temperatures at 21 European cities (°C)

Stem	Leaves
0	8 8 9
1	0 2 3 3 5 5 5 7 7
2	0 3 4 8 8 9
3	0 1 3

There are 21 data values.
The median is the 11th value, which is 17 °C.
The range is 33 – 8 = 25 °C

highest temperature (33) lowest temperature (8)

The mode is 15 °C. It occurs most often on the diagram.

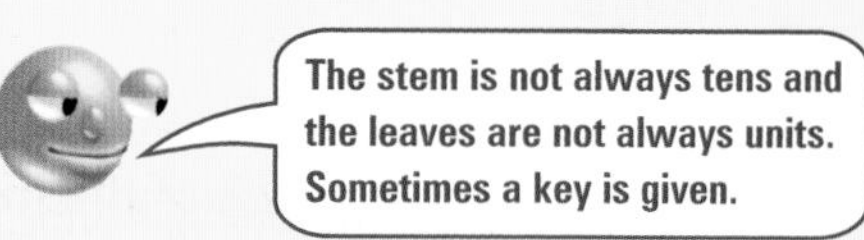

Exercise 4

1 Find the median, range and mode from these stem-and-leaf diagrams.

a **Marks out of 60**

stem = tens
leaves = units

Stem	Leaves
0	9
1	2 4 7 9
2	0 3 3 4 5 7 9
3	0 0 4 5 6 6 7 7 7 9
4	0 0 5 6 7 8 8 9
5	3 5 5

b

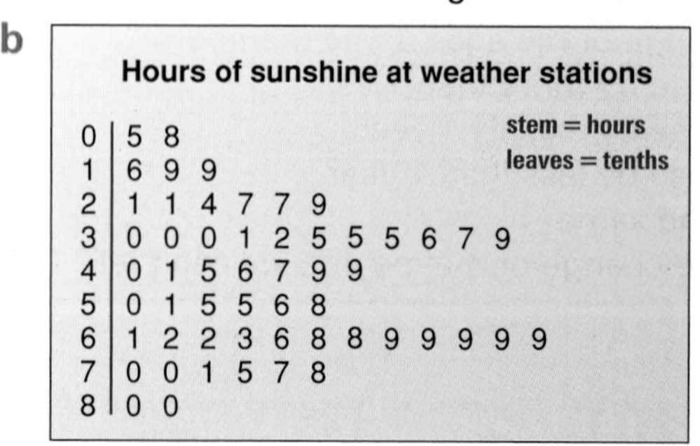

Hours of sunshine at weather stations

stem = hours
leaves = tenths

Stem	Leaves
0	5 8
1	6 9 9
2	1 1 4 7 7 9
3	0 0 0 1 2 5 5 5 6 7 9
4	0 1 5 6 7 9 9
5	0 1 5 5 6 8
6	1 2 2 3 6 8 8 9 9 9 9 9
7	0 0 1 5 7 8
8	0 0

2 Libby wanted to compare the ages of the mothers and fathers of her classmates. Libby asked each person in her class how old their mother was.
This list shows her results.
28, 31, 43, 36, 38, 37, 36, 29, 40, 36, 39, 41,
44, 30, 50, 36, 42, 43, 29, 33, 35, 36, 48, 42,
35, 32, 37, 46, 34, 30, 40, 41, 35, 39, 33

a Construct a stem-and-leaf diagram to show Libby's results.
b How many mothers were aged 30–39?
c How many were aged 40–49?
d Find the median, range and mode of the ages.

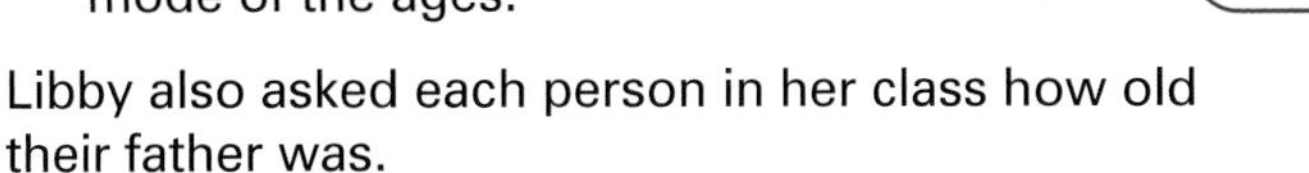

Libby also asked each person in her class how old their father was.
This stem-and-leaf diagram shows her results.
e Find the median, range and mode of the ages.
f Compare the ages of the mothers and fathers.

Age of fathers

Stem	Leaves
5	1 1 2 3 4 7
4	0 0 1 1 2 2 3 3 4 5 5 5 6 8
3	0 0 3 3 4 4 5 5 6 6 6 7 8
2	9 9

Review This stem-and-leaf diagram shows the weekly rainfall last year at a city in Europe.
Find the median, mode and range of the weekly rainfall.

Rainfall in cm stem = centimetres leaves = tenths

Stem	Leaves
3	0 1 2 3 4
2	0 0 0 0 0 1 2 3 3 5 7 7 7 7 9
1	0 0 1 1 1 1 2 2 2 3 3 3 4 4 4 4 4 5 5 6 6 7 8 8 9 9
0	6 6 8 9 9 9

Comparing data

Remember
To **compare data** we sometimes use the range and one or more of the mode, median or mean.

Example This data gives the number of hours two different brands of electric kettle lasted.

Brand A	608	635	612	585	683	697	641	610	604	664
	607	638	601	636	615	642	701	618	625	620
Brand B	437	861	735	824	632	321	532	682	614	732
	486	324	507	461	913	912	586	671	469	893

The mean of brand A = $\frac{\text{sum of data values}}{\text{number of values}} = 632{\cdot}1$
The median of brand A = 622·5
The range of brand A = 701 − 585 = 116

The mean of brand B = 629·6
The median of brand B = 623
The range of brand B = 913 − 321 = 592.

The mean of brand A is slightly higher than the mean of brand B. The median of brand A is slightly lower than that of brand B. Using only these two statistics, we could conclude that the brands are about the same. However, if we look at the range, brand B has a much higher range than brand A. This tells us that brand B has much less consistent manufacturing standards.

A buyer might choose brand A because it has a slightly higher mean and a smaller range, indicating it has more consistent standards.
A buyer might choose brand B because it has a slightly higher median and a large range. This means there is a chance the kettle might last a very long time.

Exercise 5

1 This table gives the results of the last twelve public speaking contests for Lufta and Sophie.
Only one girl can represent the school at a local public speaking competition.
Which girl would you choose? You can choose either as long as you use the results in the table to explain why.

	Mean	Median	Range
Lufta	17	16	12
Sophie	16	16	3

2 A coach must choose one of two boys to compete in the 100 m sprint. She looks at the times, in seconds, of their last five races.

Ben	12·1 s	12·0 s	12·0 s	16·8 s	12·1 s
Josh	12·3 s	12·4 s	12·4 s	12·5 s	12·4 s

a Find the mean, median, mode and range for each boy.
b Which boy do you think the coach should choose?
Explain why using your results from **a**.

3 A food scientist measured the amount of vitamin C, in milligrams, in oranges from two different countries.

Country A	92	109	114	85	85	109	98	101
	114	93	82	89	98	88	89	112
Country B	105	92	136	106	85	76	94	91
	115	82	94	120	124	90	97	117

a Find the mean, median and range for the amount of vitamin C in the oranges for each country.
b Which oranges would you buy? Explain your answer using the mean **or** median and the range. It doesn't matter which one you choose, as long as you explain why you chose it.
Explain why you chose the mean or median to compare the results.

4 Adam was doing a survey on TV viewing patterns for adults and teenagers.
He showed his results on these stem-and-leaf diagrams.

Teenage hours watched per week
stem = tens leaves = units

Stem	Leaves
4	0
3	0 1 3 5 7 7 8 9 9
2	0 3 4 4 5 6 7 8 9 9 9
1	0 2 3 4 5 5 8 9
0	0

Adult hours watched per week
stem = tens leaves = units

Stem	Leaves
4	
3	
2	0 3 5 9
1	0 1 1 1 2 2 3 3 3 4 4 4 4 5 5 6 6 7 7 8 9 9
0	8 9 9 9

a Find the mean, median, mode and range for teenagers and adults.
b Compare and contrast teenage and adult viewing habits using your results from **a**.

Review A test is done on two types of antiperspirant.
Twenty people using each brand were asked how long to the nearest hour each was effective.

Brand A	12	14	12	15	18	12	10	12	11	16
	14	14	11	12	15	17	10	14	11	15
Brand B	6	20	4	8	17	19	24	15	8	12
	18	20	13	11	9	3	20	17	12	13

a Find the mean, median and range for the number of hours it was effective for each brand.
b Which brand would you buy? Explain your answer, using the range and either the mean or median.

Practical

- Compare and contrast weather patterns in two different places.
- Compare and contrast reading habits of different age groups.
- Use the results of an experiment to find out which type of battery lasts longer.

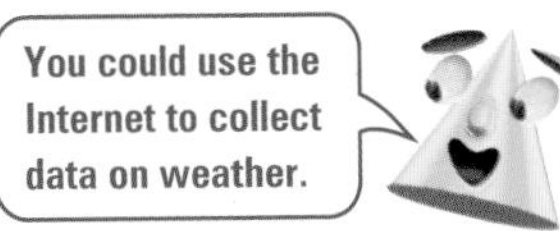

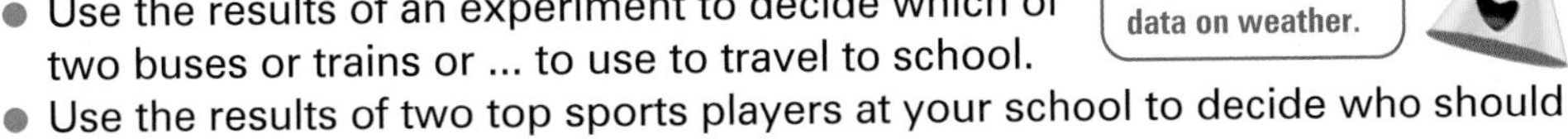

- Use the results of an experiment to decide which of two buses or trains or ... to use to travel to school.
- Use the results of two top sports players at your school to decide who should represent the school if only *one* could be chosen.

Constructing graphs and diagrams

Discussion

Look at the graphs and questions below to help your discussion.

- Data that has been collected is usually recorded on a table first. Then it is often displayed on a chart, graph or diagram.

  What does the data on a graph show us that a table does not? **Discuss.**
- Can you always write down the original data that was collected if it is displayed on a chart, graph or diagram? **Discuss.**
- What data might have been collected and put on a table before this pie chart could be drawn?

Sources of calcium in the British diet

other foods 13%
vegetables 7%
cheese 13%
cereals and cereal products 16%
meat and fish 7%
liquid milk 35%
white bread 9%

- What data was collected for this graph?

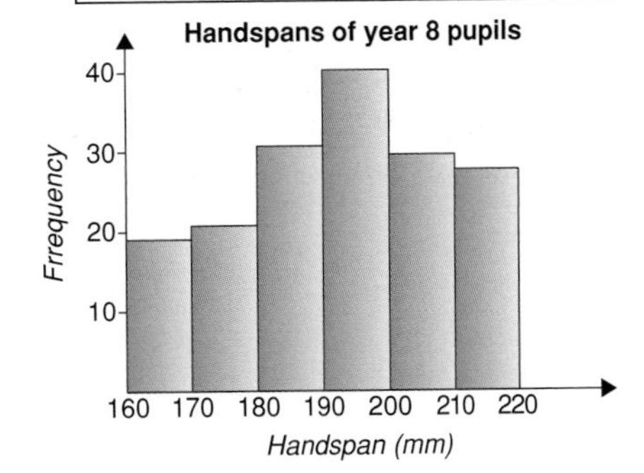

A **graph or chart** highlights features or trends that a table does not show well.
To display **categorical data** you can use a

- pictogram
- pie chart
- bar graph

Categorical data is non-numerical data.

To display ungrouped **discrete data** you can use a

- pictogram
- pie chart
- bar graph
- bar-line graph

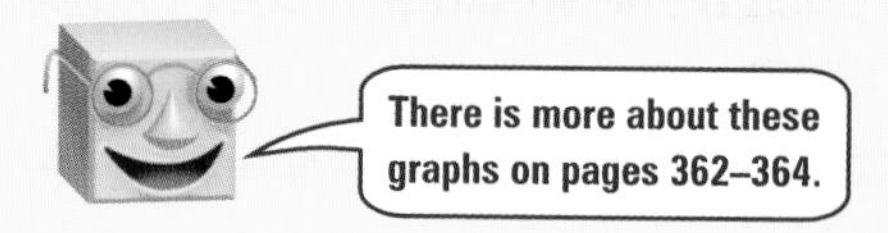

To display **grouped discrete data** use a bar chart.
To display **continuous data** use a frequency diagram.
To show **data over time** use a line graph.
More than one set of data can be displayed on bar graphs, line graphs and frequency diagrams.

Sometimes a strip bar chart like the one below is used to show information.

Example **Sources of vitamin A in British diet**

Meat	Dairy products, eggs and fats and spreads	Vegetables	Cereals	Other

About half of the strip is labelled meat. This means that about half of the vitamin A in the British diet comes from meat.
About a quarter of the strip is dairy products, eggs and fats and spreads. About a quarter of our vitamin A comes from these sources.

Line graphs are often used to compare several sets of data.

Example A study from 1986 to 1996 was carried out to test the levels, in mg per cigarette, of tar, nicotine and carbon monoxide. The table shows the results.
The results can be shown on a line graph. The graph shows the trend more clearly than the table.

Year	Tar	Nicotine	Carbon monoxide
1986	13·9	1·30	14·7
1987	13·7	1·21	14·3
1988	13·3	1·17	14·3
1989	12·9	1·17	14·2
1990	12·4	1·15	14·0
1991	11·9	1·02	13·4
1992	11·3	0·91	12·7
1993	10·9	0·86	12·3
1994	10·6	0·83	12·2
1995	10·4	0·85	12·1
1996	10·7	0·87	12·1

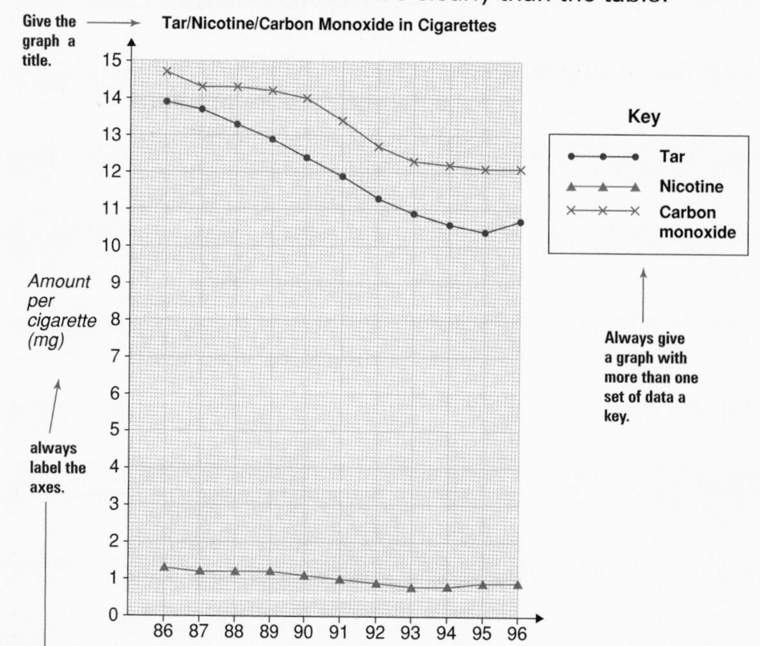

Exercise 6

1 This table gives the amount of water used yesterday by Julia's family.

Activity	Toilet flush	Personal washing	Washing clothes	Dishwashing	Garden	Cooking	Drinking
Amount of water used (ℓ)	65	55	20	15	15	10	5

a Julia started this strip bar chart.
Use a copy of it.
Finish filling it in.

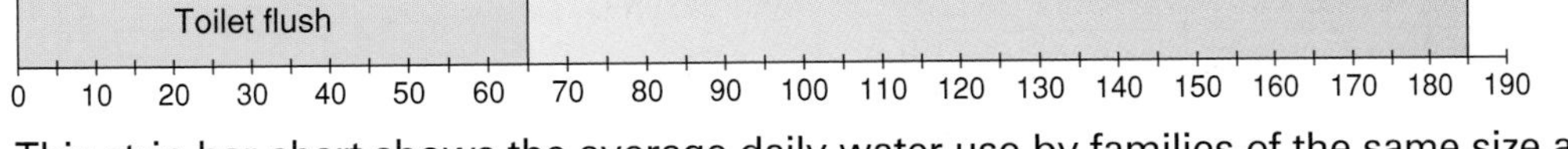

b This strip bar chart shows the average daily water use by families of the same size as Julia's.

Toilet flush | Personal washing | Washing clothes | Dishwashing | Garden | Cooking | Drinking

0 10 20 30 40 50 60 70 80 90 100 110 120 130 140 150 160 170 180

Julia's family wants to reduce the amount of water they use each day.
Use the graphs to choose an activity that could be targeted to reduce the amount of water used. Explain why you chose this.

2 This table gives the wind speed at hourly intervals at the top and bottom of a hill.

Time	6 a.m.	7 a.m.	8 a.m.	9 a.m.	10 a.m.	11 a.m.	12 p.m.	1 p.m.	2 p.m.	3 p.m.	4 p.m.	5 p.m.	6 p.m.
Windspeed (knots) Top of hill	4	4	4	8	12	16	22	24	26	27	28	23	8
Windspeed (knots) Bottom of hill	2	2	3	5	7	9	15	16	17	18	18	12	4

a Draw a line graph for this data. Have time on the horizontal axis.
Put both sets of data on the same axis. Use -■- for top of hill and -●- for bottom of hill.

b Could we use the graph to estimate wind speeds at times in between the times given?

c Compare the wind speeds at the top and bottom of the hill.

3 Riffet drew this graph to show the maximum and minimum temperatures (°C) in Madrid.

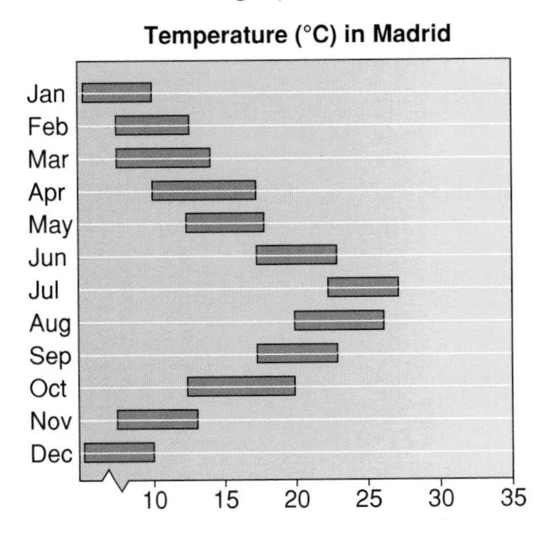

a Use the tables below to draw a similar graph to show the maximum and minimum temperatures for Sydney.

Maximum and minimum temperatures (°C) in Sydney

	Jan	Feb	Mar	Apr	May	Jun	Jul	Aug	Sep	Oct	Nov	Dec
Average minimum	18·5	18·6	17·4	14·6	11·4	9·2	7·9	8·8	10·9	13·4	15·5	17·3
Average maximum	25·7	25·6	24·6	22·2	19·2	16·7	16·0	17·5	19·7	21·9	23·5	25·1

b This table gives the mean temperatures for Sydney.
Plot this data on a line graph.

	Jan	Feb	Mar	Apr	May	Jun	Jul	Aug	Sep	Oct	Nov	Dec
Mean temperature (°C)	22·1	22·1	21·0	18·4	15·3	12·9	12·0	13·2	15·3	17·7	19·5	21·2

4 This table gives the average weekly cigarette consumption by some female smokers aged 16 and over.

	16–19	20–24	25–34	35–49	50–59	60 and over
1976	92	109	106	111	104	74
1986	75	82	98	109	97	83
1996	71	79	89	108	108	86

a Plot this data on a line graph.
b Write down two trends the graph shows.

T

5 This data shows the aircraft movements (in thousands) at five British airports in 1991, 1996 and 2001.

Aircraft movements (000s)

Airport	1991	1996	2001
Heathrow	362	428	464
Gatwick	163	212	252
Stansted	36	77	170
Manchester	124	144	181
Birmingham	66	78	112

a Use a copy of this grid.
Draw a line graph to show this data.
b Which airport had the greatest increase in aircraft movements from 1991 to 2001?
c Give two reasons why aircraft movements might have increased over this time.

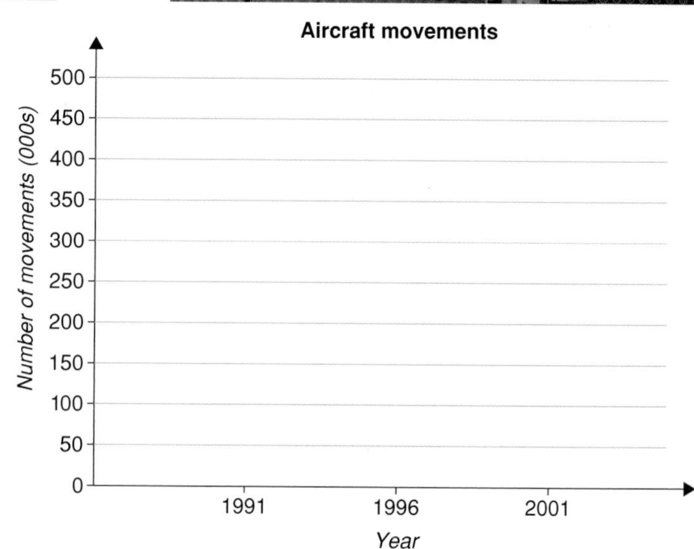

6 This data gives the number of notifications of infectious diseases in 1990 and 2000.

Year	TB	Whooping Cough	Meningitis	Malaria	Total
1990	5010	14 125	2369	1469	22 973
2000	6379	658	2251	1110	10 398

a Draw two circles of radius 3 cm.
Find the angle at the centre of each sector for each year.
Draw two pie charts to show the data in the table.

b Draw a bar chart to show the data. You will need to round the data to the nearest 100 before plotting.

c Which graph do you think gives better information about the data?

***7** Show this data on a suitable graph.
Say why you chose the type of graph you used.

Acccidental deaths: by age and gender, 2000

United Kingdom Numbers

	Males	Females
0–	282	157
15–	943	241
25–	1125	207
35–	985	264
45–	732	289
55–	658	319
65–	755	524
75 and over	2001	3551
All ages	7481	5552

Source: Office for National Statistics; General Register Office for Scotland; Northern Ireland Statistics and Research Agency

Review This table gives the ages of people at a 'Stop Smoking' seminar.

Age	Number
20–29	4
30–39	6
40–49	9
50–59	10
60+	23

Display this data on a bar chart, a pie chart and a line graph.
Choose which one you would use to put into a report on the seminar. Explain why you chose it.

Practical

You will need a temperature probe and a graphical calculator.

Use a temperature probe to get the data and draw the cooling curve for these.

a water heated to 80 °C and left to cool in a single test tube

b water heated to 80 °C and left to cool surrounded by other test tubes of water heated to 80 °C

Scatter graphs

This table gives the marks scored by pupils in a maths test and in a science test.

Maths	15	36	32	22	36	26	36	35	24	25	30	40	12	36	19
Science	21	38	31	25	32	30	29	35	27	28	32	36	10	33	26

The marks given in the table have been plotted on this **scatter graph**.

A scatter graph displays two sets of data.

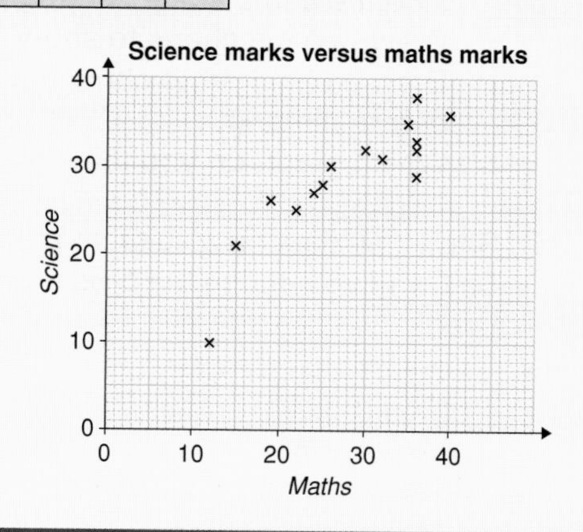

Discussion

- Does the scatter graph above show that someone who did well in maths is likely to have done well in science? **Discuss.**
  How might this information be useful? **Discuss.**

Length of shot-put throw (metres)	18·4	14·7	13·2	15·1	17·2
Time for 400 m (seconds)	62	59	63	57	63

  This table shows the results of five students who entered both the shot-put event and the 400 m event at an athletics meeting.
  These results are to be displayed on a scatter graph.
  The six members of a group each drew up a different set of axes for the scatter graph.
  These are shown below.

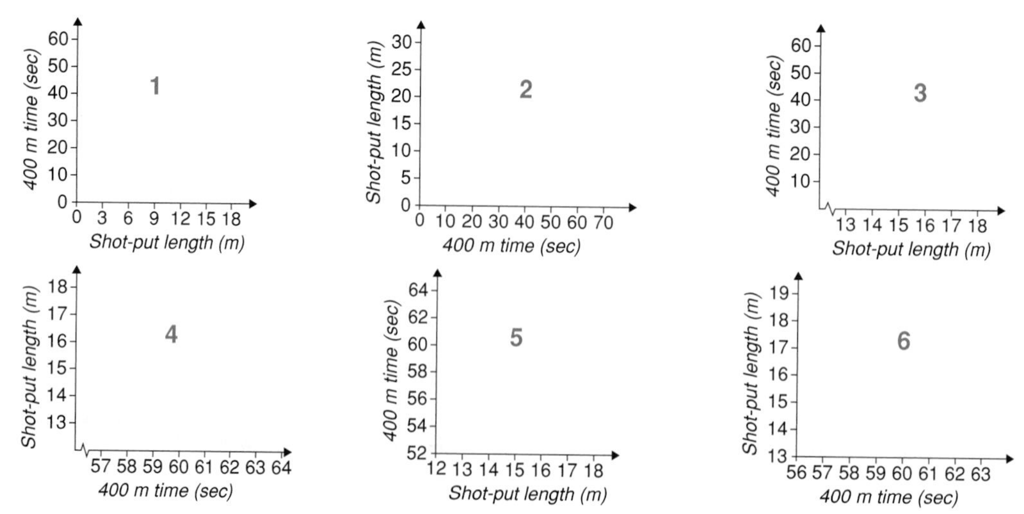

What do you think the jagged line means on the axes in **3** and **4**? **Discuss.**

Are any of these sets of axes suitable for graphing the data?
Is one of these sets of axes more suitable than the others? **Discuss.**

Shoe size	3	4	5	4	3	4	2
Dress size	8	10	9	10	12	11	8

  The data given in this table is to be displayed on a scatter graph.
  Could this data be displayed equally well on either of the two following sets of axes?

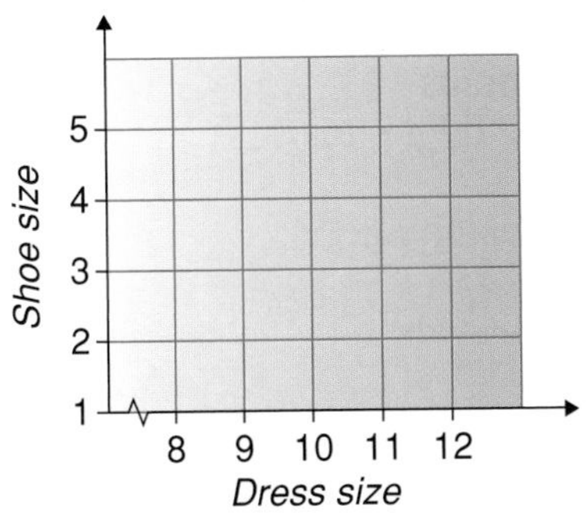

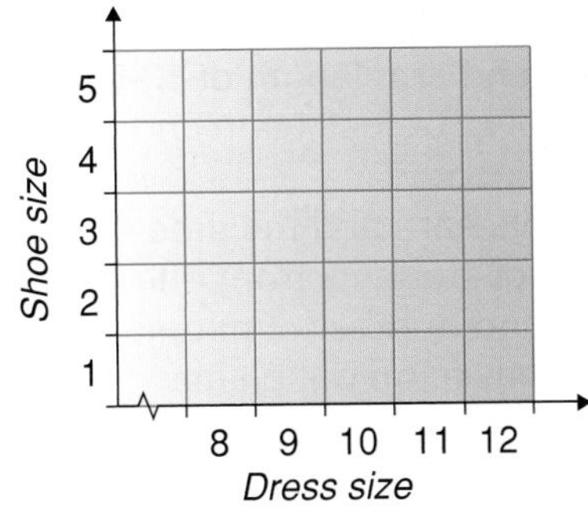

  How would you graph the second and fourth pairs of data (both 4, 10) on each of these sets of axes?
  Is this problem likely to arise when you are graphing continuous data such as in the first table of this discussion? **Discuss.**

- What do you think this scatter graph might represent? **Discuss.**
  Make up a suitable title for this graph. **Discuss.**

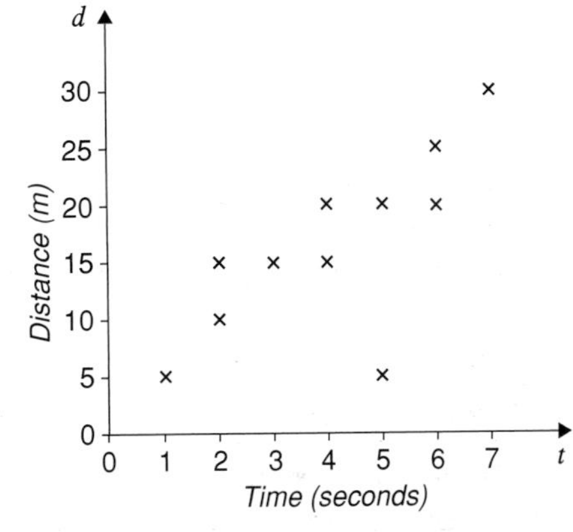

Correlation

A scatter graph sometimes shows a relationship or **correlation** between the variables.

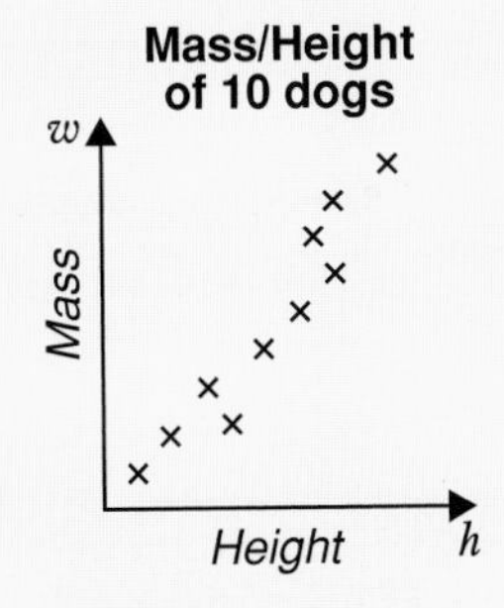

This graph shows a positive correlation between the height and mass of dogs. It shows that the taller a dog is, the heavier it is likely to be.

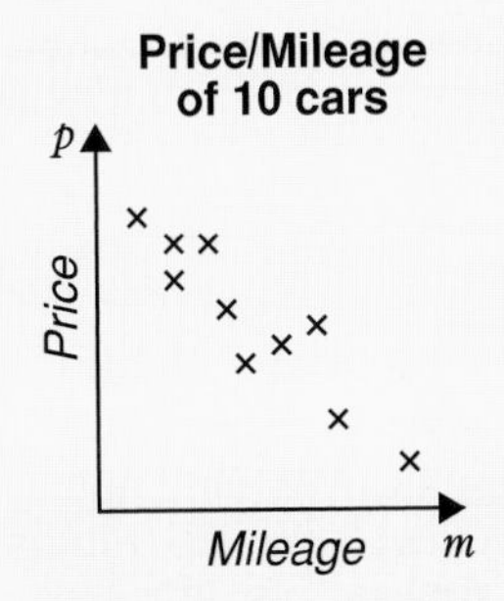

This graph shows a negative correlation between price of a car and mileage. The higher the mileage, the lower the price is likely to be.

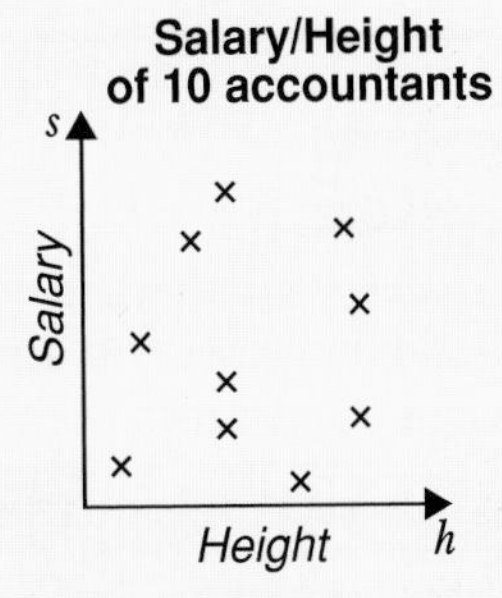

This graph shows there is no correlation between the height of an accountant and the salary earned. There doesn't seem to be any relationship between these variables.

Discussion

- Do you think there would be a positive correlation or negative correlation or no correlation between the following? **Discuss.**

  Number of pages and number of advertisements in magazines
  Number of days absent from school and marks in tests
  Circumference of head and arm length of 13-year-old students
  Number of supporters of a football team and goals scored by the team
  Years of education and income
  Amount of money people earn and the time they spend watching TV
  Maths mark and height of students from one year group
  Income of households and number of people in the households
  Height and number of leaves on pot plants
  Age and amount of pocket money received
  Length and weight of cats
  Number of rooms and number of windows in houses
  Time spent on homework and time spent watching TV
  Number of boys and number of girls in families

- Think of some pairs of variables that would be likely to have positive correlation, some that would be likely to have negative correlation and some that would be likely to have no correlation. **Discuss.**

If there is a correlation between two variables, we can use a scatter graph to estimate values.

Example
This scatter graph shows the number of worms and birds found in a garden. The greater the number of birds the fewer the number of worms.
There is a negative correlation.
We can estimate how many birds there would be if there were 25 worms. Draw a vertical line from 25 worms. Estimate from this, that between about 7 and 11 birds would be in the garden.

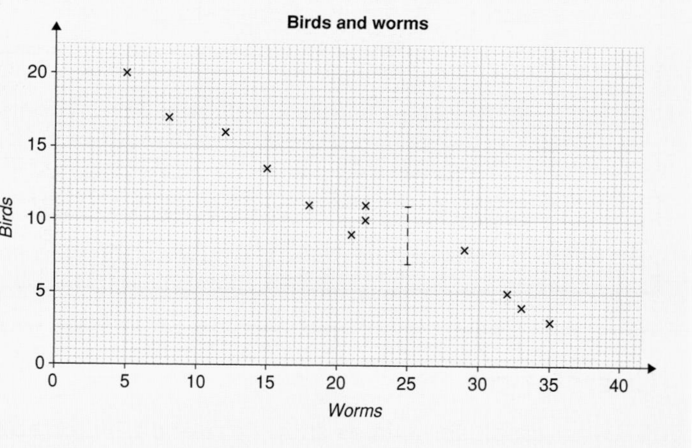
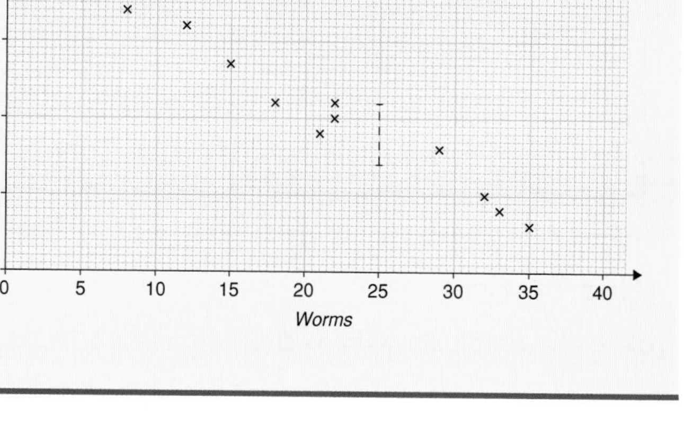

Exercise 7

1 The table shows the ages and selling prices of eight similar computers.

Age (months)	3	6	7	8	4	9	11	10
Selling price (pounds)	1120	900	820	780	980	650	560	620

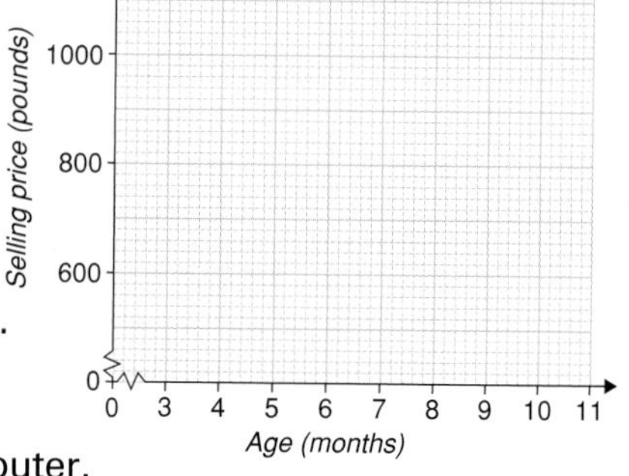

a Use a copy of this grid.
Use the information in the table to draw a scatter graph.
b Describe the relationship between the selling price of computers and their age.
c Estimate the selling price of a 5-month-old similar computer.
d A similar computer sells for £860. Estimate its age.

2 The table shows the heights of twelve people and the time it took them to eat their breakfast.

Time (minutes)	15	12	21	6	18	14	12	4	5	9	17	3
Height (cm)	163	172	184	176	168	172	184	163	181	168	185	172

a Use a copy of the grid. Draw a scatter graph of these heights and times.

b Does the scatter graph show that there is positive, negative or no correlation between height and time to eat breakfast?

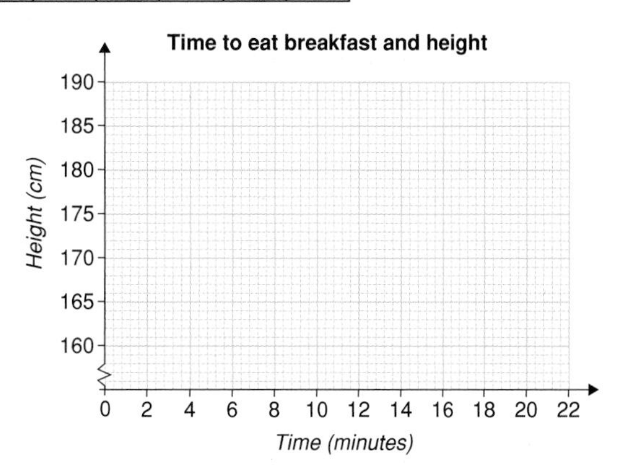

3 The table shows the time taken, to the nearest minute, by eight runners to run different distances.

Length of run (m)	2000	8000	6000	5000	3000	7000	4000	8000
Time taken (minutes)	5	22	16	14	8	20	12	21

a Use a copy of this grid. Draw a scatter graph for this data.

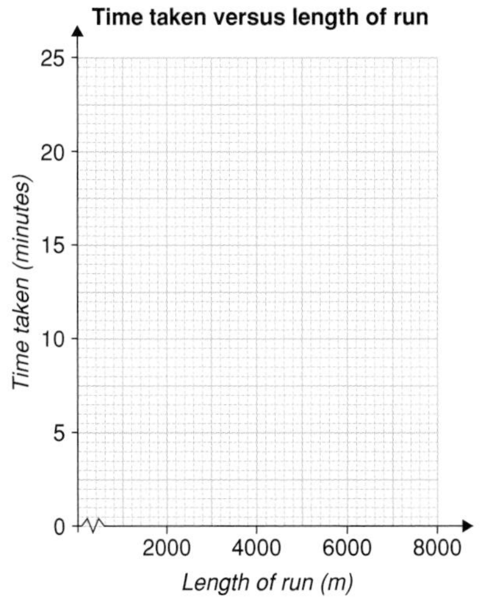

b What sort of correlation is there between time taken and length of run?

c Estimate how long it would take for a runner to run 4500 m.

4 The table below gives the time twenty people spent playing sport one weekend and the time they spent watching sport the same weekend.

Time playing (hours)	1	3	5	10	2	4	2	1	5	6	4	3	8	7	0	2	0	9	5	7
Time watching (hours)	8	6	4	2	5	5	7	9	3	3	5	7	2	3	10	8	9	1	5	2

a Draw a grid with values from 1 to 10 on the x and y axes.
Use the information in the table to draw a scatter graph.

b Does someone who plays more sport generally watch less sport?

c Estimate how much sport someone who watches $4\frac{1}{2}$ hours of sport would play.
Give a range.

5 This table shows the handspan and length of the thumb of 8 students.

Student	Ann	Roger	Meena	Jamie	Kojo	Sue	Jason	Aba
Handspan (mm)	184	196	176	203	192	196	211	205
Thumb length (mm)	55	57	51	58	62	54	62	65

a Draw a scatter graph to show the handspan and the length of the thumb of these students. Have a sensible range of values on each axis.
b Does the scatter graph show that students with longer thumbs tend to have greater handspans? Is there positive, negative or no correlation between thumb length and handspan?
c Jossie's thumb is 59 mm long. Estimate her handspan. Give a range.
d Stuart's handspan is 190 mm. Estimate the length of his thumb. Give a range.

6 Ali thinks that taller people have bigger feet. She collected this data. It shows the foot length and height of several students.

Foot length (mm)	203	207	223	196	214	239	222	230	225	231	200	238	198
Height (cm)	160	152	164	149	155	171	156	154	160	162	149	170	155

a Draw a scatter graph for the data.
b Does the scatter graph support what Ali thinks? Explain your answer.

7 Lauren thinks that as water gets hotter, more sugar will dissolve in it.
If this is true, sketch what a graph of temperature of water versus mass of sugar dissolved would look like.

8 Hannah did a survey to see if practice resulted in better shot-put distances thrown. She plotted her results on this scatter graph.

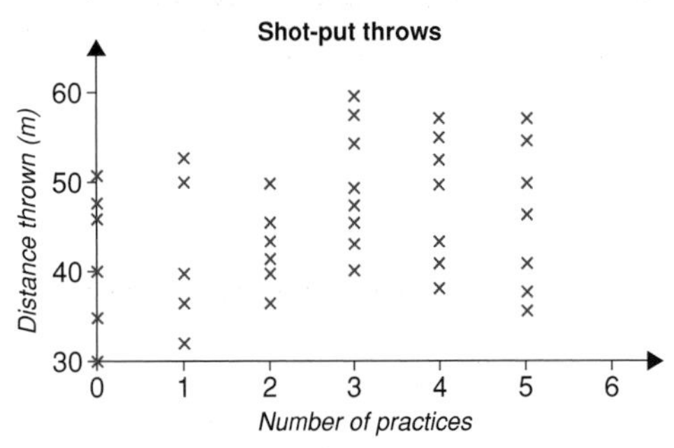

Is there enough evidence to show that as the number of practices increase from 1 to 5 the distance the shot is thrown increases? Explain.

T

Review The table lists the ages of some children and their heights in centimetres.

Age (years)	2	6	8	10	7	3	5	4	8	7	9	6
Height (cm)	80	110	122	136	114	95	104	100	126	120	130	112

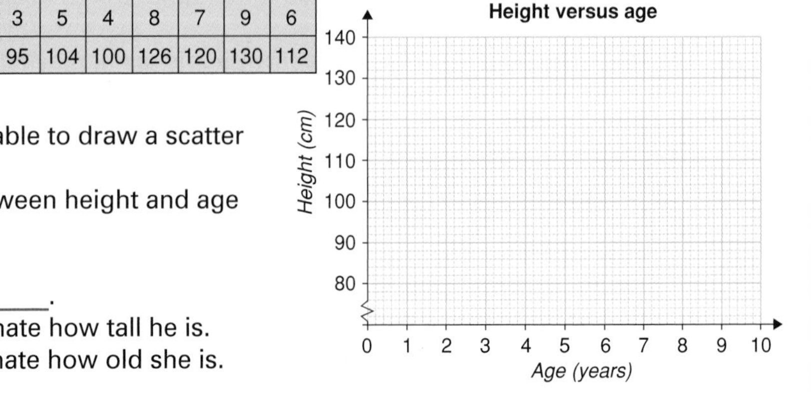

a Use a copy of the grid.
Use the information in the table to draw a scatter graph.
b What sort of correlation between height and age does the graph show?
c Copy and finish this.
Older children tend to be ______.
d Callum is $5\frac{1}{2}$ years old. Estimate how tall he is.
e Charlotte is 85 cm tall. Estimate how old she is.

Practical

You will need a spreadsheet software package.

Ask your teacher for a copy of the Scatter Graphs ICT worksheet.

Interpreting graphs and tables

Discussion

Discuss the question given under each graph or table.

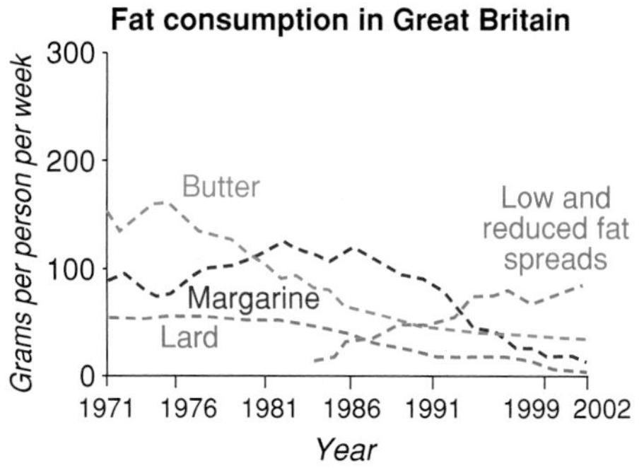

Source: National Food Survey, Ministry of Agriculture, Fisheries and Food

How has fat consumption in Britain changed over the last 30 years?

Reason why school pupils tried to give up smoking

Current smokers — *England*

Why tried to give up	Percentage Boys	Girls	Total
Worried about my health	47	55	52
Cost	28	35	32
To make me feel fitter	32	19	24
My family/friends persuaded me	12	10	11
Smoking made me smell or look nasty	7	10	9
Did not like/enjoy it	4	9	7
Other	12	13	13
Bases (=100%)	137	210	347

Percentages total more than 100 because some pupils gave more than one answer

Source: *Young Teenagers and Smoking* by Vanessa Higgins

Compare and contrast the reasons given by boys to those given by girls for trying to give up smoking.

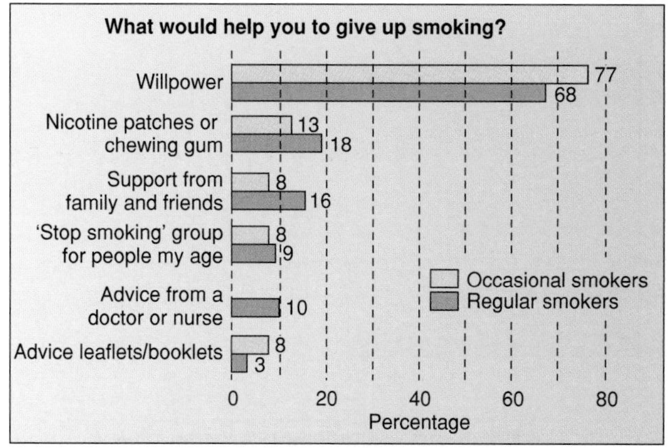

Source: *Young Teenagers and Smoking* by Vanessa Higgins

You are asked to develop a programme to help pupils stop smoking. Funds are limited. Use the graph to help choose two things your programme would focus on.
Explain why you chose these, using the graph to support your explanation.

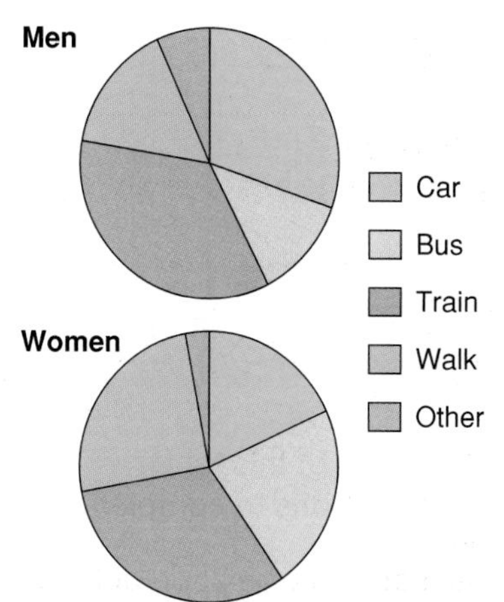

What are the differences between how men and women usually travel to work?

Exercise 8

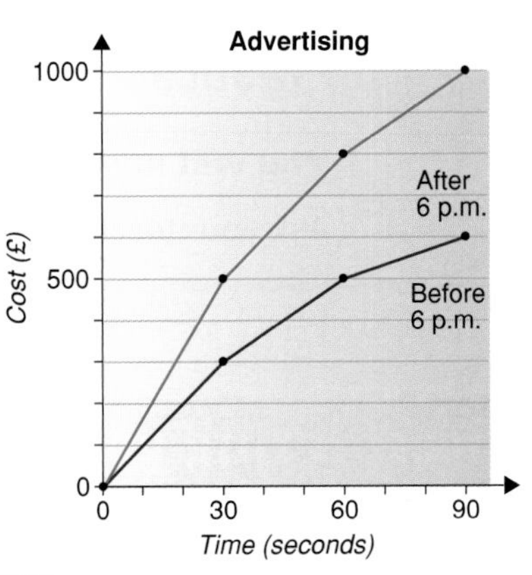

1 This chart gives the cost of buying advertising at the cinema at different times.

a An advertisement lasts 75 seconds. Use the graph to estimate how much more expensive it is to show after 6 p.m. compared with before 6 p.m.

b An advertisement is shown before 6 p.m. and again after 6 p.m. The total cost is £1050. Use the graph to estimate how long the advertisement lasted.

2 This table shows the percentage of smokers and non-smokers aged 11–17 who agreed with the statements given. Comment on the relationship between people who smoke and what they believe.

Statements	Non-smokers	Smokers
Smoking can help calm you down	24%	75%
Smokers tend to be more rebellious than people who don't smoke	32%	26%
Smokers are more boring than people who don't smoke	26%	7%
Smoking can put you in a better mood	12%	50%
Smoking can help you stay slim	12%	24%
Smoking helps give you confidence	7%	24%
Smoking can help you make friends more easily	8%	14%
Smoking makes you look more grown up	7%	13%

Source: *Young Teenagers and Smoking* by Vanessa Higgins

T

3 These two graphs convert pounds (£) to euro (€) and pounds (£) to Canadian dollars (C$).

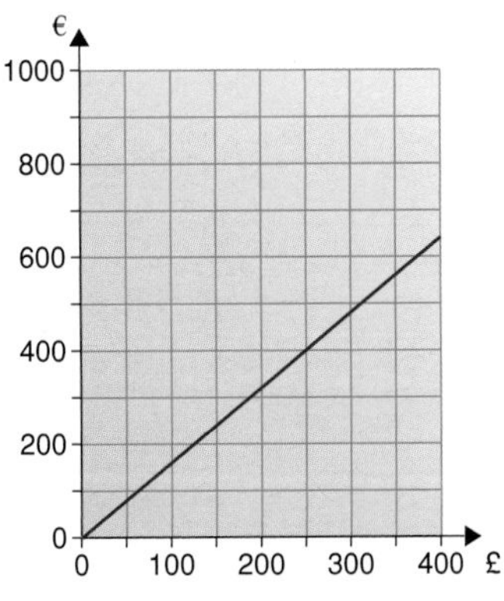

Use a copy of this table.
Use the graphs to complete it.

Number of £	Approximate number of €	Approximate number of C$
0		
200		
400		

Use the information in the table to draw a conversion graph for euro to Canadian dollars.

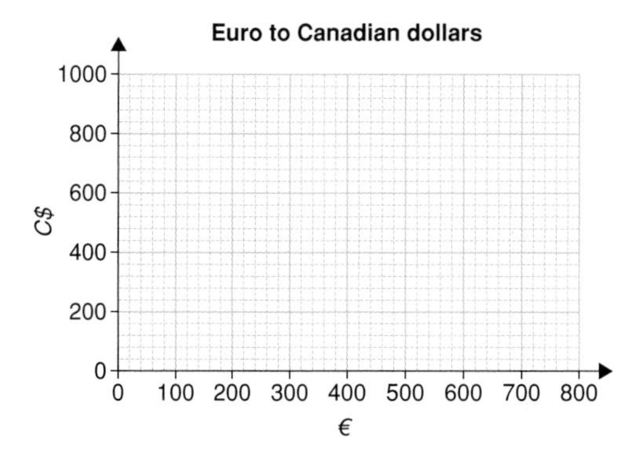

4 The swimming association wrote an article about public swimming pools.
This graph was in the article.
Use the graph to decide whether these statements from the article are true or false or you cannot tell.

a The number of pools open for more than 100 hours each week during the summer fell by more than half from 1992 to 2002.

b By the year 2006, only about 100 pools will be open for more than 100 hours each week during the summer.

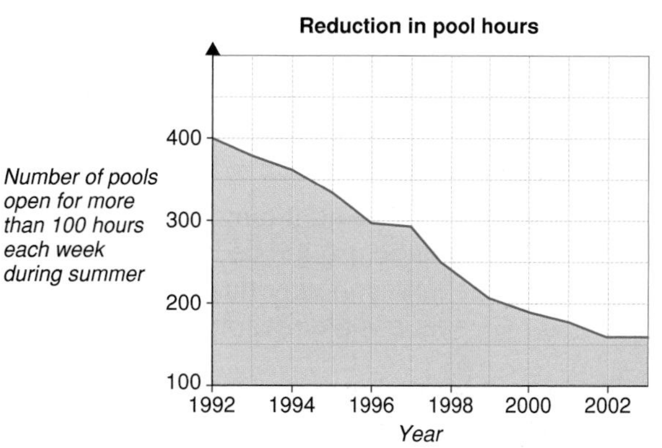

5

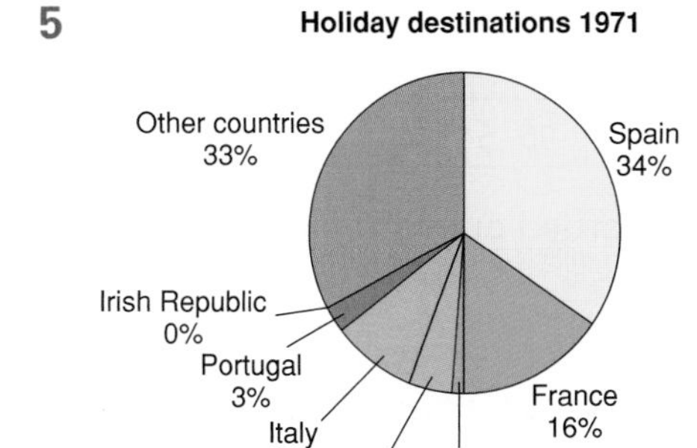

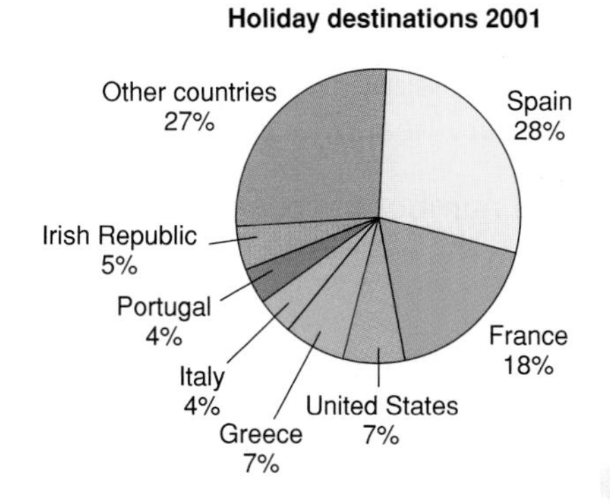

These pie charts show the holiday destinations chosen by British people in 1971 and 2001.

a Give a possible reason why a greater proportion of people holidayed in the United States in 2001 than in 1971.

b The Spanish claim that more British people visited Spain in 2001 than in 1971. Is this possible if the graphs above are correct?

6 This frequency diagram shows the high jump results for Chisnalwood school. The jumps were measured to the nearest cm and put into class intervals $70 \leqslant h < 75$, $75 \leqslant h < 80$, ...

a Jenni jumped 122 cm.
She told her mother that she came fifth in the school.
Could she be right? Explain.

*__b__ Ruben jumped 107 cm.
He said 'I was above the median'.
Explain how you can tell from the graph that he is wrong.

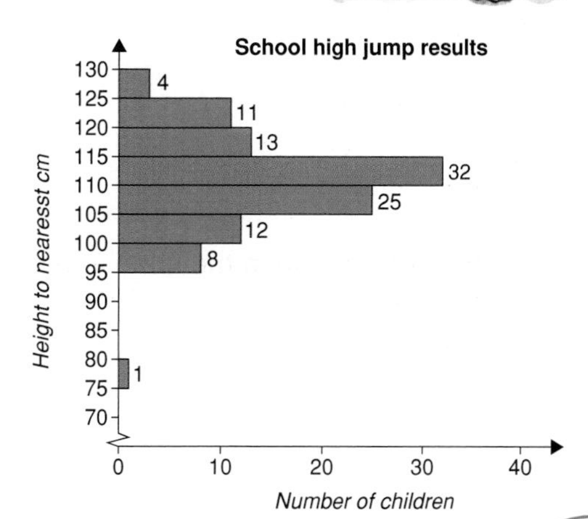

7 This graph shows cinema attendance by age.

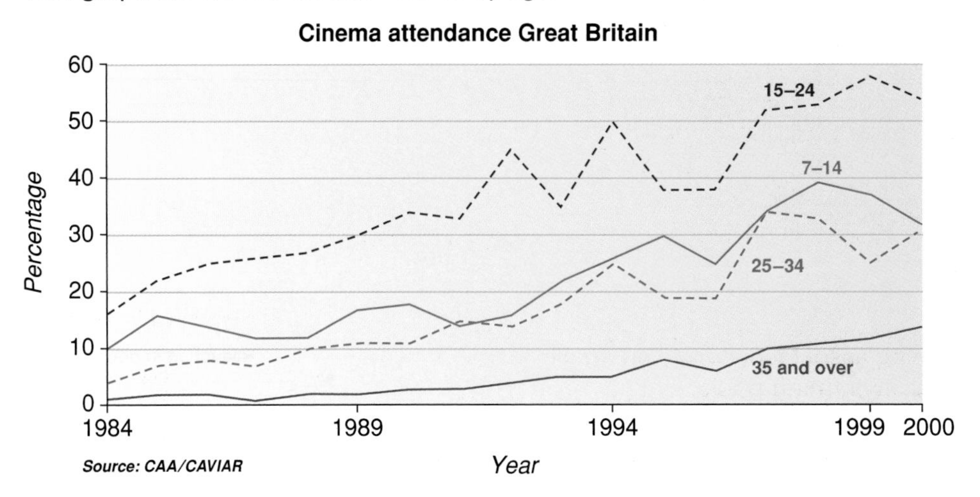

People were asked if they had gone to the cinema once or more in the last month.

a Which age group has had the greatest increase in cinema attendance since 1984? Give a reason why this might be.

b Give a possible reason why some age groups have 'peaks' in attendance.

8 A study was carried out on 100 000 people. This two-way table shows the number who died from coronary heart disease, their age and the number of cigarettes they smoked in a day.
Explain how these figures support the argument that smoking causes coronary heart disease.

	No.of cigarettes smoked per day			
Age	0	1–14	15–24	25+
45–54	118	220	368	393
55–64	531	742	819	1025

9 This graph shows the number of breath screening tests carried out each night over a 4-week period.

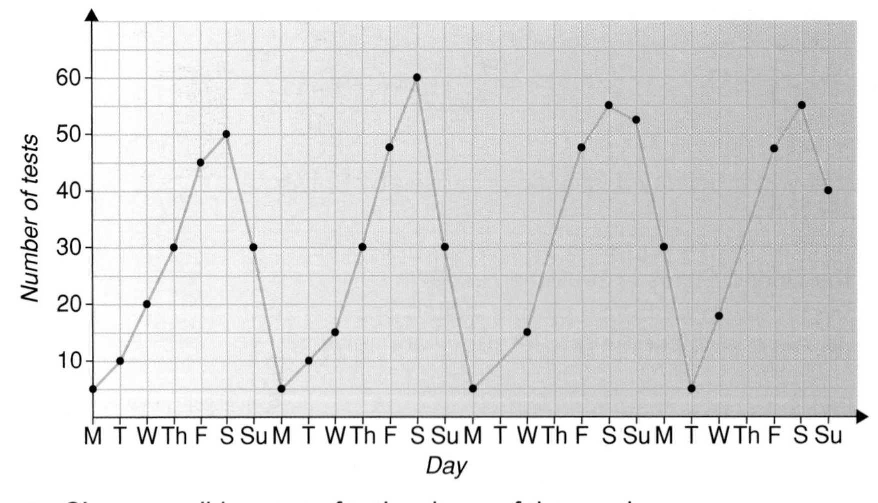

M–Monday
T–Tuesday
W–Wednesday
Th–Thursday
F–Friday
S–Saturday
Su–Sunday

a Give a possible reason for the shape of the graph.

b On the third Sunday more breath tests than usual were carried out for a Sunday night. Which of these is the most likely reason for this?

A the police had nothing else to do this day

B there are always more screening tests carried out on the third Sunday of a four-week period

C The Monday after this Sunday was a public holiday.

Review 1 Twenty pupils measured an angle. This table shows their results.

Angle measured as	Number of pupils
55°	6
124°	4
125°	6
126°	4

Use the table to decide what the angle is most likely to measure.
Give your reasons.

Review 2 This frequency diagram shows the times when goals were scored in a World Cup netball game. The times were divided into class intervals of $0 < t \leqslant 15$, $15 < t \leqslant 30$, $30 < t \leqslant 45$, ...

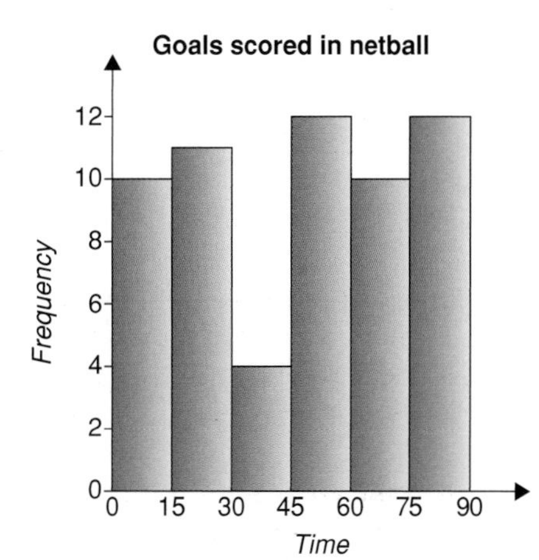

a One of the players said 'There were no goals scored in the last 10 minutes'.
Is this possible?
Explain.

b Half time is 5 minutes long. Which class interval do you think half time fell into?

c The coach thought that half the goals were scored in the first half of the game. She is wrong.
Explain how you can tell this from the graph.

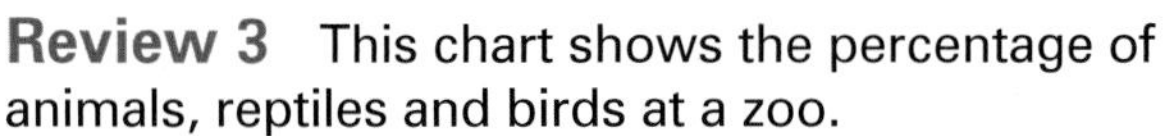

Review 3 This chart shows the percentage of animals, reptiles and birds at a zoo.

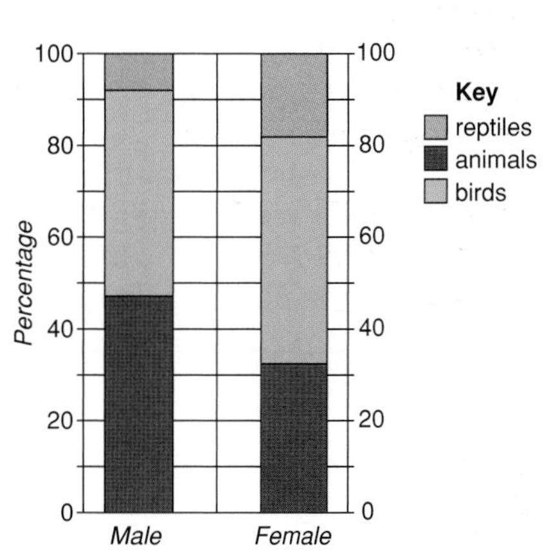

a About what percentage of females are birds?

b About what percentage of males are not birds?

c The black bars show the percentages for animals.
One green bar is taller than the other.
Does this mean that there must be more male animals than female animals at the zoo?
Explain your answer.

Misleading graphs

Often statistics or graphs are used to **mislead** us.

Example

The vertical scale not starting at zero makes the increase in male asthma sufferers look much more dramatic than it really is.
In fact it has only risen by about 20 out of 530, a percentage increase of about 3·8%.

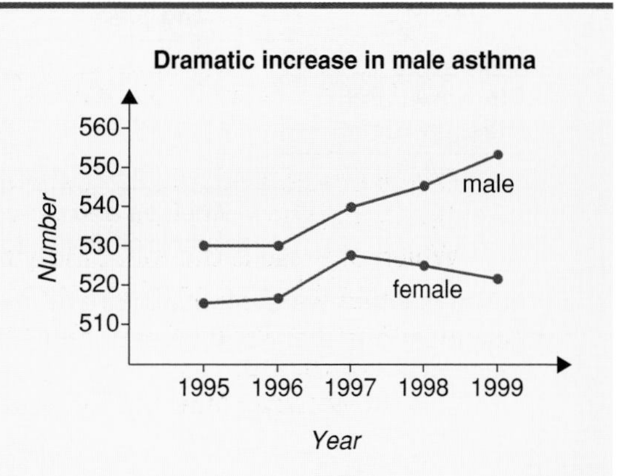

Discussion

- **Discuss** how each of these graphs is misleading.

1

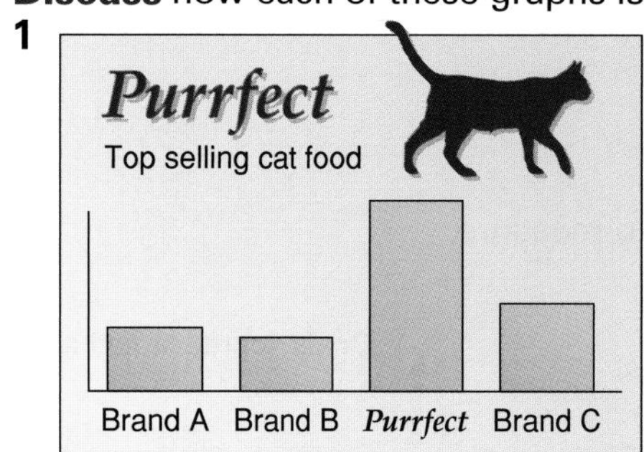

2

Disease hits old and young

Number of cases (000's)

0 1 2 3 4 5

20 25 30 35 40 45 50 55 60 100

Age (years)

3

Huge increase in population

1·5 tonnes — Year 1

1·8 tonnes — Year 2

2·2 tonnes — Year 3

4

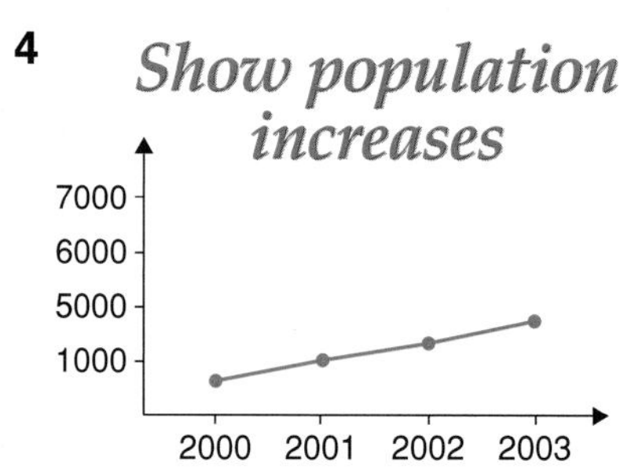

- What is misleading about these statements?
  1. Brite Colour washes your clothes 50% better.
  2. Tests prove Kittydins is best.
  3. Come to the Sunshine Resort. Annual mean rainfall only 200 mm.

Surveys

This diagram shows the **cycle for surveys**.

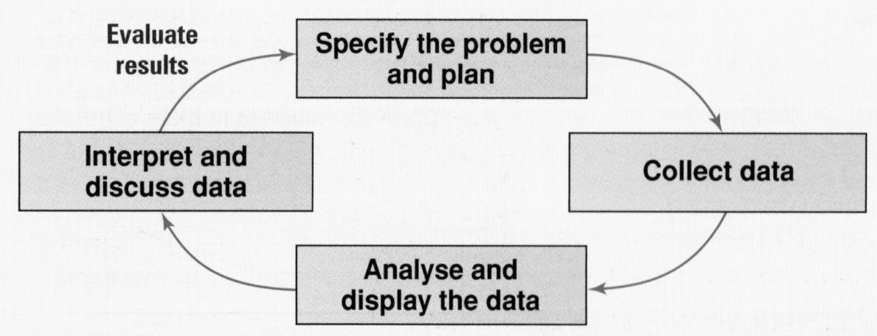

Sometimes when we evaluate the results at the end of a statistical investigation, this leads to further questions that need investigating. The cycle begins again.

When planning a survey, remember to decide on the answers to these questions:

- What do you want to find out?
- Are there any related questions?
- What might you find out?
- What data do you need to collect?
- How accurate does the data need to be?
- How will you collect the data and who from? You may choose a primary source or a secondary source.
- How many pieces of data do you need (sample size)?
- When and where will you collect the data?
- How will you display the data? Is using ICT appropriate or best?
- How will you interpret the data? Is finding the mean, median, mode or range appropriate?

Practical

- **You will need** to speak with the teacher of one of your other subjects.

1. If possible, choose something from one of your other subjects to investigate statistically. Talk to the teacher of this subject about your choice.
2. Plan your investigation/survey. Think about all the points given above.
3. Carry out the investigation/survey.
4. Display the results using ICT if appropriate.
5. Analyse the data using the mean, median, mode and range if appropriate.
6. Write a report on what you found out. Include your conclusions. Make sure your conclusions relate to the original problem you set out to investigate. In your report, write about any difficulties you had and how you solved these difficulties.

You might collect data from the Internet or other secondary source.

Use ICT to display your results and present your report.

Suggestions

- How does the weather in two different places differ?
- What factors affect the growth of plants?
- Simulate the cooling rates of penguins that huddle and those that don't.
- Compare and contrast TV viewing or radio listening habits of different age groups.
- What factors affect athlete performance on the day?
- Compare and contrast the health of two different age groups or males and females in one age group.
- What factors affect jumping or throwing distances?

Summary of key points

The **mean**, **median** and **mode** are all ways of summarising data into a single number.

The **median** is useful for comparing a set of items with a middle value.

Example About half of the goals were scored in the last 15 minutes.

The **mean** gives an idea of what would happen if there were equal shares.

The mean is affected by extreme values and does not represent the data well in this case.

Example Darren's team scored these goals in their last ten matches.

2, 1, 2, 1, 3, 1, 24, 4, 1, 3

The mean is 4·2 which is not representative of the data. They only scored more than this in one match.

The **mode** is useful for identifying the 'most popular' or 'most common'.

Example The modal shoe size of Year 8 girls is 5.

We can find the **mean** of a large data set using a calculator or spreadsheet.

We often find the **mean**, **median**, **mode** or **range** when **solving problems**.

We can find the median, range and mode from a **stem-and-leaf diagram**.

Example There are 25 data values.

The median is the 13th value, 0·8 hours.

The range is 4·0 – 0·2 = 3·8 hours.

The mode is 0·5 hours.

Time taken for lunch
stem = hours, leaves = tenths

Stem	Leaves
0	2 2 4 5 5 5 5 5 6 6 7 7 8
1	0 0 0 0 5 5 6
2	0 5
3	1 5
4	0

To **compare data** we use the range and one or more of the mean, median or mode.

Example Robert found the mean, median, mode and range of the hours of sunshine per week in two places.

Place A mean 53 range 24

Place B mean 54 range 8

The two places have about the same mean sunshine hours but place A has a much less consistent number of sunshine hours, shown by the bigger range.

Data collected on tables is often then **displayed** as one of these types of **graphs**.

Categorical data – pictogram, pie chart, bar graph

Discrete data – pictogram, pie chart, bar graph, bar-line graph

Grouped discrete – bar chart

Continuous data – frequency diagram

Data over time – line graph

A graph highlights particular features and trends that a table does not.

A **scatter graph** displays two sets of data

Example This scatter graph shows the mass of people and the mass of pizza each ate at a pizza night.

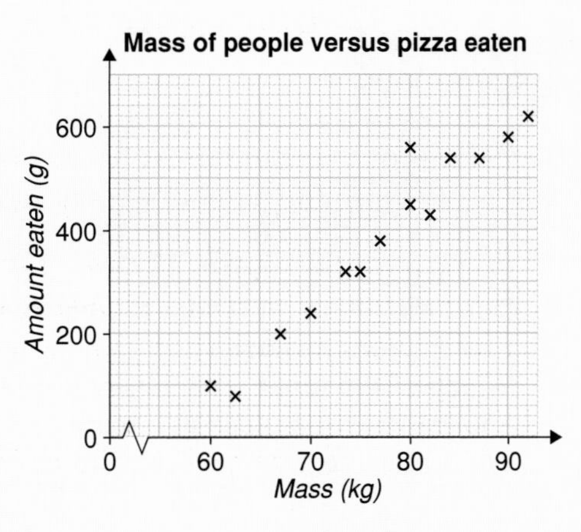

A scatter graph sometimes shows a **correlation** (relationship) between the variables.

Example

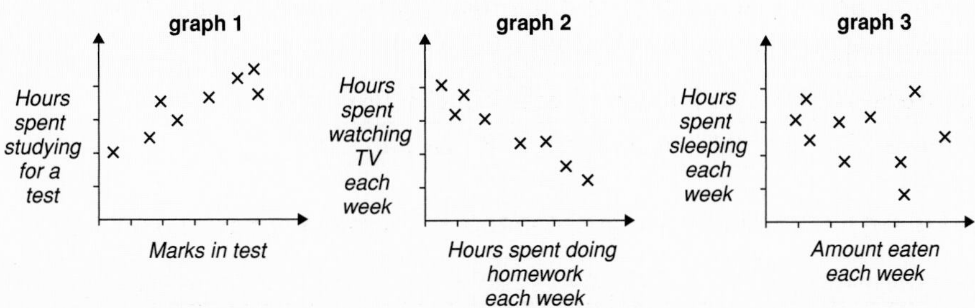

Graph 1 shows there is a **positive correlation** between the hours spent studying for a test and the marks gained in the test. As the number of hours increases so do the marks in the test.

Graph 2 shows a **negative correlation** between hours spent watching TV and hours spent doing homework. As the number of hours spent watching TV increases, the hours spent doing homework decreases.

Graph 3 shows there is no correlation between the hours spent sleeping and the amount eaten in a week.

We often use graphs and tables to help **interpret** data.

Sometimes graphs, diagrams or statements can be **misleading**.

Example

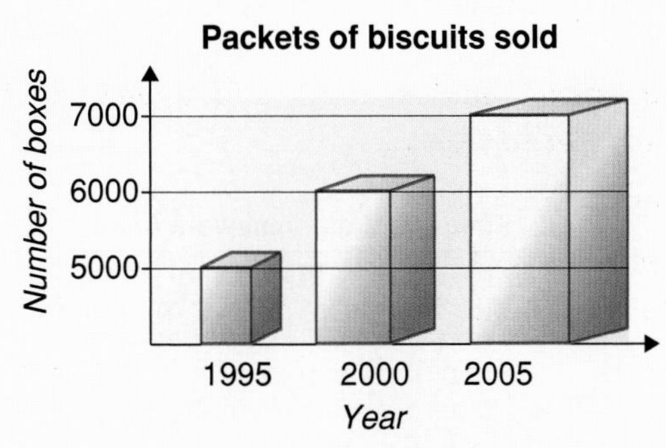

This graph is misleading because the volume of the boxes makes it look like the increase has been massive. Also the vertical scale does not start at zero, again emphasising the increase in number sold.

Test yourself

1 Rachel was goalkeeper for her hockey team. 
She wrote down the number of goals she stopped in each game.

2 4 2 6 3 3 2 10

a Find the mean, median and mode of this data.
b Does the mean represent the data well?
Explain your answer.
c Which of the mean, median or mode would Rachel be more likely to use in an application for a place in a team to represent her area.
Would this be a fair 'average' to quote?

2 Use a spreadsheet or calculator to find the mean score rolled on a dodecahedron-shaped dice. Give the answer to 2 d.p.

Score	1	2	3	4	5	6	7	8	9	10
No. of throws	26	30	37	28	32	29	33	29	35	26

3 This chart gives the marathon times of the top 100 women in a New York marathon. 

Time (hours:minutes)	2:20 ⩽ t < 2:30	2:30 ⩽ t < 2:40	2:40 ⩽ t < 2:50	2:50 ⩽ t < 3:00	3:00 ⩽ t < 3:10
Frequency	7	8	12	22	51

a What is the modal class?
b Why might it be useful to know this modal class?

4 The median score of three people in a test is 35. The mean is 38 and the range is 41. 
What are the three scores?

5 Is it possible to find a data set with five values with a mean of 4, a range of 2 and a mode of 6? 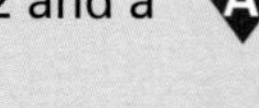

6 Write an expression for the mean of these three expressions. 

$8x + 2$ $5x - 3$ $2x - 2$

Write your expression as simply as possible.

7 Find the median, mode and range from this stem-and-leaf diagram.

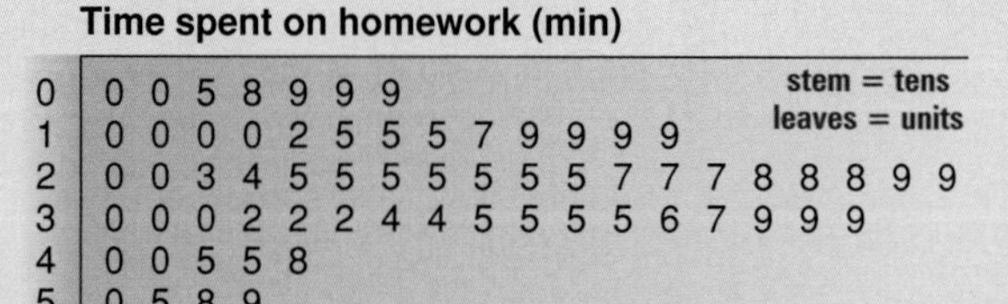

8 The school netball team has a final next week. 
The coach had to choose whether Lela or Gwen should play as goal shoot.
The following list shows their goals for their last 6 games.

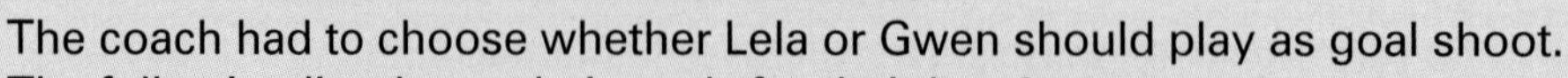

Gwen:	12	10	12	14	12	12
Lela:	10	2	4	30	2	24

a Find the mean, median and range for each girl.
b Which girl do you think the coach should choose?
Explain why using your results from **a**.

9 The average monthly rainfall in two different places is given by this table. 

	Average rainfall (mm)											
	Jan	Feb	Mar	Apr	May	Jun	Jul	Aug	Sep	Oct	Nov	Dec
New York	84	79	99	93	106	85	105	104	91	84	107	92
London	62	36	50	43	45	46	46	44	43	73	45	59

a Draw a line graph with both sets of data on the same axes.
b Do the points in between the ones plotted have meaning? Explain.
c Use your graph to compare the rainfall in New York and London.
d Could you display this data on two pie charts?
If you did, would it show the information as clearly as the line graph?

10 Jody gathered the following data from ten students in her class. Each of these students recorded the time (to the nearest hour) they spent reading and the time they spent watching TV during one weekend. 

No. of hours watching TV	10	3	8	5	1	13	6	14	12	4
No. of hours reading	2	7	4	4	10	3	7	1	1	11

a Draw a scatter graph for this data.
b Does this scatter graph show that people who watch lots of TV tend to spend less time reading?
c Amy watched 11 hours of TV that weekend. Use the graph to estimate the number of hours she spent reading.

11 Jordan thinks that people with longer arms also have longer legs.
If this were true, what would the scatter graph look like? Draw a sketch. 

12 This table gives the results for the Year 8 long jump competition at Ryan's school.

Distance jumped (d in cm)	Frequency
$180 \leq d < 190$	2
$190 \leq d < 200$	6
$200 \leq d < 210$	10
$210 \leq d < 220$	11
$220 \leq d < 230$	14
$230 \leq d < 240$	16
$240 \leq d < 250$	9
$250 \leq d < 260$	4

a Draw a frequency diagram for the data.
b Ryan jumped 243 cm. He thought he came fifth.
Is this possible? Explain your answer.
***c** The school magazine reported that over half of Year 9 students jumped further than 233 cm.
Explain how you can tell from the graph or table that this is not correct.

13 Felicity looked at this graph and said 'British people sleep far too much'.
Decide if Felicity is correct.
Explain how you decided.

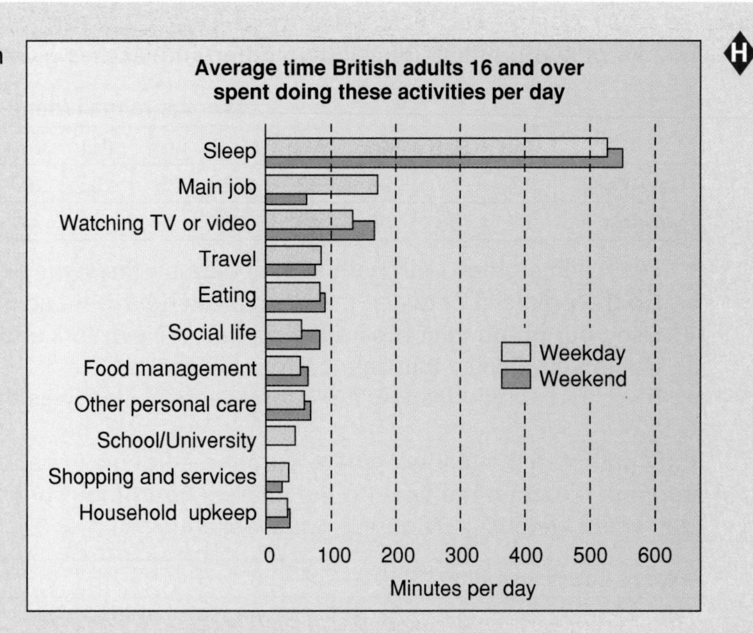

14 This graph appeared in an article claiming that we had beaten tooth decay in children.
Explain why it is misleading.

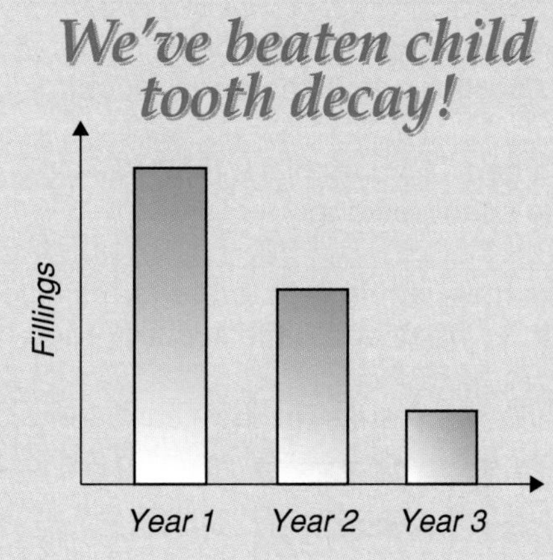

17 Probability

You need to know

 probability page 365

Key vocabulary

biased, event, experimental probability, mutually exclusive, relative frequency, sample space, theoretical probability

In a Spin

T **You will need** thin card and 2 paper fasteners (split pins).

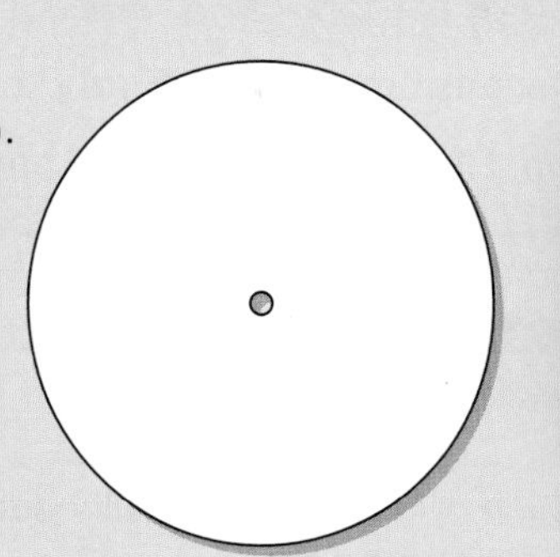

1. Design a spinner that has these probabilities:

   stopping on green $\frac{1}{4}$

   stopping on red $\frac{1}{2}$

   stopping on blue $\frac{1}{8}$

   stopping on yellow $\frac{1}{8}$

2. Design another spinner that has these probabilities:

   stopping on red $\frac{1}{5}$

   stopping on blue $\frac{1}{10}$

   stopping on green $\frac{3}{10}$

   stopping on yellow $\frac{2}{5}$

Language of probability

Discussion

An event is something that happens.

- Think of an **event** where the outcome
  a is certain **b** is impossible **c** has an even chance of happening.
  Discuss.
- Sasha and Craig are going on holiday to Majorca with their parents. Their holiday begins as follows:

  They travel by train to Gatwick Airport,
  They fly from Gatwick to Majorca,
  They hire a car and drive to the Hotel Playa.

  Think of some things that are **likely** to happen to this family as they do this.
  Think of some things that are **unlikely** to happen. **Discuss**.
- Mandy has five cards. She picks one card without looking. Is this a random event? **Discuss**.
  Is it possible to predict with certainty the outcome of a random event? **Discuss**.

Sasha has two bags of sweets.
If she takes a sweet at **random** (without looking) the outcome is **unpredictable**.
She might get a green sweet or she might get a yellow sweet.

blue bag red bag

Sasha wants a green sweet.
It is **more likely** she will get a green sweet from the blue bag than from the red bag.
There is a greater **proportion** of green sweets in the blue bag. $\frac{2}{3} > \frac{3}{5}$

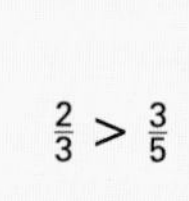

Some events are **equally likely** to happen.

Example If a dice is rolled, the outcomes 1, 2, 3, 4, 5 and 6 are all **equally likely**.

Exercise 1

1 Which of these events will have equally likely outcomes?
  a tossing a coin
  b taking a card at random from these cards
  c spinning this spinner

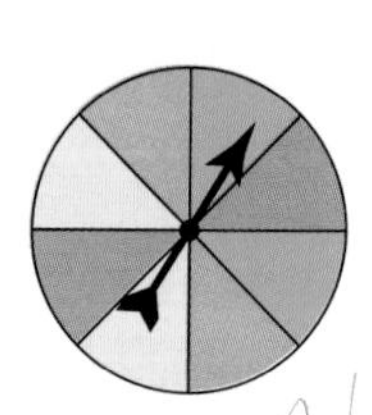

2 This table shows the colours of hats sold at a fête.
I walk around the fête. Which colour hat am I most likely to see?
Explain why using proportion.

Colour	Number
Red	14
Green	2
Yellow	7
Blue	10

3 Brookside School is selling scratch and win cards. You choose **one** square to scratch and if the square says 'win' you win £5.
On which of these cards are you most likely to get a 'win'? Explain why using proportion.

A
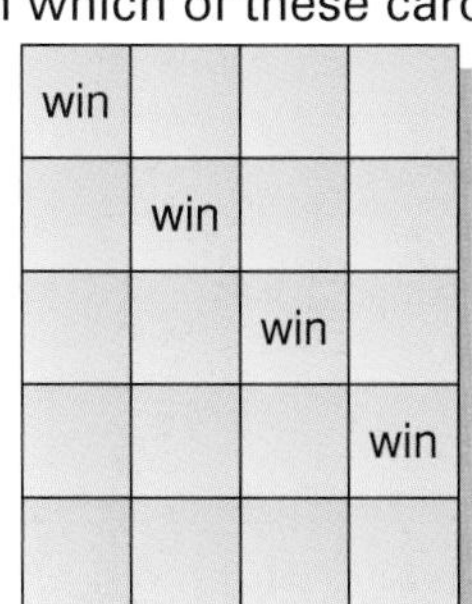

B
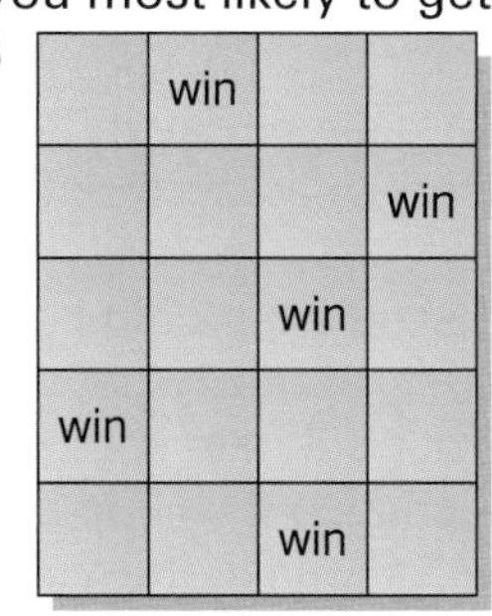

C
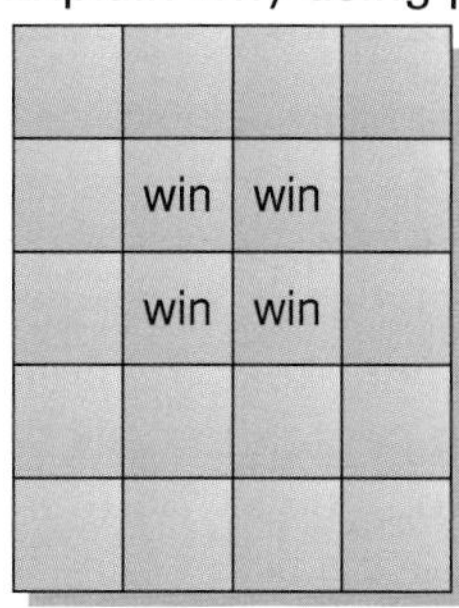

4 Two family packs of muesli bars contain different numbers of chocolate-coated and plain bars.
Pack 1 has 8 chocolate-coated and 12 plain bars.
Pack 2 has 6 chocolate-coated and 14 plain bars.
Emily only likes chocolate-coated bars. She can pick a bar at random from either pack.
Which pack should she pick from. Why?

5 Two boxes of fruit contain different numbers of apples and oranges.
Box 1 has 12 oranges and 18 apples.
Box 2 has 10 oranges and 16 apples.
Charlotte only likes oranges. She can pick a piece of fruit at random from either box.
Which box should she pick from? Why?

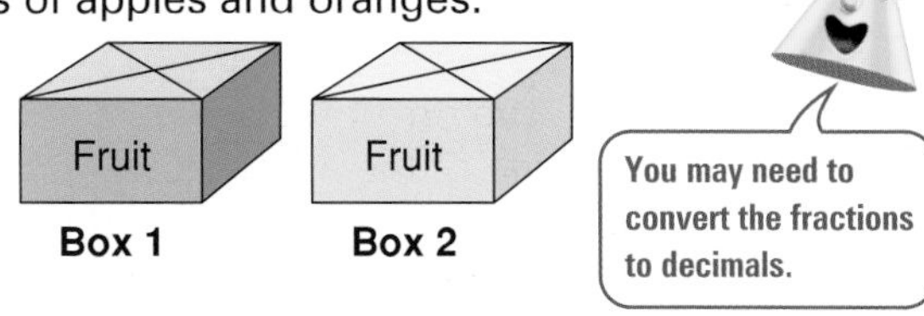

6 The diagram shows where pupils in Years 7, 8 and 9 went on their last holiday.

a A pupil from Year 7 is chosen at random.
Are they **most likely** to have holidayed in Europe, England or other?

b Repeat **a** for a pupil from Year 8.

c Repeat **a** for a pupil from Year 9.

d How many more pupils are there in Year 8 than Year 9?

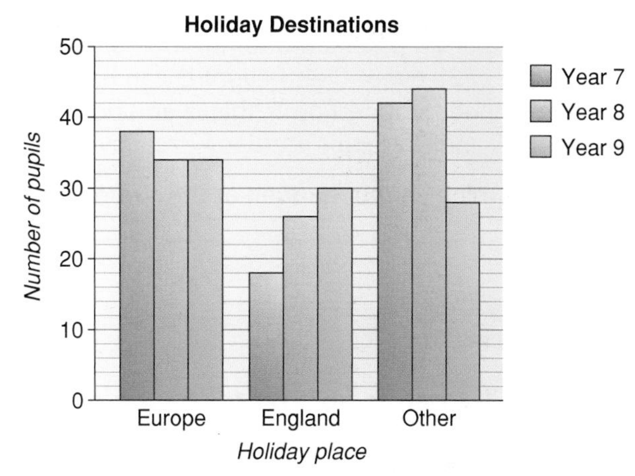

Review 1 This shows a map of Lake Deepwater.
Melanie is walking around the shores of the lake.
An emergency beacon is seen flashing in one of the squares.
The person who saw it can't tell if it is coming from land or water.
Where is it more likely to be coming from? Explain.

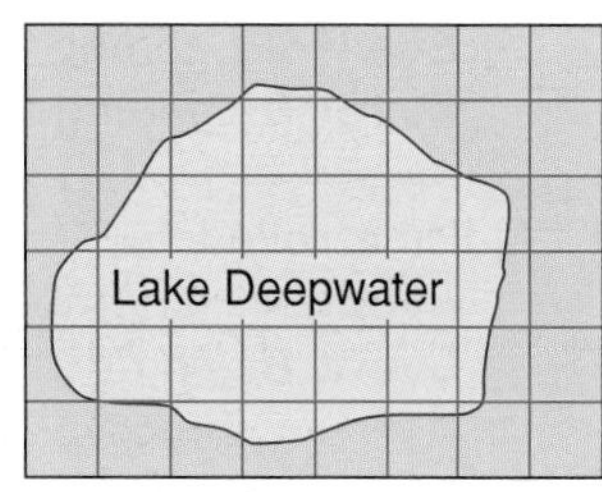

Review 2 I have two bags of counters.
Bag A contains 14 red counters and 21 yellow counters.
Bag B contains 10 red counters and 16 yellow counters.
I am going to take one counter at random from either bag A or bag B.
I want to get a red counter.
Which bag should I choose? Explain your answer.

Discussion

Discuss the truth of these statements.

a A serious road accident happens nearly every day in Britain. There are so many cars on the road today, an accident is **certain** to happen.

b For the last three days, each time I tossed a fair coin I got a head. When I toss the coin today I can't **possibly** get a head.

c The next person to walk through the school gate could be a teacher or a pupil. So there is a 50% chance it will be a pupil.

d Taylor buys a ticket each week in the National Lottery. She said 'I'm not lucky, I'll **never** win.'

e The risk of being killed in an accident is about 1 in 400 and of dying of cancer is 1 in 3. You are much more likely to die of cancer next year than in an accident.

Mutually exclusive events

This spinner could stop on 1 or 2 or 3 or 4 or 5 or 6 or 7 or 8.
It can only stop on *one* of the numbers.
If event A is 'stops on an odd number' and event B is 'stops on the 2', event A and event B cannot happen at the same time.
Event A and event B are called **mutually exclusive**.

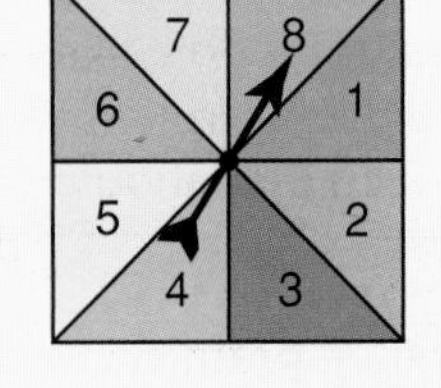

If event C is 'stops on an odd number' and event D is 'stops on a multiple of 3', event C and event D can happen at the same time.
If the spinner stopped on 3, this is an odd number **and** a multiple of 3.
Events C and D are *not* mutually exclusive.

Events which cannot happen at the same time are called **mutually exclusive**.

Exercise 2

State whether or not the following events A and B are mutually exclusive.

1 A dice is rolled.
Event A : getting an even number
Event B : getting an odd number

2 Two coins are tossed.
Event A : two heads
Event B : one head and one tail

3 A dice is rolled.
Event A : a number greater than 4
Event B : an odd number

4 Two coins are tossed.
Event A : two heads
Event B : at least one head

5 A card is drawn from a full pack.
Event A : a spade
Event B : an ace

6 A day dawns.
Event A : it is a fine day
Event B : it is windy

7 A sheep has lambs.
Event A : the lambs are twins
Event B : the lambs are triplets

8 Two students are chosen at random.
Event A : one student wears glasses
Event B : one student is tall

9 A counter is drawn from a box containing red, blue and white counters.
Event A : a red counter
Event B : not a red counter

Review 1 A dice is rolled.
Event A : a number less than 4
Event B : the number 4

Review 2 Salina goes shopping.
Event A : Salina goes shopping with Kate
Event B : Salina goes shopping with Shannon

Discussion

- Think of pairs of events that are mutually exclusive. **Discuss.**
  Think of pairs of events that are not mutually exclusive. **Discuss.**
- A card is drawn from a pack.
  Event A : an ace **Event B** : the king of spades
  What might **event C** be if the three events A, B and C are mutually exclusive? **Discuss.**
  What might **event C** be if the three events A, B and C are not mutually exclusive? **Discuss.**

Calculating probability

Remember
If the outcomes of an event are **equally likely,**

Probability of an event = $\frac{\textbf{number of favourable outcomes}}{\textbf{number of possible outcomes}}$

Example There are 50 countries represented at an export conference.
24 export food products, 14 export minerals, 6 export building products and the rest export other products.
One country is chosen at random to speak first at the conference.
The probability that this country will export food products is

probability (export food) = $\frac{24}{50}$ ← number of countries that export food (favourable outcomes) / ← total number of countries (possible outcomes)

= $\frac{12}{25}$ **or 48% or 0·48**

Probability can be given as a fraction, decimal or percentage.

An event will either happen or not happen.

Probability of an event not happening = 1 − probability of it happening

If the probability of an event happening is p, the probability of it not happening is $1 - p$.

Example If the probability of winning a game is 0·45, then the probability of not winning it is 1 − 0·45 = 0·55.

There are red, green and yellow counters in a bag. I take one at random.
It is certain that I will get red **or** green **or** yellow.
The sum of the probabilities of getting each colour must add to 1.
$p(\text{red}) + p(\text{green}) + p(\text{yellow}) = 1$.

The sum of probabilities of all the mutually exclusive outcomes of an event is 1.

Worked Example

In a game there are four types of card.
These are the probabilities of getting each card:

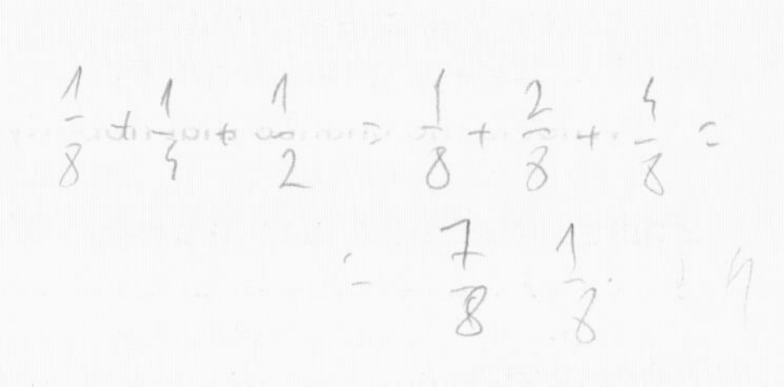

luck card	$\frac{1}{8}$
doom card	$\frac{1}{4}$
wealth card	$\frac{1}{2}$
lose money	**?**

a What is the probability of getting a lose money card?
b What is the probability of not getting a luck card?

Answer

a The four outcomes, luck card, doom card, wealth card and lose money card are mutually exclusive.
The probabilities of getting a luck card or a doom card or a wealth card or a lose money card must add to 1.

$$p(\text{luck}) + p(\text{doom}) + p(\text{wealth}) + p(\text{lose money}) = 1$$
$$\tfrac{1}{8} + \tfrac{1}{4} + \tfrac{1}{2} + \mathbf{?} = 1$$
$$\mathbf{?} = 1 - \tfrac{1}{8} - \tfrac{1}{4} - \tfrac{1}{2}$$
$$= \mathbf{\tfrac{1}{8}}$$

The probability of getting a lose money card is $\mathbf{\frac{1}{8}}$.

b $p(\text{not luck}) = 1 - p(\text{luck})$
$= 1 - \frac{1}{8}$
$= \frac{7}{8}$

The probability of not getting a luck card is $\mathbf{\frac{7}{8}}$.

Exercise 3

1. The probability of Yeorgi getting a hole in one at crazy golf is $\frac{2}{5}$. What is the probability that he will not get a hole in one?

2. Elfie has a telephone answering machine. If the probability of a caller not leaving a message on the machine is 0·3, find the probability of a caller leaving a message.

3. If the probability that a novice jumper hits the top rail is 65%, what is the probability that this jumper does not hit the top rail?

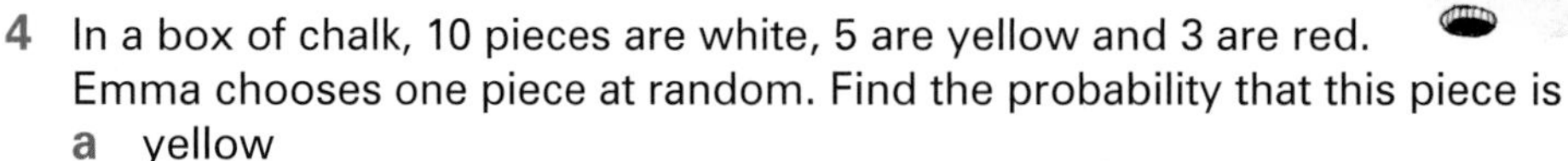

4. In a box of chalk, 10 pieces are white, 5 are yellow and 3 are red. Emma chooses one piece at random. Find the probability that this piece is
   a yellow
   b not yellow.

5. Derek has a number of different coloured felt-tip pens in a special case. There are 2 red, 2 purple, 1 green, 6 blue, 1 yellow and 2 black felt-tip pens in Derek's case. He takes one out at random. Find the probability that the one he took out is
   a blue
   b not blue.

6 There were 200 tickets sold to the school musical. They are numbered 1 to 200. One number is drawn at random for a prize.

a Brendon has number 182. What is the probability he will win?

b The Tucker family have tickets with numbers 25, 26, 27 and 28. The Billens family have tickets with numbers 120, 121, 122 and 123. Which family has the better chance of winning the prize? Why?

c Marcia buys several tickets. She works out she has a 5% chance of winning. How many tickets has she bought?

d Three people have lost their tickets and cannot claim the prize if their number is drawn. What is the chance that nobody wins?

7 There are some cubes in a bag. They are either blue or green. If a cube is taken at random, the probability it is blue is $\frac{1}{6}$.

a What is the probability it is green?

b I take a cube at random. It is blue. What is the smallest number of green cubes that could be in the bag?

c I take another cube. It is also blue. Now what is the smallest number of green cubes that could be in the bag?

d A different bag has red (R), yellow (Y) and purple (P) cubes in it. There is at least one of each of the three colours. If a cube is taken at random, the probability it will be red is $\frac{5}{8}$. There are 24 cubes in the bag. What is the greatest number of purple cubes there could be in the bag?

*__e__ Two bags, A and B, each have the same number of cubes. The probability of taking a blue cube from bag A is 0·6. The probability of taking a blue cube from bag B is 0·3. All the cubes from bags A and B are put in a new bag. What is the probability of taking a blue cube from the new bag?

8 On a game on a computer, one of four pictures appears or a 'You lose' screen appears.

picture	
p(mobile phone)	0·15
p(stereo)	0·1
p(computer)	0·05
p(head phones)	0·35

a What is the probability of getting the 'You lose' screen?

b What is the probability of not getting the mobile phone picture?

c What is the probability of getting a mobile phone, stereo or computer on the screen?

9 In a children's lucky dip at a fair, there are four possible gifts you could get. The probability of each is:

p(watch) $= \frac{1}{16}$

p(ring) $= \frac{1}{4}$

p(sticker) $=$ **?**

p(pen) $= \frac{1}{2}$

a What is the probability of getting a sticker?

b Which gift is a child most likely to get?

c What is the probability of not getting a watch?

*__d__ After many lucky dips have been bought, 4 children have got a watch. How many lucky dips do you think have been bought?

Review 1 The probability that a car calling at a service station needs oil is 12%. What is the probability that a car calling at this service station does not need oil?

Review 2 Of 50 chocolates in a box, 20 contain a nut, 16 are pure chocolate and the rest have a soft centre. One chocolate is chosen at random from this box.

Find the probability that this chocolate

a contains a nut **b** does not contain a nut **c** is soft centred

d is not pure chocolate.

Review 3 In an arcade game, one of three options appears in the final window.

The probability of each appearing is

$p(\text{win}) = \frac{1}{16}$

$p(\text{consolation}) = \frac{1}{4}$

$p(\text{lose}) = \textbf{?}$

a What is the probability of 'lose' appearing?

b Which is most likely to appear?

c What is the probability of not getting 'win'?

*****d** After many games 'win' had appeared 4 times. How many games do you think had been played?

Discussion

- Here is a spinner with five equal sections. A class is divided into two groups, A and B.
  They spin the pointer lots of times.
  If it stops on an odd number, group A get a point.
  If it stops on an even number, group B get a point.
  Is this a fair game? **Discuss.**
  What if group A get 2 points when it stops on odd and group B gets 3 points when it stops on even?
- A biased dice is one that is not equally likely to land on each of the numbers on its faces.
  What is a biased spinner? **Discuss.** Draw one.

Calculating probability by listing outcomes

Discussion

Elizabeth was tossing a coin twice.

She said 'There are three possible outcomes, so, there is a 1 in 3 chance I will get two heads.'

Craig said 'There are four possible outcomes, so there is a 1 in 4 chance you will get two heads.'

Who was right? **Discuss.**

Are all the outcomes of tossing a coin twice mutually exclusive? **Discuss.**

Tracy is vegetarian.
At Pizza Palace, there are two vegetarian pizzas, Bean Delight and Spicy Cheese.
She is going to buy pizza on Tuesday and Thursday.
All the possible mutually exclusive outcomes for what she buys could be given as a list,

Bean Delight, Bean Delight; Bean Delight, Spicy Cheese;
Spicy Cheese, Bean Delight; Spicy Cheese, Spicy Cheese.

or a table

Tuesday	Thursday
Bean Delight	Bean Delight
Bean Delight	Spicy Cheese
Spicy Cheese	Bean Delight
Spicy Cheese	Spicy Cheese

or

Bean Delight Bean Delight	Bean Delight Spicy Cheese
Spicy Cheese Bean Delight	Spicy Cheese Spicy Cheese

The set of all possible outcomes is called the **sample space**.
We can use the sample space to help calculate probability.

Worked Example
A dice and a coin are tossed together. Find the probability that we get a head and a number greater than 4.

This list of outcomes is the sample space.

Answer
All the possible outcomes can be written as a list as follows:

H1 H2 H3 H4 H5 H6 T1 T2 T3 T4 T5 T6

There are 12 possible outcomes in the sample space.
There are two favourable outcomes: H5, H6.

Then, $p(\text{H and number greater than 4}) = \frac{2}{12}$ or $\mathbf{\frac{1}{6}}$

Exercise 4

1 A fair counter is red on one side (R) and blue on the other (B). This counter and a fair dice are thrown together. List all the possible outcomes.

Remember: When a fair dice is thrown all the outcomes are equally likely.

2 What are the possible outcomes when
- **a** Josie and Charlotte choose a flavour of ice cream from vanilla, chocolate and raspberry
- **b** you choose a piece of fruit for morning tea and one for lunch from a bowl of bananas, apples and oranges
- **c** a woman has two children
- **d** you choose a card in round 1 and a card in round 2 from a pile of Jacks, Queens and Kings?

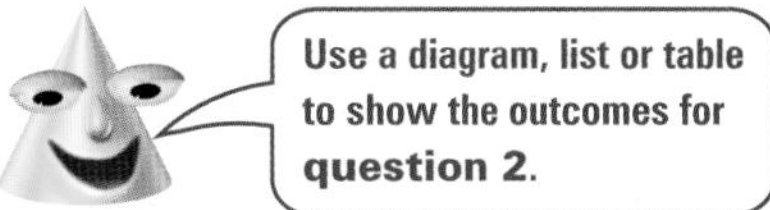

3 A coin is tossed and a dice is rolled. Use the sample space given in the worked example above to find the probability of getting
- **a** a tail and a 3
- **b** a tail and an even number
- **c** a head and a number greater than 2
- **d** a head and a multiple of 2
- **e** a tail and a number less than 7
- **f** a head and a number greater than 6.

4 Two children are playing 'paper, scissors, stone'.
They each choose one of paper, scissors or stone and at the count of three, show their choice using a hand.

a Write out the sample space for all the possible outcomes for the hands the two children show.

b What is the probability of
i both children showing the same hands
ii at least one scissors   iii no stones?

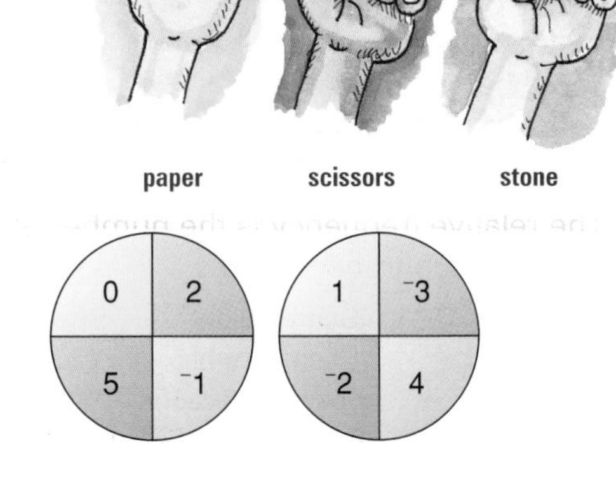

T

5 These spinners are spun at the same time.
The two numbers spun are added.

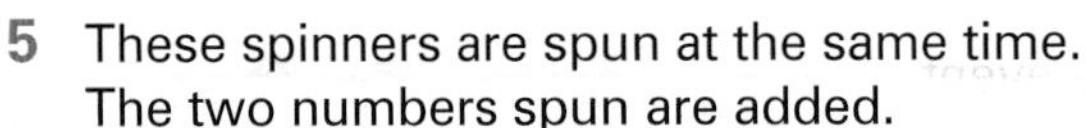

a Use a copy of this table to show the sample space for the outcomes

+	0	2	5	$^-1$
1		3	6	0
$^-3$				
$^-2$				
4				

b Find the probability of getting a total of
i 1   ii 3   iii less than 5   iv more than 6   v at least 2.

c What is the most likely total?

d What is the probability of getting 2 on one spinner and $^-3$ on the other?

e What is the probability of getting a negative total?

***6** Stella and Paolo each have three cards numbered 1, 2 and 3.
They each take one of the other person's cards.
They then add together the numbers on the four cards left.
What is the probability the total will be an odd number?

***7** Sanjay is choosing subjects for next year.
He must choose one subject from each of these blocks.

Block 1	**Block 2**	**Block 3**
French (F)	Italian (I)	Spanish (S)
Art (A)	Computers (C)	Cooking (Co)

a List all the possible combinations Sanjay could choose.

b Sanjay decided to choose his subject from each block randomly.
What is the probability his choice will
i be French, computers and cooking
ii have at least one language
iii have neither art nor cooking
iv have French and cooking but not computers?

Review A game on the computer shows two cards.
Each card is equally likely to be a cherry, a strawberry or a plum.

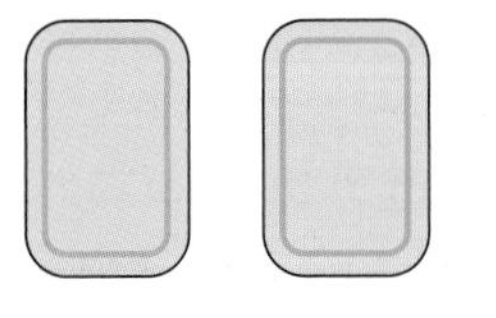

cherry
(c)

strawberry
(s)

plum
(p)

a List all the possible outcomes.

b What is the probability of getting
i two identical fruits   ii at least one strawberry   iii no cherries?

Estimating probability from relative frequency

Class 8D did an experiment to see what proportion of packets of crisps were underweight.
They weighed 50 bags of crisps. They found 17 of them weighed less than the amount stated on the packet.
The **relative frequency** of an underweight packet = $\frac{17}{50}$.

The relative frequency is the number of times the event occurs in a number of trials.

Relative frequency = $\frac{\textbf{number of times an event occurs}}{\textbf{number of trials}}$

We often use the relative frequency as an **estimate of probability**.
In the above example, the probability that an unopened bag will be underweight is estimated as $\frac{17}{50}$.

Practical

You will need a spinner like this or a similar one.
Spin the pointer twice. Add the two scores.
Do this 50 times. Record the results in a tally chart.
Draw a bar chart to show the results.
Compare your results with another group.
Are they different? Why?

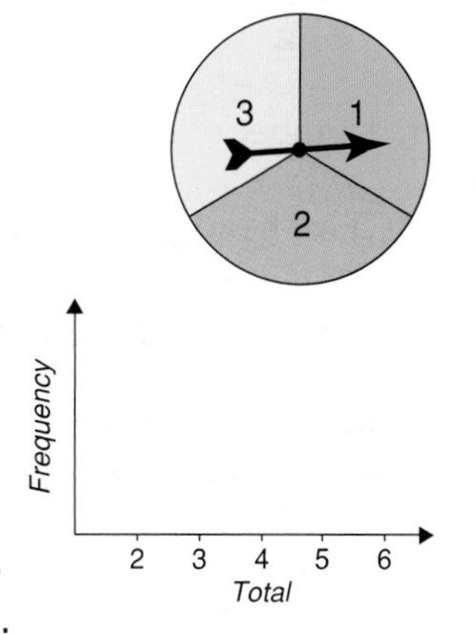

What is the relative frequency of a total of 5?

Use the relative frequency to estimate the probability of getting a total of 5 the next time you spin it twice and add the scores.

Predict what might happen if you repeat the experiment.

Repeat spinning and adding the scores another 50 times.
Combine the results together on another table.
What effect do the extra throws have on the results?

Use the combined results of your experiment to estimate the probability of getting a total of 5 the next time you spin the spinner twice and add the results.
How could you get a more accurate estimate?

If you repeated the experiment 500 times, how many times would you expect to get a total of 5?

Discussion

- If an experiment is repeated, will the outcomes be exactly the same each time? **Discuss**.
- Billie tossed a biased dice 50 times.
  He used the results of his experiment to estimate the probability of getting each number the next time he tossed the dice.

  Melanie tossed a biased dice 200 times.
  She used the results of her experiment to estimate the probability of getting each number the next time she tossed the dice.

  Who is likely to get the better estimate of the probabilities? Why? **Discuss**.

When **estimating probability** from an experimental result it is important to note that

1 when an experiment is repeated, there may be, and usually will be different outcomes
2 increasing the number of times an experiment is repeated usually leads to better estimates of probability.

Worked Example

Adam tossed two coins together a number of times. He recorded the results on the following table.

Event	Tally	Frequency
HH	||||̸ ||||̸ ||||̸	15
HT	||||̸ ||||̸ ||||̸ ||||̸ ||||	24
TH	||||̸ ||||̸ ||||̸ ||||̸ ||||̸ ||||	29
TT	||||̸ ||||̸ ||||̸ |||	18

a Find
  i the number of times Adam tossed the two coins
  ii the relative frequency of the event 'two heads'.
b Estimate the probability of getting 'one head and one tail', the next time these coins are tossed.
c What would happen if he tossed the two coins together again, the same number of times?

Answer

a i Number of tosses = 15 + 24 + 29 + 18
  = **86**
  ii 'Two heads' occurred 15 times.
  Relative frequency of 'two heads' = $\mathbf{\frac{15}{86}}$
b 'One head and one tail' occurred 53 times (24 HT and 29 TH).
  Relative frequency of 'one head and one tail' = $\frac{53}{86}$
  Probability of getting one head and one tail = $\mathbf{\frac{53}{86}}$.
c **The results would be different but probably similar.**

Exercise 5

1 This tally chart gives the number of pictures on each of the pages of a newspaper.

Number of pictures	Tally	Frequency
0	||||̸ ||||̸	10
1	||||̸ ||	7
2	||||̸ ||||	9
3					̸		6
4					̸		6
5						4
6					̸		6

a How many pages were there in this newspaper?
b Find the relative frequency of a page having just one picture.
c Estimate the probability of there being 5 pictures on a page chosen at random.
d How could you make a more accurate estimate of probability?

2 A card was drawn at random from a pack and the picture on it noted. The card was put back and the pack shuffled. This was done 250 times. The results are shown in the table.

Picture	Fruit	Animal	Car	Bird
Frequency	52	67	61	70

a Find the relative frequency of
i fruit **ii** bird **iii** animal **iv** car.
What is the sum of these relative frequencies?

b Estimate the probability of getting a bird the next time a card is drawn at random from this pack.

c If a card is drawn at random from this pack 1000 times and put back each time, how many times would you expect to get a car?

3 Some pupils tossed three coins.
This table shows their results.

Name	**No. of tosses**	**Two the same**	**All the same**
Tony	50	42	8
Huw	150	112	38
Jane	100	71	29

a Whose data is most likely to give the best estimate of the probability of getting each result? Explain.

b Use this person's results to estimate the probability of getting all the same the next time you toss three coins.
How could you improve the accuracy of this estimate?

c If the coins were tossed together 500 times, how many times would you expect to get just two the same?

4 The winning numbers in a raffle are shown below.

1076 1094 1099 1100 1110 1212 1215 1245 1256 1262 1292 1293 1299 1303 1306
1363 1365 1440 1480 1528 1544 1546 1652 1683 1729 1794 1799 1842 1892 1922
2101 2127 2186 2232 2278 2279 2284 2321 2334 2349 2384 2419 2423 2446 2461
2472 2513 2519 2520 2542 2627 2638 2663 2671 2674 2675 2732 2768 2776 2783
2794 2879 2902 2911 2913 2939 2939 2945 2963 2987 3010 3029 3080 3121 3142
3145 3148 3162 3171 3180 3189 3363 3365 3380 3425 3429 3488 3567 3612 3635
3645 3698 3718 3734 3738 3766 3766 3771 3786 3793 3801 3863 3874 3886 3954
4016 4104 4125 4141 4157 4158 4170 4171 4188 4224 4264 4274 4308 4390 4498
4504 4517 4582 4585 4614 4630 4656 4669 4694 4697

a Find the relative frequency of a winning number
i beginning with the digit 2 **ii** containing the digit 2.

b Estimate the probability of a winning number beginning with the digit 3.

5 At a fair, Tony threw a ball at numbered cans 50 times.
The highest number he could score with one ball was double forty.
Tony scored double forty 7 times.
Tony is going to have one more game.

a Estimate the probability that he will score double forty.

b Estimate the probability that he will score less than 80. Give a reason.

Review A fair dice was thrown 200 times and the results were as shown in this table.

Number on dice	1	2	3	4	5	6
Frequency	24	29	32	41	39	35

a Find the relative frequency of getting a
i 1 **ii** 2 **iii** 3 **iv** 4 **v** 5 **vi** 6
What is the sum of these relative frequencies?

b Estimate the probability that the next time this dice is thrown it will land on a 6.

c If the dice is tossed 500 times, about how many times would you expect to get a 6?

Comparing calculated probability with experimental probability

Practicals

You will need a computer software package that simulates rolling two dice and a card with a dot on one side.

2 3 4 5 6 7 8 9 10 11 12
Total

1 a Simulate rolling two dice and adding the numbers fifty times.
Draw a graph to show the results.

b Calculate the theoretical probability of each total.
Draw a graph of these probabilities.
Compare the two graphs.

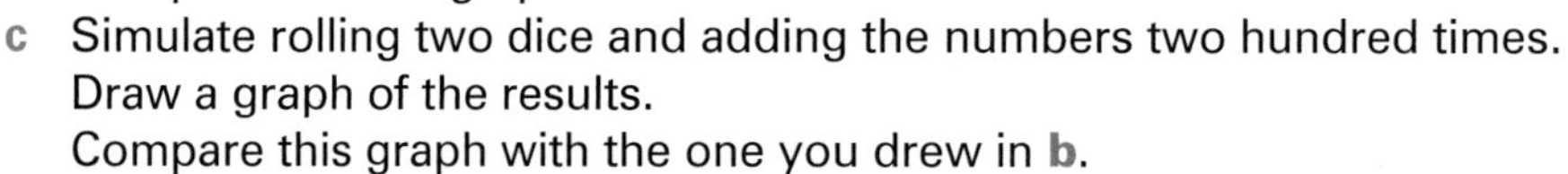

c Simulate rolling two dice and adding the numbers two hundred times.
Draw a graph of the results.
Compare this graph with the one you drew in **b**.

d Repeat **c** for five hundred trials.

e What do you notice happens as you increase the number of trials?

2 Jake thought that a buttered biscuit was more likely to land buttered side down when dropped. What is the theoretical probability of this?
Devise an experiment, using a card with a dot representing the buttered side, to test Jake's prediction. Carry out the experiment.
Compare the theoretical and experimental probabilities.

Investigation

Are You A Mind Reader?

You will need five cards with a design on each, for example:

1 Shuffle the cards and place them face down.

2 Ask someone to take one of the cards and picture the symbol on it in their mind.

3 See if you can tell them what the symbol is.

4 Repeat this with at least 20 other people.
Record your results in a table.

	Tally	Frequency
Symbol correct		
Symbol incorrect		

5 From these results, estimate the probability that the next time you try to guess the symbol you will be correct.

6 What is the theoretical probability of you guessing the correct symbol?

When we **compare experimental probability with theoretical probability**, the greater the number of trials in the experiment, the closer the experimental probability is to the theoretical probability.

Exercise 6

1 Jade and Kira were asked to toss a fair dice 90 times each and record the results.
One of them made up the results.
Which one? Explain.

	1	2	3	4	5	6
Jade	15	16	14	17	13	15
Kira	15	15	15	15	15	15

2 Design an experiment so that you could compare the theoretical probability with the experimental probability of guessing the missing digit in this combination. It could be any digit from 0 to 9.

***3** Jasmine worked out the probability of getting a total of seven when she spun this spinner twice and added the scores.

a What theoretical probability should she have got?

b She did an experiment to test the theoretical probability.
She spun the spinner twice and added the scores.
She repeated this 90 times.
She got a total of seven 33 times.
She said 'The experimental probability is not the same as the theoretical probability so I must have done something wrong.'
Is she correct? Explain.

Review Toby rolled a dice 200 times. He got a six 34 times.
Do you think the dice is biased? Explain.

Summary of key points

When an event is **random**, the outcome is **unpredictable**.

Some outcomes are **more likely** than others.

Example This spinner is more likely to stop on blue than on red when spun.

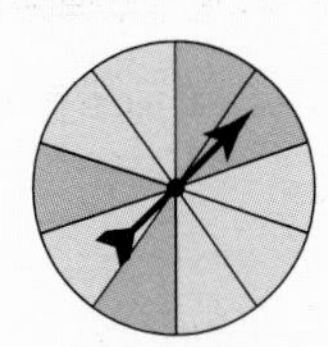

Some events have **equally likely** outcomes.

Example This spinner is equally likely to stop on red, blue, green, yellow or purple.

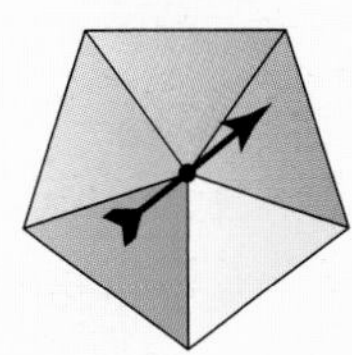

B Events that cannot happen at the same time are called **mutually exclusive**.

Example A is the event 'getting an even number'.
B is the event 'getting a 3' when a fair dice is rolled.
These events cannot happen at the same time.
They are mutually exclusive events.

If the probability of an event occurring is p, then the **probability** of it **not** occurring is $\mathbf{1} - p$.

Example If the probability of having to stop at an intersection is $\frac{1}{4}$, then the probability of not having to stop is $\frac{3}{4}$.

The sum of the probabilities of all the mutually exclusive outcomes of an event is 1.

Example There are four different colours of counters in a bag.
One is taken at random.
The probabilities of getting three of the colours are
red 0·2 green 0·25 blue 0·4
The sum of all four probabilities must equal 1.
So $p(\text{yellow}) = 1 - 0{\cdot}2 - 0{\cdot}25 - 0{\cdot}4$
$= 0{\cdot}15$
The probability of getting yellow is **0·15**.

To **calculate a probability** we often record all the possible **outcomes** using a list, diagram or table.
The set of all the possible outcomes is called the **sample space**.

Example These two spinners are spun.
The possible outcomes could be shown in a table.

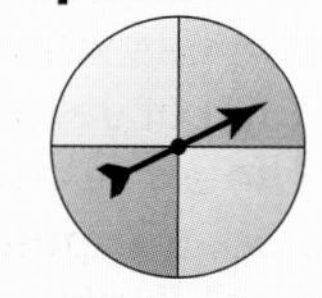

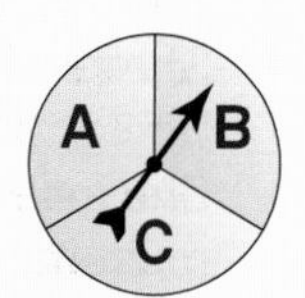

red	red	red	blue	blue	blue	purple	purple	purple	yellow	yellow	yellow
A	B	C	A	B	C	A	B	C	A	B	C

The probability of getting red A = $\frac{1}{12}$ ← **ways of getting red A** / ← **number of possible outcomes**

From the results of an experiment we can find the **relative frequency** of an event.

Relative frequency = $\dfrac{\textbf{number of times an event occurs}}{\textbf{number of trials}}$

We often use the relative frequency as an **estimate of probability**.

Example A factory tested five hundred circuit boards.
Twelve were found to have faults.
Relative frequency of a fault = $\frac{12}{500} = \frac{3}{125}$
We can use this to estimate that the probability a circuit board chosen at random will be faulty is $\frac{12}{500} = \frac{3}{125}$.

Note

1. When an experiment is repeated, there may be, and usually will be different outcomes.
2. Increasing the number of times an experiment is repeated usually leads to better estimates of probability.

When we compare **experimental probability** with **theoretical probability**, the greater the number of trials in the experiment, the closer the experimental probability is to the theoretical probability.

Test yourself

1 A prize is given if a counter, when dropped onto a board, lands on a purple square.
On which board are you most likely to win? Explain using proportion. 

A 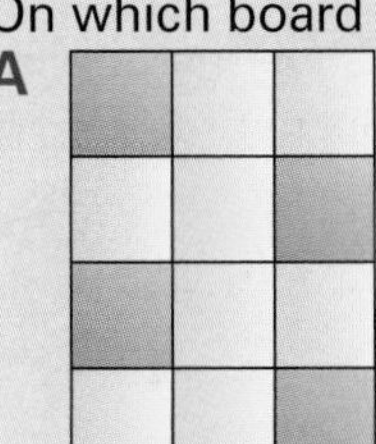 B 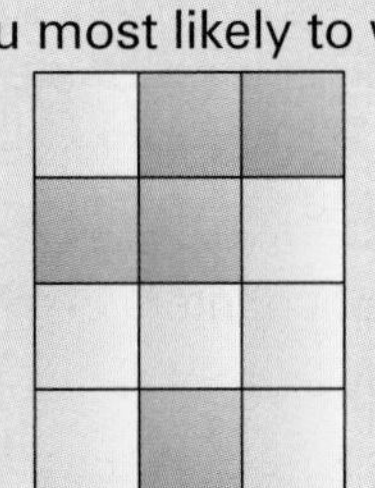 C 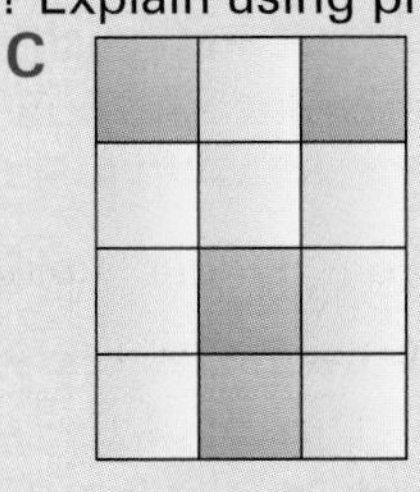 counter

2 Gareth likes white chocolates. 
He can choose a chocolate at random from one of three boxes.
Box A has 4 white and 7 milk chocolates.
Box B has 10 white and 22 milk chocolates.
Box C has 16 white and 42 milk chocolates.
From which box should Gareth take his chocolate? Explain.

3 In which of the following are the events **A** and **B** mutually exclusive? 

- **a** **Event A** : John goes to the football match.
  Event B : Jessie goes to the football match.
- **b** **Event A** : A diamond is drawn from a pack of cards.
  Event B : A picture card is drawn from a pack of cards.
- **c** **Event A** : A prime number is obtained when a dice is rolled.
  Event B : A six is obtained when a dice is rolled.

4 A cube-shaped biased dice is numbered 1, 2, 3, 4, 5 and 6. 
The probabilities of it landing on a particular number are;

$p(1) = 0{\cdot}1$
$p(2) = 0{\cdot}2$
$p(3) = 0{\cdot}15$
$p(4) = 0{\cdot}05$
$p(5) = 0{\cdot}25$

- **a** What is the probability of it landing on 6?
- **b** What is the probability of it not landing on 3 or 4?

5 The probability of a fisherman catching an under-sized lobster is 0·08. 
If under-sized lobsters are caught, they must be thrown back to sea.
Find the probability that a lobster is not thrown back to sea.

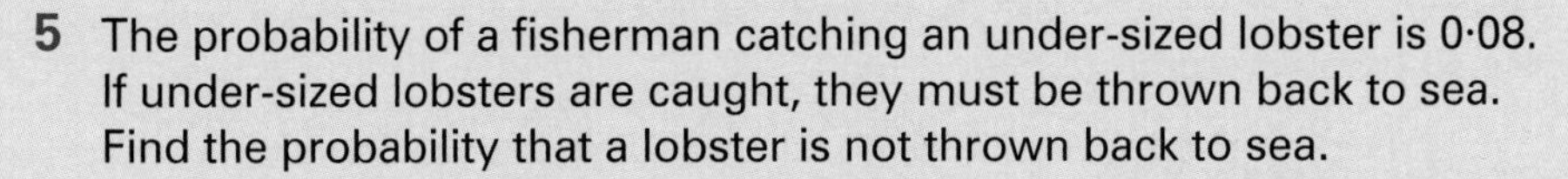

6 A pack of 26 cards is made, each with a different letter of the alphabet on it. 
What is the probability of these if one card is drawn at random?

- **a** a vowel
- **b** not a consonant
- **c** a letter between L and P inclusive
- **d** a letter from the word 'probability'
- **e** neither a vowel nor a consonant
- **f** not a letter from the word 'mathematics'.

7 Each pupil on a camp could choose two pieces of fruit, one for breakfast and one for lunch.
They could choose from bananas (b), apples (a) and oranges (o).

a List all the possible choices they could make for their two pieces.
b What is the probability that a pupil chosen at random will have
i the same fruit for breakfast and lunch **ii** at least one orange **iii** no apples?

8 The hands on these spinners are spun at the same time.
The two scores are added together.
What is the probability that the total score is negative?

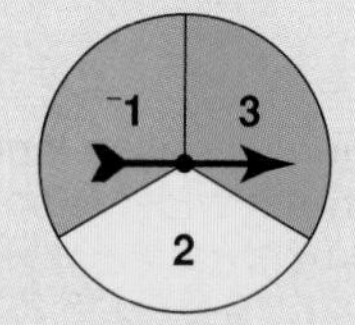

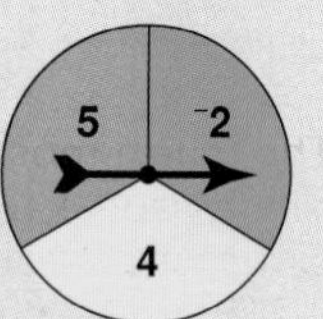

9 Three counters are put into a bag. One is green, one is red and the other is blue.
A counter is drawn at random and its colour noted. The counter is then put back in the bag.
Another counter is then drawn at random and its colour noted.
Write down the sample space for the colour of the two counters.
You could use a table or a diagram to help.
Find the probability that

a both counters are red **b** both counters are the same colour
c the counters are different colours **d** just one counter is blue
***e** at least one counter is red ***f** neither counter is green.

10 Four full packs of cards (52 cards in each) are shuffled together. A card is chosen at random, then put back and the pack shuffled again. This is done 260 times.
How many picture cards (Aces, Kings, Queens and Jacks) do you expect there to have been in the 260 cards chosen?

11 54 out of 7200 shirts sewn at the Shirt Factory last month were rejected.

a What is the relative frequency of a shirt being rejected?
b Estimate the probability that the next shirt sewn will be rejected.

12 Melanie was going to estimate the probability of a page, chosen at random, from a magazine (published 25 May) having 5 advertisements. She began by counting the advertisements on each page and recording her results on this table.

No. of advertisements	1	2	3	4	5	6	7	8 or more
Frequency								

a Explain how Melanie could continue.
b If she had chosen the 18 May publication instead of the 25 May would you expect Melanie to get the same estimate?
c Would you expect the two estimates to be quite close?

13 Annie could only remember the first three digits of her credit card pin number.
The last digit could be 0 to 9. She guesses the last digit.

a What is the theoretical probability she will be correct?
b Write down an experiment you could do to find the experimental probability.
c Is the experimental probability more likely to be closer to the theoretical probability if you do 50 trials or 200 trials?

Test yourself answers

Chapter 1 page 32

1 a 10^5 b 10^4 c 10^6 d 10^9
2 a One point four nine times ten to the power of eight.
b Two point zero times ten to the power of negative twenty-three
3 a 10^3 b 10^9 c 10^{-9} 4 C 5 a 0·57 b 47·3 c 0·835 d 68 6 £32·76
7 The fourth most widely spoken language has about 173 million speakers.
% who speak five most widely spoken is about 33%.
8 Greatest 5 349 999 999, smallest 5250 million.
9 a 76·98 b 103·00 c 0·0 d 8·0 e 160 f 9·70 g 0·10
10 a 4·34 or 4·3$\dot{3}\dot{6}$ b 62·91 c 8·76 d 9·45 e 0·00
11 a 47·8$\dot{3}$ or 47·8 b 0·1 kg 12 a 0·62 kg b 0·92 kg

Chapter 2 page 53

1 a b c d

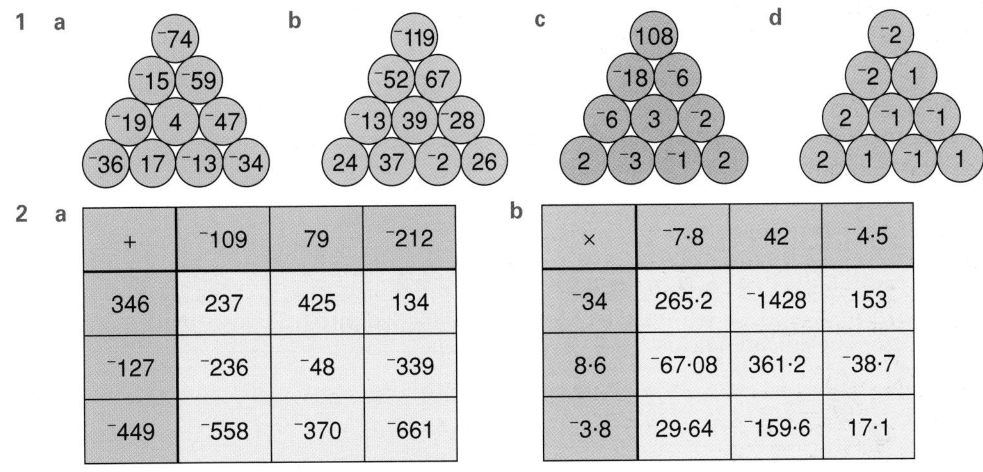

2 a

+	$^{-}109$	79	$^{-}212$
346	237	425	134
$^{-}127$	$^{-}236$	$^{-}48$	$^{-}339$
$^{-}449$	$^{-}558$	$^{-}370$	$^{-}661$

b

×	$^{-}7·8$	42	$^{-}4·5$
$^{-}34$	265·2	$^{-}1428$	153
8·6	$^{-}67·08$	361·2	$^{-}38·7$
$^{-}3·8$	29·64	$^{-}159·6$	17·1

3 a $^{-}22$ b $^{-}12$ c $^{-}34$ d 5 4 a 0 b 5 c 21 d $^{-}6$ e 26 f 33 g $^{-}21$
5 a $\frac{3}{7}$ b $\frac{4}{9}$ c $\frac{4}{7}$ d $\frac{20}{59}$ e $\frac{12}{23}$ f $\frac{4}{7}$ 6 a $\frac{37}{180}$ b $\frac{71}{120}$ c $\frac{53}{180}$ d $\frac{211}{300}$
7 a 78 125 b 7776 c 3·84 d 0·01 e $^{-}592·70$ f 1
8 a 16 b 16 c 9 d 6 e ±14 f 64 g 3
9 a 60·76 b 71·32 c 3·08 or $^{-}3·08$ d 47·62 e 2197 f $^{-}512$ g 13 h 4769
10 Negative 11 a 6·71 b 3·27 12 53 and 54 13 a 8^7 b 9^5 c 5^3 d a^8 e m^5
14 Either: Let the number be $1000u + 100t + 10t + u$.
$1000u + 100t + 10t + u = 1001u + 110t$
Both of these terms are divisible by 11 so $1001u + 110t$ is divisible by 11.
Or: $1001u + 110t = 11(91u + 10t)$ which is divisible by 11.

Chapter 3 page 71

1 a 1·23 b 0·43 c 0·014 d 4·28 2 a 7 b $^{-}0·4$ 3 1437
4 a 4 200 000 b 80 c 0·000 625 d 0·086 e 6·2 f 6·02 g $^{-}0·72$ h $^{-}0·05$ i 1·68 j 630
k 34·4 l $^{-}22·8$ m 38·4 n 50·5
5 a 1104 b 1125 6 One possible answer is 0·2 × 3 × 0·1.
7

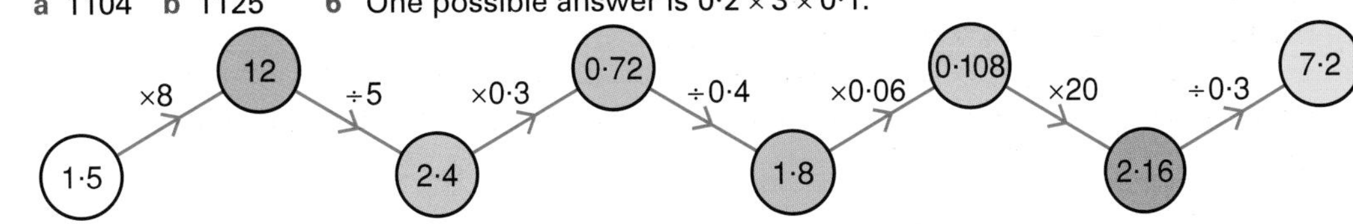

8 a 5 b 6 c 72 d 4 e 5 9 a 0·175 b 32·5% c 1150%
10 a 22·2 b 6300 c £2·55 d £2·25 11 a 952 m b 440 kg 12 150
13 a 11 cm b £7·80 c 84° d 27 e About £19·20 f About 37·5 miles g $\frac{5}{12}$ h 75

14 8·5 mℓ **15** One possible answer is

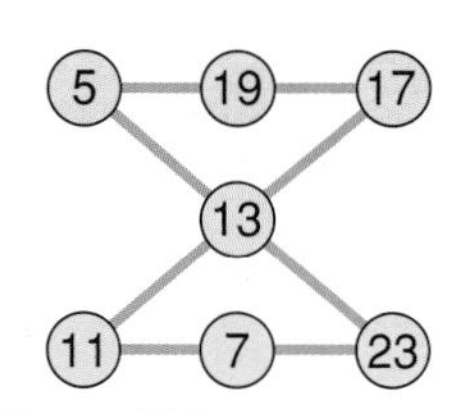

16 **a** No **b** No
17 Possible answers are: **a** About 20 **b** About 800
18 **a** About 320; 356·259 **b** About 12; 11·27 (2 d.p.)
19 **a** Area is about 450 cm^2, perimeter is about 100 cm
b Area = 446·16 cm^2, perimeter = 98·1 cm

Chapter 4 page 89

1 8·7297 kg
2 **c** and **d** because in **c** 0·26 is multiplied by a number greater than 1 and in **d** 0·26 is divided by a number less than 1.
3 **a** 42·16 **b** 175·94 **c** 4·7196 **d** 0·7638 **e** 0·042159 **4** 0·1496 m^2 **5** **a** 36·7 **b** 15·3
6 1·86 seconds **7** 36 slices **8** **a** 38·2 **b** 118·6 **c** 8778·3 **d** 11.1 **e** 344·8 **f** 24·2 **9** No
10 **a** The answer is wrong. The answer should be bigger than 5·2 because we have multiplied by a number bigger than 1.
b The answer is wrong. The answer should be bigger than 8·3 because we have divided by a number smaller than 1.
c The answer is wrong. The answer should be smaller than 89 because we have divided by a number bigger than 1.
11 A possible answer is 4·6 × 2·1 = 4·6 × 2 + 4·6 × 0·1 = 9·66. The answer is correct.
12 **a** and **d** **13** **a** 0·79 **b** 76·56 **c** 2·59 **d** $^-$0·83 **e** $^-$698 **14** **a** £1·54

Chapter 5 page 110

1 $\frac{1}{2}$ **2** $1\frac{1}{2}$ **3** **a** $\frac{1}{2}$ **b** $\frac{11}{24}$ **4** **a** 0·28 **b** $0{\cdot}\dot{8}\dot{1}$ **c** $0{\cdot}\dot{6}$ **d** $0{\cdot}\dot{4}$
5 **a** 0·55 (2 d.p.) **b** 0·76 (2 d.p.) **c** 0·5 **d** 0·01 (2 d.p.) **e** 0·02 (2 d.p.)
6 **a** $\frac{5}{9}$ **b** $\frac{7}{9}$ **c** $\frac{2}{3}$ **7** The lunchtime club **8** **a** $\frac{7}{25}, \frac{3}{10}, \frac{17}{50}, \frac{2}{5}, \frac{1}{2}$
9

$\frac{12}{20}$ $\frac{36}{40}$ $\frac{13}{10}$ $\frac{35}{25}$

0 1

10 $\frac{5}{12}$ **11** **a** $1\frac{1}{5}$ **b** $1\frac{1}{4}$ **c** $1\frac{29}{40}$ **d** $\frac{5}{24}$ **e** $\frac{7}{30}$ **f** $1\frac{3}{8}$ **g** $\frac{11}{20}$
12 **a** $\frac{1}{10}$ **b** 30 **c** $7\frac{1}{2}$ **d** $\frac{5}{6}$ **e** $1\frac{5}{7}$ **f** $\frac{1}{10}$ **g** $\frac{3}{20}$ **h** $14\frac{14}{15}$ **i** $1\frac{17}{22}$ **j** $3\frac{3}{8}$ **k** $1\frac{3}{4}$
13 **a** $\frac{13}{63}$ **b** $1\frac{11}{40}$ **c** $5\frac{47}{48}$ **d** $1\frac{23}{87}$ **14** **a** $\frac{7}{9}$ of 45 **b** 72% of 36 **15** $\frac{1}{6}$ **16** 24

Chapter 6 page 126

1 **a** 6·8 m **b** 23·4 ℓ **c** 130·5 km **d** £152·88 **2** **a** About 10 **b** About 5
3 **a** 1·275 m **b** 0·956 m to the nearest mm **4** C **5** 200 g with 25% extra at £1·64
6 18·8% (1 d.p.) **7** About 384% **8** **a** Pam **b** Tania **9** 15
10 £136·15 to the nearest penny **11** About 30 million square kilometres

Chapter 7 page 138

1 325
2 **a**

Coconut Kisses

6 egg whites
$\frac{3}{4}$ tsp vanilla
3 cups cornflakes
$\frac{3}{4}$ cup chopped nuts
$1\frac{1}{2}$ cups sugar
$2\frac{1}{4}$ cups coconut

b 100

3 32 : 14 : 45 **4** 11 : 28

5 Nightime because $\frac{1}{12\cdot5}$ is greater than $\frac{1}{13\cdot3}$. Because Nightime has a higher ratio of cotton to other materials it must have a higher proportion of cotton.

6 300 mℓ **7** 168 **8** 144 g **9** 780 mℓ **10** 29 : 20 **11** $x : 2$

12 St Christopher's because 3·3 : 1 is a lower ratio of attempted goals to goals scored than 3·8 : 1. So the proportion of goals scored must be higher for St Christopher's.

Chapter 8 page 187

1 a True b False c True d True e True f False g True

2 a $g = \frac{r}{2}$, $2g + r = p$, $p - 2g = r$, $g = \frac{p - r}{2}$

b Possible answers are: $p - r = 2g$, $p = 4g$

3 a B b C **4** a $42y$ b $32x$ c ^-12b d 24 m^2 e a^6 f $3a(g)$ g $\frac{5q^3}{3}$ h $\frac{x^2}{5}$ i $\frac{7n}{4}$ j p^4

5

$p+r$	$p-q-r$	$p+q$
$p+q-r$	p	$p-q+r$
$p-q$	$p+q+r$	$p-r$

6 a $2a + 11b$ b $6x + 3$ c $3m - 12$ d $^-3q + 3$ e $n - 1$ f $8a^2$ g $p^2 + 6p$ h $12y^2 - 2y$ i $7x + 24$ j $4p + 4$ k $12 - 4n$ l $5 - y$ m $10p + 9q$ n $14x - y$

7 $2x + 1$ and $3y + 2$

8 Possible answers are: $45 - 2h$, $5(9 - h) + 3h$, $27 + 2(9 - h)$. $5(9 - h) + 3h = 45 - 5h + 3h = 45 - 2h$ and $27 + 2(9 - h) = 27 + 18 - 2h = 45 - 2h$

All three expressions are equal to $45 - 2h$.

9 a $9(y + 2)$ b $3(2p - 3)$ c $2(3 - 4n)$ d $5(p + 3q)$ e $x(5x + 1)$ f $n(6 + n)$ g $3(a^2 + 2a + 2)$ h $x(x^2 + x + 2)$

10 ABBA = A × 1000 + B × 100 + B × 10 + A
= A × 1001 + B × 110
= 11(91A + 10B)

11(91A + 10B) is always divisible by 11.

11 a i 2·12 ii $^-$0·568 iii 1·5 iv 0·232

b i 14 ii $^-$26 iii $^-$4 iv 3·4

12 a 2500 b 36 **13** 360 **14** a 1229 kjoules b 1006·5 kjoules

15 a 3·2 b 1·3 (1 d.p.) c $T = \frac{P}{3V}$ **16** a $C = \frac{5(F - 32)}{9}$ b 37°C to the nearest degree

17 $F + V = E + 2$ **18** a 245 m^2 b 80°, 60° and 40°

19 a 1·4 b 5 c 5 d 3·5 e 8 f 2 g 2·5 h 2 i $^-$5 j ±5·2 k 4 l 0·25 m $^-$3

20 $6(n - 3) = 4n + 6$; $n = 12$

21 a

British pounds (p)	10	20	30	40	50
Euro (€)	15	30	45	60	75

b Yes, because each time we increase the pounds by £10, the number of euro increases by a constant amount.

c It is 3 : 2 for each pair.

d

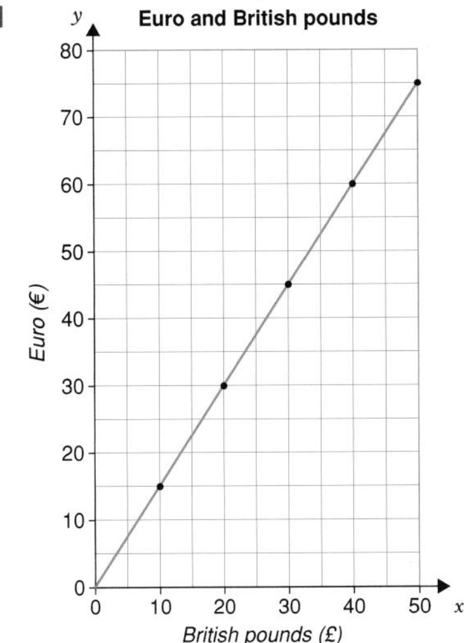

e $e = 1{\cdot}5p$ f 120 euro

22 a 14 or −14 b 8 or −8 **23** a 3.11 b 4.28 c 4·65 d 1·2 (1 d.p.) **24** 57, 58 and 59

25

x	2	3	4
y	−8	9	44

$x = 2·58$

Chapter 9 page 221

1 4, 8, 16, 32, 64 **2** a 3, 6, 12, 24, 48 b −400, 200, −100, 50, −25

3 a 3, 5, 7, 9, 11, 13 b −5, −2, 1, 4, 7, 10 c 5, 3, 1, −1, −3, −5

4 a 119, 127, 135 b 100, 81, 64 c −59, −73, −87 d $-4\frac{1}{2}$, $-5\frac{3}{4}$, −7

5 a Possible answers are: 4, 6, 10, 16, 26, 42, ... **rule** add the two previous terms **or** 4, 6, 10, 16, 24, 34, 46, ... **rule** count on in steps of 2, 4, 6, 8, 10, ...
b You need to know the rule for the sequence.

6 a 1, 5, 9, 13, 17, 21, 25, 29, 33, 37 b 70, 63, 56, 49, 42, 35, 28, 21, 14, 7 c 2·5, 5·5, 8·5, 11·5, 14·5, 17·5, 20·5, 23·5, 26·5, 29·5 d 0·3, 0·6, 0·9, 1·2, 1·5, 1·8, 2·1, 2·4, 2·7, 3 e 2, 5, 10, 17, 26, 37, 50, 65, 82, 101 f 1, 7, 17, 31, 49, 71, 97, 127, 161, 199 g −2, 1, 6, 13, 22, 33, 46, 61, 78, 97

7 a **first term** 1, **rule** add 4 b **first term** 70, **rule** subtract 7
c **first term** 2·5, **rule** add 3 d **first term** 0·3, **rule** add 0·3
e **first term** 2, **rule** add consecutive odd numbers starting at 3
f **first term** 1, **rule** add every second even number starting at 6
g **first term** −2, **rule** add consecutive odd numbers starting at 3

8 a 0·1, 0·2, 0·4, 0·8, 1·6 b $\frac{1}{2}$, 2, 8, 32, 128 c 5000, 1000, 200, 40, 8 **9** $\frac{1}{2}$, $\frac{2}{5}$, $\frac{3}{10}$

10 A possible answer is: **first term** 1·8, [0·2]

11 a Ascending numbers with a difference of 6 which are all one more than a multiple of 6. It starts at 7.
b Descending numbers with a difference of 6 which are all multiples of 6. It starts at 60.
c Ascending numbers with a difference of 6 which are all multiples of 6. It starts at 18.

12 a

5th term is 41

b

Diagram number	1	2	3	4	5
Number of dots	1	5	13	25	41
Increase		4	8	12	16

c 85
Each time a new diagram is drawn, another row of dots is added to each of the sides. Each of these added sides has one more dot than the previous sides added.
This means the number of dots **added** each time increases by 4.

13 a $T(n) = 8n + 40$ b $T(n) = 82 - 7n$ c $T(n) = 0·2n + 1·9$ d $T(n) = 7 - 9n$ **14** 24th

15 a

Input	Output
7	15
−3	−15
1·4	−1·8
$\frac{1}{2}$	$-4\frac{1}{2}$

b

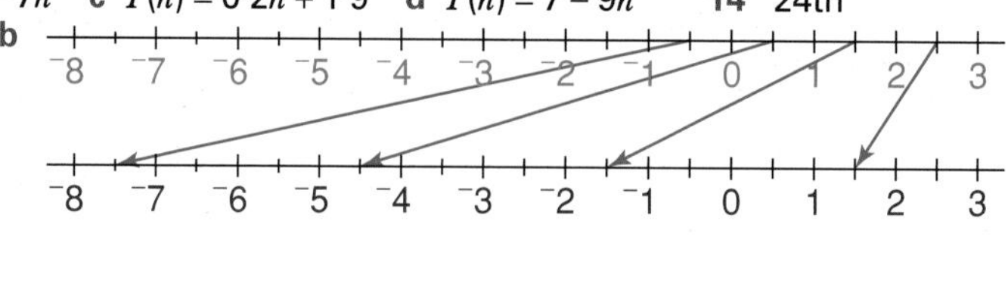

16 a The mapping lines would be parallel.
b The mapping lines would meet at the zero line.

17 $x \longrightarrow \frac{x}{2} - 1$ **18** a Subtract 1 b Multiply by 3

19 When the operations are reversed, you get a different function.
For the first function machine, the function is $y = 3(x + 4)$.
For the second function machine, the function is $y = 3x + 4$.
$y = 3(x + 4)$ is different from $y = 3x + 4$.

20 a 16, 3·5, −8 b 36, 8·2, 19·2 **21** a $x \longrightarrow \frac{x+7}{5}$ b $x \rightarrow 5x - 3$ c $x \rightarrow 4(x + 3)$

22 21·75 cm

Chapter 10 page 246

1 a $y = -2x + 3$ b $y = \frac{^{-}x}{2} + 3$ c $y = 3x - 1$

2 a

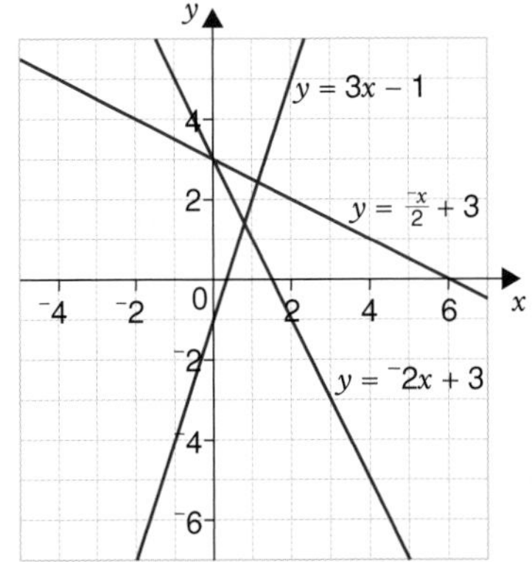

3 Yes, because $3x - 2y = 3 \times 1 - 2 \times {}^{-}4 = 3 + 8 = 11$. 4 a B b C c A d A e B f A

5 a C b D c E d A e B 6 l_1 $^{-}1$, l_2 $^{-}3$, l_3 $\frac{3}{2}$, l_4 $\frac{1}{3}$

7 a 70 cm b 80 cm c 50 cm d $3\frac{1}{2}$ years e About 5 cm

8 a $g = 0{\cdot}22\ \ell$

b

ℓ	0	10	20
g	0	2·2	4·4

c

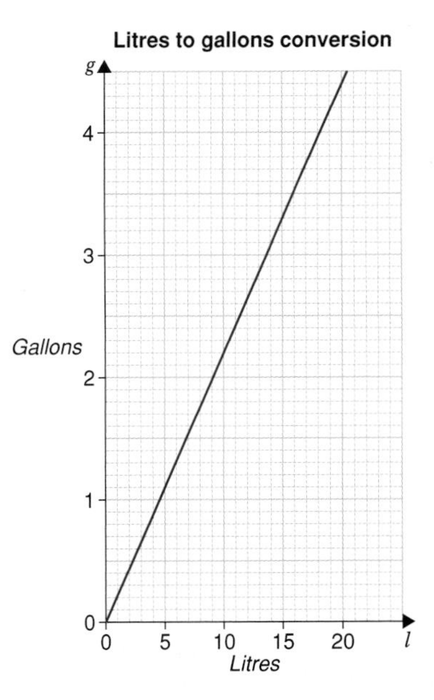

d About 3·5 gallons

e Label on smallest is 150 litres 30 gallons, label on next is 180 litres 40 gallons, label on largest is 450 litres 100 gallons

9 a About 175 m b About 70 seconds c About 85 seconds d About 50 seconds e C

10 a 2 b He went back home at a constant speed. c Afraaz

d Yes, because the line is a straight line which tells us that as time increases in equal steps, distance increases in equal steps.

11 3

12

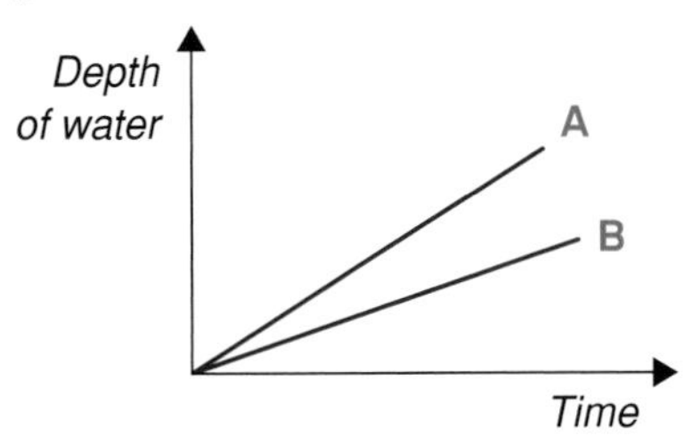

13 **a** **i**

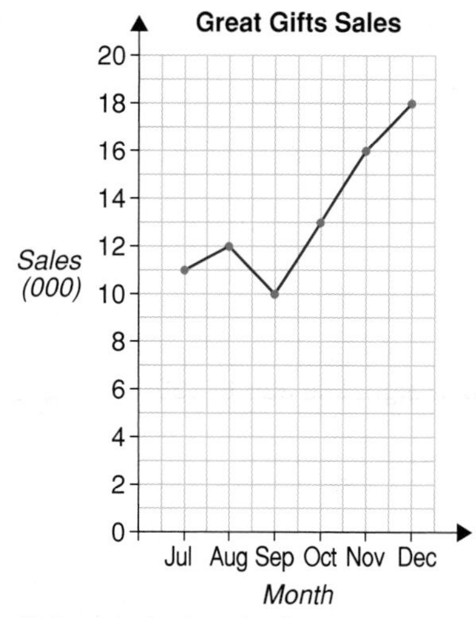

ii

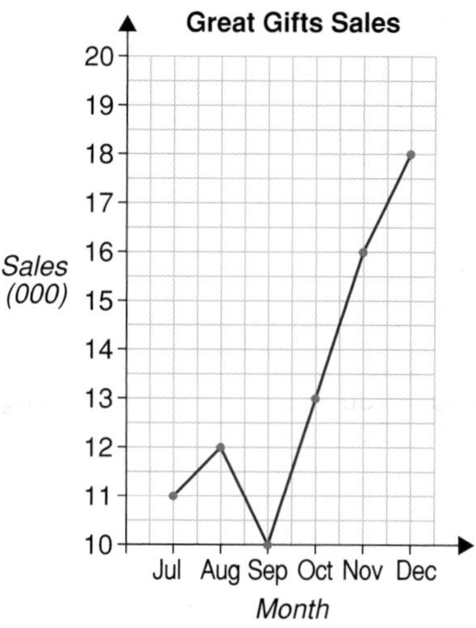

b In **ii** the vertical scale does not start at 0 and so it gives a misleading picture of the increase in sales from September to December. Also the vertical scale goes up in steps of 0·5 in graph **ii** and in steps of one in graph **i**. This makes the increase look greater on graph **ii**.

14 $2x - y + 1 = 0$

15 As equal time intervals go by, the volume increases sharply at first and then increases in decreasing steps until it stops increasing at all.

16

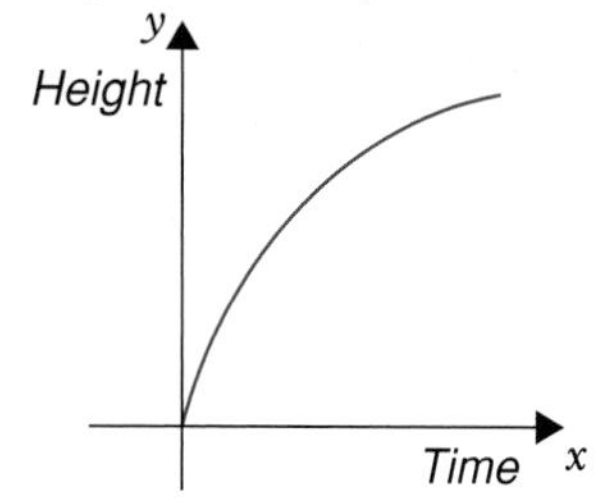

Chapter 11 page 273

1 Possible answers are:

a
$a = 70°$ alternate angles parallel lines are equal
$b = 180° - 70°$ angles in a straight line add to 180°
$= 110°$
$d = 70°$ alternate angles parallel lines are equal
$c + d = 180°$ angles on a straight line add to 180°
$c = 180° - 70°$
$= 110°$

b
$d = 25°$ alternate angles parallel lines are equal
$e + 25° + 95° = 180°$ angles in a △ add to 180°.
$e = 60°$
$f = 95°$ corresponding angles parallel lines are equal

c
$g + 47° = 88°$ vertically opposite angles are equal
$\mathbf{g = 41°}$
$x = 41°$ alternate angles parallel lines are equal
$h + x + 72° = 180°$ angles on a straight line add to 180°
$\mathbf{h = 67°}$
$i = 67°$ corresponding angles parallel lines are equal

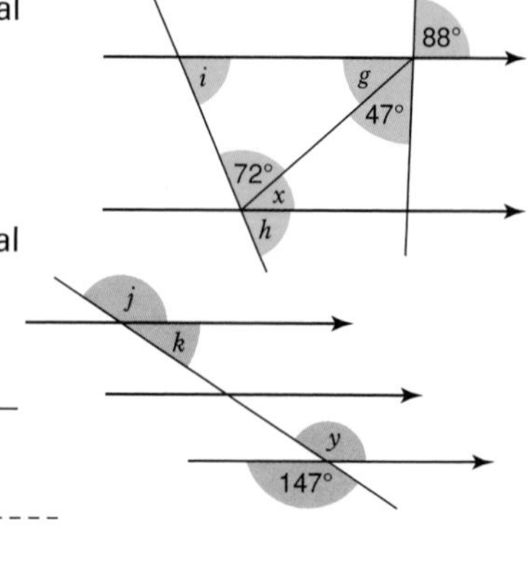

d
$y = 147°$ vertically opposite angles are equal
$j = 147°$ corresponding angles equal
$k = 180° - 147°$ angles on a straight line
$= 33°$

e
$x = 69°$ alternate angles
$y = 111°$ angles on a straight line
$z = 111°$ corresponding angles
$l = 111°$ alternate angles

2 **a** $3x + 86° + x + 42° = 180°$ angles on a straight line add to 180°

$x = 13°$

b $115° + x + 36° + 24° + 2x + 10° + 2x = 360°$ angles at a point add to 360°

$x = 35°$

c $x - 30° + x - 25° + 2x + 15° = 180°$ angles in a △ add to 180°

$x = 55°$

d $3x - 10 = 2x + 26°$ vertically opposite angles are equal

$x = 36°$

e $2x + 10° = 3x - 20°$ corresponding angles are equal

$x = 30°$

f $y = 3x + 20°$ corresponding angles are equal

$3x + 20° + x + 40° = 180°$ angles on a straight line add to 180°

$x = 30°$

g $q = 44°$ base angles isosceles △

$2p + 44° = 180°$ angles in △ add to 180°

$p = 68°$

$68° + x - 30° + 44° + 44° = 180°$ angles in △ add to 180°

$x = 54°$

3 $a = 25°$ alternate angles parallel lines are equal

$b = 30°$ alternate angles parallel lines are equal

$x = a + b$

$= 55°$

4 **a** 88° **b** 123° **c** 48° **d** 47° **e** 77·5° **f** 22·5° **g** 36°

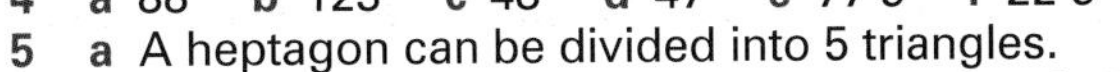

5 **a** A heptagon can be divided into 5 triangles.

The sum of the interior angles of a heptagon is $5 \times 180° = 900°$.

b At each vertex of a heptagon, the sum of the interior and exterior angle is 180°.

The sum of 7 interior and 7 exterior angles is $7 \times 180 = 1260°$.

Sum of exterior angles $= 1260° - 900°$

$= 360°$

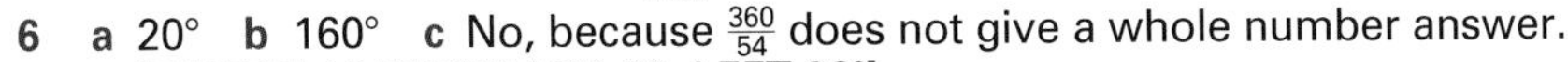

6 **a** 20° **b** 160° **c** No, because $\frac{360}{54}$ does not give a whole number answer.

7 **a** REPEAT 10 [FORWARD 20, LEFT 36°]

b REPEAT 8 [FORWARD 20, LEFT 45°]

Chapter 12 page 305

1 **a** Rectangle, rhombus **b** Rectangle, rhombus

2 AB = AC arrowhead symmetrical

$A\hat{B}C = 49°$ base angles of isosceles △ are equal

$A\hat{B}D = 49° - 27°$

$= 22°$

3

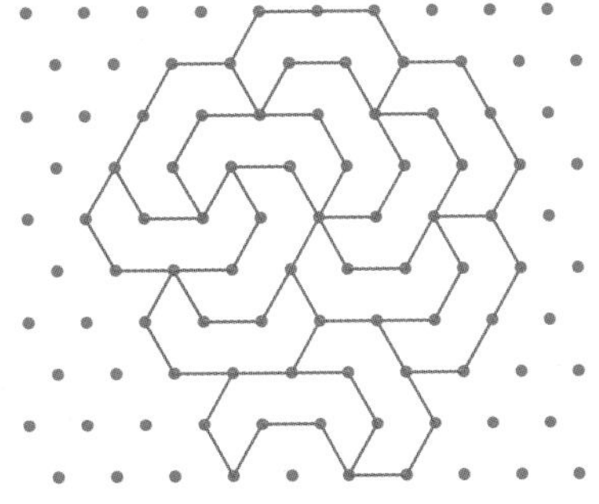

This tessellation is made by rotating the shape.

4 **a** 92° **b** 6 cm **c** B **5** **a** C **b** A **c** B **d** A **e** B

6 Check your answer using the diagram on page 304. **7** Aii, Biii, Civ, Di

8 **a**

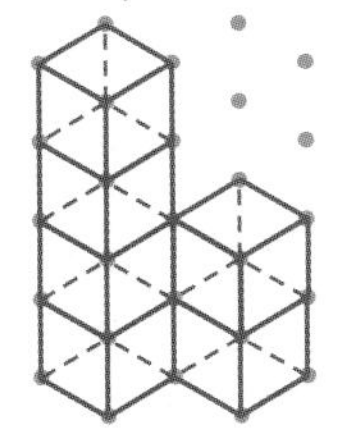

b

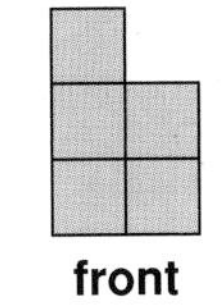

plan

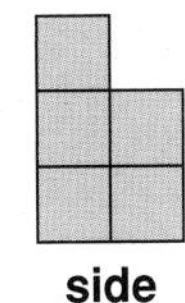

front

side

9 **a** View D **b** View C **c** View A **d** View B

10 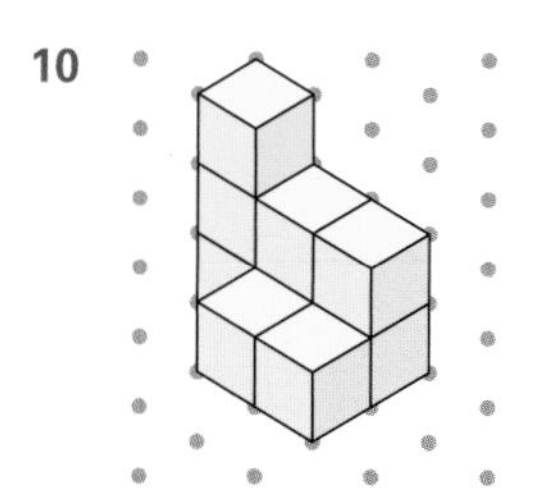

11 A net for a tetrahedron.

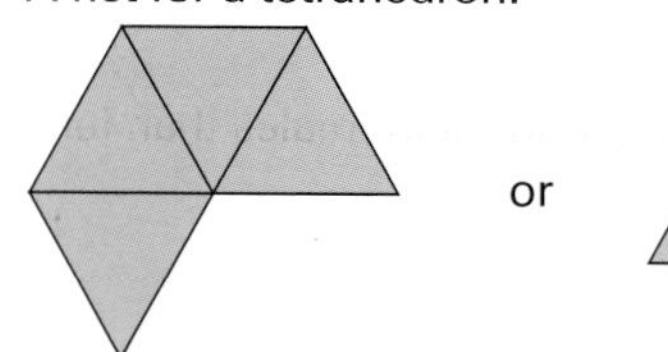

12 **a** A triangle **b** A triangle
13 10 mm
14 **a** 31° **b** 1·8 m
15 **a** A circle **b** Perpendicular bisector of the line joining the two trees
c Bisector of the angle between the two fences
16 The answer is the line $x = {}^-1{\cdot}5$.
17 A possible shape is:

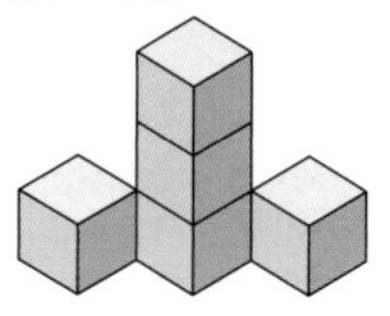

Chapter 13 page 332

1 **a** A′(2, 3), B′(5, 3), C′(1, ⁻1) **b** A′(⁻3, ⁻2), B′(⁻3, 1), C′(1, ⁻3) **c** A′(⁻2, ⁻3), B′(1, ⁻3), C′(⁻3, 1)
2 All of them are true. 3 **a** True **b** True **c** True **d** False **e** True
4 **a** Reflection in the x-axis
b A possible answer is: reflection in the x-axis and then reflection in the y-axis.
5 Isosceles triangle
6 **a** A possible answer is: reflection in the line $y = 1$ then translation by 2 units down.
b A possible answer is: rotation 180° about (0, 0) then reflection in the y-axis.
c A possible answer is: rotation 90° about (⁻5, ⁻4) then translation 3 squares right and 5 squares up.
d A possible answer is: rotation 270° about (0, 0) then reflection in the line $x = {}^-3$.
7 **a i** AB = A′B′ **ii** ∠B = ∠B′ **b i** **c i** **d** Rotation 270° about (0, 0) **e** No
8

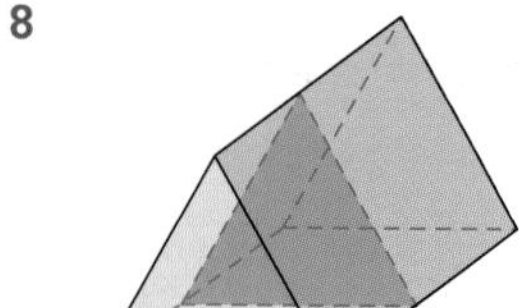 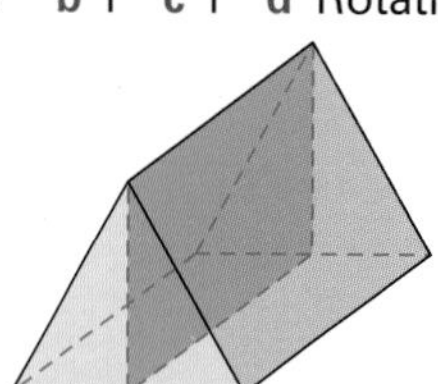

9 3 10 **a i** 8 cm **ii** 54° **b** 36 cm 11 About 10·8 m 12 **a** 40 m **b** 155 m
13 1 : 800 or 1 cm represents 8 m
14 **a** 890 cm or 8·9 m **b** 4·8 m by 5·1 m **c** 4·9 m by 3·8 m **d** 5·5 m by 7·2 m
15 **a** (3·5, 13) **b** (⁻5, 1)

Chapter 14 page 358

1 **a** 42 000 m^2 **b** 0·08 ha **c** 5100 ℓ **d** 52 mℓ **e** 5·86 ℓ **f** 4·936 m^3
g 50 000 cm^2 **h** 4000 mm^3 **i** 853·2 cm^3 **j** 840 mm^2 **k** 620 000 cm^3 **l** 528 cm^2
2 **a** 272 minutes **b** 12 hours 4 minutes **c** 9 years 4 weeks
3 **a** 18 900 cm^3 **b** 18 900 000 mm^3 **c** 18 900 mℓ **d** 18·9 ℓ
4 **a** 4440 cm^2 **b** 444 000 mm^2 5 245 kg 6 1·2 km 7 1800 cm^3
8 **a** Marcia cycled about **3·2 km** to the local shop to buy **2·4 ℓ** of milk, **2·3 kg** of apples, **90 g** of chocolate chips, **3 metres** of tape and **10 cm** of ribbon.
b Robbie lives **6·25 miles** from school. He has a dog that has a mass of **11 lb** and a rabbit that lives in a **6 foot** or **2 yard** hutch. Each day Robbie, the dog and the rabbit drink **7 pints** of milk.

9 a Depends on the individual b 8 kg ⩽ mass of 10 books ⩽ 15 kg
c 10 mℓ ⩽ capacity of egg cup ⩽ 30 mℓ
10 a 053° b 019° c 090° d 156° e 202° 11 a Pirate Ship b Roller Ride
12 a circumference = 37·7 cm (1 d.p.), area = 113·1 cm^2 (1 d.p.)
b circumference = 48·7 cm (1 d.p.), area = 188·7 cm^2 (1 d.p.)
c circumference = 51·5 cm (1 d.p.), area = 211·2 cm^2 (1 d.p.)
13 a 314·16 cm (2 d.p.) b 62·83 cm 14 a 159·27 m^2 b 65·7 m
15 a 234 cm^3 b 0·336 cm^3 c 222·75 m^3 16 b 3·34 cm^2

Chapter 15 page 384

1 a 6 b 10 c 7 d More females than males played the violin and piano. More males than females played the guitar.

2

Age	Men	Women
60–		
70–		
80–		
90+		

3 a The question does not give a great enough range of possible responses. It is possible that someone might like all except one person in their family. The answer is then 'No' but this does not give good data about attitudes to family.
b I consider I generally have a good relationship with
my mother ☐ my father ☐ sibling 1 ☐ sibling 2 ☐ sibling 3 ☐ sibling 4 ☐ sibling 5 ☐
c Mia's family is unlikely to be representative of the whole population.
Giving the questionnaire to one family is not a great enough sample to be representative.

4 A a An experiment could be carried out to collect the data or the PE department might have some.
b Between 50 and 100.
c

Number	Distance jumped (in cm) standing start	Distance jumped (in cm) 10 practice jumps	Distance jumped (in cm) 20 practice jumps	Distance jumped (in cm) with run up

B a An experiment could be carried out to collect the data or there may be data in consumer magazines or on the Internet.
b Between 8 and 12
c

Power (in watts)	Time (in minutes) to dry the same person's hair

5 Possible answers are: At what age do you think teenagers should be allowed to smoke if they choose to?
13 ☐ 14 ☐ 15 ☐ 16 ☐ 17 ☐ 18 ☐ 19 ☐ not at all ☐
Tick agree or disagree.
Teenagers should be fined for smoking. Agree ☐ Disagree ☐
Teenagers should smoke if they want to. Agree ☐ Disagree ☐
Teenagers should be educated about the dangers of smoking. Agree ☐ Disagree ☐
Teenagers who smoke should be made to visit people dying of lung cancer. Agree ☐ Disagree ☐

Chapter 16 page 414

1 a Mean = 4, median = 3, mode = 2
b No, because there are two extreme values, 6 and 10 which make it higher.
c Mean. No, because only two of the data values are greater than the mean.
2 5·51
3 a $3:00 \leqslant T < 3:10$
b One possible answer is: so that the organisers know when to have lots of time keepers on.
4 19, 35, 60 5 No 6 $5x - 1$
7 Median 27 minutes, mode 25 minutes, range 59 minutes

8 a **Gwen** mean = 12, median = 12, range = 4
Lela mean = 12, median = 7, range = 28

b The coach could choose Gwen because she has the same mean as Lela, a high median and she has more consistent scores shown by her smaller range.
The coach could choose Lela because there is a possibility she will score a lot of goals, because of her big range.

9 a

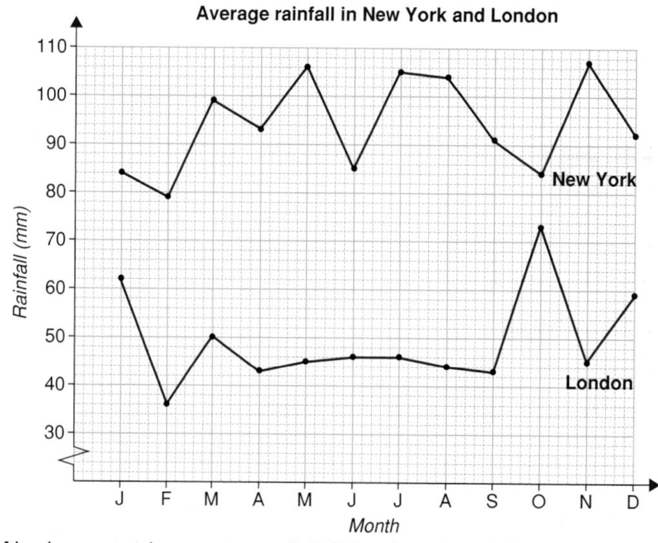

b No, because the average rainfall for that month is plotted and you cannot tell when in the month it fell so points between months are meaningless.

c The average rainfall in New York is higher than in London for every month of the year. The average rainfall in New York varies quite a lot from month to month whereas in London from April to September it is very consistent.

d Yes, but it would not show the trend as well as a line graph.

10 a

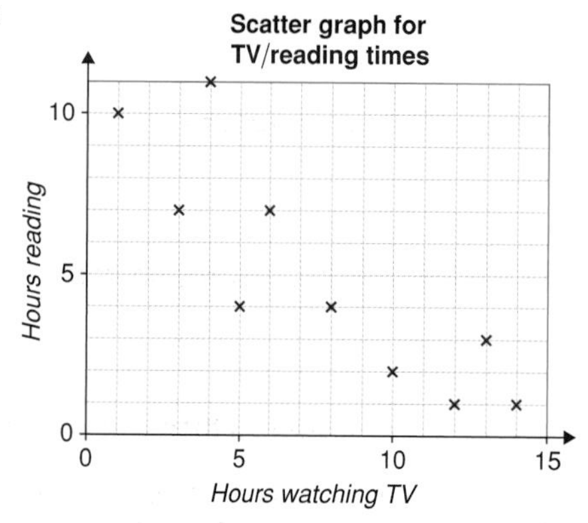

b Yes **c** Between 1 and 3 hours

11

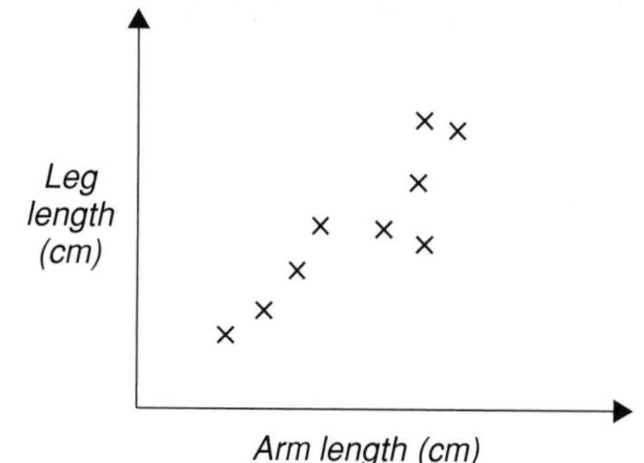

12 a

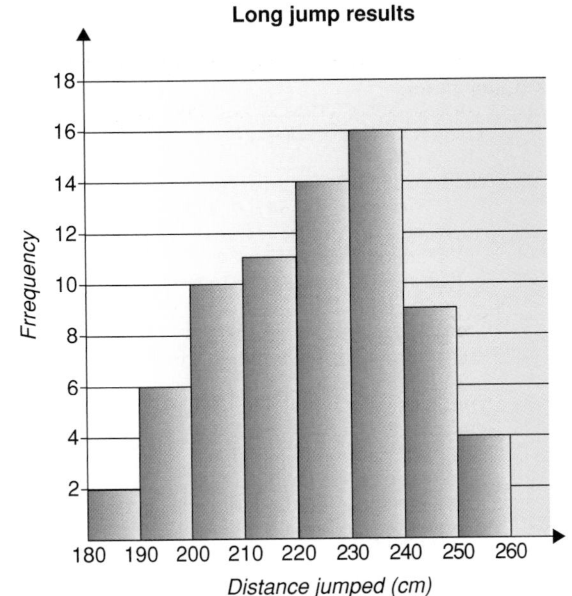

b Yes, because it is possible that all the other 8 people in the class interval $240 \leqslant d < 250$ jumped less than the 243 cm that Ryan jumped.

c There are 72 results. Over half would be more than 36.
The 36th jump is in the class interval $220 \leqslant d < 230$ which is less than the 233 cm quoted.

13 A possible answer is: On weekdays British people sleep about 520 minutes each day, which is about 8 hours 40 minutes. This is only a little more than the average 8 hours that is recommended.
At the weekend, the number of hours sleep is just over 9 hours. This is probably not excessive for the weekend.

14 There is no scale on the vertical axis. It is impossible to tell by how much tooth decay has decreased.

Chapter 17 page 433

1 Board B 2 Box A, because this has the greatest proportion of white chocolates.

3 a No b No c Yes 4 a 0·25 b 0·8 5 0·92

6 a $\frac{5}{26}$ b $\frac{5}{26}$ c $\frac{5}{26}$ d $\frac{9}{26}$ e 0 f $\frac{18}{26}$ or $\frac{9}{13}$

7 a ba, bo, bb, ab, ao, aa, ob, oa, oo b i $\frac{1}{3}$ ii $\frac{5}{9}$ iii $\frac{4}{9}$ 8 $\frac{1}{9}$

9 One possible way of showing the sample space is: green green, green red, green blue, red red, red green, red blue, blue blue, blue green, blue red
a $\frac{1}{9}$ b $\frac{1}{3}$ c $\frac{2}{3}$ d $\frac{4}{9}$ e $\frac{5}{9}$ f $\frac{4}{9}$

10 80 11 a $\frac{3}{400}$ b $\frac{3}{400}$

12 a She could calculate the relative frequency of a page having 5 advertisements then use this as an estimate of the probability. b No c Yes

13 a $\frac{1}{10}$ b One possible answer is: Put cards or counters numbered 0 to 9 in a bag. Take one at random. Record its number on a tally chart. Put it back. Do this 100 times. c 200 trials

Index